INTRODUCTORY
ALGEBRA

INSTRUCTOR'S EDITION

INTRODUCTORY ALGEBRA

Ronald Hatton

Sacramento City College

Gene R. Sellers

Sacramento City College

HBJ

Harcourt Brace Jovanovich, Publishers

and its subsidiary, Academic Press

San Diego   New York   Chicago   Austin   Washington, D.C.

London   Sydney   Tokyo   Toronto

We dedicate this book to our wives,
Susie Hatton and Linda Sellers,
whose love and encouragement
we sorely needed through
the months of toil.

ISBN: 0-15-541565-4 (Student Edition)
       0-15-541566-2 (Instructor's Edition)

Printed in the United States of America

Preface

The mathematical concepts and principles presented in this textbook can be covered in a one-semester or two-quarter course in introductory algebra. Extent and depth of coverage are among the strongest features of the book. Students who successfully complete a course using this textbook will be well prepared to continue their mathematical studies in an intermediate algebra course, a geometry course, or a course in technical or applied mathematics.

WRITING STYLE AND USE OF GRAPHICS

Throughout this textbook the aim has been to write mathematical principles as concisely as possible while maintaining clarity. A new principle is frequently introduced with a specific example to help students grasp the evolution of the principle from specific to general. The use of graphics, including lines, arrows, arcs, and marginal comments, promotes visual awareness. The book is greatly enhanced by the use of four colors, with each color having a pedagogical function. The result is a textbook that is easy to read, easy to follow, and appealing to the eye.

1. Each section begins with a list of **key topics** identifying the skills to be learned within the section. The key topics are then used to separate the section into subsections.
2. Most sections begin with an **introductory statement** that links current topics with previously learned topics or with topics to be learned in future sections.
3. **Definitions** and **important concepts** are clearly identified by color boxes. Guided solutions are also printed in color. Students have found these displays to be a successful way of identifying important items in a clear, concise, and organized manner.
4. Each section contains numerous worked examples. The solutions to these examples contain several features that enhance their effectiveness as a teaching tool: a **discussion** that sets forth a plan

of attack, **marginal comments** that help explain the solution, and a **step-by-step approach** that leaves no room for confusion.

SIX STEPS TO SUCCESS

One of the basic assumptions underlying the writing of this textbook is that the mastery of algebra is best achieved by working problems. To that end, INTRODUCTORY ALGEBRA offers a wealth of problems, with each section containing four different kinds of exercise sets. The exercise sets themselves are described in some detail below; here we offer six problem-solving steps which, if followed consistently, will maximize a student's chances for success.

1. Read the section carefully, following the problem-solving strategy in the examples.
2. Work the odd-numbered *Practice Exercises* and check the answers at the back of the book.
3. Go to the text-specific tutorial software (Hattonware) as needed for additional practice.
4. Work all 10 *Review Problems*.
5. Work all *Summary Exercises*. (A student who successfully completes all 8 for a given section has mastered that section.)
6. Work any *Supplementary Exercises* the instructor may assign.

EXERCISES

We are aware that students in a typical introductory algebra class differ greatly in their readiness. Some students need a slow-paced, low-level course and others need a moderately-paced course that includes some challenging exercises. And students who have had sufficient exposure to algebraic concepts need still more challenging exercises to keep them interested. These exercise sets will help an instructor deal with very diverse needs within the same class. Answers to all exercises can be found in the back of the Instructor's Edition.

1. **Practice Exercises** are keyed to the examples in the text and progress from easy to difficult. Each Practice Exercise set appears in a parallel odd–even format. That is, every odd-numbered exercise has a corresponding even-numbered problem that has the same level of difficulty, covers the same principle, and has a similar type of solution. A student might be required to work all the odd-numbered exercises in this set. Answers to the odd-numbered exercises can be found at the back of the student edition.

2. Beginning with Chapter 2, each section has 10 **Review Exercises**, which are intended to serve two primary purposes: they provide a continuous review of material learned in previous sections and they link related topics from different sections. For example, several review sets have each type of equation and inequality previously studied. This requires students to recognize the type of equation and then apply the correct technique to solve it. Answers to all these exercises can be found in the student edition.

3. **Supplementary Exercises** offer additional problems that are more difficult than those in the Practice Exercises, and they are not keyed to examples. These exercises frequently have guided solutions that enable students to discover new mathematical concepts. A few of the exercises in these sets cover topics that some instructors may consider too advanced for beginning students. However, these exercises are generally preceded by examples with detailed solutions. These exercises provide more flexibility in teaching a beginning algebra course. By selecting few, or none, of these exercises, the level of difficulty can be kept low. By selecting more of these exercises and fewer of the Practice Exercises, the level of difficulty can be increased. Exercises can also be used for small, study group sessions within the class; for in-class review; or for extra credit. Answers to the odd-numbered exercises can be found in the back of the student edition.

4. **Summary Exercises** covering all the concepts studied in a particular section appear, by section, at the end of each chapter. They are easily located by the color edges of the pages. There are typically 8 Summary Exercises for each section, with space between exercises for students to show their work. Furthermore, each exercise set has an answer column and a heading for name, date, and score. These exercises can be easily removed and used for homework assignments or in-class quizzes. Complete solutions to these exercises can be found in the Instructor's Edition.

5. An extensive set of **Review Exercises** can be found at the end of each chapter. These exercises include two of each type of problem found in each Practice Exercise set in each chapter. Answers to the odd-numbered exercises can be found in the back of the student edition.

WORD APPLICATIONS

In any mathematics course applied problems are a source of frustration to most students and an area of concern for all instructors. In this textbook we have improved the chances for success in this area in three ways. First, exposure to applied problems is increased by including them in more sections. Second, many applied problems in Supplementary Exercise sets have guided solutions that lead a student step-by-step to the answer. Third, the discussion component of examples gives almost every example the look of an applied problem.

EXTENDED COVERAGE

At the end of the textbook are five appendixes. Appendix A covers briefly some of the terminology used in discussing sets, Appendix B is a thorough review of fractions, and Appendix C is a review of rounding numbers. Appendixes B and C contain exercises and answers. Appendix D is a table designed to help students solve problems involving compound interest, and Appendix E provides students with blank graphs for use with problems requiring graphing.

SUPPLEMENTS FOR INSTRUCTORS

- *Instructor's Edition* provides answers to all the exercises and problems in the textbook.
- *Videotapes* provide 15 hours of lessons to complement classroom lectures.
- *Instructor's Manual with Solutions* provides detailed solutions for all exercises and problems in the textbook.
- *Computerized Test Bank (Micro-Pac Genie)* provides a test-generating and test-writing system, with graphics, that can print five types of questions and multiple versions of a given test. Also available in a hardcopy format, it accommodates 10,000 questions and creates almost any graph.

SUPPLEMENT FOR STUDENTS

A *Student Study Guide with Solutions* contains three sections for every chapter in the textbook. Section A is a summary of the topics covered in the chapter, with some additional information and all definitions and procedures. This part of the manual could be used in place of the notes students frequently take in class but seldom use. Section B contains detailed solutions for one-half of the odd-numbered Practice and Supplementary Exercise sets. The solutions are similar to those that accompany examples in the textbook. The structure of the exercise sets enables students to study a solution and then apply the technique to similar exercises. Section C is a test readiness section containing sample tests that will allow a student to determine whether he or she has the speed and skills needed to succeed in testing on a particular chapter.

COMPUTER SUPPLEMENTS FOR INSTRUCTORS AND STUDENTS

Hattonware provides a computer-assisted instructional program that gives students an opportunity to practice on an unlimited number of exercises. This software package was developed by Ronald Hatton to be used exclusively with this textbook. As a consequence, it identifies topics by section and page number. The disks are not copy-protected, and instructors are encouraged to make copies for students.

Other software tutorials available are EXPERT ALGEBRA TUTOR, MATH PAC, ALGEBRA Student Tutorial by TRUE BASIC, and BEGINNING ALGEBRA by MATH LAB.

ACKNOWLEDGMENTS

The authors would like to acknowledge several individuals who made significant contributions to this project. First, we would like to thank the following reviewers of the manuscript for suggestions that resulted in many improvements: Allan Bluman, Community College of Allegheny County; Joan Dykes, Edison Community College; Bill Foley, Southwestern Community College; Linda Holden, Indiana University; Richard Semmler, Northern Virginia Community College; and Arleen Watkins, University of Arizona.

We would also like to express our deepest appreciation to the staff at Harcourt Brace Jovanovich for their efforts in the development of this project. We especially want to thank Cindy Simpson, Judi McClellan, Don Fujimoto, Florence Kawahara, Lynne Bush, and Paulette Russo. We also want to express our appreciation to Michael Johnson, mathematics editor, and Nancy Evans, marketing manager, who worked diligently to get the support material needed for the project and who checked and rechecked to make sure the textbook covered all of the topics taught in mainstream beginning algebra courses. Without the support and encouragement of these team members, the project would not have achieved the goals we envisioned.

RONALD HATTON
GENE R. SELLERS

Contents

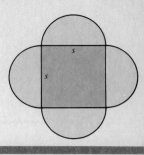

Preface   v
Introduction   xv

CHAPTER

1

Essential Topics from Prealgebra   **1**
1-1   Real Numbers and Some Related Topics   2
1-2   Addition and Subtraction of Real Numbers   13
1-3   Multiplication and Division of Real Numbers   22
1-4   Raising to Powers and Order of Operations   30
1-5   Properties of Real Numbers   37
1-6   Algebraic Expressions   45
1-7   Evaluating Algebraic Expressions   52
1-8   Word Phrases and Algebraic Expressions   60

CHAPTER

2

Linear Equations and Inequalities in One Variable   **85**
2-1   Parts of an Equation   86
2-2   Solving Linear Equations in One Variable   94
2-3   Applied Problems   101
2-4   Solving More Linear Equations in One Variable   111
2-5   Formulas and Literal Equations   120
2-6   Solving Linear Inequalities in One Variable   129

CHAPTER

3

Polynomials   **153**
3-1   Some Terminology Related to Polynomials   154
3-2   Adding and Subtracting Polynomials   161
3-3   Multiplying Polynomials by Monomials   168
3-4   Multiplying Polynomials   175
3-5   Multiplying Binomials: FOIL Method and Special Products   183
3-6   Dividing Polynomials   189

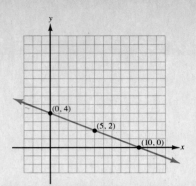
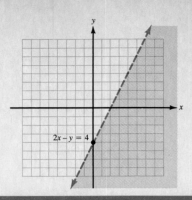

CHAPTER

4

Factoring Polynomials   **212**

4-1   Common Monomial Factors   213
4-2   Factoring a Simple Trinomial   218
4-3   Factoring by Grouping Binomials   224
4-4   Factoring a General Trinomial   228
4-5   Special Factoring Forms   236
4-6   Solving Quadratic Equations by Factoring   244
4-7   Applications of Quadratic Equations   250

CHAPTER

5

Rational Expressions   **272**

5-1   Rational Expressions, Defined and Reduced   274
5-2   Negative Integers as Exponents   283
5-3   Product and Quotient of Rational Expressions   292
5-4   Sum and Difference of Rational Expressions, Same Denominator   299
5-5   Sum and Difference of Rational Expressions, Different Denominators 306
5-6   Complex Fractions   315
5-7   Equations with Rational Expressions   324
5-8   Proportions   331
5-9   Applied Problems with Rational Expressions   339

CHAPTER

6

Linear Equations and Inequalities in Two Variables   **369**

6-1   Equations in Two Variables   371
6-2   Graphing Linear Equations   377
6-3   Slope and Intercepts of a Line   389
6-4   The Slope-Intercept Equation of a Line   398
6-5   Writing an Equation of a Line   407
6-6   Graphs of Linear Inequalities in Two Variables   414

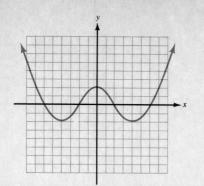
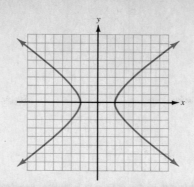
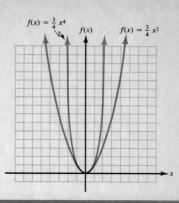

CHAPTER

7

Systems of Equations   **439**

7-1   Solutions of Linear Systems   440
7-2   The Graphical Method of Solving a System   448
7-3   The Substitution Method of Solving a System of Linear Equations
    in Two Variables   456
7-4   The Addition Method of Solving a System of Linear Equations in
    Two Variables   463
7-5   Applied Problems   471

CHAPTER

8

Radical Expressions   **495**

8-1   Square Root of a Number   496
8-2   Products and Quotients of Square Roots   505
8-3   Sums and Differences of Square Root Radicals   512
8-4   Preferred Form for a Square Root Radical   520
8-5   Equations with Square Root Radicals   527

CHAPTER

9

Quadratic Equations   **546**

9-1   Solving Quadratic Equations by Factoring   547
9-2   Solving Equations of the Form $X^2 = k$ and $k > 0$ 555
9-3   Solving Quadratic Equations by Completing the Square   560
9-4   Solving Equations with the Quadratic Formula   565
9-5   Complex Solutions of Quadratic Equations (Optional)   573
9-6   Solving Quadratic Inequalities (Optional)   579

CHAPTER

10

Relations and Functions   **601**

10-1 A General Discussion of Relations and Functions   602
10-2 The Linear and Quadratic Functions   614
10-3 Other Functions   621

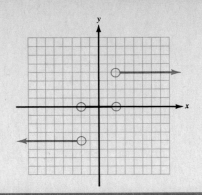 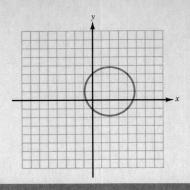

Appendix A   Sets   A-1
Appendix B   Fractions   A-5
Appendix C   Rounding a Number   A-13
Appendix D   Table: Powers of Periodic Interest Rates   A-15
Appendix E   Blank Graphs   A-17
Answers for Exercises   ANS-1
Solutions and Answers for Summary Exercises   S-1
Index   I-1

Introduction

Ridgewood Community College is located near a medium-sized city in the Midwest. The focus of area employment opportunites has been changing from industrial manufacturing to work involving electronic components and computer software. The college has been an important avenue for retraining residents so that their abilities match the requisite skills of the more technical jobs. The academic areas that have felt the impact most are mathematics and English.

Ms. Sharon Glaston has been a member of Ridgewood's mathematics department for 12 years. She recently has been studying the content of the entry-level algebra courses with three other department members to determine whether these courses meet the needs of the changing student body. The committee's preliminary report suggests that these courses should include more "problem-solving techniques" and problem sets that offer a greater variety and a greater challenge. (Many of these changes are found in this textbook.)

All chapters in this textbook begin with a short look into Ms. Glaston's Introductory Algebra classroom. These stories give an overview of the material to be covered in each chapter. The questions and answers given by class members remind us that successful learning in mathematics requires active participation on the part of the student. The exchange of ideas among teacher and students provides a learning environment that will benefit all participants.

Introduction to Schedule of Assignments

The Schedule of Assignments provides a model for assigning student homework. Although this list must be modified to fit specific courses, it is basically designed for three types of introductory algebra courses.

The Basic Course covers the standard topics taught in an introductory algebra course. Practice Exercises are assigned which are keyed to examples in the text. Summary Exercises test a student's understanding of topics covered in the Practice Exercises. Specific Review or Supplementary Exercises can be added to the Basic Course to improve students' understanding and to provide a greater challenge.

The Moderate Course adds two features to the Basic Course. Ten Review Exercises are assigned in addition to the Practice and Summary Exercises. These problems review concepts from previous chapters and highlight similarities and differences among related topics from different sections. In most cases, a few Supplementary Exercises also are assigned to improve students' understanding of the topics in the section.

The Advanced Course uses the Practice, Summary, and Review Exercises and many of the Supplementary Exercises to challenge the student and to build a firm understanding of algebra.

Schedule of Assignments

Abbreviations used in this schedule:
Prac—Practice Exercises
Rev—Ten Review Exercises
Supp—Supplementary Exercises
Summ—Summary Exercises

Day	Section	Basic Course	Moderate Course	Advanced Course
1	Intro	Read: Preface Introduction pp. 1–2	Read: Preface Introduction pp. 1–2	Read: Preface Introduction pp. 1–2
2	1-1	Prac 1–75 odd Summ 1–8 all	Prac 1–75 odd Supp 1–13 odd Summ 1–8 all	Prac 1–75 odd Supp 1–23 odd Summ 1–8 all
3	1-2	Prac 1–43 odd Summ 1–8 all	Prac 1–51 odd Supp 1–9 odd 31–35 odd Summ 1–8 all	Prac 1–51 odd Supp 1–35 odd Summ 1–8 all
4	1-3	Prac 1–67 odd Summ 1–8 all	Prac 1–67 odd Supp 1–9 odd Supp 33–39 odd Summ 1–8 all	Prac 1–67 odd Supp 1–39 odd Summ 1–8 all
5	1-4	Prac 1–43 odd Summ 1–8 all	Prac 1–43 odd Supp 1–23 odd Summ 1–8 all	Prac 1–43 odd Supp 1–39 odd Summ 1–8 all
6	1-5	Prac 1–43 odd Summ 1–15 all	Prac 1–43 odd Supp 1–19 odd Summ 1–15 all	Prac 1–43 odd Supp 1–39 odd Summ 1–15 all
7	1-6	Prac 1–57 odd Summ 1–8 all	Prac 1–57 odd Supp 1–27 odd Summ 1–8 all	Prac 1–57 odd Supp 1–37 odd Summ 1–8 all
8	1-7	Prac 1–37 odd Summ 1–7 all	Prac 1–45 odd Supp 1–9 odd Summ 1–8 all	Prac 1–45 odd Supp 1–19 odd Summ 1–8 all
9	1-8	Prac 1–33 odd Summ 1–8 all	Prac 1–33 odd Supp 1–13 odd Summ 1–8 all	Prac 1–33 odd Supp 1–21 odd Summ 1–8 all
10	Review	1–35 odd	1–35 odd (2–36 even optional)	1–36 all

Day	Section	Basic Course	Moderate Course	Advanced Course
11	[Test #1]	Read pp. 85–6	Read pp. 85–6	Read pp. 85–6
12	2-1	Prac 1–43 odd Summ 1–8 all	Prac 1–43 odd Rev 1–10 all Supp 1–11 odd Summ 1–8 all	Prac 1–43 odd Rev 1–10 all Supp 1–23 odd Summ 1–8 all
13	2-2	Prac 1–45 odd Summ 1–8 all	Prac 1–45 odd Rev 1–10 all Supp 1–27 odd Summ 1–8 all	Prac 1–45 odd Rev 1–10 all Supp 1–39 odd Summ 1–8 all
14	2-3	Prac 1–31 odd Summ 1–5 all	Prac 1–31 odd Rev 1–10 all Summ 1–5 all	Prac 1–31 odd Rev 1–10 all Supp 1–7 odd Summ 1–5 all
15	2-4	Prac 1–41 odd Summ 1–8 all	Prac 1–47 odd Rev 1–10 all Supp 1–21 odd Summ 1–8 all	Prac 1–47 odd Rev 1–10 all Supp 1–29 odd Summ 1–8 all
16	2-5	Prac 1–53 odd Summ 1–8 all	Prac 1–53 odd Rev 1–10 all Supp 1–9 odd Summ 1–8 all	Prac 1–53 odd Rev 1–10 all Supp 1–19 odd Summ 1–8 all
17	2-6	Prac 1–49 odd Summ 1–7 all	Prac 1–59 odd Rev 1–10 all Supp 1–17 odd Summ 1–8 all	Prac 1–59 odd Rev 1–10 all Supp 1–33 odd Summ 1–8 all
18	Review	1–39 odd	1–39 odd (2–40 even optional)	1–40 all
19	[Test #2]	Read p. 153	Read p. 153	Read p. 153
20	3-1	Prac 1–35 odd Supp 1 & 3 Summ 1–8 all	Prac 1–35 odd Rev 1–10 all Supp 1, 3, & 9 Summ 1–8 all	Prac 1–35 odd Rev 1–10 all Supp 1–9 odd Summ 1–8 all
21	3-2	Prac 1–49 odd Summ 1–8 all	Prac 1–49 odd Rev 1–10 all Supp 1–13 odd	Prac 1–49 odd Rev 1–10 all Supp 1–25 odd
22	3-3	Prac 1–59 odd Supp 1–7 odd Summ 1–7 all	Prac 1–65 odd Rev 1–10 all Supp 1–7 odd Summ 1–8 all	Prac 1–65 odd Rev 1–10 all Supp 1–15 odd Summ 1–8 all
23	3-4	Prac 1–41 odd Summ 1–8 all	Prac 1–41 odd Rev 1–10 all Supp 1–7 odd 17 & 19 Summ 1–10 all	Prac 1–41 odd Rev 1–10 all Supp 1–25 odd Summ 1–10 all

Day	Section	Basic Course	Moderate Course	Advanced Course
24	3-5	Prac 1–53 odd Summ 1–8 all	Prac 1–53 odd Rev 1–10 all Supp 1–15 odd Summ 1–8 all	Prac 1–53 odd Rev 1–10 all Supp 1–25 odd Summ 1–8 all
25	3-6	Prac 1–51 odd Summ 1–8 all	Prac 1–51 odd Rev 1–10 all Supp 1, 3, 11, 13 Summ 1–8 all	Prac 1–51 odd Rev 1–10 all Supp 1–13 odd Summ 1–8 all
26	Review	1–29 odd	1–29 odd (2–30 even optional)	1–30 all
27	[Test #3]	Read pp. 212–213	Read pp. 212–213	Read pp. 212–213
28	4-1	Prac 1–43 odd Summ 1–8 all	Prac 1–43 odd Rev 1–10 all Summ 1–8 all	Prac 1–43 odd Rev 1–10 all Supp 1–17 odd Summ 1–8 all
29	4-2	Prac 1–41 odd Summ 1–8 all	Prac 1–49 odd Rev 1–10 all Supp 1–11 odd Summ 1–8 all	Prac 1–49 odd Rev 1–10 all Supp 1–23 odd Summ 1–8 all
30	4-3	Prac 1–31 odd Summ 1–8 all	Prac 1–31 odd Rev 1–10 all Supp 1–9 odd Summ 1–8 all	Prac 1–31 odd Rev 1–10 all Supp 1–17 odd Summ 1–8 all
31	4-4	Prac 1–39 odd Summ 1–8 all	Prac 1–49 odd Rev 1–10 all Supp 1–9 odd 15 & 17 Summ 1–8 all	Prac 1–49 odd Rev 1–10 all Supp 1–19 odd Summ 1–8 all
32	4-5	Prac 1–49 odd Comm 1–8 all	Prac 1–49 odd Rev 1–10 all Supp 1–19 odd Summ 1–10 all	Prac 1–49 odd Rev 1–8 all Supp 1–39 odd Summ 1–10 all
33	Review 4-1 to 4-5	1–99 Every other odd	1–99 odd	1–99 odd
34	4-6	Prac 1–45 odd Summ 1–8 all	Prac 1–45 odd Rev 1–10 all Supp 1–15 odd Summ 1–8 all	Prac 1–45 odd Rev 1–10 all Supp 1–23 odd Summ 1–8 all
35	4-7	Prac 1–23 odd Summ 1–4 all	Prac 1–23 odd Rev 1–10 all Supp 1–5 odd Summ 1–4 all	Prac 1–23 odd Rev 1–10 all Supp 1–11 odd Summ 1–4 all
36	Review	1–29 odd	1–29 odd (2–30 even optional)	1–30 all

Day	Section	Basic Course	Moderate Course	Advanced Course
37	[Test #4]	Read pp. 272–3	Read pp. 272–3	Read pp. 272–3
38	5-1	Prac 1–59 odd Summ 1–8 all	Prac 1–59 odd Rev 1–10 all Supp 1–13 odd, 33 Summ 1–8 all	Prac 1–59 odd Rev 1–10 all Supp 1–33 odd Summ 1–8 all
39	5-2	Prac 1–63 odd Summ 1–7 all	Prac 1–89 odd Rev 1–10 all Supp 1–15 all Summ 1–8 all	Prac 1–89 odd Rev 1–10 all Supp 1–31 odd Summ 1–8 all
40	5-3	Prac 1–49 odd Summ 1–8 all	Prac 1–49 odd Rev 1–10 all Supp 1–17 odd Summ 1–8 all	Prac 1–49 odd Rev 1–10 all Supp 11–23 odd Summ 1–8 all
41	5-4	Prac 1–39 odd Summ 1–8 all	Prac 1–39 odd Rev 1–10 all Supp 1–11 odd Summ 1–8 all	Prac 1–39 odd Rev 1–10 all Supp 1–11 odd 19–25 odd Summ 1–8 all
42	5-5	Prac 1–51 odd Summ 1–3, 5 & 6	Prac 1–67 odd Rev 1–10 all Summ 1–8 all	Prac 1–67 odd Rev 1–10 all Supp 1–13 odd, 21 Summ 1–8 all
43	5-6	Prac 1–39 odd Summ 1–7 all	Prac 1–49 odd Rev 1–10 all Supp 1–11 odd Summ 1–8 all	Prac 1–49 odd Rev 1–10 all Supp 1–15 odd Summ 1–8 all
44	5-7	Prac 1–39 odd Summ 1–8 all	Prac 1–39 odd Rev 1–10 all Supp 1–9 odd Summ 1–8 all	Prac 1–39 odd Rev 1–10 all Supp 1–19 odd Summ 1–8 all
45	5-8	Prac 1–41 odd Summ 1–8 all	Prac 1–41 odd Rev 1–10 all Supp 1–9 odd Summ 1–8 all	Prac 1–41 odd Rev 1–10 all Supp 1–23 odd Summ 1–8 all
46	5-9	Prac 1–29 odd Summ 1–4 all	Prac 1–29 odd Rev 1–10 all Summ 1–4 all	Prac 1–29 odd Rev 1–10 all Supp 1–5 odd Summ 1–4 all
47	Review	1–47 odd (Skip 17 & 37)	1–47 odd (2–48 even optional)	1–48 all
48	[Test #5]	Read pp. 369–70	Read pp. 369–70	Read pp. 369–70
49	6-1	Prac 1–33 odd Summ 1–8 all	Prac 1–33 odd Rev 1–10 all Supp 1–19 odd Summ 1–8 all	Prac 1–33 odd Rev 1–10 all Supp 1–37 odd Summ 1–8 all

Day	Section	Basic Course	Moderate Course	Advanced Course
50	6-2	Prac 1–59 odd Summ 1–8 all	Prac 1–59 odd Rev 1–10 all Supp 1–25 odd Summ 1–8 all	Prac 1–59 odd Rev 1–10 all Supp 1–35 odd Summ 1–8 all

51	6-3	Prac 1–49 odd
		Summ 1–8 all
			Prac 1–49 odd
			Rev 1–10 all
			Supp 1–15 odd
			  29 & 31
			Summ 1–8 all
				Prac 1–49 odd
				Rev 1–10 all
				Supp 13–23 odd
				  29 & 31
				Summ 1–8 all

52	6-4	Prac 1–43 odd
		Summ 1–8 all
			Prac 1–43 odd
			Rev 1–10 all
			Supp 17–23 odd
			  33–39 odd
			Summ 1–8 all
				Prac 1–43 odd
				Rev 1–10 all
				Supp 1–23 odd
				  33–39 odd
				Summ 1–8 all

53	6-5	Prac 1–35 odd
		Summ 1–8 all
			Prac 1–35 odd
			Rev 1–10 all
			Supp 1–13 odd
			  23–27 odd
			Summ 1–8 all
				Prac 1–35 odd
				Rev 1–10 all
				Supp 1–27 odd
				  31–35 odd
				Summ 1–8 all

54	6-6	Prac 1–27 odd Summ 1–7 all	Prac 1–35 odd Rev 1–10 all Supp 5–13 odd 27–33 odd Summ 1–8 all	Prac 1–35 odd Rev 1–10 all Supp 1–31 odd Summ 1–8 all
55	Review	1–25 odd	1–27 odd (2–28 even optional)	1–28 all
56	[Test #6]	Read pp. 439–40	Read pp. 439–40	Read pp. 439–40
57	7-1	Prac 1–29 odd Summ 1–7 all	Prac 1–35 odd Rev 1–10 all Supp 1–11 odd Summ 1–8 all	Prac 1–35 odd Rev 1–10 all Supp 1–23 odd Summ 1–8 all
58	7-2	Prac 1–39 odd Summ 1–8 all	Prac 1–39 odd Rev 1–10 all Supp 5–11 odd Summ 1–8 all	Prac 1–39 odd Rev 1–10 all Supp 1–11 odd Summ 1–8 all
59	7-3	Prac 1–39 odd Summ 1–8 all	Prac 1–39 odd Rev 1–10 all Supp 13–15 odd Summ 1–8 all	Prac 1–39 odd Rev 1–10 all Supp 1–15 odd Summ 1–8 all
60	7-4	Prac 1–33 odd Summ 1–8 all	Prac 1–33 odd Rev 1–10 all Supp 1–19 odd Summ 1–8 all	Prac 1–33 odd Rev 1–10 all Supp 1–19 odd Summ 1–8 all
61	7-5	Prac 1–31 odd	Prac 1–39 odd	Prac 1–39 odd

Day	Section	Basic Course	Moderate Course	Advanced Course
61 (cont.)		Summ 1–4 all	Rev 1–10 all Summ 1–5 all	Rev 1–10 all Supp 1 & 3 Summ 1–5 all
62	Review	1–19 odd	1–21 odd (2–22 even optional)	1–22 even
63	[Test #7]	Read pp. 495–96	Read pp. 495–96	Read pp. 495–96
64	8-1	Prac 1–45 odd Summ 1–7 all	Prac 1–53 odd Rev 1–10 all Supp 1–23 odd Summ 1–8 all	Prac 1–53 odd Rev 1–10 all Supp 1–41 odd Summ 1–8 all
65	8-2	Prac 1–69 odd Summ 1–8 all	Prac 1–69 odd Rev 1–10 all Supp 1–15 odd Summ 1–8 all	Prac 1–69 odd Rev 1–10 all Supp 1–31 odd Summ 1–8 all
66	8-3	Prac 1–49 odd Summ 1–8 all	Prac 1–49 odd Rev 1–10 all Supp 1–17 odd Summ 1–8 all	Prac 1–49 odd Rev 1–10 all Supp 1–39 odd Summ 1–8 all
67	8-4	Prac 1–43 odd Summ 1–8 all	Prac 1–43 odd Rev 1–10 all Supp 1–17 odd Summ 1–8 all	Prac 1–43 odd Rev 1–10 all Supp 1–29 odd Summ 1–8 all
68	8-5	Prac 1–43 odd Summ 1–8 all	Prac 1–43 odd Rev 1–10 all Supp 15–21 odd Summ 1–8 all	Prac 1–43 odd Rev 1–10 all Supp 1–21 odd Summ 1–8 all
69	Review	1–31 odd	1–31 odd (2–32 even optional)	1–32 all
70	[Test #8]	Read p. 546	Read p. 546	Read p. 546
71	9-1	Prac 1–41 odd Summ 1–7 all	Prac 1–45 odd Rev 1–10 all Supp 1–9 odd Summ 1–8 all	Prac 1–45 odd Rev 1–10 all Supp 1–9 odd 15–21 odd Summ 1–8 all
72	9-2	Prac 1–49 odd Summ 1–8 all	Prac 1–49 odd Rev 1–10 all Supp 1–13 odd Summ 1–8 all	Prac 1–49 odd Rev 1–10 all Supp 1–19 odd Summ 1–8 all
73	9-3	Prac 1–25 odd Summ 1–6 all	Prac 1–33 odd Rev 1–10 all Supp 1–11 odd Summ 1–8 all	Prac 1–33 odd Rev 1–10 all Supp 1–11 odd 23–27 odd Summ 1–8 all

Day	Section	Basic Course	Moderate Course	Advanced Course
74	9-4	Prac 1–57 odd Summ 1–8 all	Prac 1–57 odd Rev 1–10 all Supp 13–19 odd Summ 1–8 all	Prac 1–57 odd Rev 1–10 all Supp 1–19 odd Summ 1–8 all
75	9-5 (optional)	Prac 1–49 odd Summ 1–8 all	Prac 1–49 odd Rev 1–10 all Supp 1–19 odd Summ 1–8 all	Prac 1–49 odd Rev 1–10 all Supp 1–39 odd Summ 1–8 all
76	9-6	Prac 1–23 odd Summ 1–6 all	Prac 1–29 odd Rev 1–10 all Supp 19–25 odd Summ 1–8 all	Prac 1–29 odd Rev 1–10 all Supp 1–25 odd Summ 1–8 all
77	Review	1–23 odd (29–39 odd optional)	1–27 odd (29–39 odd optional)	2–28 all (29–40 all optional)
78	[Test #9]	Read pp. 601–2	Read pp. 601–2	Read pp. 601–2
79	10-1	Prac 1–55 odd Summ 1–8 all	Prac 1–55 odd Rev 1–10 all Supp 15–25 odd Summ 1–8 all	Prac 1–55 odd Rev 1–10 all Supp 1–25 odd Summ 1–8 all
80	10-2	Prac 1–43 odd Summ 1–8 all	Prac 1–43 odd Rev 1–10 all Supp 1–11 odd Summ 1–8 all	Prac 1–43 odd Rev 1–10 all Supp 1–17 odd Summ 1–8 all
81	10-3	Prac 1–23 odd Summ 1–6 all	Prac 1–23 odd Rev 1–10 all Supp 1–7 odd Summ 1–6 all	Prac 1–23 odd Rev 1–10 all Supp 1–13 odd Summ 1–6 all
82	Review	1–23 odd	1–23 odd (2–24 even optional)	1–24 all
83	[Test #10]			

1

Essential Topics from Prealgebra

The following expressions were on the board as the members of Ms. Sharon Glaston's algebra class entered the room.

A collection of words	A collection of numbers and operations
John said Mary said hello	$3 + 6 \cdot 5 - 1$

"Good morning ladies and gentlemen," Ms. Glaston greeted the class. "Please look at the two collections written on the board. Let's consider first the five words on the left. Tell me, how would you read these words?"

Kate Derrick raised her hand and said, "I don't think you can say for certain how this should be read. Don't we need some commas, or other marks, to tell us how?"

"What you are suggesting, Kate," Ms. Glaston replied, "is that without any punctuation marks there is more than one way of reading these five words."

She then turned to the board and wrote the following:

Possibility 1.   John said, "Mary said hello."

Possibility 2.   "John," said Mary, "said hello."

Possibility 3.   John said, "Mary said hello?"

"In English," Ms. Glaston continued, "punctuation is used to help us better understand the meaning of a statement. As you can see, punctuation marks give us at least three different meanings for the five words that were written earlier. I'm sure that Kate would have no problem reading any of the three interpretations that I have suggested as Possibilities **1**, **2**, and **3**.

"Now let's turn our attention to the four numbers and three operations on the right. By carrying out the three indicated operations, the four numbers can be changed to one number. Now, can anyone tell us what the value of this number expression might be?"

Scott Cooney said, "Well, by doing the operations on my calculator I got 32."

Barbara Bulas raised her hand and said, "Ms. Glaston, I used a calculator also and got 44."

"Well, I used a paper and pencil," Eric Hauben added, "and I got 36. I guess that since Scott, Barbara, and I got three different answers for the same set of numbers and operations means the expression lacks 'punctuation marks,' just like the five words that Kate worked on."

"That's right, Eric," Ms. Glaston replied. "In mathematics, just as in English, we have punctuation marks that dictate how some numerical and algebraic expressions are to be manipulated. Furthermore, certain ordinary operations on numbers have priority over others in that they should be carried out first. We call these regulations the 'rules for order of operations.' The details of the punctuation marks and order of operations rules will be studied in this chapter."

"By the end of this chapter," Ms. Glaston continued, "we will be able to verify that the following expressions do yield the values of 44, 32, and 36 that Scott, Barbara, and Eric predicted."

She then wrote the following equations on the board:

$$(3 + 6) \cdot 5 - 1 = 44$$

$$3 + 6 \cdot 5 - 1 = 32$$

$$(3 + 6)(5 - 1) = 36$$

SECTION 1-1. Real Numbers and Some Related Topics

**KEY TOPICS
IN THIS SECTION**

1. Subsets of the real numbers

2. A number line for displaying numbers

3. Three relations for comparing numbers

4. The absolute value of a number

5. Basic operations on positive numbers

6. Multiplication and division have priority over addition and subtraction

In this first section we will review several elementary topics in mathematics. It is important that these topics be understood before starting our study in beginning algebra.

Subsets of the Real Numbers

In mathematics we find it advantageous to separate numbers into categories based on some special characteristics. For example, numbers can be separated into **positive** and **negative.** We can also separate positive numbers into **primes** and **composites.** The first set of numbers that we learn in our study of mathematics is called **natural,** or **counting numbers.**

Definition 1.1.   The natural numbers
   The **natural numbers** are

$$1, 2, 3, 4, 5, \ldots$$

The three dots after 5 indicate that the list does not end. We therefore say

"There is no greatest natural number"

or

"The number of natural numbers is infinite."

Example 1. Find the first ten multiples of 7.

Solution. **Discussion.** Multiples of a number are obtained by multiplying the number by the natural numbers. Since we want the first ten multiples of 7, we form the products of 7 with 1, 2, 3, . . . , 10. Thus, the first ten multiples are

$$1 \cdot 7, 2 \cdot 7, 3 \cdot 7, 4 \cdot 7, 5 \cdot 7, 6 \cdot 7, 7 \cdot 7, 8 \cdot 7, 9 \cdot 7, 10 \cdot 7,$$

or 7, 14, 21, 28, 35, 42, 49, 56, 63, 70.

Example 2. List the natural numbers that divide 48.

Solution. **Discussion.** By "divide," we mean the number divides 48 with a 0 remainder.

$48 \div 1 = 48$	and	$48 \div 48 = 1$	Thus 1 and 48 divide 48.
$48 \div 2 = 24$	and	$48 \div 24 = 2$	Thus 2 and 24 divide 48.
$48 \div 3 = 16$	and	$48 \div 16 = 3$	Thus 3 and 16 divide 48.
$48 \div 4 = 12$	and	$48 \div 12 = 4$	Thus 4 and 12 divide 48.
$48 \div 6 = 8$	and	$48 \div 8 = 6$	Thus 6 and 8 divide 48.

The natural numbers that divide 48 are

1, 2, 3, 4, 6, 8, 12, 16, 24, 48.

When 0 is included in the list of natural numbers, the set* of **whole numbers** is formed.

Definition 1.2. The whole numbers
The **whole numbers** are

$$0, 1, 2, 3, 4, 5, 6, \ldots$$

Numbers that are less than 0 are called **negative**. A minus bar $(-)$ is used to indicate a negative number. For example, -3 is read "negative three." The set consisting of the natural numbers, the negatives of the natural numbers, and 0 is called **integers**.

Definition 1.3. Integers
The **integers** are

$$\ldots, -5, -4, -3, -2, -1, 0, 1, 2, 3, 4, 5, \ldots$$

There is no smallest integer. That is, there are infinitely many positive and negative integers.

If a and b are integers and $b \neq 0$, then the number formed by the ratio of a to b is called a **rational number**.

* Appendix A includes some of the terminology from sets used in this text.

Definition 1.4.   Rational numbers
· The **rationals** contain any number that can be written as $\frac{a}{b}$, where a and b are integers, and $b \neq 0$.

Example 3.   Verify that each of the following numbers are rational.

   a. -2

   b. $3\frac{4}{5}$

   c.   4.73

Solution.   **Discussion.** To show that a number is rational, we write the number as a ratio of two integers. There are an infinite number of ways of accomplishing this task. Only one such ratio is needed for each number.

   a. $-2 = \dfrac{-2}{1}$ -2 and 1 are integers; thus $\frac{-2}{1}$ is rational.

   b. $3\dfrac{4}{5} = \dfrac{19}{5}$ 19 and 5 are integers; thus $\frac{19}{5}$ is rational.

   c. $4.73 = \dfrac{473}{100}$ 473 and 100 are integers; thus $\frac{473}{100}$ is rational.

When any rational number is changed to a decimal, one of two expressions will be obtained:

a. A terminating decimal   For example, 0.25 and 3.926

b. A repeating decimal   For example, 0.333 ... (repeats 3s) and 5.272727 ... (repeats 27s)

Not all decimals terminate or repeat the same block of digits. For example,

$$1.21221222122221 \ldots \quad \text{and} \quad 6.094823361475 \ldots$$

Numbers such as these cannot be written as a ratio of two integers. Therefore they are not rational numbers, and they are called **irrational.**

Definition 1.5.   Irrational numbers
   The **irrational numbers** are those that cannot be written in the form $\frac{a}{b}$, where a and b are integers and $b \neq 0$.

In a later chapter we will study numbers that are **square roots** of other numbers, such as $\sqrt{2}$. Many of these numbers, as we will learn, are irrational.

When we combine the sets of rational and irrational numbers into one set, we call them collectively **real numbers.** These are the numbers we use in our study of algebra.

Definition 1.6.   The real numbers
   The **real numbers** are numbers that are either rational or irrational.

A summary of the numbers discussed thus far is shown in Figure 1-1.

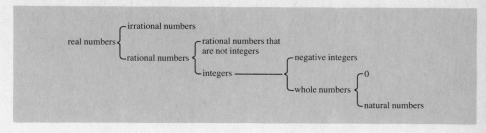

Figure 1-1. Numbers that comprise the set of real numbers.

A Number Line for Displaying Numbers

Figure 1-2. A geometric line.

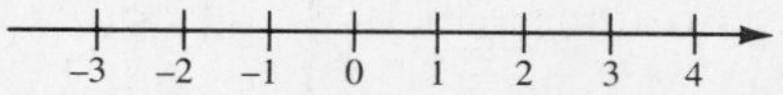

Figure 1-3. A number line.

A geometric line consists of points. The line extends infinitely far in both directions, as shown in Figure 1-2.

We can use a line to represent the real numbers. To accomplish this task we equally space several integers on the line, as shown in Figure 1-3. The point of the line paired with a number is called the **graph of the number.** The real number paired with a point is called the **coordinate of the point.** The arrow on the right end of the line shows the direction of increasing numbers.

Three Relations for Comparing Numbers

If a and b are any two real numbers, then exactly one of the following three **relations** must exist for a and b:

a. a is *less than* b, which is written $a < b$.   For example, $3 < 10$.

b. a is *equal to* b, which is written $a = b$.   For example, $\frac{28}{7} = 4$.

c. a is *greater than* b, which is written $a > b$.   For example, $9 > 5$.

When a and b are graphed on the same number line, the direction of the graph of a from the graph of b shows the relationship between the numbers a and b.

a. $a < b$ The graph of a is to the left of the graph of b.

b. $a = b$ The graph of a is the same as the graph of b.

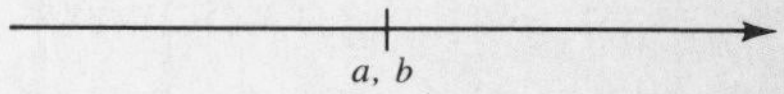

c. $a > b$ The graph of a is to the right of the graph of b.

Example 4.   Insert $<$, $=$, or $>$ between each pair of numbers to make a true statement.

a. $-3 \qquad -8$

b. $\dfrac{36}{9} \qquad \dfrac{44}{11}$

Solution.   **a.** The graph of -3 is to the right of the graph of -8; thus, $-3 > -8$.

Notice that -3 is "less negative"; therefore it is greater.

b. Since $\frac{36}{9}$ and $\frac{44}{11}$ are both equal to 4, the numbers are equal to each other, and $\frac{36}{9} = \frac{44}{11}$.

The Absolute Value of a Number

If b is any real number other than 0, then b can be thought of consisting of two parts:

1. the **magnitude** of the number

2. the **sign** of the number, either positive or negative.

Consider, for example, 10 and -10. Both numbers have a magnitude of 10, but 10 is positive while -10 is negative. Suppose these numbers are used to show a change in temperature from a current thermometer reading. The 10 would show an **increase of ten degrees,** but the -10 would show a **decrease of ten degrees.** The amounts of change in both cases would be the same, but the directions of change would be opposites. The magnitude of a number is called the **absolute value** of the number.

Definition 1.7.   The absolute value of a number

If b is a real number, and $b \neq 0$, then the **absolute value of b,** written $|b|$, is the magnitude of the number without regard to the sign.

The absolute value of a positive number is positive.

$$|8| = 8 \qquad |0.27| = 0.27 \qquad \left|\frac{3}{4}\right| = \frac{3}{4}$$

The absolute value of a negative number is positive.

$$|-8| = 8 \qquad |-0.27| = 0.27 \qquad \left|\frac{-3}{4}\right| = \frac{3}{4}$$

The absolute value of 0 is 0.

$$|0| = 0$$

In geometric terms, the absolute value of a number is the distance the graph of the number is from 0, and distances are always measured with positive numbers or 0.

Example 5.   Write parts **a**, **b**, and **c** without absolute value bars.

    a. $|3.5|$

    b. $|-2|$

    c. $-|-5|$

Solution.   **Discussion.** To illustrate the geometric interpretation of absolute value, the three numbers are graphed in Figure 1-4.

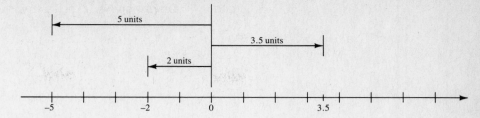

Figure 1-4. The graphs of -5, -2, and 3.5.

a. $|3.5| = 3.5$ The graph is 3.5 units from 0.

b. $|-2| = 2$ The graph is 2 units from 0.

c. $-|-5| = -5$ $|-5| = 5$ and $-|-5| = -5$

If a minus bar is written in front of a letter or symbol that represents a number, then the bar should be read "the opposite."

If b is a real number, then $-b$ is read "the opposite of b."

In general	**Examples**
(i) b and $-b$ have opposite signs.	
If b is positive, then $-b$ is negative.	10 and -10 are opposites.
If b is negative, then $-b$ is positive.	-39 and 39 are opposites.
If $b = 0$, then $-b = 0$.	0 is its own opposite.
(ii) b and $-b$ have the same absolute value.	$|10| = |-10|$
	$|-39| = |39|$
	$|0| = |-0|$

Notice in part **c** of Example 5 that $-|-5|$ can be read

"the opposite of the absolute value of -5."

Thus, we first find the absolute value of (-5), and we then take the opposite of that number.

Basic Operations on Positive Numbers

The following examples review adding, subtracting, multiplying, and dividing positive numbers. For a more thorough review of operations on common fractions see Appendix B.

Example 6. Do the indicated operations.

a. $109 - 83 + 17 - 35$

b. $\dfrac{5}{3} + \dfrac{1}{6} - \dfrac{3}{4} - \dfrac{5}{12}$

c. $0.137 + 2.4 - 0.09 - 1.025$

Solution.   **Discussion.** The expressions are simplified by adding or subtracting in order from left to right.

a. $109 - 83 + 17 - 35$

$= 26 + 17 - 35$ $\qquad$ $109 - 83 = 26$

$= 43 - 35$ $\qquad$ $26 + 17 = 43$

$= 8$

b. $\dfrac{5}{3} + \dfrac{1}{6} - \dfrac{3}{4} - \dfrac{5}{12}$

$= \dfrac{20}{12} + \dfrac{2}{12} - \dfrac{9}{12} - \dfrac{5}{12}$ $\qquad$ The least common denominator is 12.

$= \dfrac{22}{12} - \dfrac{9}{12} - \dfrac{5}{12}$ $\qquad$ $\dfrac{20}{12} + \dfrac{2}{12} = \dfrac{20 + 2}{12}$

$= \dfrac{13}{12} - \dfrac{5}{12}$ $\qquad$ $\dfrac{22}{12} - \dfrac{9}{12} = \dfrac{13}{12}$

$= \dfrac{8}{12} = \dfrac{2}{3}$ $\qquad$ $\dfrac{8}{12} = \dfrac{2 \cdot 4}{3 \cdot 4} = \dfrac{2}{3}$

c. $0.137 + 2.4 - 0.09 - 1.025$ $\qquad$ $0.137 + 2.4 = 2.537$

$= 2.537 - 0.09 - 1.025$ $\qquad$ $2.537 - 0.09 = 2.447$

$= 2.447 - 1.025$

$= 1.422$

Example 7.   Do the indicated operations.

a. $\dfrac{7}{5} \cdot \dfrac{25}{26} \cdot \dfrac{13}{14}$

b. $(15.2)(0.45)$

c. $\left(5\dfrac{1}{3}\right) \div \dfrac{10}{9}$

d. $0.735 \div 2.1$

Solution.   **a.** $\dfrac{7}{5} \cdot \dfrac{25}{26} \cdot \dfrac{13}{14} = \dfrac{7 \cdot 25 \cdot 13}{5 \cdot 26 \cdot 14}$

$= \dfrac{1 \cdot 5 \cdot 1}{1 \cdot 2 \cdot 2}$ $\begin{array}{l}\text{Divide 7 and 14 by 7.} \\ \text{Divide 25 and 5 by 5.} \\ \text{Divide 13 and 26 by 13.}\end{array}$

$= \dfrac{5}{4}$

b. $(15.2)(0.45) = 6.840$ $\qquad$ The product has three decimal places.

c. $\left(5\dfrac{1}{3}\right) \div \dfrac{10}{9} = \dfrac{16}{3} \div \dfrac{10}{9}$ Write $5\frac{1}{3}$ as $\frac{16}{3}$.

$\qquad\qquad = \dfrac{16}{3} \cdot \dfrac{9}{10}$ Multiply by the reciprocal of $\frac{10}{9}$.

$\qquad\qquad = \dfrac{8 \cdot 3}{1 \cdot 5}$ Divide 16 and 10 by 2.
Divide 9 and 3 by 3.

$\qquad\qquad = \dfrac{24}{5}$

d. $0.735 \div 2.1 = \dfrac{0.735}{2.1} \cdot \dfrac{10}{10}$

$\qquad\qquad = \dfrac{7.35}{21}$

$\qquad\qquad = 0.35$

Multiplication and Division have Priority over Addition and Subtraction

The following expression indicates one addition, one subtraction, one multiplication, and one division:

$$12 + 3 \cdot 5 - 36 \div 9$$

To simplify this expression to one number, all four operations need to be carried out. Based on the rule for order of operations the multiplication and division are carried out before the addition and subtraction. The rule for order of operations will be expanded in future sections as more operations are encountered, such as raising to powers and extracting roots.

The rule for order of operations
When addition, subtraction, multiplication and division are indicated in the same expression,

Step 1. Do the multiplications and divisions in order from left to right.

Step 2. Do the additions and subtractions in order from left to right.

Based on this rule,

$\qquad 12 + 3 \cdot 5 - 36 \div 9$

$\qquad = 12 + 15 - 4$ $\qquad 3 \cdot 5 = 15$ and $36 \div 9 = 4$

$\qquad = 27 - 4$ $\qquad 12 + 15 = 27$

$\qquad = 23$ $\qquad$ The value of the expression is 23.

Example 8.   Simplify:

$$15 - 48 \div 6 + 5 \cdot 3 \cdot 2 - 17$$

Solution. $15 - 48 \div 6 + 5 \cdot 3 \cdot 2 - 17$

$\qquad = 15 - 8 + 5 \cdot 3 \cdot 2 - 17$ Do the division first.

$\qquad = 15 - 8 + 15 \cdot 2 - 17$ Now multiply 5 by 3.

$\qquad = 15 - 8 + 30 - 17$ Do the last multiplication.

$\qquad = 7 + 30 - 17$ Replace $15 - 8$ by 7.

$\qquad = 37 - 17$ Replace $7 + 30$ by 37.

$\qquad = 20$ The value of the expression is 20.

SECTION 1-1.   Practice Exercises

In exercises **1–8**, list the first seven multiples of the given number.

[Example 1]   **1.** 2   **2.** 4

3. 3   **4.** 5

5. 9   **6.** 8

7. 15   **8.** 20

In exercises **9–16**, list the natural numbers that divide the given number.

[Example 2]   **9.** 20   **10.** 24

11. 30   **12.** 32

13. 66   **14.** 55

15. 42   **16.** 70

In exercises **17–26**, verify that each number is rational by writing it as a/b, where a and b are integers and $b \neq 0$.

[Example 3]   **17.** 12   **18.** 19

19. -3 **20.** -6

21. $5\dfrac{1}{3}$ **22.** $6\dfrac{2}{3}$

23. 0   **24.** 1

25. 7.13   **26.** 9.02

In exercises **27–38**, insert $<$, $=$, or $>$ between each pair of numbers to make a true statement.

[Example 4]   **27.** 7 -3 **28.** 12 -1

29. -21 $\quad$ -25 $\qquad$ **30.** -6 $\quad$ -2

31. 0 $\qquad$ -6 $\qquad$ **32.** 0 $\qquad$ -10

33. $\dfrac{45}{9}$ $\quad$ $\dfrac{56}{7}$ $\qquad$ **34.** $\dfrac{36}{4}$ $\quad$ $\dfrac{99}{9}$

35. $\dfrac{49}{7}$ $\quad$ $\dfrac{35}{5}$ $\qquad$ **36.** $\dfrac{121}{11}$ $\quad$ $\dfrac{55}{5}$

37. -5.3 $\quad$ -5.8 $\qquad$ **38.** -2.1 $\quad$ -2.5

In exercises **39–52**, write each number without absolute value bars.

[Example 5] **39.** $|-17|$ $\qquad$ **40.** $|-42|$

41. $|-3.2|$ $\qquad$ **42.** $|-7.8|$

43. $|0|$ $\qquad$ **44.** $-|0|$

45. $-|3|$ $\qquad$ **46.** $-|7|$

47. $-|-3|$ $\qquad$ **48.** $-|-7|$

49. $|-(-3)|$ $\qquad$ **50.** $|-(-7)|$

51. $-|-(-3)|$ $\qquad$ **52.** $-|-(-7)|$

In exercises **53–68**, perform the indicated operations.

[Example 6] **53.** $47 - 32 + 5 - 9$ $\qquad$ **54.** $81 - 54 + 18 - 21$

55. $37 - 21 + 6 - 22$ $\qquad$ **56.** $29 - 12 + 11 - 28$

57. $\dfrac{7}{8} - \dfrac{1}{4} - \dfrac{1}{2} + \dfrac{1}{8}$ $\qquad$ **58.** $\dfrac{3}{10} + \dfrac{4}{10} - \dfrac{1}{5} + \dfrac{1}{20}$

59. $3.5 - 1.23 + 5.04 - 0.92$ $\qquad$ **60.** $5.9 - 4.21 + 1.05 - 1.2$

[Example 7] **61.** $\dfrac{7}{3} \cdot \dfrac{27}{14} \cdot \dfrac{2}{5}$ $\qquad$ **62.** $\dfrac{2}{5} \cdot \dfrac{35}{10} \cdot \dfrac{3}{7}$

63. $(12.3)(7.02)$ $\qquad$ **64.** $(41.3)(9.07)$

65. $\left(7\dfrac{2}{5}\right) \div \dfrac{4}{5}$ $\qquad$ **66.** $\left(6\dfrac{3}{8}\right) \div \dfrac{3}{4}$

67. $7.434 \div 3.54$ $\qquad$ **68.** $7.623 \div 2.31$

In exercises **69–76**, simplify using the order of operations.

[Example 8] **69.** $25 + 10 - 4 \cdot 6$ $\qquad$ **70.** $45 - 13 - 16 \div 8$

71. $40 - 5 \cdot 7 + 10 \div 5$ $\qquad$ **72.** $9 \cdot 7 + 11 - 4 \cdot 2$

73. $36 \div 12 \cdot 3 + 4(2)$ $\qquad$ **74.** $8 \cdot 12 \div 4 - 7(3)$

75. $48 \cdot 2 \div 6 - 3(5) + 13$ $\qquad$ **76.** $31 - 3(5) + 72 \div 8 - 4$

SECTION 1-1.   Supplementary Exercises

In exercises **1–4**, list all of the numbers from the given sets (if any) that are

a. natural numbers

b. whole numbers

c. integers

d. rational numbers

e. irrational numbers

(Note: Some numbers may be used more than once.)

1. $\left\{-3, \dfrac{1}{2}, 0, 7\right\}$ **2.** $\left\{6, -\dfrac{3}{4}, -11, 0\right\}$

3. $\left\{1.4, -3, \sqrt{2}, \dfrac{17}{2}\right\}$ **4.** $\left\{\sqrt{5}, \dfrac{4}{3}, -1.05, 101\right\}$

In exercises **5–8**, simplify.

5. a. $|-7| - 7$ **b.** $7 - |-(-7)|$

6. a. $|-12| + 12$ **b.** $|-12| - |12|$

7. a. $|-15| - |15|$ **b.** $|-15| + |-(-15)|$

8. a. $|-(-9)| + 9$ **b.** $9 - |-9|$

In exercises **9–14**, simplify.

9. $12 \div 4 + 15 \cdot 3 - 2$ **10.** $3 \cdot 6 - 12 \div 4 + 1$

11. $13 \cdot 3 - 2(7)$ **12.** $20 \cdot 5 - 4(6)$

13. $\dfrac{12 - 2 \cdot 3 + 1}{2 \cdot 5 - 3}$ **14.** $\dfrac{18 - 2 \cdot 5 + 1}{6 + 1(3)}$

In exercises **15–22**, insert $<$ or $>$, if possible, between each pair of statements to make the entire statement true. State "cannot be determined" if that is the case.

15. Any negative number _____ zero.

16. Any positive number _____ zero.

17. Any negative number _____ any positive number.

18. Any positive number _____ any negative number.

19. Any negative improper fraction _____ any negative proper fraction.

20. Any positive improper fraction _____ any positive proper fraction.

21. Any fraction _____ any decimal.

22. Any fraction _____ any mixed number.

23. a. List the first eight multiples of 12.
  b. List the first eight multiples of 18.
  c. What multiples of 12 in part **a** are also multiples of 18 in part **b**? (Such numbers are called common multiples.)

24. a. List the first eight multiples of 21.
  b. List the first eight multiples of 28.
  c. What multiples of 21 in part **a** are also multiples of 28 in part **b**?

SECTION 1-2. Addition and Subtraction of Real Numbers

KEY TOPICS IN THIS SECTION

1. Adding integers with the same sign

2. Adding integers with different signs

3. Subtracting integers

4. Applied problems

The set of integers contains positive and negative numbers and 0. The following is a list of these numbers:

$$\ldots, -5, -4, -3, -2, -1, 0, 1, 2, 3, 4, 5, \ldots$$

In this section we will review addition and subtraction of integers. Included will be some exercises in which other rational numbers are added and subtracted.

Adding Integers with the Same Sign

To review the way in which positive and negative numbers are added, we will think in terms of money. For this exercise the following interpretations will be used:

A positive number will represent **money earned.**

A negative number will represent **money spent.**

A positive sum will represent a result in which **money remains.**

A negative sum will represent a result in which **money is owed.** Using this interpretation,

a. $15 + 12 = 27$

"$15 earned plus $12 earned means $27 remain."

b. $23 + (-9) = 14$

"$23 earned plus $9 spent means $14 remain."

c. $-30 + 22 = -8$

"$30 spent plus $22 earned means $8 owed."

d. $-13 + (-16) = -29$

"$13 spent plus $16 spent means $29 owed."

From examples **a** and **d** we see that the procedures for adding integers that are both positive or both negative is essentially the same. That is, add the absolute values of the numbers, and then give the sum the common sign. Using this interpretation the sums would be computed as follows:

Two Positive or Two Negative Numbers	Add their Absolute Values	Compute the Sums	Give to the Sum the Common Sign
a. $15 + 12$	$= \quad \lvert 15 \rvert + \lvert 12 \rvert \quad =$	27	27 (sum is positive)
d. $-13 + (-16)$	$= \lvert -13 \rvert + \lvert -16 \rvert =$	29	-29 (sum is negative)

If a and b are numbers with the same sign,

Step 1. add the absolute values of a and b.

Step 2. give to the sum the common sign:
   (i) positive if a and b are both positive
   (ii) negative if a and b are both negative

Example 1.   Simplify:

  a. $12 + 18 + 3 + 56$

  b. $-13 + (-9) + (-38) + (-7) + (-29)$

Solution

Discussion. These numerical expressions are simplified by adding the integers two at a time from left to right.

  a. $12 + 18 + 3 + 56$

$= 30 + 3 + 56$	$12 + 18 = 30$
$= 33 + 56$	$30 + 3 = 33$
$= 89$	The value of the expression is 89.

  b. $-13 + (-9) + (-38) + (-7) + (-29)$

$= -22 + (-38) + (-7) + (-29)$	$-13 + (-9) = -22$
$= -60 + (-7) + (-29)$	$-22 + (-38) = -60$
$= -67 + (-29)$	$-60 + (-7) = -67$
$= -96$	The value of the expression is -96.

Adding Integers with Different Signs

Suppose a thermometer reads $23°$ (that is, $+23$). If the temperature goes down $9°$ (that is, -9), then the thermometer would read $14°$.

$$23 + (-9) = 14 \qquad \text{Notice: } \lvert 23 \rvert - \lvert -9 \rvert = \lvert 14 \rvert.$$

Now suppose a thermometer reads $30°$ below zero (that is, -30). If the temperature goes up $22°$ (that is, $+22$), then the thermometer would read $8°$ below zero, or $-8°$.

$$-30 + 22 = -8 \qquad \text{Notice: } \lvert -30 \rvert - \lvert 22 \rvert = \lvert -8 \rvert.$$

These examples, together with examples **b** and **c** on page 13 suggest the following procedure for adding numbers with different signs:

One positive and one negative number	Subtract the smaller absolute value from the larger one	Compute the difference	Give to the difference the sign of the number with the greater absolute value
b. $23 + (-9) =$	$\lvert 23 \rvert - \lvert -9 \rvert =$	14	14 (23 has the greater absolute value.)
c. $-30 + 22 =$	$\lvert -30 \rvert - \lvert 22 \rvert =$	8	-8 (-30 has the greater absolute value.)

> **If a and b are numbers with different signs,**
>
> **Step 1.** take the absolute values of a and b.
>
> **Step 2.** find the difference between the absolute values obtained in Step 1 by subtracting the smaller number from the larger one.
>
> **Step 3.** give to the difference the sign of a or b, whichever number has the greater absolute value.

Example 2.   Simplify:

      a. $-12 + 7 + (-11) + 24 + 9$

      b. $8 + (-16 + 22) + (23 + (-41))$

Solution.   **Discussion.** In example a, we simply add the integers two at a time from left to right. In example b, we must first add the pairs of integers inside the parentheses.

    a. $\quad -12 + 7 + (-11) + 24 + 9$

$\qquad = -5 + (-11) + 24 + 9 \qquad\qquad -12 + 7 = -5$

$\qquad = -16 + 24 + 9 \qquad\qquad\qquad -5 + (-11) = -16$

$\qquad = 8 + 9 \qquad\qquad\qquad\qquad\quad -16 + 24 = 8$

$\qquad = 17 \qquad\qquad\qquad\qquad\qquad$ The value of the expression is 17.

    b. $\quad 8 + (-16 + 22) + (23 + (-41))$

$\qquad = 8 + 6 + (-18) \qquad\qquad\qquad -16 + 22 = 6$ and $23 + (-41) = -18$

$\qquad = 14 + (-18) \qquad\qquad\qquad\quad 8 + 6 = 14$

$\qquad = -4 \qquad\qquad\qquad\qquad\qquad$ The value of the expression is -4.

Subtracting Integers

When subtracting b from a, we look for a number c to add to b to get a. That is,

$$a - b = c \qquad \text{if and only if} \qquad a = b + c$$

This fact is seen when we first learn to subtract. We do not study subtraction tables, but we use the addition facts we have already learned. Examples **a–d** illustrate this idea for four pairs of positive and negative numbers.

	$a - b = c$	because	$a = b + c$
a.	$5 - 13 = -8$	because	$5 = 13 + (-8)$
b.	$-9 - 16 = -25$	because	$-9 = 16 + (-25)$
c.	$4 - (-7) = 11$	because	$4 = -7 + 11$
d.	$-16 - (-23) = 7$	because	$-16 = -23 + 7$

Consider again examples **a–d** and four additional examples **a*–d***.

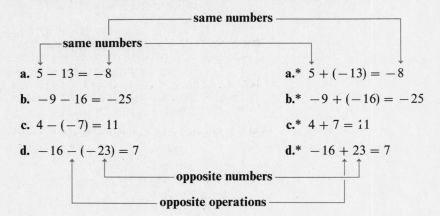

same numbers

same numbers

a. $5 - 13 = -8$ **a.*** $5 + (-13) = -8$

b. $-9 - 16 = -25$ **b.*** $-9 + (-16) = -25$

c. $4 - (-7) = 11$ **c.*** $4 + 7 = 11$

d. $-16 - (-23) = 7$ **d.*** $-16 + 23 = 7$

opposite numbers

opposite operations

In the two sets of examples, the numbers in the first columns are the same $(5, -9, 4, \text{ and } -16)$, and the answer columns are the same $(-8, -25, 11, \text{ and } 7)$. However, the operations in **a–d** are **subtractions,** and the operations in **a*** through **d*** are **additions.** Furthermore, the numbers in the middle columns are **opposites** $(13 \text{ and } -13, 16 \text{ and } -16, -7 \text{ and } 7, \text{ and } -23 \text{ and } 23)$. These comparisons suggest the following three-step procedure for subtracting integers:

To simplify $a - b$,

Step 1. change the subtraction to addition.

Step 2. change b to its opposite.

Step 3. combine $a + (-b)$ using the methods for adding positive and negative numbers.

Example 3. Simplify:

  a. $-42 - (-29)$

  b. $-17 - 22$

  c. $-10 - (-17) + 23 - 41$

Solution. **a.** $-42 - (-29)$ The given subtraction

  $= -42 + 29$ The opposite of -29 is 29.

  $= -13$ The difference is -13.

b. $\quad -17 - 22$ $\qquad$ The given subtraction

$\quad = -17 + (-22)$ $\qquad$ The opposite of 22 is -22.

$\quad = -39$ $\qquad$ Adding two negatives

c. $\quad -10 - (-17) + 23 - 41$

$\quad = 7 + 23 - 41$ $\qquad$ $-10 - (-17) = -10 + 17 = 7$

$\quad = 30 - 41$ $\qquad$ $7 + 23 = 30$

$\quad = -11$ $\qquad$ The value of the expression is -11.

Applied Problems

Many problems in a real-world setting can be solved by adding or subtracting positive and negative numbers. For example, Figure 1-5 is a diagram of a possible trek by a hiker in a mountainous region. As indicated, the hiker is at 3,500 feet when at A. The changes in elevation between the points A through E are shown in the figure.

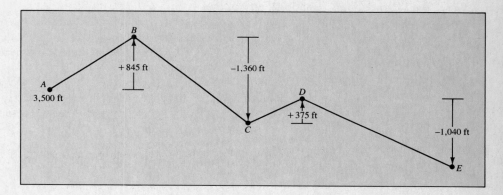

Figure 1-5. A trek of a hiker from points A through E.

To find the elevation at E requires doing the following operations:

$$3,500 + 845 + (-1,360) + 375 + (-1,040) = 2,320$$

Therefore the hiker is at an elevation of 2,320 feet when at point E.

Example 4. The closing price of one share of Megatech Electronics Incorporated stock on the first of the month was compared to the price on the first of the previous month.

January 1—$\$36\frac{3}{8}$

February 1—Up $\$4\frac{1}{4}$

March 1—Down $\$7\frac{7}{8}$

April 1—Up $\$10\frac{1}{2}$

May 1—Down $\$2\frac{1}{8}$

Find the cost of one share of this stock on May 1.

Solution. **Discussion.** Stock prices are quoted in eighths of a dollar. Therefore, any fluctuations in stock prices are given in these units.

$$\text{February 1 price} = \$36\tfrac{3}{8} + \$4\tfrac{1}{4} = \$40\tfrac{5}{8}$$

$$\text{March 1 price} = \$40\tfrac{5}{8} - \$7\tfrac{7}{8} = \$32\tfrac{6}{8} \text{ or } \$32\tfrac{3}{4}$$

$$\text{April 1 price} = \$32\tfrac{3}{4} + \$10\tfrac{1}{2} = \$43\tfrac{1}{4}$$

$$\text{May 1 price} = \$43\tfrac{1}{4} - \$2\tfrac{1}{8} = \$41\tfrac{1}{8}$$

Therefore, on May 1 the price of one share of Megatech Electronics Inc. stock was $\$41\tfrac{1}{8}$.

Calculator Assistance Supplement A

Sometimes several numbers in sequence are being operated on using a calculator. If, near the end of the list, a wrong number is keyed into the calculator, it is not necessary to start over at the beginning of the list. Most calculators have a key that will erase the number shown on the display, but not the current totals and operations pending.

The CLEAR key (labeled C) will erase the current number, but not the previous numbers or calculations. Thus, if after several calculations you enter a wrong number, press the CLEAR key, enter the correct number, and continue as though no error had been made.

The CLEAR ALL key (or ALL CLEAR, labeled AC) will erase all numbers and operations, and resets the calculator back to zero.

If you press an operation key by accident, then what should you do? For example, suppose you press the subtraction key ($-$) when you needed the addition key ($+$). Simply press the correct operation key and the calculator will automatically erase the incorrect operation and replace it with the correct one.

Calculator Assistance Supplement B

To enter a negative number into a scientific calculator, use the key marked $\boxed{+/-}$. First enter into the calculator the digits of the positive number that corresponds to the desired negative number. Now press the $\boxed{+/-}$ key. A negative sign ($-$) should appear in front of the number, making it negative. (Note: If the $\boxed{+/-}$ key is pressed a second time, it will make the number positive again.)

SECTION 1-2. Practice Exercises*

In exercises **1–40**, simplify.

[Example 1] **1.** $-3 + (-9)$ **2.** $-8 + (-5)$

* Additional problems are available in the Instructor's Manual.

    3. $-4 + (-2) + (-7)$ **4.** $-9 + (-4) + (-3)$

    5. $25 + 12 + 8$ **6.** $14 + 7 + 35$

    7. $-81 + (-3) + (-14) + (-15)$ **8.** $-23 + (-11) + (-18) + (-7)$

[Example 2]   **9.** $-17 + 22$ **10.** $32 + (-49)$

    11. $-5 + 29 + (-10)$ **12.** $-6 + 68 + (-12)$

    13. $13 + 21 + (-41)$ **14.** $17 + 29 + (-57)$

    15. $9 + (-18 + (-5)) + (-41)$ **16.** $4 + (-6 + (-13)) + (-16)$

    17. $(26 + (-56)) + (-63 + 34)$ **18.** $(48 + (-18)) + (-34 + 47)$

    19. $31 + (-67 + 67) + (-78) + 56$ **20.** $23 + (-27 + 27) + 39 + (-15)$

[Example 3]   **21.** $35 - (-21)$ **22.** $30 - (-17)$

    23. $-85 - (-34)$ **24.** $-74 - (-29)$

    25. $25 - 41$ **26.** $40 - 52$

    27. $-45 - 72$ **28.** $-103 - 84$

    29. $-12 - (-19) - 11$ **30.** $15 - 23 - (-12)$

    31. $37 - 54 - (-9)$ **32.** $-73 - 17 - (-97)$

    33. $-11 - 32 + 41 - (-17)$ **34.** $-13 - 21 + 37 - (-24)$

    35. $88 - (-31 + 12) + (31 - 24)$ **36.** $79 - (-46 + 28) + (49 - 37)$

    37. $(-7 - (-9)) + (10 - (-6)) - (-3 - 8)$

    38. $(9 - (-13)) - (-6 - (-3)) + (15 - 34)$

    39. $(17 - (42 - 51)) + (21 - (-3))$ **40.** $(25 - (38 - 49)) + (27 - (-9))$

    Given the starting elevations and trail elevation changes below, determine the final elevation.

[Example 4]   **41.** Starting elevation 1,053 feet, increase of 123 feet, decrease of 98 feet, increase of 63 feet.

    42. Starting elevation 1,231 feet, increase of 213 feet, decrease of 112 feet, increase of 78 feet.

    43. Starting elevation 7,302 feet, decrease of 623 feet, decrease of 123 feet, increase of 307 feet.

    44. Starting elevation 6,907 feet, decrease of 511 feet, decrease of 203 feet, increase of 419 feet.

    Given the starting stock prices and weekly changes, determine the final price for each stock below.

    45. Start $35\frac{1}{2}$, up $1\frac{1}{2}$, up $\frac{1}{4}$, down $\frac{3}{4}$, down $\frac{1}{2}$

    46. Start $42\frac{1}{4}$, up $\frac{1}{4}$, up $1\frac{1}{4}$, down $1\frac{3}{4}$, down $\frac{1}{2}$

    47. Start $49\frac{3}{8}$, up $\frac{1}{8}$, down $\frac{7}{8}$, up $\frac{1}{2}$, down $1\frac{1}{8}$

48. Start $53\frac{7}{8}$, down $\frac{1}{4}$, up $\frac{3}{8}$, down $1\frac{1}{8}$, up $\frac{5}{8}$

For the following starting football field positions and the results of successive plays, determine the final location of the football. Assume the ball is being advanced toward the center of the field (that is, the fifty yard line).

49. Starting location 19-yard line. A 3-yard gain, then a 5-yard loss, then a 6-yard gain.

50. Starting location 24-yard line. A 7-yard gain, then a 3-yard loss, then a 2-yard loss.

51. Starting location 7-yard line. A 9-yard gain, then a 2-yard loss, then a 5-yard gain, then a 10-yard gain, then a 6-yard loss.

52. Starting location 27-yard line. A 4-yard gain, then a 1-yard loss, then an 8-yard gain, then a 10-yard gain, then a 2-yard loss.

SECTION 1-2.   Supplementary Exercises

In exercises **1–30**, simplify.

1. $20 + 3 + (-14)$

2. $34 + 3 + (-19)$

3. $(6 - 7) + (4 - 23)$

4. $(2 - (-6)) - (15 - 19)$

5. $(6 - (-7)) + (4 - (-23))$

6. $(2 - 6) + (5 - (-19))$

7. $(7 - 4) - (3 + (-6)) + (-5 - 3)$

8. $(9 - 5) - (7 + (-11)) + (-8 - 4)$

9. $(7 - (-4)) + (3 - 6) - (-5 - (-3))$

10. $(9 - (-5)) + (7 + 11) - (-8 - (-4))$

11. $\dfrac{-4}{15} - \dfrac{7}{15}$

12. $\dfrac{-7}{15} - \dfrac{4}{15}$

13. $\dfrac{12}{23} - \dfrac{13}{23}$

14. $\dfrac{-12}{23} - \dfrac{13}{23}$

15. $\dfrac{3}{11} - \dfrac{-5}{11}$

16. $\dfrac{6}{17} - \dfrac{-4}{17}$

17. $\dfrac{1}{6} - \left(\dfrac{5}{6} + \dfrac{-7}{6}\right)$

18. $\dfrac{-2}{3} - \left(\dfrac{5}{3} - \dfrac{-1}{3}\right)$

19. $\dfrac{1}{4} - \dfrac{3}{5} - \dfrac{1}{10}$

20. $\dfrac{1}{3} - \dfrac{2}{5} + \dfrac{-1}{15}$

21. $\left(\dfrac{-2}{5} + \dfrac{1}{2}\right) - \left(\dfrac{7}{10} + \dfrac{1}{5}\right)$

22. $\left(\dfrac{5}{12} - \dfrac{1}{6}\right) - \left(\dfrac{3}{4} - \dfrac{2}{3}\right)$

23. $21.6 - (-3.7) + 6.8$

24. $31.5 - 9.2 - (-11.3)$

25. $-8.83 + 6 - (-16.4)$

26. $9.13 - (-4.2) + 10$

27. $(42.5 - 9.01) - (14.5 + (-21))$

28. $(53.2 + (-12.8)) - (23 - 31.7)$

29. $(132.61 - 93.5 + 17) - (54.1 + 18.21)$

30. $(108.7 - 45.23 - 8) - (42.7 + 63.05)$

In exercises **31–36**, answer each question as directed.

31. a. $13 - 9 = ?$
   b. $9 - 13 = ?$
   c. $13 + (-9) = ?$
   d. $-9 + 13 = ?$
   e. Are the answers to parts **a** and **b** the same?
   f. Are the answers to parts **c** and **d** the same?
   g. By writing subtraction as an addition, can the terms be interchanged without changing the answer?

32. a. $15 - 17 = ?$
   b. $17 - 15 = ?$
   c. $15 + (-17) = ?$
   d. $-17 + 15 = ?$
   e. Are the answers to parts **a** and **b** the same?
   f. Are the answers to parts **c** and **d** the same?
   g. By writing subtraction as an addition, can the terms be interchanged without changing the answer?

33. a. $20 - 7 - 4 = ?$
   b. $20 - (7 - 4) = ?$
   c. $20 + (-7) + (-4) = ?$
   d. $20 + ((-7) + (-4)) = ?$
   e. Are the answers to parts **a** and **b** the same?
   f. Are the answers to parts **c** and **d** the same?
   g. By writing a series of subtractions as additions, can the numbers be regrouped without changing the answer?

34. a. $32 - 15 - 12 = ?$
   b. $32 - (15 - 12) = ?$
   c. $32 + (-15) + (-12) = ?$
   d. $32 + ((-15) + (-12)) = ?$
   e. Are the answers to parts **a** and **b** the same?
   f. Are the answers to parts **c** and **d** the same?
   g. By writing a series of subtractions as additions, can the numbers be regrouped without changing the answer?

35. a. $0 - 8 = ?$
   b. $8 - 0 = ?$
   c. $0 - (-3) = ?$
   d. $-3 - 0 = ?$
   e. Are the answers to parts **a** and **b** the same?
   f. Are the answers to parts **c** and **d** the same?

36. a. $0 - 12 = ?$
   b. $12 - 0 = ?$
   c. $0 - (-12) = ?$
   d. $-12 - 0 = ?$
   e. Are the answers to parts **a** and **b** the same?
   f. Are the answers to parts **c** and **d** the same?

SECTION 1-3.   Multiplication and Division of Real Numbers

KEY TOPICS IN THIS SECTION

1. Multiplying and dividing integers with different signs

2. Multiplying and dividing negative integers

3. Multiplication property of 1

4. The product of -1 and b is the opposite of b.

5. Special problems involving 0

In this section we will review the rules that govern the signs of products and quotients of real numbers. We will also review the properties of 0 and 1 under the operations of multiplication.

Multiplying and Dividing Integers with Different Signs

Multiplication can be thought of in terms of repeated addition. For example,

$$5 \cdot 7 \text{ means} \quad \overbrace{\underbrace{\begin{array}{c} 5+5+5+5+5+5+5 \\ \text{or} \\ 7+7+7+7+7 \end{array}}} \quad \text{equals } 35$$

The definition of division relates it to multiplication. For example,

$$\frac{35}{5} = 7 \quad \text{because} \quad 35 = 5 \cdot 7$$

As a consequence, any rule that is given for the sign (positive or negative) of a product will also apply to a quotient.

Consider now the product of a positive and a negative integer. For example, $5(-7)$. The repeated addition interpretation can be used to verify that the product is negative.

$$5(-7) = \underbrace{(-7)+(-7)+(-7)+(-7)+(-7)}_{5(-7) \text{ means ``add } (-7) \text{ five times.''}} = -35$$

Notice, $5(-7)$ can also be written using

$-7 \cdot 5$ a raised dot.

-7×5 an $\times$ between the numbers.

$(-7)(5)$ parentheses around both numbers.

The product or quotient of numbers with different signs
If a and b are numbers, and one is positive while the other one is negative, then the product or quotient is negative.

Example 1.   Do the indicated operations.

a. $(-13)(25)$

b. $\dfrac{-140}{35}$

Solution.   **a.** $(-13)(25) = -325$ The product or quotient of numbers with unlike signs is negative.

         b. $\dfrac{-140}{35} = -4$

Example 2.   Simplify:

$$-6 + 7(-2) - \frac{45}{-9} + 19$$

Solution.   **Discussion.** Multiplications and divisions are done first in order from left to right. Then the additions and subtractions are done in order from left to right.

$$-6 + 7(-2) - \frac{45}{-9} + 19$$

$$= -6 + (-14) - (-5) + 19 \qquad 7(-2) = -14 \text{ and } \frac{45}{-9} = -5$$

$$= -20 - (-5) + 19 \qquad\qquad -6 + (-14) = -20$$

$$= -15 + 19 \qquad\qquad\qquad -20 - (-5) = -20 + 5 = -15$$

$$= 4 \qquad\qquad\qquad\qquad\quad \text{The value of the expression is 4.}$$

Multiplying and Dividing Negative Integers

Using the repeated addition interpretation of multiplication we can show that the product of any positive number and 0 is 0. For example,

$$5(0) = 0 + 0 + 0 + 0 + 0 = 0.$$

The multiplication property of 0 states that the product of any number and 0 is 0.

> **The multiplication property of zero**
> For any number b, $b \cdot 0 = 0$ and $0 \cdot b = 0$

To establish an intuitive acceptance of the rule for the product or quotient of two negative integers, we can study a pattern of products such as the following one:

$$4(-7) = -28$$

The left factor	$3(-7) = -21$	**Notice: $-21 = -28 + 7$**
is one less	$2(-7) = -14$	**Notice: $-14 = -21 + 7$**
than the one	$1(-7) = -7$	**Notice: $-7 = -14 + 7$**
on the line	$0(-7) = 0$	**Notice: $0 = -7 + 7$**
directly above it.	$-1(-7) = ?$	

To continue the pattern we should add $7 + 0$ and get 7. The number 7 is in fact the sum of $0 + 7$ and the product of -1 and -7. Thus the product (and quotient) of two integers that are both positive or both negative is positive.

The product or quotient of numbers with the same signs
   If a and b are numbers and they are both positive or both negative, then the product or quotient is positive.

Example 3.   Do the indicated operations.

   a. $(-15)(-32)$

   b. $\dfrac{-966}{-46}$

Solution.   **a.** $(-15)(-32) = 480$ The product or quotient of numbers with like signs is positive.

   b. $\dfrac{-966}{-46} = 21$

Example 4.   Simplify:

$$-13 - (-5)(-2) + \frac{-72}{9} + 17$$

Solution.

$$-13 - (-5)(-2) + \frac{-72}{9} + 17$$

$$= -13 - 10 + (-8) + 17 \qquad (-5)(-2) = 10 \text{ and } \frac{-72}{9} = -8$$

$$= -23 + (-8) + 17 \qquad \begin{aligned} -13 - 10 &= -13 + (-10) \\ &= -23 \end{aligned}$$

$$= -31 + 17 \qquad -23 + (-8) = -31$$

$$= -14 \qquad \text{The value of the expression is } -14.$$

Multiplication Property of 1

With respect to multiplication and division, 1 is a very special number.

The multiplication property of 1
   If a is any real number, then

$$a \cdot 1 = a \quad \text{and} \quad 1 \cdot a = a \quad \text{and} \quad \frac{a}{1} = a.$$

Furthermore, if $a \neq 0$, then $\frac{a}{a} = 1$.

The multiplication property of 1 is used to write a fraction as an **equivalent fraction.** Equivalent fractions have the same value but different form. Keep in mind that 1 can be written in many different ways. For example, $1 = \frac{5}{5}$ and $1 = \frac{-8}{-8}$.

Example 5. Write $\frac{3}{5}$ as an equivalent fraction with denominator 75.

Solution. **Discussion.** Since $75 = 5 \cdot 15$, multiply $\frac{3}{5}$ by $\frac{15}{15}$, a form of 1.

$$\frac{3}{5} = \frac{3}{5} \cdot \frac{15}{15} = \frac{45}{75} \qquad \text{Same fraction, different denominators.}$$

The multiplication property of 1 is also used to "reduce" fractions.

Example 6. Reduce:

a. $\dfrac{48}{120}$

b. $\dfrac{-300}{225}$

Solution. **Discussion.** Since $\frac{a}{a} = 1$, provided $a \neq 0$, we look for any number that evenly divides numerator and denominator.

a. $\dfrac{48}{120} = \dfrac{\cancel{2} \cdot \cancel{2} \cdot \cancel{2} \cdot 2 \cdot \cancel{3}}{\cancel{2} \cdot \cancel{2} \cdot \cancel{2} \cdot \cancel{3} \cdot 5} = \dfrac{2}{5}$ $\qquad \dfrac{2}{2} = 1$ and $\dfrac{3}{3} = 1$

b. $\dfrac{-300}{225} = \dfrac{-4 \cdot \cancel{75}}{3 \cdot \cancel{75}} = \dfrac{-4}{3}$ $\qquad \dfrac{75}{75} = 1$ and $\dfrac{-4}{3} \cdot 1 = \dfrac{-4}{3}$

The fact that $1 \cdot a = a$ has caused 1 to be called "the trivial factor." That is, *it is always there, so only write it when you need it.* For example,

$$\frac{5}{10} = \frac{1}{2} \text{ because } \frac{5}{10} = \frac{1 \cdot \cancel{5}}{2 \cdot \cancel{5}} = \frac{1}{2}.$$

The Product of -1 and b is the Opposite of b

The **opposite** of a real number b is the real number with the same absolute value but the opposite sign.

The opposite of 10 is -10 $\Big\langle$ $|10| = |-10|$
10 and -10 are opposite in sign.

The opposite of -6 is 6 $\Big\langle$ $|-6| = |6|$
-6 and 6 are opposite in sign.

> If b is any real number, then
>
> $$-1 \cdot b = -b.$$
>
> The product of -1 and b is the opposite of b.

This property of -1 permits us to write a negative number as the product of -1 and a positive number.

Example 7.   Reduce: $\dfrac{-12}{60}$

Solution.   **Discussion.** Write -12 as $-1 \cdot 12$, then reduce the fraction.

$$\frac{-12}{60} = \frac{-1 \cdot 12}{60} = \frac{-1 \cdot \cancel{12}}{5 \cdot \cancel{12}} = \frac{-1}{5} \qquad \frac{12}{12} = 1 \text{ and } \frac{-1}{5} \cdot 1 = \frac{-1}{5}$$

Special Problems Involving 0

Zero, like 1, has some special properties that need to be emphasized.

> **Some special properties of 0**
>   If a is any real number, then
>
> $$a + 0 = a \qquad \text{and} \qquad 0 + a = a.$$
> $$a \cdot 0 = 0 \qquad \text{and} \qquad 0 \cdot a = 0.$$
>
> Furthermore, if $a \neq 0$, then
>
> $$\frac{0}{a} = 0 \qquad \text{and} \qquad \frac{a}{0} \text{ is undefined.}$$

Division by 0 is **undefined** because when we attempt to divide by 0 we cannot find a number that fits the definition of division. Recall that

$$a \div b = c \qquad \text{if and only if} \qquad a = b \cdot c.$$

Consider now, for example, $12 \div 0$. By definition,

$$12 \div 0 = c \qquad \text{if and only if} \qquad 12 = 0 \cdot c.$$

Since $0 \cdot c = 0$ for any number c, there is no number c for which $0 \cdot c = 12$. Thus we say that the definition given to division does not apply to a divisor of 0.

Example 8.   Simplify:

    a. $-13 + 13 + 127$

    b. $-140(96)(0)$

    c. $419 \div 0$

Solution.   **a.** $-13 + 13 + 127 = 0 + 127$ $-13 + 13 = 0$

                          $= 127$ 0 plus any number is the number.

    b. $-140(96)(0) = 0$ 0 times any number is 0.

    c. $419 \div 0$ is undefined.   Division by 0 is undefined.

SECTION 1-3. Practice Exercises*

In exercises **1–44**, simplify.

[Example 1]

1. $(7)(-3)$

2. $(2)(-5)$

3. $(-19)(5)$

4. $(-8)(26)$

5. $\dfrac{45}{-5}$

6. $\dfrac{56}{-8}$

7. $\dfrac{-90}{15}$

8. $\dfrac{-80}{20}$

9. $\dfrac{156}{-52}$

10. $\dfrac{144}{-36}$

11. $(17)(-10)$

12. $(23)(-10)$

13. $(-11)(7)$

14. $(-12)(5)$

15. $(40)(-6)$

16. $(89)(-3)$

[Example 2]

17. $6 + (3)(-4) - 7$

18. $9 + (4)(-5) - 8$

19. $12 - \dfrac{-15}{3} + (4)(-2)$

20. $15 - \dfrac{-20}{4} + (7)(-3)$

21. $15 - (-3) + \dfrac{27}{-3}$

22. $13 - (-4) + \dfrac{28}{-4}$

23. $\dfrac{6 - (-5) + 4}{-21 + (2)(3)}$

24. $\dfrac{34 - (-8) - 12}{(4)(-4) - (-1)}$

[Example 3]

25. $(-4)(-7)$

26. $(-9)(-3)$

27. $(-12)(-11)$

28. $(-10)(-13)$

29. $\dfrac{-144}{-6}$

30. $\dfrac{-96}{-4}$

31. $\dfrac{-78}{-13}$

32. $\dfrac{-84}{-21}$

33. $(-68)(-12)$

34. $(-44)(-13)$

35. $(-38)(0)$

36. $(-52)(0)$

[Example 4]

37. $10 - (-3)(-2) + \dfrac{15}{-3}$

38. $14 - (-7)(-3) + \dfrac{18}{-3}$

39. $-39 + (4)(8) + \dfrac{-24}{-4}$

40. $-41 + (3)(8) + \dfrac{-36}{-4}$

41. $\dfrac{-72}{-9} - (12)(-4)$

42. $\dfrac{-56}{-8} - (13)(-5)$

* Additional problems are available in the Instructor's Manual.

43. $\dfrac{14 - (8)(-2)}{-(-3)(-4) + (-3)}$ **44.** $\dfrac{-28 - (-2)(-6)}{(3)(-6) + 8}$

In exercises **45–52**, write the fraction as an equivalent fraction with the given number as denominator.

[Example 5]   **45.** $\dfrac{3}{4}$ with denominator 48   **46.** $\dfrac{3}{5}$ with denominator 45

47. $\dfrac{-3}{7}$ with denominator 28   **48.** $\dfrac{-4}{9}$ with denominator 63

49. $\dfrac{-10}{3}$ with denominator 36   **50.** $\dfrac{-11}{4}$ with denominator 52

51. $\dfrac{17}{-13}$ with denominator 39   **52.** $\dfrac{15}{-17}$ with denominator 51

In exercises **53–60**, reduce the given fraction.

[Example 6]   **53.** $\dfrac{-5}{15}$ **54.** $\dfrac{-3}{21}$

55. $\dfrac{-24}{36}$ **56.** $\dfrac{-15}{45}$

57. $\dfrac{-60}{600}$ **58.** $\dfrac{-40}{400}$

59. $\dfrac{36}{-90}$ **60.** $\dfrac{26}{-104}$

In exercises **61–68**, simplify. Write "undefined" for any expression that does not name a number.

[Example 7]   **61.** $-3 + 3 + 7$ **62.** $6 - 6 + 10$

63. $(-47)(0)(32)$ **64.** $(41)(-2)(0)$

65. $-23 + 17 + (-17)$ **66.** $-43 + 9 + (-9)$

67. $-45 \div 0$ **68.** $-78 \div 0$

SECTION 1-3.   Supplementary Exercises

In exercises **1–32**, simplify.

1. $-7 + (-3)(10)$ **2.** $-11 + (-4)(5)$

3. $-7 + (-3)(10) \div 2$ **4.** $-11 + (-4)(5) \div 2$

5. $36 \div 9 \cdot 2$ **6.** $48 \div 12 \cdot 2$

7. $(5)(-7) \div (-35)$ **8.** $(3)(-9) \div (-27)$

9. $4 + 3(1) - 52 \div 4$

10. $13 - 5(1) - 64 \div 2$

11. $\dfrac{-2}{3} \cdot \dfrac{6}{7}$

12. $\dfrac{-2}{5} \cdot \dfrac{10}{11}$

13. $\dfrac{-7}{8} \cdot \dfrac{12}{7}$

14. $\dfrac{-3}{4} \cdot \dfrac{8}{9}$

15. $\dfrac{-3}{5} \div \dfrac{-3}{7}$

16. $\dfrac{-5}{6} \div \dfrac{-2}{3}$

17. $\dfrac{-7}{8} \div \left(\dfrac{1}{6} \div \dfrac{-1}{2} \right)$

18. $\left(\dfrac{-7}{8} \div \dfrac{1}{4} \right) \div \dfrac{-1}{2}$

19. $\dfrac{-40}{17} \div \dfrac{-20}{34} \cdot \dfrac{2}{5}$

20. $\dfrac{25}{16} \div \dfrac{-15}{32} \cdot \dfrac{-3}{2}$

21. $\dfrac{-10}{39} \cdot \dfrac{13}{8} - \dfrac{1}{6} \div \dfrac{-2}{9}$

22. $\dfrac{15}{24} \cdot \dfrac{-12}{5} - \dfrac{21}{30} \div \dfrac{-7}{15}$

23. $(3.5)(-3) + (-7.4)$

24. $(-4.1)(9) + (-4.8)$

25. $(-0.5)(32.4) - (0.2)(-1.1)$

26. $(0.1)(-14.2) - (0.5)(18.4)$

27. $(-1.06) \div 0.2 + (-1.4)(-3.1)$

28. $2.88 \div (-0.3) + (-6.2)(-4.4)$

29. $\dfrac{12}{6 - 3} + \dfrac{7}{12 - 9}$

30. $\dfrac{20}{-10 + 5} + \dfrac{8}{-12 + 8}$

31. $\dfrac{15 + (-27)}{0 + 5} - \dfrac{-12 + 7}{0 - 5}$

32. $\dfrac{18 - 14}{0 + 3} - \dfrac{19 - (-2)}{0 - 3}$

In exercises **33–39**, do as directed.

33. a. $10 \cdot 2 = ?$ **b.** $2 \cdot 10 = ?$ **c.** $10 \div 2 = ?$ **d.** $2 \div 10 = ?$
   e. Are the answers to parts **a** and **b** the same?
   f. Are the answers to parts **c** and **d** the same?

34. a. $21 \cdot 3 = ?$ **b.** $3 \cdot 21 = ?$ **c.** $21 \div 3 = ?$ **d.** $3 \div 21 = ?$
   e. Are the answers to parts **a** and **b** the same?
   f. Are the answers to parts **c** and **d** the same?

35. a. $(36 \div 6) \div 3 = ?$ **b.** $36 \div (6 \div 3) = ?$
   c. Are the answers to parts **a** and **b** the same?

36. a. $(100 \div 10) \div 2 = ?$
   b. $100 \div (10 \div 2) = ?$
   c. Are the answers to parts **a** and **b** the same?

37. a. $(17)(0) = ?$ **b.** $(0)(17) = ?$ **c.** $(25)(0) = ?$
   d. $17 \cdot \dfrac{1}{17} = ?$ **e.** $\dfrac{1}{17} \cdot 17 = ?$ **f.** $9 \cdot \dfrac{1}{9} = ?$
   g. Are the answers to parts **a**, **b**, and **c** the same?
   h. Are the answers to parts **d**, **e**, and **f** the same?

38. a. $(13)(0) = ?$ **b.** $(0)(13) = ?$ **c.** $(21)(0) = ?$
   d. $12 \cdot \dfrac{1}{12} = ?$ **e.** $\dfrac{1}{12} \cdot 12 = ?$ **f.** $5 \cdot \dfrac{1}{5} = ?$

g. Are the answers to parts **a**, **b**, and **c** the same?

h. Are the answers to parts **d**, **e**, and **f** the same?

39. Based on your answers to exercises **37** and **38**, answer the following:

a. Zero multiplied by any number is _____.

b. Any nonzero number multiplied by one over the number is _____.

SECTION 1-4.   Raising to Powers and Order of Operations

KEY TOPICS IN THIS SECTION

1. A definition of a positive integer exponent

2. Determining whether b^n is positive or negative

3. Expanding the rule for order of operations

4. Commonly used grouping symbols

5. Simplifying expressions with grouping symbols

Many numerical expressions have operations other than addition, subtraction, multiplication, and division. One such operation is **raising a number to a power.** Furthermore, symbols that group two or more numbers are also used in numerical expressions. In this section the methods used to simplify such expressions will be studied.

A Definition of a Positive Integer Exponent

Previously we noted that multiplication is a shorthand way of showing an indicated sum of several terms of the same number. For example,

$$\underbrace{3 + 3 + 3 + 3 + 3}_{\textbf{5 terms of 3}} \text{ can be written as } 3 \cdot \underset{\textbf{5 is a factor}}{5}$$

Frequently we need to indicate a product of several factors of the same number. Such an expression can be written with fewer symbols by using an **exponent.** For example,

$$\underbrace{3 \cdot 3 \cdot 3 \cdot 3 \cdot 3}_{\textbf{5 factors of 3}} \text{ can be written as } 3^{\underset{\textbf{5 is an exponent}}{5}}$$

A definition of b^n

If b is any number and n is a positive integer, then

$b^1 = b$ If n is 1, then it is usually not written.

$b^n = \underbrace{b \cdot b \cdot b \cdot \,\cdots\, \cdot b}_{n \text{ factors}}$ Read b^n as "b to the nth power."

In the expression b^n, b is the **base,** n is the **exponent,** and b^n is called a **power term.**

Example 1. Find the value of the following.

    a. 13^2

    b. $(-5)^3$

    c. $\left(\dfrac{2}{3}\right)^4$

    d. -6^2

Solution.   **a.** $13^2 = 13 \cdot 13 = 169$ The base is 13.

An exponent of 2 is read "squared." Thus 13^2 is read "thirteen squared."

    b. $(-5)^3 = (-5)(-5)(-5) = -125$ The base is -5.

An exponent of 3 is read "cubed." Thus $(-5)^3$ is read "negative five cubed."

    c. $\left(\dfrac{2}{3}\right)^4 = \dfrac{2}{3} \cdot \dfrac{2}{3} \cdot \dfrac{2}{3} \cdot \dfrac{2}{3} = \dfrac{16}{81}$ The base is $\frac{2}{3}$.

This power term is read "two-thirds to the fourth power."

    d. $-6^2 = -(6)(6) = -36$ The base is 6.

The expression -6^2 has two operations:

    Operation 1. Square 6. $6 \cdot 6 = 36$

    Operation 2. Take the opposite of the power. $-(36) = -36$

Determining Whether b^n is Positive or Negative

Consider examples **a–d**:

a. $(-2)^2 = 4$ 2 is even and the product is positive.

b. $(-2)^3 = -8$ 3 is odd and the product is negative.

c. $(-2)^4 = 16$ 4 is even and the product is positive.

d. $(-2)^5 = -32$ 5 is odd and the product is negative.

Based on examples **a–d**, the following general statements can be made:

The sign of b^n, where n is one of the numbers 1, 2, 3, 4, . . .
(i) If $b > 0$, then   **The product of only positive numbers is positive.**
    $b^n > 0$ for all n
(ii) If $b < 0$, then
    $b^n > 0$ for n equal to $2, 4, 6, . . .$ **A negative number raised to an even power is positive.**
    $b^n < 0$ for n equal to $1, 3, 5, . . .$ **A negative number raised to an odd power is negative.**

Example 2.   Without evaluating, determine the sign of the product.

a. $(-7)^5$

b. $(-15)^4$

c. -3^6

Solution.   a. The base is negative $(-7 < 0)$ and the exponent is an **odd number.** Therefore the product is negative: $(-7)^5 < 0$.

b. The base is negative $(-15 < 0)$ and the exponent is an **even number.** Therefore the product is positive: $(-15)^4 > 0$.

c. The base is positive $(3 > 0)$; therefore the power is positive. The opposite sign will make the final answer negative: $-3^6 < 0$.

Expanding the Rule for Order of Operations

Section 1-1 stated a rule for order of operations. The rule indicated that multiplications and divisions in any expression must be computed (from left to right) prior to any additions or subtractions. The following expanded rule covers expressions that also have power terms.

Expanded rule for order of operations
  Unless grouping symbols indicate otherwise,

Step 1. find the power of any term in the expression.

Step 2. do the multiplications and divisions in the order in which they occur, from left to right.

Step 3. do the additions and subtractions in the order in which they occur, from left to right.

Example 3.   Simplify:

$$5 \cdot 2^2 + 39 \div 13 + 3^3 - 10 \cdot 6$$

Solution.

$5 \cdot 2^2 + 39 \div 13 + 3^3 - 10 \cdot 6$	The given expression
$= 5 \cdot 4 + 39 \div 13 + 27 - 10 \cdot 6$	$2^2 = 4$ and $3^3 = 27$
$= 20 + 39 \div 13 + 27 - 10 \cdot 6$	$5 \cdot 4 = 20$
$= 20 + 3 + 27 - 10 \cdot 6$	$39 \div 13 = 3$
$= 20 + 3 + 27 - 60$	$10 \cdot 6 = 60$
$= 23 + 27 - 60$	$20 + 3 = 23$
$= 50 - 60$	$23 + 27 = 50$
$= -10$	The value of the expression is -10.

In the last step of Example 3, $50 - 60 = 50 + (-60) = -10$. Such a subtraction may be also done mentally, without actually changing the subtraction to addition.

Commonly Used Grouping Symbols

In mathematics an expression may include so-called **grouping symbols** that can be used to link two or more numbers together. Using these symbols gives the impression that the numbers inside the grouping symbol are to be treated as one term. To illustrate this feature of grouping symbols, parentheses are used to change the number of terms in the following expression.

$$9 + 6 - 10 + 3 \qquad \text{has four terms: 9, 6, 10 and 3}$$

$$9 + 6 - (10 + 3) \qquad \text{has three terms: 9, 6, and } (10 + 3)$$

$$(9 + 6) - (10 + 3) \qquad \text{has two terms: } (9 + 6) \text{ and } (10 + 3)$$

There are four commonly used grouping symbols:

Name	Symbol		Example
Parentheses	(	)	$8 - (4 + 9)$
Brackets	[	]	$-13 + [5^2 - 7] + 3$
Braces	{	}	$\{17 - (9 + 4)\} - 10$
Bar	——————		$\dfrac{3 \cdot 8}{10 + 2}$

Simplifying Expressions with Grouping Symbols

It is often necessary to simplify an expression containing one or more grouping symbols. To simplify such an expression the rule for order of operations must be followed. However, **any grouped expression must be simplified first.** Furthermore, if one grouping symbol occurs within a second set, then the **innermost grouped expression must be simplified first.**

Example 4. Simplify:

$$\frac{3 \cdot 5 + 2^4}{7^2 - 14} + \frac{3^3 + 4 \cdot 3}{13 \cdot 3 - 2^2}$$

Solution. **Discussion.** The fraction bars are grouping symbols. Therefore the numerators and denominators must be simplified first.

$$\frac{3 \cdot 5 + 2^4}{7^2 - 14} + \frac{3^3 + 4 \cdot 3}{13 \cdot 3 - 2^2}$$

$$= \frac{3 \cdot 5 + 16}{49 - 14} + \frac{27 + 4 \cdot 3}{13 \cdot 3 - 4} \qquad \text{Do all the raising to powers first.}$$

$$= \frac{15 + 16}{49 - 14} + \frac{27 + 12}{39 - 4} \qquad \text{Now do any multiplications or divisions.}$$

$$= \frac{31}{35} + \frac{39}{35} \qquad \text{Next do the indicated additions and subtractions.}$$

$$= \frac{70}{35} \qquad \frac{a}{c} + \frac{b}{c} = \frac{a + b}{c}$$

$$= 2 \qquad \text{The value of the expression is 2.}$$

Example 5.   Simplify:

$$12 - \{-3 \cdot 5 + 2[2^2 - 3(10 - 13)]\}$$

Solution.   **Discussion.** The expression grouped by parentheses is the innermost one and is simplified first. The expression grouped with brackets is inside the braces and is therefore simplified next. The expression inside the braces is simplified last.

$$12 - \{-3 \cdot 5 + 2[2^2 - 3(10 - 13)]\}$$

$= 12 - \{-3 \cdot 5 + 2[2^2 - 3(-3)]\}$ Replace $10 - 13$ with -3.

$= 12 - \{-3 \cdot 5 + 2[13]\}$ Replace $2^2 - 3(-3)$ with 13.

$= 12 - \{11\}$ Replace $-3 \cdot 5 + 2[13]$ with 11.

$= 1$ The value of the expression is 1.

Calculator Assistance Supplement C

Order of Operations

The following expression, without parentheses, must be simplified using the order of operations.

$$3 + 4^2 \div 2 - 5$$

Some of the less expensive calculators will perform operations as they are keyed into the machine. That is, they will calculate values out of order with the agreed-upon order of operations. The scientific calculator, on the other hand, will allow a user to key in the expression as it is written. The machine will perform the operations according to the order of operations. For example,

Press $\boxed{3}$.	Display shows	3.
Press $\boxed{+}$.	Display shows	3.
Press $\boxed{4}$.	Display shows	4.
Press $\boxed{x^2}$.	Display shows	16.
Press $\boxed{\div}$.	Display shows	16.
Press $\boxed{2}$.	Display shows	2.
Press $\boxed{-}$.	Display shows	11.
Press $\boxed{5}$.	Display shows	5.
Press $\boxed{=}$.	Display shows	6.

Thus the correct simplification for this expression is 6.

Calculator Assistance Supplement D

Raising to a Power

Most calculators have a key that is marked $\boxed{x^2}$. When a number shows on the display and this key is pressed, the display will change to the square of the number. For example,

the display shows $\qquad$ 21

you press $\boxed{x^2}$

the display now shows $\qquad$ 441

thus $21^2 = 441$

So-called scientific calculators also have a key that is marked x^y or y^x. With a proper sequence of key strokes, a number can be raised to any real number power. For example, to find 2^{15} ($= 2 \cdot 2 \cdot 2 \ldots 2$, fifteen factors),

Press $\boxed{2}$.	Display shows	2.
Press $\boxed{x^y}$.	Display shows	2.
Enter $\boxed{15}$.	Display shows	15.
Press $\boxed{=}$.	Display shows	32,768.

Thus, $2^{15} = 32,768$.

SECTION 1-4.   Practice Exercises

In exercises **1–10**, find the value.

[Example 1]

1. 2^5

2. 2^7

3. $(-4)^2$

4. $(-2)^4$

5. $(-3)^3$

6. $(-5)^3$

7. $\left(\dfrac{2}{5}\right)^3$

8. $\left(\dfrac{3}{4}\right)^3$

9. -3^2

10. -4^2

In exercises **11–20**, without evaluating, determine the sign of the product.

[Example 2]

11. $(-7)^{11}$

12. $(-9)^{13}$

13. $(-6)^{22}$

14. $(-7)^{24}$

15. $(-52)^{17}$

16. $(-47)^{15}$

17. -5^3

18. -7^5

19. -5^4

20. -7^6

In exercises **21–44**, simplify according to the rule for order of operations.

[Example 3]   **21.** $5 + 2 \cdot 3^2$ **22.** $7 + 3 \cdot 2^3$

23. $7(-2)^3 - 6(3)^4$ **24.** $3(-5)^2 - 4^4$

25. $4(-3)^3 + 6(7) + 5^2$ **26.** $19 - 7 + 6^2 - 2^6 + 11$

27. $6 \cdot 5^2 - 76 \div 2^2 - 3^4$ **28.** $3 \cdot 6^2 - 72 \div 2^3 - 7^2$

[Example 4]   **29.** $\dfrac{3 \cdot 6 - 4}{2 - 9}$ **30.** $\dfrac{5 \cdot 6 - 8}{6 + 5}$

31. $\dfrac{9 + 2^2 \cdot 5}{11 - 10 \cdot 2^2}$ **32.** $\dfrac{7 + 3^2 \cdot 6}{3 - 4^3}$

33. $\dfrac{15}{1 - 4} + \dfrac{1 + 4}{2^3 - 3}$ **34.** $\dfrac{21}{-9 + 2} + \dfrac{7 + 2}{3^2}$

35. $\dfrac{33 + 3^2}{7} - \dfrac{2^3 - 63}{11}$ **36.** $\dfrac{22 + 2^3}{5} - \dfrac{(-3)^2 - 48}{13}$

[Example 5]   **37.** $18 - 2(6 + 8)$ **38.** $10 - 3(12 - 3)$

39. $6^2 + 7(15 - 4)$ **40.** $10^2 - 5(23 - 16)$

41. $-3(10 + 4) - 5(13 - 6)$ **42.** $-4(2 - 11) - 3(17 - 10)$

43. $38 - 7[20 - (12 + 5)]$ **44.** $-45 + 6[[18 - 7) - 3]$

SECTION 1-4.   Supplementary Exercises

In exercises **1–34**, simplify.

1. $4 + 3(1 - 5) \div 6$ **2.** $13 - 5(3 - 7) \div 10$

3. $[13 - 2(7 - 3)] + 3(9 - 11)$ **4.** $73 + (-23 + 2(-25)) + 12$

5. $2^3 \cdot 3^2 - 5 \cdot 6$ **6.** $2^4 \cdot 3^2 - 7(4)$

7. $(-2)^4 \div 8 - 5 \cdot 3 + 6 \cdot 5$ **8.** $(-3)^3 \div 3 - 7 \cdot 2 + 10$

9. $(24 \div (12 \cdot 2))(42 - (-3))$ **10.** $(12 - (-6))(35 \div (5 \cdot 7))$

11. $50 - 3\{5 + 2[8 + 4(-4)]\}$ **12.** $10 - 2\{5 - 3[10 + 4(6 - 9)] + 8\}$

13. $[5(13 - 4) - 7(19 - 15)] + 4^3$ **14.** $[9(17 - 5) - 3(27 - 23)] + 5^2$

15. $\dfrac{1}{8} \div \dfrac{1}{4} + \dfrac{-1}{2}$ **16.** $\dfrac{1}{12} \div \dfrac{1}{2} - \dfrac{1}{6}$

17. $\left(\dfrac{-3}{5}\right)^2 \cdot \dfrac{5}{6}$ **18.** $\left(\dfrac{-2}{3}\right)^2 \cdot \dfrac{3}{4}$

19. $\left(\dfrac{3}{4}\right)^2 + \dfrac{-1}{4} \cdot \dfrac{7}{4}$ **20.** $\left(\dfrac{4}{5}\right)^2 + \dfrac{1}{5} \cdot \dfrac{-9}{5}$

21. $3 - \left(\dfrac{-1}{4} + 5\left(\dfrac{-3}{4}\right)\right)$

22. $7 - \left(\dfrac{-1}{3} + 4\left(\dfrac{-2}{3}\right)\right)$

23. $\dfrac{-10}{9} \cdot \dfrac{-3}{4} + (-3) \div \left(\dfrac{6}{5}\right)^2$

24. $\dfrac{-7}{2} \cdot \dfrac{3}{8} - 6 \div \left(\dfrac{-4}{5}\right)^2$

25. $4.6 + 2.3(-5.4) - 8.6$

26. $3.1 - 9.2(5.3) + (-4.8)$

27. $6.5(2.3 - 1.2)^2 - (4.8 - 7.1)$

28. $2.5(8.3 - 7.4)^2 - (2.1 - (-3.9))$

29. $[8 \div 0.4(-20)]^2 \div (-10)^3$

30. $[4 \div 0.5(-8)]^2 \div (-2)^3$

31. $(2^4 - 3^2)^2 - 4[6^2 - 7 \cdot 5]^3$

32. $(3^3 - 5^2)^2 - 5[7^2 - 5(10)]^3$

33. $\left(\dfrac{7 \cdot 5 - 8}{3^2}\right)^2 + \dfrac{5^3 + 1}{5 + 4}$

34. $\left(\dfrac{13 \cdot 5 + 15}{5^2 - 5}\right)^2 + \dfrac{3^4 + 3^2}{2^4 - 6}$

In exercises **35–40**, insert parentheses in the expression where necessary so that when simplified it will yield the given number.

Example: $80 \div 2 + 8 \rightarrow 8$

Solution: $80 \div (2 + 8) = 8$

35. $5 + (-3)(2) + 4 \rightarrow 8$

36. $5 + (-3)(2) + 4 \rightarrow 12$

37. $(-3)(6) + 4(-2) \rightarrow 6$

38. $(-3)(6) + 4(-2) \rightarrow 60$

39. $-5 + 4(-9) + 6 \div (-2) \rightarrow 6$

40. $-5 + 4(-9) + 6 \div (-2) \rightarrow 10$

SECTION 1-5.   Properties of Real Numbers

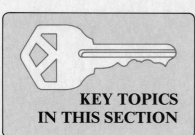

**KEY TOPICS
IN THIS SECTION**

1. The commutative property of addition and multiplication

2. The associative property of addition and multiplication

3. Identity elements for addition and multiplication

4. Inverse elements for addition and multiplication

5. The distributive property of multiplication over addition

Every game has a set of rules that specify what actions can or cannot be taken within the limits of the game. If a rule is broken, then most games have a set of penalties that may be assessed for each violation. Mathematics is not a game; however, there are rules that must be followed when mathematical principles are being applied. If any of these rules are violated, then the penalty will probably be a wrong answer to the problem being solved. In this section we will study some of the basic rules of mathematics.

The Commutative Property of Addition and Multiplication

Frequently we take actions in pairs. That is, one action (call it x) is immediately followed by another action (call it y). If the symbol "$*$" is used to stand for "followed by," then the two actions can be written symbolically as

$$x * y \qquad \text{"action } x \text{ followed by action } y\text{"}$$

Sometimes $x * y$ and $y * x$ yield the same result. Sometimes $x * y$ and $y * x$ yield entirely different results. To illustrate the first instance, consider the following actions:

Action x	**Action y**	**$x * y$ or $y * x$**
$\left(\begin{array}{l}\text{Put on your} \\ \text{right sock.}\end{array}\right)$	$\left(\begin{array}{l}\text{Put on your} \\ \text{left sock.}\end{array}\right)$	$\left(\begin{array}{l}\text{Both feet have} \\ \text{socks on them.}\end{array}\right)$

To illustrate the second kind, consider the following actions:

Action x	**Action y**	**$x * y$**	**$y * x$**
$\left(\begin{array}{l}\text{Put on your} \\ \text{right sock.}\end{array}\right)$	$\left(\begin{array}{l}\text{Put on your} \\ \text{right shoe.}\end{array}\right)$	$\left(\begin{array}{l}\text{Shoe is correctly} \\ \text{placed over sock.}\end{array}\right)$	$\left(\begin{array}{l}\text{Sock is incorrectly} \\ \text{placed over shoe.}\end{array}\right)$

The first illustration, in which $x * y$ and $y * x$ yielded the same result, is an example of the **commutative property.** Addition and multiplication are commutative operations.

The commutative property of addition and multiplication
If a and b are real numbers, then

$$a + b = b + a \quad \text{and} \quad a \cdot b = b \cdot a.$$

The commutative property permits the order in which two numbers are added or multiplied to be reversed without changing the sum or product.

$$-26 + 15 = -11 \quad \text{and} \quad 15 + (-26) = -11$$

$$(-26)(15) = -390 \quad \text{and} \quad (15)(-26) = -390$$

Subtraction and division are examples of operations that are not commutative.

$$-26 - 15 = -41 \quad \text{but} \quad 15 - (-26) = 41 \quad \text{and} \quad -41 \neq 41$$

$$-8 \div 2 = -4 \quad \text{but} \quad 2 \div (-8) = \frac{-1}{4} \quad \text{and} \quad -4 \neq \frac{-1}{4}$$

Example 1.   Use the commutative property to rewrite the following addition and multiplication.

   a. $-8 + (-12)$

   b. $8(-12)$

Solution.   **a.** $-8 + (-12) = -12 + (-8)$ $a + b = b + a$

   b. $8(-12) = -12(8)$ $a \cdot b = b \cdot a$

The Associative Property of Addition and Multiplication

Frequently we may do a task that involves three activities. It may make no difference which activities are carried out first, second, or third. Suppose x is one of the activities, y is another activity, and z is a third. If the grouping of the activities has no consequence on the final result, then

$$(x * y) * z \text{ would be the same as } x * (y * z).$$

To illustrate, suppose the task is eating a breakfast, where

x is the activity of drinking a glass of fruit juice.

y is the activity of eating a bowl of cereal.

z is the activity of eating an english muffin.

The sequence $(x * y) * z$ would indicate the following:

1. drinking the fruit juice while eating the bowl of cereal

2. then eating the english muffin

The sequence $x * (y * z)$ would indicate

1. drinking the fruit juice alone

2. then eating the cereal and english muffin together

Either sequence has essentially the same result. This example illustrates the **associative property.** Addition and multiplication are associative operations.

The associative property of addition and multiplication
If a, b, and c are real numbers, then

$$a + (b + c) = (a + b) + c \qquad \text{and} \qquad a \cdot (b \cdot c) = (a \cdot b) \cdot c$$

Addition and multiplication are **binary operations**—only two numbers at a time can be added or multiplied. Thus, where three or more numbers are to be added or multiplied, we add (or multiply) two of them. Then the result is operated on by the third number. The associative property permits the grouping to be done in any order without affecting the results.

Example 2.   Use the associative property to do the following operations.

$$\textbf{a. } \frac{3}{7} + \left(\frac{4}{7} + 8 \right)$$

$$\textbf{b. } \frac{9}{5} \cdot \left(\frac{10}{3} \cdot 4 \right)$$

Solution. $\textbf{a. } \dfrac{3}{7} + \left(\dfrac{4}{7} + 8 \right) = \left(\dfrac{3}{7} + \dfrac{4}{7} \right) + 8 \qquad a + (b + c) = (a + b) + c$

$$= \frac{7}{7} + 8 \qquad \frac{a}{c} + \frac{b}{c} = \frac{a + b}{c}$$

$$= 1 + 8 \qquad \frac{a}{a} = 1, \text{ provided } a \neq 0$$

$$= 9$$

$\textbf{b. } \dfrac{9}{5} \cdot \left(\dfrac{10}{3} \cdot 4 \right) = \left(\dfrac{9}{5} \cdot \dfrac{10}{3} \right) \cdot 4 \qquad$ Divide the 9 by 3, and the 10 by 5.

$$= (3 \cdot 2) \cdot 4$$

$$= 6 \cdot 4$$

$$= 24$$

Identity Elements for Addition and Multiplication

As previously noted, 0 and 1 have special properties under the operations of addition and multiplication. Two of these properties result in giving them the names of **identity elements.**

The identity elements for addition and multiplication
  If a is any real number, then

$$a + 0 = 0 + a = a, \text{ and } 0 \text{ is the identity element for addition.}$$

$$a \cdot 1 = 1 \cdot a = a, \text{ and } 1 \text{ is the identity element for multiplication.}$$

Examples **a–d** illustrate the properties of the identity elements.

a. $8 + 0 = 0 + 8 = 8$ **b.** $-10 + 0 = 0 + (-10) = -10$

c. $8 \cdot 1 = 1 \cdot 8 = 8$ **d.** $-10 \cdot 1 = 1 \cdot (-10) = -10$

Inverse Elements for Addition and Multiplication

If a is any real number, then there is a unique (only one) real number to add to a to get 0. This number we have previously called **the opposite of a**, and symbolized by $-a$. This number is also called **the additive inverse of a**.

If a is any real number and $a \neq 0$, then there is a unique (only one) real number to multiply by a to get 1. This number we have previously called **the reciprocal of a** and represented by $\frac{1}{a}$. This number is also called **the multiplicative inverse of a**.

The additive and multiplicative inverses of a
  The additive inverse of a is $-a$, such that

$$a + (-a) = -a + a = 0 \qquad \text{The identity element for addition}$$

If $a \neq 0$, then the multiplication inverse of a is $\frac{1}{a}$, such that

$$a \cdot \frac{1}{a} = \frac{1}{a} \cdot a = 1 \qquad \text{The identity element for multiplication}$$

Example 3. For the following numbers state

  (i) the additive inverse.

  (ii) the multiplicative inverse.

  a. $\dfrac{-3}{4}$

  b. $2\dfrac{1}{3}$

  c. 0

Solution. **a.** **(i)** The additive inverse is $\dfrac{3}{4}$. $\dfrac{-3}{4} + \dfrac{3}{4} = 0$

  (ii) The multiplicative inverse is $\dfrac{-4}{3}$. $\dfrac{-3}{4} \cdot \dfrac{-4}{3} = \dfrac{12}{12} = 1$

b. First change the mixed number to an improper fraction:

$$2\frac{1}{3} = \frac{7}{3}$$

   (i) The additive inverse is $\dfrac{-7}{3}$. $\dfrac{7}{3} + \dfrac{-7}{3} = 0$

   (ii) The multiplicative inverse is $\dfrac{3}{7}$. $\dfrac{7}{3} \cdot \dfrac{3}{7} = \dfrac{21}{21} = 1$

c. (i) The additive inverse of 0 is 0.

   (ii) 0 has no multiplicative inverse, because $\frac{1}{0}$ is not defined.

The Distributive Property of Multiplication Over Addition

Whenever an addition and a multiplication are both indicated in an expression, the rule for order of operation states that the multiplication must be carried out first. The only exception to this rule is when the addition is inside a grouping symbol and the multiplication is outside. The distributive property of multiplication over addition states the two ways in which an expression can be written—in one form the multiplication must be carried out first, and in the other the addition must be carried out first.

The distributive property of multiplication over addition
  If a, b, and c are real numbers, then

add first   **multiply first**

$$a(b + c) = a \cdot b + a \cdot c$$

multiply the sum by a   **add the products**

Some expression are more easily simplified by using the distributive property to change the given form to the other form in the equation.

Example 4.   Simplify: $36\left(\dfrac{5}{9} + \dfrac{3}{4}\right)$.

Solution.   **Discussion.** The given form of this expression requires adding the unlike fractions first. However, 36 is a multiple of both 9 and 4. Therefore, multiplying both fractions first by 36 will yield products that are whole numbers.

$$36\left(\frac{5}{9} + \frac{3}{4}\right) = 36 \cdot \frac{5}{9} + 36 \cdot \frac{3}{4} \qquad a(b + c) = a \cdot b + a \cdot c$$

$$= 20 + 27 \qquad\qquad \text{Do the multiplications.}$$

$$= 47 \qquad\qquad\quad \text{Add the products.}$$

Example 5.   Simplify:

   a. $(2.54)(63) + (2.54)(37)$

   b. $\dfrac{-5}{8} \cdot 17 + \dfrac{-5}{8} \cdot 39$

Solution.   **Discussion.** The given forms of both expressions require doing the multiplications first. However, 2.54 is a common factor in a, and $\frac{-5}{8}$ is a common factor in b. By writing these common factors outside parentheses, the additions would then be done first. In both cases, the consequence of these actions makes the simplification easier.

a. $(2.54)(63) + (2.54)(37)$

$= 2.54(63 + 37)$ $a \cdot b + a \cdot c = a(b + c)$

$= 2.54(100)$ Do the addition.

$= 254$ Do the multiplication.

b. $\dfrac{-5}{8} \cdot 17 + \dfrac{-5}{8} \cdot 39$

$= \dfrac{-5}{8}(17 + 39)$ $a \cdot b + a \cdot c = a(b + c)$

$= \dfrac{-5}{8}(56)$ Do the addition.

$= -35$ Do the multiplication.

The distributive property can be extended to sums of more than two numbers. Another form of the property distributes multiplication over subtraction.

If a, b, c, and d are real numbers, then

$$a(b + c + d) = a \cdot b + a \cdot c + a \cdot d$$

$$a(b - c) = a \cdot b - a \cdot c$$

$$a(b + c - d) = a \cdot b + a \cdot c - a \cdot d$$

For example,

$$24\left(\frac{5}{6} + \frac{7}{12} - \frac{3}{8}\right) = 24 \cdot \frac{5}{6} + 24 \cdot \frac{7}{12} - 24 \cdot \frac{3}{8}$$

$$= 20 + 14 - 9$$

$$= 25$$

SECTION 1-5.   Practice Exercises

In exercises **1–12**, use the commutative property to rewrite the following additions and multiplications.

[Example 1]   **1.** $32 + (-5)$ **2.** $19 + (-9)$

3. $(21)(-3)$ **4.** $(42)(-8)$

5. $\dfrac{1}{2} + \dfrac{3}{4}$ **6.** $\dfrac{1}{9} + \dfrac{7}{8}$

7. $\dfrac{-3}{7} \cdot \dfrac{2}{5}$ **8.** $\dfrac{-4}{9} \cdot \dfrac{5}{6}$

9. $\square + 0$ **10.** $\triangle + *$

11. $(\square)(0)$ **12.** $(\triangle)(*)$

In exercises **13–20**, simplify parts a and b. Notice the associative property relationship between part a and part b.

[Example 2] **13. a.** $(4 + 10) + 7$ **b.** $4 + (10 + 7)$

14. a. $(9 + 6) + 5$ **b.** $9 + (6 + 5)$

15. a. $(6 \cdot 5)3$ **b.** $6(5 \cdot 3)$

16. a. $(7 \cdot 2)4$ **b.** $7(2 \cdot 4)$

17. a. $(-72 + 21) + 43$ **b.** $-72 + (21 + 43)$

18. a. $(-35 + 49) + 83$ **b.** $-35 + (49 + 83)$

19. a. $\left(\dfrac{5}{6} \cdot \dfrac{3}{10}\right) \cdot 4$ **b.** $\dfrac{5}{6} \cdot \left(\dfrac{3}{10} \cdot 4\right)$

20. a. $\left(\dfrac{7}{8} \cdot \dfrac{4}{9}\right) 18$ **b.** $\dfrac{7}{8}\left(\dfrac{4}{9} \cdot 18\right)$

In exercises **21–28**, state

a. the additive inverse and

b. the multiplicative inverse

[Example 3] **21.** -8 **22.** -5

23. $\dfrac{2}{3}$ **24.** $\dfrac{4}{5}$

25. $3\dfrac{1}{2}$ **26.** $2\dfrac{1}{3}$

27. -4.2 **28.** -3.1

In exercises **29–36**, use the distributive property to simplify by changing the order of operation to multiplication first.

[Example 4] **29.** $\dfrac{7}{12}(48 + 36)$ **30.** $\dfrac{5}{8}(72 + 96)$

31. $12\left(\dfrac{1}{6} + \dfrac{2}{3}\right)$ **32.** $15\left(\dfrac{1}{5} + \dfrac{2}{3}\right)$

33. $-21\left(\dfrac{3}{7} + \dfrac{4}{3}\right)$ **34.** $-12\left(\dfrac{1}{6} + \dfrac{5}{4}\right)$

35. 3.2(4.1 + 3.2)   **36.** 2.8(6.3 + 2.4)

In exercises **37–44**, use the distributive property to simplify by changing the order of operation to addition first.

[Example 5] **37.** 8(18) + 8(12)   **38.** 6(29) + 6(11)

39. 14(36) − 14(16)   **40.** 18(54) − 18(44)

41. (−5)(74.8) + (−5)(25.2)   **42.** −8(41.6) + (−8)(8.4)

43. 73.1(58) + 73.1(42)   **44.** 58.5(62) + 58.5(38)

SECTION 1-5. Supplementary Exercises

In exercises **1–8**, determine the value that x must be for the equations to be true.

1. $p + x = p$ **2.** $x + q = q$

3. $a \cdot x = a$ **4.** $x \cdot b = b$

5. $\dfrac{x}{a} \cdot a = 1, a \neq 0$ **6.** $p \cdot \dfrac{x}{p} = 1, p \neq 0$

7. $-a + x = -a$ **8.** $x - z = -z$

In exercises **9–14**, determine if the following actions are necessarily commutative.

9. a. Put cereal in a bowl.
   b. Put milk in the same bowl.

10. a. Add salt.
    b. Add pepper.

11. a. Go to bed.
    b. Go to sleep.

12. a. Fill the swimming pool.
    b. Go for a swim.

13. a. Put ice in a glass.
    b. Put ice tea in a glass.

14. a. Put ham on a slice of bread.
    b. Put cheese on the same slice of bread.

In exercises **15–20**, determine if the following actions are necessarily associative.

15. Buy lettuce, buy tomatoes, buy salad dressing for a salad.

16. Decide on what dress to wear, decide on what shoes to wear, decide on what coat to wear.

17. Fill the tank with gas, check the oil, check the air in the tires.

18. Wash the clothes, clean the kitchen, shop for food.

19. Listen to the lecture, read the book, do the homework.

20. Put a letter in an envelope, seal the envelope, mail the envelope.

In exercises **21–30**, simplify by multiplying first, then adding.

21. $\dfrac{1}{2}(4 + 2 + 8)$ **22.** $\dfrac{1}{3}(9 + 3 + 6)$

23. $\dfrac{1}{6}(12 - 48)$ **24.** $\dfrac{1}{9}(45 - 81)$

25. $\dfrac{3}{4}(12 + 4 - 20)$ **26.** $\dfrac{2}{7}(14 - 7 + 28)$

27. $24\left(\dfrac{1}{8} + \dfrac{5}{6} - \dfrac{1}{4}\right)$ **28.** $28\left(\dfrac{1}{7} + \dfrac{3}{4} - \dfrac{1}{2}\right)$

29. $\dfrac{1}{2}(8.2 + 6.8 - 4.4)$ **30.** $\dfrac{1}{2}(10.6 - 8.4 + 2.8)$

In exercises **31–40**, identify the property illustrated.

31. $7 + 3 = 3 + 7$ **32.** $14 + 1 = 1 + 14$

33. $12 + 0 = 12$ **34.** $0 + 42 = 42$

35. $\dfrac{2}{7} \cdot 1 = \dfrac{2}{7}$ **36.** $\dfrac{3}{8} = \dfrac{3}{8} \cdot 1$

37. $3\left(\dfrac{7}{15} \cdot 4\right) = \left(3 \cdot \dfrac{7}{15}\right)4$ **38.** $(9 \cdot 8)\dfrac{3}{4} = 9\left(8 \cdot \dfrac{3}{4}\right)$

39. $\dfrac{1}{2}\left(3 - \dfrac{1}{4}\right) = \dfrac{1}{2}(3) - \dfrac{1}{2} \cdot \dfrac{1}{4}$ **40.** $\dfrac{2}{3}\left(\dfrac{1}{5} - 7\right) = \dfrac{2}{3} \cdot \dfrac{1}{5} - \dfrac{2}{3} \cdot 7$

SECTION 1-6.   Algebraic Expressions

**KEY TOPICS
IN THIS SECTION**

1. Parts of an algebraic expression

2. Some terminology associated with algebraic expressions

3. A definition of like terms

4. Adding or subtracting like terms

5. Simplifying expressions with grouping symbols

In the past few sections we have simplified **numerical expressions.** The process required carrying out all the operations indicated in the expression. The final form was a single number. In algebra we need to work with expressions that also include symbols that stand for numbers, but not always one specific number. Such symbols are called **variables.** Expressions that include variables are called **algebraic.**

In this section we begin to examine algebraic expressions and some of the ways in which they can be simplified.

Parts of an Algebraic Expression

To help us recognize an algebraic expression, the elements that can be included in such an expression are listed below.

Parts of an algebraic expression

1. **Real numbers,** such as -3, 10, $\frac{3}{5}$, and 2.54.

2. **Variables** that stand for numbers, such as x, y, t and k.

3. **Operations,** such as addition, multiplication, and raising to powers.

4. **Grouping symbols,** such as parentheses and bars.

An expression is frequently referred to by the variable or variables in the expression. Examples **a–d** give four illustrations.

a. $x^2 + 5x + 6$ An expression in x

b. $2t^4 + t^2$ An expression in t

c. $m^2 - 7mn + 10n^2$ An expression in m and n

d. $3xy + 2xz - 5yz + 1$ An expression in x, y, and z

Some Terminology Associated with Algebraic Expressions

Several words or phrases are used in connection with algebraic expressions. It is important that such terminology be understood.

A **term** of an expression is either a single number or a product of a number and one or more variables. The variable or variables may have exponents.

In any expression, terms are separated from each other by $+$ and $-$ signs.

Algebraic Expression	Number of Terms	Name for Expression
$-73k^3$	one	monomial
$16t^2 - 25$	two	binomial
$4x^2 - 12xy + 9y^2$	three	trinomial
$16 - m^2 + 10mn - 25n^2$	four	expression of four terms

When two or more terms are grouped by a grouping symbol, then the grouped expression is counted as one term because the grouping symbol is used to indicate that everything within the symbol is to be treated as a single unit.

Example 1. How many terms are in the following expression?

$$5(x^2 + y^2) - 2(x + y) + 7$$

Solution.

$5(x^2 + y^2)$

$2(x + y)$ ——— are single terms

7

Thus, as written, the expression has three terms.

> Any number, variable, product of variables, or product of number and variables that evenly divides a term is a **factor** of that term.

Recall that terms are added or subtracted. By contrast, factors are multiplied or divided. To illustrate the difference, consider $2x$ and $5y$:

$2x \cdot 5y = 10xy$ ⟨ $2x$ is a *factor* of $10xy$.

$5y$ is a *factor* of $10xy$.

$2x + 5y$ or $2x - 5y$ ⟨ $2x$ is a *term* of either expression.

$5y$ is a *term* of either expression.

> If the product of a and b is an algebraic term, then
>
> a is the **coefficient** of b, and
>
> b is the **coefficient** of a.

If a is the **number factor** of an algebraic term, then it is called the **numerical coefficient**. If b is the variable factor of an algebraic term, then it is called the **literal coefficient**. If a given term does not have a numerical coefficient, then the coefficient is either

1 (such as $x^2 = 1 \cdot x^2$), or -1 (such as $-t^3 = -1 \cdot t^3$).

Example 2. Identify the

(i) numerical coefficient.

(ii) literal coefficient.

a. $-6x^2$

b. $\dfrac{3t}{4}$

c. m^2n^3

Solution. **Discussion.** The numerical coefficient is the number factor. The literal coefficient is the variable factor.

a. (i) The numerical coefficient is -6.
(ii) The literal coefficient is x^2.

b. (i) The numerical coefficient is $\frac{3}{4}$.
(ii) The literal coefficient is t.

c. (i) The numerical coefficient is 1.
(ii) The literal coefficient is m^2n^3.

It is frequently possible to write a term of an algebraic expression in many different ways. To illustrate, consider a term such as $10xyz$. This term could also be written as

$$x10yz \quad \text{or} \quad y10xz \quad \text{or} \quad yz10x \quad \text{and so on.}$$

The following statement identifies what is considered to be the **preferred way** to write a term.

The conventional form for writing a term

1. Write the numerical coefficient on the left.

2. Write the literal coefficient in alphabetical order.

Example 3. Write in conventional form $v(-10)w^3u^2$.

Solution. In conventional form

$$v(-10)w^3u^2 \text{ is written } -10u^2vw^3.$$

Number factor first $\uparrow$ $\uparrow$ Variables in alphabetical order

A Definition of Like Terms

A numerical expression such as

$$2 \cdot 5 + 3 \cdot 8$$

can be simplified by doing the indicated operations.

$2 \cdot 5 + 3 \cdot 8$	The given expression
$= 10 + 24$	Do the multiplications first.
$= 34$	Now do the addition.

An algebraic expression such as

$$2x + 3y$$

cannot be simplified, because the multiplications (2 times x and 3 times y) must be carried out before the addition. Since x and y are variables, we cannot replace $2x$ and $3y$ with products until x and y are first replaced by numbers. However, an algebraic expression such as

$$2x + 3x$$

can be simplified. As will be seen, the distributive property can be used to change the sequence in which the multiplication and addition need to be carried out. Terms such as $2x$ and $3x$ which can be combined (that is, added or subtracted) are called **like terms.**

A definition of like terms

Terms that have the same literal coefficients are called **like terms.** That is, the variables in like terms are exactly the same, and a given variable has the same exponent.

Examples **a** and **b** illustrate like terms.

a. $5t^2$, $-2t^2$, and t^2 Literal coefficient is t^2.

b. $\dfrac{-2}{3}xyz$, $\dfrac{4}{5}xyz$, and $\dfrac{7}{10}xyz$ Literal coefficient is xyz.

Examples **c** and **d** illustrate terms that are not alike.

c. $5t^3$, $-2t^2$, and t Literal coefficients are t^3, t^2, and t.

d. $\dfrac{-2}{3}x^2yz$ and $\dfrac{4}{5}xy^2z$ Literal coefficients are x^2yz and xy^2z.

Adding or Subtracting Like Terms

Like terms can be added or subtracted. The distributive property of multiplication over addition is the tool that permits us to do these operations.

$$5t^2 - 2t^2 + t^2 \qquad \text{The given expression}$$

$$= (5 - 2 + 1)t^2 \qquad \text{Use the distributive property.}$$

$$= 4t^2 \qquad \text{Simplify within the parentheses.}$$

This example illustrates the fact that combining like terms can be accomplished by merely adding or subtracting (as indicated) the numerical coefficients.

Example 4. Simplify.

$$7mn - 5mn - 9mn + 10mn$$

Solution.

$$7mn - 5mn - 9mn + 10mn \qquad \text{The given expression}$$

$$= (7 - 5 - 9 + 10)mn \qquad \text{Use the distributive property.}$$

$$= 3mn \qquad \text{Simplify within the parentheses.}$$

Simplifying Expressions with Grouping Symbols

When two or more expressions are added or subtracted, they can frequently be simplified by combining any like terms. The associative and commutative properties of addition can be used to group any like terms together.

Example 5. Simplify:

$$(3x^2 + 7x + 5) + (x^2 - 10x + 3)$$

Solution. **Discussion.** A preliminary step would be to remove the parentheses, and then regroup any like terms in the expression.

$$(3x^2 + 7x + 5) + (x^2 - 10x + 3)$$

$$= 3x^2 + 7x + 5 + x^2 - 10x + 3$$

$$= 3x^2 + 7x + 5 + x^2 + (-10x) + 3$$

$$= (3x^2 + x^2) + (7x + (-10x)) + (5 + 3)$$

$$= 4x^2 + (-3x) + 8$$

$$= 4x^2 - 3x + 8$$

Example 6.   Simplify:

$$2(t^2 - 3t - 5) + 5(t^2 + t + 3) - 3(2t^2 - t + 1)$$

Solution.   **Discussion.** Use the distributive property to multiply the trinomials by the constants outside the parentheses. In the case of the last expression (that is, $-3(2t^2 - t + 1)$), distribute a (-3) to each term.

$$2(t^2 - 3t - 5) + 5(t^2 + t + 3) - 3(2t^2 - t + 1)$$

$$= 2t^2 - 6t - 10 + 5t^2 + 5t + 15 - 6t^2 + 3t - 3$$

$$= (2t^2 + 5t^2 - 6t^2) + (-6t + 5t + 3t) + (-10 + 15 - 3)$$

$$= t^2 + 2t + 2$$

SECTION 1-6.   Practice Exercises

In exercises **1–12**, indicate the number of terms in each expression.

[Example 1]   **1.** $-5a^2b$ **2.** $-3ab^3$

3. $3t^2 - 5t + 1$ **4.** $7x^2 - 3x - 5$

5. $x^2 - 25$ **6.** $a^2 - 49$

7. $(x^2 + 1)^2 + 7x$ **8.** $(3a + 2)^2 - 9$

9. $6x^2 - 13y^2 + 7z^2 + 1$ **10.** $-5x^2 + 4y^2 - z^2 + 9$

11. $(x^2 + 6x + 9) - 16$ **12.** $(y^2 - 10y + 25) - 100$

In exercises **13–22**,

a. identify the numerical coefficient.

b. identify the literal coefficient.

[Example 2]   **13.** $23x^2$ **14.** $45y^3$

15. $7x^2y$ **16.** $4a^2b$

17. $\dfrac{-3p^2q^2}{2}$ **18.** $\dfrac{-7xy}{10}$

19. a^2b **20.** x^2y

21. $0.3mn^2$ **22.** $0.8t^2w$

In exercises **23–30**, write each term in conventional form.

[Example 3]   **23.** $z5x^2$ **24.** p^28q

25. $b^2(-6)ac$ **26.** $q^2(-9)pr$

27. $(x + y)^4(-10)$ **28.** $(a - b)^5(-3)$

29. $-vw^2(23)u$ **30.** $-rs(13)q^2$

In exercises **31–58**, simplify by adding like terms.

[Example 4]   **31.** $12x + 7x$ **32.** $35y + 23y$

33. $13a^2 - 5a^2$ **34.** $17b^2 - 20b^2$

35. $xyz + 9xyz$ **36.** $ab + 5ab$

37. $22a^2b - a^2b$ **38.** $43xy^2 - xy^2$

39. $14a^2 - 8a^2 - 3a^2$ **40.** $18k^3 - 9k^3 - 7k^3$

41. $5xyz - 15xyz + 10xyz$ **42.** $-17m^2n + 11m^2n + 6m^2n$

[Example 5]   **43.** $(5a + 8b) + (-a + 3b)$ **44.** $(14c - 9d) + (-4d + 8c)$

45. $(25h^2 - 10k^2 + 6) + (9h^2 + k^2 - 3)$ **46.** $(7x^2 + 33y^2 + 1) + (3x^2 - y^2 - 8)$

47. $(1 - y) + (-5 + 8y)$ **48.** $(18 - 30x) + (21 + x)$

49. $(a^3 - 5a^2 + 3a) + (7a^2 - a)$ **50.** $(5y^3 + 3y^2 - 2y) + (8y^3 + y^2)$

[Example 6]   **51.** $12 + 2(5x + 1)$ **52.** $9y + 3(y + 6)$

53. $(3b - 4) + 5(2b - 3)$ **54.** $7(3 - 2c) + 2(8 + c)$

55. $3(a^2 + 5a - 4) - 7(2a^2 + 4a + 1)$ **56.** $5(m^2 - 2m + 4) - 8(3m^2 + 2m + 2)$

57. $5(-3v^2 + 8v + 10) - 2(-5v^2 + 6v) + 4(v^2 + 9v + 3)$

58. $6(-2w^2 - 4w + 5) - 3(4w^2 + w) + 5(w^2 - 3w + 2)$

SECTION 1-6.   Supplementary Exercises

In exercises **1–4**, indicate the number of terms.

1. a. $x^2 + 10x + 5$ **b.** $x^2 + (10x + 5)$
   c. $(x^2 + 10x) + 5$ **d.** $(x^2 + 10x + 5)$

2. a. $t^2 + 5t + 1$ **b.** $(t^2 + 5t) + 1$
   c. $t^2 + (5t + 1)$ **d.** $(t^2 + 5t + 1)$

3. a. $x^2 + 6xy + 9y^2 - 25$ **b.** $(x^2 + 6xy + 9y^2) - 25$
   c. $(x^2 + 6xy) + 9y^2 - 25$ **d.** $(x^2 + 6xy) + (9y^2 - 25)$

4. a. $y^2 + 10y + 25 - z^2$ **b.** $(y^2 + 10y) + (25 - z^2)$
   c. $(y^2 + 10y + 25) - z^2$ **d.** $(y^2 + 10y) + 25 - z^2$

In exercises **5–12**, identify the numerical coefficient of

a. the term with exponent two.

b. the term with exponent one.

5. $17x^2 + 6x + 3$ **6.** $12a^2 - 5a + 2$

7. $5 - 6t - t^2$ **8.** $3 - 4w - w^2$

9. $49b^2 - 1$ **10.** $25x^2 - 9$

11. $-4(a + b)^2 + 3(a + b) + 1$ 12. $-9(m - n)^2 - 2(m - n) + 5$

In exercises **13–28**, simplify.

13. $37 + 13xy + 13 - 3xy$ 14. $20 - xy - 10 - 9xy$

15. $5(2y^3 + 1) - 2(2 + y^3)$ 16. $6(3x^2 + 1) - 3(4 + x^2)$

17. $3b^2 - 2b - 1 + 2b - 4 + b^2$ 18. $b^2 - 5b + 6 - b + 3b^2 - 6$

19. $12(2x - 3y) - 8(3x - 2y)$ 20. $-4(x + y) + 9(x - y)$

21. $(12x + y) + 8(y - x) - 7(2x - y)$ 22. $(a - 3b) - 2(2a - b) + 3(5a - 4b)$

23. $8m - 2[6 + 3(m - 4)]$ 24. $8n + 4[10 - 5(n + 3)]$

25. $-5[2(a - 3b) - 6(2a + b)]$ 26. $-6[3(2c + d) - (10c + 3d)]$

27. $4\{3(m - 2n) - 2[m + 4(m - 3n)]\}$ 28. $-3\{2(m + n) - [6m - 2(3m - n)]\}$

In exercises **29–36**, find the value of k so that expression **a** simplifies to expression **b**.

29. **a.** $-7x + 15x + kx$ **b.** $17x$

30. **a.** $5y - 11y + ky$ **b.** $-4y$

31. **a.** $3mn + kmn - 19mn$ **b.** $-5mn$

32. **a.** $-8pq + kpq + 12pq$ **b.** $7pq$

33. **a.** $3x^2 + 6x - kx^2 + x$ **b.** $-4x^2 + 7x$

34. **a.** $4z^2 - 3z + kz^2 + 5z$ **b.** $-3z^2 + 2z$

35. **a.** $7m^2 + 3m - 5m^2 + km$ **b.** $2m^2 + 11m$

36. **a.** $-4p^2 + kp + 3p^2 + 6p$ **b.** $-p^2 - 2p$

In exercises **37** and **38**, find the values of h, i, and j so that expression **a** simplifies to expression **b**.

37. **a.** $5x^2 + ix - 7 + hx^2 + 6x + j$ **b.** $-3x^2 - 7$

38. **a.** $-11y^2 + iy - 12 + hy^2 - 2y + j$ **b.** $5y^2 - 2y$

SECTION 1-7. Evaluating Algebraic Expressions

**KEY TOPICS
IN THIS SECTION**

1. Changing an algebraic expression to a numerical expression

2. Evaluating algebraic expressions with grouping symbols

3. Some applied problems

In an algebraic expression the variable (or variables) usually stands for more than just one number. As a consequence, we sometimes call an algebraic expression a **variable expression.** This name is appropriate because we can **vary the value** of the expression by replacing all variables by different numbers. Once we have made the replacements, we have changed the algebraic expression to a numerical expression. The indicated operations can now be carried out and the expression can be simplified to one number. In this section we will study the process of changing algebraic expressions to numerical expressions, and then simplifying them.

Changing an Algebraic Expression to a Numerical Expression

An expression such as $2t^2 + 9t - 10$ is algebraic because it contains the variable t. Unless some restriction is stated, it is assumed that t represents any real number. We frequently refer to the set of numbers that a variable represents as the **replacement set.** As we will see later in this section, some applied problems will involve algebraic expressions in which the replacement set is a subset of the real numbers.

Example 1. Evaluate $2t^2 + 9t - 10$ for $t = 7$.

Solution.   **Discussion.** The evaluation consists of two parts. First replace t with 7 in the given expression. Then carry out the indicated operations to simplify the numerical expression.

$$2(7)^2 + 9(7) - 10 \qquad \text{Replace } t \text{ with 7.}$$

$$= 2(49) + 9(7) - 10 \qquad \text{Raise to powers first.}$$

$$= 98 + 63 - 10 \qquad \text{Next, multiply.}$$

$$= 161 - 10 \qquad \text{Addition is on the left.}$$

$$= 151 \qquad \text{Do the subtraction last.}$$

Thus, $2t^2 + 9t - 10$ has the value 151 when $t = 7$.

When an expression has more than one variable, then all variables must be replaced by numbers to change it to a numerical expression.

Example 2. Evaluate $4m^2 - 8mn - 9n^2$ for $m = 5$ and $n = -4$.

Solution. $4(5)^2 - 8(5)(-4) - 9(-4)^2 \qquad \text{Replace } m \text{ by 5 and } n \text{ by } -4.$

$$= 4(25) - 8(5)(-4) - 9(16) \qquad \text{Do the powers first.}$$

$$= 100 - (-160) - 144 \qquad \text{Now do the multiplications.}$$

$$= 260 - 144 \qquad 100 - (-160) = 100 + 160$$

$$= 116$$

Thus, $4m^2 - 8mn - 9n^2$ has the value 116 when $m = 5$ and $n = -4$.

Evaluating Algebraic Expressions with Grouping Symbols

If an algebraic expression with grouping symbols is to be evaluated after replacement for the variable(s), then the grouped portion of the resulting numerical expression must be simplified first.

Example 3.   Evaluate $3[3k^2 - 9(k + 8)]$ for $k = -4$.

Solution.   **Discussion.** After replacing k with (-4), do the operation within the parentheses. Then do the operations inside the brackets.

$$3[3(-4)^2 - 9(-4 + 8)] \qquad \text{Replace } k \text{ with } -4.$$

$$= 3[3(-4)^2 - 9(4)] \qquad -4 + 8 = 4$$

$$= 3[3(16) - 9(4)] \qquad (-4)^2 = 16$$

$$= 3[48 - 36] \qquad \text{Do both multiplications.}$$

$$= 3[12] \qquad \text{Subtract inside the brackets.}$$

$$= 36$$

The value of the expression is 36 when $k = -4$.

Example 4.   Evaluate $\dfrac{5x + 3y}{x^2 + y^2} + \dfrac{3x - 2y - z}{z^2}$ for $x = 3$, $y = -4$ and $z = 5$.

Solution.   **Discussion.** Keep in mind that the bar is a grouping symbol. Therefore the operations above and below the bars must be carried out first.

$$\frac{5(3) + 3(-4)}{3^2 + (-4)^2} + \frac{3(3) - 2(-4) - 5}{5^2} \qquad \begin{array}{l}\text{Replace } x \text{ with 3, } y \text{ with} \\ -4 \text{ and } z \text{ with 5.}\end{array}$$

$$= \frac{15 + (-12)}{9 + 16} + \frac{9 - (-8) - 5}{25}$$

$$= \frac{3}{25} + \frac{12}{25}$$

$$= \frac{15}{25} = \frac{3}{5}$$

Thus the value of the expression is $\frac{3}{5}$ when $x = 3$, $y = -4$, and $z = 5$.

Some Applied Problems

Many algebraic expressions contain variables that represent quantities with physical applications. Examples 5–7 illustrate three such expressions.

Example 5.   A four-sided figure with exactly two parallel sides is called a **trapezoid** (shown in Figure 1-6). The lengths of the parallel sides are a and b. The distance between these sides is h, which stands for **height**. The area of the trapezoid is obtained by evaluating the expression $\frac{1}{2}h(a + b)$. Find the area of a trapezoid in which $a = 15$ meters, $b = 20$ meters, and $h = 8$ meters. The abbreviation for meters is m.

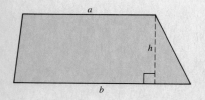

Figure 1-6. A trapezoid.

Solution.   **Discussion.** To find the area, simply evaluate $\frac{1}{2}h(a + b)$ for $h = 8$, $a = 15$, and $b = 20$.

$$\frac{1}{2} \cdot 8m(15m + 20m)$$

$$= \frac{1}{2} \cdot 8m(35m) \qquad\qquad 15m + 20m = 35m$$

$$= 4m(35m) \qquad\qquad \frac{1}{2} \cdot 8m = 4m$$

$$= 140m^2 \qquad\qquad 4 \cdot 35 = 140 \text{ and } m \cdot m = m^2$$

Thus the area of the trapezoid is 140 square meters (written m^2).

When money is invested in certain types of funds, the interest earned is **compounded.** Basically *compound interest* means that interest earned during one period will earn interest over the subsequent periods. The amount returned to the investor can be found by evaluating the expression $P(1 + r)^{nt}$. In this expression P is the **principal** (or amount put into the fund), r is **rate** of interest per period, n is the **number of periods** in one year that interest is compounded, and t is the **time** in years that the money is left in the fund. The following table identifies the values of n for various compounding periods.

Compounding Period	Value of n
Annually	1
Semiannually	2
Quarterly	4
Monthly	12

The annual interest rate is usually stated for accounts that earn or charge compound interest. To find the periodic interest rate in the formula, the annual interest rate is divided by the number of compounding periods per year. To illustrate, suppose the annual interest rate is 12% ($= 0.12$).

Annual Interest Rate	Interest is Compounded	Periodic Interest Rate
12% $= 0.12$	annually ($n = 1$)	$\dfrac{0.12}{1} = 0.12$
12% $= 0.12$	semiannually ($n = 2$)	$\dfrac{0.12}{2} = 0.06$
12% $= 0.12$	quarterly ($n = 4$)	$\dfrac{0.12}{4} = 0.03$
12% $= 0.12$	monthly ($n = 12$)	$\dfrac{0.12}{12} = 0.01$

Appendix D contains values for $(1 + r)^{nt}$ for several values of r and nt. A portion of this table is shown in Figure 1-7.

To illustrate how to use this table, suppose an amount of money (P) is invested in an account with a periodic interest rate of 4% ($r = 0.04$), and interest is compounded semiannually ($n = 2$). Furthermore, the money is invested for three years ($t = 3$). For this example,

$$(1 + r)^{nt} \text{ becomes } (1 + 0.04)^{2(3)} = 1.04^6$$

nt	$(1.01)^{nt}$	$(1.02)^{nt}$	$(1.03)^{nt}$	$(1.04)^{nt}$	$(1.05)^{nt}$
1	1.01000000	1.02000000	1.03000000	1.04000000	1.05000000
2	1.02010000	1.04040000	1.06090000	1.08160000	1.10250000
3	1.03030100	1.06120800	1.09272700	1.12486400	1.15762500
4	1.04060401	1.08243216	1.12550881	1.16985856	1.21550625
5	1.05101005	1.10408080	1.15927407	1.21665290	1.27628156
6	1.06152015	1.12616241	1.19405229	1.26531901	1.34009563
7	1.07213535	1.14868566	1.22987386	1.31593177	1.40710042
8	1.08285670	1.17165937	1.26677007	1.36856904	1.47745544
9	1.09368527	1.19509256	1.30477318	1.42331180	1.55132821
10	1.10462212	1.21899441	1.34391637	1.48024427	1.62889461

Figure 1-7. A portion of Appendix D.

In Figure 1-7 locate 6 in the column headed *nt*. Now move to the right in this row to the entry under the column headed 1.04. The entry is 1.26531901. Thus,

$$1.04^6 = 1.26531901, \text{ to eight decimal places.}$$

For any amount P invested in this fund, the amount at the end of the three years will be $P \cdot 1.26531901$. That is, the original amount P will have increased by approximately 26.5% (the first three digits after the 1 in the table factor).

Example 6.   Sarah Johnson put $7,500 in an investment fund that compounds interest semiannually for a period of ten years. The periodic interest rate is 5% $(=0.05)$. Compute, to the nearest cent, the amount Sarah will get back at the end of the ten years.

Solution.   **Discussion.** To find the amount, evaluate $P(1 + r)^{nt}$ for $P = \$7,500$, $r = 0.05$, $n = 2$, and $t = 10$.

$$\$7,500(1 + 0.05)^{2(10)}$$

$$= \$7,500(1.05)^{20} \qquad 1 + 0.05 = 1.05$$

$$= \$7,500(2.65329767) \qquad \text{from Appendix Table 1}$$

$$= \$19,899.73253 \qquad \text{the exact value}$$

$$= \$19,899.73, \text{ to the nearest cent.}$$

When a solid object is dropped near the earth's surface, it will fall a **distance,** d, in a specified **time,** t. The distance can be approximated with the expression $4.9t^2$ where d is measured in meters and t is measured in seconds. For example, suppose a steel ball is dropped from the bridge in Figure 1-8. The distance from the bridge to the water can be estimated by knowing the time it takes the dropped ball to hit the water. The expression $4.9t^2$ would then be evaluated for the value of t.

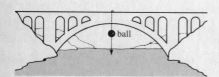

Figure 1-8. A bridge over a deep canyon.

Example 7.   If the steel ball in Figure 1-8 takes eight seconds to reach the water, how high is the bridge from the water to the nearest meter?

Solution.   Evaluating the expression $4.9t^2$ for $t = 8$ seconds,

$$4.9(8)^2 \qquad \text{Replace } t \text{ by 8.}$$

$$= 4.9(64) \qquad \text{Raise to powers first.}$$

$$= 313.6 \text{ or about 314 meters.}$$

SECTION 1-7.   Practice Exercises

In exercises **1–4**, evaluate for $x = 7$.

[Example 1]   **1.** $3x + 2$ **2.** $4x - 10$

3. $2x^2 - 15$ **4.** $3x^2 - 9$

In exercises **5–10**, evaluate for $a = -4$.

5. $5a + 9$ **6.** $2a - 11$

7. $a^2 - 9$ **8.** $4a^2 - 25$

9. $2a^2 - a - 3$ **10.** $5a^2 - 6a + 1$

In exercises **11–14**, evaluate for $x = 3$ and $y = -5$.

[Example 2]   **11.** $5x + 2y$ **12.** $-4x + 3y$

13. $2x^2 + 7xy$ **14.** $5y^2 + 9xy$

In exercises **15–20**, evaluate for $a = -2$ and $b = 4$.

15. $2a - 3b$ **16.** $8a - 9b$

17. $10a^2 - 20b^2$ **18.** $4a^2 - 3b^2$

19. $a^2b^2 + 3ab - 10$ **20.** $2a^2b^2 - ab + 25$

In exercises **21–26**, evaluate for $x = 2$, $y = -3$, and $z = 4$.

[Example 3]   **21.** $2[x - (2y + z)]$ **22.** $3[2y + (x + 2z)]$

23. $-2[3z + 5(x + y)]$ **24.** $-5[3x - 7(3y + 2z)]$

25. $5(x - y) - 9(x + z)$ **26.** $-6(5x + y) - 7(4y + z)$

In exercises **27–32**, evaluate for the given values.

[Example 4]   **27.** $\dfrac{xy + x^2}{xy + y^2}$ for $x = 8$, $y = -2$

28. $\dfrac{x^3 + y^3}{x + y}$ for $x = -5$, $y = 2$

29. $\dfrac{x^2 y}{x + y}$ for $x = 10$, $y = 15$

30. $\dfrac{m^3 - n^3}{2m - n}$ for $m = 6$, $n = 5$

31. $\dfrac{x^2 - y^2}{xy} + \dfrac{2x - z}{2z}$ for $x = 4$, $y = 5$, $z = 10$

32. $\dfrac{n^2 + p^2}{m^2} + \dfrac{mn}{4p^2}$ for $m = -8$, $n = 3$, $p = 4$

In exercises **33–38**, find the area of the trapezoid with the given measurements.

[Example 5]   **33.** $h = 5$ inches (in), $a = 4$ in, $b = 8$ in

34. $h = 8$ in, $a = 10$ in, $b = 5$ in

35. $a = 13$ centimeters (cm), $b = 10$ cm, $h = 6$ cm

36. $a = 15$ cm, $b = 12$ cm, $h = 10$ cm

37. $a = 75$ millimeters (mm), $h = 40$ mm, $b = 60$ mm

38. $a = 42$ mm, $h = 52$ mm, $b = 31$ mm

In exercises **39–42**, find the answers to the compound interest problems.

[Example 6]   **39.** Find the amount of money available if \$2,000 is invested at 8%, compounded quarterly, and left in the investment for nine years.

40. Find the amount of money available if \$1,000 is invested at 6%, compounded semiannually for five years.

41. A \$3,500 investment is left in an account for four years at 6% compounded semiannually.
a. How much money is in the fund at the end of four years?
b. How much interest was earned?

42. \$25,000 is invested at 12% compounded monthly for two years.
a. How much money is in the fund at the end of the two years?
b. How much interest was earned?

In exercises **43–46**, answer the following distance problems.

[Example 7]   **43.** If an object was dropped from the top of the Washington Monument in Washington, D.C., it would take about 6 seconds to hit the ground. About how many meters high is the monument?

44. If a crescent wrench was dropped from a tower in a chemical plant and took 4 seconds to hit the ground, how tall is the tower?

45. A rock takes 4.5 seconds to reach the valley at the foot of a high cliff. How high is the cliff above the valley?

46. If it takes 2.5 seconds for a pebble to fall from the top of a building to the ground, how tall is the building?

SECTION 1-7.   Supplementary Exercises

In exercises **1–10**, evaluate the expression for the given values.

1. $(2x - 2)^2 + 5$ $x = 7$

2. $(2p - 5)^2 - 33$ $p = 4$

3. $-5(4a + 3b)$ $a = -0.2, b = 1.4$

4. $-3(5a + 2b)$ $a = -0.3, b = 2.1$

5. $4a^2 - b^2$ $a = -2, b = 4$

SECTION 1-1.   Summary Exercises

For exercises **1–4**, solve each problem.

1. List the first five multiples of 7.

1. _____

2. List the natural numbers that divide 36.

2. _____

3. Write in $\frac{a}{b}$ form the rational number 7.3.

3. _____

4. Insert $<$, $=$, or $>$ between each pair of numbers to make the statement true.

4. a. _____

a. -23 11

b. -32 -34

b. _____

c. $3\frac{1}{2}$ $3\frac{4}{5}$

c. _____

67

For problems **5–8**, simplify.

5. $43 - 21 - 12 + 17$

5. _____

6. $\dfrac{3}{4} \div \dfrac{7}{8} \cdot \dfrac{14}{15}$

6. _____

7. $9 \cdot 8 - 36 \div 9 - 8$

7. _____

8. $1 + 8 \cdot 16 + 25 \div 5 - 9$

8. _____

Name _____

Date _____

Score _____

SECTION 1-2. Summary Exercises

Answer

For exercises **1–7**, simplify.

1. $-3 + 7 + (-15)$

1. _____

2. $12 - (-15) - 11$

2. _____

3. $(42 - 39) + (62 - 75)$

3. _____

4. $(31 - (-12)) - (45 + (-23))$

4. _____

5. $-32 + 73 - (92 - (43 + 6))$

5. _____

6. $-(21 + (-6)) + (39 - (28 - (-11)))$

6. _____

7. $34 + (17 + (35 - (6 + (-12)))) - 30$

7. _____

8. Susan started a diet at a weight of 163 pounds. At the end of each week her weight had changed as follows: loss of 4 pounds, loss of 5 pounds, gain of 2 pounds, loss of 4 pounds and a gain of 1 pound. What was her final weight?

8. _____

SECTION 1-3. Summary Exercises

Name

Date

Score

Answer

For exercises **1–7**, simplify.

1. $(-42)(-21)$

1. _____

2. $-84 \div 12$

2. _____

3. $(-180) \div (-3) + (8)(-7)$

3. _____

4. $-17 + 3(15) - (-4)$

4. _____

5. $\dfrac{-41 + 45(-2) + 14}{-5 - (11)(-4)}$

5. _____

6. $14 - 17 \cdot 2 + 8 - 3(-9)$

6. _____

7. $(75)(-5) \div 15(5)$

7. _____

8. Write $\dfrac{-7}{22}$ as an equivalent fraction with a denominator of 176.

8. _____

SECTION 1-4. Summary Exercises

In exercises **1–8**, simplify.

1. $4 + 6 \cdot 2 - 3^3$

1. _____

2. $2(10 - 3) - 4(2^2 - 5)$

2. _____

3. $12 + 3 \cdot 5 - 2^4$

3. _____

4. $\dfrac{2 \cdot 8^2 + 2^2 \cdot 8}{2 \cdot 8}$

4. _____

5. $\dfrac{9 \cdot 3 - 4^2 + 2}{3^2 + 2^2}$

5. _____

6. $\{(4 + 8)^2 \div (4)(-2)\} + (7)(-3)$

6. _____

7. $[-32 \div (4)(-8)][93 - (-7)] + ((2)(8) - 4^2)$

7. _____

8. $-3[2(15 - 2^3) - (5^2 - 13)]$

8. _____

SECTION 1-5. Summary Exercises

Answer

In exercises **1** and **2**, use the commutative property to rewrite each expression.

1. $-8.5 + 10.7$

1. _____

2. $\dfrac{7}{10} \cdot \dfrac{-8}{3}$

2. _____

In exercises **3** and **4**, use the associative property to rewrite each expression.

3. $\left(-10 \cdot \dfrac{4}{9}\right) \cdot 27$

3. _____

4. $-26 + (26 + 89)$

4. _____

5. State

    a. the additive inverse of -7.4

5. **a.** _____

    b. the multiplicative inverse of 4.7

    b. _____

In exercises **6** and **7**, use the distributive property to

a. change the order of operations.

b. simplify the changed expression.

6. $45\left(\dfrac{-7}{9} + \dfrac{2}{5}\right)$

6. **a.** _____

    b. _____

7. $0.49(62) + 0.49(38)$

In exercises **8–15**, identify completely by name the property illustrated.

8. $(4 + 9) + 0 = 4 + 9$

8. _____

9. $-15 \cdot 1 = -15$

9. _____

10. $2.8 + (1.2 + 7.0) = (2.8 + 1.2) + 7.0$

10. _____

11. $2.8\left(\dfrac{1}{2} + \dfrac{3}{7}\right) = 2.8\left(\dfrac{1}{2}\right) + 2.8\left(\dfrac{3}{7}\right)$

11. _____

12. $\dfrac{4}{3} \cdot \left(\dfrac{1}{2} \cdot \dfrac{3}{7}\right) = \left(\dfrac{4}{3} \cdot \dfrac{1}{2}\right) \cdot \dfrac{3}{7}$

12. _____

13. $\dfrac{-10}{9} + \dfrac{10}{9} = 0$

13. _____

14. $\left(2\dfrac{1}{3}\right)\left(\dfrac{3}{7}\right) = 1$

14. _____

15. $53(19) + 53(81) = 53(19 + 81)$

15. _____

SECTION 1-6. Summary Exercises

1. Identify for the following terms

 (i) the numerical coefficient

(ii) the literal coefficient

 a. $-6x^2y$

1. **a.** _____

_____

 b. $\dfrac{3a^4}{5}$

b. _____

_____

 c. pq

c. _____

_____

 In exercises **2–8**, simplify.

2. $5x - 4x + 7x$

2. _____

3. $-6m + 4 + 12m - 11$

3. _____

4. $(b^2 - 3b + 7) + (-12b^2 - b + 15)$

4. _____

5. $6x^2y - 9xy^2 + x^2y + 11xy^2$

5. _____

6. $5(-6x + 4) + 7(x - 7)$

6. _____

7. $8(x^2 + 7xy + y^2) - 2(3x^2 - xy + 6y^2)$

7. _____

8. $3(-2c^2d + 4cd^2 - 3cd) - 5(5c^2d - 7cd^2 + cd)$

8. _____

SECTION 1-7. Summary Exercises

Answer

In exercises **1–6**, evaluate the expressions for the given values.

1. $5y^2 - 15$ and $y = 0$

1. _____

2. $a^2 - 3a + 6$ and $a = -3$

2. _____

3. $-9p + 3q$ and $p = 6$, $q = -5$

3. _____

4. $(a^2 - 9)^2 + (a^2 - 9) + 4$ and $a = 4$

4. _____

5. $11[2z + 3(5 - 2z)]$ and $z = -6$

5. _____

6. $\dfrac{x^2 - 5xy}{3x - 15y}$ and $x = -12, y = 3$

6. _____

7. a. Sketch a trapezoid.

7. a. _____

b. Find the area if the parallel sides are 6 inches and 9 inches and the height is 4 inches.

b. _____

8. How much money would be in a fund if $500 was invested at 8% compounded quarterly for two years?

8. _____

SECTION 1-8. Summary Exercises

In exercises **1–8**, write each word phrase as an algebraic expression.

1. The sum of x and 3 is decreased by the sum of y and 5.

1. _____

2. The product of 2 and x is increased by 11.

2. _____

3. 3 more than x is multiplied by 6 less than y.

3. _____

4. k reduced by 1 is divided by the square of x.

4. _____

5. *y* minus 6 is multiplied by one-half of *x*.

5. _____

6. The cube of the sum of *a* and *b* is divided by the square of *c*.

6. _____

7. The sum of *k* squared and 3 is divided by 2.

7. _____

8. The quotient of 3 and *x* is subtracted from the product of 3 and *x*.

8. _____

CHAPTER 1. Review Exercises

In exercises **1–5**, solve each problem.

1. Write $3\frac{1}{5}$ in rational number form.

2. List the natural numbers that divide 12.

3. Write the following without absolute value bars:
   a. $|-19|$ **b.** $-|-19|$ **c.** $-|19|$

4. Simplify $12 - 2(6) + 20 \div 2$

5. Insert $<$, $=$, or $>$ between each pair of numbers to make the statement true.

   a. $\dfrac{3}{4} \quad \dfrac{4}{3}$ **b.** $\dfrac{-7}{8} \quad \dfrac{-2}{3}$

In exercises **6–13**, simplify.

6. $5 + (-7) + (-10) + 4$

7. $5 - (-7) - (-10) - 4$

8. $-12 + (-31) - 14 - (-7)$

9. $5xy + 12xy - 20xy$

10. $(-7)(-8) + (4)(-6)$

11. $\dfrac{-36}{9} + (4)(3)$

12. $\dfrac{14 - 30 - (-1)}{(-2)(-3) - 1}$

13. $(-32)(17)(-6)(0)$

14. Find the missing number needed to make the two fractions equal.

   a. $\dfrac{-4}{7} = \dfrac{?}{35}$ **b.** $\dfrac{-5}{8} = \dfrac{?}{-56}$

In exercises **15–19**, simplify.

15. $3 + 5 + 2 \cdot 6 - (-4)$

16. $5 \cdot 6 - 2 \cdot 3 + 2^2 \cdot 3^2$

17. $-3[2(15 - 2^3) - (5^2 - 13)]$

18. $\dfrac{2(10 - 13) - 3(14 - 19)}{9 - 6}$

19. $(3^2 - 5)^2 - (30 - 2^5)^3$

In exercises **20–24**, identify the properties illustrated as either commutative, associative, or distributive.

20. $12 \cdot 8 = 8 \cdot 12$

21. $5(x + 3) = 5(x) + 5(3)$

22. $(x + 2) + y = y + (x + 2)$

23. $a + (b + 4) = (a + b) + 4$

24. $(x + y)z = x(z) + y(z)$

25. **a.** State the additive inverse of -12.
    b. State the multiplicative inverse of -12.

In exercises **26–28**, simplify.

26. $3x^2 + 7x - 9x^2 + 2$

27. $2(ab + 3a) - (6ab - 2b)$

28. $-5m^2n + 13mn^2 + 7m^2n - mn^2$

29. State the number of terms in each expression.
   a. $-4x^2y + 32xy - z$
   b. $(a + b)^3 - 27$
   c. $-132x^2yz^4$

In exercises **30–33**, evaluate for the given values.

30. $3x - x^2$ and $x = -5$

31. $4x - 2y + 3$ and $x = 0, y = -8$

32. $\dfrac{a^2 - ab + 6b^2}{a + 4b}$ and $a = 5, b = 2$

33. $m^3 + 9m^2 + 27m + 27$ and $m = -2$

In exercises **34–36**, write each word phrase as an algebraic expression.

34. The product of 25 and x is increased by the cube of x.

35. The 4th power of y is reduced by the quotient of x and 4.

36. The square of the difference of 10 and a is multiplied by the sum of c and d.

2

Linear Equations
and Inequalities
in One Variable

When Ms. Glaston arrived for class one morning, she placed two small paper bags on the desk. A large "A" was painted on one bag and a large "B" was painted on the other. (See Figure 2-1.)

Figure 2-1. Two bags containing an unknown number of jelly beans.

"In each of these bags I have a certain number of jelly beans," Ms. Glaston said. "Of interest at this time is the number of jelly beans in the two bags. The simplest action to take to determine these numbers would be to remove the jelly beans from the bags and count them. Suppose, however, that this action for some reason is not possible. Instead, assume that some information is given that makes it possible for us to analytically determine these numbers."

She then placed an acetate sheet on the overhead projector and displayed the following paragraph on the screen:

> Bag B has 33 more jelly beans than Bag A. Furthermore, if 8 is added to five times the number of jelly beans in Bag A, then the sum is the same as when 7 is substracted from three times the number of jelly beans in Bag B. How many jelly beans are in Bags A and B?

"Ms. Glaston," Bob Gazdacko said, "that paragraph reads like a riddle. There's no way I could plow through those words to get an answer to your question."

"OK Bob," Ms. Glaston replied, "suppose the word phrases are difficult to interpret. What can we do to get around the problem?"

"I have an idea," Shannon Schanze said. "We can change some of those words to mathematical symbols. For example, 'added' can be replaced by a plus sign and 'subtracted' can be replaced by a minus sign. We could do a lot more if we used a variable for something in the problem."

"A good suggestion, Shannon," Ms. Glaston replied. "Suppose we do use a variable for some unknown quantity. What variable might we use and what would the variable represent?"

"Use x," Marty Rosen replied quickly. "In algebra we always use x to represent anything that is unknown."

"Come on, Marty," Annie Todd said. "We can be a little more creative than that! Since we're looking for the **number** of jelly beans in the bags, let's use n to represent that number."

"But there are two bags, Annie," Shannon replied. "Maybe we should let n represent the number of jelly beans in just one of the bags."

"Okay, let's follow up on these suggestions," Ms. Glaston said. "The first sentence in the paragraph states that Bag B has 33 more jelly beans than Bag A. If we let n represent the number in Bag A, then $n + 33$ could represent the number in Bag B." She then turned to the board and wrote

Let n represent the number of jelly beans in Bag A.

Then $n + 33$ represents the number of jelly beans in Bag B.

"The phrase '8 added to five times the number in Bag A' can be written '$5n + 8$'," Shannon suggested.

"And the phrase '7 is subtracted from three times the number in Bag B' can be written '$3(n + 33) - 7$'," Annie said.

"Finally," Ms. Glaston said, "the phrase 'is the same as' can be replaced by an equal sign." She then wrote the following equation on the board:

$$5n + 8 = 3(n + 33) - 7$$

"This is an example of an **equation**," Ms. Glaston continued. "It is an accurate description in mathematical symbols of the content of the problem stated in words. The number replacement for n that makes a true equality answers the question of how many jelly beans are in Bag A. The same replacement for n in $n + 33$ will tell us how many jelly beans are in Bag B. The process of determining what number will accomplish these tasks is called 'solving the equation.' In today's lesson we will begin the study of equations like the one written on the board. After we learn how to solve equations, we'll figure out just how many jelly beans there are in these bags. To check our results we will then actually count the jelly beans. As a final task, we will then eat every last one of them!"

SECTION 2-1.   Parts of an Equation

KEY TOPICS IN THIS SECTION

1. A definition of an equation

2. A solution, or root of an equation

3. Equivalent equations

4. Algebraic expressions versus equations

This section starts our study of equations. First we define an equation, and then we identify the parts that comprise it. We then study the meaning of a solution of an equation. Finally, we state the conditions under which two or more equations are equivalent.

A Definition of an Equation

The first step in studying equations is to learn to recognize one. Definition 2.1 identifies the essential elements in an equation.

> **Definition 2.1. Equation**
> An equation is a mathematical sentence in which the verb is **equals** (represented by =). The sentence asserts that the expressions to the left and right of the equal sign name the same number.

Before we begin our study of equations, we should point out that the equal sign is also used to compare two number expressions. Examples **a** and **b** illustrate two such uses of the equal sign.

a. $3 + 5(2) = 6(4) - 11$

b. $7(4) - 36 \div 2 = 3^2 + 2$

Since the expressions in examples **a** and **b** are numerical, they can be simplified to a single number. As a consequence, the equation can be determined to be *true* (both expressions do name the same number), or *not true* (both expressions do not name the same number).

Example 1.   Determine whether the following equations are true or not true.

a. $3 + 5(2) = 6(4) - 11$

b. $7(4) - 36 \div 2 = 3^2 + 2$

Solution.   **Discussion.** Simplify the expressions on both sides of the equal sign. If they name the same number, the equation is true. If not, the equation is not true.

a. Simplify the left side.

$$3 + 5(2) = 3 + 10 \qquad \text{Multiply before adding.}$$

$$= 13 \qquad \text{The left side is 13.}$$

Simplify the right side.

$$6(4) - 11 = 24 - 11 \qquad \text{Multiply before subtracting.}$$

$$= 13 \qquad \text{The right side is 13.}$$

Since both expressions simplify to 13, the equation is true.

b. Simplify the left side.   Simplify the right side.

$$7(4) - 36 \div 2 \qquad\qquad 3^2 + 2$$

$$= 28 - 18 \qquad\qquad\quad = 9 + 2$$

$$= 10 \qquad\qquad\qquad\quad = 11$$

Since the two expressions simplify to different numbers, the equation is not true.

Equations may include one or more **algebraic expressions.** Since a variable does not usually represent exactly one number, such an equation cannot be classified as true or not true. This type of equation is called **conditional.** For example,

$$10x - 7 = 14 + 3x$$

is a **conditional equation.** This equation becomes true if x is replaced by an appropriate number. The equation is not true if x is replaced by other numbers.

Example 2.   Determine whether the following numbers make the equation

$$10x - 7 = 14 + 3x$$

true or not true.

a. 5

b. 3

Solution.

a. $10(5) - 7 = 14 + 3(5)$ Replace x with 5.

$50 - 7 = 14 + 15$ Multiply first.

$43 = 29$ This equation is not true.

b. $10(3) - 7 = 14 + 3(3)$ Replace x with 3.

$30 - 7 = 14 + 9$ Multiply first.

$23 = 23$ This equation is true.

The equation is false when x is replaced with 5, but it is true when x is replaced with 3.

A Solution, or Root of an Equation

In Example 2 the conditional equation $10x - 7 = 14 + 3x$ is true when x is replaced with **3.** The number 3 is called a **solution** of the equation.

Definition 2.2.   A solution of a conditional equation
   A **solution** of a conditional equation in one variable is a number that makes the equation **true** when the variable is replaced with that number. A solution is also called a **root** of the equation.

Example 3.   Determine which of the numbers $-3, -2, 2,$ or 3 are roots of the following equation:

$$3t = 18 - 3t^2$$

Solution.   **Discussion.** Replace t with each of these numbers to determine which, if any, make the equation true.

Replace t by -3	Replace t by -2
$3(-3) = 18 - 3(-3)^2$	$3(-2) = 18 - 3(-2)^2$
$3(-3) = 18 - 3(9)$	$3(-2) = 18 - 3(4)$
$-9 = 18 - 27$	$-6 = 18 - 12$
$-9 = -9$, true	$-6 = 6$, false
Thus -3 is a root.	Thus -2 is not a root.

Replace t by 2	Replace t by 3
$3(2) = 18 - 3(2)^2$	$3(3) = 18 - 3(3)^2$
$3(2) = 18 - 3(4)$	$3(3) = 18 - 3(9)$
$6 = 18 - 12$	$9 = 18 - 27$
$6 = 6$, true	$9 = -9$, false
Thus 2 is a root.	Thus 3 is not a root.

Therefore -3 and 2 are roots, but -2 and 3 are not.

Equivalent Equations

A major task in this course will be to learn procedures whereby given conditional equations can be changed to forms in which any roots can be easily found. Furthermore, the procedures will use properties in mathematics that alter the forms of the equations, but not any roots of the equations. We call equations that have the same solutions **equivalent**.

Definition 2.3. Equivalent equations
If two or more equations have exactly the same solutions, then they are **equivalent**.

Example 4. Are equations **a** and **b** equivalent, given that both have exactly one solution?

a. $6k + 9 = 4k + 10$

b. $k = \dfrac{1}{2}$

Solution. **Discussion.** The root of the **b** equation is $\frac{1}{2}$. If $\frac{1}{2}$ is also a root of the **a** equation, then **a** and **b** are equivalent. If $\frac{1}{2}$ is not a root of **a**, then **a** and **b** are not equivalent.

Replace k with $\frac{1}{2}$ in equation **a**.

$$6 \cdot \frac{1}{2} + 9 = 4 \cdot \frac{1}{2} + 10$$

$3 + 9 = 2 + 10$ Multiply first on both sides.

$12 = 12$, true   Now add on both sides.

Since $\frac{1}{2}$ is a root of both equations they are equivalent.

Algebraic Expressions Versus Equations

Consider the following:

a. $7(m - 3) - 3(2m - 5) - 4(m - 6)$

b. $7(m - 3) - 3(2m - 5) = 4(m - 6)$

Just glancing at the symbols in parts **a** and **b** may lead one to conclude that they are the same. However, by examining them more closely we can see that part

b has an equal sign, but part **a** does not. As a consequence, we would call part **a** an **expression,** but part **b** is an **equation.**

In a language class we are taught to distinguish between a **phrase** and a **sentence.** Similarly, in a mathematics class we are taught to distinguish between *expressions* and *equations.* We must keep in mind that expressions merely represent numbers. We can change the form of an expression by doing certain operations, such as combining like terms within the expression. But the numbers that the expression represents remain the same.

By contrast, an equation is a statement that is either true or false. If the equation is conditional, then replacement of the variable or variables by certain numbers will make the equation either true or false. The goal we attempt to reach when working with equations is quite different from our goal when working with expressions.

Example 5.   **a.** Simplify: $7(m - 3) - 3(2m - 5) - 4(m - 6)$

b. Are equations **A** and **B** equivalent, given that they both have exactly one root?
A. $7(m - 3) - 3(2m - 5) = 4(m - 6)$
B. $m = 6$

Solution.   **a.** $7(m - 3) - 3(2m - 5) - 4(m - 6)$

$= 7m - 21 - 6m + 15 - 4m + 24$ Remove parentheses.

$= -3m + 18 \text{ or } 18 - 3m$ Combine like terms.

b. The root of equation **B** is 6. Replacing m by 6 in equation **A,**

$7(6 - 3) - 3(2(6) - 5) = 4(6 - 6)$

$7(3) - 3(7) = 4(0)$ Simplify within parentheses.

$21 - 21 = 0, \text{ true}$

Since 6 is a solution of both equations, they are equivalent.

SECTION 2-1.   Practice Exercises

In exercises **1–8,** determine whether the equations are true or not true.

[Example 1]   **1.** $2(17) + 1 = 5 + 6(5)$

2. $5(8) + 2 = 6 + 6(6)$

3. $7(-3) + (-5)(-4) = -8(-7) - 11(5)$

4. $9(-6) - (10)(-6) = (-6)(4) - 9(-3)$

5. $\dfrac{5}{3} \cdot \dfrac{-2}{7} + \dfrac{1}{3} = \dfrac{1}{7} \cdot \dfrac{3}{7} - \dfrac{10}{49}$

6. $\dfrac{3}{4} - \dfrac{1}{3} \cdot \dfrac{5}{2} = \dfrac{-1}{6} \cdot \dfrac{7}{2} + \dfrac{1}{2}$

7. $4.9 \div 0.1 + (-2.0) = (0.5)(132) - 19$

8. $-7.86 \div (0.2 - 0.8) + 0.9 = (1.7 + 8.3)28 \div 20$

In exercises **9–16**, determine whether the given number makes the equation true or false.

[Example 2]   **9.** $9(x - 1) + 1 = 3x + 16$ and $x = 4$

10. $5(x - 2) - 2 = 3(20 - x)$ and $x = 9$

11. $6(x + 3) = 7 - 4(x + 5)$ and $x = -3$

12. $2(2x + 5) = 10 - (x + 40)$ and $x = -8$

13. $6x + 12 = 15 - 3x$ and $x = \dfrac{1}{3}$

14. $15x + 7 = 5x + 20$ and $x = \dfrac{7}{5}$

15. $2(4x - 1) = 1 - (4x + 9)$ and $x = \dfrac{-1}{2}$

16. $5(3x + 1) = 3x - 10$ and $x = \dfrac{-4}{3}$

In exercises **17–30**, determine which of the given numbers are roots of the stated equations.

[Example 3]   **17.** $8x + 3 = 19$; 0, 1, 2, 3

18. $20 - 7y = 6$; 0, 1, 2, 3

19. $9a - 8 = 2 - a$; 0, 1, 2, 3

20. $33 - 10x = 5x + 3$; 0, 1, 2, 3

21. $2 - 3x = 8$; $-3, -2, -1, 0$

22. $9 - 4y = 17$; $-3, -2, -1, 0$

23. $2(2m + 1) = 4$; $0, \dfrac{1}{2}, \dfrac{1}{4}$

24. $2(4x - 1) = 2$; $0, \dfrac{1}{2}, \dfrac{1}{4}$

25. $n^2 - 6 = 4n + 6$; 2, 4, 6

26. $3n^2 - 2 = 7n - 4$; 2, 4, 6

27. $x^2 + 4 = 5x$; 1, 2, 3, 4

28. $y^2 + 8 = 6y$; 1, 2, 3, 4

29. $a^2 + a = 2$; $-2, -1, 0, 1, 2$

30. $b^2 - 2b = 8$; $-4, -2, 0, 2, 4$

In exercises **31–38**, determine whether equations **a** and **b** are equivalent, given that they have exactly one solution.

[Example 4]   **31. a.** $17x + 3 = 12 + 8x$ **b.** $x = 1$

32. a. $10y - 14 = 36 - 15y$ **b.** $y = 2$

33. a. $6(z + 5) = 4 - 3z$ **b.** $z = -3$

34. a. $3b + 17 = 12 - 2b$ **b.** $b = -5$

35. a. $3(2m - 1) = 7m + 4$ **b.** $m = -7$

36. a. $2(2a + 3) = 6a - 6$ **b.** $a = 6$

37. a. $30n + 1 = 10n + 10$ **b.** $n = \dfrac{1}{2}$

38. a. $1 - 5x = 15x - 2$ **b.** $x = \dfrac{1}{5}$

In exercises **39–44**,

a. simplify the expression

b. determine if the equations are equivalent, given that both equations have exactly one solution

[Example 5] **39. a.** $2(x + 2) - 3x - 7$ **b.** $2(x + 2) = 3x - 7$
$$x = 11$$

40. a. $10x + 9 + 12x + 1$ **b.** $10x + 9 = 12x + 1$
$$x = 4$$

41. a. $6a + 9 - 10a - 2$ **b.** $6a + 9 = 10a + 2$
$$a = 7$$

42. a. $25 - 3b - 9b + 11$ **b.** $25 - 3b = 9b - 11$
$$b = 3$$

43. a. $8c - 7 + 2(6c - 5)$ **b.** $8c - 7 = 2(6c - 5)$
$$c = \dfrac{3}{4}$$

44. a. $6(m + 1) + 12m + 2$ **b.** $6(m + 1) = 12m + 1$
$$m = \dfrac{2}{3}$$

SECTION 2-1.   Ten Review Exercises

In exercises **1–10**, solve each problem.

1. Simplify: $2 \cdot 5^2 - 7 \cdot 5 - 15$

2. Evaluate for $t = 5$.
$2t^2 - 7t - 15$

3. Is 5 a root of $2t^2 - 7t = 15$?

4. Is $\frac{-3}{2}$ a root of $2t^2 = 7t + 15$?

5. Simplify: $2(3(-2) + 8) - 5(-2 + 4)$

6. Evaluate for $k = -2$.
   $2(3k + 8) - 5(k + 4)$

7. Simplify: $2(3k + 8) - 5(k + 4)$

8. Evaluate the expression from exercise 7 for $k = -2$.

9. Is -2 a root of $2(3k + 8) = 5(k + 4)$?

10. Are $2(3k + 8) = 5(k + 4)$ and $k = -2$ equivalent equations?

SECTION 2-1.   Supplementary Exercises

In exercises **1–6**, determine whether the equations are true or not true.

1. $(-4)^2 - 3(-5) = 3(2 - (-3)) + (-1)^2$

2. $(-1)^3 + 2(3)^3 = -5(7 - (-2) + 2^3)$

3. $(6 - 2)^2(-8 + 6) = (-5)^2(-2) - (-3)(6)$

4. $(9 - 2)(5 - 3)^2 = (6 - (-1))(3^2 - 5)$

5. $4(2)^3 - 3(2)^2 + 6 = 2^2 + 10(2) + 2$

6. $9(3)^2 - 3 = 3^3 + 8(3) + (5^2 + 2)$

In exercises **7–12**, determine which of the given numbers are roots of the stated equations.

7. $3 + 12a = 0; \dfrac{-1}{4}, \dfrac{-1}{2}, 0$

8. $8 + 8a = 6; \dfrac{-1}{4}, \dfrac{-1}{2}, 0$

9. $c^2 = 3c + 18; -6, -3, 0, 3, 6$

10. $3m = 18 - 3m^2; -3, -2, 0, 2, 3$

11. $x^4 + x^3 = 4x^2 + 4x; -2, -1, 0, 1, 2$

12. $x^4 + 2x^3 = x^2 + 2x; -2, -1, 0, 1, 2$

In exercises **13–18**, determine whether equations **a** and **b** are equivalent, given that they have exactly one solution.

13. **a.** $4(2y - 1) + 7 = 9y - 5$ **b.** $y = 8$

14. **a.** $6a + 9 = 17a - 68$ **b.** $a = 7$

15. **a.** $4y + 3 = -2y + 15$ **b.** $y = 2$

16. **a.** $6 - 5z = -8z - 6$ **b.** $z = -4$

17. a. $5(t + 1) = 5t + 1$ **b.** $t = 2$

18. a. $2(w - 2) = 2w - 2$ **b.** $w = -3$

In exercises **19–24**, answer parts **a**, **b**, and **c**.

19. All the roots for equations **a** and **b** are in the set $\{-2, -1, 0, 1, 2\}$.
   a. Find the root(s) of $x^2 - 4 = 0$.
   b. Find the root(s) of $(x - 2)(x + 2) = 0$.
   c. Are the equations equivalent?

20. All the roots for equations **a** and **b** are in the set $\{-3, -1, 0, 1, 3\}$.
   a. Find the root(s) of $x^2 - 9 = 0$.
   b. Find the root(s) of $(x - 3)(x + 3) = 0$.
   c. Are the equations equivalent?

21. All the roots for equations **a** and **b** are in the set $\{-10, -5, 0, 5, 10\}$.
   a. Find the root(s) of $x^2 - 100 = 0$.
   b. Find the root(s) of $x - 10 = 0$.
   c. Are the equations equivalent?

22. All the roots for equations **a** and **b** are in the set $\{-4, -2, 0, 2, \ 4\}$.
   a. Find the root(s) of $x^2 - 16 = 0$.
   b. Find the root(s) of $x - 4 = 0$.
   c. Are the equations equivalent?

23. The cost of a bottle is always $1 more than the cost of the cork.
   a. If the possible costs of the cork are $1, 75¢, 50¢, or 5¢ then what are the corresponding costs of the bottle?
   b. Compute the corresponding possible costs of the bottle and the cork together.
   c. If the cost of the bottle and cork together is $1.50, what is the cost of each item?

24. Tom weighs 10 pounds more than Ramon.
   a. If the possible weights for Tom are 120 pounds, 135 pounds, 155 pounds, or 180 pounds, what are the corresponding weights of Ramon?
   b. Compute the corresponding possible weights of Tom and Ramon together.
   c. If the combined weight of Tom and Ramon is 290 pounds, what is the weight of each man?

SECTION 2-2.   Solving Linear Equations in One Variable

KEY TOPICS IN THIS SECTION

1. A definition of a linear equation in one variable

2. What it means to solve an equation

3. Solving equations with the addition property of equality

4. Solving equations with the multiplication property of equality

5. Solving equations using both properties

In this section we will study in more detail how to find the roots of an equation. We will use the linear equation in one variable to study this problem.

A Definition of a Linear Equation in One Variable

In a beginning algebra course we learn procedures for finding the roots of several different kinds of equations. As we will see, each kind of equation has a recommended procedure for finding the roots. As a consequence, a general form is given for each kind of equation so we know which procedure to use. Definition 2.4 gives the general form of a linear equation in one variable.

Definition 2.4.   A linear equation in x
   A **linear equation** in the variable x is one that can be written in the form

$$ax + b = c$$

where a, b, and c are real numbers, and $a \neq 0$.

Examples **a–c** identify three equations that can be changed to the form of the equation in Definition 2.4.

Given form	The $ax + b = c$ form
a. $2(x + 3) = 9$	$2x + 6 = 9$
b. $7x - 5 = 4x + 22$	$3x - 5 = 22$
c. $5(x - 4) + 3(x + 3) = 1$	$8x - 11 = 1$

What it Means to Solve an Equation

Consider equations **1.** and **2.**

1. $5(x + 3) - 12 = 22 - 3(x - 7)$

2. $\qquad\qquad x = 5$

The solution to **2** is easily seen to be 5. When x is replaced by 5 in equation **1** the true equality $28 = 28$ is obtained. The process of changing an equation such as **1** to an equivalent equation such as **2** is called **solving an equation.**

   To **solve a linear equation in x** means to change, if possible, the given equation to an equivalent equation with the form

$$x = k \qquad \text{or} \qquad k = x$$

where k is a real number.

If the equation $x = k$ or $k = x$ is equivalent to the given equation, then k is a root of that equation.

Solving Equations with the Addition Property of Equality

Two properties of equality are useful in solving equations. The first is the **addition property,** which states that the same quantity (positive or negative) can be added to both sides of an equation, and the resulting equation will be equivalent.

Addition property of equality

If $a = b$, and c is any quantity, then

$$a + c = b + c$$

Figure 2-2 illustrates what the addition property of equality means. The scale shown is level, indicating that the quantity of material on the two pans weigh the same amount. If exactly the same amount of material is added to (or taken away from) both pans, then the scale will remain in balance.

Examples 1 and 2 illustrate equations that can be solved by using the addition property of equality. Remember, the object of solving an equation is to change it to the form $x = k$ or $k = x$.

Figure 2-2. A balanced scale.

Example 1.   Solve and check:

a. $t + 9 = 26$

b. $12 + 5y = 6y$

Solution.   **a. Discussion.** The equation has one variable term and two constant terms. We need to combine the constants on the side of the equation opposite to the variable term.

$t + 9 = 26$	The given equation
$t + 9 + (-9) = 26 + (-9)$	Add (-9) to both sides.
$t + 0 = 17$	$9 + (-9) = 0$
$t = 17$	
Check: $17 + 9 = 26$	Replace t by 17.
$26 = 26$, true	The root is 17.

b. Discussion. This equation has one constant term and two variable terms. We need to combine the variable terms on the side of the equation opposite to the constant term.

$12 + 5y = 6y$	The given equation
$12 + 5y + (-5y) = 6y + (-5y)$	Add $-5y$ to both sides.
$12 + 0y = y$	$6y + (-5y) = 1y$ or y
$12 = y$	The root is 12.

The check is similar to part **a** and is omitted.

In Example 2 the addition property is used twice.

Example 2.   Solve and check. $5 - 9z = -8 - 10z$

Solution.   **Discussion.** To combine the z terms on the left side, we can add $10z$ to both sides of the equation. Then, to combine the constant terms on the right side, we can add (-5) to both sides of the equation.

$$5 - 9z = -8 - 10z \qquad \text{The given equation}$$

$$5 - 9z + 10z = -8 - 10z + 10z \qquad \text{Add } 10z \text{ to both sides.}$$

$$5 + z = -8 \qquad -10z + 10z = 0$$

$$5 + z + (-5) = -8 + (-5) \qquad \text{Add } (-5) \text{ to both sides.}$$

$$z = -13 \qquad 5 + (-5) = 0$$

$$\text{Check: } 5 - 9(-13) = -8 - 10(-13) \qquad \text{Replace } z \text{ with } (-13).$$

$$5 - (-117) = -8 - (-130) \qquad \text{Multiply first.}$$

$$122 = 122, \text{ true} \qquad \text{The root is } -13.$$

Solving Equations with the Multiplication Property of Equality

The second property of equality useful in solving equations is the multiplication property. This property states that the same nonzero quantity (positive or negative) can be multiplied to both sides of an equation, and the resulting equation will be equivalent.

Multiplication property of equality
  If $a = b$, and c does not equal zero ($c \neq 0$), then

$$a \cdot c = b \cdot c.$$

Example 3 illustrates equations that can be solved by using the multiplication property of equality. Note that in the equations $x = k$ or $k = x$, the coefficient of x is 1. (Since $1 \cdot x = x$, we do not write the 1.) Therefore, when the coefficient of the variable is not 1, the multiplication property of equality is used to change it to 1.

Example 3. Solve:

  a. $12a = -60$

  b. $\dfrac{-3b}{2} = 9$

Solution.   **a. Discussion.** The coefficient of a is 12. To change it to 1 we multiply both sides by $\frac{1}{12}$.

$$12a = -60 \qquad \text{The given equation}$$

$$\frac{1}{12}(12a) = \frac{1}{12}(-60) \qquad \text{Multiply both sides by } \tfrac{1}{12}.$$

$$a = -5 \qquad \frac{1}{12} \cdot 12 = 1 \text{ and } 1 \cdot a = a$$

The root is -5.   The check is omitted.

  b. Discussion. The coefficient of b is $\frac{-3}{2}$. To change it to 1 we multiply both sides by $\frac{-2}{3}$. (Notice that $\frac{-2}{3}$ is the **reciprocal** of $\frac{-3}{2}$.)

$$\frac{-3b}{2} = 9 \qquad \text{The given equation}$$

$$\frac{-2}{3} \cdot \frac{-3b}{2} = \frac{-2}{3} \cdot 9 \qquad \text{Multiply both sides by } \tfrac{-2}{3}.$$

$$b = -6 \qquad \frac{-2}{3} \cdot \frac{-3}{2} = 1 \text{ and } 1 \cdot b = b$$

The root is -6.   The check is omitted.

Solving Equations Using Both Properties

In Example 4 the equation has one variable term and one constant term on each side of the equation. To solve this equation we first use the addition property of equality to combine the variable terms on one side and the constant terms on the other. (Choose the side of the equation that will result in a positive number coefficient for the variable.) Second, we use the multiplication property of equality to change, if necessary, the coefficient to one.

Example 4. Solve:

$$13t + 9 = 19t - 27$$

Solution.   **Discussion.** A $13t$ is on the left side, and a $19t$ is on the right side. If we add $-13t$ to both sides, a positive 6 will be the coefficient.

$$13t + 9 = 19t - 27 \qquad \text{The given equation}$$

$$-13t + 13t + 9 = -13t + 19t - 27 \qquad \text{Add } -13t \text{ to both sides.}$$

$$9 = 6t - 27 \qquad \text{Simplify both sides.}$$

$$9 + 27 = 6t - 27 + 27 \qquad \text{Add 27 to both sides.}$$

$$36 = 6t \qquad \text{Simplify both sides.}$$

$$\frac{1}{6} \cdot 36 = \frac{1}{6} \cdot 6t \qquad \text{Multiply both sides by } \tfrac{1}{6}.$$

$$6 = t \qquad \text{Simplify the products.}$$

The root is 6.   The check is omitted.

SECTION 2-2. Practice Exercises

In exercises **1–20**, solve by using the addition property of equality.

[Examples 1 and 2]

1. $x - 3 = 7$ **2.** $y - 8 = 0$

3. $z - 10 = -5$ **4.** $a - 1 = -17$

5. $7x = 3 + 6x$ **6.** $4y = 3y + 7$

7. $8 - 3z = -2z$ **8.** $4b + 9 = 5b$

9. $x + 13 = 5$

10. $y + 4 = -15$

11. $-11 + 3z = 4z$

12. $z - 2 = 2z$

13. $5w - 7 = 4w - 3$

14. $9x - 5 = 8x + 1$

15. $12x - 4 = 5 + 13x$

16. $4p - 10 = 3 + 5p$

17. $3 - 18z = -17z + 1$

18. $5 - 9y = -8y + 4$

19. $z + 3 = 2z - 10$

20. $4 + w = 2w + 9$

In exercises **21–36**, solve using the multiplication property of equality.

[Example 3]   21. $3c = 27$

22. $7d = 28$

23. $-8x = -40$

24. $-13z = -39$

25. $72 = -6y$

26. $80 = -5y$

27. $\frac{1}{5}x = 15$

28. $\frac{1}{3}y = 30$

29. $\frac{1}{7}t = -5$

30. $\frac{1}{9}w = -4$

31. $-21 = \frac{-1}{8}t$

32. $-36 = \frac{-1}{4}w$

33. $\frac{2}{3}w = 20$

34. $\frac{3}{4}z = 15$

35. $48 = \frac{-3a}{8}$

36. $36 = \frac{-2b}{9}$

In exercises **37–46**, solve and check.

[Example 4]   37. $3y = y + 10$

38. $7y = 2y + 30$

39. $18 - a = 8a$

40. $27 + 2a = 5a$

41. $t = 9t + 160$

42. $x = 3x + 16$

43. $17t + 3 = 13t - 1$

44. $6t - 3 = 8t + 5$

45. $7a - 8 = 2 - 3a$

46. $7a + 8 = 8 - 2a$

SECTION 2-2.   Ten Review Exercises

In exercises **1–10**, solve each problem.

1. Evaluate: $(5 \cdot 4 - 3) - (2 \cdot 4 + 9)$

2. Simplify: $(5x - 3) - (2x + 9)$

3. Solve: $5x - 3 = 2x + 9$

4. Are $5x - 3 = 2x + 9$ and $x = 4$ equivalent equations?

5. Is 4 a root of $5x - 3 = 2x + 9$?

6. Evaluate: $3[3(-5) - 2] - [2(-5) - 41]$

7. Simplify: $3(3y - 2) - (2y - 41)$

8. Solve: $3(3y - 2) = 2y - 41$

9. Are $9y - 6 = 2y - 41$ and $y = -3$ equivalent equations?

10. Is 5 a root of $9y - 6 = 2y - 41$?

SECTION 2-2.   Supplementary Exercises

In exercises **1–28**, solve and check.

1. $p + 6 = 6$ **2.** $7 + x = 7$

3. $6x = 5x$ **4.** $3y = 4y$

5. $x - \dfrac{1}{4} = \dfrac{3}{4}$ **6.** $y + \dfrac{3}{8} = \dfrac{7}{8}$

7. $-12y + \dfrac{5}{8} = -11y$ **8.** $-13w = \dfrac{1}{10} - 14w$

9. $\dfrac{7}{4}w + 3 = \dfrac{3}{4}w - 5$ **10.** $\dfrac{8}{3}t - 2 = \dfrac{5}{3}t + 6$

11. $\dfrac{1}{9}x = 0$ **12.** $\dfrac{1}{8}y = 0$

13. $-x = 65$ **14.** $-y = 32$

15. $-1.5y = 2.25$ **16.** $-1.2x = 1.44$

17. $28 = 4 - 4w$ **18.** $34 = 4 - 6b$

19. $23 = \dfrac{3a}{4} - 7$ **20.** $15 = -11 + \dfrac{2a}{3}$

21. $\dfrac{-2}{3}d = 0$ **22.** $\dfrac{-4}{5}p = 0$

23. $w + \dfrac{5}{7} = \dfrac{-2}{7}$ **24.** $z + \dfrac{2}{3} = \dfrac{-4}{3}$

25. $3.2 + x = 5.1$ **26.** $7.8 + m = 4.7$

27. $10 = 4 + \dfrac{x}{5}$ **28.** $1 = \dfrac{x}{2} - 2$

In exercises **29–36**, solve each equation for x.

29. $7x - 5a = 6x - 5a - 3$ **30.** $9x - 13q = 8x - 13q - 6$

31. $10.7x - 10 = 3.2 + 9.7x$

32. $8.9x - 11 = 7.2 + 7.9x$

33. $4.9 = 2.8 - 0.3x$

34. $13.5 = 1.2x + 3.9$

35. $\dfrac{4}{3}x = \dfrac{2}{3}x + 4$

36. $\dfrac{9}{4}x = \dfrac{3}{4}x - 12$

In exercises **37–40**, solve each problem.

37. The area of a rectangle is the product of the length and the width.
  a. Find the length if the area is 40 square inches and the width is 5 inches.
  b. Find the width if the area is 65 square centimeters and the length is 13 centimeters.

38. The distance a car travels is the product of the rate and the time.
  a. If a car travels 440 miles at 55 mph, find the time.
  b. If the car travels 240 miles in 5 hours, find the rate.

39. The simple interest earned on an investment is the product of the amount invested, the rate at which interest is earned, and the time in years the money is invested.
  a. Find the amount invested if the interest is $205, the rate is 0.05, and the time is two years.
  b. Find the number of years needed to earn $3,600 if the amount invested is $15,000 and the rate is 0.06.

40. The volume of a rectangular solid is the product of the length, the width, and the height.
  a. If the volume is 770 cubic inches, the length is 11 inches, and the width is 10 inches, find the height.
  b. If the volume is 288 cubic feet, the length is 9 feet, and the height is 4 feet, find the width.

SECTION 2-3. Applied Problems

**KEY TOPICS
IN THIS SECTION**

1. A recommended format for solving applied problems

2. Number problems

3. Mixture problems

4. Geometry problems

5. Motion problems

6. Age problems

7. Supplement: How to find the equation

Knowing how to solve equations can be a valuable skill for finding answers to problems that arise in many commonplace areas. A major obstacle to using the process is the requirement that the problem be correctly phrased in mathematical symbols. To gain some practice in learning this skill, we will use a few types of problems that are traditionally used in beginning algebra courses.

A Recommended Format for Solving Applied Problems

A fundamental ingredient of the successful completion of any task is a good plan of attack. Solving a problem by using a mathematical equation is an excellent example of a task that is more readily solved with a well-organized procedure. The five steps in the following outline provide a recommended procedure.

A recommended procedure for solving an applied problem

Step 1. Assign a variable to the unknown quantity in the problem.

Step 2. Write an equation that accurately describes the relationship expressed in the problem.

Step 3. Solve the equation for the variable.

Step 4. Check the root of the equation in the problem.

Step 5. Answer the question stated in the problem.

Step 2 is the most difficult one. Practice appears to be one of the best ways to improve one's skill in this area. The supplement to this section sets forth the "Guess and Test" problem-solving technique as a more direct way of finding an equation.

Number Problems

The simplest of the applied problems is the **number problem,** in which some information is usually given about an unknown number. The information can be used to write an equation in which the variable stands for the unknown number.

Example 1.   When three times a number is increased by 15, the result is 54. Find the number.

Solution.   **Discussion.** Since we are trying to find a number, the letter n will be used for the variable.

Step 1. Let n stand for the unknown number.

Step 2. "Three times the number" is written $3n$.
"Three times the number is increased by 15" is written $3n + 15$.
"The result is" is written $=$.

$$3n + 15 = 54 \qquad \text{An equation for the problem}$$

Step 3. $3n + 15 + (-15) = 54 + (-15)$ Add (-15) to both sides.

$$3n = 39 \qquad \text{Simplify both sides.}$$

$$\frac{1}{3} \cdot 3n = \frac{1}{3} \cdot 39 \qquad \text{Multiply both sides by } \tfrac{1}{3}.$$

$$n = 13 \qquad \text{Simplify both sides.}$$

Step 4. "Three times the number" becomes $3 \cdot 13 = 39$.
"Increased by 15" becomes $39 + 15 = 54$.
"The result is 54" is true.

Step 5. The number is 13.

Mixture Problems

Another type of applied problem is the **mixture problem.** In a typical mixture problem two or more items are combined in some way. Frequently a mixture problem has two statements regarding the items being combined:

1. the **quantities** of the items combined

2. the **values** of the items combined.

Example 2. Karen Hayashi is a secretary for Granite Construction Company. She just returned from the post office where she bought 25¢ and 45¢ stamps. The number of 25¢ stamps Karen bought was twice the number of 45¢ stamps. The total cost of the stamps was $33.25. How many of each kind of stamp did Karen buy?

Solution. Let x stand for the number of 45¢ stamps. Then $2x$ would stand for the number of 25¢ stamps. The value of the x stamps at 45¢ each is the product of x and 0.45. Similarly, the value of the $2x$ stamps at 25¢ each is $2x(0.25)$.

Items	Quantity	Value
25¢ stamps	$2x$	$2x(0.25)$
45¢ stamps	x	$x(0.45)$

$$2x(0.25) \quad + \quad x(0.45) \quad = \quad 33.25$$

$$\left(\begin{matrix} \text{Value of} \\ \text{25¢ stamps} \end{matrix} \right) + \left(\begin{matrix} \text{Value of} \\ \text{45¢ stamps} \end{matrix} \right) = \text{Total value}$$

$$0.5x + 0.45x = 33.25$$

$$0.95x = 33.25$$

$$x = 35 \text{ of the 45¢ stamps}$$

$$2x = 70 \text{ of the 25¢ stamps}$$

Check: $70(0.25) = \$17.50$ for the 25¢ stamps
$\underline{35(0.45) = \$15.75}$ for the 45¢ stamps
$\$33.25$ for both

Thus, Karen bought 35 stamps of the 45¢ denomination and 70 stamps of the 25¢ denomination.

Geometry Problems

Many applied problems in algebra involve geometric figures. These figures are displayed on the inside cover of this text, together with formulas related to these figures.

Example 3.   Cindy Simpson has a vegetable garden in the shape of a rectangle behind her backyard. The length of the garden is 23 meters. If the distance around the edge of the garden is 76 meters, find the width.

Solution.   **Discussion.** It is advisable to draw a figure when attempting to solve any geometry problem. The information regarding the figure can then be displayed. For example, the given rectangle is shown in Figure 2-3.

Step 1. Let w stand for the width.

Step 2. "The distance around the edge of the garden" is the perimeter of the rectangle. The formula that relates the perimeter of a rectangle to its length and width is

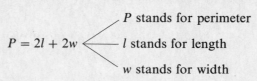

$$P = 2l + 2w$$

P stands for perimeter
l stands for length
w stands for width

With $P = 76$, $l = 23$, and $w =$ the width,

$76 = 2(23) + 2w$ An equation for the problem

Step 3. $76 = 46 + 2w$ Do the indicated multiplication.

$76 + (-46) = 46 + 2w + (-46)$ Add -46 to both sides.

$30 = 2w$ Simplify both sides.

$\frac{1}{2} \cdot 30 = \frac{1}{2} \cdot 2w$ Multiply both sides by $\frac{1}{2}$.

$15 = w$ Simplify both sides.

Step 4. Two times the length is $2(23) = 46$

Two times the width is $2(15) = 30$

The sum is 76, the given perimeter.

Step 5. The width of Cindy's garden is 15 meters.

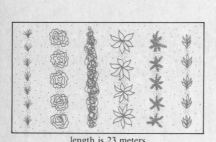

length is 23 meters

Figure 2-3. Cindy's garden.

Motion Problems

If an object is moving with **uniform motion** (that is, at some constant speed), then three quantities related to that motion can be identified.

Distance: *How far* the object moves from some point A to some point B, measured in feet, miles, kilometers.

Rate: The *amount of distance* the object moves in *some unit of time,* measured in feet per second (fps), miles per hour (mph), and kilometers per hour (kph). Uniform motion implies the rate is constant (does not increase or decrease).

Time: The *amount of time* the motion is observed or measured, measured in seconds, minutes, hours.

The formula that relates these three quantities is

$$d = rt$$

distance ↑ ↑↑rate × time

Example 4. At 8:00 A.M. Clair leaves Phoenix heading east at 55 mph driving an 18-wheeler with a Peterbilt cab. At the same time and from the same place, Linda leaves in a red-and-white '57 Corvette heading west. At noon Phoenix time they both stop for lunch. If at that time they are approximately 500 miles apart, at what rate had Linda been driving?

Solution. **Discussion.** A table and a line diagram are frequently useful aids in solving motion problems.

Phoenix 8:00 A.M.

noon   noon

← Linda's Corvette | Clair's truck →

←————————————— 500 miles —————————————→

Vehicle	Rate	Time	Distance
truck	55 mph	4 hours	(55 mph)(4 hours)
car	r	4 hours	$4 \cdot r$ hours

Let r stand for the rate of the Corvette.

$$\begin{pmatrix} \text{Total} \\ \text{distance} \end{pmatrix} = \begin{pmatrix} \text{Truck's} \\ \text{distance} \end{pmatrix} + \begin{pmatrix} \text{Car's} \\ \text{distance} \end{pmatrix}$$

$$500 \text{ miles} = 55 \frac{\text{miles}}{\text{hour}} \cdot 4 \text{ hours} + r \cdot 4 \text{ hours}$$

$$500 \text{ miles} = 220 \text{ miles} + 4r \text{ hours}$$

$$280 \text{ miles} = 4r \text{ hours}$$

$$70 \frac{\text{miles}}{\text{hour}} = r$$

Thus, Linda had been driving at a rate of 70 mph.

Age Problems

Age problems usually compare the ages of two or more persons, places, or things. Some age problems will compare the ages at different times, such as years in the past, or years in the future. In either type of problem, a table is usually helpful for identifying the relationship.

Example 5. Jamison is Tom's older son. Based on their current ages, Tom is five years more than four times Jamison's age. The sum of their current ages is 40 years. Find their current ages.

Solution.   Let a stand for Jamison's current age.

Then $4a + 5$ stands for Tom's current age.

Items	Current ages
Jamison	a
Tom	$4a + 5$

"The sum of their current ages is 40 years."

$$a + (4a + 5) = 40$$

$$5a + 5 = 40$$

$$5a = 35$$

$$a = 7$$

$$4a + 5 = 4(7) + 5 = 33$$

Check: $7 + 33 = 40$, true

Thus, Jamison is 7 and Tom is 33.

Supplement: How to Find the Equation

The most difficult part in solving an applied problem is writing an equation that accurately describes the problem. A problem-solving technique that is frequently helpful in writing an equation is the "Guess and Test" method. This method suggests that you arbitrarily pick any number to be the answer. You then test the selected number in the problem. The number will probably not be the correct answer, but the process of checking the number will generate an equation that may be the one needed to solve the problem.

The following problem is used to illustrate this method:

The length of a rectangle is 3 feet more than the width. The perimeter of the rectangle is 58 feet. Find the width and length of the rectangle.

The Guess and Test method would use the following procedure: The perimeter of a rectangle is the sum of two times the length and two times the width. (That is, $P = 2l + 2w$.) We therefore set up a table with the following headings:

Width	Length	$2 \cdot$ Width	$2 \cdot$ Length	Sum $= 58$?

Guess a number for the width. Suppose the number guessed is 5.

Width	Length	$2 \cdot$ Width	$2 \cdot$ Length	Sum $= 58$?
5				

If the width is 5, then the length is $5 + 3$, since the length is 3 feet more than the width. Now fill in the remaining columns in the table.

Width	Length	$2 \cdot$ Width	$2 \cdot$ Length	Sum	$= 58$?
5	$5 + 3$	$2 \cdot 5$	$2(5 + 3)$	$2 \cdot 5 + 2(5 + 3)$	
				$= 10 + 16 = 26$	No

Since $26 \neq 58$, the width is not 5 feet. Also, since $26 < 58$, our next guess will be greater than 5 feet, say 10 feet.

Width	Length	2 · Width	2 · Length	Sum	= 58?
5	5 + 3	2 · 5	2(5 + 3)	2 · 5 + 2(5 + 3) = 26	No
10	10 + 3	2 · 10	2(10 + 3)	2 · 10 + 2(10 + 3)	
				= 20 + 26 = 46	No

Let's now guess that the width is x feet:

Width	Length	2 · Width	2 · Length	Sum	= 58?
5	5 + 3	2 · 5	2(5 + 3)	2 · 5 + 2(5 + 3) = 26	No
10	10 + 3	2 · 10	2(10 + 3)	2 · 10 + 2(10 + 3) = 46	No
x	$x + 3$	$2x$	$2(x + 3)$	$2x + 2(x + 3) = ?$	?

The indicated sum in terms of x can now be set equal to 58, the desired perimeter. The root of this equation is the unknown width.

$$2x + 2(x + 3) = 58$$

SECTION 2-3. Practice Exercises

In exercises **1–32**, use the five-step method to solve each problem.

[Example 1] **1.** Find the number if 8 more than 3 times the number is 53.

2. Find the number if, when 5 times the number is decreased by 15, the result is 100.

3. Find the number when the sum of 4 times the number and 32 is -40.

4. Find the number when the difference between one-fourth the number and 10 is -14.

5. When 4 times a number is decreased by 15, the result is the same as the sum of 3 times the number and 6. Find the number.

6. When 7 times a number is decreased by 10, the result is the same as the sum of 5 times the number and 20. Find the number.

7. The difference between 100 and 8 times a number is 12 more than 3 times the number. Find the number.

8. The difference between 20 and 5 times a number is 104 more than 2 times the number. Find the number.

[Example 2] **9.** The ten members of the pep band went for a quick lunch at the hamburger stand on 54th Street. Each had a 79¢ soft drink and a hamburger. If the total bill, without any tax, was $23.80, what was the cost of a hamburger?

10. All five of the 10th Street Dance Group had a sandwich and cup of coffee at the corner diner. The coffee was 95¢ a cup and the total bill, without tax or tip, was $17.50. What was the cost of a sandwich?

11. Sandy bought three sweatshirts and a pair of running shoes. The cost of the shoes turned out to be four times the cost of a sweatshirt. If she paid $84 for all of the items, what was the cost of the running shoes?

12. Billy Bovane left the corner store with three small candy bars and two large candy bars. If the large bars cost twice as much as the small ones, and if he paid $2.45, what is the cost of a large candy bar?

13. Friends bought Craig Brooks three different types of stamps for his collection. They gave him ten common stamps, three foreign stamps, and one rare stamp. A foreign stamp costs three times as much as a common one, and the rare stamp costs seven times as much as a common one. If his friends paid $10.14, what is the cost of each type of stamp?

14. Sheila Kearney bought three types of used books at a Cincinnati book swap. The second type of book was twice as expensive as the first type, and the third type was three times as expensive as the first type. If Sheila bought four of each type of book and spent $6, how much did each type of book cost?

[Example 3]   **15.** The foil found on a gum wrapper has a length of 92 millimeters. The perimeter of this rectangular shape is 280 millimeters. What is the width?

16. Rodney uses a rectangular-shaped mounting board to display his photographs. The board has a height of 14 inches and a perimeter of 50 inches. What is the board's width?

17. A rectangular flower bed has a perimeter of 24 feet. If the length is twice the width, what are the dimensions of the flower bed?

18. A picture frame is three times as tall as it is wide. If the perimeter is 40 inches, what are the dimensions of this rectangular frame?

19. The longest side of a triangle is 4 centimeters more than the length of its shortest side, and the third side is 1 centimeter more than the length of the shortest side. Find the lengths of the sides if the perimeter of the triangle is 56 centimeters.

20. Two sides of a triangle are the same length, but the third side is 3 inches longer than either of the other two sides. Find the lengths of the sides if the perimeter of the triangle is 39 inches.

[Example 4]   **21.** Bob Smith and Marty Simmons leave a sales meeting at the same time and start to drive home. Bob heads west at a rate of 55 mph. Marty heads east, and after two hours of travel the cars are 190 miles apart. What is the rate of Marty's car?

22. Jean and Judy bicycle in opposite directions home from the park. Jean can average 12 mph on her bike, and after two hours the women are 46 miles apart. What is the rate of Judy's bike?

23. Dave and Ken leave the barn together on a trail with their horses. Ken decides to go at a faster pace. If Ken's horse travels twice as fast as Dave's horse, and after two hours of travel Ken is 6 miles ahead of Dave, how fast is Dave traveling?

24. After eating breakfast at a highway restaurant, the Hendersons and the Loftans continue their trip together. The Hendersons travel at about 53 mph, and at the end of three hours they are 12 miles ahead of the Loftans. How fast are the Loftans traveling?

25. A car and a truck leave the same rest stop at the same time heading in the same direction. The truck travels at $\frac{4}{5}$ the average speed of the car. After an hour of travel the two vehicles are 11 miles apart. At what rate is the truck traveling?

26. At 1:15 P.M. a train leaves downtown Winterford on the northbound track. A Dodge van leaves the station at the same time but heads south. If the train travels at $\frac{7}{8}$ the speed of the van and in an hour the two are 120 miles apart, at what rate is the train traveling?

[Example 5]   **27.** Gary is four years older than Tammy. Their combined age is 52. How old is each?

28. Mattie's car is nine years older than Dale's pickup truck. The total of their vehicles' ages is 37. How old is each vehicle?

29. The age of Sam's cat and the age of his dog differ by four years. The cat is three times as old as the dog. How old is the cat?

30. George's oldest daughter, Julia, is twice as old as his youngest daughter, Nicole. Julia is seven years older than Nicole. How old is Julia?

31. Ricky's hamster, Gus, is two months more than twice as old as his new hamster, Gert. If Gus is nine months old, how old is Gert?

32. Sara's favorite pet goose is three months more than twice as old as her pet gander. If the goose is 21 months old, how old is the gander?

SECTION 2-3.   Ten Review Exercises

In exercises **1–4**, evaluate the following equation for the indicated values.

$$5(2a + 1) - (12a + 5) + 2a$$

1. $a = 7$　　　　　　　　　　　**2.** $a = -4$

3. $a = \dfrac{-1}{2}$　　　　　　　　**4.** $a = \dfrac{3}{2}$

5. Simplify. $5(2a + 1) - (12a + 5) + 2a$

6. Is 7 a root of $5(2a + 1) = (12a + 5) - 2a$?

7. Is -4 a root of $5(2a + 1) = (12a + 5) - 2a$?

8. Is $\dfrac{-1}{2}$ a root of $5(2a + 1) = (12a + 5) - 2a$?

Exercises **9** and **10** refer to equations **a** and **b**.

a. $5(z - 3) = 8z - 3(z + 5)$　　　　**b.** $z = 6$

9. Is 6 a root of both **a** and **b**?

10. Are **a** and **b** equivalent equations? Justify your answer.

SECTION 2-3.   Supplementary Exercises

In exercises **1–8**, solve each problem.

1. John Gerrity is 23 years old. His wife, Mary Hanson-Gerrity, is 27 years old. John's father is four years more than twice as old as John. Mary's father is four years less than twice as old as Mary.
   a. How old is John's father?
   b. How old is Mary's father?
   c. Find the difference in the fathers' ages.

2. Skip Aboujoud has a three-year-old cat called Omolene. He also has a five-year-old dog called Beta. Omolene's mother is one year less than three times Omolene's age. Beta's mother is twice as old as Beta.
   a. How old is Omolene's mother?
   b. How old is Beta's mother?
   c. Find the difference in the mothers' ages.

3. On a back trail to Bear Lake, a Ford truck averages 10 mph and a Toyota truck averages 8 mph. They both travel for two hours on the trail and start at the same time. However, when they begin, the Toyota is 4 miles ahead of the Ford.
   a. Find the distance covered by the Ford in the two hours.
   b. Find the distance covered by the Toyota in the two hours.
   c. At the end of the two hours, how far apart are the trucks?

4. A Cessna and a Hawk private plane leave Reno, Nevada, at the same time heading in opposite directions. The pilots of the two planes want to be 850 miles apart in two hours. Suppose the Cessna averages 215 mph and the Hawk averages 190 mph.
   a. Find the distance covered by the Cessna in the two hours.
   b. Find the distance covered by the Hawk in the two hours.
   c. Find the difference between the distances actually covered and the desired 850-mile goal.

5. The Falcons and Mustangs were playing in a Friday-night football game at Dewey Hall Football Field. The following series of plays occurred during the game:
   i. The Falcons took possession on the Mustang's 30-yard line.
   ii. On the first play the Falcons gained 8 yards.
   iii. On each of the next two plays they lost 2 yards.
   iv. On fourth down they threw an incomplete pass for no gain or loss.
   v. The Mustangs took possession and ran three plays that gained 2 yards each.
   vi. On fourth down the Mustangs completed a pass that gained 7 yards.
      a. Compute the net gain for the Falcons on their series of plays.
      b. Compute the net gain for the Mustangs on their series of plays.
      c. On what yard line was the ball located after the last play?

6. On Saturday morning Scout Troops 111 and 66 left Silver Lake Camp Ground. Troop 111 headed west and Troup 66 headed east. Relative to the elevation at Silver Lake, their altitudes changed as follows:
   i. Troop 111 climbed 250 feet, then went down 125 feet, and then hiked up again 425 feet.
   ii. Troop 66 climbed 175 feet, then went down 50 feet, and then hiked up again 400 feet.

   a. Compute the net change in elevation for Troop 111.
   b. Compute the net change in elevation for Troop 66.
   c. Compute the difference in the net changes.

7. Jean Spencer is given the task of spending as much as possible of the $10 she was given on stamps.
   a. If she buys sixteen 25¢ stamps, what is the maximum number of 40¢ stamps she can also buy?
   b. If she buys ten 40¢ stamps, what is the maximum number of 25¢ stamps she can also buy?
   c. If she buys the same number of 25¢ and 40¢ stamps, how many of each can she buy?
   d. If she buys twice as many 25¢ stamps as 40¢ stamps, how many of each can she buy?
   e. If she buys two more 25¢ stamps than 40¢ stamps, how many of each can she buy?

8. The perimeter of a triangle is 100 feet.
   a. If the length of one side is 30 feet and the other two are equal in length, state the lengths of all three sides.
   b. If the length of the longest side is 15 feet more than the shortest side, and the middle-sized side is 10 feet more than the shortest side, find the lengths of all three sides.
   c. If side two is 10 feet longer than side one, and side three is the same length as side one, find the lengths of all three sides.
   d. If the lengths of the three sides are equal, and only integers are possible numbers used for the lengths, what are the lengths of the sides given that the perimeter cannot be more than 100 feet?

SECTION 2-4. Solving More Linear Equations in One Variable

KEY TOPICS IN THIS SECTION

1. Solving equations with more than two steps

2. Solving equations with grouping symbols

3. Solving equations that have no roots

4. Solving equations that have all real numbers for roots

5. Applied problems

Many linear equations in one variable are written with expressions that need to be simplified before the properties of equality can be used to solve them. The expressions may also include grouping symbols, such as parentheses. In general, it is necessary to remove such symbols before the equation can be solved. In this section we will study the procedure for solving such equations.

Solving Equations with More Than Two Steps

The following equation can be solved by using only the addition and multiplication properties of equality:

$$10x - 3 = 7x + 15 \qquad \text{The given equation}$$

$$10x - 3 + (-7x) = 7x + 15 + (-7x) \qquad \text{Add } -7x \text{ to both sides.}$$

$$3x - 3 = 15$$ Combine like terms on both sides.

$$3x - 3 + 3 = 15 + 3$$ Add 3 to both sides.

$$3x = 18$$ Combine like terms on both sides.

$$3x \cdot \frac{1}{3} = 18 \cdot \frac{1}{3}$$ Multiply both sides by $\frac{1}{3}$.

$$x = 6$$ Simplify the indicated products.

The root of the equation is 6.

Consider now the following equation:

$$16x + 11 - 6x - 14 = 8 - x + 2 + 8x + 5$$

The expressions on both sides of this equation have more than one variable and one constant term. Therefore the expressions should first be simplified by combining like terms.

Left side	Right side
$16x + 11 - 6x - 14$	$8 - x + 2 + 8x + 5$
$= 10x + 11 - 14$	$= 8 + 7x + 2 + 5$
$= 10x - 3$	$= 15 + 7x$

Now setting the simplified expressions equal to each other,

$$10x - 3 = 15 + 7x.$$

This is the equation that was previously solved. Thus the root is 6.

Example 1. Solve:

$$9 + 3t - 4 = 34 - 2t - 5 + 9t$$

Solution. **Discussion.** First simplify the expressions on both sides. Then use the addition and multiplication properties of equality to solve the simplified equation.

$$9 + 3t - 4 = 34 - 2t - 5 + 9t$$

$$3t + 5 = 29 + 7t$$ Combine like terms on both sides.

$$5 = 29 + 4t$$ Add $-3t$ to both sides.

$$-24 = 4t$$ Add -29 to both sides.

$$-6 = t$$ Multiply both sides by $\frac{1}{4}$.

Check: $9 + 3(-6) - 4 = 34 - 2(-6) - 5 + 9(-6)$

$$9 + (-18) - 4 = 34 - (-12) - 5 + (-54)$$

$$-9 - 4 = 46 - 5 + (-54)$$

$$-13 = 41 + (-54)$$

$$-13 = -13, \text{ true}$$

The resulting number statement is true; therefore the root is -6.

Solving Equations with Grouping Symbols

A linear equation in one variable may have one or more grouping symbols on one or both sides of the equation. These grouping symbols must be removed in order to simplify the expressions on both sides of the equation. The distributive property of multiplication over addition is used to remove such grouping symbols.

Example 2. Solve:

$$y - 5(2y + 1) = 4 - 2(2y - 3)$$

Solution.   **Discussion.** To remove the parentheses on the left side, think of $y - 5(2y + 1)$ as $y + (-5)(2y + 1)$ and distribute a (-5) to the $2y$ and 1. Similarly on the right side, think of

$$4 - 2(2y - 3) \text{ as } 4 + (-2)(2y - 3)$$

and distribute a (-2) to terms inside the parentheses.

$y - 5(2y + 1) = 4 - 2(2y - 3)$	
$y - 10y - 5 = 4 - 4y + 6$	$x(y + z) = x \cdot y + x \cdot z$
$-9y - 5 = 10 - 4y$	Combine like terms
$-5 = 10 + 5y$	Add $9y$ to both sides.
$-15 = 5y$	Add -10 to both sides.
$-3 = y$	Multiply both sides by $\frac{1}{5}$.

The check is omitted and the root is -3.

Example 3. Solve:

$$5(k - 2) = 4k + 2[5k - 3(k + 1)]$$

Solution.

$5(k - 2) = 4k + 2[5k - 3k - 3]$	$-3(k + 1) = -3k - 3$
$5(k - 2) = 4k + 2[2k - 3]$	Simplify within the brackets.
$5k - 10 = 4k + 4k - 6$	Remove grouping symbols.
$5k - 10 = 8k - 6$	Combine like terms on right side.
$-10 = 3k - 6$	Add $-5k$ to both sides.
$-4 = 3k$	Add 6 to both sides.
$\dfrac{-4}{3} = k$	Multiply both sides by $\frac{1}{3}$.

The check is omitted and the root is $\frac{-4}{3}$.

Solving Equations That Have No Roots

Not all equations have roots. For example, consider the following equation:

$$x + 2 = x + 5$$

The left side is an expression in which 2 is added to some number x. The right side is an expression in which 5 is added to the same number x. The equal sign indicates the two sums are the same. Since there is no number whose sums with 2 and 5 can be the same, *we say the equation has no root.*

Notice what happens when we attempt to solve this equation.

$$x + 2 = x + 5 \qquad \text{The given equation}$$

$$2 = 5, \text{false} \qquad \text{Add } -x \text{ to both sides.}$$

This result will always happen for equations that have no roots.

When solving a linear equation in one variable
   If the variable terms combine to 0 and the resulting number statement is false, then the equation has no root.

Example 4. Solve:

$$4b + 13 - b = 24 + 3b - 10$$

Solution. $\qquad 3b + 13 = 14 + 3b \qquad \text{Combine like terms.}$

$$13 = 14, \text{false} \qquad \text{Add } -3b \text{ to both sides.}$$

The b terms combined to 0, and a false number statement resulted. There is no root.

Solving Equations That Have All Real Numbers for Roots

Some equations have all real numbers for roots. For example, consider the following equation:

$$3(t + 2) = 3t + 6$$

Notice what happens when we attempt to solve this equation.

$$3(t + 2) = 3t + 6 \qquad \text{The given equation}$$

$$3t + 6 = 3t + 6 \qquad \text{Remove parentheses.}$$

$$6 = 6, \text{true} \qquad \text{Add } -3t \text{ to both sides.}$$

A similar result will always happen for equations that have all real numbers as roots. That is, a true number statement is obtained when t is replaced by any real number. For example, suppose we replace t by 5:

$$3(5 + 2) = 3 \cdot 5 + 6$$

$$3(7) = 15 + 6$$

$$21 = 21, \text{true}$$

Any real number replacement for t would also yield a true statement.

When solving a linear equation in one variable
   If the variable terms combine to 0 and the resulting number statement is true, then all real numbers are solutions, denoted by R.

Example 5. Solve:

$$10(y - 5) - 2(y - 9) = 8y - 32$$

Solution.

$10(y - 5) - 2(y - 9) = 8y - 32$	The given equation
$10y - 50 - 2y + 18 = 8y - 32$	Remove parentheses.
$8y - 32 = 8y - 32$	Combine like terms.
$-32 = -32$, true	Add $-8y$ to both sides.

The y terms combined to 0 and the resulting number statement is true. Thus all real numbers are solutions and we write R.

Applied Problems

The equations used to solve some applied problems require expressions written inside grouping symbols. One such type of problem is called the **consecutive integer problem.** Recall that the integers are

$$\ldots, -2, -1, 0, 1, 2, \ldots$$

If i is an integer, then

(i) i and $i + 1$ are consecutive integers.   **3 and 4 are consecutive integers.**

(ii) i and $i + 2$ are consecutive even integers, if i is even.   **6 and 8 are consecutive even integers.**

(iii) i and $i + 2$ are consecutive odd integers, if i is odd.   **11 and 13 are consecutive odd integers.**

Notice that both the even integers and odd integers are two units apart. Thus, adding 2 to an even integer (0, 2, 4, 6, ...) yields the next even integer. Likewise, adding 2 to an odd integer (1, 3, 5, 7, ...) yields the next odd integer.

Example 6. Three times the sum of an integer and 8 is equal to two more than four times the next consecutive integer. Find both integers.

Solution. Let $i = $ the first integer.
$i + 1 = $ the next consecutive integer.
"Three times the sum of an integer and 8" is written $3(i + 8)$.
"Two more than four times the next consecutive integer" is written $2 + 4(i + 1)$.
Thus, an equation for the problem is

$$3(i + 8) = 2 + 4(i + 1)$$

$3i + 24 = 2 + 4i + 4$	Remove parentheses.
$3i + 24 = 6 + 4i$	Combine like terms.
$24 = 6 + i$	Add $-3i$ to both sides.
$18 = i$	Add -6 to both sides.
$19 = i + 1$	Add 1 to both sides.

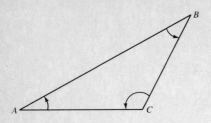

Figure 2-4. In any triangle,
$A + B + C = 180°$.

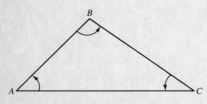

Figure 2-5. A triangle.

Check: If 18 is the first integer, then $3(18 + 8) = 78$.
If 19 is the next consecutive integer, then $2 + 4(19) = 78$.
Since $78 = 78$, the integers are 18 and 19.

In Figure 2-4 a triangle is drawn. The letters A, B, and C represent the points where the sides of the triangle join. These letters can also be used to represent the measures of the angles of the triangle. It can be shown that for any triangle.

$A + B + C = 180°$.   The sum of the measures of the angles of any triangle is 180° (read "180 degrees").

Example 7.   For the triangle in Figure 2-5, B is 5° less than two times C, and A is 25° more than C. Find the measures of the three angles of this triangle.

Solution.   **Discussion.** C is the key to finding the values of A and B, since both A and B can be determined once C is known.

Let $m = C$	C is the measure in degrees.
$2m - 5° = B$	B is 5° less than two times C.
$m + 25° = A$	A is 25° more than C.
$(m + 25°) + (2m - 5°) + m = 180°$	$A + B + C = 180°$.
$4m + 20° = 180°$	Combine like terms.
$4m = 160°$	Add $-20°$ to both sides.
$m = 40°$	Multiply both sides by $\frac{1}{4}$.
$2m - 5° = 75°$	
$m + 25° = 65°$	

Check: $65° + 75° + 40° = 180°$

$180° = 180°$, true

Thus A is 65°, B is 75°, and C is 40°.

SECTION 2-4.   Practice Exercises

In exercises **1–26**, solve and check.

[Example 1]   **1.** $5y - 2y = 2 + y + 8$ **2.** $10y - 3y = 14 + 2y + 16$

3. $6 + 11z + 42 = 13z - 10z$ **4.** $29 + 5z + 4 = 4z - 10z$

5. $11t + 3 + 6t = 15t - 1 - 2t$ **6.** $8t - 3 - 2t = 7 + 7t + t - 2$

7. $2a + 10 - a - 2 + 6a = 15 - 3a - 13$

8. $a + 1 + 3a + 7 = 13 - 2a - 2$

[Examples 2 and 3]   **9.** $3(x + 2) = 15$ **10.** $5(x - 1) = -15$

11. $5y = 2(y + 5)$ **12.** $8x = 2(x + 8)$

13. $6(x + 2) = 9 + 4(2x + 1)$ **14.** $4(x - 3) = 5(x - 4) + 5x$

15. $17(a + 3) - 4(a - 10) = 13$ **16.** $10(y - 5) - 2(y - 9) = -28$

17. $4(2a - 3) - 80 = 5a - 5(3a + 4)$ **18.** $2(5b + 4) + 19 = 4b - 3(2b + 11)$

19. $9x + 4(2x + 1) = 3(3x + 5) - 6(2 - x)$

20. $7(2y - 1) - 2(y + 4) = 10(y - 1) - 8$

21. $7z + 2 = 2[3(z - 2) - 2z] + 4$

22. $3z - 5 = 1 - 3[2(z + 1) - 3]$

23. $4x + 2[3 + 2(3x - 5)] = 2x - 5(3 - 2x)$

24. $6 - 7(x - 2) = 4 - [4x - 2(10 - 3x)]$

25. $2[2(2y - 1) - 3] = 3[5(y + 2) - 2(2y + 7)]$

26. $4 - 2[16 - 5(y + 2)] = 3(2y + 1) + 7$

In exercises **27–34**, find any roots.

[Examples 4 and 5] **27.** $3(x - 4) = 5x - 4 - 2x$ **28.** $2x + 10 + x = 3(4 + x)$

29. $5x + 7 = 7(x + 1) - 2x$ **30.** $10x - 12 = 4(x - 3) + 6x$

31. $3w - 2(w + 3) = w + 6$ **32.** $9t - (8 - t) = 10t + 8$

33. $6 - 4y = -2(2y - 3)$ **34.** $2x + 12 - 8x = -6(x - 2)$

In exercises **35–42**, solve each integer problem.

[Example 6] **35.** The sum of three consecutive integers is 36. Find the three integers.

36. The sum of three consecutive integers is 66. Find the three integers.

37. The sum of three consecutive odd integers is 219. Find the three odd integers.

38. The sum of three consecutive odd integers is 117. Find the three odd integers.

39. If the smaller of two consecutive integers is added to three times the larger, the resulting sum is 43. Find the integers.

40. If five times the smaller of two consecutive integers is added to three times the larger integer, the result is 59. Find both integers.

41. The sum of the first and twice the second of three consecutive odd integers is 18 more than the third. Find the integers.

42. The sum of the smallest and twice the largest of three consecutive even integers is 58 more than the second even integer. Find the integers.

In exercises **43–48**, A, B, and C are the measures of the angles of triangle ABC. In each exercise find A, B, and C.

[Example 7] **43.** In triangle ABC, B is twice the sum of A and $5°$, and C is four times the difference between A and $10°$.

44. In triangle ABC, A is one-half C, and B is one-third the sum of C and $45°$.

45. In triangle ABC, B is $5°$ less than twice A, and C is five times the difference between A and $3°$.

46. In triangle ABC, A is $11°$ more than one-half B, and C is $15°$ more than twice B.

47. In triangle ABC, A is $6°$ more than four times B, and C is $6°$ more than B.

48. In triangle ABC, A is $3°$ less than three-halves B, and C is three times the sum of B and $6°$.

SECTION 2-4.   Ten Review Exercises

In exercises **1–10**, solve each problem.

1. Write the following as an algebraic expression: Eight more than three times a number n

2. Write the following as an algebraic expression: Twelve less than five times a number n

3. Evaluate: $3n + 8$ for $n = 10$

4. Evaluate: $5n - 12$ for $n = 10$

5. If 8 is added to the product of some number and 3, the result is the same as when 12 is subtracted from the product of five times the same number. Find the number.

6. Write the following as an algebraic expression: The sum of two times a number n and 4 is decreased by 10.

7. Evaluate: $(2n + 4) - 10$ for $n = 3$

8. Simplify: $(2n + 4) - 10$

9. If 4 is added to twice a number, the result is 10. Find the number.

10. Is 3 a root of the equation $2n + 4 = 10$?

SECTION 2-4.   Supplementary Exercises

In exercises **1–18**, find any solutions.

1. $10 = 4(2 - a)$

2. $8 = 3(2 - a)$

3. $6(z - 2) = (11 - z)3$

4. $5(9 + z) = (z + 1)(-3)$

5. $2(3x + 1) + 2(x - 1) = 4(2x + 1) - 3$

6. $13(x + 3) - 21 = 10x + 3(x + 5)$

7. $18 - 4x = (21 - 2x)2 - 24$

8. $12x = (5 - 6x)(-2) + 10$

9. $4(5z + 6) - 2(2z + 8) = 6z - 2(3 - 2z) + 18$

10. $13z - 7(z - 2) = 14 - 2(z + 5)$

11. $-(4x + 2) - (-3x - 5) = 3$ **12.** $-(6k - 5) - (-5k + 8) = -4$

13. $2[x + 3(x - 1)] = 18$ **14.** $3[2(x - 5) - 10] = -25 - x$

15. $-(3x - 2x^2) + 2(x + 5) = 2(x^2 - 3x)$

16. $5(x + 3) - (x^2 + 2x) = (3x - x^2) - (1 - 4x)$

17. $4x + 2[3 - (3x^2 - 5x)] = 2x + 3(3x - 2x^2)$

18. $6x^2 - 7(x - 6) = 4 - [4x - 2(10 + 3x^2)]$

In exercises **19–22**, find the indicated integers.

19. The sum of three consecutive even integers is 0.

20. The sum of three consecutive odd integers is -3.

21. Twice the smaller of three consecutive even integers is six less than the sum of the two larger integers.

22. The sum of the smallest and twice the largest of three consecutive odd integers is two more than three times the second integer.

In exercises **23–31**, solve each problem.

23. The length of a rectangular room is 2 feet more than the width. The difference between four times the width and 4 is equal to three times the length. Find the length and width of the room.

24. When a rectangular dining room table is extended to its largest size, the length is three times the width. Twice the sum of the width and 10 inches is equal to the difference between the length and 20 inches. Find the length and width of the table when it is extended to its largest size.

25. A bank has some nickels and dimes worth $2.75. The number of nickels is five fewer than twice the number of dimes. How many of each kind of coin are in the bank?

26. A collection of old pennies and nickels has a face value of $6. The number of pennies is 40 more than three times the number of nickels. How many of each kind of coin are in the collection?

27. A clerk bought $18 worth of stamps for the office in 1¢, 15¢, and 25¢ denominations. The mixture had five more 1¢ stamps than 25¢ stamps. The number of 15¢ stamps was five more than four times the number of 25¢ stamps. How many of each kind of stamp did he buy?

28. A box consists of three kinds of soup. Kind A costs 23¢ a can, kind B costs 32¢ a can, and kind C costs 48¢ a can. The number of cans of kind A in the mixture is two fewer than twice the number of cans of kind B. The number of cans of kind C is seven fewer than the number of cans of kind B. The value of the soup is $11.30. How many cans of each kind are in the box?

29. A shawl is shaped like a triangle. Two edges of the shawl are equal in length. The longest edge is 40 inches less than twice the length of one of the equal edges. What are the dimensions of the shawl if the distance around the edge is 184 inches?

30. A portion of the supporting structure of an ice arena is constructed of steel beams. The beams join to form a triangle. The longest beam is 10 feet shorter than twice the length of the shortest beam. The intermediate beam is 5 feet longer than the shortest beam, and the sum of the lengths of the three beams is 95 feet. Find the length of each beam.

31. A four-sided figure is labeled *ABCD*. A line is drawn from *A* to *C* forming two triangles that have the common side *AC*.
  a. What is the sum of the six angles in the two triangles?
  b. Suppose the measures of the six angles are related as follows:
    1. The two smallest angles are equal in measure, namely $x°$ each.
    2. The measure of the next largest angle is $x° + 5°$.
    3. The measure of the next largest angle is $x° + 10°$.
    4. The measure of the next largest angle is $x° + 20°$.
    5. The measure of the largest angle is $x° + 25°$.
    Find the measures of all six angles.

SECTION 2-5.   Formulas and Literal Equations

KEY TOPICS IN THIS SECTION

1. The definition of a formula

2. Evaluating a formula

3. Solving a formula for a particular variable

4. The definition of a literal equation

5. Solving a literal equation for an indicated variable

6. Evaluating a literal equation

Not all equations have just one variable. Many equations have two or more variables. Some of these equations are called **formulas,** and others are simply called **literal equations.** In this section we will study both formulas and literal equations.

The Definition of a Formula

Two scales are commonly used to measure temperature. In the United States the Fahrenheit scale (F°) is most commonly used. If the temperature predicted for tomorrow is 95°, we know it will be hot.

In other parts of the world the Celsius scale (C°) is used. If the predicted high temperature in Paris, France, is 35°, the person living there knows it will be hot. As seen in Figure 2-6, 95° F and 35° C are the same.

The way in which F° and C° are related to each other can be stated as an equation. The equation is an example of a formula.

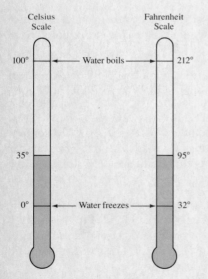

Celsius Scale   Fahrenheit Scale

100° ← Water boils → 212°

35°   95°

0° ← Water freezes → 32°

Figure 2-6. The Celsius and Fahrenheit scales for measuring temperatures.

Definition 2.5.   Formula
  A **formula** is an equation that always represents a true relationship between two or more quantities in the world.

Actually, three equations can be used as formulas for F° and C°:

a. $5\,F° - 9\,C° = 160°$ A general statement of how F° and C° are related

b. $C° = \dfrac{5}{9}(F - 32°)$ The equation is solved for C° in terms of F°.

c. $F° = \dfrac{9}{5}C° + 32°$ The equation is solved for F° in terms of C°.

To show that all three equations are equivalent, we will use the 35° Celsius and 95° Fahrenheit values in equations **a**, **b**, and **c**.

a. $5(95°) - 9(35°) = 160$

$\qquad 475° - 315° = 160°$

$\qquad\qquad 160° = 160°,\ \text{true}$

b. $35° = \dfrac{5}{9}(95° - 32°)$

$\quad\ 35° = \dfrac{5}{9}(63°)$

$\quad\ 35° = 35°,\ \text{true}$

c. $95° = \dfrac{9}{5} \cdot 35° + 32°$

$\quad\ 95° = 63° + 32°$

$\quad\ 95° = 95°,\ \text{true}$

Evaluating a Formula

Suppose the values of all the variables in a formula are known except one. The value of the unknown variable can be found by **evaluating the formula.**

Example 1.   Ellen Wynn lives in Katonah, New York. She needs to make a business trip of approximately 476 miles. She plans to drive her car and wants to reach her destination at about 4:00 P.M. Since the trip will be primarily on interstate highways, she thinks she will be able to average 56 mph. What time should Ellen leave home to reach her destination at the indicated time?

Solution.   **Discussion.** The formula for a distance (d), rate (r), and time (t) problem is $d = r \cdot t$. From the given information,

$d = 476$ miles

$r = 56$ miles per hour (*mph*)

t is unknown

$476 = 56t \qquad\quad d = r \cdot t$

$\ \ 8.5 = t \qquad\qquad$ Divide both sides by 56.

It will take Ellen about 8.5 hours to make the trip.

4:00 P.M. − 8.5 hours is 7:30 A.M.

Thus Ellen should leave home at about 7:30 in the morning.

Example 2.   Bill Weaver wants to build a flower bed in the shape of the triangle shown in Figure 2-7. He wants the area to be about 70 square meters (m^2). The base is 14 meters (m), and Bill needs to know h.

Solution.   **Discussion.** The formula for the area of a triangle is $A = \frac{1}{2}b \cdot h$, where A is **area,** b is **base,** and h is **height.** From the given information,

$$A = 70m^2$$

$$b = 14m$$

h is unknown

$$70m^2 = \frac{1}{2} \cdot 14m \cdot h \qquad A = \frac{1}{2}bh$$

$$70m^2 = 7m \cdot h \qquad \frac{1}{2} \cdot 14m = 7m$$

$$\frac{70m^2}{7m} = h \qquad \text{Divide both sides by } 7m.$$

$$10m = h$$

Thus the height of the flower bed is about 10 meters.

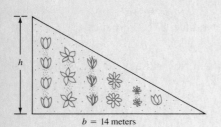

h

$b = 14$ meters

Figure 2-7. A flower bed.

Solving a Formula for a Particular Variable

Examples **a, b,** and **c** on page 121 illustrate three different forms of the formula that relates the Celsius and Fahrenheit temperature scales. In example **b** the formula is solved for C° in terms of F°. The instructions to "solve the equation for F° in terms of C°" would yield the equation in example **c.**

Example 3.   Solve $V = \frac{1}{3}\pi r^2 h$ for h.

Solution.   **Discussion.** This formula is used to compute the **volume** (V) of a right circular cone, such as the one shown in Figure 2-8.

r stands for **radius** of the base.

h stands for **height** of the cone.

π is an irrational number **constant.**

$$V = \frac{1}{3}\pi r^2 h \qquad \text{The given equation}$$

$$3V = \pi r^2 h \qquad \text{Multiply both sides by 3.}$$

$$\frac{3V}{\pi r^2} = h \qquad \text{Divide both sides by } \pi r^2.$$

Figure 2-8. A right circular cone.

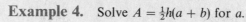

Thus, $h = \dfrac{3V}{\pi r^2}$ is the volume formula solved for h in terms of V and r.

Example 4. Solve $A = \frac{1}{2}h(a + b)$ for a.

Solution. **Discussion.** This formula is used to compute the **area** (A) of a trapezoid, such as the one shown in Figure 2-9.

h stands for the **height.**

a stands for the **length of one base.**

b stands for the **length of the other base.**

$$A = \frac{1}{2}h(a + b)$$ The given equation

$$2A = h(a + b)$$ Multiply both sides by 2.

$$2A = ah + bh$$ Use the distributive property.

$$2A - bh = ah$$ Subtract bh from both sides.

$$\frac{2A - bh}{h} = a$$ Divide both sides by h.

Thus, $a = \dfrac{2A - bh}{h}$ is the area formula solved for a in terms of A, b, and h.

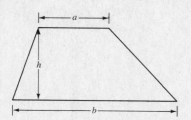

Figure 2-9. A trapezoid.

The Definition of a Literal Equation

In mathematics we frequently study equations with more than one variable in which the variables are not related to specific objects in the world. Such equations are not formulas, but are called **literal equations.**

Definition 2.6.   Literal equations
  A **literal equation** is one that contains more than one variable.

Examples **a–d** illustrate four literal equations.

a. $5x + 3y = 15$

b. $2x - y + 3z = 9$

c. $y = 2x^2 + 8x + 11$

d. $z = 9x^2 - 16y^2$

Solving a Literal Equation for an Indicated Variable

Equation **a** states that a relationship exists between the variables x and y. We say that the relationship is an **implied one,** because the equation is not solved for either variable. We can make the relationship an **explicit one** by isolating one of the variables on one side of the equation as we did in the formulas in Examples 3 and 4.

part a solved for y in terms of x   **part a solved for x in terms of y**

$$3y = 15 - 5x \qquad\qquad 5x = 15 - 3y$$

$$y = \frac{15 - 5x}{3} \qquad\qquad x = \frac{15 - 3y}{5}$$

Example 5.   Solve $2x - y + 3z = 9$ for z.

Solution.

$2x - y + 3z = 9$	The given equation
$3z = 9 - 2x + y$	Isolate the $3z$ term.
$z = \dfrac{9 - 2x + y}{3}$	Divide both sides by 3.

The equation is solved for z in terms of x and y.

Example 6.   Solve $5x + 2 = ax + b$ for x.

Solution.   **Discussion.** This equation has two x terms, which are first isolated on one side of the equation. The distributive property is then used to change the two terms to only one term with a factor of x.

$5x + 2 = ax + b$	The given equation
$5x - ax = b - 2$	Subtract ax and 2 from both sides.
$x(5 - a) = b - 2$	Write $5x - ax$ as $x(5 - a)$.
$x = \dfrac{b - 2}{5 - a}$	Divide both sides by $5 - a$.

The equation is solved for x in terms of b and a.

Evaluating a Literal Equation

The literal equation $5x + 3y = 15$ becomes a true number statement when x and y are replaced by 6 and -5 respectively:

$5(6) + 3(-5) = 15$	Replace x by 6 and y by -5.
$30 + (-15) = 15$	Multiply first.
$15 = 15$	A true number statement

To find other pairs of numbers that make the equation true, it would be easier to use one of the equivalent equations that is solved for either x or y. For example,

$$y = \tfrac{15 - 5x}{3} \text{ is equivalent to } 5x + 3y = 15.$$

In this form of the equation, any replacement for x will yield an expression that can be simplified to a single number. This number would be the replacement for y which when paired with the replacement for x would make the equation true.

Example 7.   Given $x - 3y + 2z = 8$

a. Solve for z in terms of x and y.

b. Find the value of z when $x = -6$ and $y = 2$.

Solution.   **a.** $x - 3y + 2z = 8$

$$2z = 8 - x + 3y$$

$$z = \frac{8 - x + 3y}{2}$$

The equation is solved for z in terms of x and y.

b. $z = \dfrac{8 - (-6) + 3(2)}{2}$

$$= \frac{8 - (-6) + 6}{2}$$

$$= \frac{20}{2}$$

$$= 10$$

Thus $z = 10$, when $x = -6$ and $y = 2$.

SECTION 2-5.   Practice Exercises

In exercises **1–8**, use the formula $d = r \cdot t$.

[Example 1]   **1. a.** If a car averages 52 mph, find the time needed to travel 312 miles.
    b. If the driver of the car wants to arrive at 2:00 P.M., approximately what time should she leave?

2. a. If a truck averages 48 mph, find the time needed to travel 336 miles.
    b. If the driver of the truck wants to arrive at 7:30 P.M., approximately what time should he leave?

3. a. Karen Shepherd owns a late-model, blue Ford Mustang convertible. She averages 62 mph on a trip to see her parents. If the trip takes 2.5 hours, how far apart in miles do Karen and her parents live?
    b. If Karen could average 70 mph, how much time could she cut off the 2.5 hours to the nearest minute?

4. a. Scott Cooney uses a '72 Ford pickup to commute to work. If he averages 58 mph and the commute time is three-quarters of an hour, find the distance from home to work.
    b. On the days that Scott carpools with Elizabeth, he gets to work in 0.60 hours. Find the average speed that Elizabeth drives.

5. Al and Eve Mae make an annual trip with their Venture motor home. The trip is 546 miles and the estimated travel time is 12 hours. Find the mph rate for this trip.

6. Rick and Nancy DeTar took their tent trailer on a camping trip to Big Sur, a distance of 84 miles from their home. The round trip time was 3.5 hours. Find the mph rate for this trip.

7. Monica needs to travel to Waterloo, a distance of 1105 miles, and she needs to be there by 2:00 P.M. She will fly in her private plane, which averages 260 mph. Approximately what time should she leave to meet her schedule?

8. Carol needs to get to work this morning by 7:45 A.M. The crosstown route permits her to average only 24 mph. If the distance is 18 miles, what time should she leave home?

In exercises **9–14**, use the stated formulas.

[Example 2]

9. George Fong is going to build a patio in his backyard. If the base of the triangular-shaped patio is 10 feet and the area is 50 feet^2, then what should be the height of the triangle? (Use $A = \frac{1}{2}bh$.)

10. The Sunset Memorial Park has a triangular-shaped lawn with an area of 300 feet^2. If the base side of the lawn is 25 feet, then what is the height? (Use $A = \frac{1}{2}bh$.)

11. A circular play area is outlined on a playground. What is the area enclosed if the radius is 5 feet? (Use $A = \pi r^2$ where $\pi \approx 3.14$.)

12. A circular design is to be constructed on a football field. What is the area of the design if the radius is 8 feet? (Use $A = \pi r^2$ where $\pi \approx 3.14$.)

13. A rectangular dog run has a perimeter of 34 meters and a width of 7 meters. What is its length? ($P = 2w + 2l$.)

14. A picture has a width of 24 inches and a perimeter of 120 inches. What is the length of the picture? (Use $P = 2w + 2l$.)

In exercises **15–34**, solve each formula for the indicated variable.

[Examples 3 and 4]

In **15** and **16**, use $P = a + b + c$ (perimeter of a triangle).

15. Solve for a.   16. Solve for b.

In **17** and **18**, use $P = na$ (perimeter of a regular polygon).

17. Solve for n.   18. Solve for a.

In **19** and **20**, use $C = 2\pi r$ (circumference of a circle).

19. Solve for r.   20. Solve for π.

In exercises **21** and **22**, use $A = \frac{1}{2}PS + B$ (surface area of a pyramid).

21. Solve for P.   22. Solve for S.

In exercises **23** and **24**, use $A = Ph + 2B$ (regular prism).

23. Solve for h.   24. Solve for B.

In **25** and **26**, use $P = 2l + 2w$ (perimeter of a rectangle).

25. Solve for l.   26. Solve for w.

In **27** and **28**, use $P = 2(a + b)$ (perimeter of a parallelogram).

27. Solve for a.   **28.** Solve for b.

In **29** and **30**, use $S = \frac{1}{2}(a + b + c)$ (semiperimeter).

29. Solve for c.   **30.** Solve for b.

In **31** and **32**, use $A = \pi(R + r)s$ (lateral area of a frustrum of a cone).

31. Solve for R.   **32.** Solve for r.

In **33** and **34**, use $A = \pi R(R + s)$ (surface area of a cone).

33. Solve for s.   **34.** Solve for π.

In exercises **35–42**, solve for the indicated variable.

[Example 5]   In **35** and **36**, use $9x + 4y = 36$.

35. Solve for y.   **36.** Solve for x.

In **37** and **38**, use $5p + 9q - 2r = 7$.

37. Solve for q.   **38.** Solve for r.

In **39** and **40**, use $ax^2 + bx + c = 10$.

39. Solve for a.   **40.** Solve for b.

In **41** and **42**, use $d = at^2 + vt$.

41. Solve for v.   **42.** Solve for a.

In exercises **43–48**, solve for x.

[Example 6] **43.** $4x + 3 = cx - 2$ **44.** $9x + 5 = cx - 4$

45. $ax - 13 = -5x + 12$ **46.** $dx - 1 = -2x + 19$

47. $x - 3 = bx - 14$ **48.** $x - 7 = bx - 5$

In exercises **49–54**, answer each part.

[Example 7] **49.** Given $3x + y = 6$
   a. Solve for y.   **b.** Find y when $x = 5$.

50. Given $5x + y = 15$
   a. Solve for y.   **b.** Find y when $x = 4$.

51. Given $2x + 3y + 4z = 16$
   a. Solve for y.   **b.** Find y when $x = 2$ and $z = 3$.

52. Given $5x + 2y + z = 12$
   a. Solve for x.   **b.** Find x when $y = 2$ and $z = 3$.

53. Given $P = a + 2b + c$
   a. Solve for b.   **b.** Find b when $P = 48$, $a = 12$, and $c = 20$.

54. Given $p = 2l + 2w + 2h$
   a. Solve for h.   **b.** Find h when $l = 3$, $w = 5$ and $p = 28$.

SECTION 2-5.   Ten Review Exercises

In exercises **1–4**, solve each problem.

1. Simplify: $3(6) + 5(-2)$

2. Simplify: $3(-4) + 5(4)$

3. Evaluate: $3x + 5y$ for $x = -9$ and $y = 7$

4. Evaluate: $3x + 5y$ for $x = 16$ and $y = -8$

In exercises **5** and **6**, use the following equation: $y = \dfrac{8 - 3x}{5}$.

5. Find the value of y when $x = 31$.

6. Find the value of y when $x = 0$.

In exercises **7** and **8**, solve each problem.

7. Simplify: $4(-4)^2 + 6(-4) - 7$

8. Simplify: $4\left(\dfrac{1}{2}\right)^2 + 6\left(\dfrac{1}{2}\right) - 7$

In exercises **9** and **10**, use the following equation: $b = 4a^2 + 6a - 7$.

9. Find the value of b when $a = 0$.

10. Find the value of b when $a = 0.2$.

SECTION 2-5.   Supplementary Exercises

In exercises **1** and **2**, use $A + B + C + D = 360°$ (interior angles of a quadrilateral).

1. Solve for D.   **2.** Solve for B.

In exercises **3** and **4**, use $A = 2\pi Rh$ (lateral surface of a circular cylinder).

3. Solve for R.   **4.** Solve for h.

In exercises **5** and **6**, use $t = a + (n - 1)d$ (term of a sequence).

5. Solve for n.   **6.** Solve for d.

In exercises **7–10**, solve each problem.

7. If a train travels at 52 mph,
   a. How far does it travel in two hours?
   b. How far does it travel in five hours?
   c. How far did it travel between the second and fifth hours?

8. Use $I = PRt$ (simple interest formula).
   a. Solve for t.   **b.** Find t if $P = \$1500$, $R = 0.06$, and $I = \$180$.

9. Solve for x: $3(6x + 5) - ax = 27$

10. Solve for y: $py + 2(y + 3) = 10$

In exercises **11** and **12**, use $\frac{a}{b} = \frac{c}{d}$ (proportion).

11. Solve for a.   **12.** Solve for c.

In exercises **13** and **14**, use $D = b^2 - 4ac$ (discriminant).

13. Solve for a.   **14.** Solve for c.

In exercises **15** and **16**, solve each problem.

15. Given $8x - 2y = 4$
  a. Solve for y.   **b.** Find y if $x = \frac{1}{2}$.

16. Given $x - 3y = 3$
  a. Solve for y.   **b.** Find y if $x = -3$.

In exercises **17–20**, write an equation for finding A, the area of each figure. Do not multiply out or simplify any expressions.

17.
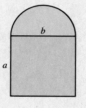

18.
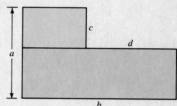

19.
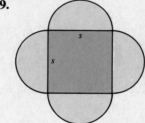

20.
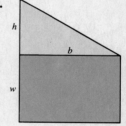

SECTION 2-6. Solving Linear Inequalities in One Variable

KEY TOPICS IN THIS SECTION

1. A definition of a linear inequality in one variable

2. Solving a linear inequality in one variable

3. The addition property of inequality

4. The multiplication property of inequality

5. Compound inequalities in one variable

6. Some applied problems

Sometimes problems are encountered that cannot be written as equations. Examples **a** through **c** are word statements that can be represented mathematically as an **inequality**, not an **equality**.

a. "In this city you should expect to pay at least $75,000 for a home."

b. "The mpg rating of this model car is less than 28."

c. "The employees of this company earn between $7.50 and $12.75 per hour."

The phrases "at least," "less than," and "between" are symbolized mathematically by an inequality. In this section we will learn how to solve linear inequalities in one variable.

A Definition of a Linear Inequality in One Variable

In Section 1-1 we are told that only three symbols are needed to compare any two real numbers. Specifically, if m and n are real numbers, then

1. $m < n$ which is read "m is less than n"

2. $m = n$ which is read "m is equal to n"

3. $m > n$ which is read "m is greater than n"

If m or n, or both m and n, are algebraic expressions, statements **1** and **3** are called inequalities. The following definition identifies the general form of a linear inequality in one variable.

Definition 2.7.   A linear inequality in x
A linear inequality in the variable x can be written in the form

$$ax + b < c \qquad \text{or} \qquad ax + b > c$$

where a, b and c are real numbers, and $a \neq 0$.

Notice that the forms of linear inequalities in x are the same as the linear equation in x, except the relations are $<$ or $>$ instead of $=$. Examples **a** and **a***, **b** and **b***, and **c** and **c*** illustrate this comparison.

Linear equation in x	Linear inequality in x
a. $4x - 5 = 3$	**a*.** $4x - 5 < 3$
b. $3x - 1 = x + 11$	**b*.** $3x - 1 > x + 11$
c. $2(x - 2) = 5(x + 3) - 7$	**c*.** $2(x - 2) > 5(x + 3) - 7$

Solving a Linear Inequality in One Variable

To solve a linear inequality in x means finding all real numbers that make the inequality true. In general, the number of solutions for an inequality is infinite. To illustrate, consider the inequality $x < 2$ and a partial listing of the solutions:

Inequality in x	Partial listing of solutions
$x < 2$	$-983, \ -10, \ -5.6, \ \dfrac{-3}{4}, \ 0, \ \dfrac{9}{80}, \ 1.5, \ 1.999$

We cannot list all the solutions for $x < 2$ in the same sense that we can for an equation, such as $x = 2$. We therefore agree to identify only the number that separates the real numbers that are solutions from those that are not solutions. Since 2 is the number that separates solutions from nonsolutions for the inequality $x < 2$, we write $x < 2$ to show that any number less than 2 is a solution.

Solving a linear inequality in x means to change, if possible, a given inequality to one of the following forms:

(i) $x < k$ **(ii)** $x > k$ **(iii)** $k < x$ **(iv)** $k > x$

where k is a real number that separates solutions from nonsolutions.

The correspondence between real numbers and points of a line gives a graphic representation of numbers. We use this correspondence to graph the solutions of an inequality in one variable.

Example 1.   Graph the solutions of the following inequalities.

$\qquad$ **a.** $x < -3$

$\qquad$ **b.** $x > 5$

Solution.   **Discussion.** To graph the solutions of $x < k$ or $x > k$, we locate two points on a line and label them 0 and k. We put a hollow dot around the point labeled k to show that it is not a solution. Then we darken the line to the left or right of k to indicate the solutions.

$\qquad$ **a.** $x < -3$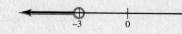

$\qquad$ **b.** $x > 5$

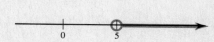

The Addition Property of Inequality

The addition property of inequality, like the addition property of equality, gives us a tool to relocate terms from one side of the inequality to the other side.

Addition property of inequality
$\quad$ If $a < b$ and c is any quantity, then

$$a + c < b + c.$$

The addition property of inequality is valid when $<$ is replaced by $>$.

Example 2.   Solve and graph.

$\qquad$ $8 + x < 2$

Solution.   **Discussion.** There are two constant terms in the inequality (8 on the left side and 2 on the right side). By adding (-8) to both sides of the inequality we can combine the constant terms on the right side.

$$8 + x < 2 \qquad \text{The given inequality}$$

$$8 + x + (-8) < 2 + (-8) \qquad \text{Add } (-8) \text{ to both sides.}$$

$$x < -6 \qquad 8 + (-8) = 0 \text{ and } 0 + x = x$$

Graph of the solutions

In an equation the left and right sides can be interchanged and the equations are still equivalent. (Such an interchanging is an example of the **symmetric property of equality**.)

The given equation	The same equations with left and right sides interchanged
$w = 10$	$10 = w$
$-3 = t$	$t = -3$

The left and right sides of an inequality can be interchanged *provided the sense of the order of the inequality is also interchanged.* (Such an interchanging of sides and change of order is an example of the **antisymmetric property of inequality**.)

The given inequality	The inequality with sides interchanged and order reversed	In general terms
$10 > w$	$w < 10$	If $a > b$, then $b < a$
$-4 < t$	$t > -4$	If $a < b$, then $b > a$

When an inequality is solved, the final form should be written with the variable on the left side. When the variable is on the right side, an incorrect graph may be drawn.

Example 3.   Solve and graph.

$$5t - 9 < 6t - 5$$

Solution. $\quad 5t - 9 < 6t - 5 \qquad$ The given inequality

$$-9 < t - 5 \qquad \text{Add } -5t \text{ to both sides.}$$

$$-4 < t \qquad \text{Add 5 to both sides.}$$

$$t > -4 \qquad \text{If } a < b, \text{ then } b > a.$$

Graph of the solutions

The Multiplication Property of Inequality

The multiplication property of inequality, like the multiplication property of equality, gives us a tool to change the coefficient of the variable from some number other than one to the number one. However, there are two parts to this property. When an inequality is multiplied by a positive number, the sense of the order is unchanged. If an inequality is multiplied by a negative number, the sense of the order is reversed.

A given inequality	Multiply by 4	Multiply by -4
$3 < 7$	$12 < 28$	$-12 > -28$
$-2 < 5$	$-8 < 20$	$8 > -20$
$-8 < -6$	$-32 < -24$	$32 > 24$

The order stays the same

The order in each case must be reversed.

Multiplication property of inequality
If $a < b$ and c is a positive number ($c > 0$), then $ac < bc$.
If $a < b$ and c is a negative number ($c < 0$), then $ac > bc$.

Example 4.   Solve and graph.

a. $2z - 5 < 3$

b. $3 - 5t < 18$

Solution.   **Discussion.** First use the addition property to isolate the variable terms on one side of the inequality. Then use the multiplication property to change the coefficient to one.

a. $2z - 5 < 3$ .... The given inequality

$2z < 8$ .... Add 5 to both sides.

$\dfrac{1}{2} \cdot 2z < \dfrac{1}{2} \cdot 8$ .... Multiply both sides by $\frac{1}{2}$.

$z < 4$ .... $\dfrac{1}{2} \cdot 2 = 1$ and $1 \cdot z = z$

Graph of solutions

b. $3 - 5t < 18$ .... The given inequality

$-5t < 15$ .... Add (-3) to both sides.

$\dfrac{-1}{5}(-5t) > \dfrac{-1}{5}(15)$ .... Multiply by $\frac{-1}{5}$ and reverse the order of the inequality.

$t > -3$ .... $\dfrac{-1}{5}(-5) = 1$ and $1 \cdot t = t$

Graph of solutions

Compound Inequalities in One Variable

Frequently the equality relation is included with a "less than" or "greater than" relation. The result is a **compound inequality.**

> **Definition of a compound inequality in the variable x**
> A compound inequality in the variable x can be written in the form
>
> $ax + b \leq c$,   which means $ax + b < c$ or $ax + b = c$
>
> $ax + b \geq c$,   which means $ax + b > c$ or $ax + b = c$
>
> where a, b, and c are real numbers and $a \neq 0$.

A compound inequality is solved when it is written in one of the following forms:

$$x \leq k \qquad x \geq k \qquad k \leq x \qquad k \geq x$$

When we graph the solutions of such an inequality we put a solid dot at k. The solid dot indicates that k is a solution.

Example 5. Solve and graph.

$$2(5z - 4) \leq 6 - 8(1 - 2z)$$

Solution.

$2(5z - 4) \leq 6 - 8(1 - 2z)$	The given inequality
$10z - 8 \leq 6 - 8 + 16z$	Remove the parentheses.
$10z - 8 \leq 16z - 2$	Write $-2 + 16z$ as $16z - 2$.
$-8 \leq 6z - 2$	Keep the coefficient of z positive.
$-6 \leq 6z$	Add 2 to both sides.
$-1 \leq z$	Multiply both sides by $\frac{1}{6}$.
$z \geq -1$	

If $a \leq b$, then $b \geq a$.

Graph of the solutions

Some Applied Problems

Example 6 illustrates a number problem whose solutions are written in terms of an inequality.

Example 6. The sum of two times a number and 9 is less than 5. Find the possible values for the number.

Solution.   Let $n = $ the number
"Two times a number" is written $2n$.
"The sum of two times a number and 9" is written $2n + 9$.
"The sum is less than 5" is written $2n + 9 < 5$.
An inequality of the problem is

$$2n + 9 < 5$$
$$2n < -4$$
$$n < -2$$

n is any number less than -2.

SECTION 2-6. Practice Exercises

In exercises **1–10**, graph the solutions.

[Example 1]

1. $x < 4$

2. $x < 2$

3. $y > -3$

4. $y > -1$

5. $z < -2$

6. $z < -5$

7. $x < 0$

8. $x > 0$

9. $x < \dfrac{3}{2}$

10. $x < \dfrac{5}{2}$

In exercises **11–42**, solve and graph.

[Examples 2–4]

11. $x + 3 < 8$

12. $x + 2 < 10$

13. $y - 5 > -2$

14. $z - 3 < -1$

15. $z + 4 < 2$

16. $p + 7 > 3$

17. $a - 7 > -12$

18. $t - 9 > -15$

19. $4x + 6 < 5x$

20. $3q + 2 < 4q$

21. $6(p - 1) > 7p$

22. $5(x - 1) > 6x$

23. $5(x - 3) < 6(x + 1)$

24. $9(y - 1) > 10(y - 2)$

25. $-3(4 - w) < 3w + 3 + w$

26. $-2(3 - t) < -4t + 1 + 7t$

27. $\dfrac{3b}{5} > -6$

28. $\dfrac{4d}{3} < -12$

29. $\dfrac{-x}{3} < 18$

30. $\dfrac{-y}{5} > 20$

31. $-3z > -21$

32. $-4z < -8$

33. $5 - 2t < 9$

34. $10 - 3t > -8$

35. $3k + 10 > 4$

36. $5k - 10 < 11 - 2k$

37. $6 - 5b > 20 + 2b$

38. $8b - 13 < 19 + 4b$

39. $2(x - 1) - 1 < 9(x + 2) + 7$

40. $3(x + 4) + 5 > 2x + 7(x - 3) + 8$

41. $5(t + 2) - 4(t + 1) < 3(t + 3) + 1$

42. $9(t - 2) - (t - 5) < 3(t + 7) - (3t + 2)$

In exercises **43–50**, solve and graph.

[Example 5]

43. $6z - 5 \le 2z + 19$

44. $10a - 4 \ge 7a - 1$

45. $7(z - 4) \le 5(4 - z)$

46. $13(a + 2) \le 3(a - 5) + 1$

47. $8c - (c - 5) \le c + 17$

48. $5x - (5 - x) \ge 10 + x$

49. $5x \le 10 + 3(2x + 4)$

50. $3(b + 5) \ge 7(b - 3) + 2b$

In exercises **51–60**, solve each problem.

[Example 6]

51. The sum of six and five times a number is always less than or equal to 1. Find all such numbers.

52. The difference between three times a number and 5 is always more than the difference between 3 and the number. Find all such numbers.

53. Two times the sum of a number and 9 is always more than the difference between 3 and the number. Find all such numbers.

54. Four times the difference between a number and 2 is always greater than or equal to the sum of two times the number and 4. Find all such numbers.

55. When the sum of three times a number and 4 is multiplied by two, the product is always less than 14. Find all such numbers.

56. The length of a rectangular picture frame is equal to 1 inch more than twice the width. If the perimeter of the frame is to be less than 38 inches, find the possible values for the width of the frame. (Recall: $P = 2l + 2w$.)

57. The length of a flower box is six times its width. If the perimeter of the box must be less than or equal to 112 inches, find the possible values for the width of the box.

58. The perimeter of a rectangular-shaped wall hanging must be less than 216 inches. If the width is equal to one-half the sum of the length and 36 inches, find the possible values for the length of the wall hanging.

59. On the first two tests Steve Koch got scores of 77 and 82. What can his final test score be so that he gets a total of more than 210 points on the three tests?

60. At the last football game, John Hiffman, a running back for the Grizzly Flat High School team, ran for 27 yards in the first quarter, 39 yards in the second quarter, and 11 yards in the third quarter. Find the number of yards John must run in the fourth quarter to gain more than 100 yards total for the game.

SECTION 2-6.   Ten Review Exercises

In exercises **1–6**, solve each problem.

1. Simplify: $4(2k - 1) + 10(k - 5)$

2. Solve: $4(2k - 1) + 10(k - 5) = 0$

3. Solve and graph: $4(2k - 1) + 10(k - 5) < 0$

4. Simplify: $6(y + 1) + 5 - 7(3y - 4) + 6$

5. Solve: $6(y + 1) + 5 = 7(3y - 4) - 6$

6. Solve and graph: $6(y + 1) + 5 \leq 7(3y - 4) - 6$

In exercises **7–10**, simplify.

7. $5^2 - 2^3 + 72 \div 6(10 - 2 \cdot 3)$

8. $2^2 \cdot 5^2 - 5(23 - 2^4) - 4 \cdot 3$

9. $[5(13 - 2^2) - 7(19 - 5 \cdot 3)] + 4^3$

10. $5[3 + 7(10 - 13) - 2 \cdot 3] + 2^5 \cdot 3$

SECTION 2-6.   Supplementary Exercises

In exercises **1–18**, solve and graph.

1. $6(z - 3) \geq 4(3 - z)$ **2.** $8(y + 2) \geq 3y + 1$

3. $\dfrac{4}{7}z > \dfrac{-2}{5}$ **4.** $\dfrac{5}{4}t < \dfrac{-3}{2}$

5. $4(n - 1) > 7n + 8$ **6.** $2(z + 3) < 5z + 9$

7. $-3z \leq 0$ **8.** $-2p \geq 0$

9. $\dfrac{4}{3}p + 32 > 0$ **10.** $\dfrac{9}{2}p + 27 < 0$

11. $-3(2x - 8) < 2(x + 14)$ **12.** $4x - (x - 3) \leq -3(2x - 7)$

13. $3y - 2(8y - 11) > 6 - (2y + 6)$ **14.** $6 - (3y + 5) > 4 - (2y + 7)$

15. $\dfrac{2x}{3} - \dfrac{5}{3} \geq \dfrac{3x}{2} + \dfrac{5}{2}$ **16.** $\dfrac{5x}{4} - \dfrac{1}{4} \leq \dfrac{6x}{5} + \dfrac{1}{5}$

17. $1.1x - 0.2 < 1.0 - 0.4x$ **18.** $0.8y + 1.3 > 0.2y + 3.1$

In **19–22**, graph on a number line the numbers indicated by x.

19. x is negative.   **20.** x is positive.

21. x is nonnegative.   **22.** x is nonpositive.

An inequality such as

$$-2 < x < 4$$

is a **compound inequality** that is satisfied by any number between -2 and 4. A graph of these numbers is shown on the following number line:

In exercises **23–30**, graph the solutions.

23. $-5 < x < 2$ **24.** $-1 < x < 6$

25. $2 < x < 7$ **26.** $4 < x < 10$

27. $-9 < x < -3$ **28.** $-13 < x < -8$

29. $-4 \leq x \leq 3$ **30.** $-2 \leq x \leq 5$

In exercises **31–34**, solve and graph.

Example. $-2 < x + 5 < 4$

Solution.

$-2 < x + 5 < 4$	The given inequality
$-2 - 5 < x + 5 - 5 < 4 - 5$	Subtract 5 from each part.
$-7 < x < -1$	Simplify each part.

Graph of solutions

31. $-4 < x + 2 < 5$ **32.** $-1 < x + 3 < 7$

33. $-5 < 2x - 3 < 1$ **34.** $-3 < 3x - 6 < 0$

SECTION 2-1. Summary Exercises

Name _____

Date _____

Score _____

Answer

1. Is the equation below true or not true?
$$5(-2)^2 - 3(-2) = (-4)(-1) + (-3)(10)$$

1. _____

In exercises **2–4**, determine whether the given number is a root.

2. $6x - 15 = -23 + 2x$ and $x = -3$

2. _____

3. $4x - 5 - x = 13 - 2x - 3$ and $x = 3$

3. _____

4. $2m^2 - 20 = -2m^2 + 5$ and $m = \dfrac{5}{2}$

4. _____

In exercises **5** and **6**, determine which of the given numbers are roots.

5. $28x + 20 = 4x - 76;$ $-6, -4, -2, 0$

5. _____

6. $x^2 + 3x = x + 3;$ $-3, -1, 1, 3$

6. _____

In exercises **7** and **8**, solve each problem.

7. Are equations **a** and **b** equivalent, given that they both have exactly one root?
   a. $9z - 36 = 3z - 18$ **b.** $z = 3$

7. _____

8. a. Simplify.
   b. Determine whether equations A and B are equivalent, given that they both have exactly one root.
   a. $5(x + 1) - 4x + x - 5$

8. **a.** _____

   b. A. $5(x + 1) - 4x = x - 5$

      B. $x = -2$

b. _____

SECTION 2-2.   Summary Exercises

In exercises **1–8**, solve and check.

1. $7w + 5 = 68$

1. _____

2. $7t + 13 = 6t - 5$

2. _____

3. $8a = -56$

3. _____

4. $\dfrac{-2}{3} w = 36$

4. _____

5. $19 = 5x - 11$

5. _____

6. $5z - 17 = 9z + 19$

6. _____

7. $11 + 14a = 2a - 10$

7. _____

8. $\dfrac{4x}{3} + 13 = 21$

8. _____

SECTION 2-3. Summary Exercises

In exercises **1–5**, solve each problem.

1. If four times a number is increased by 30, the result is the same as the difference between three times the number and -25. Find the number.

1. _____

2. Debbie bought two different types of coffee mugs to give away as gifts. The larger mugs cost $3.50 each, and the smaller mugs cost $2.25 each. If she bought three more small mugs than large mugs and spent a total of $35.50, how many of each size of mug did she buy?

2. _____

3. The length of a rectangle is one less than three times the width. If the perimeter is 158 centimeters, find the length and width.

3. _____

4. A car and a truck leave a rest area at the same time heading in the same direction. The car travels at an average rate of 60 mph. If the two vehicles are 12 miles apart after four hours, find the rate at which the truck is traveling.

4. _____

5. Kerry's blue bike is six years older than her white bike. Her blue bike is also four times as old as her white bike. How old are the bikes?

5. _____

SECTION 2-4. Summary Exercises

In exercises **1–8**, solve and check.

1. $3y - 5 = 7y + 7$

1. _____

2. $6(x - 2) + 21 = 5(x - 11)$

2. _____

3. $-4w + 6 = -2(-3 + 2w)$

3. _____

4. $17y - 7(y + 2) = 2(4y - 3) - 7$

4. _____

5. $5p + 4 = 7(p + 1) - 2p$

5. _____

6. $3(2y - 4) = 0$

6. _____

7. If two times the smallest of three consecutive odd integers is subtracted from three times the largest, the difference is -3. Find all three odd integers.

7. _____

8. A, B, and C are the measures of the angles of a triangle. B is $10°$ less than two times A. C is two times the sum of A and $5°$. Find A, B, and C.

8. _____

SECTION 2-5.   Summary Exercises

In exercises **1–8**, solve each problem.

1. a. Fred Klampett will drive a school bus with students to a science fair 126 miles away. If Fred can average 42 mph, how long will the trip take?

1. a. _____

   b. If Fred needs to get the students to the fair by 11:00 A.M., at approximately what time should he leave?

b. _____

2. One of the parking areas at The Arden Mall has the shape of a rectangle. If the perimeter of this area is 106 yards, and it is 31 yards long, find its width. (Use $P = 2l + 2w$.)

2. _____

3. A football field has an area of 57,600 ft^2. If the length of the field is 300 feet, find the width. (Use $A = lw$.)

3. _____

4. Solve $PV = nRT$ for R (ideal gas law).

4. _____

5. Solve $F = \dfrac{w(R - r)}{2R}$ for w (differential pulley).

5. _____

6. a. Solve $3x - 2y + 5z = 8$ for z.

6. a. _____

    b. Find z for $x = -5$ and $y = 6$.

b. _____

7. a. Solve $P = 4l + 4w + 4h$ for w.

7. a. _____

    b. Find w if $P = 60$, $l = 7$, and $h = 3$.

b. _____

8. Solve for y.
    $3(y + 2) = ay - 10$

8. _____

SECTION 2-6. Summary Exercises

Name

Date

Score

Answer

In exercises **1–7**, solve and graph.

1. $3t - 9 \leq 36$

1. _____

——————————————————▶

2. $-5x + 10 > 25$

2. _____

——————————————————▶

3. $3(x + 7) \geq 4x - 3$

3. _____

——————————————————▶

4. $\dfrac{-3}{2}w - 5 < 4$

4. _____

——————————————————▶

5. $-5y - 2 < 32 + 12y$

5. _____

_____→

6. $2(3x + 1) - (5x + .4) > 2$

6. _____

_____→

7. $3x - 1 \leq x - (3x - 14)$

7. _____

_____→

8. The product of a number and 5 is less than or equal to the number increased by 8. Find the possible numbers.

8. _____

CHAPTER 2.   Review Exercises

In exercises **1–4**, determine which of the given numbers are solutions.

1. $3x + 8 = 3 - 2x$; $-1, 0, 1, 2$

2. $3(y - 5) = 2(4y + 5)$; $-7, -5, 0$

3. $a^2 + 5 = 6a$; $1, 2, 3, 4, 5$

4. $z^2 + 12 = 7z$; $2, 3, 4, 5$

In exercises **5** and **6**, determine whether equations **a** and **b** are equivalent, given that they both have exactly one solution.

5. **a.** $4(x - 2) = 13 - 3x$ **b.** $x = 3$

6. **a.** $2(1 + 3y) = 9y + 3$ **b.** $y = \dfrac{1}{3}$

In exercises **7–16**, solve and check.

7. $x - 15 = 3$ 8. $y + 42 = -7$

9. $10y = -18$ 10. $-5a = 55$

11. $\dfrac{b}{10} = 12$ 12. $\dfrac{x}{3} = 17$

13. $3a + 12 = 2a$ 14. $7p - 3 = 6p$

15. $15w + 18 = 11w + 10$ 16. $7t + 20 = 5t + 14$

In exercises **17–22**, solve each problem.

17. The difference between three times a number and 12 is the number itself. Find the number.

18. The difference between 56 and four times a number is 28. Find the number.

19. A child's bank contains three times as many nickles as dimes. The value of the coins is $1.50. How many nickles are in the bank?

20. If twice as many 25¢ stamps as 45¢ stamps were purchased and if the total cost of the stamps was $9.50, how many of each was purchased?

21. A triangle has for the lengths of its sides three consecutive integers. If the distance around the triangle is 48 centimeters, what is the length of the longest side?

22. A triangle's second side is twice as long as its first side, and its third side two and one-half times as long as its first side. If the perimeter is 33 centimeters, how long is the third side?

In exercises **23–26**, solve and check.

23. $3(3y + 6) = 17 + 6y$ 24. $9 + 6(16 - a) = 3(3a - 5)$

25. $5(8b + 6) - 33 = 3(12 - 4b)$ 26. $3(m - 1) = 2(4m + 3) - 4$

In exercises **27** and **28**, solve each problem.

27. The sum of the first and two times the third of three consecutive integers is 67. Find all three integers.

28. When two times the first of three consecutive integers is added to the third, the sum is 38 less than four times the middle integer. Find all three integers.

In exercises **29–32**, solve for the indicated variable.

29. $2x + 3y = 0$ for x

30. $S = V + at$ for a

31. $F = \dfrac{kAv}{l}$ for v

32. $3x^2 + 4y - 1 = 0$ for y

In exercises **33** and **34**,

a. solve for y.

b. find y when $x = -15$.

33. $x - 5y = 10$

34. $4x + 2y = 8$

In exercises **35–40**, solve and graph.

35. $4x + 1 > 5x - 3$

36. $8y + 3 \le 9y + 10$

37. $4y + 15 \ge 3 - 2y$

38. $8t + 2 < 5t - 10$

39. $13(y + 3) - 3 \le 3(y + 2)$

40. $7(3 - x) > 5(x - 7) - 4$

3
Polynomials

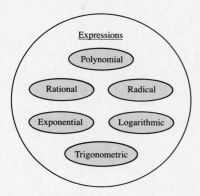

Figure 3-1. Some of the categories of mathematical expressions.

"In Chapter 2 we worked on solving linear equations in one variable," Ms. Glaston began. "The task associated with solving equations is to find any roots, or solutions. That is, *find any real numbers that make a given equation a true number statement.* In Chapter 3 we have distinctly different tasks to work on. Specifically, we will begin an extensive study of expressions called **polynomials.**"

She then turned on the projector to show the diagram in Figure 3-1.

"Recall that real numbers can be separated into categories, or sets, based on clearly defined characteristics," Ms. Glaston continued. "In the diagram on the screen the circle represents all the different categories, or sets, of mathematical expressions. Like real numbers, expressions can be separated into categories based on clearly defined characteristics. Within the circle I have listed some of the expressions that are studied in mathematics. As you can see, the polynomial expression is included in the diagram."

Bill O'Quinn raised his hand and asked, "Ms. Glaston, what exactly is the difference between an equation and an expression?"

"Well, Bill," Ms. Glaston replied, "an equation is a mathematical sentence. We look for number replacements for the variable or variables in the equation that will make the statement true. Expressions are combinations of numbers, variables, and operations that represent numbers. A specific value of an expression is obtained when the variable or variables are replaced by numbers, and the resulting numerical expression is simplified. In general, *equations are solved,* but *expressions are manipulated, simplified,* or *written in different forms.*"

She then turned to the board and wrote the following examples:

Polynomials	**Non-Polynomials**
a. $4t^2 - 9$	**d.** $k^2 + \dfrac{1}{k}$
b. $2u^4 + 3u^2 - 5$	**e.** $p^2q^{-1} - 4p^{-1}q^2 + 9$
c. $\dfrac{1}{4}x^3 + \dfrac{2}{3}x^2y - \dfrac{3}{5}xy^2 + y^2$	**f.** $\sqrt{-4z^3} - 1$

"In the chapter we are about to study," Ms. Glaston said, "we will learn a definition that shows that the expressions in examples **a, b,** and **c** are polynomials. The definition will also show why the expressions in examples **d, e,** and **f** are not polynomials. Throughout this chapter we will study methods for adding, subtracting, multiplying, and dividing expressions such as those in examples **a, b,** and **c.**"

SECTION 3-1.   Some Terminology Related to Polynomials

KEY TOPICS IN THIS SECTION

1. A definition of a polynomial in x

2. Polynomials in more than one variable

3. The degree of a polynomial in x

4. The degree of a polynomial in more than one variable

5. Writing polynomials in ascending or descending powers

Polynomials are expressions that can be used to describe many quantities in the world. For example, polynomials can be used to show how the areas and volumes of objects are related to their dimensions. Other polynomials describe relationships found in business, medicine, transportation, and electronics. Some of the terminology associated with polynomials will be given in this section.

A Definition of a Polynomial in x

Polynomials are frequently identified by the variable (or variables) found in the expression. For example, we may refer to a polynomial as

"a polynomial in t."

"a polynomial in x and y."

"a polynomial in a, b, and c."

Definition 3.1 identifies the form of every term of a polynomial in x.

Definition 3.1.   A polynomial in x
The form of every term of a polynomial in x is

$$ax^n$$

where a is a real number and n is a positive integer or 0. The a is called the **coefficient** of x^n.

Examples **a–e** are five polynomials in x. Notice that each term has the form ax^n given in Definition 3.1. The corresponding name of each polynomial is based on the number of terms in each expression.

Polynomial in x	Name
a. $-8x^3$	Monomial
b. $9x^2 - 1$	Binomial
c. $2x^2 - x - 3$	Trinomial
d. $kx^2 + 2kx - x - 2k$; k is a real number	Polynomial of four terms
e. $81x^4 + 108x^3 + 54x^2 + 12x + 1$	Polynomial of five terms

The expressions in examples **f–i** would not qualify as terms of a polynomial.

Expression	Not a Term of a Polynomial in x
f. $\dfrac{5}{x^3}$	The variable is written in the denominator.
g. $5x^{-3}$	The exponent on x is negative.
h. $5x^{2/3}$	The exponent on x is not an integer.
i. $\sqrt{-9x^3}$	As will be seen in Section 8-1, $\sqrt{-9}$ is not a real number.

Polynomials in More Than One Variable

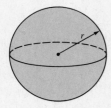

Figure 3-2. A sphere with radius r.

Polynomials can be used to describe the areas and volumes of many objects in terms of the dimensions of the object. Sometimes the polynomials require only one variable, but sometimes they require more.

In Figure 3-2 the **sphere** (shaped like a ball) has a radius of r units. The **surface area** and **volume** can be described as monomials in r.

	Monomial in r	If $r = 3$ inches (in)
Surface area	$4\pi r^2$	surface area is $4\pi\,(3\text{ in})^2 = 36\pi\text{ in}^2$
Volume	$\dfrac{4}{3}\pi r^3$	volume is $\frac{4}{3}\pi\,(3\text{ in})^3 = 36\pi\text{ in}^3$

In Figure 3-3 the **rectangular solid** (shaped like a box) has length x units, width y units, and height z units. The **surface area** and **volume** of this object require polynomials in x, y, and z.

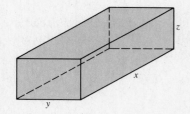

Figure 3-3. A rectangular solid.

	Polynomial in x, y, and z	If $x = 5$ inches $y = 4$ inches and $z = 3$ inches
Surface area	$2xy + 2xz + 2yz$	$2(5)(4) + 2(5)(3) + 2(4)(3)$ $= 40 + 30 + 24$ $= 94\text{ in}^2$
Volume	xyz	$(5\text{ in})(4\text{ in})(3\text{ in}) = 60\text{ in}^3$

The Degree of a Polynomial in x

Polynomial expressions are used to write equations. We will study examples of such equations in later chapters. As an illustration, consider the following:

$x^2 - 5x + 6$ is a polynomial expression consisting of 3 terms.

$x^2 - 5x + 6 = 0$ is an example of a polynomial equation.

The number of possible roots, or solutions, of a polynomial equation is linked to **the degree of the polynomial expression** in the equation. Definition 3.2 identifies what is meant by the degree of a monomial, and a polynomial of more than one term.

Definition 3.2.   Degree of a polynomial in x

(i) The degree of ax^n is n, the exponent on x.

(ii) The degree of a polynomial in x is the highest degree term in the polynomial.

If the term in the polynomial is a nonzero constant, then the degree of the term is 0. For example, the degree of 3, -10, $\frac{2}{3}$, and $\sqrt{2}$ is 0. If 0 occurs in a polynomial, we say that the term has no degree.

Example 1.   For the polynomial $7x^3 + 2x^5 - x + 9$,

      a. state the degree of each term.

      b. state the degree of the polynomial.

Solution.   **a.** The degree of $7x^3$ is 3.   The exponent is 3.

      The degree of $2x^5$ is 5.   The exponent is 5.

      The degree of x is 1. x can be written x^1.

      The degree of 9 is 0.   The degree of a nonzero constant is 0.

      b. The degree of $7x^3 + 2x^5 - x + 9$ is 5.

The Degree of a Polynomial in More Than One Variable

If a polynomial contains more than one variable, then the degree of each term is **the sum of the exponents of the variables in that term.** The degree of the polynomial is then the highest degree term in the polynomial.

Example 2.   If x, y, and z are variables, state for the polynomial $9x^2y^2 - 30xyz + 25z^2 - 16$

      a. the degree of each term.

      b. the degree of the polynomial.

Solution.   **Discussion.** Since x, y, and z are variables, the degree of each term is the sum of the exponents on these variables in that term.

      a. The degree of $9x^2y^2$ is $2 + 2 = 4$.
      The degree of $30xyz$ is $1 + 1 + 1 = 3$.
      The degree of $25z^2$ is 2.
      The degree of 16 is 0.

      b. The degree of $9x^2y^2 - 30xyz + 25z^2 - 16$ is 4.

Writing Polynomials in Ascending or Descending Powers

Mathematics is a highly structured body of knowledge. Built into this structure are "preferred forms" for writing things such as answers, expressions, and equations. Many of these preferred forms have reasons for their existence. However, many are the result of conventions that have been passed on from generation to generation.

Polynomials should be written in one of two preferred forms: **ascending powers** or **descending powers** of some variable in the polynomial.

Ascending and descending powers forms for writing a polynomial in x

(i) The terms of a polynomial in x are written in **ascending powers** when the lowest-degree term is written first (at the extreme left), and the degree of each successive term is more than the degree of the term on the left.

(ii) The terms of a polynomial in x are written in **descending powers** when the highest-degree term is written first (at the extreme left), and the degree of each successive term is less than the degree of the term on the left.

Example 3.   Write in ascending powers

$$x^2 - 2x^3 - 5x^4 + 7x + 1.$$

Solution.   **Discussion.** The constant term 1 has degree 0 and is written first. The $7x$ term has degree 1 and is written next. The $5x^4$ term has degree 4 and is the last term written.

In ascending powers, the polynomial is written as follows:
$1 + 7x + x^2 - 2x^3 - 5x^4$

Example 4.   Write in descending powers of a

$$6a^2b^2 - 4ab^3 - 4a^3b + b^4 + a^4.$$

Solution.   **Discussion.** The polynomial is to be written in descending powers of a. Therefore the b is treated as a constant, and the exponents on the factors of b are ignored in determining the order in which the terms are written.

In descending powers of a, the polynomial is written as follows:
$a^4 - 4a^3b + 6a^2b^2 - 4ab^3 + b^4$

Calculator Supplement

Using a Calculator to Evaluate Polynomials

Frequently it is necessary to find the value of a polynomial for a given value of the variable. If the polynomial has more than one variable, values must be given for each of the variables. The instructions "Evaluate" are usually given for such an activity. A calculator with a **memory function** is a valuable tool for doing these evaluations.

Consider, for example, the following trinomial in x:

$$x^4 - 6x^2 + 7$$

Based on the rule for order of operations, whenever x is replaced by some number, then the powers (x^4 and x^2) would be simplified first, the product ($6 \cdot x^2$) would be done next, and the subtraction and addition (in that order) would be done last. The following set of instructions can be followed to evaluate this trinomial for $x = 3.5$ using a calculator.

Steps		Display Shows
Enter	3.5	3.5
Press	M in*	3.5
Press	x^y	3.5
Enter	4	4.
Press	−	150.0625
Enter	6	6.
Press	×	6.
Press	MR**	3.5
Press	x^2	12.25
Press	+	76.5625
Enter	7	7.
Press	=	83.5625

Thus $x^4 - 6x^2 + 7$ has the value 83.5625 when x is replaced by 3.5.

SECTION 3-1. Practice Exercises

In exercises **1–8**,

a. state the degree of each term.

b. state the degree of the polynomial.

c. state the name of the polynomial.

[Example 1]
1. $3x^5 + x^3 - 1$ **2.** $10 - x^2 - 4x^4$

3. $-6x^4 + 6x$ **4.** $7x^5 - 3x$

5. $9x^4 - 6x^8 + 3x^2 + 2$ **6.** $11x^7 - 8x^9 - x + 1$

7. $x^2 - 3 + 6x^7 - 2x^5 + 8x^3$ **8.** $9 - 4x^3 + 7x^9 + 3x^{10} + x^2$

In exercises **9–18**,

a. state the degree of each term.

b. state the degree of the polynomial.

[Example 2]
9. $a^2b^2 - 5ab + 6$ **10.** $a^3b^3 + 9ab - 5$

11. $5u^2v^3 - 3u^2v$ **12.** $-8x^2y + 5x^2y^4$

13. $5r^3s^2t + 2rs^2t^2 - r^2s^3t^4$ **14.** $r^3s^3 + r^2s^4t - 3rs^2t^2$

15. $u^6 - 4u^4v + 6u^2v^2 - v^3$ **16.** $9 - 2u^3v^2 + 5uv^4 - v^6$

17. $a^4b^2 - 3a^3b^3 + 8a^2b^4$ **18.** $x^3y^2 + 4x^2y^3 - xy^4$

* The "M in" is a **memory in** key that enters the number on the display into the calculator's memory. The "stored" number can then be used repeatedly as needed.

** The "MR" is a **memory recall** key that retrieves the stored number from the calculator's memory. The stored number can be recalled as many times as needed, but it will be replaced by a new number when the "M in" key is pressed.

In exercises **19–24**, write in ascending powers.

[Example 3]   **19.** $3x + 4 + x^2$ **20.** $7x + 6x^2 + 1$

21. $8a^3 + 10 - 4a^2 + 2a$ **22.** $27 + a^2 - 3a - 9a^3$

23. $7b^5 - 2b^{10} + 4b^3 + b$ **24.** $3b^3 - 7b^5 - b^8 + 3b$

In exercises **25–30**, write in descending powers.

25. $64 + y^2 - 16y$ **26.** $121y^2 + 1 + 22y$

27. $8m^3 + 1 - 4m^5 - m$ **28.** $3n^2 - 5 - 6n^5 + n$

29. $11z^3 + 8z^5 - 3z + 6z^2$ **30.** $14z^4 + z^7 - 8z + 3z^3$

In exercises **31–36**, write in descending powers of the stated variable.

[Example 4]   **31.** $10y^2 + x^2 - 7xy$ **a.** x **b.** y

32. $-12xy + 4x^2 + 9y^2$ **a.** x **b.** y

33. $5 - 3ab - a^2 + 5b^2$ **a.** a **b.** b

34. $12 - mn + n^2 - m^2$ **a.** m **b.** n

35. $5r^3s^2 - r^4st^2 - 8s^4t^5 + 2r^5t^3$ **a.** r **b.** s **c.** t

36. $6a^2b^2c^3 - 4ab^3 - 4a^3bc + b^4c^2$ **a.** a **b.** b **c.** c

SECTION 3-1.   Ten Review Exercises

In exercises **1–10**, solve each problem.

1. Write the following as an algebraic expression:

Five times a number n is increased by two.

2. Evaluate the expression of exercise **1** for $n = -5$.

3. Write the following as an algebraic expression:

Eight is subtracted from three times a number n.

4. Evaluate the expression of exercise **3** for $n = -5$.

5. If five times a number is increased by two, the result is the same as eight less than three times the number. Find the number.

6. Evaluate for $z = \frac{-1}{2}$: $3(5z + 7) - 3$

7. Evaluate for $z = \frac{-1}{2}$: $3 + 5(1 - z)$

8. Solve and check: $3(5z + 7) - 3 = 3 + 5(1 - z)$

9. Solve and graph: $3(5z + 7) - 3 < 3 + 5(1 - z)$

10. Solve and graph: $3(5z + 7) - 5(1 - z) \geq 6$

SECTION 3-1.   Supplementary Exercises

In exercises **1–4**, state the value of k so that the term will have the stated degree.

1. $7x^k y^2$
   a. 5th degree
   b. 3rd degree

2. $-4x^k y^3$
   a. 5th degree
   b. 4th degree

3. $10x^2 y^k z$
   a. 5th degree
   b. 7th degree

4. $12xy^k z^3$
   a. 6th degree
   b. 10th degree

In exercises **5–8**, answer parts **a–c**.

5. Given $x^2 + 5y^2 + 8 - 3xy$
   a. Write in descending powers of x.
   b. Identify the non-x terms.
   c. Which non-x term should be written first for preferred form?

6. Given $x^2 + 5y^2 + 8 - 3xy$
   a. Write in descending powers of y.
   b. Identify the non-y terms.
   c. Which non-y term should be written first for preferred form?

7. Given $a^3 b^3 + 3a^2 b + 2ab^2 + 6a^2 + 5b^2 - 9$
   a. Write in descending powers of a.
   b. Identify the a^2 terms.
   c. Which a^2 term should be written first?

8. Given $a^3 b^3 - 3a^2 b + 2ab^2 + 6a^2 + 5b^2 - 9$
   a. Write in descending powers of b.
   b. Identify the b^2 terms.
   c. Which b^2 term should be written first?

In exercises **9** and **10**, the polynomials are written in descending powers with respect to x. Answer parts **a–d** for each question.

9. $-4x^k y^2 + 5x^m y - x^n + 4$ is a polynomial of degree 5.
   a. What is the value of k?
   b. What is the largest possible value of m?
   c. What is the smallest possible value of m?
   d. What is the smallest possible value of n?

10. $-4x^k y^2 + 5x^m y - x^n + 4$ is a polynomial of degree 7.
    a. What is the value of k?
    b. What is the largest possible value of m?
    c. What is the smallest possible value of m?
    d. What is the smallest possible value of n?

In exercises **11–14**, use paper and pencil or a calculator to evaluate the given polynomials for the stated values.

11. $2t^3 - t^2 + 5t - 8$
    a. $t = 3$
    b. $t = -2$
    c. $t = 0$

12. $1 + t - 2t^2 - 5t^3$
    a. $t = 2$
    b. $t = -3$
    c. $t = 0$

13. $3x^2 - 4xy + 10y^2$
   a. $x = 2$ and $y = -2$
   b. $x = -5$ and $y = 3$
   c. $x = 7$ and $y = 0$

14. $2y^2 - 7xy - 5x^2$
   a. $x = 3$ and $y = -3$
   b. $x = -4$ and $y = 6$
   c. $x = 0$ and $y = 10$

SECTION 3-2. Adding and Subtracting Polynomials

KEY TOPICS IN THIS SECTION

1. What it means to add or subtract polynomials

2. Adding polynomials

3. Subtracting polynomials

4. Combined operations

In this section we will learn how polynomials can be added or subtracted.

What it Means to Add or Subtract Polynomials

Consider the following polynomials in x, which have been labeled P and Q:

$$P: 3x^2 - 4x + 5 \quad \text{and} \quad Q: 2x^2 + x - 9$$

If x is replaced by any number, then P and Q can each be simplified to a single number. The exact values for P and Q depend on the number replacement for x. For example, if x is replaced by 3,

P becomes $3(3^2) - 4(3) + 5$ Q becomes $2(3^2) + 3 - 9$

           $= 27 - 12 + 5$ $= 18 + 3 - 9$

           $= 20$ $= 12$

Suppose now that we want to find $P + Q$, or $P - Q$; that is, the sum or difference of the polynomials.

$$P + Q \text{ is written as } (3x^2 - 4x + 5) + (2x^2 + x - 9).$$
$$P - Q \text{ is written as } (3x^2 - 4x + 5) - (2x^2 + x - 9).$$

The indicated sum and difference cannot be found in the way a sum or difference of numbers are computed. However, some of the properties of real numbers that are reviewed in Chapter 1 can be used to write the sum and difference with fewer symbols. For example, in $P + Q$,

$$3x^2 + 2x^2 = (3 + 2)x^2 = 5x^2$$

$$-4x + x = (-4 + 1)x = -3x$$

$$5 + (-9) = -4$$

As a consequence,

$$(3x^2 - 4x + 5) + (2x^2 + x - 9) \text{ can be written as } 5x^2 - 3x - 4.$$

Notice, if x is replaced by 3 in all three polynomials,

$3x^2 - 4x + 5$ becomes 20,
$2x^2 + x - 9$ becomes 12, and $20 + 12 = 32$. $\Big\}$ $5x^2 - 3x - 4$ becomes 32.

In fact, $5x^2 - 3x - 4$ always yields the same number as $P + Q$ for any number replacement for x. We therefore write the following equation to show the sum of P and Q:

$$(3x^2 - 4x + 5) + (2x^2 + x - 9) = 5x^2 - 3x - 4$$

Adding Polynomials

To simplify the sum of P and Q above, the coefficients of the x^2 and x terms respectively were added, as were the constant terms. *Terms in which the exponents of the variable factors are exactly the same are called like terms.*

> To simplify a sum of two or more polynomials, add the coefficients of any like terms in the polynomials.

There are basically two ways of adding the coefficients of any like terms when polynomials are added. The two procedures are shown in Examples **1** and **2**.

Example 1. Add:

$$(10xy + 4x^2 - 7y^2) + (2y^2 - 3x^2 - 9xy) + (2x^2 - xy + 3y^2)$$

Solution. **Discussion.** In general, a sum can be more easily found by first re-writing each polynomial in descending powers of the same variable. This first method of finding the sum uses arcs to link any like terms in the polynomials.

$$(10xy + 4x^2 - 7y^2) + (2y^2 - 3x^2 - 9xy) + (2x^2 - xy + 3y^2)$$
$$= (4x^2 + 10xy - 7y^2) + (-3x^2 - 9xy + 2y^2) + (2x^2 - xy + 3y^2)$$

$$4x^2 - 3x^2 + 2x^2 = (4 - 3 + 2)x^2 = 3x^2$$
$$= 3x^2 - 2y^2 \quad 10xy - 9xy - xy = (10 - 9 - 1)xy = 0xy$$
$$-7y^2 + 2y^2 + 3y^2 = (-7 + 2 + 3)y^2 = -2y^2$$

Since $0xy = 0$, the xy term is not written.

Example 2. Add:

$$(-ab - bc + 9) + (4ac - 5 - 2bc) + (2ab + 3ac - 1)$$

Solution. **Discussion.** Another method writes the polynomials in a vertical format in which the like terms are aligned in columns. For these polynomials, the columns from left to right will contain the ab terms, the ac terms, the bc terms, and the constant terms respectively.

$-ab$		$-bc + 9$		Leave a space for the missing ac term.
	$4ac$	$-2bc - 5$		Align the bc and constant terms.
$2ab$	$+3ac$	-1		Align the ac and constant terms.
ab	$+7ac$	$-3bc + 3$		Add the coefficients of the like terms.

Subtracting Polynomials

Numbers are subtracted by changing the subtraction to an addition.

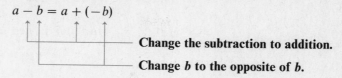

$$a - b = a + (-b)$$

Change the subtraction to addition.

Change *b* to the opposite of *b*.

The subtraction of two polynomials can similarly be changed to an addition.

Definition 3.3. The subtraction of polynomials

If P and Q are polynomials, then

$$P - Q = P + (-Q)$$

where $-Q$ is the opposite of Q.

Consider the P and Q identified earlier in the section:

$$P: 3x^2 - 4x + 5 \quad \text{and} \quad Q: 2x^2 + x - 9$$

$P - Q$ becomes $(3x^2 - 4x + 5) - (2x^2 + x - 9)$.

To simplify this subtraction we add the opposite of Q to P. The question is, "What is the form of the opposite of Q?"

A recommended procedure for finding the opposite of a polynomial is to use the distributive property to multiply each term of the polynomial by -1. Examples **a–c** illustrate this procedure for several polynomials.

Q	$-Q$, the opposite of Q
a. $2x^2 + x - 9$	$-1 \cdot (2x^2 + x - 9) = -2x^2 - x + 9$
b. $3a^2 - 2ab + 5$	$-1 \cdot (3a^2 - 2ab + 5) = -3a^2 + 2ab - 5$
c. $-2xy + xz - 3yz - 1$	$-1 \cdot (-2xy + xz - 3yz - 1) = 2xy - xz + 3yz + 1$

Once the polynomial being subtracted has been changed to its opposite, the subtraction is changed to an addition. The indicated sum is then simplified as in Example 1 or Example 2.

Example 3. Simplify:

$$(t^5 + 3t + 7 - 3t^3) - (4 - t^5 + 3t - 5t^3)$$

Solution. **Discussion.** As with addition, the polynomials should first be written in descending powers. Then write both polynomials without parentheses and change the appropriate polynomial to its opposite.

$$(t^5 + 3t + 7 - 3t^3) - (4 - t^5 + 3t - 5t^3)$$

$$= (t^5 - 3t^3 + 3t + 7) - (-t^5 - 5t^3 + 3t + 4)$$

$$= t^5 - 3t^3 + 3t + 7 + t^5 + 5t^3 - 3t - 4$$

$$= 2t^5 + 2t^3 + 0t + 3 \qquad 3t - 3t = 0t$$

$$= 2t^5 + 2t^3 + 3 \qquad \text{Delete the 0 term.}$$

Combined Operations

In Examples 4 and 5, more than one operation must be performed to get a simplified form for an answer.

Example 4. Do the indicated operations.

$$(7 - 4z^2 - 12z) - (2z^2 + 10 - 9z) + (9z^2 + 5z - 1)$$

Solution.

$$(7 - 4z^2 - 12z) - (2z^2 + 10 - 9z) + (9z^2 + 5z - 1)$$

$$= (-4z^2 - 12z + 7) - (2z^2 - 9z + 10) + (9z^2 + 5z - 1)$$

$$= -4z^2 - 12z + 7 - 2z^2 + 9z - 10 + 9z^2 + 5z - 1$$

$$= 3z^2 + 2z - 4$$

Thus the simplified answer is $3z^2 + 2z - 4$.

Example 5. Do the indicated operations.

$$3(2k^3 + 5k - 1) - 2(k^2 + 8k - 4)$$

Solution. **Discussion.** Use the distributive property to remove the parentheses. Then combine any like terms.

$$3(2k^3 + 5k - 1) - 2(k^2 + 8k - 4)$$

$$= 6k^3 + 15k - 3 - 2k^2 - 16k + 8$$

$$= 6k^3 - 2k^2 - k + 5$$

Thus the simplified answer is $6k^3 - 2k^2 - k + 5$.

SECTION 3-2. Practice Exercises

In exercises **1–18**, add.

[Examples 1 and 2]

1. $(3x^2 + 7) + (4x^2 - 5)$ **2.** $(5x^2 - 3) + (8x^2 + 1)$

3. $(z^2 + 6z - 8) + (9z^2 - 7z + 1)$ **4.** $(2z^2 + 3z - 13) + (z^2 + z + 10)$

5. $(y^3 + 5y + 10) + (5 - 5y - 2y^3)$ **6.** $(6 + 5y - y^2) + (3y^2 - y - 6)$

7. $(2a^4 - 3a^3 + a) + (5a^3 - 4a - a^4)$

8. $(-a^2 + a^3 + 3a^4) + (2a^4 - a^2 - 8a^3)$

9. $(3a^2 + ab - 2b^2) + (2b^2 - a^2 + 2ab)$

10. $(6a^2 - 5b^2 - 8ab) + (5b^2 + 4ab - a^2)$

11. $(8m^3 + m^2 - 4) + (3 - 2m - m^2)$ **12.** $(1 - 4m + 4m^2) + (m^3 + 4m - 3)$

13. $(x^3 + 1) + (3x^2 - 3) + (2x^3 - x^2)$ **14.** $(7 - 5x) + (x^2 - 4) + (9x - 3x^2)$

15. $(4xy - 2xz - 7) + (5yz + 3xz + 4) + (3 - xy - 2yz)$

16. $(10 - xy + 3yz) + (3xy - 6) + (8xz - yz - 2xy)$

17. $(5ab - 3ac + 1) + (4 - 2bc + ac) + (3bc - 8 + ab)$

18. $(-ab - bc + 9) + (-2bc - 5 + 4ac) + (-1 + 2ab + 3ac)$

In exercises **19–26**, find the indicated opposites.

[Examples a–c]   **19.** $-(x^2 - 3x + 4)$ **20.** $-(x^2 - 5x + 10)$

21. $-(-a^2 + 5a + 7)$ **22.** $-(-a^2 + a + 6)$

23. $-(10a^2 + ab - 2b^2 - 13)$ **24.** $-(5a^2 - 3ab + b^2 + 4)$

25. $-(a^3 - b^3)$ **26.** $-(a^5 - b^4)$

In exercises **27–34**, subtract.

[Example 3]   **27.** $(2y + 1) - (4y + 3)$ **28.** $(9 - 4y) - (4 + y)$

29. $(3z^2 + z - 6) - (z^2 - 2z - 3)$ **30.** $(7 - 12z - 4z^2) - (10 - 9z - 2z^2)$

31. $(3a^4 - 5a^3 + 2a^2 + 6a) - (-5a^3 - 2a^2 - 6a + 4)$

32. $(8a^4 + 2a^3 - a^2 + 9a) - (-2a^3 - a^2 - 9a + 1)$

33. $(2xz + 3yz - 4xy) - (3yz + 2xz + 4xy)$

34. $(5yz - 7xy - 3xz) - (3yz - 3xz - xy)$

In exercises **35–50**, do the indicated operations.

[Examples 4 and 5]   **35.** $(x^2 + 6x - 3) + (2x^2 - 5x + 1) - (3x^2 + 9x - 5)$

36. $(x^2 - x + 2) + (3x^2 + 4x - 5) - (4x^2 - x - 2)$

37. $(a^3 + 4a^2 + 5a) - (5a^3 - 4a^2 + a) + (a^3 + 3a^2 - 4a)$

38. $(a^4 - 3a^2 + 6) - (2a^4 + 5a - 12) + (3a^2 + a - 5)$

39. $(3a^2 - 2ab + b^2) + (a^2 + ab - b^2) - (2a^2 - 3ab - 2b^2)$

40. $(7a^2 + ab + 2b^2) + (a^2 - ab + 3b^2) - (6a^2 + 4ab - 7b^2)$

41. $4(2x^2 - 3x + 5) - 2(x^2 + 4x - 7)$ **42.** $3(x^2 + 5x - 1) - 4(2x^2 - 3x + 1)$

43. $5(t^3 - 3t^2 + 2t) - 6(2t^3 + 2t^2 - t)$ **44.** $6(t^3 + 3t^2 - 2t) - 2(3t^3 - 4t^2 - 8t)$

45. $-(a^5 - 2a^3 + a) - 3(a^5 + 6a^4 - a)$

46. $-(2a^4 + a^3 + 3a) - 4(a^4 - 6a^2 + a)$

47. $\frac{1}{2}(18k^3 - 10k) + \frac{2}{3}(6k^2 - 9k) - \frac{3}{4}(12k^3 + 20k^2 - 16k)$

48. $\frac{1}{3}(15k^2 + 6k - 12) - \frac{5}{4}(4k^2 - 12k + 20) + \frac{1}{6}(12k^2 - 30k + 180)$

49. $20\left(\frac{3}{10}m^2 - \frac{4}{5}mn + \frac{1}{4}n^2\right) + 35\left(\frac{1}{7}m^2 + \frac{2}{5}mn - \frac{3}{5}n^2\right)$

50. $24\left(\frac{3}{8}m^2 + \frac{5}{12}mn - \frac{7}{6}n^2\right) - 30\left(\frac{1}{6}m^2 + \frac{3}{10}mn - \frac{7}{15}n^2\right)$

SECTION 3-2.   Ten Review Exercises

In exercises **1–8**, solve each problem.

1. Simplify: $2(3(5)^2 + 5 - 7) - 3(5^2 + 5 - 8)$

2. Evaluate: $2(3y^2 + y - 7) - 3(y^2 + y - 8)$ for $y = 5$

3. Simplify: $2(3y^2 + y - 7) - 3(y^2 + y - 8)$

4. Evaluate the expression of exercise **3** for $y = 5$.

5. Simplify: $(8(2)^2 + 6(2)(-1) + (-1)^2) + (5(2)^2 - 3(2)(-1) + 3(-1)^2)$

6. Evaluate for $x = 2$ and $y = -1$: $(8x^2 + 6xy + y^2) + (5x^2 - 3xy + 3y^2)$

7. Simplify: $(8x^2 + 6xy + y^2) + (5x^2 - 3xy + 3y^2)$

8. Evaluate the expression of exercise **7** for $x = 2$ and $y = -1$.

In exercises **9** and **10**, use the following polynomial:

$$6k^3 + k^6 + 3k + 1 - 9k^2 - 7k^4$$

9. Write the polynomial in descending powers.

10. State the degree of the polynomial.

SECTION 3-2.   Supplementary Exercises

In exercises **1–10**, simplify.

1. $(5x^3 - 3x + 3) - (4x^2 + 9) + 3(x - x^3 + 2)$

2. $(8x^3 + 7x^2 - 1) - (8x - 3) + 4(1 - 2x^3 + 2x)$

3. $(9y^2 - 6y + 3) - (10y^2 - 4y - 3) - (7 - 2y - y^2)$

4. $(8x^2 + x - 5) - (12x^2 + 3x - 6) - (1 + 5x - 4x^2)$

5. Subtract $(2x^2 + 3x - 3)$ from the sum of $(5x^2 - 2x + 3)$ and $(-4x^2 - 6x + 8)$.

6. Subtract $(6z^2 + 3z + 4)$ from the sum of $(-2z^2 + z + 8)$ and $(8z^2 + 2z - 4)$.

7. Subtract $(5y + 7)$ from the difference between $(9y - 3)$ and $(y + 1)$.

8. Subtract $(8t + 1)$ from the difference between $(12t - 3)$ and $(7t + 9)$.

9. $(a^2 - 3a + 4) - (a^2 + 5a - (7 + a^2))$

10. $(c^2 + 13) - (c^2 + 7c - (9 - c^2))$

In exercises **11–14**, subtract the bottom polynomial from the top polynomial.

11. $\begin{array}{r} 3x^3 + 7x^2 - 8x + 2 \\ (-)\ 4x^3 - 8x^2 + 2x - 5 \\ \hline \end{array}$ **12.** $\begin{array}{r} 5x^3 - 3x^2 - 5x + 4 \\ (-)\ 2x^3 - \ x^2 + 4x - 5 \\ \hline \end{array}$

13. $\quad\begin{array}{l}4a^2 - 8ab + 2b^2 \\ (-)\ \underline{\ a^2 - 8ab - 3b^2}\end{array}$ 14. $\quad\begin{array}{l}7u^2 + 5uv - \ v^2 \\ (-)\ \underline{-5u^2 + 5uv + 6v^2}\end{array}$

In exercises **15** and **16**, simplify.

15. $(9.3b^2 + 8.1b - 0.05) - (7.2b^2 + 6b - 1.3)$

16. $(5.1b^2 - 4.8b - 1.07) - (6b^2 - 2.9b + 2.8)$

In exercises **17–24**, simplify.

17. $\dfrac{1}{5}(40z^4 + 45z^2) - \dfrac{1}{3}(24z^4 - 12z) - \dfrac{3}{5}(15z^2 + 10z)$

18. $\dfrac{2}{3}(12z^3 - 9z^2) + \dfrac{1}{5}(20z^2 + 25z) - \dfrac{1}{4}(32z^3 + 20z)$

19. $\dfrac{4}{7}(49x^2 - 7x + 21) - \dfrac{2}{3}(42x^2 - 6x + 30)$

20. $\dfrac{5}{6}(36x^2 + 12x - 24) - \dfrac{2}{5}(15x^2 + 25x - 50)$

21. $28\left(\dfrac{2}{7}p^2 - \dfrac{1}{14}pq + \dfrac{3}{4}q^2\right) + 30\left(\dfrac{1}{6}p^2 + \dfrac{1}{15}pq - \dfrac{7}{10}q^2\right)$

22. $32\left(\dfrac{1}{8}p^2 + \dfrac{3}{4}pq - \dfrac{3}{16}q^2\right) + 15\left(p^2 - \dfrac{2}{3}pq + \dfrac{2}{5}q^2\right)$

23. $24\left(\dfrac{2}{3}m^2n + \dfrac{5}{12}mn + \dfrac{3}{8}n^2\right) + 40\left(\dfrac{3}{10}m^2n - \dfrac{1}{4}mn - \dfrac{2}{5}n^2\right)$

24. $12\left(\dfrac{1}{6}m^2n - \dfrac{3}{4}mn - \dfrac{2}{3}n^2\right) + 18\left(\dfrac{2}{3}m^2n + \dfrac{1}{2}mn + \dfrac{4}{9}n^2\right)$

In exercises **25** and **26**, use paper and pencil or a calculator.

a. Evaluate P for the given value of the variable.

b. Evaluate Q for the given value of the variable.

c. Find the polynomial expression for $P + Q$.

d. Evaluate the polynomial from part **c** and compare it with the sum of the numbers obtained for parts **a** and **b**.

e. Find the polynomial expression for $P - Q$.

f. Evaluate the polynomial from part **e** and compare it with the difference in the numbers obtained in parts **a** and **b**.

25. $P: 2t^2 + 8 - 9t$ and $Q: 4t - 6 + t^2$, and $t = -5$

26. $P: 4t - 3t^2 + 9$ and $Q: 4t^2 - 10 - 5t$, and $t = 3$

SECTION 3-3.   Multiplying Polynomials by Monomials

KEY TOPICS IN THIS SECTION

1. Product of powers, same base, property of exponents

2. Multiplying monomials

3. Power of a power, same base, property of exponents

4. Power of a product, same exponent, property of exponents

5. Power of a quotient, same exponent, property of exponents

6. Multiplying polynomials by monomials

   In the past section we added and subtracted polynomials. In this section we begin to study the procedure for multiplying polynomials. To use this procedure we will need to establish some properties of exponents.

Product of Powers, Same Base, Property of Exponents

   Multiplying polynomials frequently involves multiplying variable factors with exponents. For example, it is quite likely that multiplications like the following could result from multiplying two polynomials:

$$t^4 \cdot t^3 \qquad \text{and} \qquad (x^3y^2)(xy^3)$$

The definition of an exponent can be used to write these products with each variable written only once.

$$t^4 \cdot t^3 \text{ means } \underbrace{(t \cdot t \cdot t \cdot t)}_{4 \text{ factors of } t}\underbrace{(t \cdot t \cdot t)}_{3 \text{ factors of } t}$$

Now (4 factors of t)(3 factors of t) is the same as $4 + 3 = 7$ factors of t. Thus,

$$t^4 \cdot t^3 = t^{4+3} = t^7$$

— **The t is written once.**

— **The exponent 7 is the sum of 3 and 4.**

Similarly,

$$(x^3y^2)(xy^3) = (x^3 \cdot x)(y^2 \cdot y^3) = x^4y^5$$

— **Bases are used once.**

— **Exponents are added.**

The product of powers, same base, property of exponents
  If b is a real number, and m and n are positive integers, then

$$b^m \cdot b^n = b^{m+n}$$

— **Write the base once.**

— **Add the exponents.**

Example 1.   Simplify:

a. $k^2 \cdot k^4 \cdot k$

b. $(xy^2z^3)(x^2y^2z^2)$

Solution.   **a.** $k^2 \cdot k^4 \cdot k = k^{2+4+1}$ $b^m \cdot b^n = b^{m+n}$

$\qquad\qquad\qquad = k^7$ Do the indicated additions.

b. $(xy^2z^3)(x^2y^2z^2)$

$\qquad = (x \cdot x^2)(y^2 \cdot y^2)(z^3 \cdot z^2)$ Regroup like bases.

$\qquad = x^3y^4z^5$ Add exponents on like bases.

Examples **a–c** illustrate expressions in which the product of powers, same base property is used incorrectly, or the property is used when it should not be.

a. $2^5 \cdot 2^3 \neq 4^8$ **Do not multiply the bases.**
That is, $2^5 \cdot 2^3 = 2^8$

b. $2^3 \cdot 3^2 \neq 6^5$ **The bases are not alike; do not apply the property.**
$(2^3 \cdot 3^2 = 8 \cdot 9 = 72)$

c. $5^3 + 5^2 \neq 5^5$ **The property does not apply.**
$(5^3 + 5^2 = 125 + 25 = 150)$

Multiplying Monomials

The product of powers, same base, property of exponents permits one to write $t^4 \cdot t^3$ as t^7. Consider now the following indicated product:

$$(-5t^4)(9t^3)$$

Multiplication is commutative and associative; therefore the factors can be regrouped as follows:

$$(-5t^4)(9t^3) = (-5 \cdot 9)(t^4 \cdot t^3) = -45t^7$$

Thus the product can be written as a single monomial by multiplying the coefficients and adding the exponents on the common base.

Example 2.   Simplify: $(4xy)(-7x^2)(3y^3)$

Solution.   **Discussion.** The indicated products should be rewritten by grouping the coefficients and the like bases. Then multiply the coefficients and add the exponents on the like bases.

$\qquad (4xy)(-7x^2)(3y^3)$

$\qquad = (4 \cdot (-7) \cdot 3)(x \cdot x^2)(y \cdot y^3)$ Regrouping

$\qquad = -84x^3y^4$ Simplifying

Power of a Power, Same Base, Property of Exponents

There are times when power terms, such as t^4 and x^3, are raised to powers. For example,

$\qquad (t^4)^5$ means that t^4 is raised to the power of five

$\qquad (x^3)^2$ means that x^3 is squared

If the definition of an exponent is applied to the outside exponents, we get the following:

$$(t^4)^5 = \underbrace{t^4 \cdot t^4 \cdot t^4 \cdot t^4 \cdot t^4}_{\text{5 factors of } t^4} \qquad (x^3)^2 = \underbrace{x^3 \cdot x^3}_{\text{2 factors of } x^3}$$

$$= t^{20} \longleftarrow \text{Adding exponents} \longrightarrow = x^6$$

Since $4 \cdot 5 = 20$ (the exponent on t), and $3 \cdot 2 = 6$ (the exponent on x), the following property of exponents is a general statement of these examples.

> **Power of a power, same base, property of exponents**
> If b is a real number, and m and n are positive integers, then
>
> $$(b^m)^n = b^{mn} \qquad \begin{array}{l} \text{Write the base once.} \\ \text{Multiply the exponents.} \end{array}$$

Example 3. Simplify:

  a. $(p^2)^5(q^8)^2$

  b. $(-3a^2b)(a^3)^2(b^2)^4$

Solution. **a.** $(p^2)^5(q^8)^2$ The given expression

   $= p^{10}q^{16}$ Multiply exponents.

  b. $(-3a^2b)(a^3)^2(b^2)^4$

   $= (-3a^2b)(a^6)(b^8)$ $(a^3)^2 = a^6$ and $(b^2)^4 = b^8$

   $= -3(a^2 \cdot a^6)(b \cdot b^8)$ Regrouping

   $= -3a^8b^9$ Adding exponents on same bases

Power of a Product, Same Exponent, Property of Exponents

The $(b^m)^n = b^{mn}$ property of exponents can be used to simplify a power of a single base. But sometimes, in multiplying polynomials, products such as $5t$ and $-3xy$ are raised to powers. For example,

$(5t)^3$ means $5t$ is raised to the power of three

$(-3xy)^4$ means $-3xy$ is raised to the power of four

If the definition of an exponent is applied to the exponents 3 and 4, we get the following:

$$(5t)^3 = \underbrace{(5t)(5t)(5t)}_{\text{3 factors of } 5t} \qquad (-3xy)^4 = \underbrace{(-3xy)(-3xy)(-3xy)(-3xy)}_{\text{4 factors of } -3xy}$$

By regrouping factors, each factor can be written once with a single exponent.

$$(5t)(5t)(5t) = 5^3 \cdot t^3 \text{ or } 125t^3$$

$$(-3xy)(-3xy)(-3xy)(-3xy) = (-3)^4 \cdot x^4 \cdot y^4 = 81x^4y^4$$

In both examples the simplified expression has each factor with the exponent that was outside the parentheses of the given expressions.

$$(5t)^3 = 5^3 \cdot t^3$$

The exponent 3 is written on both the 5 and the t.

$$(-3xy)^4 = (-3)^4 \cdot x^4 \cdot y^4$$

The exponent 4 is written on the (-3), the x, and the y.

The following property of exponents is a general statement of these examples.

Power of a product, same exponent, property of exponents
If a and b are real numbers, and n is a positive integer, then

$$(ab)^n = a^n b^n$$

Write each base once.

Write the exponent on each base.

Example 4.   Simplify:

a. $(4p^2 q)^3$

b. $(5y)^2(-2y^2)^3$

Solution.   **a.** $(4p^2q)^3$ The given expression

$= 4^3(p^2)^3 q^3$ Raise each factor to the power 3.

$= 64p^6 q^3$ $(p^2)^3 = p^{2 \cdot 3} = p^6$

b. $(5y)^2(-2y^2)^3$

$= 5^2 \cdot y^2 \cdot (-2)^3 \cdot (y^2)^3$ $(ab)^n = a^n \cdot b^n$

$= 5^2 \cdot y^2 \cdot (-2)^3 \cdot y^6$ Replace $(y^2)^3$ with y^6.

$= 25y^2(-8)y^6$ $5^2 = 25$, and $(-2)^3 = -8$

$= -200y^8$ $y^2 \cdot y^6 = y^8$

Power of a Quotient, Same Exponent, Property of Exponents

The **power of a product** property of exponents is used to simplify expressions such as $(5t)^3$ and $(-3xy)^4$. The **power of a quotient** property of exponents can similarly be used to simplify expressions such as $\left(\dfrac{t}{5}\right)^3$ and $\left(\dfrac{-xy}{3}\right)^4$.

$$\left(\frac{t}{5}\right)^3 = \left(\frac{t}{5}\right)\left(\frac{t}{5}\right)\left(\frac{t}{5}\right) = \frac{t \cdot t \cdot t}{5 \cdot 5 \cdot 5} = \frac{t^3}{5^3} \text{ or } \frac{t^3}{125}$$

Writing $-xy$ as $-1 \cdot xy$, then

$$\left(\frac{-xy}{3}\right)^4 = \left(\frac{-1 \cdot xy}{3}\right)^4 = \frac{(-1)^4 \cdot x^4 \cdot y^4}{3^4} = \frac{1 \cdot x^4 y^4}{81} \text{ or } \frac{x^4 y^4}{81}$$

The following property of exponents is a general statement of these examples.

Power of a quotient, same exponent, property of exponents
If a and b are real numbers ($b \neq 0$), and n is a positive integer, then

$$\left(\frac{a}{b}\right)^n = \frac{a^n}{b^n}$$

Write each base once.

Write the exponent on each base.

Example 5.   Simplify: $\left(\dfrac{5t}{2}\right)^2 \left(\dfrac{t}{3}\right)^3$

Solution.

$$\left(\dfrac{5t}{2}\right)^2 \left(\dfrac{t}{3}\right)^3$$

$$= \dfrac{(5t)^2}{2^2} \cdot \dfrac{t^3}{3^3} \qquad \left(\dfrac{a}{b}\right)^n = \dfrac{a^n}{b^n}$$

$$= \dfrac{5^2 \cdot t^2}{4} \cdot \dfrac{t^3}{27} \qquad (ab)^n = a^n \cdot b^n \text{ on the left expression}$$

$$= \dfrac{25t^2 \cdot t^3}{108} \qquad \dfrac{\text{Multiply numerators.}}{\text{Multiply denominators.}}$$

$$= \dfrac{25t^5}{108} \qquad t^2 \cdot t^3 = t^{2+3} = t^5$$

Multiplying Polynomials by Monomials

The distributive property of multiplication over addition can be used to simplify an indicated product of a polynomial and a monomial. The result is that each term of the polynomial is multiplied by the monomial. Then the products of the monomials can be simplified using the properties of exponents.

Example 6.   Simplify: $3xy(7x + 5y - 1)$

Solution.

$$3xy(7x + 5y - 1)$$

$$= 3xy(7x) + 3xy(5y) - 3xy(1) \qquad \text{Use the distributive property.}$$

$$= 21x^2y + 15xy^2 - 3xy \qquad \text{Multiply coefficients; add exponents on like bases.}$$

Example 7.   Simplify: $(12t^3 + 45t^2 - 3t - 93)\dfrac{t^2}{3}$

Solution.

$$(12t^3 + 45t^2 - 3t - 93)\dfrac{t^2}{3}$$

$$= 12t^3 \cdot \dfrac{t^2}{3} + 45t^2 \cdot \dfrac{t^2}{3} - 3t \cdot \dfrac{t^2}{3} - 93 \cdot \dfrac{t^2}{3}$$

$$= 4t^5 + 15t^4 - t^3 - 31t^2$$

SECTION 3-3.   Practice Exercises

In exercises **1–66**, simplify.

[Example 1]   **1.** $x^5 \cdot x^7$ **2.** $x^4 \cdot x^6$

3. $t^2 \cdot t^5 \cdot t^3$ **4.** $t^3 t^4 t^2$

5. $(a^4b)(a^2b^3)$

6. $(a^3b)(a^4b^3)$

7. $(w^7y^5z^5)(w^6y^3z)$

8. $(w^8y^5z^6)(w^2y^4z^2)$

9. $(xy^2)(x^3y^2)(xy)$

10. $(x^2y)(xy^2)(x^3y^2)$

[Example 2]　11. $(7a^2b)(-2ab)$

12. $(5a^3b)(-3ab^2)$

13. $(-4m^4n^4)(-3mn^3)$

14. $(-5m^2n^2)(-6mn^3)$

15. $(-10ab^3)(-8a^5b^3)(2ab^3)$

16. $(-5a^3b^3)(7a^3b^2)(2a^2b)$

17. $(-15b^5a^2)(b^2a^3)(4b)$

18. $(-13b^2a^7)(a^5b^8)(3a)$

[Example 3]　19. $(a^3)^4(b^2)^3$

20. $(a^2)^4(b^3)^5$

21. $(x^3)(y^5)^3$

22. $(x^4)^5(y)^4$

23. $(p^3)^4(q^7)^2(r^6)^4$

24. $(p^3)^5(q^7)^3(r^8)^2$

25. $(-5a^2b)(a^4)^3(b^2)^5$

26. $(-2ab^3)(a^3)^2(b^4)^5$

[Example 4]　27. $(5p)^3$

28. $(7p)^2$

29. $(-2ab)^5$

30. $(-10ab)^4$

31. $(3x^4)^2$

32. $(5x^3)^3$

33. $(4a^2b^5c)^3$

34. $(3ab^4c^2)^4$

35. $(-2xy^2)^5$

36. $(-3xy^4)^3$

37. $(2x^3)^3(-3x^2)^2$

38. $(4x^2)^2(-2x^3)^3$

39. $(-5a^2bc^2)^3(-3ab^2)^2$

40. $(-6ab^2c^2)^2(-3a^4b)^3$

[Example 5]　41. $\left(\dfrac{k}{2}\right)^5$

42. $\left(\dfrac{k}{10}\right)^3$

43. $\left(\dfrac{-st}{5}\right)^4$

44. $\left(\dfrac{-st}{3}\right)^5$

45. $\left(\dfrac{3p}{2}\right)^3$

46. $\left(\dfrac{2p}{5}\right)^2$

47. $\left(\dfrac{x^2y}{z^3}\right)^4$

48. $\left(\dfrac{xy^3}{z^2}\right)^3$

49. $\left(\dfrac{2p^2}{q}\right)^2\left(\dfrac{p^3}{q^2}\right)^3$

50. $\left(\dfrac{3p}{q^3}\right)^2\left(\dfrac{p^3}{q}\right)^5$

51. $\left(\dfrac{ab^2c^3}{10}\right)^4\left(\dfrac{a^4c}{2}\right)^3$

52. $\left(\dfrac{ab^3c^2}{4}\right)^3\left(\dfrac{a^2b^4}{3}\right)^2$

[Example 6]　53. $3uv(2u+5v)$

54. $-7uv(6u-9v)$

55. $2x^2y(6x^2-3xy+5xy^2)$

56. $3xy^2(2x^2+7xy-9xy^2)$

57. $-5st(-3s^2t^3+6st-4t^3)$

58. $-6st(-4s^2t-7st+5t^2)$

59. $(7m^5-4m^3n+mn^2-6n^3)3m^3n$

60. $(2m^4+4m^2n^2-6n^4-7n^6)5m^2n^3$

[Example 7]　61. $\dfrac{p^3}{5}(15p^3-25p^2+70p-10)$

62. $\dfrac{p^4}{4}(8p^5 + 4p^3 - 28p + 40)$

63. $(18a^3b^4 - 12a^2b^2 - 21a^2b + 27ab)\left(\dfrac{-a^2b}{3}\right)$

64. $(2a^4b^4 + 10a^3b^2 - 20a^2 - 14ab)\left(\dfrac{-ab^3}{2}\right)$

65. $\dfrac{x^3y^3}{6}(-48x^3 + 87xy^3 - 72y^4)$

66. $\dfrac{x^2y^2}{9}(-45x^5 - 60xy^2 + 108y^4)$

SECTION 3-3.   Ten Review Exercises

In exercises **1–4**, solve each problem.

1. Multiply: $5t^2(t - 2)$

2. Multiply: $-t(3 + 10t - 5t^2)$

3. Add the products from exercises **1** and **2**.

4. Subtract the product of exercise **2** from the product of exercise **1**.

In exercises **5–10**, use the following polynomial:

$$3a^2b^2 - 5b^3 + 2a^5 - a^3b$$

5. Write the polynomial in descending powers of a.

6. Write the polynomial in descending powers of b.

7. State the degree of the polynomial.

8. State the degree of the term with a b^2 factor.

9. Write the opposite of the polynomial in exercise **5**.

10. Identify the polynomial by name.

SECTION 3-3.   Supplementary Exercises

In exercises **1–8**, simplify each part.

1. a. $2^3 \cdot 2^2$　　　　　　　　　　　　　　**2. a.** $3^2 \cdot 3^3$
　　b. $2^3 + 2^2$　　　　　　　　　　　　　　　**b.** $3^2 + 3^3$
　　c. $3^2 \cdot 2^2$　　　　　　　　　　　　　　**c.** $2^3 \cdot 3^3$

3. a. $2^4 \cdot (-2)^3$　　　　　　　　　　　　**4. a.** $3^2 \cdot (-3)^3$
　　b. $2^4 + (-2)^3$　　　　　　　　　　　　　**b.** $3^2 + (-3)^3$
　　c. $2^4 - (-2)^3$　　　　　　　　　　　　　**c.** $3^2 - (-3)^3$

5. a. $8x^2 + 5x^2$
  b. $(8x)^2 + (5x)^2$
  c. $(8x)^2 \cdot (5x)^2$

6. a. $3s^2 + 2s^2$
  b. $(3s)^2 + (2s)^2$
  c. $(3s)^2 \cdot (2s)^2$

7. a. $(-12p)(11p)$
  b. $-12p + 11p$

8. a. $(3q^4)(5q^4)$
  b. $3q^4 + 5q^4$

In exercises **9–16**, assume that the variable exponents represent positive integers and simplify.

9. $2^x \cdot 2^{3x}$

10. $5^{2p} \cdot 5^p$

11. $(a^2)^n$

12. $(b^3)^{2n}$

13. $(r^2 s^3)^t$

14. $(a^3 b^2)^{2t}$

15. $\left(\dfrac{p^{2x}}{q^y} \right)^z$, and $q \neq 0$

16. $\left(\dfrac{m^x}{n^{2y}} \right)^{3z}$, and $n \neq 0$

**KEY TOPICS
IN THIS SECTION**

SECTION 3-4.   Multiplying Polynomials

1. Multiplying binomials

2. Multiplying polynomials with a horizontal format

3. Multiplying polynomials with a vertical format

4. Some applications

In Section 3-3 we used the distributive property of multiplication over addition (and subtraction) to multiply a polynomial by a monomial. If both polynomials being multiplied are at least binomials, then the distributive property can be used more than once to find the product.

Multiplying Binomials

Consider the following indicated product of two binomials:

$$(2t^2 + 3)(5t + 2)$$

To see how the distributive property can be used to write this product as a single polynomial, we can temporarily replace one of the binomials by a single letter. For example, let P stand for $2t^2 + 3$. Then $(2t^2 + 3)(5t + 2)$ can be written as $P(5t + 2)$. Now use the distributive property to multiply both $5t$ and 2 by P.

$P(5t + 2)$

$= P(5t) + P(2)$ The distributive property

$= (2t^2 + 3)(5t) + (2t^2 + 3)(2)$ Replace P by $2t^2 + 3$.

$= (2t^2)(5t) + 3(5t) + (2t^2)2 + 3(2)$ The distributive property again

$= 10t^3 + 15t + 4t^2 + 6$ Multiplying the monomials

$= 10t^3 + 4t^2 + 15t + 6$ In descending powers of t

This example shows that to multiply two polynomials, we must multiply each term of the first polynomial by each term of the second one. This procedure follows the one we use to multiply whole numbers of two or more digits. To illustrate this fact, consider finding the product of $(203)(52)$. If we write these numbers as $(200 + 3)(50 + 2)$, then

$$(200 + 3)(50 + 2) = 10{,}000 + 400 + 150 + 6$$
$$= 10{,}556$$

Replace t with 10 in $(2t^2 + 3)$, $(5t + 2)$, and $10t^3 + 4t^2 + 15t + 6$:

$2t^2 + 3$ becomes $2(100) + 3 = 200 + 3$ or 203.

$5t + 2$ becomes $5(10) + 2 = 50 + 2$ or 52.

$10t^3 + 4t^2 + 15t + 6$ becomes $10{,}000 + 400 + 150 + 6$, or $10{,}556$.

Thus, when t is replaced with 10, the product of $(2t^2 + 3)$ and $(5t + 2)$ has the value of the product of 203 and 52. The two multiplications have striking similarities.

Multiplying polynomials is really no more difficult than multiplying several pairs of monomials. The major obstacle to finding a correct answer is overlooking one or more pairs of monomials that should be multiplied, or making an error in signs. The next example illustrates a potential problem when a subtraction is indicated in one of the binomials.

Example 1.   Multiply: $(3x + 5y)(2x - 7y)$

Solution.   **Discussion.** To guard against an error in signs, it is sometimes helpful to write $2x - 7y$ as $2x + (-7y)$. That is, change the subtraction to an addition using the definition

$$a - b = a + (-b).$$

After the products of the pairs of monomials have been simplified, use the definition again to change any negative coefficients to positive numbers.

$(3x + 5y)(2x - 7y)$	
$= (3x + 5y)(2x + (-7y))$	Change the subtraction to addition.
$= (3x + 5y)(2x) + (3x + 5y)(-7y)$	Distribute $(3x + 5y)$.
$= (3x)(2x) + (5y)(2x) + (3x)(-7y) + (5y)(-7y)$	Distribute $2x$ and $-7y$.
$= 6x^2 + 10xy + (-21xy) + (-35y^2)$	Multiply the monomials.
$= 6x^2 + (-11xy) + (-35y^2)$	$10xy + (-21xy) = -11xy$
$= 6x^2 - 11xy - 35y^2$	Change back to subtractions.

Multiplying Polynomials with a Horizontal Format

The procedure illustrated in Example 1 is called a **horizontal format.** This procedure uses the distributive property as many times as is necessary to distribute each term of the first polynomial to each term of the second polynomial. It is called a *horizontal format* because the work is always carried out line by line.

Example 2. $(2k - 5)(3k^2 + k - 2)$

Solution. **Discussion.** As suggested in Example 1, the two subtractions will first be changed to additions, to safeguard against any possible errors in signs.

$$(2k - 5)(3k^2 + k - 2)$$

$$= (2k + (-5))(3k^2 + k + (-2))$$

$$= (2k + (-5))(3k^2) + (2k + (-5))k + (2k + (-5))(-2)$$

$$= (2k)(3k^2) + (-5)(3k^2) + (2k)k + (-5)k + (2k)(-2) + (-5)(-2)$$

$$= 6k^3 + (-15k^2) + 2k^2 + (-5k) + (-4k) + 10$$

$$= 6k^3 + (-13k^2) + (-9k) + 10$$

$$= 6k^3 - 13k^2 - 9k + 10$$

Example 3. Multiply: $(3 + 5y^2 - y)(y - 2 + 2y^2)$

Solution. **Discussion.** The first step will be to write the trinomials in descending powers.

$$(5y^2 - y + 3)(2y^2 + y - 2)$$

$$= (5y^2 + (-y) + 3)(2y^2) + (5y^2 + (-y) + 3)y + (5y^2 + (-y) + 3)(-2)$$

$$= 10y^4 + (-2y^3) + 6y^2 + 5y^3 + (-y^2) + 3y + (-10y^2) + 2y + (-6)$$

$$= 10y^4 + 3y^3 + (-5y^2) + 5y + (-6)$$

$$= 10y^4 + 3y^3 - 5y^2 + 5y - 6$$

Multiplying Polynomials with a Vertical Format

A **vertical format** can also be used to multiply two polynomials. This format organizes the work up and down, rather than from left to right. When using such a format, we write one polynomial on a line under the other one. (Usually, if one polynomial has fewer terms, it is written on the second line.) The products of each term of the bottom polynomial and each term of the top polynomial are aligned in columns of like terms. The like terms in each column are then combined to yield the polynomial that is the product.

Example 4. Multiply: $(a + 3a^2 - 2)(3a^2 - 2 - a)$

Solution. **Discussion.** Both trinomials are first written in descending powers.

$$
\begin{array}{ll}
\begin{aligned}
3a^2 + a - 2 \\
3a^2 - a - 2 \\
\hline
9a^4 + 3a^3 - 6a^2 \\
\quad\;\; - 3a^3 - a^2 + 2a \\
\quad\quad\quad\;\; - 6a^2 - 2a + 4 \\
\hline
9a^4 \quad\quad\quad - 13a^2 \quad\quad + 4
\end{aligned}
&
\begin{aligned}
&\text{Put first polynomial on top line.} \\
&\text{Put second polynomial on bottom line.} \\
&3a^2(3a^2 + a - 2) = 9a^4 + 3a^3 - 6a^2 \\
&(-a)(3a^2 + a - 2) = -3a^3 - a^2 + 2a \\
&(-2)(3a^2 + a - 2) = -6a^2 - 2a + 4 \\
&3a^3 - 3a^3 = 0 \text{ and } 2a - 2a = 0
\end{aligned}
\end{array}
$$

Thus, $(3a^2 + a - 2)(3a^2 - a - 2) = 9a^4 - 13a^2 + 4$.

Example 5.   Multiply: $(2z^3 + 3z - 1)(2z^2 + z)$

Solution.   **Discussion.** The first polynomial is missing a z^2 term, so we will leave a space when we list this polynomial on the top line. Then, as the products are formed, we will have a space for any missing terms that might occur.

$2z^3 \qquad + 3z - 1$	Leave a space for a missing z^2 term.
$2z^2 + \; z$	Align on the far left.
$\overline{4z^5 \qquad + 6z^3 - 2z^2}$	Leave a space for a missing z^4 term.
$\quad 2z^4 \qquad + 3z^2 - z$	Align the z^2 terms.
$\overline{4z^5 + 2z^4 + 6z^3 + \; z^2 - z}$	Combine like terms.

Thus $(2z^3 + 3z - 1)(2z^2 + z) = 4z^5 + 2z^4 + 6z^3 + z^2 - z.$

Some Applications

Examples 6 and 7 illustrate how the products of polynomials are related to the areas and volumes of objects, when the dimensions of objects are written as polynomials.

Example 6.   The Tipton Sign Company makes signs for many purposes. The one shown in Figure 3-4 is an example of a Series R sign. This series offers rectangular-shaped signs in which the length is always 6 inches more than the width.

  a. Let w stand for the width of a Series R sign. Write an expression for the corresponding length.

  b. Write a polynomial that expresses the area of the sign.

  c. If a given sign in the series has a width of 15 inches, find the area of that sign.

Figure 3-4. A residence sign.

THE JONES FAMILY

9301 HARBOR BLVD.

Solution.   **a.** The length is always 6 inches more than the width. Thus the length is $(w + 6)$ inches.

  b. The area of a rectangle is the product of the width and length. Thus the area is $w(w + 6) = w^2 + 6w$.

  c. If $w = 15$ inches, then

$$\text{Area} = 15^2 + 6(15)$$

$$= 225 + 6(15)$$

$$= 225 + 90$$

$$= 315 \text{ square inches.}$$

Example 7.   The Masek Tropical Fish Company constructs fish tanks. The one shown in Figure 3-5 is a Model 72 J tank. This tank is one in an A25SK series in which the length is 8 inches more than 3 times the width, and the height is 10 inches more than the width.
  Let w stand for the width of an A25SK series tank.

  a. Write a polynomial for the length.

  b. Write a polynomial for the height.

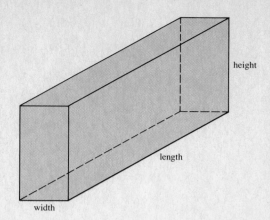

Figure 3-5. A Model 72 J fish tank.

  c. Write a polynomial for the volume.

  d. Find the volume, given the width is 10 inches.

Solution. **a.** The length is 8 inches more than 3 times the width. Thus the length is $(3w + 8)$ inches.

  b. The height is 10 inches more than the width. Thus the height is $(w + 10)$ inches.

  c. The volume of the tank is the product of the width times the length times the height.

$$\text{Volume is } w(3w + 8)(w + 10)$$

$$= (3w^2 + 8w)(w + 10)$$

$$= 3w^3 + 30w^2 + 8w^2 + 80w$$

$$= 3w^3 + 38w^2 + 80w$$

  d. If $w = 10$ inches, then

$$\text{Volume} = 3(10^3) + 38(10^2) + 80(10)$$

$$= 3{,}000 + 3{,}800 + 800$$

$$= 7{,}600 \text{ in}^3$$

As a point of interest, there are 12^3 cubic inches in a cubic foot. This tank has $7{,}600 \div 1{,}728$, or about 4.4, cubic feet of space in it. One cubic foot of water weighs about 62.4 pounds. Therefore this tank, when filled with water, will have approximately 275 pounds of water.

SECTION 3-4. Practice Exercises

In exercises **1–24**, multiply.

[Examples 1–3]

1. $(3a + 2)(2a + 1)$ **2.** $(5a + 3)(a + 2)$

3. $(5s - 3)(6s + 5)$ **4.** $(2t + 9)(3t - 2)$

5. $(4m + 5n)(2m - 3n)$ **6.** $(7x - 3y)(2x + y)$

7. $(2a - b)(a - 5b)$

8. $(6c - 7d)(3c - 2d)$

9. $(x + 3)(2x^2 + 1)$

10. $(4x + 1)(x^2 + 1)$

11. $(2y + 3)(y^2 + 3y + 3)$

12. $(9x + 4)(2x^2 + x + 5)$

13. $(x + 2y)(x^2 - xy + y^2)$

14. $(3a + b)(a^2 - ab + b^2)$

15. $(2a - b)(2a^2 - 3ab + b^2)$

16. $(a + 3b)(a^2 - 6ab - 9b^2)$

17. $(x^2 + x + 1)(x^2 + x + 1)$

18. $(a^2 - a + 1)(a^2 - a + 1)$

19. $(y^2 - 3y + 1)(3y^2 + 2y + 1)$

20. $(5y^2 - y + 3)(2y^2 + y - 2)$

21. $(a^3 + a^2 - a)(2a^2 - a + 3)$

22. $(2b^3 - b + 2)(5b - 1 + 3b^2)$

23. $(1 - 2x - 3x^2)(3x^2 + 2x - 5)$

24. $(2y^2 - 1 + y)(2y^2 - y - 3)$

In exercises **25–36**, multiply using a vertical format.

[Examples 4 and 5]

25. $(5y - 3)(7y + 4)$

26. $(3w + 7)(2w - 3)$

27. $(2z - 5)(z^2 + 3z + 3)$

28. $(9y - 8)(2y^2 + y + 5)$

29. $(4p^2 - p - 8)(3p^2 + 2p - 1)$

30. $(7m^2 + 4m - 3)(2m^2 - m + 4)$

31. $(4c^3 - 5c^2 + 3)(2c + 9)$

32. $(3q^3 + 5q^2 - 2)(5q + 2)$

33. $(a^3 - 5a)(a^2 - a + 7)$

34. $(m^3 + 4m)(m^2 + 4m - 3)$

35. $(n^3 + 8n + 6)(n^2 + 2n - 3)$

36. $(a^3 - 3a + 1)(a^2 - 5a + 2)$

In exercises **37–40**, answer parts **a–c**.

[Example 6]

37. A rectangular picture frame is adjusted so that the length is always 3 inches more than the width.
   a. If x represents the width, write an expression for the length.
   b. Write a polynomial in x for the area inclosed by the frame.
   c. If a frame has a width of 12 inches, use the polynomial to find the area.

38. Tulip bulbs are planted in a rectangular array with ten more on the longer side than on the shorter side.
   a. If the number of bulbs on the short side is s, write an expression for the number of bulbs on the long side.
   b. Write a polynomial in s for the total number of bulbs planted.
   c. If the short side has 21 bulbs, use the polynomial to find the total number planted.

39. A marching band uses a rectangular formation in which the length has two more people than the width.
   a. Let n represent the number of people in the width of the formation. Write an expression for the number of people in the length.
   b. Write a polynomial in n for the total number of people in the formation.
   c. If the formation is eight people wide, use the polynomial to find the total number of people in the formation.

40. A landscape architect wants to cover an outside play area with square tiles. The rectangular area requires 18 more tiles on the long side than on the short side.
   a. Let t be the number of tiles on the short side. Write an expression for the number of tiles on the long side.

b. Write a polynomial in t for the total number of tiles in the play area.

c. Use the polynomial to find the number of tiles if the short side has 35 tiles.

In exercises **41** and **42**, answer parts **a–d** based on the information below.

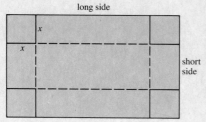

A flat piece of cardboard has squares cut out of each corner, and the remaining parts are folded up to form a box that is open at the top. (See figure.) The cardboard has a short side and a long side. The height of the box becomes x inches when the cardboard is folded. The width of the box is the short side of the cardboard minus two times the height. The length of the box is the long side of the cardboard minus two times the height.

41. Suppose the short side of the cardboard is 30 centimeters and the long side is 40 centimeters.

a. Write an expression for the width of the box.

b. Write an expression for the length.

c. Write a polynomial to find the volume.

d. What is the volume of the box if the cutout square is 2 centimeters by 2 centimeters?

42. Answer parts **a–d** of problem **41** above if the short side is 10 centimeters and the long side is 15 centimeters.

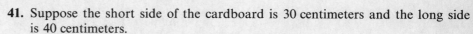

SECTION 3-4. Ten Review Exercises

In exercises **1–6**, solve each problem.

1. State the degree of $2t^3 - 5t^2$.

2. State the degree of $3t^2 + 7$.

3. Find the sum of the binomials in exercises **1** and **2**.

4. State the degree of the sum in exercise **3**.

5. Find the product of the binomials in exercises **1** and **2**.

6. State the degree of the product in exercise **5**.

In exercises **7–10**, solve and check any solutions.

7. $0 = 11 - 2(5a + 2) - 18a$

8. $76 + 9b + 15(b - 1) = 49$

9. $3c - 10 = 4 + 3(c + 2)$

10. $10(2n - 3) + 8(2n + 1) = 2$

SECTION 3-4. Supplementary Exercises

In exercises **1–16**, multiply.

1. $(7x + 3y)^2$ **2.** $(5x - 6y)^2$

3. $(a - 2b + 3)^2$　　　　　　　　**4.** $(x + y - 3)^2$

5. a. $(2t)(5t)$　　　　　　　　　　**6. a.** $(6x)(3x)$
  b. $(2t)(5 + t)$　　　　　　　　　　**b.** $6x(3 + x)$
  c. $(2 + t)(5 + t)$　　　　　　　　　**c.** $(6 + x)(3 + x)$

7. a. $(3y)(7y)$　　　　　　　　　　**8. a.** $(10y)(4y)$
  b. $3y(7 - y)$　　　　　　　　　　**b.** $10y(4 - y)$
  c. $(3 - y)(7 - y)$　　　　　　　　**c.** $(10 - y)(4 - y)$

9. $\left(\dfrac{1}{2}t + \dfrac{3}{4}\right)\left(\dfrac{1}{4}t^2 - \dfrac{1}{3}t + \dfrac{5}{6}\right)$　　　**10.** $\left(\dfrac{1}{2}x + \dfrac{1}{3}\right)\left(\dfrac{2}{9}x^2 + \dfrac{5}{6}x + \dfrac{2}{3}\right)$

11. $(2.3u - 0.5)(0.7u^2 + 1.2u - 3.1)$　　**12.** $(1.5u + 3.6)(2.6u^2 - 0.8u + 5.5)$

13. $(a - b)(a^2 + ab + b^2)$　　　　　**14.** $(a + b)(a^2 - ab + b^2)$

15. $(a - 2)(a^3 + 2a^2 + 4a + 8)$　　　**16.** $(a - 3)(a^3 + 3a^2 + 9a + 27)$

The product of two polynomials can be interpreted as an expression that represents the area of a rectangle. For example,

Length · Width =　　Area

$$(2x + 3)(x + 1) = 2x^2 + 5x + 3$$

To illustrate, suppose the length and width of a rectangle are $(2x + 3)$ and $(x + 1)$. The area of the figure can be found by adding the areas of each part.

In exercises **17–20**, find the products by first finding the areas of each part.

17. $(3x + 5)(x + 7) =$　　　　　　**18.** $(x + 4)(2x + 3) =$

19. $(x + 4)(x + 2y + 3) =$　　　　**20.** $(x + 7)(2x + y + 5) =$

In exercises **21–26**, simplify using the order of operations.

21. $3x(x + 5) - 5(x^2 + 3x)$　　　　**22.** $4x(x - 10) + 2(x^2 + 20x)$

23. $(x + 7)(x - 2) + (x - 6)(x + 1)$　　**24.** $(x - 5)(x - 4) + (x + 10)(x - 1)$

25. $(3x^2 + 4)(6x - 2) - (9x - 3)(2x^2 + 1)$

26. $(4x^2 + 5)(6x + 5) - (3x - 2)(8x^2 + 10)$

SECTION 3-5.   Multiplying Binomials: FOIL Method and Special Products

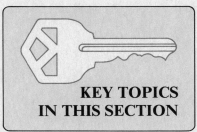

KEY TOPICS IN THIS SECTION

1. Multiplying binomials using the FOIL method

2. Multiplying a sum and difference of the same terms

3. Squaring a sum or difference of terms

The study of polynomials in beginning and intermediate algebra is dominated by binomials, trinomials, and polynomials of four terms. Often two binomials are multiplied to form another binomial, a trinomial, or a polynomial of four terms. As we will see in the next chapter, binomials, trinomials, and polynomials of four terms are frequently written as a product of two binomials. As a consequence, we need a technique that will enable us to form the product of two binomials as quickly and as easily as possible. In this section we will study some techniques that will achieve this goal.

Multiplying Binomials Using the FOIL Method

The word **FOIL** is a reminder that four products are formed whenever two binomials are multiplied. Each of the letters in FOIL stands for the locations of the monomials when the binomials are written next to each other. To illustrate, consider the following pair of binomials:

First terms in the binomials

Last terms in the binomials

$(3t + 7)(5t + 9)$

Inner terms in the array

Outer terms in the array

Using the first letters in First, Last, Inner, and Outer, the word *FOIL* is formed. A series of four arcs can be drawn as each product is computed. We recommend that the arcs be actually drawn as shown in Figure 3-6.

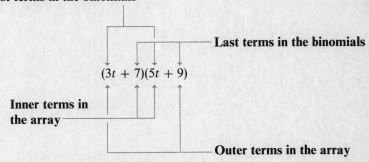

F stands for $(3t)(5t) = 15t^2$.

O stands for $(3t)(9) = 27t$.

I stands for $(7)(5t) = 35t$.

L stands for $(7)(9) = 63$.

Figure 3-6. F, O, I, and L for $(3t + 7)(5t + 9)$

Thus,

$$(3t + 7)(5t + 9) = 15t^2 + 27t + 35t + 63$$
$$= 15t^2 + 62t + 63.$$

Example 1.   Multiply: $(2y^2 - 3)(5y + 4)$

Solution.   **Discussion.** In the following display, the arcs are numbered 1–4 to show the sequence in which the products are taken. The products of the monomials are computed mentally, and only the answers are written.

$$(2y^2 - 3)(5y + 4) = 10y^3 + 8y^2 - 15y - 12$$

Example 2.   Multiply: $(7ab - 8)(3 - 4ab)$

Solution.   **Discussion.** The two binomials are not written in the same order. Although the multiplication can be carried out as they are written, it is recommended that one of them be changed to the form of the other. Arbitrarily we choose to write $3 - 4ab$ as $-4ab + 3$.

$$(7ab - 8)(-4ab + 3) = -28a^2b^2 + 21ab + 32ab - 24$$

$$= -28a^2b^2 + 53ab - 24$$

Multiplying a Sum and Difference of the Same Terms

There is something special about the following pair of binomials:

$$(7t + 5)(7t - 5)$$

Same first terms, namely $7t$ Same last terms, namely 5
First binomial adds 5 to $7t$ Second binomial subtracts 5 from $7t$

A product of two binomials that has all four of these features is called **the product of the sum and difference of terms.** When the FOIL method is used to multiply these binomials, the outer and inner products combine to zero.

$$(7t + 5)(7t - 5) = (7t)^2 - 35t + 35t - 5^2$$

$$= (7t)^2 - 5^2 \qquad -35t + 35t = 0$$

Notice that the operational sign between the indicated squares of $7t$ and 5 is subtraction. As a consequence, we say that the product of the sum and difference of terms is the difference of squares. The simplified form of the product is $49t^2 - 25$.

> **The product of the sum and difference of terms**
>
> $$(x + y)(x - y) = x^2 - y^2$$

Example 3.   Multiply: $(12m + 9n)(12m - 9n)$

Solution.   **Discussion.** Recognize that the two binomials have the form $(x + y)(x - y)$ with $x = 12m$ and $y = 9n$. The product is the difference between the squares of $12m$ and $9n$.

$$(12m + 9n)(12m - 9n) = (12m)^2 - (9n)^2$$
$$= 144m^2 - 81n^2$$

Example 4.   Multiply: $(10k^3 + 13)(10k^3 - 13)$

Solution.   **Discussion.** The indicated product has the form $(x + y)(x - y)$ with $x = 10k^3$ and $y = 13$.

$$(10k^3 + 13)(10k^3 - 13) = (10k^3)^2 - 13^2$$
$$= 100k^6 - 169$$

Squaring a Sum or Difference of Terms

The following indicated products are special in that the binomials being multiplied are identically the same. As a consequence, the products can be written as a square of one of the binomials.

$(7t + 5)(7t + 5)$ can be written $(7t + 5)^2$.

$(3x - 4y)(3x - 4y)$ can be written $(3x - 4y)^2$.

When the FOIL method is used to multiply these binomials, the F and L products are the squares of the terms in the binomials. The O and I products are exactly the same.

$$(7t + 5)(7t + 5) = (7t)^2 + 35t + 35t + 5^2$$
$$= (7t)^2 + 2(35t) + 5^2$$

$$(3x - 4y)(3x - 4y) = (3x)^2 - 12xy - 12xy + (4y)^2$$
$$= (3x)^2 - 2(12xy) + (4y)^2$$

The simplified forms of the products are $49t^2 + 70t + 25$ and $9x^2 - 24xy + 16y^2$. These expressions are called **perfect square trinomials** because they can be written as **squares of binomials**.

> **The square of a binomial is a perfect square trinomial**
>
> $$(x + y)^2 = x^2 + 2xy + y^2$$
>
> $$(x - y)^2 = x^2 - 2xy + y^2$$

Example 5.   Multiply:

a. $(7r + 11s)^2$

b. $(m^2 - 5n)^2$

Solution.   **Discussion.** Both **a** and **b** are squares of binomials. To write the product we square the first term, double the product of the terms, and square the second term. The signs of the first and last terms of the trinomial will be positive. The sign of the middle term will be the same as the sign of the binomial.

a. $(7r + 11s)^2 = (7r)^2 + 2(7r)(11s) + (11s)^2$

$$= 49r^2 + 154rs + 121s^2$$

b. $(m^2 - 5n)^2 = (m^2)^2 - 2(m^2)(5n) + (5n)^2$

$$= m^4 - 10m^2n + 25n^2$$

The square of a binomial always yields a trinomial. That is,

$$(x + y)^2 = x^2 + 2xy + y^2$$

Square of a binomial ⎯⎯⎯⎯   ⎯⎯ **Perfect square trinomial**

An error that many students make is to write

$$(x + y)^2 = x^2 + y^2.$$ The middle term is missing.

The area of a square provides a geometric picture of the square of a binomial that may help in preventing this mistake. Figure 3-7 shows a square. The lengths of the sides are, as shown, $x + y$. The area of a square is (length of side)^2, which for this square is

$$(x + y)^2.$$

The horizontal and vertical lines drawn inside the square divide the area into four pieces, labeled A_1, A_2, A_3, and A_4.

A_1 is a square whose area is $x \cdot x = x^2$.

A_2 is a rectangle whose area is $y \cdot x = xy$.

A_3 is a rectangle whose area is $y \cdot x = xy$.

A_4 is a square whose area is $y \cdot y = y^2$.

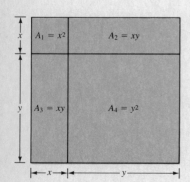

Figure 3-7. The area of the square is $(x + y)^2 = x^2 + 2xy + y^2$.

Thus, the total area is the sum of the four pieces, and

$$\text{Area} = x^2 + 2xy + y^2.$$

SECTION 3-5. Practice Exercises

In exercises **1–24**, multiply using FOIL.

[Examples 1 and 2]

1. $(x + 5)(x + 3)$ **2.** $(x + 1)(x + 5)$

3. $(z + 4)(3z + 2)$ **4.** $(z + 5)(2z + 5)$

5. $(4a + 3)(2a + 7)$ **6.** $(5a + 6)(3a + 2)$

7. $(y + 3)(y - 7)$ **8.** $(y + 5)(y - 9)$

9. $(2a + 7)(4a - 3)$ **10.** $(9a - 2)(2a + 3)$

11. $(5y - 2)(3y - 8)$ **12.** $(4y - 3)(6y - 7)$

13. $(3x^2 + 4)(6x - 2)$ **14.** $(4x^2 + 5)(7x + 1)$

15. $(a^2b^2 + 3)(ab - 4)$ **16.** $(a^2b^2 - 5)(2ab + 1)$

17. $(4a - 3)(2 - a)$ **18.** $(6a - 1)(3 - 2a)$

19. $(x^2 + 3)(4 - x)$ **20.** $(1 + 3n^2)(3n - 5)$

21. $(a + 2b)(3b^2 - a^2)$ **22.** $(2a + b)(2b^2 - 3a^2)$

23. $(z^2 + 4)(5 - 2z^2)$ **24.** $(w^2 - 6)(9 + 4w^2)$

In exercises **25–54**, multiply.

[Examples 3–5]

25. $(y + 4)(y - 4)$ **26.** $(x + 6)(x - 6)$

27. $(3x - 1)(3x + 1)$ **28.** $(4y + 1)(4y - 1)$

29. $(5z - 2)(5z + 2)$ **30.** $(9b - 4)(9b + 4)$

31. $(7m + 10)(7m - 10)$ **32.** $(4x + 3)(4x - 3)$

33. $(8c + 11d)(8c - 11d)$ **34.** $(2c + 3d)(2c - 3d)$

35. $(4x^3 + 3)(4x^3 - 3)$ **36.** $(7m^3 + 10)(7m^3 - 10)$

37. $(3y^2 - 5z^2)(3y^2 + 5z^2)$ **38.** $(2p^2 - 3q^2)(2p^2 + 3q^2)$

39. $(8 + 3ab)(8 - 3ab)$ **40.** $(6 + 7xy)(6 - 7xy)$

41. $(11t^5 - 7)(11t^5 + 7)$ **42.** $(13s^6 - 5)(13s^6 + 5)$

43. $(x + 5)^2$ **44.** $(m + 3)^2$

45. $(2a + 3)^2$ **46.** $(10c + 9)^2$

47. $(6y - z)^2$ **48.** $(5w - 4x)^2$

49. $(4y^3 - 3z)^2$ **50.** $(6m^3 - n)^2$

51. $(a^2 + 3b^2)^2$

52. $(2c^2 + 5d^2)^2$

53. $(7mn^2 - 10)^2$

54. $(12x^2y^3 - 7)^2$

SECTION 3-5.   Ten Review Exercises

In exercises **1–7**, solve each problem.

1. Evaluate: $(3m + 5)(2m - 1)$ for $m = -7$.

2. Multiply: $(3m + 5)(2m - 1)$.

3. Evaluate the product of exercise **2** for $m = -7$.

4. Evaluate: $(6m + 2)(m + 1)$ for $m = -7$.

5. Multiply: $(6m + 2)(m + 1)$.

6. Evaluate the product of exercise **5** for $m = -7$.

7. Solve: $(3m + 5)(2m - 1) = (6m + 2)(m + 1)$.

In exercises **8–10**, simplify.

8. $u^2(2u^3)(5u)$

9. $\left(\dfrac{-a^2b^3}{3}\right)^4$

10. $(-5r^2s)(2rs^2)^2$

SECTION 3-5.   Supplementary Exercises

In exercises **1–24**, multiply.

1. $\left(\dfrac{1}{2}x + \dfrac{1}{3}\right)\left(\dfrac{1}{2}x + \dfrac{1}{3}\right)$

2. $\left(\dfrac{1}{3}y + \dfrac{1}{5}\right)\left(\dfrac{1}{3}y + \dfrac{1}{5}\right)$

3. $\left(3z - \dfrac{1}{3}\right)\left(6z + \dfrac{1}{3}\right)$

4. $\left(4a + \dfrac{1}{4}\right)\left(8a - \dfrac{1}{4}\right)$

5. $(0.2b - 0.5)(0.4b + 0.2)$

6. $(0.3w + 0.1)(0.6w - 0.3)$

7. $(1.2m - 2.1)^2$

8. $(1.3x - 2.5)^2$

9. $(x - 3)(x + 3)(x^2 + 9)$

10. $(b - 2)(b + 2)(b^2 + 4)$

11. $(6xy + 5)(6xy + 5)$

12. $(9xy + 2)(9xy + 2)$

13. $(8y - 5)(-5 + 8y)$

14. $(7p - 3)(-3 + 7p)$

15. $(-3 - a)(6 - 3a)$

16. $(-4 - x)(5 - 2x)$

17. $2t(3t + 1)(5t - 4)$

18. $7t^2(2t - 3)(t + 6)$

19. $-k(2k + 9)^2$

20. $-3k(5k - 1)^2$

21. $(2b + 1)^3$

22. $(3b - 2)^3$

23. $(a + 10)^2(a - 10)^2$

24. $(3a + 5)^2(3a - 5)^2$

In exercises **25** and **26**, answer parts **a–e**.

25. If $a = 4$ and $b = 7$,
  a. Find $(a + b)(a + b)$.
  b. Find $a^2 + b^2$.
  c. Find $(a + b)^2$.
  d. Find $a^2 + 2ab + b^2$.
  e. Which answers are the same?

26. If $x = 3$ and $y = 5$,
  a. Find $(x + y)(x + y)$.
  b. Find $x^2 + y^2$.
  c. Find $(x + y)^2$.
  d. Find $x^2 + 2xy + y^2$.
  e. Which answers are the same?

In exercises **27–30**, answer each part.

27. a. Find the product of 22 and 18.
  b. Evaluate $x - 2$ if $x = 20$.
  c. Evaluate $x + 2$ if $x = 20$.
  d. Multiply $(x - 2)(x + 2)$.
  e. Evaluate the polynomial in part **d** for $x = 20$.
  f. Compare the answers in parts **a** and **e**.
  g. Write a polynomial in x to find $(41)(39)$ using $x = 40$.

28. a. Find the product of 33 and 27.
  b. Evaluate $x - 3$ if $x = 30$.
  c. Evaluate $x + 3$ if $x = 30$.
  d. Multiply $(x - 3)(x + 3)$.
  e. Evaluate the polynomial in part **d** for $x = 30$.
  f. Compare the answers in parts **a** and **e**.
  g. Write a polynomial in x to find $(52)(48)$ using $x = 50$.

29. a. Find 23^2.
  b. Find $(x + 3)$ for $x = 20$.
  c. Multiply $(x + 3)^2$.
  d. Evaluate the polynomial in part **c** for $x = 20$.
  e. Compare the answers to parts **a** and **d**.
  f. Write a polynomial in x to find 42^2 by using $x = 40$.

30. a. Find 32^2.
  b. Find $(x + 2)$ for $x = 30$.
  c. Multiply $(x + 2)^2$.
  d. Evaluate the polynomial in part **c** for $x = 30$.
  e. Compare the answers to parts **a** and **d**.
  f. Write a polynomial in x to find 65^2 by using $x = 60$.

SECTION 3-6.   Dividing Polynomials

**KEY TOPICS
IN THIS SECTION**

1. Quotient of powers, same base, property of exponents

2. Zero as an exponent

3. Dividing monomials

4. Dividing a polynomial by a monomial

5. Dividing polynomials

In earlier sections of this chapter we added, subtracted, multiplied, and raised to powers polynomials. In this section we will study the procedure for dividing polynomials. To use the procedure we need to establish another property of exponents.

Quotient of Powers, Same Base, Property of Exponents

Dividing polynomials frequently involves dividing variable factors with exponents. For example, it is quite likely that divisions like the following could result from dividing polynomials:

$$\frac{t^5}{t^3}, t \neq 0 \qquad \text{and} \qquad \frac{x^4 y^3}{x^2 y^2}, x \neq 0 \text{ and } y \neq 0$$

The definition of an exponent can be used to write these quotients with each variable written only once.

$$\overbrace{}^{\text{5 factors of } t}$$
$$\frac{t^5}{t^3} = \underbrace{\frac{\cancel{t} \cdot \cancel{t} \cdot \cancel{t} \cdot t \cdot t}{\cancel{t} \cdot \cancel{t} \cdot \cancel{t}}}_{\text{3 factors of } t} = t^{5-3} = t^2$$

The t is written once.

The exponent 2 is the difference between 5 and 3.

Since $t \neq 0$, three factors of t on the top and bottom can be changed to one, and that factor does not need to be written. Similarly,

$$\frac{x^4 y^3}{x^2 y^2} = x^{4-2} \cdot y^{3-2} = x^2 y$$

Bases are used once.

Exponents are subtracted.

> **The quotient of powers, same base, property of exponents**
> If b is a nonzero real number ($b \neq 0$), and m and n are positive integers with $m > n$, then
>
> $$\frac{b^m}{b^n} = b^{m-n}$$
>
> **Write the base once.**
>
> **Subtract the exponents.**

Example 1.   Simplify:

a. $\dfrac{3^4 k^8}{3^2 k^5}$

b. $\dfrac{(r^2 s)(r^3 s^5)}{r^4 s^3}$

Solution.   a. $\dfrac{3^4 k^8}{3^2 k^5} = 3^{4-2} k^{8-5}$ $\dfrac{b^m}{b^n} = b^{m-n}$

$= 3^2 k^3$ or $9k^3$

b. $\dfrac{(r^2 s)(r^3 s^5)}{r^4 s^3} = \dfrac{r^5 s^6}{r^4 s^3}$ $r^{2+3} = r^5$ and $s^{1+5} = s^6$

$= rs^3$ $r^{5-4} = r$ and $s^{6-3} = s^3$

The quotient of powers property of exponents stated here is not complete. We need to consider two other possibilities for exponents m and n:

Possibility 1. $m = n$, thus $m - n = 0$ A zero exponent

Possibility 2. $m < n$, thus $m - n < 0$ A negative integer exponent

The answer to the zero exponent possibility is discussed below. The answer to the negative integer exponent is covered in Chapter 5, when we study expressions that are called *rational*.

Zero as an Exponent

When polynomials are divided, we frequently have a division in which the exponents on one or more variable factors are the same. For example,

$$\frac{t^3}{t^3}, t \neq 0 \quad \text{and} \quad \frac{x^2 y}{x^2 y}, x \neq 0 \text{ and } y \neq 0$$

If the quotient of powers property of exponents is applied to these expressions, a zero is obtained for an exponent.

$$\frac{t^3}{t^3} = t^{3-3} = t^0 \quad \text{and} \quad \frac{x^2 y}{x^2 y} = x^{2-2} \cdot y^{1-1} = x^0 \cdot y^0$$

An exponent as a symbol specifies that the exponent must be a positive integer $(1, 2, 3, 4, \ldots)$. Therefore an interpretation of 0 as an exponent must be given as a definition.

Definition 3.4. Zero as an exponent
If $b \neq 0$, then $b^0 = 1$.

To demonstrate that Definition 3.4 agrees with other facts we learned in mathematics, consider the following series of equalities:

$$\frac{3^4}{3^4} = \frac{3 \cdot 3 \cdot 3 \cdot 3}{3 \cdot 3 \cdot 3 \cdot 3} = \frac{3}{3} \cdot \frac{3}{3} \cdot \frac{3}{3} \cdot \frac{3}{3} = 1 \cdot 1 \cdot 1 \cdot 1 = 1$$

$$\frac{3^4}{3^4} = 3^{4-4} = 3^0 = 1$$

In either case, whether we write out the factors and then reduce, or use Definition 3-4 and get zero as an exponent, we get a result of one. Furthermore, this result would be obtained for any nonzero base b and any positive integer m.

If $b = 0$, Definition 3.4 does not apply. Therefore we say that 0^0 is not defined. To simplify any future problems involving division by variables, let it be understood that each variable represents only nonzero numbers.

Example 2. Simplify:

a. $\dfrac{r^5 s^2 t^3}{r^2 s^2 t^2}$

b. $\dfrac{(xy^2)^2(x^3 y^5)}{(xy^3)^3}$

Solution.   a. $\dfrac{r^5 s^2 t^3}{r^2 s^2 t^2} = r^{5-2} \cdot s^{2-2} \cdot t^{3-2}$

$$= r^3 s^0 t$$

$$= r^3 t \qquad\qquad s^0 \text{ is replaced by 1.}$$

b. $\dfrac{(xy^2)^2 (x^3 y^5)}{(xy^3)^3} = \dfrac{(x^2 y^4)(x^3 y^5)}{x^3 y^9}$ $\quad\begin{aligned}(xy^2)^2 &= x^2 y^4 \\ (xy^3)^3 &= x^3 y^9\end{aligned}$

$$= \dfrac{x^5 y^9}{x^3 y^9} \qquad\qquad x^2 \cdot x^3 = x^5 \text{ and } y^4 \cdot y^5 = y^9$$

$$= x^2 y^0 \qquad\qquad x^{5-3} = x^2 \text{ and } y^{9-9} = y^0$$

$$= x^2 \qquad\qquad y^0 = 1 \text{ and } 1 \cdot x^2 = x^2$$

Dividing Monomials

With the quotient of powers property of exponents we can simplify an indicated quotient of two monomials. The following two-step procedure can be used to accomplish this task.

To simplify an indicated quotient of two monomials,

Step 1. Completely reduce the numerical coefficients.

Step 2. Subtract the exponents on each variable that is contained in both monomials.

Example 3.   Divide:

a. $\dfrac{72 r^5 s^2}{9 r^2 s^2}$

b. $\dfrac{54 t^4 u^3 v}{36 tv}$

Solution.   **Discussion.** To demonstrate the two-step procedure, each quotient is computed in two separate steps.

a. $\qquad\qquad \dfrac{72 r^5 s^2}{9 r^2 s^2}$

Step 1. $= \dfrac{8 r^5 s^2}{r^2 s^2} \qquad\qquad \dfrac{72}{9} = 8$

Step 2. $= 8 r^3 \qquad\qquad \dfrac{r^5}{r^2} = r^{5-2} = r^3 \text{ and } \dfrac{s^2}{s^2} = s^0 = 1$

b. $\qquad\qquad \dfrac{54 t^4 u^3 v}{36 tv}$

Step 1. $= \dfrac{3 t^4 u^3 v}{2 tv} \qquad\qquad \begin{aligned}54 \div 18 &= 3 \\ 36 \div 18 &= 2\end{aligned}$

Step 2. $= \dfrac{3t^3u^3}{2}$ $\dfrac{t^4}{t} = t^{4-1} = t^3$ and $\dfrac{v}{v} = v^0 = 1$

Dividing a Polynomial by a Monomial

The formula for adding fractions with a common denominator is well known. Specifically, if $c \neq 0$, then

$$\frac{a}{c} + \frac{b}{c} = \frac{a+b}{c}.$$

This formula can be used to change a division of a polynomial by a monomial to one that contains only monomials divided by monomials. To illustrate this fact, consider the following division of a binomial by a monomial:

$$\frac{96t^4 + 42t^2}{6t^2}$$

This division can be rewritten with both terms of the binomial over the monomial denominator.

$$\frac{96t^4 + 42t^2}{6t^2} = \frac{96t^4}{6t^2} + \frac{42t^2}{6t^2}$$

$$= 16t^2 + 7$$

Example 4. Divide: $\dfrac{28m^6n + 20m^4n^2 - 12m^2n^3}{4m^2n}$

Solution. **Discussion.** Write each term of the trinomial over the monomial denominator.

$$\frac{28m^6n + 20m^4n^2 - 12m^2n^3}{4m^2n} \qquad \frac{A+B-C}{D} = \frac{A}{D} + \frac{B}{D} - \frac{C}{D}$$

$$= \frac{28m^6n}{4m^2n} + \frac{20m^4n^2}{4m^2n} - \frac{12m^2n^3}{4m^2n}$$

$$= 7m^4 + 5m^2n - 3n^2$$

Dividing Polynomials

The following expression shows a division of a polynomial of 4 terms divided by a binomial:

$$\frac{15x^3 - 26x^2 + 29x - 28}{3x - 4}$$

If the division of a polynomial by a polynomial has a divisor with two or more terms, a long division format is recommended. The procedure is illustrated in Example 5, using the polynomials above.

Example 5. Divide: $\dfrac{15x^3 - 26x^2 + 29x - 28}{3x - 4}$

Solution. **Discussion.** The long division uses the $\overline{)}$ symbol, the same one that can be used to divide numbers with paper and pencil.

The steps in the procedure are identified in the following problem.

1. Compute $\dfrac{15x^3}{3x} = 5x^2$.

2. Write the $5x^2$ above the $15x^3$.

3. Compute $5x^2(3x - 4) = 15x^3 - 20x^2$ and subtract from $15x^3 - 26x^2$ to get $-6x^2$. (Bring down the $29x - 28$.)

4. Compute $\dfrac{-6x^2}{3x} = -2x$.

5. Write the $-2x$ next to the $5x^2$.

6. Compute $-2x(3x - 4) = -6x^2 + 8x$ and subtract from $-6x^2 + 29x$ to get $21x$. (Bring down the -28.)

7. Compute $\dfrac{21x}{3x} = 7$.

8. Write $+7$ next to the $2x$.

9. Compute $7(3x - 4) = 21x - 28$ and subtract from $21x - 28$ to get 0. Thus,
$$\frac{15x^3 - 26x^2 + 29x - 28}{3x - 4} = 5x^2 - 2x + 7.$$

$$
\begin{array}{r}
5x^2 - 2x + 7 \\
3x - 4 \,\overline{)\,15x^3 - 26x^2 + 29x - 28} \\
(-)\ 15x^3 - 20x^2 \\
\hline
-6x^2 + 29x - 28 \\
(-)\ -6x^2 + 8x \\
\hline
21x - 28 \\
(-)\ 21x - 28 \\
\hline
0
\end{array}
$$

Notice that the first three steps are repeated in the next three steps, and then in the last three steps.

1. Divide the first term of the dividend by the first term of the divisor.

2. Write the quotient as the first term of the answer.

3. Multiply the divisor by the monomial obtained in 1 and subtract the product from the dividend. (Bring down the remaining terms of the dividend and continue.)

Example 6.   Divide: $\dfrac{25t^2 - 18 + 10t^4 - 17t^3 + 21t}{5t + 4}$

Solution.   **Discussion.** The dividend is not written in ascending or descending powers of t. It is usually easier to divide when both dividend and divisor are written in descending powers.

$$
\begin{array}{r}
2t^3 - 5t^2 + 9t - 3 \\
5t + 4 \,\overline{)\,10t^4 - 17t^3 + 25t^2 + 21t - 18} \\
(-)\ 10t^4 + 8t^3 \\
\hline
-25t^3 + 25t^2 + 21t - 18 \\
(-)\ -25t^3 - 20t^2 \\
\hline
45t^2 + 21t - 18 \\
(-)\ 45t^2 + 36t \\
\hline
-15t - 18 \\
(-)\ -15t - 12 \\
\hline
-6
\end{array}
$$

$\dfrac{10t^4}{5t} = 2t^3$

$2t^3(5t + 4) = 10t^4 + 8t^3$

$\dfrac{-25t^3}{5t} = -5t^2$

$-5t^2(5t + 4) = -25t^3 - 20t^2$

$\dfrac{45t^2}{5t} = 9t$

$9t(5t + 4) = 45t^2 + 36t$

$\dfrac{-15t}{5t} = -3$

$-3(5t + 4) = -15t - 12$

The degree of -6 is zero and the degree of $5t + 4$ is one. Since the degree of the divisor is greater than the degree of the last remainder, the dividing stops, and we write $\dfrac{-6}{5t + 4}$ as a remainder. Thus the quotient with remainder is

$$2t^3 - 5t^2 + 9t - 3 + \frac{-6}{5t + 4}.$$

Frequently a dividend polynomial will not list all the terms whose degrees are less than the degree of the polynomial. (Such a polynomial is not complete.) For example,

$$\frac{4x^4 - 101x^2 + 25}{2x - 1}$$

The degree of the polynomial is 4.
There are no x^3 and x terms.

When such a division is carried out, it is common practice to insert the missing terms with a 0 coefficient.

Example 7.   Divide: $\dfrac{4x^4 - 101x^2 + 25}{2x - 1}$

Solution.   **Discussion.** As recommended, we insert $0x^3$ and $0x$ terms when writing the dividend polynomial.

$$
\begin{array}{r}
2x^3 + x^2 - 50x - 25 \\
2x - 1 \overline{)\, 4x^4 + 0x^3 - 101x^2 + 0x + 25} \\
(-)\ \underline{4x^4 - 2x^3} \\
2x^3 - 101x^2 + 0x + 25 \\
(-)\ \underline{2x^3 - x^2} \\
-100x^2 + 0x + 25 \\
(-)\ \underline{-100x^2 + 50x} \\
-50x + 25 \\
(-)\ \underline{-50x + 25} \\
0
\end{array}
$$

$\dfrac{4x^4}{2x} = 2x^3$

$0x^3 - (-2x^3) = 2x^3$

$\dfrac{2x^3}{2x} = x^2$

$\dfrac{-100x^2}{2x} = -50x$

$0x - 50x = -50x$

$\dfrac{-50x}{2x} = -25$

Thus the quotient is $2x^3 + x^2 - 50x - 25$, and the remainder is 0.

SECTION 3-6.   Practice Exercises

In exercises **1–32**, divide. Assume all variables are nonzero.

[Examples 1–4]

1. $\dfrac{x^{10}}{x^6}$

2. $\dfrac{3x^{13}}{x^7}$

3. $\dfrac{5^3 p^4}{5^2 p}$

4. $\dfrac{7^4 p^3}{7^3 p^2}$

5. $\dfrac{(a^2 b)(a^3 b^2)}{a^2 b^2}$

6. $\dfrac{(a^3 b^2)(ab^2)}{a^2 b^2}$

7. $\dfrac{(x^6 y^7 z^3)(x^2 y^4 z^2)}{xy^3 z^2}$

8. $\dfrac{(x^4 y^3 z^6)(x^3 y^3 z^4)}{xy^2 z^3}$

9. $\dfrac{15p^3q^5r^2}{5pq^3r^2}$

10. $\dfrac{-36p^8q^{10}r^5}{12p^2q^{10}r^3}$

11. $\dfrac{3^8s^4t^7}{3^5s^4t^6}$

12. $\dfrac{11^7s^8t^5}{11^5s^2t^4}$

13. $\dfrac{(10x^2y)^2(x^4y^2)}{(5xy^2)^2}$

14. $\dfrac{(-x^3y^2)^2(36xy^3)}{(2x^2y)^2}$

15. $\dfrac{(2m^2n)^2(3m^3n^4)^2}{(m^2n)^3}$

16. $\dfrac{(-3m^3n^2)^2(2m^2n)^3}{(-6m^2n^3)^2}$

17. $\dfrac{-56a^7b^4}{8a^5b^3}$

18. $\dfrac{-42a^{12}b^8c^4}{7a^{12}b^6c^3}$

19. $\dfrac{78x^{10}y^3z^2}{13x^8y^3z}$

20. $\dfrac{120x^5y^5z^8}{8x^5y^4z^5}$

21. $\dfrac{-44m^6n^3}{6m^4n^3}$

22. $\dfrac{-64m^4n^3}{12m^4n}$

23. $\dfrac{25p^5q^3r^2}{10p^3q^3}$

24. $\dfrac{28p^7q^6r^5}{18p^7q^2}$

25. $\dfrac{9x^4 + 6x}{3x}$

26. $\dfrac{40x^4 + 10x^2}{10x^2}$

27. $\dfrac{36y^3 - 100y^5}{4y^3}$

28. $\dfrac{15y^6 - 10y^4}{5y^2}$

29. $\dfrac{8a^4 + 16a^3 + 8a^2}{8a^2}$

30. $\dfrac{24a^5 + 16a^4 + 4a^3}{4a^3}$

31. $\dfrac{-2b^7 - 10b^5 + 6b^3}{-2b^3}$

32. $\dfrac{-5b^5 + 15b^3 - 10b}{-5b}$

In exercises **33–54**, divide using a long division format.

[Examples 5–7]

33. $(x^2 + 5x + 6) \div (x + 2)$

34. $(x^2 - x - 6) \div (x - 3)$

35. $(4a^2 - 9a - 9) \div (a - 3)$

36. $(5a^2 + 7a - 6) \div (a + 2)$

37. $(6y^3 - 5y^2 - 3y + 2) \div (3y + 2)$

38. $(6y^3 - y^2 - 5y + 2) \div (2y - 1)$

39. $(2m^3 - 3m^2n + 4mn^2 + 3n^3) \div (2m + n)$

40. $(30m^3 + m^2n - 17mn^2 + 2n^3) \div (3m - 2n)$

41. $(4 - 24a + 27a^2) \div (9a - 2)$

42. $(48a^2 - 1 + 8a) \div (12a - 1)$

43. $(5s - 4s^2 + 4s^3 - 8) \div (2s - 1)$

44. $(10 - 6s + 3s^3 - 4s^2) \div (s - 2)$

45. $(8y^2 + 3y^3 - 7 + y) \div (y + 2)$

46. $(6y^3 - 4y + 6 - 13y^2) \div (2y - 3)$

47. $(4x^3 + 17x - 12x^2 - 14) \div (2x - 3)$

48. $(9x^3 + 20 - 5x + 3x^2) \div (3x + 5)$

49. $(u^4 - 20u^2 + 64) \div (u - 4)$

50. $(81u^4 + 72u^2 - 48) \div (3u + 2)$

51. $(64x^6 + 16x^3 + 1) \div (2x + 1)$

52. $(64x^6 - 16x^3 + 1) \div (2x - 1)$

53. $(a^4 - 81) \div (a + 3)$

54. $(a^4 - 625) \div (a - 5)$

SECTION 3-6. Ten Review Exercises

In exercises **1–5**, solve each problem.

1. Multiply: $(x^2 - x - 6)(2x + 1)$.

2. Multiply: $[(x + 2)(x - 3)](2x + 1)$.

3. Divide: $(2x^3 - x^2 - 13x - 6)$ by $(2x + 1)$.

4. Divide the quotient of exercise **3** by $(x - 3)$.

5. Use the results of exercises **3** and **4** to write

$$(2x^3 - x^2 - 13x - 6)$$

as an indicated product of three binomials.

In exercises **6–10**, find the indicated products.

6. $(7u + 3v)(3u - 7v)$
7. $(7u + 3v)(7u - 3v)$

8. $(7u + 3v)^2$
9. $(3u - 7v)^2$

10. $(3u + 2v)(3u - 2v)(9u^2 + 4v^2)$.

SECTION 3-6. Supplementary Exercises

In exercises **1–4**, simplify each quotient.

1. a. $\dfrac{x^3 + 5x^2 + 4x}{4x}$

    b. $\dfrac{x^3 + 5x^2 + 4x}{x + 4}$

2. a. $\dfrac{x^3 + 5x^2 + 6x}{2x}$

    b. $\dfrac{x^3 + 5x^2 + 6x}{x + 2}$

3. a. $\dfrac{x^4 + 6x^3 - 27x^2}{-3x}$

    b. $\dfrac{x^4 + 6x^3 - 27x^2}{x - 3}$

4. a. $\dfrac{x^4 + 3x^3 - 10x^2}{-2x}$

    b. $\dfrac{x^4 + 3x^3 - 10x^2}{x - 2}$

In exercises **5–10**, divide.

5. $(y^4 - 3y^2 + 4y - 3) \div (y^2 - y + 1)$

6. $(2y^4 + 3y^3 + 8y^2 + 8) \div (y^2 + 2y + 4)$

7. $(4b^5 - 4b^4 - b + 1) \div (2b^2 + 1)$

8. $(b^4 + 2b^3 + 10b^2 + 10b + 25) \div (b^2 + 5)$

9. $(6a^4 + 7a^3 + 2a^2 + 16a - 5) \div (2a^2 - a + 3)$

10. $(10a^4 + 3a^3 - 3a - 3) \div (2a^2 + a + 1)$

In exercises **11–14**, answer each part.

11. a. Evaluate $9x^2$ for $x = 2$.
  b. Evaluate $3x$ for $x = 2$.
  c. Find the quotient of the values in parts **a** and **b**.

  d. Simplify: $\dfrac{9x^2}{3x}$.

  e. Evaluate part **d** for $x = 2$.
  f. Compare the results of parts **c** and **e**.

12. a. Evaluate $15a^2$ for $a = 3$.
  b. Evaluate $5a$ for $a = 3$.
  c. Find the quotient of the values in parts **a** and **b**.

  d. Simplify: $\dfrac{15a^2}{5a}$.

  e. Evaluate part **d** for $a = 3$.
  f. Compare the results of parts **c** and **e**.

13. Given the expression

$$\frac{x^2 + 5x + 6}{x + 3}$$

  a. Evaluate for $x = 10$.
  b. Use long division to find the indicated quotient.
  c. Evaluate part **b** for $x = 10$.
  d. Compare the answers to parts **a** and **c**.

14. Given the expression

$$\frac{x^2 + 6x + 8}{x + 4}$$

  a. Evaluate for $x = 10$.
  b. Use long division to find the indicated quotient.
  c. Evaluate part **b** for $x = 10$.
  d. Compare the answers to parts **a** and **c**.

In exercises **15–20**, simplify using the order of operations.

15. $\dfrac{25x^2 + 10x}{5x} + \dfrac{35x^3 - 14x^2}{7x^2}$

16. $\dfrac{36x^2 - 28x}{4x} + \dfrac{33x^3 + 21x^2}{3x^2}$

17. $\dfrac{a^2b^2c - ab^2c}{abc} - \dfrac{a^4b^3c^2 - a^3b^3c^2}{a^3b^2c^2}$

18. $\dfrac{a^4b^5c^2 - a^4b^3c^2}{a^4b^3c} - \dfrac{a^6bc^2 + a^5b^3c^3}{a^5bc^2}$

19. $\dfrac{y^2 - 4y - 45}{y + 5} + \dfrac{y^2 + 11y + 18}{y + 2}$

20. $\dfrac{y^2 - 10y + 21}{y - 7} + \dfrac{y^2 + 8y + 15}{y + 5}$

SECTION 3-1.   Summary Exercises

Name _____

Date _____

Score

1. What are the names of the following polynomials?

    a. $32x^2 - 6x$

    b. $4x^2y^2 + x^2y - 6x + 5y + 7$

1. a. _____

   b. _____

In exercises **2** and **3**,

a. state the degree of each term.

b. state the degree of the polynomial.

2. $4x^2 - 2xy + y^2 - 9$

2. a. _____

   b. _____

3. $-9r^2s^3 + 3rs^2 + 5s - 1$

3. a. _____

   b. _____

In exercises **4–6**, write in descending powers.

4. $3b^2 - 18b^3 - 27b$

4. _____

5. $-3x^4 - x^2 + 7x^5 + 8 + 6x^3$

5. _____

6. $60a^2b + 6a^3 - 25 + 150ab^2$

   a. For the variable a

   b. For the variable b

6. **a.** _____

   b. _____

In exercises **7** and **8**, answer each question.

7. The term $6x^3y^k$ is a term of degree seven. What is the value of k?

7. _____

8. The polynomial $x^2y^2 - 4x^3y^k - 6xy^3$ is a 5th degree polynomial.

   a. Can the value of k be found?

   b. If so, what is the value?

8. **a.** _____

   b. _____

SECTION 3-2. Summary Exercises

Answer

In exercises **1–8**, simplify.

1. $(6x^2 + 3x - 10) + (14 + 7x - 12x^2)$

1. _____

2. $(-7x^2y + 3xy - 4xy^2) + (9xy - xy^2 + 10x^2y)$

2. _____

3. $(7z^3 + 3z^2 + 5) + (-7z^3 + 4z - 5)$

3. _____

4. $(4p^2 + 8p - 7) - (p^2 - 2p + 6)$

4. _____

5. $(a^3 - b^3) + (a^2 + ab + b^2) - (a^3 + ab)$

5. _____

6. $(m^2n^2 + 4m^2n - 3mn^2 - 6) - (5m^2n^2 + m^2n - 3mn^2 + 10)$

6. _____

7. $(y^2 + 6y + 9) + (2y^2 - 11y + 5) - (3y^2 - 5y + 12)$

7. _____

8. $-3(t^2 - 4t + 8) + 2(5t^2 + t - 3)$

8. _____

SECTION 3-3. Summary Exercises

In exercises **1–8**, simplify.

1. $w^4 \cdot w \cdot w^5$

1. _____

2. $(-6x^2y)(-12x^3y^2)$

2. _____

3. $(p^3)^5(q^2)^3$

3. _____

4. $(-5a^4b)^3$

4. _____

5. $(-5a^2b)(a^4b)^3(-3ab^2)^2$

5. _____

6. $\left(\dfrac{2m^4n}{p^3}\right)^4$

6. _____

7. $(z^2 - 11z + 10)(3z)$

7. _____

8. $\left(\dfrac{-cd^2}{3}\right)(33c^2 + 24cd - 12d^2)$

8. _____

SECTION 3-4. Summary Exercises

In exercises **1–7**, multiply.

1. $(x + 7)(3x + 5)$

1. _____

2. $(4x - 9)(2x + 4)$

2. _____

3. $(5a^2 + 2)(3a - 1)$

3. _____

4. $(x^2 - 2x + 5)(3x + 2)$

4. _____

5. $(z^2 - 4z + 5)(z - 3)$

5. _____

6. $(x^2 + 3x - 1)(x^2 - 3x + 1)$

6. _____

7. $(7w^3 + 3w + 2)(2w - 5)$

7. _____

8. The length of a rectangle is 13 centimeters more than the width.

   a. Write a polynomial in x for finding the area of the rectangle.

8. **a.** _____

   b. Find the area if the width is 11 centimeters.

     b. _____

   c. Write a polynomial in x for finding the perimeter of the rectangle.

     c. _____

   d. Find the perimeter if the width is 11 centimeters.

     d. _____

SECTION 3-5. Summary Exercises

Answer

In exercises **1–8**, multiply.

1. $(x + 10)(x + 8)$

1. _____

2. $(3x + 8)(2x - 5)$

2. _____

3. $(p^2 + 2q)(p + q)$

3. _____

4. $(3a + 7)(5 - 2a)$

4. _____

5. $(3y^2 - 4)(3y^2 + 4)$

5. _____

6. $(5x - 2)(2 + 5x)$

6. _____

7. $(7x - 3)^2$

7. _____

8. $3t(7t - 15)(4t + 9)$

8. _____

Name _____

Date _____

Score _____

SECTION 3-6. Summary Exercises

In exercises **1–8**, simplify.

Answer

1. $\dfrac{6^4 a^3 b^5}{6^3 a b^2}$

1. _____

2. $\dfrac{(x^2 y^3)^3 (x y^2)}{x^3 y^3}$

2. _____

3. $\dfrac{m^8 n^4 p^{10}}{m^5 n^4 p}$

3. _____

4. $\dfrac{96 s^9 t^3 u^5}{30 s^3 t^2}$

4. _____

5. $\dfrac{35a^6 - 21a^3 + 49a^2}{7a^2}$

5. _____

6. $(x^2 - 13x + 30) \div (x - 3)$

6. _____

7. $\dfrac{(4y^4 + 20y^3 - y^2 - 5y + 3)}{5 + y}$

7. _____

8. $(x^3 - 64) \div (x - 4)$

8. _____

CHAPTER 3.   Review Exercises

In exercises **1–4**, use polynomials P and Q.

$$P: 3x^2 + 7 - 8x + 5x^5$$

$$Q: 3m^2n - n^3 + m^3 + 3mn^2$$

1. State the degree of P.

2. State the degree of Q.

3. Write P in descending powers.

4. Write Q in descending powers of m.

5. If $-9x^2y^k$ has degree 6, what is the value of k?

6. If $13a^2b^3c^k$ has degree 6, what is the value of k?

In exercises **7–30**, do the indicated operations.

7. $(2x^2 - 3x + 1) + (x^2 + 4x - 3)$

8. $(5x^2 - y^2 - 3xy) - (5xy - x^2 - y^2)$

9. $3(t^5 - t^4 + 3t^3 - 2t^2 + t - 1) - 2(t^5 - 2t^4 + 5t^3 - 3t^2 - t + 2)$

10. $5(t^4 + 2 - t^2) - 7(3 + t^4 - 2t^2) + 10(1 - t^2 + t^4)$

11. $-3u^2(2u^3)(5u)$

12. $(-2u^2v)^3$

13. $\left(\dfrac{-a^2b^3}{3}\right)^4$

14. $3a^2b(2a^2 + ab - 5b^2)$

15. $(3z + 4)(5z^2 - 1)$

16. $(3z + 4)(4z^2 + 3z - 2)$

17. $(3m + 4n)(m + 2n)$

18. $(m - 5n)(8m + 7n)$

19. $(3k - 2)(3k + 2)$

20. $(k^2 + 3j)(k^2 - 3j)$

21. $(13t - 4)^2$

22. $(10t^2 + 7)^2$

23. $\dfrac{-120a^4b^2}{15a^2b^2}$

24. $\dfrac{(2a^2b^3)^2(-3a^3b)^3}{(-6a^2b^2)^2}$

25. $\dfrac{24y^4 - 18y^2}{6y^2}$

26. $\dfrac{8y^3 - 25y^2 + 5y}{10y}$

27. $(10b^4 + 7b^3 - 10b^2 - 3b - 9) \div (2b + 3)$

28. $(18b^5 + 3b^4 + 4b^2 - 8b - 4) \div (3b^2 - 2)$

29. $(2x + 7)^2 - (5x - 1)(3x + 4) + 11(x^2 - x - 5)$

30. $(15x - 4)(15x + 4) - (10x - 9)(23x + 1) - 5(40x - 2 - x^2)$

4

Factoring Polynomials

"Good morning," Ms. Glaston greeted the class. "Recall that earlier we learned how to solve linear equations in one variable. For example, we know how to solve equations like **1** and **2**."

She then wrote on the board

1. $2x - 5 = 0$ **2.** $x + 4 = 0$

Paula Radosevich raised her hand and asked, "Isn't the root of equation **1**, $\frac{5}{2}$ and the root of equation **2**, -4?"

"Right, Paula," Ms. Glaston replied, "so the roots of **1** and **2** when taken together are $\frac{5}{2}$ and -4. Now let's see what happens when we multiply these equations."

She then turned to the board and wrote the following series of equations:

$$(2x - 5)(x + 4) = 0$$

$$2x^2 + 8x - 5x - 20 = 0$$

3. $2x^2 + 3x - 20 = 0$

"Suppose I asked you to solve the equation I labeled **3**," Ms. Glaston continued. "How would you find any roots of the equation?"

After a brief silence Thom Heinz said, "Ms. Glaston, this isn't a linear equation, because it has a $2x^2$ term. Doesn't that mean we can't solve it like the ones in Chapter 2?"

"You're right, Thom, we can't use the procedure for solving linear equations in one variable," Ms. Glaston replied, "but is it possible that we might already know the number, or numbers, that make the equation true?"

"Could $\frac{5}{2}$ and -4 be roots of this equation?," Stacey Matthews said.

"Let's find out, Stacey," Ms. Glaston replied. She then turned to the board and wrote

$$\text{If } x = \frac{5}{2}, \text{ then } 2\left(\frac{5}{2}\right)^2 + 3\left(\frac{5}{2}\right) - 20 = 0$$

$$2 \cdot \frac{25}{4} + 3 \cdot \frac{5}{2} - 20 = 0$$

$$\frac{25}{2} + \frac{15}{2} - 20 = 0$$

$$\frac{40}{2} - 20 = 0$$

$$20 - 20 = 0, \text{ true}$$

$$\text{If } x = -4, \text{ then } 2(-4)^2 + 3(-4) - 20 = 0$$

$$2(16) + 3(-4) - 20 = 0$$

$$32 + (-12) - 20 = 0$$

$$20 - 20 = 0, \text{ true}$$

"As you can see," Ms. Glaston continued, "Stacey's guess was correct. Now, does this fact suggest a way in which we might solve equation **3**?"

"I have an idea," Steve Koch said. "We multiplied $2x - 5$ by $x + 4$ to get $2x^2 + 3x - 20$, the left side of equation **3**. Now we just showed that the roots of equations **1** and **2** are also roots of **3**. Maybe we should just reverse the process by changing $2x^2 + 3x - 20$ back to $(2x - 5)(x + 4)$. Does that make any sense?"

"It not only makes sense, Steve," Ms. Glaston replied, "but it is a method that is frequently used to solve equations like **3**. As you pointed out, the process requires writing the trinomial as an indicated product of two binomials. This process is called **factoring.** In this chapter we will learn techniques for writing various kinds of polynomials in **factored form.** One purpose for doing this, as was demonstrated with equations **1**, **2**, and **3**, is to give us a technique for solving some equations that have a polynomial of degree two. In future chapters we will use the factoring of polynomials for purposes other than solving equations."

SECTION 4-1.   Common Monomial Factors

**KEY TOPICS
IN THIS SECTION**

1. The factored form of a polynomial

2. A common factor that is an integer

3. A common factor that is a power of a variable

4. A common factor that is a product of an integer and variable or variables

We begin the study of writing polynomials in factored form by considering polynomials in which each term has a **common factor.** The common factor can be an integer, a power of a variable, or a product of an integer and variables.

The Factored Form of a Polynomial

The primary tool for multiplying polynomials, or writing them in factored form, is the distributive property of multiplication over addition. For example, expressions **a** and **b** are two different forms of the same expression.

a. $5t + 20$ **A polynomial of two terms**

b. $5(t + 4)$ **A polynomial of one term, but two factors**

Expression **b** can be changed to the form in **a** by multiplying both terms inside the parentheses by the 5 outside the parentheses. Expression **a** can be changed to the form in **b** by dividing both terms by the common factor 5. The quotients are written inside the parentheses, and the 5 is written outside the parentheses. Expression **a** can be called the **multiplied form,** and expression **b** can be called the **factored form.**

> The process of changing a polynomial from the multiplied form to the factored form is called **factoring.**

A Common Factor that is an Integer

If each term of a polynomial can be evenly divided by some nonzero number, then the polynomial can be written in factored form by removing that common number factor.

Example 1. Factor: $18x^2 - 30x + 42$

Solution. **Discussion.** The polynomial has three terms. The number parts of these terms are 18, 30, and 42. Each of these numbers is evenly divisible by 1, 2, 3, and 6. The largest of the common factors is 6; therefore 6 is called the **greatest common factor (gcf).** The polynomial is written in factored form by removing the factor 6.

$$18x^2 - 30x + 42$$

$$= 6(3x^2) - 6(5x) + 6(7) \qquad \text{Write each term with a factor of 6.}$$

$$= 6(3x^2 - 5x + 7) \qquad \text{Use the distributive property.}$$

The discussion portion of Example 1 pointed out that 1, 2, 3, and 6 are all common factors of $18x^2 - 30x + 42$. As a consequence, each of the following expressions is a factored form of the polynomial.

$$1(18x^2 - 30x + 42)$$

$$2(9x^2 - 15x + 21)$$

$$3(6x^2 - 10x + 14) \qquad \text{Each expression consists of one term, but two factors.}$$

$$6(3x^2 - 5x + 7)$$

The goal in removing a common factor is to extract the greatest common factor from the polynomial.

> If a polynomial has terms with common factors, then the **gcf (greatest common factor)** should be removed.

A Common Factor that is a Power of a Variable

If every term of a polynomial contains at least one factor of a variable, then the highest degree of that variable contained in every term can be factored.

Example 2. Factor: $2t^5 - 3t^4 + t^3$

Solution. **Discussion.** The first term $(2t^5)$ has five factors of t, the second term $(3t^4)$ has four factors, and the third term (t^3) has three factors. Thus, every term has at least three factors of t. These three factors of t can be removed.

$$2t^5 - 3t^4 + t^3$$

$$= 2t^2(t^3) - 3t(t^3) + 1 \cdot (t^3) \qquad \text{Write } t^3 \text{ as } 1 \cdot (t^3).$$

$$= t^3(2t^2 - 3t + 1) \qquad \text{Use the distributive property.}$$

Example 3. Factor and write in descending powers of u.

$$u^2v^3 - 5u^3v^2 - 3u^2v + 2u^4v$$

Solution. **Discussion.** Every term has at least two factors of u and one factor of v.

$$u^2v^3 - 5u^3v^2 - 3u^2v + 2u^4v$$

$$= v^2(u^2v) - 5uv(u^2v) - 3(u^2v) + 2u^2(u^2v)$$

$$= u^2v(v^2 - 5uv - 3 + 2u^2) \qquad \text{Use the distributive property.}$$

$$= u^2v(2u^2 - 5uv + v^2 - 3) \qquad \text{In descending powers of } u$$

A Common Factor that is a Product of an Integer and Variable or Variables

Frequently the common factor in a polynomial is a product of a number and powers of one or more variables. The gcf of both numbers and variables should be removed.

Example 4. Factor: $16x^4y - 48x^3y^2 + 8x^2y^3$

Solution. **Discussion.** Consider separately the number, x, and y factors of each term.

The gcf of 16, 48, and 8 is 8.
The gcf of x^4, x^3, and x^2 is x^2.
The gcf of y, y^2, and y^3 is y.
The gcf for the polynomial is $8x^2y$.

$$16x^4y - 48x^3y^2 + 8x^2y^3$$

$$= 8x^2y(2x^2) - 8x^2y(6xy) + 8x^2y(y^2)$$

$$= 8x^2y(2x^2 - 6xy + y^2)$$

Example 5. Factor and write in descending powers.

$$27t^4 - 9t^6 + 18t^7 - 54t^3 + 45t^5$$

Solution. The gcf of 27, 9, 18, 54, and 45 is 9.
The gcf of t^4, t^6, t^7, t^3, and t^5 is t^3.
The gcf of the polynomial is $9t^3$.

$$27t^4 - 9t^6 + 18t^7 - 54t^3 + 45t^5$$

$$= 9t^3(3t) - 9t^3(t^3) + 9t^3(2t^4) - 9t^3(6) + 9t^3(5t^2)$$

$$= 9t^3(3t - t^3 + 2t^4 - 6 + 5t^2) \qquad \text{In factored form}$$

$$= 9t^3(2t^4 - t^3 + 5t^2 + 3t - 6) \qquad \text{In descending powers}$$

It is good mathematical procedure to always check one's results. The results of removing a common factor can be quickly checked by mentally multiplying each term of the grouped expression by the common factor. Such a check will also verify that the operational signs within the grouped expression are correct.

SECTION 4-1.   Practice Exercises

In exercises **1–44**, write in factored form.

[Example 1]   **1.** $24 + 36x$

2. $44x^2 - 12$

3. $8m - 12n$

4. $40m + 35n$

5. $54a + 30b - 2$

6. $36a - 42b + 3$

7. $8x^2 + 120x + 24$

8. $10x^2 + 250x - 90$

9. $9z^2 + 36z - 99$

10. $80z^2 - 16z + 144$

11. $11a^2 - 66b^2$

12. $49x^2 - 14y^2$

13. $100a^2 - 200ab + 10b^2$

14. $24a^3 - 12a^2b + 6ab^2 - 96b^3$

[Example 2]   **15.** $5x^2 + 3x$

16. $4x^3 + 9x^2$

17. $7y^3 - 2y^2 + 13y$

18. $y^4 - 10y^3 + 3y^2$

19. $5b^3 + b^5 - 4b^7 - 2b^9$

20. $2b^4 - 7b^5 - 3b^6 - b^7$

21. $a^8 - a^6 + a^4 - 3a^2$

22. $5a^9 + 2a^7 - a^5 - 4a^3$

[Example 3]   **23.** $5a^4b + 7a^3b^2 + a^2b^3$

24. $3a^3b^2 + 5a^2b^3 + 3ab^4$

25. $10x^4y^2 - x^3y^3 - x^2y^4$

26. $2x^5y^3 + 3x^4y^4 - x^3y^5$

27. $m^8n^2 - 5m^6n^4 + 6m^4n^6 - 2m^2n^8$

28. $m^9n^3 + m^7n^5 - m^5n^7 + 3m^3n^9$

29. $u^2vw - uv^2w + uvw^2$

30. $u^3v^2w^2 + 2u^2v^3w^2 - 3u^2v^2w^3$

[Example 4]   **31.** $4x^2 + 6x$

32. $9x^3 + 12x^2$

33. $15y^4 - 20y^3 + 35y^2$

34. $25y^5 - 35y^4 + 60y^3$

35. $75x^3y^2 + 27x^2y^3 - 3xy^4$

36. $4x^4y^3 - 144x^3y^4 + 36x^2y^5$

37. $4t^3u - 28t^2u^2 - 40tu^3 + 100tu$

38. $6t^4u^2 - 4t^3u^3 - 2t^2u^4 + 30t^2u^2$

[Example 5]   **39.** $16x^3 - 40x^4 + 36x^2 + 8x^5 - 12x$

40. $25x^2 + 10x^4 - 35x^3 - 15x$

41. $-42z + 36z^5 - 66z^7 + 24z^3$

42. $9z^4 - 18z^2 + 12z^6 - 27z^8$

43. $10t^6 - 70t^3 + 30t^4 - 50t^5 + 10t^7$

44. $36t^5 - 81t^3 - 18t^7 + 54t^4 + 72t^6$

SECTION 4-1.   Ten Review Exercises

In exercises **1** and **2**, multiply.

1. $7t^2(3t^3 - 5t^2 + 9)$

2. $-3ab(5a^2 - 4ab + b^2)$

In exercises **3** and **4**, factor and write the polynomial in descending powers.

3. $63t^2 + 21t^5 - 35t^4$

4. $12a^2b^2 - 15a^3b - 3ab^3$

In exercises **5** and **6**, multiply.

5. $(7k - 3)(2k - 5)$

6. $(2x + 1)(x^2 - 3x + 5)$

In exercises **7** and **8**, use the results of exercises **5** and **6** to write the polynomials in factored form.

7. $14k^2 - 41k + 15$

8. $2x^3 - 5x^2 + 7x + 5$

In exercises **9** and **10**, evaluate the expressions for $a = 5$ and $b = -2$.

9. $3a^2 + 13ab + 14b^2$

10. $(3a + 7b)(a + 2b)$

SECTION 4-1.   Supplementary Exercises

In exercises **1–18**, factor. Write answers in descending powers of one variable.

1. $5x^2y + 35x^3y - 9x^4y$

2. $21x^2y - 7x^2y^2 - 6x^2y^3$

3. $13a^2 + 65a^3 - 91a^4$

4. $34a^3 - 85a^4 + 68a^5$

5. $6a^2x - 20b^2x + 8c^2x$

6. $9at + 15bt - 21ct$

7. $\dfrac{x^2}{2} - \dfrac{x}{4}$

8. $\dfrac{y^2}{9} - \dfrac{y}{3}$

9. $\dfrac{2}{9}z^2 - \dfrac{10}{27}z^3 + \dfrac{8}{9}z^4$

10. $\dfrac{1}{25}z^2 + \dfrac{1}{50}z^3 - \dfrac{3}{100}z^4$

11. $0.6a^2 - 0.8ab + 1.4b^2$

12. $0.8a^3b - 0.4a^2b^2 - 2.8ab^3$

13. $0.05x^5 + 0.25x^3 - 0.35x$

14. $0.06x^6 + 0.42x^4 - 0.72x^2$

15. $-2x^3 - 8x^2 - 4x$

16. $-16x^4 - 48x^3 - 24x^2$

17. $-y^2 - 4y - 2$

18. $-y^5 - 3y^4 - 5y^3$

In exercises **19–22**, answer each part.

19. a. Is 5 a root of $x^2 - 5x = 0$?
   b. Write $x^2 - 5x$ in factored form.
   c. Replace $x^2 - 5x$ in the equation in part **a** with the factored form in part **b**.
   d. Is 5 a root of the equation in part **c**?
   e. Looking at the equation in part **c**, do you see another root for the equation?

20. a. Is -7 a root of $y^2 + 7y = 0$?
   b. Write $y^2 + 7y$ in factored form.
   c. Replace $y^2 + 7y$ in the equation in part **a** with the factored form in part **b**.
   d. Is -7 a root of the equation in part **c**?
   e. Looking at the equation in part **c**, do you see another root for the equation?

21. a. Is zero a root of $5k^2 - 3k = 0$?
   b. Factor the left side of the equation.
   c. Is zero a root of the equation in part **b**?
   d. State another root of the equation.

22. a. Is zero a root of $10s^2 - 3s = 0$?
   b. Factor the left side of the equation.
   c. Is zero a root of the equation in part **b**?
   d. State another root of the equation.

SECTION 4-2.   Factoring a Simple Trinomial

**KEY TOPICS
IN THIS SECTION**

1. A definition of a simple trinomial

2. The general factored form of a simple trinomial

3. Simple trinomials with a common factor

4. Simple trinomials that cannot be factored by using only integers

As we mentioned previously, much of our work on polynomials is dominated by binomials and trinomials. As evidence of this fact, several techniques are given for factoring only binomials and trinomials. In this section we will learn the first of these techniques, which can be used to factor **simple trinomials.**

A Definition of a Simple Trinomial

Any trinomial of degree two in x can be written in the general form

$$ax^2 + bx + c$$

where a, b, and c are nonzero real numbers. The number a is called the **leading coefficient,** because it is the coefficient of the x^2 term. If a is equal to 1, the trinomial is called **simple.**

> **Definition 4.1.   A simple trinomial in x**
> A **simple trinomial** of degree two in x is one that can be written in the form
>
> $$x^2 + bx + c.$$

The General Factored Form of a Simple Trinomial

To factor a simple trinomial, we need to find numbers p and q, such that

$$x^2 + bx + c = (x + p)(x + q)$$

To simplify the study of how to find p and q, we will agree to restrict our search to only integers. That is, p and q must be numbers from the following set:

$$\{\ldots, -4, -3, -2, -1, 0, 1, 2, 3, 4, \ldots\}$$

If for a given simple trinomial there do not exist integers p and q such that $(x + p)(x + q) = x^2 + bx + c$, we will say the trinomial *cannot be factored*. After we learn how to determine the appropriate integers that will factor a simple trinomial, we will consider why some trinomials cannot be factored using only integers for p and q.

Examples **a–d**, are four simple trinomials and the corresponding factored forms. The factored forms can be multiplied using FOIL to verify that their multiplied forms are, in each case, the given trinomials.

Simple trinomials	Factored forms	Values of $p \cdot q$	Values of $p + q$
a. $t^2 + 9t + 20$ $= (t + 5)(t + 4)$		$p \cdot q = 20$	$p + q = 9$
b. $y^2 - 15y + 56$ $= (y - 7)(y - 8)$		$p \cdot q = 56$	$p + q = -15$
c. $z^2 + 5z - 24$ $= (z - 3)(z + 8)$		$p \cdot q = -24$	$p + q = 5$
d. $u^2 - 13u - 48$ $= (u - 16)(u + 3)$		$p \cdot q = -48$	$p + q = -13$

These examples illustrate the following four-step procedure for factoring a simple trinomial.

To factor a trinomial of the form $x^2 + bx + c$,

Step 1. If necessary, write the trinomial in descending powers with a leading coefficient of 1.

Step 2. Consider integers p and q such that $p \cdot q = c$.

Step 3. Of the pairs of integers listed in Step 2, find the pair for which $p + q = b$ (if one exists).

Step 4. Use the pair from Step 3 to write

$$x^2 + bx + c = (x + p)(x + q).$$

Example 1. Factor: $t^2 + 16t + 48$

Solution. **Discussion.** For the given trinomial

$$p \cdot q = 48 \quad \text{and} \quad p + q = 16$$

Factors of 48	Corresponding sum
$1 \cdot 48 = 48$	$1 + 48 = 49$, reject
$2 \cdot 24 = 48$	$2 + 24 = 26$, reject
$3 \cdot 16 = 48$	$3 + 16 = 19$, reject
$4 \cdot 12 = 48$	$4 + 12 = 16$, accept

Therefore p is 4 and q is 12.

$$t^2 + 16t + 48 = (t + 4)(t + 12)$$

Example 2. Factor: $k^2 - 4k - 45$

Solution. **Discussion.** For the given trinomial

$$p \cdot q = -45 \quad \text{and} \quad p + q = -4$$

Factors of (-45)	Corresponding sum
$1(-45) = -45$	$1 + (-45) = -44$, reject
$3(-15) = -45$	$3 + (-15) = -12$, reject
$5(-9) = -45$	$5 + (-9) = -4$, accept

Therefore, one of the factors is 5 and the other one is -9:

$$k^2 - 4k - 45 = (k - 9)(k + 5)$$

Example 3. Factor: $8mn - 105n^2 + m^2$

Solution. **Discussion.** The given trinomial is not written in descending powers of either m or n. Since the coefficient of m^2 is 1, the trinomial will qualify as simple when it is written in descending powers of m.

$$m^2 + 8mn - 105n^2$$

Factors of (-105)	Corresponding sum
$105(-1) = -105$	$105 + (-1) = 104$, reject
$35(-3) = -105$	$35 + (-3) = 32$, reject
$21(-5) = -105$	$21 + (-5) = 16$, reject
$15(-7) = -105$	$15 + (-7) = 8$, accept

Therefore, $m^2 + 8mn - 105n^2 = (m + 15n)(m - 7n)$.

Simple Trinomials with a Common Factor

When a trinomial does not have a leading coefficient that is 1, it is not simple according to Definition 4.1. However, if the leading coefficient is a common factor of all three terms, then the trinomial can be changed to one that is simple.

Example 4. Factor: $7u^2 + 560 - 126u$

Solution. **Discussion.** The leading coefficient of this trinomial is 7, and the trinomial is not written in descending powers. However, 7 is a common factor.

$$7u^2 + 560 - 126u = 7(u^2 - 18u + 80)$$

Factor a 7 and write in descending powers.

Since $p \cdot q$ is positive, but $p + q$ is negative, we look for two *negative numbers* for p and q. By trial and error,

$$(-8)(-10) = 80 \quad \text{and} \quad (-8) + (-10) = -18.$$

Thus, $7(u^2 - 18u + 80) = 7(u - 8)(u - 10)$

Notice the final factored polynomial must also include the common factor 7.

Example 5. Factor: $-3a^3b + 12a^2b^2 + 96ab^3$

Solution. **Discussion.** Each term of this trinomial has a common factor of ab. Each term also has a common factor of 3 or -3. When we remove the common factor of -3, the leading coefficient can be changed to a positive one.

$$-3a^3b + 12a^2b^2 + 96ab^3 = -3ab(a^2 - 4ab - 32b^2)$$

By trial and error,

$$4(-8) = -32 \quad \text{and} \quad 4 + (-8) = -4.$$

Thus, $-3ab(a^2 - 4ab - 32b^2) = -3ab(a + 4b)(a - 8b)$.

Simple Trinomials that Cannot be Factored by Using Only Integers

Consider again the general form of the simple trinomial:

$$x^2 + bx + c$$

To write this trinomial in factored form, we have determined integers p and q such that

$$p \cdot q = c \quad \text{and} \quad p + q = b.$$

By insisting that p and q be only integers, simple trinomials can be found for which such numbers do not exist. For trinomials like this, we simply write "Cannot be factored."

Example 6.   Factor: $t^2 - 3t - 30$

Solution.

Factors of (-30)	Corresponding sum of factors
$1(-30) = -30$	$1 + (-30) = -29$, reject
$2(-15) = -30$	$2 + (-15) = -13$, reject
$3(-10) = -30$	$3 + (-10) = -7$, reject
$5(-6) = -30$	$5 + (-6) = -1$, reject

None of the possible integral factors of (-30) have a sum of -3. As a consequence, the given trinomial cannot be factored using only integers, and we write "Cannot be factored."

SECTION 4-2.   Practice Exercises

In exercises **1–42**, factor.

[Examples 1–3]

1. $y^2 + 18y + 77$

2. $y^2 + 8y + 15$

3. $x^2 - 7x + 10$

4. $x^2 - 13x + 42$

5. $a^2 + 21a + 54$

6. $a^2 + 23a + 130$

7. $b^2 + 8b - 33$

8. $b^2 + b - 56$

9. $t^2 - 17t - 38$

10. $t^2 - 10t - 39$

11. $z^2 - 12z + 32$

12. $z^2 - 7z + 12$

13. $k^2 - 6k - 72$

14. $k^2 - 8k - 65$

15. $u^2 + 10u - 56$

16. $u^2 + 4u - 77$

17. $m^2 + 23m - 50$

18. $m^2 + 21m - 100$

19. $y^2 - 7yz - 30z^2$

20. $y^2 - 7yz + 12z^2$

21. $a^2 + 17ab + 66b^2$

22. $a^2 + 17ab - 60b^2$

23. $s^2 + 28t^2 - 16st$

24. $s^2 - 48t^2 + 2st$

25. $p^2 + 15pq + 50q^2$

26. $p^2 + 20pq + 64q^2$

[Example 4]   27. $3x^2 + 24x + 36$

28. $4x^2 + 44x + 40$

29. $2z^2 - 20z - 78$

30. $4z^2 - 40z + 64$

31. $5a^2 + 40ab - 240b^2$

32. $3a^2 - 48ab + 144b^2$

33. $10m^2 - 640 - 120m$

34. $12m^2 + 252 - 120m$

[Example 5]   35. $2x^3 - 2x^2 - 112x$

36. $3x^3 + 3x^2 - 270x$

37. $-3y^4 + 15y^3 - 12y^2$

38. $-2x^6 - 12x^5 - 16x^4$

39. $2u^3v + 2u^2v^2 - 4uv^3$

40. $3u^3v + 6u^2v^2 - 45uv^3$

41. $-4x^4y^2 + 40x^3y^3 + 96x^2y^4$

42. $-5x^5y^3 + 35x^4y^4 - 50x^3y^5$

In exercises **43–50**, determine whether the polynomial can be factored and then write it in factored form. If for a given value it cannot be factored, write "Cannot be factored."

[Example 6]   43. $x^2 + kx + 12$
   a. $k = 7$
   b. $k = 8$
   c. $k = 10$

44. $x^2 + kx + 15$
   a. $k = 8$
   b. $k = 10$
   c. $k = 16$

45. $t^2 + kt - 20$
   a. $k = -1$
   b. $k = -2$
   c. $k = -19$

46. $t^2 + kt - 30$
   a. $k = -7$
   b. $k = -13$
   c. $k = -11$

47. $a^2 + ka - 42$
   a. $k = 19$
   b. $k = -41$
   c. $k = 0$

48. $a^2 + ka - 72$
   a. $k = 14$
   b. $k = -71$
   c. $k = 0$

49. $b^2 + kb + 99$
   a. $k = 20$
   b. $k = -20$
   c. $k = -2$

50. $b^2 + kb + 64$
   a. $k = 16$
   b. $k = -16$
   c. $k = 63$

SECTION 4-2.   Ten Review Exercises

In exercises **1–4**, use the rule for order of operations to simplify each expression.

1. $2^2 + \dfrac{36}{3} - 3^3$

2. $1 - 2^3 + 7 \cdot 3^2 - 5 \cdot 2$

3. $\dfrac{-2(13 - 2 \cdot 5) + 3(19 - 7 \cdot 2)}{3^2 - 3 \cdot 2}$ 4. $-3[2(15 - 2^3) + (13 - 5^2)]$

In exercises **5** and **6**,

a. use the distributive property to rewrite each expression.

b. simplify the expression obtained in part **a.**

5. $112\left(\dfrac{5}{8} - \dfrac{2}{7}\right)$ 6. $\dfrac{5}{2}(24 - 16)$

In exercises **7–10**, simplify.

7. $(2u^3vw^2)^3$ 8. $\left(\dfrac{-u^2v}{2}\right)^2 \left(\dfrac{-uv}{3}\right)^3$

9. $(-3a^2b)(2ab^2)^2$ 10. $(-a^3b^2)^2(3ab^2)(2a^2b)$

SECTION 4-2. Supplementary Exercises

In exercises **1–12**, factor if possible. If the trinomial can not be factored, write "Cannot be factored."

1. $3x + x^2 - 40$ 2. $63 + y^2 - 16y$

3. $y^2 + 3y + 10$ 4. $y^2 - 5y - 15$

5. $5x^2 - 65xy - 150y^2$ 6. $4a^2 - 40ab + 64b^2$

7. $6z^3 - 42z^2 + 36z$ 8. $5z^3 + 45z^2 + 100z$

9. $4b^3 - 56b^2 - 288b$ 10. $6b^3 + 138b^2 + 792b$

11. $r^2 - 14s^2 - 13rs$ 12. $18pq + p^2 + 65q^2$

In exercises **13–20**, find all integer values of k that will make it possible to write the trinomial as a product of two binomials. In each case, identify the value of k and write the resulting trinomial in factored form.

13. $x^2 + kx + 12$ 14. $x^2 + kx - 12$

15. $y^2 + ky - 9$ 16. $y^2 + ky - 25$

17. $z^2 + kz - 33$ 18. $z^2 + kz - 55$

19. $a^2 + ka + 4$ 20. $a^2 + ka + 9$

To factor $t^4 + 9t^2 + 20$, temporarily replace t^2 by k and t^4 by k^2. Therefore, $t^4 + 9t^2 + 20$ becomes $k^2 + 9k + 20$. This trinomial $k^2 + 9k + 20$ can be factored as $(k + 5)(k + 4)$. Replacing k by t^2, $t^4 + 9t^2 + 20$ can be written $(t^2 + 5)(t^2 + 4)$.

In exercises **21–24**, factor.

21. $t^4 + 18t^2 + 77$; replace t^2 by k.   22. $t^4 + 8t^2 + 15$; replace t^2 by k.

23. $z^6 - 7z^3 + 10$; replace z^3 by k.   24. $z^6 - 13z^3 + 42$; replace z^3 by k.

SECTION 4-3. Factoring by Grouping Binomials

**KEY TOPICS
IN THIS SECTION**

1. Another look at FOIL

2. Changing the order to find the common binomial factor

3. Factoring a negative common factor

The trinomials in examples **a** and **b** are not simple because the leading coefficients are not one. In example **a** the leading coefficient is 6, and in example **b** it is 5.

a. $6t^2 + 19t + 15$ **b.** $5k^2 + 17k - 12$

These trinomials are called **general**. One technique for factoring them is **factoring by grouping.** We will study this method in this section, and then we will apply it to general trinomials in the next section.

Another Look at FOIL

To show why the factor-by-grouping technique works, let's first look closely at the product of two binomials obtained by using FOIL.

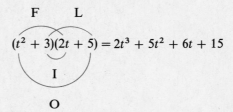

$$F \qquad L$$
$$(t^2 + 3)(2t + 5) = 2t^3 + 5t^2 + 6t + 15$$
$$I$$
$$O$$

The multiplication yielded four terms. Suppose now that the task is to put the four terms back in the factored form; that is, suppose we reverse the multiplication that was performed on the two binomials.

Taken as a group of four terms, the polynomial has no common factor. However, when taken as two groups of two terms each, both pairs have common factors.

Write $2t^3 + 5t^2 + 6t + 15$ as $(2t^3 + 5t^2) + (6t + 15)$.

The first binomial has a common factor t^2, and the second one has a common factor of 3.

$(2t^3 + 5t^2) + (6t + 15)$ The grouped binomials

$= t^2(2t + 5) + 3(2t + 5)$ Remove the common factors.

common factor

Notice that $(2t + 5)$ is now a common factor of the two terms. This binomial can also be removed as a common factor.

$t^2(2t + 5) + 3(2t + 5)$ $(2t + 5)$ is a common factor.

$= (t^2 + 3)(2t + 5)$ Remove the common factor.

It is sometimes difficult to see the binomial $(2t + 5)$ as being a common factor. To help see the process, we can temporarily replace $(2t + 5)$ with a single variable, such as k. The common factor k can then be removed, and as a final step, the variable k is replaced with $(2t + 5)$.

$$t^2(2t + 5) + 3(2t + 5)$$

becomes $\quad t^2 \cdot k + 3 \cdot k \qquad\qquad$ Replace $(2t + 5)$ with k.

$\qquad\quad = (t^2 + 3)k \qquad\qquad$ Factor the k.

becomes $\quad (t^2 + 3)(2t + 5) \qquad\quad$ Replace k with $(2t + 5)$.

Example 1. Factor: $14xy + 63x + 6y + 27$

Solution. **Discussion.** Taken in the order in which the polynomial is written, the first two terms have a common factor of $7x$ and the second pair has a common factor of 3.

$\quad (14xy + 63x) + (6y + 27) \qquad$ Grouped as two binomials

$= 7x(2y + 9) + 3(2y + 9) \qquad$ Remove the common factors.

$= 7xk + 3k \qquad\qquad\qquad$ Replace $(2y + 9)$ with k.

$= (7x + 3)k \qquad\qquad\qquad$ Factor the k.

$= (7x + 3)(2y + 9) \qquad\qquad$ Replace k with $(2y + 9)$.

Thus, $14xy + 63x + 6y + 27 = (7x + 3)(2y + 9)$.

Changing the Order to Find the Common Binomial Factor

The order in which the terms of a polynomial are given may have to be changed to factor it by grouping.

Example 2. Factor: $8x^2 + 3a + 2ax + 12x$

Solution. **Discussion.** As written, the first two terms $(8x^2 + 3a)$ have no common factor. If we change the order of the two middle terms, the first two terms will have a common factor, $2x$.

$\quad 8x^2 + 3a + 2ax + 12x$

$= 8x^2 + 2ax + 3a + 12x \qquad$ Interchange middle terms.

$= 2x(4x + a) + 3(a + 4x) \qquad$ Factor the $2x$ and 3.

$= 2x(4x + a) + 3(4x + a) \qquad$ Write $a + 4x$ as $4x + a$.

$= 2x \cdot k + 3 \cdot k \qquad\qquad\quad$ Replace $(4x + a)$ with k.

$= (2x + 3)k \qquad\qquad\qquad$ Factor the k.

$= (2x + 3)(4x + a) \qquad\qquad$ Replace k with $(4x + a)$.

Thus, $8x^2 + 3a + 2ax + 12x = (2x + 3)(4x + a)$.

Factoring a Negative Common Factor

The factor-by-grouping technique requires that the same binomial be obtained when the common factors are removed from the grouped terms. With this goal in mind, we sometimes must decide whether to remove a positive or negative factor from one of the grouped pairs. To illustrate, consider the following:

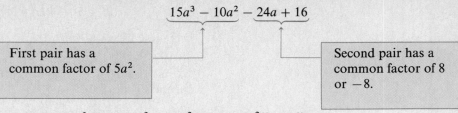

Factoring a $5a^2$ from $15a^3 - 10a^2$, we get $5a^2(3a - 2)$.
Factoring an 8 from $-24a + 16$, we get $8(-3a + 2)$.
Factoring a -8 from $-24a + 16$, we get $-8(3a - 2)$.

When we factor the -8, we get the same binomial that results from factoring a $5a^2$ from the first pair.

$$15a^3 - 10a^2 - 24a + 16$$

$$= 5a^2(3a - 2) - 8(3a - 2) \qquad \text{Remove the common factors.}$$

$$= (5a^2 - 8)(3a - 2) \qquad \text{Factor the } (3a - 2).$$

Example 3.   Factor: $4m^3 + 3m^2n - 8mn - 6n^2$

Solution.   **Discussion.** The first two terms have a common factor of m^2. The last two terms have a common factor of $2n$ or $-2n$. To get the same binomial, factor the $-2n$.

$$4m^3 + 3m^2n - 8mn - 6n^2$$

$$= m^2(4m + 3n) - 2n(4m + 3n) \qquad \text{Remove the common factors.}$$

$$= (m^2 - 2n)(4m + 3n) \qquad \text{Factor the } (4m + 3n).$$

Thus, $4m^3 + 3m^2n - 8mn - 6n^2 = (m^2 - 2n)(4m + 3n)$.

Example 4.   Factor: $8z^4 + 30z - 10z^3 - 24z^2$

Solution.   **Discussion.** When factoring any polynomial, two general principles should always be applied:

1. Look for a common factor first.

2. Write the polynomial in descending powers of one of the variables.

These principles are illustrated in the given polynomial.

$$8z^4 + 30z - 10z^3 - 24z^2$$

$$= 2z(4z^3 + 15 - 5z^2 - 12z) \qquad \text{A common factor of } 2z$$

$$= 2z(4z^3 - 5z^2 - 12z + 15) \qquad \text{Write in descending powers.}$$

$$= 2z[(4z^3 - 5z^2) + (-12z + 15)] \qquad \text{Group in two binomials.}$$

$$= 2z[z^2(4z - 5) - 3(4z - 5)] \qquad \text{Factor } z^2 \text{ and } -3.$$

$$= 2z(z^2 - 3)(4z - 5) \qquad \text{Factor the } (4z - 5).$$

Thus, $8z^4 + 30z - 10z^3 - 24z^2 = 2z(z^2 - 3)(4z - 5)$.

SECTION 4-3.   Practice Exercises

In exercises **1–32**, factor.

[Example 1]   **1.** $6y^3 + y^2 + 18y + 3$ **2.** $3y^3 + 27y^2 + 5y + 45$

3. $24ab + 42a + 20b + 35$ **4.** $18ab + 30a + 21b + 35$

5. $5x^2y + xy + 20xz + 4z$ **6.** $9x^2y + 63xy + 2xz + 14z$

7. $30t^5 + 48t^3 + 5t^2 + 8$ **8.** $16t^5 + 24t^4 + 2t + 3$

[Example 2]   **9.** $14u^2 + 24v + 16uv + 21u$ **10.** $18u^2 + 5uv + 5v + 18u$

11. $9ac + 40bd + 15bc + 24ad$ **12.** $8ac + 21bd + 56ad + 3bc$

13. $14mn + 28 + 49m + 8n$ **14.** $28 + 2mn + 7m + 8n$

15. $21 + 16xy + 12y + 28x$ **16.** $2xy + 27 + 6x + 9y$

[Example 3]   **17.** $12z^3 - 4z^2 - 9z + 3$ **18.** $10z^3 - 20z^2 - 7z + 14$

19. $16s^2 + 8s - 22st - 11t$ **20.** $3s^2 + 6s - 13st - 26t$

21. $32ac - 8ad - 12bc + 3bd$ **22.** $24ac - 16ad - 3bc + 2bd$

23. $3r^2 + 21rt - 5r - 35t$ **24.** $30r^2 + 25rt - 36r - 30t$

[Example 4]   **25.** $180x^4 + 12x^3 + 90x^2 + 6x$ **26.** $18x^5 + 36x^4 + 10x^3 + 20x^2$

27. $16x^2y^2 - 8xy^2 + 48x^2y - 24xy$ **28.** $24x^2y^2 + 24x^2y - 48xy^2 - 48xy$

29. $96a^2b^2c^2 + 128a^2bc^2 + 240ab^2c^2 + 320abc^2$

30. $72a^2b + 24a^2d + 6abc + 2acd$

31. $2x^2y + 6xy - 14xy^2 - 42y^2$ **32.** $32x^2z + 32xz - 2xyz - 2yz$

SECTION 4-3.   Ten Review Exercises

In exercises **1–5**, solve each problem.

1. Simplify: $4k + 25 - 3(4k - 5)$.

2. Evaluate the expression in exercise **1** for $k = 5$.

3. Solve and check: $4k + 25 = 3(4k - 5)$.

4. Simplify: $4(a^2 + 5) - 7(a^2 - 2) - 3(2a + 6) + 6(a - 3 + 2a^2)$.

5. Evaluate the expression in exercise **4** for $a = 3$.

In exercises **6–10**, do the indicated operations.

6. $(5x + 3)(2x - 5)$ **7.** $(5x + 3)(5x - 3)$

8. $(5x + 3)^2$ **9.** $-3x(2x - 5)^2$

10. $(4x^5 - 4x^4 - x + 1) \div (2x^2 + 1)$

SECTION 4-3. Supplementary Exercises

In exercises **1–10**, factor by grouping. Do not combine the middle terms before factoring.

1. $8x^2 + 24x + x + 3$ **2.** $20x^2 + 15x + 28x + 21$

3. $24y^2 + 56y + 3y + 7$ **4.** $20y^2 + 24y + 25y + 30$

5. $6z^2 + 7z - 6z - 7$ **6.** $14z^2 + 35z - 2z - 5$

7. $5a^2 - 20a - 7a + 28$ **8.** $6a^2 - 8a - 3a + 4$

9. $14a^2 - 56ab - 3ab + 12b^2$ **10.** $6a^2 - 24ab - ab + 4b^2$

In exercises **11–18**, factor by grouping the first three terms and the last three terms.

11. $x^2 + 2xz + 5x + xy + 2yz + 5y$ **12.** $x^2 + xy + 3x + ax + ay + 3a$

13. $x^2 + 3xy + x + 3xz + 9yz + 3z$ **14.** $x^2 + 4xy + 5x + xz + 4yz + 5z$

15. $m^2 + 2mn + 6mp - 4m - 8n - 24p$ **16.** $m^2 + 4mn + 5mp - 5m - 20n - 25p$

17. $4st + 16t^2 + 24t - su - 6u - 4tu$ **18.** $4st + 12t^2 + 8t - 15tu - 10u - 5su$

In exercises **19–22**, answer each part.

19. a. Evaluate $2t^3 + 7t^2 - 16t - 56$ for $t = -3$.
   b. Factor: $2t^3 + 7t^2 - 16t - 56$
   c. Evaluate the factored form of part **b** for $t = -3$.

20. a. Evaluate $15t^3 + 27t^2 - 40t - 72$ for $t = -2$.
   b. Factor: $15t^3 + 27t^2 - 40t - 72$
   c. Evaluate the factored form of part **b** for $t = -2$.

21. a. Evaluate $12ac + 8ad - 9bc - 6bd$ for $a = \frac{1}{2}$, $b = \frac{1}{3}$, $c = \frac{-2}{3}$, and $d = \frac{-1}{2}$.
   b. Factor: $12ac + 8ad - 9bc - 6bd$
   c. Evaluate the factored form of part **b** for $a = \frac{1}{2}$, $b = \frac{1}{3}$, $c = \frac{-2}{3}$, and $d = \frac{-1}{2}$.

22. a. Evaluate $10ac - 25ad + 4bc - 10bd$ for $a = \frac{3}{5}$, $b = \frac{-1}{2}$, $c = \frac{-3}{2}$, $d = \frac{2}{5}$.
   b. Factor: $10ac - 25ad + 4bc - 10bd$
   c. Evaluate the factored form of part **b** for $a = \frac{3}{5}$, $b = \frac{-1}{2}$, $c = \frac{-3}{2}$, and $d = \frac{2}{5}$.

SECTION 4-4. Factoring a General Trinomial

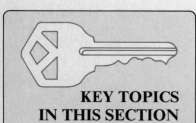

KEY TOPICS IN THIS SECTION

1. A definition of a general trinomial

2. Factoring a general trinomial by the grouping method

3. Using the product of a and c to find the replacement for b

4. General trinomials with a common factor

5. The trial-and-error method for factoring general trinomials

The grouping technique for factoring polynomials of four terms can also be used to factor trinomials. To use this method, we must rewrite the trinomial as an equivalent expression of four terms. The steps that can be used to change the trinomial to four terms will be studied in this section.

A Definition of a General Trinomial

Any trinomial of degree two in x can be written in the general form

$$ax^2 + bx + c$$

where a, b, and c are nonzero numbers. If the leading coefficient a is not 1, then the trinomial is called **general.**

> **Definition 4.2. A general trinomial in x**
>   A general trinomial of degree two in x is one that can be written in the form
>
> $$ax^2 + bx + c$$
>
> where a, b, and c are nonzero numbers, and $a \neq 1$.

Factoring a General Trinomial by the Grouping Method

Examples **a** and **b** were given in Section 4-3 to illustrate two general trinomials.

a. $6t^2 + 19t + 15$ **b.** $5k^2 + 17k - 12$

Let's focus our attention on the trinomial in example **a**. Suppose we wrote $19t$, the middle term, as follows:

$$9t + 10t$$

When $19t$ is replaced with $9t + 10t$, the trinomial is changed to a polynomial of four terms.

As a trinomial	As a polynomial of four terms
$6t^2 + 19t + 15$	$6t^2 + 9t + 10t + 15$

Now that we have four terms, the grouping method can be used to attempt to factor the trinomial.

$$6t^2 + 9t + 10t + 15 \qquad \text{Four-term polynomial}$$

$$= 3t(2t + 3) + 5(2t + 3) \qquad \text{Remove the common factors.}$$

$$= (3t + 5)(2t + 3) \qquad \text{Factor the } (2t + 3).$$

Thus, $6t^2 + 19t + 15 = (3t + 5)(2t + 3)$.

   There is a way of using the product of the leading coefficient and the constant term in the trinomial to determine the terms with which to replace the middle term of the trinomial. This technique will be shown later in this section.

Example 1.   Factor $5k^2 + 17k - 12$ by writing $17k$ as $20k - 3k$.

Solution.   **Discussion.** When $17k$ is written as $20k - 3k$, the following four-term polynomial is obtained:

$$5k^2 + 20k - 3k - 12$$

The common factor in the second pair of terms is either 3 or -3. The -3 is factored to obtain the same binomial that we get when $5k$ is factored from the first pair of terms.

$$5k^2 + 20k - 3k - 12$$

$$= 5k(k + 4) - 3(k + 4) \qquad \text{Remove the common factors.}$$

$$= (5k - 3)(k + 4) \qquad \text{Factor the } (k + 4).$$

Thus, $5k^2 + 17k - 12 = (5k - 3)(k + 4)$.

Using the Product of a and c to Find the Replacement for b

In Example **1**, the trinomial is

$$5k^2 + 17k - 12.$$

Therefore, $a = 5$, $b = 17$, and $c = -12$. In the example, $17k$ was replaced by

$$20k - 3k. \qquad \boxed{17 = 20 + (-3)}$$

Note the following:

1. $20(-3) = -60$, and -60 is the product of a and c.

2. $20 + (-3) = 17$, and 17 is the value of b.

In general, the product of a (the leading coefficient), and c (the constant term) can be used to find the terms with which to replace the middle term of the trinomial. The details are given in the following four-step procedure.

To factor $ax^2 + bx + c$ by the grouping method,

Step 1. Form the product of a and c.

Step 2. If they exist, find integers p and q such that
   i. $p \cdot q = a \cdot c$ and   **ii.** $p + q = b$.

Step 3. Write $ax^2 + bx + c$ as $ax^2 + px + qx + c$.

Step 4. Factor the four-term polynomial by the grouping method.

Example 2.   Factor: $4y^2 - 16y + 15$

Solution.   **Discussion.** With $a = 4$ and $c = 15$, $a \cdot c = 4 \cdot 15 = 60$. The product is positive; therefore p and q are either both positive or both negative. Since $b = -16$, and $p + q$ must equal b, we may conclude that p and q are both negative integers.

Step 1. With $a = 4$ and $c = 15$

$$a \cdot c = 4 \cdot 15 = 60 \qquad \text{Therefore, } p \cdot q = 60 \text{ also.}$$

Step 2. We now consider **negative integers** that are factors of 60.

$p \cdot q = 60$	Corresponding $p + q$
$(-1)(-60) = 60$	$-1 + (-60) = -61$, reject
$(-2)(-30) = 60$	$-2 + (-30) = -32$, reject
$(-3)(-20) = 60$	$-3 + (-20) = -23$, reject
$(-4)(-15) = 60$	$-4 + (-15) = -19$, reject
$(-5)(-12) = 60$	$-5 + (-12) = -17$, reject
$(-6)(-10) = 60$	$-6 + (-10) = -16$, accept

Step 3. Write $4y^2 - 16y + 15$ as $4y^2 - 6y - 10y + 15$.

Step 4. $4y^2 - 6y - 10y + 15$

$= 2y(2y - 3) - 5(2y - 3)$ Remove common factors.

$= (2y - 5)(2y - 3)$ Factor the $(2y - 3)$.

Thus, $4y^2 - 16y + 15 = (2y - 5)(2y - 3)$.

If instead of $-6y - 10y$ we had replaced $-16y$ with $-10y - 6y$, the results would have been the same.

$4y^2 - 10y - 6y + 15$

$= 2y(2y - 5) - 3(2y - 5)$ Remove the common factors.

$= (2y - 3)(2y - 5)$ Factor the $(2y - 5)$.

With practice, it will not be necessary to check all the possible factors of ac to find the appropriate integers p and q. In Example 2, $p \cdot q = -60$ and $p + q = -15$. Since we determined that p and q were both negative, we may conclude that in absolute value both p and q must be less than 15. As a consequence, the first four choices for p and q in Step **2** of the last example were unnecessary.

Example 3.   Factor: $10m^2 - 17mn - 20n^2$

Solution.   **Discussion.** With $a = 10$ and $c = -20$, $a \cdot c = 10(-20) = -200$. The product is negative; therefore p and q are opposite in signs. Since $b = -17$, the negative integer in absolute value is larger than the positive integer. We might therefore look for negative factors of -200 near -20 or less.

Step 1. As indicated previously, $a \cdot c = -200$.

Step 2.

$p \cdot q = -200$	Corresponding $p + q$
$-20 \cdot 10 = -200$	$-20 + 10 = -10$, reject
$-25 \cdot 8 = -200$	$-25 + 8 = -17$, accept

Step 3. Write $10m^2 - 17mn - 20n^2$ as

$10m^2 - 25mn + 8mn - 20n^2$.

Step 4. $10m^2 - 25mn + 8mn - 20n^2$

$= 5m(2m - 5n) + 4n(2m - 5n)$

$= (5m + 4n)(2m - 5n)$

Thus, $10m^2 - 17mn - 20n^2 = (5m + 4n)(2m - 5n)$.

General Trinomials with a Common Factor

Even when attempting to factor a general trinomial, be sure first to check whether the trinomial has a common factor.

Example 4.   Factor: $35z^3 - 20z^2 + 10z^4$

Solution.   **Discussion.** For this trinomial, notice two things:

1. Each term has a factor $5z^2$.

2. The trinomial is not written in descending powers of z.

Before beginning the four-step procedure, remove the common factor and write the resulting trinomial in descending powers.

$$35z^3 - 20^2 + 10z^4 = 5z^2(2z^2 + 7z - 4)$$

Step 1. With $a = 2$ and $c = -4$, $a \cdot c = 2(-4) = -8$.

Step 2. $p \cdot q = -8$ **Corresponding $p + q$**

$-1 \cdot 8 = -8$ $-1 + 8 = 7$, accept

Step 3. Write $5z^2(2z^2 + 7z - 4)$ as

$$5z^2(2z^2 - z + 8z - 4).$$

Step 4. $5z^2(2z^2 - z + 8z - 4)$

$$= 5z^2[z(2z - 1) + 4(2z - 1)]$$

$$= 5z^2(z + 4)(2z - 1)$$

Thus, $35z^3 - 20z^2 + 10z^4 = 5z^2(z + 4)(2z - 1)$.

The Trial-and-Error Method for Factoring General Trinomials

If a general trinomial can be factored using only integers, then the grouping method can be used to factor it. However, for some trinomials, the trial-and-error method may be more efficient. This is particularly true for trinomials in which the number of possible factors of a (the leading coefficient) and c (the constant term) are few.

To factor $ax^2 + bx + c$ by the trial-and-error method,

Step 1. List the possible factors of a.

Step 2. List the possible factors of c.

Step 3. Using the possible factors of a and c, find the correct combination (if one exists) that yields b, the coefficient of the x term.

Example 5.   Factor: $3x^2 - 8x + 5$

Solution.   **Discussion.** Notice that both 3 and 5 are prime numbers. As a consequence, the only possible factors for both are 1 and the numbers themselves. Also notice that the signs in the trinomial indicate that the signs in the factored form are both minus.

Step 1. The factors of 3 are 1 and 3.

Step 2. The factors of 5 are 1 and 5.

Step 3. Use FOIL to determine the correct combinations.

$$3x^2 \qquad 5 \qquad\qquad 3x^2 \qquad 5$$

$$(3x - 1)(x - 5) \qquad (3x - 5)(x - 1)$$

$$-x \qquad\qquad -5x$$

$$\frac{-15x}{-16x, \text{ reject}} \qquad \frac{-3x}{-8x, \text{ accept}}$$

Thus, $3x^2 - 8x + 5 = (3x - 5)(x - 1)$.

Example 6. Factor: $6 - 7t^2 - 11t$

Solution. **Discussion.** The given trinomial is not written in descending powers of t. Furthermore, when the trinomial is written in descending powers, the leading coefficient will be -7, a *negative number*. In general, it is easier to factor trinomials with positive leading coefficients. To obtain this goal we first factor a (-1) from the trinomial.

$$6 - 7t^2 - 11t$$

$$= -1 \cdot (7t^2 + 11t - 6)$$

Step 1. The factors of 7 are 1 and 7.

Step 2. The factors of 6 are 1 and 6, and 2 and 3.

Step 3. Using trial-and-error, the following pair of factors are obtained:

$$7t^2 \qquad -6$$

$$-1 \cdot (7t - 3)(t + 2)$$

$$-3t$$

$$\frac{+14t}{+11t, \text{ accept}}$$

Thus, $6 - 7t^2 - 11t = -1 \cdot (7t - 3)(t + 2)$.

SECTION 4-4. Practice Exercises

In exercises **1–40**, factor completely.

[Example 1] **1.** $12x^2 + 31x + 7$ (Write $31x$ as $3x + 28x$.)

2. $5x^2 + 36x + 7$ (Write $36x$ as $x + 35x$.)

3. $8y^2 - 26y + 15$ (Write $-26y$ as $-6y - 20y$.)

4. $12y^2 - 29y + 15$ (Write $-29y$ as $-20y - 9y$.)

5. $20r^2 - 41r - 9$ (Write $-41r$ as $-45r + 4r$.)

6. $8r^2 - 22r - 63$ (Write $-22r$ as $-36r + 14r$.)

7. $15s^2 + 8s - 63$ (Write $8s$ as $-27s + 35s$.)

8. $15s^2 - 32s - 7$ (Write $-32s$ as $3s - 35s$.)

[Example 2]   **9.** $3a^2 - 11a + 10$ **10.** $2a^2 - 9a + 10$

11. $3b^2 - 7b + 2$ **12.** $4b^2 - 27b + 18$

13. $5t^2 - 36t + 7$ **14.** $2t^2 - 15t + 7$

15. $6u^2 - 11u + 3$ **16.** $9u^2 - 18u + 5$

[Example 3]   **17.** $6m^2 + m - 35$ **18.** $5m^2 - 7m - 6$

19. $3n^2 + 7n - 10$ **20.** $5n^2 + 28n - 12$

21. $30w^2 - 19w - 4$ **22.** $8w^2 - 26w - 15$

23. $4y^2 - 4y - 15$ **24.** $5y^2 - 6y - 8$

[Example 4]   **25.** $12a^2 + 22a - 20$ **26.** $12a^2 - 75a + 18$

27. $18b^3 - 21b^2 - 15b$ **28.** $16b^3 + 52b^2 + 30b$

29. $-15y^3 - 36y^2 + 6y^4$ **30.** $15y^3 + 40y - 50y^2$

31. $12x^5 - 18x^3 - 6x^4$ **32.** $-15x^3 + 9x^2 + 6x^4$

33. $6u^2 + uv - 2v^2$ **34.** $5u^2 - 21uv + 4v^2$

35. $12a^2 + 5ab - 2b^2$ **36.** $10a^2 - 11ab + 3b^2$

37. $20r^2 + s^2 + 9rs$ **38.** $18r^2 + s^2 - 9rs$

39. $3xy - 54y^2 + 2x^2$ **40.** $7xy - 40y^2 + 3x^2$

In exercises **41–50**, factor by trial-and-error method.

[Examples 5 and 6]   **41.** $2t^2 - 11t - 6$ **42.** $3t^2 + 11t - 4$

43. $10u^2 - u - 2$ **44.** $12u^2 - 16u - 3$

45. $6v^2 + 19v + 10$ **46.** $8v^2 + 14v + 3$

47. $15y^2 + 2 - 11y$ **48.** $14y^2 + 6 - 25y$

49. $3a^3b + 7a^2b^2 - 10ab^3$ **50.** $5a^3b - 12ab^3 + 28a^2b^2$

SECTION 4-4.   Ten Review Exercises

In exercises **1–10**, solve each problem.

1. Evaluate for $t = 5$: $5(t - 2) - 2 + 3(t + 20)$

2. Simplify: $5(t - 2) - 2 + 3(t + 20)$

3. Evaluate the expression obtained in exercise **2** for $t = 5$.

4. Solve: $5(t - 2) = 2 - 3(t + 20)$

5. Solve and graph: $5(t - 2) + 3(t + 20) < 2$

6. Multiply: $(8k + 3)(3k - 5)$

7. Multiply: $4k(6k - 8)$

8. Simplify: $(8k + 3)(3k - 5) - 4k(6k - 8)$

9. Evaluate the expression in exercise **8** for $k = 10$.

10. Solve: $(8k + 3)(3k - 5) = 4k(6k - 8)$

SECTION 4-4. Supplementary Exercises

In exercises **1–10**, factor completely.

1. $4y^2 + 25y - 56$ **2.** $6y^2 + y - 12$

3. $10a^2 - 51a + 27$ **4.** $5a^2 - 36a + 36$

5. $5b^2 + 44b - 60$ **6.** $5b^2 + 23b - 42$

7. $4t^2 + 8t - 45$ **8.** $2t^2 - 13t - 24$

9. $13mn - 12n^2 + 4m^2$ **10.** $6m^2 - 10n^2 - 7mn$

In exercises **11–14**, answer parts **a–d** for each trinomial written in the form $ax^2 + bx + c$.

a. Compute $a \cdot c$.

b. List all the pairs of integer factors of $a \cdot c$.

c. List the sum of the pairs found in part **b**.

d. Based on the answers to parts **b** and **c**, what conclusion can be made regarding the factorability of the trinomial using only integers?

11. $3x^2 - 9x + 8$ **12.** $4x^2 - 7x + 9$

13. $7a^2 + 7a - 2$ **14.** $6a^2 + 5a - 3$

Suppose $3x^2 + 17x + 10$ represents the area of a rectangle. The area can be written as a product of the length and width as shown in the following figure:

	c	d
a	$3x^2$	$15x$
b	$2x$	10

$3x^2 + 15x + 2x + 10$

$= 3x^2 + 17x + 10$

In the diagram, let a, b, c, and d represent the lengths and widths of the four rectangles as shown. The following products can be written

$$a \cdot c = 3x^2 \qquad a \cdot d = 15x \qquad b \cdot c = 2x \qquad b \cdot d = 10$$

Notice $3x$ is the common factor of Row 1; thus $a = 3x$.

2 is the common factor of Row 2; thus $b = 2$.

x is the common factor of Column 1; thus $c = x$.

5 is the common factor of Column 2; thus $d = 5$.

The rectangle can be displayed as follows:

	x	5
$3x$	$3x^2$	$15x$
2	$2x$	10

As a consequence,

$3x^2 + 17x + 10$

$= (3x + 2)(x + 5)$

In exercises **15–20**,

a. State the values of a, b, c, and d.

b. Write the trinomial in factored form.

15. $2x^2 + 15x + 18$

	c	d
a	$2x^2$	$12x$
b	$3x$	18

16. $3x^2 + 13x + 4$

	c	d
a	$3x^2$	x
b	$12x$	4

17. $3x^2 + 16x + 20$

	c	d
a	$3x^2$	$10x$
b	$6x$	20

18. $3x^2 + 29x + 56$

	c	d
a	$3x^2$	$21x$
b	$8x$	56

19. $2x^2 + 15x + 28$

	c	d
a	$2x^2$	$8x$
b	$7x$	28

20. $3x^2 + 20x + 32$

	c	d
a	$3x^2$	$8x$
b	$12x$	32

SECTION 4-5.   Special Factoring Forms

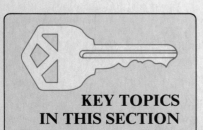

**KEY TOPICS
IN THIS SECTION**

1. Factoring a difference of two squares

2. Factoring a perfect square trinomial

3. Factoring a sum or difference of cubes

Some polynomials possess special characteristics that considerably reduce the work needed to factor them. The challenge is to know specifically what characteristics are needed for a polynomial to be classified "special." The task is then knowing how to apply the **general form** to the given polynomial to factor it. In this section we will study how this can be done.

Factoring a Difference of Two Squares

The binomials in examples **a** and **b** have the form called the **sum and difference of terms.**

a. $(5t + 9)(5t - 9)$ **b.** $(3m + 2n)(3m - 2n)$

The sum and difference of 5t and 9   **The sum and difference of 3m and 2n**

When binomials with this form are multiplied using FOIL, the O and I terms always combine to zero. The simplified product is a binomial that is **the difference of the squares of the terms.**

a. $(5t + 9)(5t - 9)$ **b.** $(3m + 2n)(3m - 2n)$

$= (5t)^2 - 9^2$ $= (3m)^2 - (2n)^2$

$= 25t^2 - 81$ $= 9m^2 - 4n^2$

If we are faced with the task of factoring a binomial that can be written as a difference of two squares, we simply reverse the steps used above.

To factor a binomial of the form $x^2 - y^2$

$$x^2 - y^2 = (x + y)(x - y)$$

As examples:

$$t^2 - 9 = (t + 3)(t - 3)$$

$$4q^2 - 1 = (2q + 1)(2q - 1)$$

Example 1.   Factor: $16u^2 - 121$

Solution.   **Discussion.** First write the given binomial in the form $x^2 - y^2$. Then apply the factored form in the equation above.

$16u^2 - 121$ The given binomial

$= (4u)^2 - 11^2$ Written in the form $x^2 - y^2$

$= (4u + 11)(4u - 11)$ $x^2 - y^2 = (x + y)(x - y)$

Thus, $16u^2 - 121 = (4u + 11)(4u - 11)$.

To help identify numbers that are squares of integers, the following table lists the squares of integers 1 through 25.

The squares of integers 1 through 25

$1^2 = 1$	$2^2 = 4$	$3^2 = 9$	$4^2 = 16$	$5^2 = 25$
$6^2 = 36$	$7^2 = 49$	$8^2 = 64$	$9^2 = 81$	$10^2 = 100$
$11^2 = 121$	$12^2 = 144$	$13^2 = 169$	$14^2 = 196$	$15^2 = 225$
$16^2 = 256$	$17^2 = 289$	$18^2 = 324$	$19^2 = 361$	$20^2 = 400$
$21^2 = 441$	$22^2 = 484$	$23^2 = 529$	$24^2 = 576$	$25^2 = 625$

Example 2. Factor: $75a^3b - 147ab^3$

Solution. **Discussion.** Checking for a common factor, we see that $3ab$ is a factor of both terms. Removing the common factor yields a binomial that has the form $x^2 - y^2$.

$$75a^3b - 147ab^3 \qquad \text{The given binomial}$$

$$= 3ab(25a^2 - 49b^2) \qquad \text{Remove the common factor.}$$

$$= 3ab[(5a)^2 - (7b)^2] \qquad \text{Written in the form } x^2 - y^2$$

$$= 3ab(5a + 7b)(5a - 7b) \qquad x^2 - y^2 = (x + y)(x - y)$$

Thus, $75a^3b - 147ab^3 = 3ab(5a + 7b)(5a - 7b)$.

Factoring a Perfect Square Trinomial

The binomials in examples **a** and **b** have the forms called **the square of a sum** and **the square of a difference** respectively.

a. $(5t + 9)^2$ **b.** $(3m - 2n)^2$

The square of the sum of 5t and 9.   **The square of the difference of 3m and 2n.**

When binomials with this form are multiplied using FOIL, the O and I terms are always the same and combine to form two times that term. The simplified product is a trinomial that is called a **perfect square trinomial.**

a. $(5t + 9)^2$ **b.** $(3m - 2n)^2$

$$= (5t)^2 + 2(5t)(9) + 9^2 \qquad\qquad = (3m)^2 - 2(3m)(2n) + (2n)^2$$

$$= 25t^2 + 90t + 81 \qquad\qquad\qquad = 9m^2 - 12mn + 4n^2$$

Perfect square trinomial

If we are faced with the task of factoring a trinomial that can be written as a square of a binomial, we simply reverse the steps used above.

To factor a perfect square trinomial

$$x^2 + 2xy + y^2 = (x + y)(x + y) \qquad \text{or} \qquad (x + y)^2$$

$$x^2 - 2xy + y^2 = (x - y)(x - y) \qquad \text{or} \qquad (x - y)^2$$

As examples:

$$t^2 + 8t + 16 = (t + 4)(t + 4) = (t + 4)^2$$

$$4q^2 - 4q + 1 = (2q - 1)(2q - 1) = (2q - 1)^2$$

Example 3. Factor: $16u^2 + 40u + 25$

Solution. **Discussion.** Write the given trinomial in the form $x^2 + 2xy + y^2$. Then apply the factored form given in the equation above.

$$16u^2 + 40u + 25 \qquad \text{The given trinomial}$$

$$= (4u)^2 + 2(4u)(5) + 5^2 \qquad \text{Written in the } x^2 + 2xy + y^2 \text{ form}$$

$$= (4u + 5)^2 \qquad \text{Factored as the square of } (4u + 5)$$

Example 4. Factor: $-200a^2 + 120ab - 18b^2$

Solution. **Discussion.** The given trinomial cannot be written in the form of a perfect square trinomial. However, each term has a common factor of 2 or (-2). To give both the a^2 and b^2 terms positive coefficients, we factor a (-2).

$$-200a^2 + 120ab - 18b^2 \qquad \text{The given trinomial}$$

$$= -2(100a^2 - 60ab + 9b^2) \qquad \text{Remove the common} \\ \text{factor of } -2.$$

$$= -2[(10a)^2 - 2(10a)(3b) + (3b)^2] \qquad \text{Write as } x^2 - 2xy + y^2.$$

$$= -2(10a - 3b)^2 \qquad \text{Factored as } (10a - 3b)^2$$

Thus, $-200a^2 + 120ab - 18b^2 = -2(10a - 3b)^2$.

Factoring a Sum or Difference of Cubes

The products of the polynomials in examples **a** and **b** yield binomials. As will be seen, the binomials are respectively the sum and difference of cubed terms.

a. $(2t + 1)(4t^2 - 2t + 1)$ **b.** $(3m - 2n)(9m^2 + 6mn + 4n^2)$

a. $4t^2 - 2t + 1$
$2t + 1$
$\overline{8t^3 - 4t^2 + 2t}$
$4t^2 - 2t + 1$
$\overline{8t^3 + 1}$
$\downarrow \downarrow$
$(2t)^3$ **sum of cubes** 1^3

b. $9m^2 + 6mn + 4n^2$
$3m - 2n$
$\overline{27m^3 + 18m^2n + 12mn^2}$
$ - 18m^2n - 12mn^2 - 8n^3$
$\overline{27m^3 - 8n^3}$
$\downarrow \downarrow$
$(3m)^3$ **difference of cubes** $(2n)^3$

Suppose we are faced with the task of factoring a binomial that can be written as a sum of cubes (example **a**) or a difference of cubes (example **b**). The factored forms would consist of a binomial and a trinomial. The terms of these polynomials would have the pattern given in the following equations.

To factor a binomial of the form $x^3 + y^3$ or $x^3 - y^3$

$$x^3 + y^3 = (x + y)(x^2 - xy + y^2)$$

$$x^3 - y^3 = (x - y)(x^2 + xy + y^2)$$

As examples:

$$t^3 + 125 = (t)^3 + (5)^3 = (t + 5)(t^2 - 5t + 25)$$

$$q^3 - 64 = (q)^3 - (4)^3 = (q - 4)(q^2 + 4q + 16)$$

To help identify numbers that are cubes of integers, the following table lists the cubes of integers 1 through 10.

The cubes of integers 1 through 10

$1^3 = 1$	$2^3 = 8$	$3^3 = 27$	$4^3 = 64$	$5^3 = 125$
$6^3 = 216$	$7^3 = 343$	$8^3 = 512$	$9^3 = 729$	$10^3 = 1000$

Example 5.   Factor: $216z^3 + 125$

Solution.   **Discussion.** First write the given binomial as the sum of two cubes. Then use the factored form for $x^3 + y^3$ in the equation above.

$216z^3 + 123$	The given binomial
$= (6z)^3 + 5^3$	Written in the $x^3 + y^3$ form
$= (6z + 5)((6z)^2 - (6z)(5) + 5^2)$	$x^3 + y^3 = (x + y)(x^2 - xy + y^2)$
$= (6z + 5)(36z^2 - 30z + 25)$	Simplify the trinomial.

Notice that once we verify the binomial as an $x^3 + y^3$ polynomial, we obtain the factored form by following the pattern in the formula. We know the factored form is correct because if the multiplication is carried out, only the sum of the two cubed terms will be obtained.

Example 6.   Factor: $54u^4v - 16uv^7$

Solution.   **Discussion.** After the common factor is removed, the given binomial has the form $x^3 - y^3$.

$54u^4v - 16uv^7$	The given binomial
$= 2uv(27u^3 - 8v^6)$	Remove the common factor $2uv$.
$= 2uv[(3u)^3 - (2v^2)^3]$	Written in the $x^3 - y^3$ form
$= 2uv(3u - 2v^2)((3u)^2 + (3u)(2v^2) + (2v^2)^2)$	
$= 2uv(3u - 2v^2)(9u^2 + 6uv^2 + 4v^4)$	

In factored form, $54u^4v - 16uv^7 = 2uv(3u - 2v^2)(9u^2 + 6uv^2 + 4v^4)$

SECTION 4-5.   Practice Exercises

In exercises **1–50**, factor completely.

[Example 1]   **1.** $16x^2 - 9$ **2.** $4x^2 - 49$

3. $36y^2 - 1$ **4.** $49y^2 - 1$

5. $81u^2 - 100$ **6.** $25u^2 - 121$

7. $4p^2 - 81$ **8.** $64p^2 - 25$

[Example 2]   **9.** $9a^2 - 16b^2$ **10.** $4a^2 - 25b^2$

11. $50m^2 - 98n^2$ **12.** $72m^2 - 2n^2$

13. $121x^2y - 196y^3$ **14.** $169x^2y^2 - 64y^4$

15. $18p^3q - 162pq^3$ **16.** $50p^3q - 722pq^3$

[Example 3]   **17.** $x^2 + 8x + 16$ **18.** $x^2 + 20x + 100$

19. $y^2 - 14y + 49$ **20.** $y^2 - 12y + 36$

21. $9m^2 + 42m + 49$ **22.** $81m^2 + 36m + 4$

23. $49a^2 - 42a + 9$ **24.** $64a^2 - 80a + 25$

25. $4m^2 - 12mn + 9n^2$ **26.** $25m^2 + 10mn + n^2$

[Example 4] **27.** $4k^2 - 24k + 36$ **28.** $20k^2 - 20k + 5$

29. $24x^2 - 120xy + 150y^2$ **30.** $36x^2 + 24xy + 4y^2$

31. $-10 + 120a - 360a^2$ **32.** $-36 + 180a - 225a^2$

33. $-36u^3v - 132u^2v^2 - 121uv^3$ **34.** $-4u^3v + 28u^2v^2 - 49uv^3$

[Example 5] **35.** $x^3 - 8$ **36.** $125x^3 - 27$

37. $27y^3 + 8$ **38.** $y^3 + 343$

39. $8t^3 - 125$ **40.** $27t^3 - 64$

41. $216w^3 + 1$ **42.** $729w^3 + 1,000$

[Example 6] **43.** $4x^3 - 500$ **44.** $5x^3 - 320$

45. $x^4 + 27xy^3$ **46.** $2y^5 + 432y^2z^3$

47. $16t^4s + 250ts^4$ **48.** $54t^4s^2 + 128ts^5$

49. $40x^6 - 5y^6$ **50.** $250x^9 - 2y^3$

SECTION 4-5. Ten Review Exercises

In exercises **1–7**, solve each problem.

1. Multiply: $(5k^2)(4k)$ **2.** Multiply: $5k^2(k + 4)$

3. Multiply: $(k^2 + 5)(k + 4)$ **4.** Multiply: $(7t)^2$

5. Multiply: $(t + 7)^2$ **6.** Multiply: $(3x^2)(-3x^2)$

7. Multiply: $(x^2 + 3)(x^2 - 3)$

In exercises **8–10**, factor completely.

8. $64y^2 - 1$ **9.** $64y^3 - 1$

10. $64y^2 - 16y + 1$

SECTION 4-5. Supplementary Exercises

In exercises **1–26**, factor completely.

1. $16x^4 - 9$ **2.** $4x^4 - 49$

3. $289u^2 - 34u + 1$ **4.** $225u^2 + 30u + 1$

5. $81 - 100b^2$

6. $225 - 16b^2$

7. $36a^2b^2 - 1$

8. $49a^2b^2 - 100$

9. $a^2 + 10ab + 25b^2$

10. $4a^2 - 4ab + b^2$

11. $81u^6 - 100$

12. $25u^6 - 121$

13. $16v^5 - 40v^4 + 25v^3$

14. $64v^5 + 16v^4 + v^3$

15. $192x^4y - 3x^2y$

16. $245x^3y^2 - 5xy^2$

17. $27m^3 - 125n^3$

18. $8m^3 - 343n^3$

19. $192a^4 + 81a$

20. $686a + 1024a^4$

21. $\dfrac{t^2}{4} - \dfrac{1}{9}$

22. $\dfrac{9t^2}{25} - \dfrac{1}{36}$

23. $\dfrac{u^2}{16} - \dfrac{4v^2}{9}$

24. $\dfrac{4u^2}{49} - \dfrac{v^2}{100}$

25. $0.09w^2 - 0.25$

26. $0.04w^2 - 0.81$

Example. Factor: $16t^4 - 1$

Solution. **Discussion.** Notice that $16t^4$ can be written as $(4t^2)^2$. Thus, $16t^4 - 1$ can be written as $(4t^2)^2 - 1^2$, a difference of two squares.

$$16t^4 - 1$$
$$= (4t^2 + 1)(4t^2 - 1) \qquad x^2 - y^2 = (x + y)(x - y)$$
$$= (4t^2 + 1)(2t + 1)(2t - 1) \qquad \text{Factor } 4t^2 - 1 \text{ as } (2t)^2 - 1^2$$

In exercises **27–34**, factor completely.

27. $k^4 - 81$ (Hint: Factor twice.)

28. $k^4 - 16$ (Hint: Factor twice.)

29. $p^4 - 625$

30. $p^4 - 256$

31. $81q^4 - 16$

32. $625q^4 - 256$

33. $x^4 - y^4$

34. $x^4y^4 - z^4$

In exercises **35–40**, find the values of k that will make each trinomial a perfect square.

35. $9x^2 + kx + 1$

36. $25x^2 + kxy + 36y^2$

37. $4y^2 - 20y + k$

38. $81y^2 + 18yz + kz^2$

39. $kt^2 + 60t + 9$

40. $kt^2 - 130tu + 25u^2$

In exercises **41–46**,

a. Identify the factor or sign that makes the trinomial not a perfect square.

b. Change the factor or sign to make the trinomial a perfect square.

c. Factor the expression obtained in part **b**.

41. $16m^2 - 40m - 25$

42. $9m^2 + 12m - 4$

43. $7n^2 - 140n + 100$

44. $36n^2 + 60n + 5$

45. $121a^2 + 33ab + 9b^2$

46. $225a^2 - 15ab + b^2$

Review Exercises for Sections 4-1 through 4-5

In exercises **1–100**, factor completely.

1. $3x^3 + 15x$

2. $6x^3 - 10x^2$

3. $2a^2b - 2ab^2 + 6ab$

4. $6a^3b + 3a^2b^2 - 3a^2b$

5. $x^2 - 36$

6. $9x^2 - 25$

7. $125y^3 - 1$

8. $8y^3 - 27z^3$

9. $9a^2 + 24a + 16$

10. $25a^2 - 20a + 4$

11. $x^2 - 8xy + 16y^2$

12. $4x^2 - 20xy + 25y^2$

13. $m^2 + 8mn + 15n^2$

14. $m^2 - mn - 2n^2$

15. $k^3 + 7k^2 - 8k - 56$

16. $k^4 - 4k^3 - 6k + 24$

17. $2x^2 - 5x + 3$

18. $3x^2 - 13x - 10$

19. $3y^2 + 2y - 5$

20. $2y^2 - 17y + 8$

21. $6x^2 + 6xy + 30y^2$

22. $x^3y - x^2y^2 - 7xy^3$

23. $100z^2 - 9$

24. $16 - 25z^2$

25. $4a^2 - 28a + 49$

26. $a^2 + 20a + 100$

27. $b^2 + 7b - 18$

28. $b^2 - 11b + 28$

29. $5c^2 + 13c + 6$

30. $6c^2 - 5c + 1$

31. $6m^3 + 750$

32. $12m^3n + 96n^4$

33. $6ab - 45 - 27a + 10b$

34. $14a^2 + 7at - 4t - 8a$

35. $4x^2 + 28xy + 49y^2$

36. $25x^2 - 30xy + 9y^2$

37. $a^2 + 6a - 40$

38. $a^2 - 11a + 30$

39. $6m^2 - 47m - 8$

40. $4m^2 + 11m + 6$

41. $121a^2 - 25b^2$

42. $169a^2 - 4b^2$

43. $8x^2 + 8x + 24$

44. $12x^3 + 6x^2 + 6x$

45. $25y^2 - 60y + 36$

46. $4y^2 + 36y + 81$

47. $3z^2 - 10z - 8$

48. $6z^2 + 13z + 6$

49. $x^2 - 2xy - 15y^2$

50. $x^2 + 9xy + 14y^2$

51. $4a^4 - 256a$

52. $20a^5 + 160a^2$

53. $4a^2 + 12a + 9$

54. $4a^2 - 12a + 9$

55. $4a^2 - 5a - 9$ 56. $4a^2 - 13a + 9$

57. $4a^2 - 16a - 9$ 58. $4a^2 - 9a - 9$

59. $4a^2 + 37a + 9$ 60. $4a^2 - 35a - 9$

61. $18y^3 + 8y - 27y^2 - 12$ 62. $5y^2 - 40 - 8y + y^3$

63. $9y^2 + 15y + 4$ 64. $9y^2 - 9y - 4$

65. $9y^2 + 20y + 4$ 66. $9y^2 - 16y - 4$

67. $9y^2 - 12y + 4$ 68. $9y^2 + 12y + 4$

69. $9y^2 + 5y - 4$ 70. $9y^2 - 37y + 4$

71. $2x^2 - 72y^2z^2$ 72. $12x^2 - 75y^4$

73. $5a^3 + 5{,}000b^3c^3$ 74. $2a^4 + 54ab^6$

75. $3x^2 - 30xy + 75y^2$ 76. $20x^2 + 60xy + 45y^2$

77. $2m^3n + 8m^2n^2 + 8mn^3$ 78. $12m^4 - 36m^3n + 27m^2n^2$

79. $5k^3 + 15k^2 + 10k$ 80. $7k^5 + 14k^4 - 105k^3$

81. $9a^3 + 6a^2b + ab^2$ 82. $4a^3b - 12a^2b^2 + 9ab^3$

83. $36x^3 - 25x$ 84. $25x^3y - 49xy^3$

85. $3y^4 + 3y^3 + 9y^2$ 86. $5x^3y + 5x^2y^2 + 5xy^3$

87. $z^3 - 3z^2 - 70z$ 88. $2z^4 + 5z^3 - 3z^2$

89. $2m^3 - 14m^2 + 20m$ 90. $3m^3 + 9m^2 - 54m$

91. $18k^3 + 24k^2 + 9k + 12$ 92. $75k^3 - 120k^2 + 25k - 40$

93. $80x^3 - 45x$ 94. $12x^4 - 147x^2$

95. $4y^3 + 24y^2 - 28y$ 96. $12y^3 + 4y^2 - 56y$

97. $4z^4 - 22z^3 - 12z^2$ 98. $9z^4 - 33z^3 - 12z^2$

99. $40x^3y - 120x^2y^2 + 90xy^3$ 100. $32x^3y - 72xy^3$

SECTION 4-6.   Solving Quadratic Equations by Factoring

**KEY TOPICS
IN THIS SECTION**

1. A definition of a quadratic equation

2. The zero-product property

3. Solving a quadratic equation using the zero-product property

4. Quadratic equations that do not have real numbers as solutions

Factoring has many uses in mathematics, including solving certain types of **quadratic equations.** This section gives the general form of a quadratic equation, and a property of multiplication that can be used to solve certain types of such equations.

A Definition of a Quadratic Equation

In the introduction to this chapter, Ms. Glaston wrote the following equation on the board:

$$2x^2 + 3x - 20 = 0$$

This equation is not a **linear equation in x** because it does not fit the definition, namely

$$ax + b = c.$$

Equation **3** has a $2x^2$ term, and the linear equation does not have a term of degree two (also called a **squared term**). Equation **3** is an example of equations called **quadratic in x.**

Definition 4.3.   A quadratic equation in x
  A **quadratic equation in x** is one that can be written in the form

$$ax^2 + bx + c = 0 \qquad \text{or} \qquad 0 = ax^2 + bx + c$$

where a, b, and c are real numbers and $a \neq 0$.

The characteristic of the equation in Definition 4.3 that makes it quadratic is the term of degree two (that is, ax^2). Furthermore, this equation has no term of degree more than two. Finally, it may or may not have terms of degree one and zero (that is, the bx and c terms respectively).

The equation in the definition has the polynomial on one side of the equation and a zero on the other side. This is called the **standard form of the equation.** The definition states that quadratic equations "can be written" in this form. Examples **a** and **b** change quadratic equations in t and z to the standard form.

a. $\qquad\qquad t = 40 - 6t^2$ $\qquad$ **b.** $64 = 9z^2$ $\qquad\qquad$ The given equations

$\quad 6t^2 + t - 40 = 0$ $\qquad\qquad\qquad\quad 0 = 9z^2 - 64$ $\qquad\qquad$ In standard form

The Zero-Product Property

To solve a quadratic equation means to find numbers that make the equation true. Such numbers are called **solutions** (or **roots**) of the equation. Recall that Ms. Glaston showed that $\frac{5}{2}$ and -4 are solutions of $2x^2 + 3x - 20 = 0$. In this section we will solve quadratic equations that have only real numbers for solutions. However, we will also look at quadratic equations that do not have real numbers for solutions. In a later chapter we will study numbers that are roots of such equations.

The technique used to solve quadratic equations in this section uses **the zero-product property.**

Zero-product property
  If $x \cdot y = 0$, then $x = 0$ or $y = 0$, or both x and y are 0.

The statement of the zero-product property shows that a product of zero is possible only if at least one of the factors is zero. In other words, *a product of two nonzero numbers can never be zero.*

Example 1.   Find the values of t that make the following equations true.

  a. $5t = 0$

  b. $(t + 2)(t - 3) = 0$

Solution.   **Discussion.** In both **a** and **b** we see an indicated product of two factors that are equal to zero. Based on the zero-product property, at least one of the factors must be zero.

  a. If $5t = 0$, then t must be 0.
  Notice, if $t = 0$, then $5t$ becomes $5(0)$, and $5(0) = 0$.

  Thus, 0 is a solution of the equation.

  b. If $(t + 2)(t - 3) = 0$, then either $t + 2 = 0$ or $t - 3 = 0$.
  Furthermore, if $t + 2 = 0$ then $t = -2$. If $t - 3 = 0$ then $t = 3$.

  Check: If $t = -2$, then $(-2 + 2)(-2 - 3) = 0$

$$0(-5) = 0, \text{ true}$$

  If $t = 3$, then $(3 + 2)(3 - 3) = 0$

$$5(0) = 0, \text{ true}$$

  Thus, -2 and 3 are solutions of the equation.

Solving a Quadratic Equation Using the Zero-Product Property

The following five steps provide a recommended procedure for using the zero-product property to solve a quadratic equation.

To solve a quadratic equation using the zero-product property

Step 1. If necessary, write the given equation in standard form.

Step 2. Factor the polynomial.

Step 3. Use the zero-product property to set both factors equal to zero.

Step 4. Solve both equations obtained in Step 3.

Step 5. Check the roots obtained in Step 4 in the given equation.

Example 2.   Solve and check: $3x + 28 = x^2$

Solution.   **Step 1.** $3x + 28 = x^2$ The given equation

$$0 = x^2 - 3x - 28$$ Subtract $3x$ and 28 from both sides.

  Step 2. $0 = (x - 7)(x + 4)$ Factor the trinomial.

  Step 3. $x - 7 = 0$ or $x + 4 = 0$ Use the zero-product property.

  Step 4. $x = 7$ $x = -4$ Solve both equations.

Step 5. If $x = 7$, then $3(7) + 28 = 7^2$

$$21 + 28 = 49, \text{ true}$$

If $x = -4$, then $3(-4) + 28 = (-4)^2$

$$-12 + 28 = 16, \text{ true}$$

Thus, 7 and -4 are both solutions of the given equation.

Example 3. Solve and check: $3z^2 = 12 + 16z$

Solution.

Step 1. $3z^2 = 12 + 16z$	The given equation
$3z^2 - 16z - 12 = 0$	Subtract 12 and 16z from both sides.
Step 2. $(3z + 2)(z - 6) = 0$	Factor the trinomial.
Step 3. $3z + 2 = 0$ or $z - 6 = 0$	Use the zero-product property.
Step 4. $\quad 3z = -2 \qquad\qquad z = 6$	Solve both equations.
$\qquad z = \dfrac{-2}{3}$	

Step 5. If $z = \dfrac{-2}{3}$, then $3\left(\dfrac{-2}{3}\right)^2 = 12 + 16\left(\dfrac{-2}{3}\right)$

$$\dfrac{4}{3} = \dfrac{36}{3} + \dfrac{-32}{3}, \text{ true}$$

If $z = 6$, then $3(6)^2 = 12 + 16(6)$

$$108 = 12 + 96, \text{ true}$$

Thus, $\dfrac{-2}{3}$ and 6 are both solutions of the given equation.

If the polynomial in a quadratic equation is a perfect square trinomial, then the same root is obtained from the two equations of Step **3**. This single root is frequently called a **double root.**

Example 4. Solve and check: $20a - 4a^2 = 25$

Solution.

Step 1. $20a - 4a^2 = 25$

$$0 = 4a^2 - 20a + 25$$

Step 2. $\qquad 0 = (2a - 5)(2a - 5)$

Step 3. $2a - 5 = 0 \qquad$ or $\qquad 2a - 5 = 0$

Step 4. $\quad 2a = 5 \qquad\qquad\qquad 2a = 5$

$$a = \dfrac{5}{2} \qquad\qquad\qquad a = \dfrac{5}{2}$$

Step 5. The check is omitted. The solution of the equation is the double root $\frac{5}{2}$.

Quadratic Equations that Do Not Have Real Numbers as Solutions

Consider the following quadratic equation:

$$t^2 + 9 = 0$$

The binomial on the left side consists of the sum of two terms, namely the square of some number t and the positive number 9. Since the square of any real number is positive or zero, there does not exist a real number t whose square plus 9 can equal zero. For such an equation we write "No real roots." In a later chapter we will define a set of numbers that will provide solutions for equations such as this one.

SECTION 4-6. Practice Exercises

In exercises **1–46**, solve and check.

[Examples 1–4]

1. $10x = 0$

2. $-4x = 0$

3. $y(y - 3) = 0$

4. $3y(y - 2) = 0$

5. $(z + 1)(z - 5) = 0$

6. $(z + 7)(z - 3) = 0$

7. $(5a - 2)(a + 6) = 0$

8. $(3a - 5)(a - 4) = 0$

9. $(2b + 3)(2b - 1) = 0$

10. $(3b - 2)(4b + 1) = 0$

11. $t^2 - 6t + 8 = 0$

12. $t^2 + 9t + 18 = 0$

13. $u^2 + 3u - 54 = 0$

14. $u^2 - 5u - 50 = 0$

15. $w^2 - w = 56$

16. $37w = -36 - w^2$

17. $k^2 = 100$

18. $k^2 = 49$

19. $m^2 - 7m = 0$

20. $4m^2 + 8m = 0$

21. $n^2 - 27n + 50 = 0$

22. $n^2 - 20n + 64 = 0$

23. $2x^2 - x = 6$

24. $3x^2 = 8x - 4$

25. $0 = 4y^2 - 7y + 3$

26. $0 = 3y^2 - 8y - 3$

27. $3z^2 + 25 = 20z$

28. $3z^2 + 7z = -4$

29. $4m^2 = 13m - 10$

30. $20 = 13m - 2m^2$

31. $9n - 5n^2 = 0$

32. $3n^2 = -7n$

33. $x^2 - 14x + 49 = 0$

34. $x^2 - 16x + 64 = 0$

35. $y^2 + 10y = -25$

36. $y^2 = -12y - 36$

37. $12a + 9 = -4a^2$

38. $30a + 25 = -9a^2$

39. $36b^2 + 1 = 12b$

40. $16b^2 + 49 = 56b$

41. $25 = 4w^2$

42. $100 = 9w^2$

43. $6t^2 = 18t$ **44.** $-4t^2 = -44t$

45. $16x^2 = 1$ **46.** $1 = 49x^2$

SECTION 4-6.   Ten Review Exercises

In exercises **1–10**, solve each problem.

1. Divide $(6t^3 + 29t^2 - 7t - 10)$ by $(3t - 2)$.

2. Divide the quotient of exercise **1** by $(2t + 1)$.

3. Multiply: $(3t - 2)(2t + 1)$.

4. Multiply the product of exercise **3** by $(t + 5)$.

5. Use the results of exercises **3** and **4** to write $(6t^3 + 29t^2 - 7t - 10)$ in completely factored form.

6. Evaluate $6t^3 + 29t^2 - 7t - 10$ for $t = 2$.

7. Evaluate the factored form of exercise **5** for $t = 2$.

8. Evaluate $6t^3 + 29t^2 - 7t - 10$ for $t = -5$.

9. Evaluate the factored form of exercise **5** for $t = -5$.

10. State the solutions of $6t^3 + 29t^2 - 7t - 10 = 0$.

SECTION 4-6.   Supplementary Exercises

In exercises **1–16**, solve and check. If the quadratic does not factor over the integers, write, "No rational number solutions."

1. $21t^2 = 65t - 24$ **2.** $45t^2 = 45 - 56t$

3. $u^2 + 49 = 0$ **4.** $25u^2 + 16 = 0$

5. $3v^2 = 15v$ **6.** $2v^2 = -13v$

7. $100w^2 - 1 = 0$ **8.** $64w^2 - 9 = 0$

9. $x^2 + x - 10 = 0$ **10.** $x^2 - 3x + 8 = 0$

11. $(5x - 4)^2 = 0$ **12.** $(3x + 2)^2 = 0$

13. $16 - 9y^2 = 0$ **14.** $121 - 4y^2 = 0$

15. $91 + 6z - z^2 = 0$ **16.** $27z - 180 - z^2 = 0$

In exercises **17–24**, solve.

17. $(x + 2)(x - 1)(x + 3) = 0$ **18.** $(x - 5)(x + 4)(x - 3) = 0$

19. $3y(y + 6)(y - 7) = 0$ **20.** $-2y(y - 10)(y + 8) = 0$

21. $(3z + 4)(z^2 - 9) = 0$ **22.** $(2z - 3)(4z^2 - 25) = 0$

23. $(t + 4)(t^2 + 100) = 0$ **24.** $(t + 3)(4t^2 + 25) = 0$

In exercises **25–28**, answer parts **a–c**.

a. Factor the left side of the equation.

b. Set each factor equal to zero.

c. Find any rational number roots of the two equations.

25. $x^3 - 8 = 0$ **26.** $x^3 + 27 = 0$

27. $8y^3 + 125 = 0$ **28.** $27y^3 - 1000 = 0$

In exercises **29–32**, solve.

29. $(x + 4)^2 + (x + 5)^2 = (x + 15)(x + 4) - 13$

30. $(x - 1)^2 + (x + 3)^2 = 90 - 2x^2$

31. $y^2 = y(10 - y) - y(y + 8)$

32. $(2y + 1)^2 - (y + 2)^2 = y^2 + 5(y - 1)$

SECTION 4-7.   Applications of Quadratic Equations

**KEY TOPICS
IN THIS SECTION**

1. Solving number problems

2. Solving right-triangle problems

3. Solving areas-of-rectangles problems

4. Solving areas-of-triangles problems

A quadratic equation can frequently be used to solve a problem stated in words. Some problems are related to the physical world, and other problems state relationships between numbers.

In this section, we will solve word problems that require quadratic equations. The general procedure for solving these problems is similar to the one used for solving word problems in Chapter 2. Should a formula be needed to solve a problem, it will be stated in the problem. A listing of many such formulas is given on the inside back cover of the text.

Solving Number Problems

The number problems studied in this section include a statement about two numbers and their products. Quadratic equations are used to solve such problems. These equations can be solved by factoring and using the zero-product property.

Example 1.   The sum of two integers is 15, and their product is 54. Find the numbers.

Solution.   **Discussion.** The sum of the integers is 15. If x and y are used to represent the numbers, then

$$x + y = 15.$$

However, we need an equation in only one variable. Solving the equation for y, we get

$$y = 15 - x.$$

Thus, if x is used to represent one of the numbers, then $15 - x$ can be used to represent the other one. Let x be one of the integers, and let $15 - x$ be the other.

$x(15 - x) = 54$	The product is 54.
$15x - x^2 = 54$	Use the distributive property.
$0 = x^2 - 15x + 54$	Write in standard form.
$0 = (x - 6)(x - 9)$	Factor the trinomial.
$x - 6 = 0 \quad \text{or} \quad x - 9 = 0$	Use the zero-product property.
$x = 6 \quad \text{or} \qquad x = 9$	Solve for x.

Case 1. If $x = 6$, then $15 - x$ becomes $15 - 6 = 9$.

Case 2. If $x = 9$, then $15 - x$ becomes $15 - 9 = 6$.

Thus, the integers are 6 and 9.

Notice that $6 + 9 = 15$ and $6 \cdot 9 = 54$ are checks of the solutions.

Solving Right-Triangle Problems

The Pythagorean theorem is one of the most widely known theorems of geometry. The theorem is a statement about the lengths of the sides of any right triangle.

In Figure 4-1, right triangle ABC is shown. Sides AC and BC, called the **legs of the triangle,** form right angle C. Side AB, called the **hypotenuse,** is opposite the right angle and is the longest side of the triangle. As shown in the figure, the lengths of the legs are a and b, and the length of the hypotenuse is c.

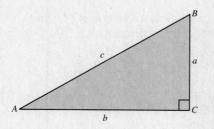

Figure 4-1. Right triangle ABC.

Pythagorean theorem

If a and b are the lengths of the legs of a right triangle, and c is the length of the hypotenuse, then

$$a^2 + b^2 = c^2.$$

Example 2.   In right triangle ABC (Figure 4-2), one leg is 7 meters longer than the other leg. The length of the hypotenuse is 13 meters. Find the lengths of the two legs.

Solution.   **Discussion.** In any problem involving a geometric figure, a figure should be drawn. The given information should then be displayed on the figure.

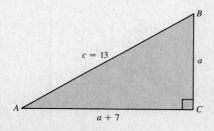

Figure 4-2. Triangle ABC.

Let a be the length of the shorter leg; then $a + 7$ is the length of the longer leg.

$$a^2 + (a + 7)^2 = 13^2 \qquad \text{The Pythagorean theorem}$$

$$a^2 + a^2 + 14a + 49 = 169 \qquad (x + y)^2 = x^2 + 2xy + y^2$$

$$2a^2 + 14a - 120 = 0 \qquad \text{Write in standard form.}$$

$$2(a^2 + 7a - 60) = 0 \qquad \text{A common factor of 2}$$

$$2(a + 12)(a - 5) = 0 \qquad \text{Factor the simple trinomial.}$$

$$2 = 0, \text{false} \qquad \text{or} \qquad a + 12 = 0 \qquad \text{or} \qquad a - 5 = 0$$

$$a = -12, \text{reject} \qquad a = 5$$

The zero-product property asserts that one of the three factors $(2, (a + 12)$, or $(a - 5))$ must equal 0 for the product to be 0. Setting 2 equal to 0 leads to the false statement $2 = 0$. Thus 2 is not a root. Setting $a + 12$ equal to 0 leads to the equation $a = -12$. Since a represents the length of a leg of the triangle, the -12 is rejected. (Lengths are not measured with negative numbers.) The 5 is accepted as a possible root.

If $a = 5$, then $a + 7 = 12$. Thus the sides of the triangle are 5 meters, 12 meters, and 13 meters.

$$5^2 + 12^2 = 13^2$$

$$25 + 144 = 169, \text{true}$$

Solving Areas-of-Rectangles Problems

A rectangle is a four-sided figure in which opposite sides are equal in length and parallel, and the sides meet to form right angles. In Figure 4-3 the longer side of the rectangle is called the **length** (represented by l), and the shorter side is called the **width** (represented by w).

For any rectangle, the area (represented by A) can be found by finding the product of the length and the width.

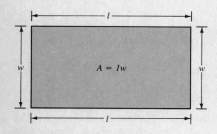

Figure 4-3. Rectangle with length l and width w.

Example 3.   Nancy Peck has a garden shaped like a rectangle. The length is 15 feet more than the width. The area of the garden is 700 square feet. Find the length and width of the garden.

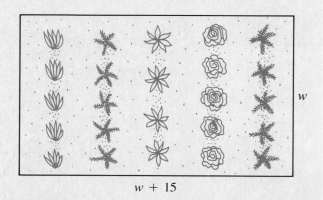

$w + 15$

Figure 4-4. The garden.

Solution.   **Discussion.** The garden is shown in Figure 4-4. If w represents the width, then $w + 15$ represents the length. The area is given as 700 square feet.

$$w(w + 15) = 700 \qquad \text{width} \cdot \text{length} = \text{area}$$

$$w^2 + 15w = 700$$

$$w^2 + 15w - 700 = 0 \qquad \text{Write in standard form.}$$

$$(w + 35)(w - 20) = 0 \qquad \text{Factor the trinomial.}$$

$$w + 35 = 0 \quad \text{or} \quad w - 20 = 0 \qquad \text{The zero-product property}$$

$$w = -35, \text{reject} \qquad w = 20 \qquad \text{Reject the negative root.}$$

If the width is 20 feet, then the length is $20 + 15 = 35$ feet. Notice, (20 feet)(35 feet) = 700 square feet, check.

Solving Areas-of-Triangles Problems

A triangle has three sides. The **area** (represented by A) of a triangle can be computed by knowing the length of one side, called the **base** (represented by b), and the distance from this side to the opposite angle, called the **height** (represented by h). These symbols are shown in Figure 4-5. The area is $\frac{1}{2}$ the product of the base times the height.

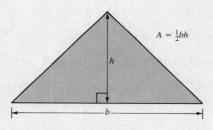

$A = \frac{1}{2}bh$

Figure 4-5. $A = \frac{1}{2}bh$

Example 4.   A table has the shape of a triangle. The distance from a corner of the table to the longest side is one foot less than the length of that side. The surface area of the table is 10 square feet. Find the length of the longest side of the table and the distance from that side to the opposite corner.

Solution.   **Discussion.** A picture of the table is shown in Figure 4-6. The longest side is the base of the triangle and has length b. The distance from that side to the opposite angle is $b - 1$ (that is, one foot less in length).

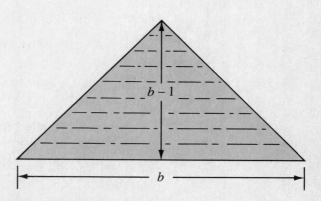

Figure 4-6. The table.

$$\frac{1}{2}b(b-1) = 10 \qquad\qquad \frac{1}{2}bh = A$$

$$b^2 - b = 20 \qquad\qquad \text{Multiply both sides by 2.}$$

$$b^2 - b - 20 = 0 \qquad\qquad \text{Write in standard form.}$$

$$(b-5)(b+4) = 0 \qquad\qquad \text{Factor the trinomial.}$$

$$b - 5 = 0 \quad \text{or} \quad b + 4 = 0 \qquad \text{The zero-product property}$$

$$b = 5 \qquad\qquad b = -4, \text{reject} \qquad \text{Reject the negative root.}$$

If the base is 5 feet, then the height is $5 - 1 = 4$ feet. $\frac{1}{2}(5)(4) = 5(2) = 10$ square feet, check.

SECTION 4-7.   Practice Exercises

In exercises **1–6**, find the numbers.

[Example 1] **1.** The sum of two integers is 17, and their product is 60. Find the numbers.

2. The sum of two integers is -22, and their product is 85. Find the numbers.

3. The sum of two integers is 3, and their product is -70. Find the numbers.

4. The difference between two integers is 5, and their product is 36. Find the numbers.

5. The difference between two integers is -15, and their product is -56. Find the numbers.

6. The difference between two integers is -15, and their product is 54. Find the numbers.

In exercises **7–12**, use the Pythagorean theorem to find the lengths of the legs of right triangle *ABC* with the stated conditions.

[Example 2] **7.** One leg is 2 feet longer than the other leg, and the hypotenuse is 10 feet long.

8. One leg is 14 inches longer than the other leg, and the hypotenuse is 26 inches long.

9. The sum of the lengths of the legs is 23 meters, and the hypotenuse is 17 meters long.

10. The difference in lengths between the longer and shorter legs is 14 inches, and the hypotenuse is 34 inches long.

11. The longer leg is 17 yards longer than the shorter leg, and the hypotenuse is 25 yards long.

12. The longer leg is 6 feet more than 3 times the length of the shorter leg, and the hypotenuse is 50 feet long.

In exercises **13–18**, use the equation $A = lw$ to find the length and width of the rectangles.

[Example 3] **13.** The length of the rectangle is 4 inches more than the width, and the area is 96 square inches.

14. The length of the rectangle is twice the width, and the area is 162 square meters.

15. The length of the rectangle is 1 foot more than twice the width, and the area is 300 square feet.

16. The area is 150 square centimeters, and the width is 5 centimeters shorter than the length.

17. The area is 85 square centimeters, and the length is 2 centimeters more than three times the width of the rectangle.

18. The area is 260 square yards, and the length is 6 yards less than twice the width.

In exercises **19–24**, use the equation $A = \frac{1}{2}bh$ to find the base and height of the triangles.

[Example 4] **19.** The height of the triangle is 3 centimeters less than the base, and the area is 35 square centimeters.

20. The base is 2 feet more than three times the height, and the area is 28 square feet.

21. The height is 3 meters less than the base, and the area is 90 square meters.

22. The area is 200 square yards, and the base is 7 yards less than twice the height.

23. The area is 96 square inches, and the base is three times the height.

24. The area is 12 square miles, and the base is 2 miles more than twice the height.

SECTION 4-7. Ten Review Exercises

In exercises **1–6**, solve and check the roots.

1. $7x + 17 = -23 - 3x$

2. $4(y + 6) + 1 = 3(4y - 3) - 6$

3. $x(2x - 3) = 2(x + 7) - 2(2 - x^2)$

4. $x(x - 1) = 2(x + 5)$

5. $5(3q - 1) + 1 = 4(2q + 3) + 7(q - 1)$

6. $2(3m + 2) - 2(m - 2) = 4(m + 2)$

7. Solve $15x + 7y = 12$ for y.

8. Solve and graph: $3(4a + 2) < 12 - 2(4a - 3)$

9. In Macy's advertisement, a popular brand of bath towels was featured on sale. Dorothy Johnson bought three of the bath towels and two of the hand towels and paid $22.97. If one bath towel costs 59¢ more than two times the cost of one hand towel, find the cost of one bath towel and one hand towel.

10. A truck left a truck stop heading east at a speed of 60 mph. One hour later a car left the same truck stop heading west at a speed of 70 mph. In how many hours after the truck left the truck stop would the vehicles be 450 miles apart?

SECTION 4-7.   Supplementary Exercises

In exercises **1–12**, solve.

1. The sum of the squares of two consecutive integers is 113. Find the integers.

2. The sum of the squares of two consecutive integers is 265. Find the integers.

3. The sum of the squares of two consecutive odd integers is 202. Find the integers.

4. The sum of the squares of two consecutive even integers is 244. Find the integers.

5. The product of the first and third of three consecutive integers is one less than the square of 12. Find the three integers.

6. The product of the first and third of three consecutive integers is one less than the square of 15. Find the three integers.

7. The sum of the squares of the first two of three consecutive integers is 12 more than the square of the third.
   a. Find the three integers.
   b. Is the smallest integer odd or even?

8. The sum of the squares of the first and third of three consecutive integers is 83 more than the square of the second.
   a. Find the three integers.
   b. Is the smallest integer odd or even?

9. For three consecutive odd integers, twice the product of the first and second is 27 more than the product of the second and third. Find the three integers.

10. For three consecutive even integers, twice the product of the first and second is 16 more than the product of the second and third. Find the integers.

11. Square A has sides that are 5 feet longer than the sides of square B. Twice the area of B is 14 square feet less than the area of A. Find the lengths of the sides of both squares.

12. Square A has sides that are 5 centimeters longer than the sides of square B. Three times the area of B is 134 square centimeters less than twice the area of A. Find the lengths of the sides of both squares.

SECTION 4-1. Summary Exercises

Answer

In exercises **1–8**, factor.

1. $12x^2 + 6x - 18$

1. _____

2. $5z^4 - 3z^3 + 4z^2$

2. _____

3. $8a^3b^2 - 3a^2b^2 + 5a^2b^3$

3. _____

4. $15x^3 - 20x^2 + 10x$

4. _____

5. $30a^2 - 96a + 6a^3 + 18a^4$

5. _____

6. $-5x^2y + 45x^2y^2 - 25x^2y^3$

6. _____

7. $20j^4k^3 - 24j^3k^2 + 36j^2k$

7. _____

8. $-12s^3 - 48s^4 + 18s^5 + 6s - 24s^2$

8. _____

SECTION 4-2.   Summary Exercises

In exercises **1–8**, factor if possible.

1. $x^2 + 3x - 40$

1. _____

2. $z^2 - 23z - 50$

2. _____

3. $y^2 + 17y + 16$

3. _____

4. $x^2 - 14x + 7$

4. _____

5. $m^2 + 12mn + 27n^2$

5. _____

6. $y^3 - 17y^2 + 42y$

6. _____

7. $-a^3b + 15a^2b - 26ab$

7. _____

8. $2x^3 + 40x^2 + 128x$

8. _____

SECTION 4-3. Summary Exercises

In exercises **1–8**, factor.

1. $16s^3 + 14s^2 + 24s + 21$

1. _____

2. $14xy + 4x + 49y + 14$

2. _____

3. $36x^2 + 10y + 24x + 15xy$

3. _____

4. $3x + 42xy + 7y + 18x^2$

4. _____

5. $8xy + 6x - 28y - 21$

5. _____

6. $2ab - 4ac - 7b + 14c$

6. _____

7. $60x^3 + 10x^2 - 18xy - 3y$

7. _____

8. $40p^2r^2 + 30p^2r + 48pqr^2 + 36pqr$

8. _____

SECTION 4-4.   Summary Exercises

Answer

In exercises **1–8**, factor completely.

1. $16w^2 - 24w - 27$
(Write $-24w$ as $-36w + 12w$.)

1. _____

2. $15x^2 + 32x - 7$
(Write $32x$ as $-3x + 35x$.)

2. _____

3. $2a^2 - 9a + 7$

3. _____

4. $15y^2 - 28y + 12$

4. _____

5. $2t^2 - 3t - 27$

5. _____

6. $14uv - 15v^2 + 8u^2$

6. _____

7. $10a^3 + 26a^2 - 12a$

7. _____

8. $-38m^3 - 20m^2 + 24m^4$

8. _____

SECTION 4-5. Summary Exercises

Answer

In exercises **1–8**, factor completely.

1. $4x^2 - 121$

1. _____

2. $64t^2 - 81$

2. _____

3. $a^2 - 30a + 225$

3. _____

4. $9x^2 + 84x + 196$

4. _____

5. $12a^2 - 12ab + 3b^2$

5. _____

6. $32a^2 + 48ab + 18b^2$

6. _____

7. $8x^3 + 27$

7. _____

8. $625a^3b - 5b^4$

8. _____

SECTION 4-6. Summary Exercises

Answer

In exercises **1–8**, solve.

1. $(x - 8)(3x + 7) = 0$

1. _____

2. $4x(x + 20) = 0$

2. _____

3. $a^2 - 7a - 30 = 0$

3. _____

4. $y^2 = 5y$

4. _____

5. $6b^2 + 5b = -1$

5. _____

6. $8k^2 + 1 = 6k$

6. _____

7. $9z^2 - 16 = 0$

7. _____

8. $4w^2 + 81 = 36w$

8. _____

SECTION 4-7. Summary Exercises

Answer

In exercises **1–4**, solve.

1. The product of two integers is 84, and the sum of the numbers is 19. Find both integers.

1. _____

2. The hypotenuse of a right triangle is 30 centimeters. One leg is 6 centimeters longer than the other. Find the lengths of both legs.

2. a. Shorter leg: _____

 b. Longer leg: _____

3. The length of a rectangle is 12 inches longer than the width. The area is 160 square inches. Find the length and the width of the rectangle.

3. a. Width: _____

b. Length: _____

4. The height of a triangle is one-half the length of the base. The area is 25 square meters. Find the lengths of the base and the height.

4. a. Base: _____

b. Height: _____

CHAPTER 4.   Review Exercises

In exercises **1–6**, factor completely.

1. $8m^3 - 16m^2 + 32m$

2. $36m^3n^2 + 9m^2n^3 - 63mn^4$

3. $x^2 - 10x - 24$

4. $x^2 - 126 - 11x$

5. $a^4 + 7a^3 - 144a^2$

6. $-75a^2 - 280a - 5a^3$

In exercises **7** and **8**, factor by grouping.

7. $6t^2 + 3t + 2tu + u$

8. $10tu + 15t - 2u - 3$

In exercises **9–20**, factor completely.

9. $5y^2 - 7y - 6$

10. $12y^2 - 29y + 15$

11. $9k^2 - 6k - 35$

12. $9k^2 - 45k + 50$

13. $18x^2 + 24xy - 10y^2$

14. $49x^2 - y^2$

15. $49m^2 - 42mn + 9n^2$

16. $343m^3 + 125n^3$

17. $192p^3q - 3pq^3$

18. $60pq + 225q^2 + 4p^2$

19. $30uv + 75 + 3u^2v^2$

20. $50uv - 15u^3v^3 + 65u^2v^2$

In exercises **21–26**, solve and check.

21. $(3x + 5)(2x - 5) = 0$

22. $x^2 + 6x = 27$

23. $144y^2 = 25$

24. $y = 3 - 10y^2$

25. $5a^2 - 28 = 13a$

26. $0 = 2a^2 - 10a$

In exercises **27–30**, solve.

27. The sum of two integers is -3, and their product is -40. Find the numbers.

28. One leg of a right triangle is 5 meters longer than the shorter leg. The hypotenuse is 25 meters long. Find the length of each leg of the triangle.

29. The length of a rectangle is 7 feet less than twice the width. The area is 400 square feet. Find the length and width of the rectangle.

30. The base of a triangle is 1 centimeter less than three times the height. The area is 70 square centimeters. Find the length of the base and the height of the triangle.

5

Rational Expressions

The display shown below was on the board as the algebra class arrived this morning.

Rational Numbers and Operations

a. $\dfrac{18}{30}$ and $\dfrac{48}{36}$

b. $\dfrac{9}{4} \cdot \dfrac{5}{8}$

c. $\dfrac{2}{3} \div \dfrac{11}{7}$

d. $\dfrac{1}{4} + \dfrac{9}{4}$ and $\dfrac{7}{5} - \dfrac{3}{5}$

e. $\dfrac{4}{3} + \dfrac{1}{4}$ and $\dfrac{3}{8} - \dfrac{5}{6}$

f. $\left(2\dfrac{1}{4}\right) \div \left(5\dfrac{1}{3}\right)$

Rational Expressions and Operations

a.* $\dfrac{2x^2 + 3x - 9}{x^2 + 5x + 6}$ and $\dfrac{4x^2 + 4x}{8x^2 - 12x}$

b.* $\dfrac{3x}{x + 1} \cdot \dfrac{x + 2}{x^2 - 1}$

c.* $\dfrac{x - 1}{x} \div \dfrac{3x + 2}{2x + 1}$

d.* $\dfrac{x - 2}{x + 1} + \dfrac{3x}{x + 1}$ and $\dfrac{2x + 1}{2x - 1} - \dfrac{x}{2x - 1}$

e.* $\dfrac{x + 1}{x} + \dfrac{x - 2}{x + 1}$ and $\dfrac{x}{x^2 - 1} - \dfrac{x + 2}{x^2 - x}$

f.* $\dfrac{x - 1 + \dfrac{1}{x + 1}}{x + 2 + \dfrac{1}{x}}$

"On the board," Ms. Glaston began, "you see some familiar expressions in examples **a** through **f**. Next to these examples are expressions that are probably new. The expressions in **a** through **f** are all **rational numbers**. The expressions in **a*** through **f*** are called **rational expressions**. To begin, let's have Jim Medley remind us of the definition of a rational number."

Jim thought for a moment and said, "I think a rational number has something to do with integers. Is a rational number an integer over an integer?"

"That's very close Jim," Ms. Glaston replied. "A rational number is one that can be written as a ratio of integers, provided the denominator is not zero." She then turned to the board and wrote the following sentence:

A rational number can be written in the form $\frac{a}{b}$
where a and b are integers, and $b \neq 0$.

"Consider now only the expressions in examples **a** through **f**," Ms. Glaston continued. "What instructions might accompany a set of problems like these?"

"The fractions in example **a** need to be reduced," Sheila Kearney said. "The first one can be reduced because 18 and 30 have a common factor of 6. The 48 and 36 in the second fraction have a common factor of 12."

"In examples **b** through **e**, the fractions need to be multiplied, divided, added, or subtracted," John Parker added. But in **e** we first need to find a common denominator."

"Aren't the numbers in **f** mixed numbers?" Mike Queen asked. "If so, aren't they out of place in this list of numbers?"

"Good question, Mike," Ms. Glaston replied. "Recall that mixed numbers can be changed to the form of an integer over an integer. For example, $2\frac{1}{4}$ can be written as $\frac{9}{4}$. So we can say that mixed numbers are rational numbers."

"Why did you put examples **a*** through **f*** next to the ones in **a** through **f**?" Cathy Lankenau asked. "I don't see any similarity at all between the two sets of examples."

"Well, Cathy," Ms. Glaston replied, "what do you see when you look at the expressions in **a*** through **f***?"

"I see a bunch of polynomials in x," Cathy said. "But there's only one problem. In most of the examples there's a polynomial divided by a polynomial."

"That's good, Cathy," Ms. Glaston replied. "As we will learn in the chapter we begin studying today, the ratio of two polynomials can be called a **rational expression,** or simply an **algebraic fraction.** The purpose of the parallel examples is to illustrate that operating on rational expressions is very similar to operating with rational numbers. In fact, if all the numbers in the polynomials are integers, as in these examples, and x is replaced by an integer, then the rational expressions simplify to rational numbers."

She then turned to the board and wrote as follows:

Replace x by 3 in $\dfrac{2x^2 + 3x - 9}{x^2 + 5x + 6}$.

$$\frac{2(3)^2 + 3(3) - 9}{3^2 + 5(3) + 6} = \frac{2(9) + 3(3) - 9}{9 + 5(3) + 6}$$

$$= \frac{18 + 9 - 9}{9 + 15 + 6}$$

$$= \frac{18}{30}, \text{ the first number in example } \mathbf{a}$$

"In fact," Ms. Glaston continued, "the rational expressions in the * examples yield the rational numbers in the corresponding lettered examples to the left when x is replaced by 3. I set up this parallelism to emphasize from the beginning that many of the formulas for operating on rational expressions are similar to those we use when we operate on rational numbers. By keeping this fact in mind, we can more readily understand how to simplify the operations on rational expressions. If there are no further questions, let's begin by looking at Section 5-1."

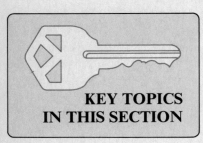

**KEY TOPICS
IN THIS SECTION**

SECTION 5-1. Rational Expressions, Defined and Reduced

1. A definition of a rational expression

2. Evaluating rational expressions

3. Finding restricted values for rational expressions

4. Reducing rational expressions

Chapter 3 contained the general form of expressions called *polynomials*. A rational expression is one in which a polynomial is divided by a polynomial. As with a rational number, the denominator of a rational expression cannot be zero. Thus, it may be necessary to restrict any variables so that the denominator polynomial cannot be zero. And as with rational numbers, rational expressions are usually written in **reduced form.** That is, the numerator and denominator of a given rational number or rational expression has no common factors other than 1.

A Definition of a Rational Expression

Definition 5.1 gives the general form of a rational expression.

> **Definition 5.1.   A rational expression in x**
> If P and Q are polynomials in x and $Q \neq 0$, then $\frac{P}{Q}$ is a **rational expression** in x.

A rational expression, like a polynomial, is frequently referred to by the variables in the expression. Examples **a–c** illustrate three such rational expressions.

Rational expressions	Name
a. $\dfrac{t}{t^2 + 1}$	A rational expression in t
b. $\dfrac{x + y}{x^2 - 2xy + y^2}$	A rational expression in x and y
c. $\dfrac{2abc}{a - b + 3c}$	A rational expression in a, b, and c

Evaluating Rational Expressions

A given rational number names exactly one real number. By contrast, a rational expression can represent an infinite number of real numbers. If all the numbers in the numerator and denominator polynomials are integers, and the variables in the polynomials are replaced by integers, then the resulting number will be rational.

Example 1.   Evaluate:

$$\textbf{a.}\ \ \frac{2x + 3}{x - 5} \text{ for } x = 4$$

$$\textbf{b.}\ \ \frac{2m + 5n}{4m - n} \text{ for } m = 3 \text{ and } n = -2$$

Solution.   **Discussion.** To find values for the expressions, replace the variables by the stated values. Then observe the rule for order of operations and recognize the fraction bar as a grouping symbol.

a. $\dfrac{2(4) + 3}{4 - 5}$ Replace x by 4.

$= \dfrac{8 + 3}{-1}$ Multiply in numerator before adding.

$= -11$ $\frac{11}{-1}$ can be written as -11.

b. $\dfrac{2(3) + 5(-2)}{4(3) - (-2)}$ Replace m by 3 and n by -2.

$= \dfrac{6 + (-10)}{12 - (-2)}$ Multiply first.

$= \dfrac{-4}{14}$ Add or subtract as indicated.

$= \dfrac{-2}{7}$ Reduce the fraction.

Example 2.   Evaluate $\dfrac{2t^2 - 7t - 15}{3t^2 - 17t + 10}$ for $t = -2$.

Solution. $\dfrac{2(-2)^2 - 7(-2) - 15}{3(-2)^2 - 17(-2) + 10}$ Replace t by (-2).

$= \dfrac{2(4) - 7(-2) - 15}{3(4) - 17(-2) + 10}$ Raise to powers first.

$= \dfrac{8 - (-14) - 15}{12 - (-34) + 10}$ Now do the multiplications.

$= \dfrac{7}{56}$ Add or subtract from left to right.

$= \dfrac{1}{8}$ Reduce the fraction.

Thus, the given expression has the value $\frac{1}{8}$ when t is replaced by (-2). Notice that $\frac{1}{8}$ is a rational number.

Example 3.   Evaluate $\dfrac{3u + 8v - 3w}{u^2 - 2v^2 + w^2}$ for $u = 2$, $v = -3$, and $w = 5$.

Solution. $\dfrac{3(2) + 8(-3) - 3(5)}{2^2 - 2(-3)^2 + 5^2}$ Replace u by 2, v by -3, and w by 5.

$= \dfrac{3(2) + 8(-3) - 3(5)}{4 - 2(9) + 25}$ Do the powers first.

$= \dfrac{6 + (-24) - 15}{4 - 18 + 25}$ Do the multiplications next.

$= \dfrac{-33}{11}$ Do the additions and subtractions in order from left to right.

$= -3$ The value of the expression is -3.

Finding Restricted Values for Rational Expressions

Consider the rational expressions in examples **a** and **b**.

a. $\dfrac{3y + 2}{y + 3}$ **b.** $\dfrac{-10}{k^2 - 4k - 32}$

In example **a**, if y is replaced by (-3), the expression becomes $\dfrac{3(-3) + 2}{-3 + 3} = \dfrac{-7}{0}$, which is an undefined expression. In example **b**, if k is replaced by (-4) or 8, the expression becomes

$$\frac{-10}{(-4)^2 - 4(-4) - 32} = \frac{-10}{0} \quad \text{and} \quad \frac{-10}{8^2 - 4(8) - 32} = \frac{-10}{0}.$$

In all three cases, the resulting expressions have zero denominators. Since division by zero is undefined, these rational expressions must be **restricted**, to prevent them from taking on those values that make the denominator polynomials zero. The restrictions can be written as follows:

a. $\dfrac{3y + 2}{y + 3}$ and $y \neq -3$ **b.** $\dfrac{-10}{k^2 - 4k - 32}$ and $k \neq -4$ or 8

If P and Q are polynomials, then to find any restricted values for the rational expression $\frac{P}{Q}$,

Step 1. Set $Q = 0$.

Step 2. Find any real roots of the equation.

Step 3. State the roots of the equation as restricted values.

Example 4.   Find any restricted values of $\dfrac{m - 3}{3m - 1}$.

Solution.   **Step 1.** $3m - 1 = 0$ Set $Q = 0$.

Step 2. $3m = 1$ Solve for m.

$$m = \frac{1}{3}$$ The root is $\frac{1}{3}$.

Step 3. $\dfrac{m - 3}{3m - 1}$ and $m \neq \dfrac{1}{3}$.

Not all rational expressions have restricted values. Examples **c** and **d** are two such rational expressions.

c. $\dfrac{t^2 - 8t}{10}$ The denominator is the nonzero constant 10.

d. $\dfrac{x^2 - 25}{x^2 + 4}$ $x^2 + 4$ is greater than 0 for all replacements of x.

Reducing Rational Expressions

Examples **e** and **f** are rational expressions that are written in factored form.

e. $\dfrac{6x^2 + 3x}{3x^2 + 6x} = \dfrac{3x(2x + 1)}{3x(x + 2)}$ $3x$ is a factor of both numerator and denominator.

f. $\dfrac{2t^2 - 7t - 15}{3t^2 - 17t + 10} = \dfrac{(2t + 3)(t - 5)}{(3t - 2)(t - 5)}$ $t - 5$ is a factor of both numerator and denominator.

If $x \neq 0$, then example **e** can be written in **reduced form** as $\dfrac{2x + 1}{x + 2}$. If $t \neq 5$, then example **f** can be written in reduced form as $\dfrac{2t + 3}{3t - 2}$. For any replacements in the given expressions or reduced forms (other than restricted values), the two expressions will yield the same numbers.

If P, Q, and R are polynomials with $Q \neq 0$ and $R \neq 0$, then

$$\frac{P \cdot R}{Q \cdot R} = \frac{P}{Q}$$ The common factor R is changed to one.

and $\dfrac{P}{Q}$ is called the **reduced form** of $\dfrac{P \cdot R}{Q \cdot R}$.

It is important to note that R is a *factor* and not a *term*. Remember, factors are *multiplied*, terms are *added* and *subtracted*.

Example 5.   Write in reduced form.

a. $\dfrac{z^2 + 3z - 70}{z^2 - 14z + 49}$ **b.** $\dfrac{2a - 3}{3 - 2a}$

Solution.   **a.** $\dfrac{z^2 + 3z - 70}{z^2 - 14z + 49}$ The given expression.

$= \dfrac{(z + 10)(z - 7)}{(z - 7)(z - 7)}$ Factor numerator and denominator.

$= \dfrac{z + 10}{z - 7}$ Change the common factor to 1, but do not write it.

b. Discussion. The terms on both top and bottom are $2a$ and 3. However, the signs are not the same, because on the top 3 is subtracted from $2a$, but on the bottom $2a$ is subtracted from 3. The signs can be made to be the same by factoring a (-1) in the numerator.

$\dfrac{2a - 3}{3 - 2a}$ The given expression

$= \dfrac{-1 \cdot (3 - 2a)}{(3 - 2a)}$ Factor a (-1) on top.

$= -1$ Reduce the fraction.

Calculator Assistance Supplement

The Use of the Parentheses on a Calculator

A scientific calculator is designed to follow the order of operations. To illustrate, consider the following numerical expression:

$$\frac{15 - 3}{4 + 2}$$

The order of operations requires computing $15 - 3$ first, then $4 + 2$, and then the division. Thus,

$$\frac{15 - 3}{4 + 2} = \frac{12}{6} = 2.$$

To key this problem into a calculator requires parentheses. That is, enter

$$(15 - 3) \div (4 + 2) = .$$

Without parentheses the calcualtor would read the following:

$$15 - 3 \div 4 + 2 = ,$$

and the division would be computed first.

The following steps illustrate how to evaluate the rational expression $\dfrac{x^2 + 1}{x^2 - 2x + 1}$ for $x = 1.2$.

Step	Display Shows
Press $($	[01 0.*
Enter 1.2	1.2
Press $M\ in$	1.2
Press x^2**	1.44
Press $+$	1.44
Enter 1	1.
Press $)$	2.44
Press $\div$	2.44
Press $($	[01 0.
Press MR	1.2
Press x^2	1.44
Press $-$	1.44
Enter 2	2.
Press $\times$	2.
Press MR	1.2
Press $+$	-0.96
Enter 1	1.
Press $)$	0.04
Press $=$	61.

The expression has the value 61 when $x = 1.2$.

* The display may vary for different brands of calculators.

** The x^2 key may require the use of a second function or inverse key.

SECTION 5-1.   Practice Exercises

In exercises **1–24**, evaluate for the stated value of the variable or variables.

[Examples 1, 2, and 3]

1. $\dfrac{x}{x+1}$ and $x=4$

2. $\dfrac{3x}{x-1}$ and $x=7$

3. $\dfrac{y+9}{y}$ and $y=3$

4. $\dfrac{y-10}{y}$ and $y=5$

5. $\dfrac{a}{a^2+1}$ and $a=-1$

6. $\dfrac{a}{a^2-1}$ and $a=-2$

7. $\dfrac{2y^2+5}{2y+1}$ and $y=5$

8. $\dfrac{2y+1}{y^2-2}$ and $y=-4$

9. $\dfrac{z^3-8}{z-2}$ and $z=-3$

10. $\dfrac{z^3-27}{z+3}$ and $z=-2$

11. $\dfrac{x+3y}{3x+y}$ and $x=2,\ y=3$

12. $\dfrac{2x-y}{x-2y}$ and $x=3,\ y=2$

13. $\dfrac{11x^2-9y^2}{9x^2-6xy+y^2}$ and $x=3,\ y=2$

14. $\dfrac{4x^2+y^2}{4x^2-4xy+y^2}$ and $x=2,\ y=3$

15. $\dfrac{x^2-y^2}{y^2-x^2}$ and $x=-5,\ y=7$

16. $\dfrac{y^2-x^2}{x^2-y^2}$ and $x=4,\ y=-9$

17. $\dfrac{x+y}{z^2}$ and $x=15,\ y=-3,\ z=2$

18. $\dfrac{xy}{x^2-z^2}$ and $x=4,\ y=12,\ z=-2$

19. $\dfrac{2a+b+c}{2c-2b-2a}$ and $a=-2,\ b=4,\ c=3$

20. $\dfrac{2a-4b+2c}{-a+2b-c}$ and $a=-5,\ b=-6,\ c=2$

21. $\dfrac{p^2+2pq+2r}{p^2+2r}$ and $p=4,\ q=3,\ r=-9$

22. $\dfrac{p-2qr+r^2}{p-r^2}$ and $p=4,\ q=5,\ r=-4$

23. $\dfrac{3xz-yz}{12x^2+6x-4xy-2y}$ and $x=-2,\ y=3,\ z=6$

24. $\dfrac{xz-2yz}{3x^2-x+6xy+z}$ and $x=9,\ y=-5,\ z=-2$

In exercises **25–40**, find any restricted values.

[Example 4]

25. $\dfrac{2x-3}{2x}$

26. $\dfrac{4x+1}{5x}$

27. $\dfrac{4y-3}{y-6}$

28. $\dfrac{6y-5}{y+5}$

29. $\dfrac{t}{3t + 5}$

30. $\dfrac{t - 6}{2t - 9}$

31. $\dfrac{z + 11}{z(z - 3)}$

32. $\dfrac{z + 7}{z(z + 8)}$

33. $\dfrac{2b - 3}{b^2 + 9b + 20}$

34. $\dfrac{4b - 9}{b^2 - b - 20}$

35. $\dfrac{x + 9}{4x^2 - 81}$

36. $\dfrac{2x + 15}{9x^2 - 25}$

37. $\dfrac{2a^2 - 1}{2a^2 + 1}$

38. $\dfrac{a^2 + 4}{9 + 4a^2}$

39. $\dfrac{1}{6m^2 - 7mn - 3n^2}$

40. $\dfrac{1}{10m^2 + 13mn - 3n^2}$

In exercises **41–60**, reduce. Assume no denominator is zero.

[Example 5] 41. $\dfrac{2y^2 + y}{3y}$

42. $\dfrac{5y^2 + 10y}{15y}$

43. $\dfrac{a^2 - 4}{a + 2}$

44. $\dfrac{a^2 - 25}{a - 5}$

45. $\dfrac{x^2 - x - 12}{x^2 - 8x + 16}$

46. $\dfrac{x^2 - 4x + 4}{x^2 - x - 2}$

47. $\dfrac{3x - 5}{5 - 3x}$

48. $\dfrac{10 - 9x}{9x - 10}$

49. $\dfrac{3y^2 - 5y - 2}{y^2 + y - 6}$

50. $\dfrac{y^2 + 8y + 15}{5y^2 + 16y + 3}$

51. $\dfrac{3a^2 - 11a - 20}{4a^2 - 17a - 15}$

52. $\dfrac{2a^2 - 15a + 7}{a^2 - 9a + 14}$

53. $\dfrac{3a^3 + 6a^2}{9a^3}$

54. $\dfrac{15a^5}{5a^3 - 25a^2}$

55. $\dfrac{t^3 - 27}{t^2 - 9}$

56. $\dfrac{8t^3 + 125}{4t^2 + 20t + 25}$

57. $\dfrac{-x - y - 5}{x + y + 5}$

58. $\dfrac{3 + x - 2y}{2y - x - 3}$

59. $\dfrac{9 - z^2}{2z^2 - z - 15}$

60. $\dfrac{25 - 10z + z^2}{3z^2 - 11z - 20}$

SECTION 5-1. Ten Review Exercises

In exercises **1** and **2**, simplify.

1. $3(4a - 2b - 1) - 5(2a - b + 2) - 2(a - 2b - 7)$

2. $-2[3t - 2(1 - 2t)] + 3(10t - 1)$

In exercises **3** and **4**, factor.

3. $6x^3 + 10x - 35 - 21x^2$ **4.** $8x^3 - 343$

In exercises **5–8**, solve each problem.

5. Reduce: $\dfrac{8x^3 - 343}{6x^3 + 10x - 35 - 21x^2}$ **6.** Multiply: $(6y^2 + 2y - 5)(3y - 1)$

7. Divide: $(18y^3 - 17y + 5) \div (3y - 1)$

8. Use the results of exercises **6** and **7** to reduce $\dfrac{18y^3 - 17y + 5}{3y - 1}$.

In exercises **9** and **10**, simplify.

9. $-2k^2(3k^3)(5k)^2$ **10.** $\dfrac{(-8xy^2)(x^2y)^3}{(2xy)^5}$

SECTION 5-1. Supplementary Exercises

In exercises **1–14**, find any restricted values.

1. $\dfrac{t^2 - 1}{t^2 + 1}$ **2.** $\dfrac{t^2 - 25}{t^2 + 25}$

3. $\dfrac{2a + 3}{18}$ **4.** $\dfrac{7a - 1}{12}$

5. $\dfrac{x + 1}{x + 1}$ **6.** $\dfrac{x + 3}{x + 3}$

7. $\dfrac{(x + 5)(x - 1)}{x(x + 3)(2x - 1)}$ **8.** $\dfrac{(x + 2)(x - 3)}{x(3x + 1)(x - 4)}$

9. $\dfrac{b^2 + 4b + 3}{b^2 + 10b}$ **10.** $\dfrac{b^2 - 10b + 25}{b^2 - 2b}$

11. $\dfrac{x^3}{x^3 + 8}$ **12.** $\dfrac{x^4 - 16}{27x^3 - 1}$

13. $\dfrac{5}{x - y}$ **14.** $\dfrac{x + 2y}{2x + y}$

In exercises **15–24**, reduce.

15. $\dfrac{y^2 - 6y + 9}{9 - y^2}$ **16.** $\dfrac{y^2 - 10y + 25}{25 - y^2}$

17. $\dfrac{a^3 - 8}{a^2 - 4}$ **18.** $\dfrac{a^3 - 27}{a^2 + 3a + 9}$

19. $\dfrac{a^2 + 10ab + 25b^2}{a^2 - 25b^2}$

20. $\dfrac{16a^2 - 25b^2}{16a^2 + 40ab + 25b^2}$

21. $\dfrac{4b^3 + 8b^2 - 12b}{2b^3 + 10b^2 + 12b}$

22. $\dfrac{6b^3 + 18b^2 - 60b}{3b^3 + 24b^2 + 45b}$

23. $\dfrac{t^4 - 81}{t^2 + 9}$

24. $\dfrac{t^4 - 16}{t^2 + 4}$

In exercises **25–28**,

1. Evaluate **a** for $t = 3$.

2. Evaluate **b** for $t = 3$.

3. Compare the results of parts **1** and **2**. Are the rational expressions the same?

25. a. $\dfrac{3t^2 - 4}{t^2 + 4}$ **b.** $\dfrac{3t^2 - 1}{t^2 + 1}$ **26. a.** $\dfrac{t^2 - 4}{8}$ **b.** $\dfrac{t^2 - 1}{2}$

27. a. $\dfrac{2t^2 + 3}{4t + 1}$ **b.** $\dfrac{t^2 + 3}{2t + 1}$ **28. a.** $\dfrac{2t^2 + 3}{4}$ **b.** $\dfrac{t^2 + 3}{2}$

In exercises **29–32**,

a. Evaluate the rational expression for $a = 2$.

b. Reduce the rational expression.

c. Evaluate the reduced expression for $a = 2$.

d. Compare the results in parts **a** and **c**. Do the rational expressions yield the same value?

29. $\dfrac{a^2 - 3a}{2a - 6}$

30. $\dfrac{6a + 6}{a^2 + a}$

31. $\dfrac{a^2 + 2a - 15}{a^2 - 8a + 15}$

32. $\dfrac{a^2 + 4a - 21}{a^2 - 4a + 3}$

In exercises **33** and **34**, work parts **a–d**.

33. a. Simplify: $\dfrac{6 + 7 + 8}{9 + 10 + 11}$

b. Simplify: $\dfrac{4 + 5 + 6}{7 + 8 + 9}$

c. Factor and reduce: $\dfrac{x + (x + 1) + (x + 2)}{(x + 3) + (x + 4) + (x + 5)}$

d. Without working the problem, what must the simplified form of the following fraction be?

$$\dfrac{120 + 121 + 122}{123 + 124 + 125}$$

34. a. Simplify: $\dfrac{9 + 11}{12 + 14}$

b. Simplify: $\dfrac{6 + 8}{9 + 11}$

c. Factor and reduce: $\dfrac{(x) + (x + 2)}{(x + 3) + (x + 5)}$

d. Without working the problem, what must the simplified form of the following fraction be?

$$\frac{192 + 194}{195 + 197}$$

SECTION 5-2.   Negative Integers as Exponents

KEY TOPICS IN THIS SECTION

1. A review of zero as an exponent

2. A definition of a negative integer exponent

3. Three formulas related to negative integer exponents

4. A review of the five properties of exponents

5. Scientific notation as an application of negative exponents

Exponents have been used to write indicated products of the same factor (called the **base**), using fewer symbols. As examples,

$5 \cdot 5 \cdot 5 \cdot 5 \cdot 5 \cdot 5 \cdot 5$ can be written as 5^7.

$x \cdot x \cdot y \cdot y \cdot y \cdot y \cdot y$ can be written as $x^2 y^5$.

In both examples the exponents are positive integers. Recall that the positive integers are $1, 2, 3, 4, \ldots$. In this section we will use negative integers $(-1, -2, -3, -4, \ldots)$ as exponents.

A Review of Zero as an Exponent

In Chapter 3 the following definition was given:

Definition 3.4.   Zero as an exponent
If $b \neq 0$, then $b^0 = 1$.

Example 1.   Simplify:

     a. 10^0 **b.** $(-8)^0$ **c.** -3^0

Solution.   **a.** $10^0 = 1$ By definition 3.3, $b^0 = 1$, if $b \neq 0$.

          b. $(-8)^0 = 1$ A negative number to the 0 power is 1.

          c. $-3^0 = -1$ The base is 3, not -3.

In part **c** of Example 1, the expression indicates two operations:

Operation 1. Raise 3 (the base) to the 0 power.

Operation 2. Take the opposite of the result of Operation 1.

Since $3^0 = 1$, and the opposite of 1 is -1,

$$-3^0 = -(1) = -1.$$

A Definition of a Negative Integer Exponent

The following definition now extends exponents to negative integers.

Definition 5.2.   Negative integers as exponents
If n is a positive integer (1, 2, 3, . . .) and $b \neq 0$, then

$$b^{-n} = \frac{1}{b^n}.$$

Notice that if $n > 0$, then $-n < 0$.

The equation in Definition 5.2 states that a negative integer exponent can be changed to a positive integer exponent by writing one over the number. You may recall that if b is not zero, then $\frac{1}{b}$ is called the **reciprocal** of b. Thus you may think of b^{-n} as a way of writing the reciprocal of b^n.

Example 2.   Write parts **a** and **b** with only positive exponents.

   a. 3^{-4}

   b. $-x^{-2}$, and $x \neq 0$

Solution.   **a.** $3^{-4} = \frac{1}{3^4}$ Write as 1 over 3^4.

   $\qquad\qquad = \frac{1}{81}$ Compute the indicated power.

   b. Write $-x^{-2}$ as $-1 \cdot x^{-2}$.   The opposite of $x^{-2} = -1 \cdot x^{-2}$

   $\qquad -1 \cdot x^{-2} = \frac{-1}{x^2}$ Write as -1 over x^2.

Example 3.   Write parts **a** and **b** with only positive exponents.

   a. $10t^{-1}$, and $t \neq 0$

   b. $2a^{-3}b$, and $a \neq 0$

Solution.   **a.** In $10t^{-1}$, the base of the -1 exponent is t, not $10t$. To change the exponent to a positive number, we write the reciprocal of only the t, not $10t$.

   $\qquad 10t^{-1} = \frac{10}{t}$ Write t^{-1} as $\frac{1}{t}$.

b. In $2a^{-3}b$, the base of the -3 exponent is only the a. To change this exponent to a positive number, we write the reciprocal of only the a.

$$2a^{-3}b = 2 \cdot a^{-3} \cdot b \qquad \text{Write as three separate factors.}$$

$$= 2 \cdot \frac{1}{a^3} \cdot b \qquad \text{Write } a^{-3} \text{ as } \frac{1}{a^3}.$$

$$= \frac{2b}{a^3} \qquad \text{Write as a single ratio.}$$

In Examples 2 and 3 we stated that variables with negative exponents are not zero. For simplicity, let us assume that all variables in this section represent non-zero real numbers.

Three Formulas Related to Negative Integer Exponents

In example **a** below, the factor t^{-3} is in the denominator of the expression. The equation in Definition 5.2 is used to change the exponent to a positive number.

a. $\dfrac{1}{t^{-3}} = \dfrac{1}{\dfrac{1}{t^3}} \qquad \text{Write } t^{-3} \text{ as } \dfrac{1}{t^3}.$

$$= \frac{1}{\dfrac{1}{t^3}} \cdot \frac{t^3}{t^3} \qquad \text{Multiply top and bottom by } t^3.$$

$$= \frac{t^3}{1} \qquad \frac{1}{t^3} \cdot t^3 \text{ in the denominator is } 1.$$

$$= t^3 \qquad \text{The } t^{-3} \text{ on the bottom is } t^3 \text{ on the top.}$$

In example **b** the ratio $\frac{2}{u}$ has a negative exponent. We use the equation in Definition 5.2 to change the exponent to a positive number.

b. $\left(\dfrac{2}{u}\right)^{-4} = \dfrac{1}{\left(\dfrac{2}{u}\right)^4} \qquad b^{-n} = \dfrac{1}{b^n}, \text{ and } b \text{ represents } \left(\dfrac{2}{u}\right).$

$$= \frac{1}{\left(\dfrac{2}{u}\right)^4} \cdot \frac{\left(\dfrac{u}{2}\right)^4}{\left(\dfrac{u}{2}\right)^4} \qquad \text{Multiply top and bottom by } \left(\dfrac{u}{2}\right)^4.$$

$$= \frac{\left(\dfrac{u}{2}\right)^4}{1} \qquad \left(\dfrac{2}{u}\right)^4 \cdot \left(\dfrac{u}{2}\right)^4 \text{ in the denominator is } 1.$$

$$= \left(\dfrac{u}{2}\right)^4 \qquad \text{The result is the reciprocal of } \dfrac{2}{u}.$$

The equation in Definition 5.2 and the results of examples **a** and **b** are stated below as three formulas involving negative exponents.

General equations	**Some examples**
a and b are not zero, and n is a positive integer.	All variables are nonzero real numbers.

1. $b^{-n} = \dfrac{1}{b^n}$ $\qquad\qquad$ $t^{-2} = \dfrac{1}{t^2}$ and $-u^{-1} = \dfrac{-1}{u}$

2. $\dfrac{1}{b^{-n}} = b^n$ $\qquad\qquad$ $\dfrac{1}{x^{-3}} = x^3$ and $\dfrac{-1}{y^{-5}} = -y^5$

3. $\left(\dfrac{a}{b}\right)^{-n} = \left(\dfrac{b}{a}\right)^n$ $\qquad$ $\left(\dfrac{5}{p}\right)^{-1} = \dfrac{p}{5}$ and $\left(\dfrac{-3}{q}\right)^{-2} = \left(\dfrac{-q}{3}\right)^2$

Example 4. Write parts **a** and **b** with only positive exponents.

$\qquad$ **a.** $\dfrac{2m^{-2}}{n^{-1}}$

$\qquad$ **b.** $\left(\dfrac{-10}{mn}\right)^{-3}$

Solution. $\quad$ **a.** $\dfrac{2m^{-2}}{n^{-1}} = 2 \cdot m^{-2} \cdot \dfrac{1}{n^{-1}}$ $\qquad$ Write as three separate factors.

$\qquad\qquad\qquad = 2 \cdot \dfrac{1}{m^2} \cdot n$ $\qquad$ $m^{-2} = \dfrac{1}{m^2}$ and $\dfrac{1}{n^{-1}} = n$.

$\qquad\qquad\qquad = \dfrac{2n}{m^2}$ $\qquad\qquad$ Write as a single ratio.

$\qquad$ **b.** $\left(\dfrac{-10}{mn}\right)^{-3} = \left(\dfrac{-mn}{10}\right)^3$ $\qquad$ Keep the minus bar on top.

A Review of the Five Properties of Exponents

The five properties of exponents stated in Chapter 3 for positive integer exponents are also valid for zero and negative integer exponents.

Five properties of exponents
$\quad$ Let a and b be nonzero real numbers and let m and n be integers.

Property 1. $a^m \cdot a^n = a^{m+n}$ $\qquad$ $t^5 \cdot t^{-2} = t^{5+(-2)} = t^3$

Property 2. $(a^m)^n = a^{mn}$ $\qquad\quad$ $(t^{-2})^{-3} = t^{(-2)(-3)} = t^6$

Property 3. $\dfrac{a^m}{a^n} = a^{m-n}$ $\qquad\quad$ $\dfrac{t^2}{t^{-1}} = t^{2-(-1)} = t^3$

Property 4. $(ab)^m = a^m \cdot b^m$ $\qquad$ $(3t)^{-2} = 3^{-2} \cdot t^{-2}$

Property 5. $\left(\dfrac{a}{b}\right)^m = \dfrac{a^m}{b^m}$ $\qquad\quad$ $\left(\dfrac{-2}{t}\right)^{-3} = \dfrac{(-2)^{-3}}{t^{-3}}$

Example 5. Simplify and write the answers with only positive exponents.

    a. $2^3 \cdot 2^{-5}$

    b. $\dfrac{t^3}{t^{-1}}$

Solution.   **a.** $2^3 \cdot 2^{-5} = 2^{3+(-5)}$ Add exponents.

                $= 2^{-2}$ Simplify.

                $= \dfrac{1}{4}$ $2^{-2} = \dfrac{1}{2^2}$, or $\dfrac{1}{4}$

    b. Discussion. To get a positive exponent, we will subtract the smaller exponent from the larger one.

        $\dfrac{t^3}{t^{-1}} = t^{3-(-1)}$ Subtract -1 from 3.

            $= t^4$ Simplify.

Example 6. Simplify and write the answers with only positive exponents.

    a. $(u^{-2})^{-1}$

    b. $(2^3)^{-2}$

Solution.   **a.** $(u^{-2})^{-1} = u^{(-2)(-1)}$ Multiply the exponents.

                $= u^2$ Simplify the indicated product.

    b. $(2^3)^{-2} = 2^{3(-2)}$ Multiply the exponents.

              $= 2^{-6}$ Simplify the indicated product.

              $= \dfrac{1}{2^6}$ or $\dfrac{1}{64}$ Write the fraction with a positive exponent.

Example 7. Simplify and write answers with only positive exponents.

    a. $(3k^{-1})^{-2}$

    b. $\left(\dfrac{10}{z^{-2}}\right)^3$

Solution.   **a.** $(3k^{-1})^{-2} = 3^{-2}(k^{-1})^{-2}$ Raise both factors to the -2 power.

                $= 3^{-2}k^2$ $(k^{-1})^{-2} = k^{(-1)(-2)} = k^2$

                $= \dfrac{k^2}{9}$ $3^{-2} = \dfrac{1}{3^2} = \dfrac{1}{9}$

    b. $\left(\dfrac{10}{z^{-2}}\right)^3 = \dfrac{10^3}{(z^{-2})^3}$ Raise both factors to the power 3.

              $= \dfrac{1,000}{z^{-6}}$ $(z^{-2})^3 = z^{(-2)(3)} = z^{-6}$

              $= 1,000z^6$ $\dfrac{1}{z^{-6}} = z^6$

Scientific Notation as an Application of Negative Exponents

In science and engineering we frequently encounter numbers that are very large or very small. Examples **a** through **d** illustrate four such numbers.

a. 300,000,000 meters per second   The speed of light

b. 67,300,000 miles   The average distance from Venus to the sun

c. 0.00000025 meters   The diameter of a smallpox virus

d. 0.0000000875 meters   The wavelength of light

Because of the difficulty of working with such large or small numbers written in **ordinary notation,** we frequently write such numbers in **scientific notation.**

Writing a number in scientific notation

A number N written in ordinary notation is changed to **scientific notation** when it is written as an indicated product of

(i) a number d, where $1 \leq d < 10$, and

(ii) a power of 10.

We change a number from ordinary notation to scientific notation by moving the decimal point until exactly one nonzero digit is to the left of it. We accomplish this by multiplying or dividing the number by the appropriate power of 10.

Recall that a power of 10 agrees with the number of 0s in the number.

$$100 = 10^2 \qquad \textbf{Two 0s in 100, and 2 is the exponent on 10.}$$

$$10,000,000 = 10^7 \qquad \textbf{Seven 0s in 10,000,000, and 7 is the exponent on 10.}$$

Example 8.   Write in scientific notation.

a. 7,500

b. 0.00009

Solution.   **a. Discussion.** To move the decimal point between the 7 and the 5, we need to divide by 1,000 (that is, three places). To keep the value of the number the same, we also multiply by 1,000.

$$\frac{7,500}{1,000} \cdot 1,000 = 7.5 \times 1,000$$

$$= 7.5 \times 10^3$$

b. Discussion. To move the decimal point to the right of the 9, we need to multiply by 100,000 (that is, five places). To keep the value of the number the same, we also divide by 100,000.

$$0.00009 \cdot \frac{100,000}{100,000} = \frac{9}{100,000}$$

$$= \frac{9}{10^5}$$

$$= 9 \times 10^{-5}$$

Example 9. Write in ordinary notation.

    a. 5.28×10^{-4}

    b. 6.3×10^6

Solution.   **a. Discussion.** The power on 10 is -4. Thus, we move the decimal point four places to the left.
$5.28 \times 10^{-4} = 0.000528$

     b. Discussion. The power on 10 is six. Thus, we move the decimal point six places to the right.
$6.3 \times 10^6 = 6,300,000$

SECTION 5-2. Practice Exercises

In exercises **1–8**, simplify.

[Example 1]

1. 17^0 **2.** 93^0 **3.** $(-8)^0$

4. $(-12)^0$ **5.** -21^0 **6.** -49^0

7. $\left(\dfrac{3}{4}\right)^0$ **8.** $\dfrac{3^0}{4}$

In exercises **9–42**, write with only positive exponents.

[Example 2]

9. 2^{-4} **10.** 5^{-3} **11.** $(-8)^{-2}$

12. $(-4)^{-2}$ **13.** -3^{-4} **14.** -6^{-2}

15. a^{-8} **16.** a^{-11} **17.** $-p^{-6}$

18. $-p^{-4}$ **19.** -7^{-1} **20.** -6^{-1}

[Example 3]

21. $10p^{-3}$ **22.** $17p^{-5}$

23. $4m^2n^{-2}$ **24.** $3m^4n^{-3}$

25. $a^{-5}b^{-2}$ **26.** $a^{-3}b^{-7}$

27. $5^{-1}xy^{-3}$ **28.** $2^{-1}x^{-2}y^3$

29. $7^2s^5t^{-8}$ **30.** $5^2s^{-4}t^3$

[Example 4]

31. $\dfrac{1}{7^{-2}}$ **32.** $\dfrac{1}{11^{-2}}$ **33.** $\dfrac{1}{q^{-7}}$

34. $\dfrac{1}{q^{-3}}$ **35.** $\dfrac{4x^{-2}}{y^{-3}}$ **36.** $\dfrac{5x^{-1}}{y^{-4}}$

37. $\left(\dfrac{2}{5}\right)^{-1}$ **38.** $\left(\dfrac{8}{3}\right)^{-1}$ **39.** $\left(\dfrac{3}{x}\right)^{-2}$

40. $\left(\dfrac{x}{6}\right)^{-2}$ **41.** $\left(\dfrac{5a}{2b^2}\right)^{-3}$ **42.** $\left(\dfrac{3a^3}{4b}\right)^{-2}$

In exercises **43–64**, simplify and write answers with only positive exponents.

[Examples 5–7]

43. $x^5 \cdot x^{-3}$ **44.** $x^{-4} \cdot x^2$

45. $(y^{-2})^3$

46. $(y^{-3})^{-1}$

47. $\dfrac{z^{-6}}{z^{-4}}$

48. $\dfrac{z^{-2}}{z^4}$

49. $(a^{-1}b^2)^{-3}$

50. $(a^{-3}b^2)^{-2}$

51. $\left(\dfrac{p^3}{q^{-1}}\right)^{-4}$

52. $\left(\dfrac{p^{-5}}{q^{-2}}\right)^{-3}$

53. $5^{-9} \cdot 5^7$

54. $5^6 \cdot 5^{-5}$

55. $(2t^{-2})^{-3}$

56. $(2t^{-4})^{-2}$

57. $\dfrac{m^{-4}n^{-3}}{m^{-4}n^{-5}}$

58. $\dfrac{m^2n^{-3}}{m^{-3}n^{-3}}$

59. $\left(\dfrac{5s^2}{t^{-3}}\right)^2$

60. $\left(\dfrac{2s^{-1}}{t^4}\right)^3$

61. $-5a^{-3}(2a^{-2})^{-2}$

62. $(-3t)^{-2}(2t^{-1})^{-1}$

63. $\dfrac{(6x^{-2}y)^{-1}}{(-2^{-1}xy^{-1})^2}$

64. $\dfrac{(-3x^{-1}y)(2x^2y^3)^{-1}}{(6x^{-2}y^{-1})^2}$

In exercises **65–78**, write in scientific notation.

[Example 8] **65.** 540,000

66. 65,000

67. 9,100,000

68. 16,000,000

69. 0.0008

70. 0.00006

71. 0.000064

72. 0.0000052

73. 861,000,000

74. 3,650,000

75. 0.000807

76. 0.00000421

77. 28,500

78. 725,000

In exercises **79–90**, write in ordinary notation.

[Example 9] **79.** 7.4×10

80. 3.8×10^2

81. 4.3×10^6

82. 5.8×10^4

83. 1.7×10^{-4}

84. 9.1×10^{-3}

85. 8.7×10^{-6}

86. 5.7×10^{-5}

87. 1.15×10^8

88. 7.96×10^7

89. 5.53×10^{-9}

90. 6.81×10^{-6}

SECTION 5-2.   Ten Review Exercises

In exercises **1–4**, do the indicated operations.

1. $3^4 - 5 \cdot 2^4$

2. $-3 \cdot 5 + 7 \cdot 2 - 39 \div 13$

3. $-4(3^2 - 2^4) - (5^2 + 7 \cdot 2)$ **4.** $5(6^2 - 7 \cdot 3) + (3 \cdot 2 - 19)$

In exercises **5** and **6**, use the distributive property to simplify.

5. $48\left(\dfrac{5}{12} - \dfrac{7}{8}\right)$ **6.** $37 \cdot 43 + 37 \cdot 57$

In exercises **7–10**, solve each problem.

7. Factor: $3k^2 - 10k - 8$ **8.** Evaluate for $k = \frac{-2}{3}$: $3k^2 - 10k - 8$

9. Solve: $3k^2 = 8 + 10k$ **10.** Reduce: $\dfrac{2k - 8}{3k^2 - 10k - 8}$

SECTION 5-2. Supplementary Exercises

Example. $(8.2 \times 10^{15})(2.0 \times 10^{-3})$

Solution. $(8.2 \times 10^{15})(2.0 \times 10^{-3})$

$= 8.2 \times 10^{15} \times 2.0 \times 10^{-3}$

$= 8.2 \times 2.0 \times 10^{15} \times 10^{-3}$

$= (8.2 \times 2.0) \times (10^{15} \times 10^{-3})$

$= 16.4 \quad \times \quad 10^{15+(-3)}$

$= 16.4 \quad \times \quad 10^{12}$

$= 1.64 \quad \times \quad 10^{13}$

In exercises **1–8**, find the product.

1. $(3.3 \times 10^{12})(2.1 \times 10^{-4})$ **2.** $(1.9 \times 10^{11})(3.2 \times 10^{-3})$

3. $(8.2 \times 10^{-5})(2.0 \times 10^8)$ **4.** $(1.4 \times 10^{-7})(3.1 \times 10^{12})$

5. $(7.4 \times 10^{-3})(2.4 \times 10^{-7})$ **6.** $(8.6 \times 10^{-6})(7.2 \times 10^{-5})$

7. $(4.01 \times 10^7)(2.2 \times 10^{11})$ **8.** $(5.04 \times 10^9)(1.2 \times 10^6)$

Example. $\dfrac{1.2 \times 10^9}{2.0 \times 10^5}$

Solution. $\dfrac{1.2 \times 10^9}{2.0 \times 10^5}$

$= \dfrac{1.2}{2.0} \times \dfrac{10^9}{10^5}$

$= 0.6 \times 10^{9-5}$

$= 0.6 \times 10^4$

$= 6.0 \times 10^3$

In exercises **9–16**, find the quotient.

9. $\dfrac{2.53 \times 10^{12}}{2.3 \times 10^{7}}$

10. $\dfrac{6.82 \times 10^{8}}{3.1 \times 10^{2}}$

11. $\dfrac{6.12 \times 10^{15}}{1.2 \times 10^{11}}$

12. $\dfrac{5.4 \times 10^{13}}{1.8 \times 10^{4}}$

13. $\dfrac{3.5 \times 10^{-6}}{1.4 \times 10^{-3}}$

14. $\dfrac{8.06 \times 10^{-11}}{2.6 \times 10^{-5}}$

15. $\dfrac{1.23 \times 10^{5}}{4.1 \times 10^{-4}}$

16. $\dfrac{9.02 \times 10^{6}}{4.1 \times 10^{-7}}$

In exercises **17–20**, simplify.

17. $\dfrac{(2.52 \times 10^{6})(3.1 \times 10^{-2})}{1.2 \times 10^{9}}$

18. $\dfrac{(4.16 \times 10^{11})(2.3 \times 10^{-4})}{3.2 \times 10^{2}}$

19. $\dfrac{(9.52 \times 10^{-6})(1.3 \times 10^{15})}{2.8 \times 10^{3}}$

20. $\dfrac{(3.52 \times 10^{-8})(1.9 \times 10^{15})}{2.2 \times 10^{4}}$

In exercises **21–32**, find any restricted values.

21. $5x^{-1}y$

22. $6x^{2}y^{-1}$

23. $10p^{-2}q^{-3}$

24. $7p^{-1}q^{-5}$

25. $2^{-3}mn^{4}$

26. $3^{-2}m^{2}n$

27. $(x + 2)^{-1}$

28. $(x - 5)^{-1}$

29. $(y + 10)(y - 6)^{-1}$

30. $(y + 8)^{-1}(y - 7)$

31. $z^{-2}(z + 1)(z - 6)^{-2}$

32. $z^{-1}(z - 5)(z + 10)^{-2}$

SECTION 5-3.   Product and Quotient of Rational Expressions

**KEY TOPICS
IN THIS SECTION**

1. Multiplying rational expressions

2. Dividing rational expressions

3. Combined operations

Rational expressions, like rational numbers, can be multiplied and divided. In this section, both operations are studied. To simplify the study, let us understand that all operations are being carried out on rational expressions in which the variables have been restricted so that no denominator is zero.

Multiplying Rational Expressions

The following expression is an indicated product of two rational expressions:

$$\frac{t - 3}{t + 5} \cdot \frac{t^{2} + 3t - 10}{t^{2} - 9}$$

The four polynomials $(t - 3)$, $(t^2 + 3t - 10)$, $(t + 5)$, and $(t^2 - 9)$ represent real numbers. Therefore, we can write the given indicated product as a single rational expression.

$$\frac{t - 3}{t + 5} \cdot \frac{t^2 + 3t - 10}{t^2 - 9} \text{ can be written as } \frac{(t - 3)(t^2 + 3t - 10)}{(t + 5)(t^2 - 9)}.$$

We certainly have the skills to multiply the two polynomials on the top, as well as the two on the bottom. However, should we want to evaluate this expression for some value of t (say $t = 4$), it would be easier to use the factored forms rather than the multiplied forms.

$$\frac{(t - 3)(t^2 + 3t - 10)}{(t + 5)(t^2 - 9)}$$ The given product

$$= \frac{(t - 3)(t - 2)(t + 5)}{(t + 5)(t - 3)(t + 3)}$$ Factoring $(t^2 + 3t - 10)$ and $(t^2 - 9)$

$$= \frac{t - 2}{t + 3}$$ Remove the common factors $(t - 3)$ and $(t + 5)$.

By completely factoring the numerator and denominator, we were able to reduce the expression to a simpler form. It should be reasonably apparent that the reduced form would be the easiest one to use for evaluating for $t = 4$.

$$\frac{4 - 2}{4 + 3}$$ Replace t by 4.

$$= \frac{2}{7}$$ Do the indicated operations, using the rule for order of operations.

If P, Q, R, and S are polynomials where $Q \neq 0$ and $S \neq 0$, then

$$\frac{P}{Q} \cdot \frac{R}{S} \text{ can be written as } \frac{P \cdot R}{Q \cdot S}.$$

If possible, factor and reduce.

Example 1. Multiply:

a. $\dfrac{10p^3}{9q^2} \cdot \dfrac{21q^5}{5p}$

b. $\dfrac{2x^2 - xy - y^2}{4x^2 + 2xy} \cdot \dfrac{2x^2 - 4xy}{x^2 - xy - 2y^2}$

Solution. a. $\dfrac{10p^3}{9q^2} \cdot \dfrac{21q^5}{5p}$ The given expressions

$$= \frac{10p^3 \cdot 21q^5}{9q^2 \cdot 5p}$$ $\dfrac{P}{Q} \cdot \dfrac{R}{S} = \dfrac{P \cdot R}{Q \cdot S}$

$$= \frac{14p^2q^3}{3}$$ Remove the common factors 5, 3, p, and q^2.

b. $\dfrac{2x^2 - xy - y^2}{4x^2 + 2xy} \cdot \dfrac{2x^2 - 4xy}{x^2 - xy - 2y^2}$

$= \dfrac{(2x + y)(x - y)2x(x - 2y)}{2x(2x + y)(x - 2y)(x + y)}$

$= \dfrac{x - y}{x + y}$ Remove the common factors $2x$,
     $(2x + y)$, and $(x - 2y)$.

Dividing Rational Expressions

The following is an indicated quotient of two rational expressions:

$$\frac{15t - 6t^2}{t^2 + 20t + 100} \div \frac{2t - 5}{t + 10}$$

The definition of division can be used to write this expression in terms of multiplication. To do this we multiply the first fraction by the *reciprocal* of the second one. The reciprocal of a rational expression is obtained in the same way as the reciprocal of a rational number.

If a and b are nonzero integers, then the reciprocal of $\dfrac{a}{b}$ is $\dfrac{b}{a}$.

The reciprocal of $\dfrac{5}{9}$ is $\dfrac{9}{5}$.

The reciprocal of $\dfrac{-4}{7}$ is $\dfrac{-7}{4}$.

If P and Q are nonzero polynomials, then the reciprocal of $\dfrac{P}{Q}$ is $\dfrac{Q}{P}$.

The reciprocal of $\dfrac{5u}{9v}$ is $\dfrac{9v}{5u}$.

The reciprocal of $\dfrac{2t - 5}{t + 10}$ is $\dfrac{t + 10}{2t - 5}$.

Using the reciprocal of $\dfrac{2t - 5}{t + 10}$,

$\dfrac{15t - 6t^2}{t^2 + 20t + 100} \div \dfrac{2t - 5}{t + 10}$ can be written $\dfrac{15t - 6t^2}{t^2 + 20t + 100} \cdot \dfrac{t + 10}{2t - 5}$.

This operation can be simplified as a product.

$\dfrac{15t - 6t^2}{t^2 + 20t + 100} \cdot \dfrac{t + 10}{2t - 5}$

$= \dfrac{(15t - 6t^2)(t + 10)}{(t^2 + 20t + 100)(2t - 5)}$ $\dfrac{P}{Q} \cdot \dfrac{R}{S} = \dfrac{P \cdot R}{Q \cdot S}$

$= \dfrac{-3t(2t - 5)(t + 10)}{(t + 10)(t + 10)(2t - 5)}$ Factor numerator and denominator.

$= \dfrac{-3t}{t + 10}$ Remove the common factors of
     $(2t - 5)$ and $(t + 10)$.

Notice that a $(-3t)$ was factored from $15t - 6t^2$ and not a $3t$. When the $(-3t)$ was factored, the signs of the resulting binomial were the same as the $(2t - 5)$ in the denominator.

$$15t - 6t^2 \begin{cases} \text{Factor } 3t \text{ yields } 3t(5 - 2t). \textbf{ Wrong signs} \\ \text{Factor } (-3t) \text{ yields } (-3t)(2t - 5). \textbf{ Correct signs} \end{cases}$$

If P, Q, R, and S are polynomials and Q, R, and S are not zero, then

$$\frac{P}{Q} \div \frac{R}{S} = \frac{P}{Q} \cdot \frac{S}{R} = \frac{P \cdot S}{Q \cdot R}.$$

If possible, factor and reduce.

Example 2. Divide:

a. $\dfrac{25k^2 - 1}{18k^3} \div \dfrac{5k - 1}{9k}$

b. $\dfrac{4u - 5v}{u^3 + 125v^3} \div \dfrac{16u^2 - 40uv + 25v^2}{u^2 - 5uv + 25v^2}$

Solution.

a. $\dfrac{25k^2 - 1}{18k^3} \div \dfrac{5k - 1}{9k}$ The given expressions

$$= \frac{25k^2 - 1}{18k^3} \cdot \frac{9k}{5k - 1} \qquad \frac{P}{Q} \div \frac{R}{S} = \frac{P}{Q} \cdot \frac{S}{R}$$

$$= \frac{(25k^2 - 1)9k}{18k^3(5k - 1)} \qquad \frac{P}{Q} \cdot \frac{S}{R} = \frac{P \cdot S}{Q \cdot R}$$

$$= \frac{9k(5k + 1)(5k - 1)}{18k^3(5k - 1)} \qquad \text{Factor } 25k^2 - 1.$$

$$= \frac{5k + 1}{2k^2} \qquad\qquad \begin{array}{l}\text{Remove the common factors} \\ \text{of } 9k \text{ and } (5k - 1).\end{array}$$

b. $\dfrac{4u - 5v}{u^3 + 125v^3} \div \dfrac{16u^2 - 40uv + 25v^2}{u^2 - 5uv + 25v^2}$

$$= \frac{4u - 5v}{u^3 + 125v^3} \cdot \frac{u^2 - 5uv + 25v^2}{16u^2 - 40uv + 25v^2}$$

$$= \frac{(4u - 5v)(u^2 - 5uv + 25v^2)}{(u^3 + 125v^3)(16u^2 - 40uv + 25v^2)}$$

$$= \frac{(4u - 5v)(u^2 - 5uv + 25v^2)}{(u + 5v)(u^2 - 5uv + 25v^2)(4u - 5v)(4u - 5v)}$$

$$= \frac{1}{(u + 5v)(4u - 5v)}$$

Combined Operations

Multiplication and division have equal priority in the rule for order of operations. As a consequence, it is the operation to the left that is simplified before the one on the right.

Example 3. Do the indicated operations.

$$\left(\frac{a^2 + 4ab - 5b^2}{4a^2 - b^2} \div \frac{ab - b^2}{3a^2b - 2ab^2}\right)\left(\frac{4a - 2b}{3a^2 + 15ab}\right)$$

Solution.　**Discussion.** The division is to the left of the multiplication; therefore this operation should be simplified first. However, once the division is changed to multiplication by using the reciprocal, we can simplify both products at the same time. Such a procedure can be used because multiplication is commutative and associative.

$$\left(\frac{a^2 + 4ab - 5b^2}{4a^2 - b^2} \div \frac{ab - b^2}{3a^2b - 2ab^2}\right)\left(\frac{4a - 2b}{3a^2 + 15ab}\right)$$

$$= \frac{a^2 + 4ab - 5b^2}{4a^2 - b^2} \cdot \frac{3a^2b - 2ab^2}{ab - b^2} \cdot \frac{4a - 2b}{3a^2 + 15ab}$$

$$= \frac{(a + 5b)(a - b)(ab)(3a - 2b)(2)(2a - b)}{(2a + b)(2a - b)(b)(a - b)(3a)(a + 5b)}$$

$$= \frac{2(3a - 2b)}{3(2a + b)} \qquad \begin{array}{l}\text{Remove the common factors:}\\ (a + 5b), (a - b), a, b, \text{ and } (2a - b)\end{array}$$

SECTION 5-3.　Practice Exercises

In exercises **1–20**, multiply and reduce.

[Example 1]　**1.** $\dfrac{6}{t^3} \cdot \dfrac{t^2}{2}$ 　　　　　　**2.** $\dfrac{t^4}{10} \cdot \dfrac{15}{t^2}$

3. $\dfrac{16x^3}{3y^3} \cdot \dfrac{9y^4}{4x}$ 　　　　**4.** $\dfrac{10x^5}{13y} \cdot \dfrac{39y^3}{15x^4}$

5. $\dfrac{3 - p}{4p} \cdot \dfrac{20p^3}{p^2 - 9}$ 　　　**6.** $\dfrac{5p^2}{49 - p^2} \cdot \dfrac{2p - 14}{25p}$

7. $\dfrac{4a^3}{27} \cdot \dfrac{9}{a^3 - a^2}$ 　　　**8.** $\dfrac{3a - 6}{5a^3} \cdot \dfrac{5a}{3}$

9. $\dfrac{a^2 - b^2}{3b^2} \cdot \dfrac{9b}{6a^2 + 6ab}$ 　　**10.** $\dfrac{a + b}{3ab} \cdot \dfrac{3a^2b - 3ab^2}{a^2 + 2ab + b^2}$

11. $\dfrac{4x^2 - 1}{5x^2 + 15x} \cdot \dfrac{x + 3}{2x - 1}$ 　　**12.** $\dfrac{x^3}{2x^2 + 5x + 3} \cdot \dfrac{6x^2 - x - 15}{3x^2 - 5x}$

13. $\dfrac{x^2 + x - 6}{x^2 + 7x + 12} \cdot \dfrac{x^2 + 3x - 4}{x^2 - 4}$ 　　**14.** $\dfrac{x^2 + 2x - 15}{x^2 - 9} \cdot \dfrac{x^2 - 3x - 18}{x^2 - 7x + 6}$

15. $\dfrac{9x^2 + 12xy + 4y^2}{9x^2y + 6xy^2} \cdot \dfrac{9x^2y - 6xy^2}{3x - 2y}$ 　　**16.** $\dfrac{4x^2 - 12xy + 9y^2}{10x^2y - 15xy^2} \cdot \dfrac{5x^2y + 5xy^2}{x + y}$

17. $\dfrac{t^3 - 125}{4t^2 + 6t} \cdot \dfrac{2t + 3}{t^2 - 10t + 25}$ 　　**18.** $\dfrac{t + 5}{25t^2 - 4} \cdot \dfrac{125t^3 + 8}{7t^2 + 35t}$

19. $\dfrac{a^2b - ab^2}{3a^2 + 15a - 2ab - 10b} \cdot \dfrac{a + 5}{a - b}$

20. $\dfrac{a^3 + a^2b + ab^2 + b^3}{9a^2 - b^2} \cdot \dfrac{3a + b}{a^2 + b^2}$

In exercises **21–40**, divide and reduce.

[Example 2]

21. $\dfrac{15z^2}{8} \div \dfrac{z^3}{2}$

22. $\dfrac{z^3}{12} \div \dfrac{z}{18}$

23. $\dfrac{42}{x^2y^4} \div \dfrac{6}{x^4y}$

24. $\dfrac{x^2y^3}{150} \div \dfrac{x^4y^2}{90}$

25. $\dfrac{6a^2}{a^2 - 1} \div \dfrac{3a}{a + 1}$

26. $\dfrac{a^2 - 4}{15a^3} \div \dfrac{a - 2}{5a^2}$

27. $\dfrac{9 - u^2}{9u^4} \div \dfrac{2u - 6}{3u^3}$

28. $\dfrac{70u^3}{u^2 - 10u + 25} \div \dfrac{14u^2}{5 - u}$

29. $\dfrac{a^2 + 2ab + b^2}{4a^2b + 2ab^2} \div \dfrac{a + b}{2a + b}$

30. $\dfrac{a^2 + 2ab + b^2}{a^2 - 2ab + b^2} \div \dfrac{a + b}{a - b}$

31. $\dfrac{8x^3 + 4x^2}{x^2 - 3x + 2} \div \dfrac{16x^2 + 8x}{x^2 - 4}$

32. $\dfrac{6x^2 - 3x}{x^2 + 6x + 5} \div \dfrac{4x^2 - 1}{2x^4 + 10x^3}$

33. $\dfrac{x^2 - y^2}{x^2 - 2xy + y^2} \div \dfrac{x^2 + 2xy + y^2}{x^2 - y^2}$

34. $\dfrac{x^2 - 1}{y^2 + 2y + 1} \div \dfrac{x^2 - 2x + 1}{y^2 - 1}$

35. $\dfrac{u^3 - 8v^3}{8u^3v^3} \div \dfrac{u - 2v}{4uv}$

36. $\dfrac{u^3 + 27v^3}{3uv} \div \dfrac{u^2 - 3uv + 9v^2}{9u^2v^2}$

37. $\dfrac{2t^2 + t - 15}{5t} \div (5 - 2t)$

38. $\dfrac{2t^2 + 7t - 4}{2t} \div (t + 4)$

39. $\dfrac{9m^4 - 1}{10m} \div \dfrac{3m^3 + m - 21m^2 - 7}{5m^2 - 35m}$

40. $\dfrac{6mn}{3m^2 + 3mn + m + n} \div \dfrac{6m^2n - 6mn^2}{3m + 1}$

In exercises **41–50**, do the indicated operations.

[Example 3]

41. $\left(\dfrac{12t^2u}{25v} \div \dfrac{6u^3}{5v^2}\right) \cdot \dfrac{10uv}{t^3}$

42. $\left(\dfrac{36uv}{7t^4} \div \dfrac{4v^3}{21t}\right) \cdot \dfrac{t^3u^2}{6v}$

43. $\left(\dfrac{x^2 - 9}{6x - 3} \cdot \dfrac{5x}{x + 3}\right) \div \dfrac{5x^2}{2x - 1}$

44. $\left(\dfrac{5x^2 + 2x}{x + 5} \cdot \dfrac{x^2 + 10x + 25}{15x^4 + 6x^3}\right) \div \dfrac{5}{3x}$

45. $\left(\dfrac{3y}{6y^2 - 2y} \div \dfrac{3y^2 - 9y}{2y^2 + 4y}\right) \cdot \dfrac{9y^2 - 1}{y + 2}$

46. $\left(\dfrac{10y^2 - 5y}{y^3 - 2y^2} \div \dfrac{4y^2 - 1}{2y}\right) \cdot \dfrac{y^2 - 4y + 4}{10}$

47. $\left(\dfrac{a^2 + 2ab + b^2}{9a^2} \div \dfrac{5b}{3a^2 - 2ab}\right) \cdot \dfrac{15b}{3a^2 + ab - 2b^2}$

48. $\left(\dfrac{5ab}{a^2 - 2ab + b^2} \div \dfrac{10a + 15b}{a^2 - b^2}\right) \cdot \dfrac{2a^2b + 3ab^2}{a^2b + ab^2}$

49. $\left(\dfrac{6n^2 + 13n + 6}{4n^2 - 9} \div \dfrac{6n^2 + n - 2}{4n^2 - 1}\right) \cdot \dfrac{3 - 2n}{2n + 1}$

50. $\left(\dfrac{2n^2 - 3n - 20}{2n^2 - 7n - 30} \div \dfrac{2n^2 - 5n - 12}{4n^2 + 12n + 9}\right) \cdot \dfrac{6 - n}{2n + 3}$

SECTION 5-3.   Ten Review Exercises

In exercises **1–4**, solve and check.

1. $5(a - 2) + 2(a + 6) = -3a - 68$ **2.** $12 - 3(4b + 2) = 2(4b - 3)$

3. $9x^2 = 4$ **4.** $5y^2 - 28 = 13y$

5. Solve and graph
$2(z - 4) + 6 \le 5z - 2(6 - z)$.

In exercises **6–10**, do the indicated operations.

6. $2(3x^3 + 5x - 7) + 3(2x^2 - 7x + 10) - 6(x^3 + x^2 - 2x + 3)$

7. $(3t - 5u)^2$ **8.** $(5m + 2)(2m^2 - 4m + 1)$

9. $\dfrac{18u^3v - 42u^2v^2 + 24uv^3}{6uv}$ **10.** $(6k^3 - 13k^2 + 13k + 6) \div (3k + 1)$

SECTION 5-3.   Supplementary Exercises

In exercises **1–10**,

a. evaluate the product or quotient for $x = 3$.

b. multiply or divide as indicated and reduce.

c. evaluate the expression found in part **b** for $x = 3$.

1. $\dfrac{1}{x - 1} \cdot \dfrac{3x - 3}{5}$ **2.** $\dfrac{1}{2x - 2} \cdot \dfrac{x - 1}{7}$

3. $\dfrac{x - 1}{x} \cdot \dfrac{x^2}{x^2 - 2x + 1}$ **4.** $\dfrac{2x + 4}{3x} \cdot \dfrac{2x^2}{x^2 + 4x + 4}$

5. $\dfrac{x + 1}{2x} \div \dfrac{x^2 + 2x + 1}{x}$ **6.** $\dfrac{4x + 2}{x^2 + 2x + 1} \div \dfrac{2x + 1}{x + 1}$

7. $(6x + 6) \div \dfrac{3x + 3}{x}$ **8.** $(2x + 6) \div \dfrac{x^2 + 2x - 3}{x^2 - x}$

9. $\dfrac{7x + 7}{2x + 6} \div \dfrac{7}{x + 3}$ **10.** $\dfrac{x + 3}{3x + 3} \div \dfrac{1}{2x + 2}$

In exercises **11–18**, simplify.

11. $\dfrac{16x - 24}{14x + 7} \cdot \dfrac{42x + 21}{24 - 16x}$ **12.** $\dfrac{15x - 60}{27x - 54} \cdot \dfrac{6 - 3x}{5x - 20}$

13. $\dfrac{s^4t^3}{s^2 - 3s - 28} \cdot \dfrac{49 - s^2}{st^4}$ **14.** $\dfrac{s^6t^3}{s^2 - 13s + 30} \cdot \dfrac{100 - s^2}{s^6t^2}$

15. $\dfrac{a^2 - a - 2}{a^2 - 7a + 10} \div \dfrac{a^2 - 3a - 4}{40 - 3a - a^2}$ **16.** $\dfrac{a^2 - 11a + 28}{a^2 - a - 42} \div \dfrac{8 + 2a - a^2}{a^2 + 7a + 10}$

17. $\dfrac{3x^3 + 4x^2}{5xy - 3y} \div \dfrac{3x^2 + 4x}{3y^3 - 5xy^3}$ 18. $\dfrac{2xy^2 + y^2}{5x^2 - 2x^3} \div \dfrac{2xy + y}{2x^2 - 5x}$

In exercises **19–24**, find the value of k in the expression on the left that causes it to reduce to the form of the expression on the right.

19. $\dfrac{x + 5}{3x} \cdot \dfrac{12x^2}{x^2 + kx + 10}$ becomes $\dfrac{4x}{x + 2}$.

20. $\dfrac{x}{x^2 + kx + 15} \cdot \dfrac{2x + 10}{x^2}$ becomes $\dfrac{2}{x(x + 3)}$.

21. $\dfrac{4a + 8}{a^2 - 25} \cdot \dfrac{a - 5}{5a + k}$ becomes $\dfrac{4}{5(a + 5)}$.

22. $\dfrac{3a - k}{a^2 - 9} \cdot \dfrac{a + 3}{a^2 - 2a}$ becomes $\dfrac{3}{a(a - 3)}$.

23. $\dfrac{b^2 + 20b + 99}{b + 9} \cdot \dfrac{b + 7}{b^2 + kb + 11}$ becomes $\dfrac{b + 7}{b + 1}$.

24. $\dfrac{b^2 + 19b + 84}{b - 3} \cdot \dfrac{b^2 - 9}{b^2 + kb + 36}$ becomes $b + 7$.

SECTION 5-4. Sum and Difference of Rational Expressions, Same Denominator

KEY TOPICS IN THIS SECTION

1. Rational expressions with the same denominator

2. Adding and subtracting rational expressions with the same denominator

3. Combined operations

Rational expressions, like rational numbers, can be added and subtracted. In this section both operations are studied for rational expressions that have the same denominator.

Rational Expressions with the Same Denominator

Examples **a** and **b** are indicated sums of two rational expressions that have the same denominator $15x$:

a. $\dfrac{45x^3}{15x} + \dfrac{75x^2}{15x}$ **b.** $\dfrac{3x + 7}{15x} + \dfrac{2x + 3}{15x}$

According to the rule for the order of operations, division has priority over addition. Since the numerators of the rational expressions in example **a** can be divided by $15x$ with a remainder zero, this expression can be simplified as follows:

a. $\quad \dfrac{45x^3}{15x} + \dfrac{75x^2}{15x}$ \qquad The given expression

$\quad = 3x^2 + 5x$ \qquad Simplify the divisions.

In example **b** the numerators cannot be divided by $15x$ with a remainder zero. However, by using the definition of division and the distributive property of multiplication over addition, the adding of the numerators can be given priority over the divisions. Therefore, this expression could be simplified as follows:

b. $\dfrac{3x + 7}{15x} + \dfrac{2x + 3}{15x}$ The given expression

$= (3x + 7) \cdot \dfrac{1}{15x} + (2x + 3) \cdot \dfrac{1}{15x}$ $a \div b = a \cdot \dfrac{1}{b}$

$= [(3x + 7) + (2x + 3)] \cdot \dfrac{1}{15x}$ Factor $\frac{1}{15x}$ as a common factor.

$= \dfrac{(3x + 7) + (2x + 3)}{15x}$ $a \cdot \dfrac{1}{b} = a \div b$, or $\dfrac{a}{b}$

$= \dfrac{5x + 10}{15x}$ $3x + 2x = 5x$ and $7 + 3 = 10$

$= \dfrac{5(x + 2)}{15x}$ Factor a 5 in the numerator.

$= \dfrac{x + 2}{3x}$ Reduce the fraction.

The fourth line of the above process shows an indicated sum of the numerators of the two expressions written over the common denominator $15x$. In common practice we skip lines two and three of the procedure and go directly to line four. The following are general statements of this fact for indicated sums and differences of rational expressions with the same nonzero denominators.

If P, Q, and R are polynomials and $Q \neq 0$, then

$$\dfrac{P}{Q} + \dfrac{R}{Q} \text{ can be written as } \dfrac{P + R}{Q}.$$

$$\dfrac{P}{Q} - \dfrac{R}{Q} \text{ can be written as } \dfrac{P - R}{Q}.$$

If possible, simplify the numerator, factor, and reduce.

Adding and Subtracting Rational Expressions with the Same Denominator

In the following examples, the denominators of the rational expressions being added or subtracted are the same.

Example 1.   Add or subtract, as indicated.

a. $\dfrac{17}{13t} + \dfrac{9}{13t}$

b. $\dfrac{31}{10mn} - \dfrac{7}{10mn}$

Solution.   **a.** $\dfrac{17}{13t} + \dfrac{9}{13t}$ The given expression

$= \dfrac{17 + 9}{13t}$ $\dfrac{P}{Q} + \dfrac{R}{Q} = \dfrac{P + R}{Q}$

$= \dfrac{26}{13t}$ Simplify the numerator.

$= \dfrac{2}{t}$ Reduce the fraction.

b. $\dfrac{31}{10mn} - \dfrac{7}{10mn}$ The given expression

$= \dfrac{31 - 7}{10mn}$ $\dfrac{P}{Q} - \dfrac{R}{Q} = \dfrac{P - R}{Q}$

$= \dfrac{24}{10mn}$ Simplify the numerator.

$= \dfrac{12}{5mn}$ Reduce the fraction.

Not all numerators can be simplified, nor can all fractions be reduced as the ones in Example 1. However, we should always simplify as much as possible, and reduce any fraction that can be reduced.

Example 2.   Add or subtract as indicated.

   a. $\dfrac{7u}{12v} + \dfrac{5}{12v}$ **b.** $\dfrac{7u - 1}{12v} - \dfrac{3u + 2}{12v}$

Solution.   **a.** $\dfrac{7u}{12v} + \dfrac{5}{12v}$ The given expression

$= \dfrac{7u + 5}{12v}$ Cannot be simplified or reduced.

b. $\dfrac{7u - 1}{12v} - \dfrac{3u + 2}{12v}$ The given expression

$= \dfrac{(7u - 1) - (3u + 2)}{12v}$ Keep the numerators grouped.

$= \dfrac{7u - 1 - 3u - 2}{12v}$ $-(3u + 2) = -1 \cdot (3u + 2)$
                         $= -3u - 2$

$= \dfrac{4u - 3}{12v}$ Simplify the numerator.

Since the numerator of this last expression cannot be factored, the fraction cannot be reduced.

Example 3.   Do the indicated operations.

$$\frac{2y^2}{3y^2 + 5y - 2} - \frac{y^2 + 4y}{3y^2 + 5y - 2} + \frac{3(y - 2)}{3y^2 + 5y - 2}$$

Solution.   **Discussion.** Each of the three fractions has the common denominator $3y^2 + 5y - 2$. As a consequence we can write the sum or difference (as indicated), over the common denominator.

$$\frac{2y^2}{3y^2 + 5y - 2} - \frac{y^2 + 4y}{3y^2 + 5y - 2} + \frac{3(y-2)}{3y^2 + 5y - 2}$$

$$= \frac{2y^2 - (y^2 + 4y) + 3(y-2)}{3y^2 + 5y - 2}$$

$$= \frac{2y^2 - y^2 - 4y + 3y - 6}{3y^2 + 5y - 2}$$

$$= \frac{y^2 - y - 6}{3y^2 + 5y - 2} \qquad \text{Combine like terms.}$$

$$= \frac{(y-3)(y+2)}{(3y-1)(y+2)} \qquad \text{Factor top and bottom.}$$

$$= \frac{y-3}{3y-1} \qquad \text{Remove the common factor } (y+2).$$

Combined Operations

In Example 4 the rule for order of operations must be followed to simplify the expression.

Example 4.   Do the indicated operations.

$$\frac{t^2}{t^2 - t - 2} - \frac{t^2 - 2t}{t^2 + 2t + 1} \div \frac{t^2 - 4t + 4}{3t + 3} + \frac{2}{t^2 - t - 2}$$

Solution.   **Discussion.** This expression indicates three operations on four rational expressions. The division in the middle has the highest priority. The quotient from this division will be completely simplified before the subtraction and addition are attempted.

$$\frac{t^2}{t^2 - t - 2} - \frac{t^2 - 2t}{t^2 + 2t + 1} \div \frac{t^2 - 4t + 4}{3t + 3} + \frac{2}{t^2 - t - 2}$$

$$= \frac{t^2}{t^2 - t - 2} - \frac{(t^2 - 2t)(3t + 3)}{(t^2 + 2t + 1)(t^2 - 4t + 4)} + \frac{2}{t^2 - t - 2}$$

$$= \frac{t^2}{t^2 - t - 2} - \frac{t(t - 2)(3)(t + 1)}{(t + 1)(t + 1)(t - 2)(t - 2)} + \frac{2}{t^2 - t - 2}$$

$$= \frac{t^2}{(t + 1)(t - 2)} - \frac{3t}{(t + 1)(t - 2)} + \frac{2}{(t + 1)(t - 2)}$$

$$= \frac{t^2 - 3t + 2}{(t + 1)(t - 2)}$$

$$= \frac{(t - 1)(t - 2)}{(t + 1)(t - 2)}$$

$$= \frac{t - 1}{t + 1}$$

Example 4 illustrates vividly that simplifying an expression means writing that expression with fewer symbols. The values the expression yields in the simplified form are the same as the values obtained in the unsimplified form. (We must, of course, exclude restricted values.) For example, suppose we replace t by 5 in the given expression. We can choose to simplify the following:

$$\frac{5^2}{5^2 - 5 - 2} - \frac{5^2 - 2(5)}{5^2 + 2(5) + 1} \div \frac{5^2 - 4(5) + 4}{3(5) + 3} + \frac{2}{5^2 - 5 - 2}$$

or choose to simplify the following:

$$\frac{5-1}{5+1}.$$

Both expressions yield the same number, namely $\frac{2}{3}$.

SECTION 5-4. Practice Exercises

In exercises **1–18**, simplify.

[Examples 1 and 2]

1. $\dfrac{7}{12y} - \dfrac{1}{12y}$ **2.** $\dfrac{17}{3y} - \dfrac{2}{3y}$

3. $\dfrac{31}{15a^2} + \dfrac{4}{15a^2}$ **4.** $\dfrac{19}{21a^2} + \dfrac{5}{21a^2}$

5. $\dfrac{8x}{3y} + \dfrac{7x}{3y}$ **6.** $\dfrac{11x}{4y} + \dfrac{5x}{4y}$

7. $\dfrac{32}{9ab} - \dfrac{5}{9ab}$ **8.** $\dfrac{26}{3ab} - \dfrac{2}{3ab}$

9. $\dfrac{8}{5x^2y} - \dfrac{43}{5x^2y}$ **10.** $\dfrac{3}{2xy^2} - \dfrac{31}{2xy^2}$

11. $\dfrac{10u}{11w} + \dfrac{5}{11w}$ **12.** $\dfrac{8u}{3w} + \dfrac{4}{3w}$

13. $\dfrac{3x}{8y^2} - \dfrac{12}{8y^2}$ **14.** $\dfrac{4x}{7y^2} - \dfrac{20}{7y^2}$

15. $\dfrac{3p+2}{3p} + \dfrac{4p-1}{3p}$ **16.** $\dfrac{5p-8}{2p} + \dfrac{p+3}{2p}$

17. $\dfrac{6m-5n}{7mn} - \dfrac{m-4n}{7mn}$ **18.** $\dfrac{10m+3n}{5mn} - \dfrac{4m-n}{5mn}$

In exercises **19–34**, simplify.

[Example 3]

19. $\dfrac{3x}{x+1} + \dfrac{3}{x+1}$ **20.** $\dfrac{5x}{x-3} - \dfrac{15}{x-3}$

21. $\dfrac{10b}{b-1} - \dfrac{10}{b-1}$ **22.** $\dfrac{4b}{2b-1} - \dfrac{2}{2b-1}$

23. $\dfrac{7a+3}{2(a+1)} + \dfrac{3a-3}{2(a+1)}$

24. $\dfrac{8a-1}{3(a-1)} + \dfrac{4a+1}{3(a-1)}$

25. $\dfrac{3a}{4a^2-1} + \dfrac{a+2}{4a^2-1}$

26. $\dfrac{7a}{a^2-4} - \dfrac{4a+6}{a^2-4}$

27. $\dfrac{b}{b^2-3b+2} - \dfrac{2-b^2}{b^2-3b+2}$

28. $\dfrac{b^2}{b^2+3b+2} - \dfrac{2-b}{b^2+3b+2}$

29. $\dfrac{3z(z+2)}{6z^2-5z-6} + \dfrac{2z+4}{6z^2-5z-6}$

30. $\dfrac{4z^2}{8z^2-2z-1} - \dfrac{4z-1}{8z^2-2z-1}$

31. $\dfrac{a-3}{a^2+2a-3} - \dfrac{a^2-3a}{a^2+2a-3} + \dfrac{2a^2+6}{a^2+2a-3}$

32. $\dfrac{a^2}{a^2-8a+12} + \dfrac{a-1}{a^2-8a+12} - \dfrac{2a+1}{a^2-8a+12}$

33. $\dfrac{2z^2+3z-21}{z^2-2z-63} - \dfrac{z^2-3z}{z^2-2z-63} + \dfrac{7-z}{z^2-2z-63}$

34. $\dfrac{2z^2-4}{z^2-9z+20} - \dfrac{4z^2+2z}{z^2-9z+20} + \dfrac{2z^2+3z}{z^2-9z+20}$

In exercises **35–40**, simplify.

[Example 4]　**35.** $\dfrac{2x}{x^2-1} \cdot \dfrac{x-1}{x+2} + \dfrac{2}{x^2+3x+2}$

36. $\dfrac{2x}{4x^2-1} \cdot \dfrac{2x+1}{x-3} - \dfrac{6}{2x^2-7x+3}$

37. $\dfrac{5y}{y^2+4y+4} \div \dfrac{15}{y+2} + \dfrac{2}{3y+6}$

38. $\dfrac{3y}{y^2-6y+9} \div \dfrac{4y}{y-3} - \dfrac{y}{4y-12}$

39. $\dfrac{8}{15xy} + \dfrac{2x-2y}{3x^2+3xy} \cdot \dfrac{x+y}{5xy-5y^2} - \dfrac{4}{15xy}$

40. $\dfrac{x+3}{24xy} + \dfrac{3x-y}{6x^2+18xy} \cdot \dfrac{x+3y}{12xy-4y^2} - \dfrac{x+1}{24xy}$

SECTION 5-4.　Ten Review Exercises

In exercises **1–8**, solve each problem.

1. Multiply: $(4t-3)(t+6)$

2. Reduce: $\dfrac{4t^2+21t-18}{4t-3}$

3. Divide: $(4t^2+21t-18) \div (4t-3)$

4. Evaluate $4t^2+21t-18$ for $t=-5$.

5. Evaluate $21t-18+4t^2$ for $t=\frac{1}{2}$.

6. Solve: $4t^2=18-21t$

7. Simplify: $\dfrac{4t^2}{4t-3} - \dfrac{18}{4t-3} + \dfrac{21t}{4t-3}$

8. Is -5 a root of $4t^2+21t-18=0$?

In exercises **9** and **10**, factor completely.

9. $-3m^3n + 48m^2n^2 - 192mn^3$

10. $180u^3v - 5uv^3$

SECTION 5-4.   Supplementary Exercises

In exercises **1–6**, simplify parts **a** and **b**. (Note that in part **b** division can be done first.)

1. a. $\dfrac{3}{5x} + \dfrac{7}{5x}$ **b.** $\dfrac{15x^2}{5x} + \dfrac{35x}{5x}$ **2. a.** $\dfrac{12}{5x} - \dfrac{2}{5x}$ **b.** $\dfrac{60x^2}{5x} - \dfrac{10x}{5x}$

3. a. $\dfrac{7}{3y} - \dfrac{4}{3y}$ **b.** $\dfrac{21y^3}{3y} - \dfrac{12y^2}{3y}$ **4. a.** $\dfrac{7}{4y} + \dfrac{5}{4y}$ **b.** $\dfrac{28y^4}{4y} + \dfrac{20y^2}{4y}$

5. a. $\dfrac{2x+1}{3x} + \dfrac{x+5}{3x}$ **b.** $\dfrac{3x^2+6x}{3x} + \dfrac{9x^2-3x}{3x}$

6. a. $\dfrac{8x-14}{5x} - \dfrac{3x+1}{5x}$ **b.** $\dfrac{40x^2-70x}{5x} - \dfrac{15x^2+5x}{5x}$

In exercises **7–12**,

a. evaluate both expressions for $x = 3$; then add or subtract as indicated.

b. add or subtract (as indicated) the expressions and reduce.

c. evaluate the reduced expression in part **b** for $x = 3$ and compare with part **a**.

7. $\dfrac{x+3}{x+2} + \dfrac{x+1}{x+2}$

8. $\dfrac{3x-2}{x-1} - \dfrac{1}{x-1}$

9. $\dfrac{3x+4}{2x} - \dfrac{x+2}{2x}$

10. $\dfrac{x+2}{6x} - \dfrac{3x-2}{6x}$

11. $\dfrac{6x+5}{4(x+2)} - \dfrac{2x+1}{4(x+2)}$

12. $\dfrac{6x+3}{4(x+2)} - \dfrac{2x+5}{4(x+2)}$

In exercises **13–18**, simplify.

13. $\dfrac{a^3}{a+b} + \dfrac{b^3}{a+b}$

14. $\dfrac{a^3}{a-b} - \dfrac{b^3}{a-b}$

15. $\dfrac{x^2}{x^3+8} - \dfrac{2x-4}{x^3+8}$

16. $\dfrac{x^2}{x^3-27} + \dfrac{3x+9}{x^3-27}$

17. $\dfrac{-5(x-5)}{125+x^3} + \dfrac{x^2}{125+x^3}$

18. $\dfrac{4(x+4)}{64-x^3} + \dfrac{x^2}{64-x^3}$

In exercises **19–26**, find the values of k so that the denominators will be the same.

19. $\dfrac{1}{(x+5)(x-3)}, \dfrac{1}{x^2+kx-15}$

20. $\dfrac{1}{(x+4)(x-7)}, \dfrac{1}{x^2+kx-28}$

21. $\dfrac{1}{5x^4y^3}, \dfrac{1}{5x^ky^3}$

22. $\dfrac{1}{8x^3y^5}, \dfrac{1}{8x^ky^5}$

23. $\dfrac{1}{(8x^6y^3)}, \dfrac{1}{(2x^2y)^k}$

24. $\dfrac{1}{16x^4y^8}, \dfrac{1}{(2xy^2)^k}$

25. $\dfrac{1}{(x+3)(x-3)}, \dfrac{1}{x^2+kx-9}$

26. $\dfrac{1}{(x-5)(x+5)}, \dfrac{1}{x^2+kx-25}$

SECTION 5-5.   Sum and Difference of Rational Expressions, Different Denominators

KEY TOPICS IN THIS SECTION

1. Rational expressions with different denominators

2. The least common multiple (LCM) of two or more polynomials

3. Writing a rational expression as an equivalent expression with an LCM as denominator

4. Adding and subtracting rational expressions with different denominators

5. Combined operations

In Section 5-4 we added and subtracted rational expressions with the same denominator. In this section we will study a procedure for adding and subtracting rational expressions with different denominators.

Rational Expressions with Different Denominators

Examples **a** and **b** are indicated sums of two rational expressions with different denominators, namely $8x^2$ and $12x^3$:

a. $\dfrac{72x^5}{8x^2} + \dfrac{60x^4}{12x^3}$ **b.** $\dfrac{2x+1}{8x^2} + \dfrac{x-1}{12x^3}$

The rule for order of operations gives division priority over addition. In example **a** the numerators are evenly divisible by the denominators; therefore the expression can be simplified as follows:

a. $\dfrac{72x^5}{8x^2} + \dfrac{60x^4}{12x^3}$ The given expression

$= 9x^3 + 5x$ Simplify the divisions.

In example **b** the numerators cannot be divided by the denominators with a remainder zero. Furthermore, the two denominators are not the same. Therefore, the distributive property of multiplication over addition cannot be used to change the order of the indicated operations.

$$\frac{2x+1}{8x^2} + \frac{x-1}{12x^3}$$ The given expression

$$= (2x+1) \cdot \frac{1}{8x^2} + (x-1) \cdot \frac{1}{12x^3}$$ $a \div b = a \cdot \dfrac{1}{b}$

$\underset{\textbf{Not the same}}{\underbrace{\qquad\qquad}}$ Cannot be factored

To simplify this sum we must first change both expressions to ones that have the same denominator. Furthermore, we still want the new expressions to have the same values for all suitable replacements of x. To make this change we use the multiplication property of one.

For any a, $a \cdot 1 = a$ and $1 \cdot a = a$.

Although an infinite number of expressions could be used as a common denominator, the work is easiest when the *least common multiple* of $8x^2$ and $12x^3$ is used.

The Least Common Multiple (LCM) of Two or More Polynomials

The following definition tells the meaning of the least common multiple of two or more polynomials.

Definition 5.3. The least common multiple of Q and S

If Q and S are polynomials, then the least common multiple of Q and S is the polynomial with the *fewest factors* that Q and S divide with a remainder of zero.

Based on Definition 5.3, the LCM of $8x^2$ and $12x^3$ is the polynomial T, such that $8x^2$ and $12x^3$ both divide T with a remainder zero. That is, T is evenly divisible by $8x^2$ and $12x^3$.

For the coefficients 8 and 12, the LCM is 24
$$\frac{24}{8} = 3$$
$$\frac{24}{12} = 2$$

For the variable parts x^2 and x^3, the LCM is x^3
$$\frac{x^3}{x^2} = x$$
$$\frac{x^3}{x^3} = 1$$

Thus T is the polynomial $24x^3$. That is, $24x^3$ has the fewest factors that $8x^2$ and $12x^3$ can divide evenly.

Example 1. Find the LCM.

  a. $10a^2b$ and $15ab^3$

  b. $u^2 - 7u + 10$ and $2u^2 - 9u - 5$

Solution. **a.** The LCM of 10 and 15 is 30.
  The LCM of a^2 and a is a^2.
  The LCM of b and b^3 is b^3.
  Thus, the LCM is $30a^2b^3$.

  b. Discussion. To find the LCM of the two trinomials we must first write them in factored form.

  $$u^2 - 7u + 10 = (u - 2)(u - 5)$$

  $$2u^2 - 9u - 5 = (2u + 1)(u - 5)$$

  Thus, the LCM is $(u - 2)(u - 5)(2u + 1)$.

Example 2. Find the LCM.

  $3t^2 - 12$, $2t^2 + 8t + 8$, and $6t - 12$

Solution. **Discussion.** To extend Definition 5.3 to three polynomials, we must find a polynomial that contains the fewest factors that all three polynomials divide with a remainder zero.

$$3t^2 - 12 = 3(t^2 - 4) = 3(t + 2)(t - 2)$$

$$2t^2 + 8t + 8 = 2(t^2 + 4t + 4) = 2(t + 2)^2$$

$$6t - 12 = 6(t - 2)$$

The LCM of 2, 3, and 6 is 6.
The LCM must contain one factor of $(t - 2)$, but two factors of $(t + 2)$.
Therefore the LCM is $6(t - 2)(t + 2)^2$.

Writing a Rational Expression as an Equivalent Expression with an LCM as Denominator

As previously determined, the LCM of $8x^2$ and $12x^3$ is $24x^3$. We can therefore write the rational expressions of example **b** as **equivalent expressions** with denominator $24x^3$. By *equivalent expressions* we mean that the numbers the expressions represent are not changed, only the forms of the expressions. The multiplication property of one is the reason the expressions are equivalent.

$$\frac{2x + 1}{8x^2} \qquad\qquad \frac{x - 1}{12x^3} \qquad\qquad \text{The given expressions}$$

$$= \frac{2x + 1}{8x^2} \cdot \frac{3x}{3x} \qquad = \frac{x - 1}{12x^3} \cdot \frac{2}{2} \qquad \frac{3x}{3x} = \frac{2}{2} = 1 \text{ and } a \cdot 1 = a$$

$$= \frac{3x(2x + 1)}{24x^3} \qquad = \frac{2(x - 1)}{24x^3} \qquad \begin{array}{l}\text{Equivalent fractions,}\\\text{LCM as denominator}\end{array}$$

Example 3.   For the fractions $\dfrac{5}{18r^2s}$, $\dfrac{11}{24rt}$, and $\dfrac{1}{9st}$,

   a. determine the LCM of the denominators.

   b. write as equivalent fractions with the LCM as denominator.

Solution.   **a.** The LCM of 18, 24, and 9 is 72.
The LCM of r^2s, rt, and st is r^2st.
Thus, the LCM of the denominators is $72r^2st$.

   b. Discussion. To determine what to multiply each fraction by to get the LCM as a denominator, divide the LCM by each denominator.

$$\frac{72r^2st}{18r^2s} = 4t \qquad \frac{72r^2st}{24rt} = 3rs \qquad \frac{72r^2st}{9st} = 8r^2$$

$$\left.\begin{array}{l}\dfrac{5}{18r^2s} \cdot \dfrac{4t}{4t} = \dfrac{20t}{72r^2st}\\[2mm]\dfrac{11}{24rt} \cdot \dfrac{3rs}{3rs} = \dfrac{33rs}{72r^2st}\\[2mm]\dfrac{1}{9st} \cdot \dfrac{8r^2}{8r^2} = \dfrac{8r^2}{72r^2st}\end{array}\right\} \quad \begin{array}{c}\textbf{Equivalent fractions}\\\textbf{but}\\\textbf{LCM as denominators}\end{array}$$

Notice that the procedure in Example 3 is the reverse of that used in reducing fractions. Instead of removing common factors, we put common factors into the numerator and denominator. This process is referred to as **building up fractions.**

Adding and Subtracting Rational Expressions with Different Denominators

Prior to Example 3 we wrote the fractions of example **b** as equivalent fractions with $24x^3$ as a common denominator. These fractions can now be added using the procedure learned in the last section.

b. $\dfrac{2x + 1}{8x^2} + \dfrac{x - 1}{12x^3}$ The given fractions

$= \dfrac{3x(2x + 1)}{24x^3} + \dfrac{2(x - 1)}{24x^3}$ Equivalent fractions with $24x^3$ as the common denominator.

$= \dfrac{3x(2x + 1) + 2(x - 1)}{24x^3}$ $\dfrac{P}{Q} + \dfrac{R}{Q} = \dfrac{P + R}{Q}$

$= \dfrac{6x^2 + 3x + 2x - 2}{24x^3}$ Remove parentheses in numerator.

$= \dfrac{6x^2 + 5x - 2}{24x^3}$ Simplify the numerator.

Example 4.   Do the indicated operations.

a. $\dfrac{3x + 2}{10x^2y} - \dfrac{7y - 1}{15xy^2}$

b. $\dfrac{1}{3t} + \dfrac{5}{3t + 6}$

Solution.   **a.** $\dfrac{3x + 2}{10x^2y} - \dfrac{7y - 1}{15xy^2}$ The given expressions

$= \dfrac{3x + 2}{10x^2y} \cdot \dfrac{3y}{3y} - \dfrac{7y - 1}{15xy^2} \cdot \dfrac{2x}{2x}$ The LCM of denominators is $30x^2y^2$.

$= \dfrac{3y(3x + 2) - 2x(7y - 1)}{30x^2y^2}$ Combine the numerators over the common denominator.

$= \dfrac{9xy + 6y - 14xy + 2x}{30x^2y^2}$ Remove the parentheses.

$= \dfrac{2x + 6y - 5xy}{30x^2y^2}$ Combine like terms.

b. $\dfrac{1}{3t} + \dfrac{5}{3(t + 2)}$

$= \dfrac{1}{3t}\left(\dfrac{t + 2}{t + 2}\right) + \dfrac{5}{3(t + 2)}\left(\dfrac{t}{t}\right)$

$= \dfrac{t + 2 + 5t}{3t(t + 2)}$

$= \dfrac{6t + 2}{3t(t + 2)}$ The fraction will not reduce.

Example 5.   Do the indicated operations.

$$\frac{2t}{3t + 9} + \frac{5}{2t - 4} - \frac{10t + 55}{6t^2 + 6t - 36}$$

Solution.

$$\frac{2t}{3t + 9} + \frac{5}{2t - 4} - \frac{10t + 55}{6t^2 + 6t - 36}$$

$$= \frac{2t}{3(t + 3)} + \frac{5}{2(t - 2)} - \frac{10t + 55}{6(t + 3)(t - 2)}$$

(The LCM of $3(t + 3)$, $2(t - 2)$, and $6(t + 3)(t - 2)$ is $6(t + 3)(t - 2)$.)

$$= \frac{2t}{3(t + 3)} \cdot \frac{2(t - 2)}{2(t - 2)} + \frac{5}{2(t - 2)} \cdot \frac{3(t + 3)}{3(t + 3)} - \frac{10t + 55}{6(t + 3)(t - 2)}$$

$$= \frac{4t(t - 2) + 15(t + 3) - (10t + 55)}{6(t + 3)(t - 2)}$$

$$= \frac{4t^2 - 8t + 15t + 45 - 10t - 55}{6(t + 3)(t - 2)}$$

$$= \frac{4t^2 - 3t - 10}{6(t + 3)(t - 2)}$$

$$= \frac{(4t + 5)(t - 2)}{6(t + 3)(t - 2)}$$

$$= \frac{4t + 5}{6(t + 3)}$$

Example 6.   Do the indicated operations.

$$\frac{-3}{m^2 - 9} - \frac{2}{3 - m}$$

Solution.   **Discussion.** The denominator of the left expression factors as $(m + 3)(m - 3)$. The binomial in the right expression can be changed to $m - 3$ by multiplying numerator and denominator by (-1).

$$\frac{2}{3 - m} = \frac{2}{3 - m} \cdot \frac{-1}{-1} = \frac{-2}{m - 3}$$

Now the LCM is $(m + 3)(m - 3)$.

$$\frac{-3}{(m + 3)(m - 3)} - \frac{-2}{m - 3} \cdot \frac{m + 3}{m + 3}$$

$$= \frac{-3 - (-2(m + 3))}{(m + 3)(m - 3)}$$

$$= \frac{-3 + 2m + 6}{(m + 3)(m - 3)}$$

$$= \frac{2m + 3}{(m + 3)(m - 3)}$$

Combined Operations

When addition, subtraction, multiplication, and division are indicated in the same expression, the rule for order of operations must be followed to simplify the expression.

Example 7. Do the indicated operations.

$$\frac{x+3}{8x^2y} - \frac{3x-y}{4x^2+12xy} \cdot \frac{x+3y}{18xy-6y^2} - \frac{y+1}{12xy^2}$$

Solution.   **Discussion.** The multiplication must be simplified before finding the LCM of the three denominators.

$$\frac{x+3}{8x^2y} - \frac{(3x-y)(x+3y)}{(4x^2+12xy)(18xy-6y^2)} - \frac{y+1}{12xy^2}$$

$$= \frac{x+3}{8x^2y} - \frac{(3x-y)(x+3y)}{4x(x+3y)(6y)(3x-y)} - \frac{y+1}{12xy^2}$$

$$= \frac{x+3}{8x^2y} - \frac{1}{24xy} - \frac{y+1}{12xy^2}$$

(The LCM of $8x^2y$, $24xy$, and $12xy^2$ is $24x^2y^2$.)

$$= \frac{x+3}{8x^2y} \cdot \frac{3y}{3y} - \frac{1}{24xy} \cdot \frac{xy}{xy} - \frac{y+1}{12xy^2} \cdot \frac{2x}{2x}$$

$$= \frac{3y(x+3) - xy - 2x(y+1)}{24x^2y^2}$$

$$= \frac{3xy + 9y - xy - 2xy - 2x}{24x^2y^2}$$

$$= \frac{9y - 2x}{24x^2y^2}$$

SECTION 5-5. Practice Exercises

In exercises **1–24**, find the least common multiple.

[Examples 1 and 2]

1. $3x$ and $4x$

2. $5x$ and $7x$

3. $6y$ and $8y^2$

4. $6y^3$ and $4y^2$

5. $6a$ and $9b$

6. $10a$ and $8b$

7. $5a^2b$ and $2ab^3$

8. $3ab^2$ and $5a^3b$

9. $9y$ and $3y^2 - 3y$

10. $10y$ and $2y^2 + 4y$

11. $6a^3$ and $2a^3 - 2a^2$

12. $9a^2$ and $3a^2 + 6a$

13. $a^2 - b^2$ and $a^2 - 2ab + b^2$

14. $4a^2 - 9b^2$ and $4a^2 + 12ab + 9b^2$

15. $x^2 - xy - 6y^2$ and $2x^2 + 8xy + 8y^2$

16. $2x^2 + 4xy - 6y^2$ and $3x^2 + 18xy + 27y^2$

17. $3m^2 - 15m + 18$ and $9m^2 - 36m + 36$

18. $40m^2 + 40m + 10$ and $20m^2 - 5$

19. $3x^2 - 12$, $x^2 + 5x + 6$, and $3x^2 + 3x - 18$

20. $x^2 + 8x + 15$, $2x^2 - 18$, and $2x^2 + 4x - 30$

21. $3x^2 + 6x + 3$, $x^2 + 5x + 4$, and $2x + 8$

22. $5x^2 + 20x + 20$, $3x + 6$, and $x^2 - x - 6$

23. $8y^3 - 1$, $8y^2 + 4y + 2$, and $4y^2 - 1$

24. $y^3 + z^3$, $y^2 - z^2$, and $2y^2 - 2yz + 2z^2$

In exercises **25–36**,

a. determine the LCM.

b. write each expression as an equivalent fraction with the LCM as the denominator.

[Example 3] 25. $\dfrac{7}{4x^2y}$ and $\dfrac{5}{2xy^2}$ 26. $\dfrac{3}{4xy}$ and $\dfrac{11}{2x^2y}$

27. $\dfrac{5}{2a + 10}$ and $\dfrac{4}{3a + 15}$ 28. $\dfrac{8}{5a - 5}$ and $\dfrac{7}{2a - 2}$

29. $\dfrac{a}{3a - 6}$ and $\dfrac{2a}{a^2 - 4}$ 30. $\dfrac{3}{2a - 10}$ and $\dfrac{a}{a^2 - 10a + 25}$

31. $\dfrac{2x}{x^2 - 9}$ and $\dfrac{x + 2}{6x^2 - 17x - 3}$ 32. $\dfrac{x - 3}{2x^2 - 10x + 8}$ and $\dfrac{x}{x^2 - 8x + 16}$

33. $\dfrac{mn}{m^3 + 8n^3}$ and $\dfrac{1}{m^2 - 2mn + 4n^2}$ 34. $\dfrac{3mn}{64m^3 - n^3}$ and $\dfrac{2}{4m - n}$

35. $\dfrac{1}{6u^2v + 12uv^2}$, $\dfrac{1}{12u^3 + 24u^2v}$, and $\dfrac{1}{8uv^2 + 16v^3}$

36. $\dfrac{1}{10u^2 - 30uv}$, $\dfrac{1}{15u - 45v}$, and $\dfrac{1}{6u^4 - 18u^3v}$

In exercises **37–60**, add or subtract as indicated.

[Examples 4 and 5] 37. $\dfrac{3}{2a^2b} + \dfrac{2}{2ab^2}$ 38. $\dfrac{1}{3ab^2} - \dfrac{2}{5a^2b}$

39. $\dfrac{6}{5x} + \dfrac{7}{10x^2}$ 40. $\dfrac{7}{2x^2} - \dfrac{6}{5x}$

41. $\dfrac{2}{3m - 6} - \dfrac{1}{2m - 4}$ 42. $\dfrac{1}{6m + 6} + \dfrac{5}{8m + 8}$

43. $\dfrac{3}{4n} - \dfrac{1}{2n + 2}$ 44. $\dfrac{2}{3n - 3} + \dfrac{5}{6n}$

45. $\dfrac{1}{b^2 - 4} - \dfrac{1}{b^2 - 4b + 4}$

46. $\dfrac{2b}{b^2 - 4} - \dfrac{3}{b^2 - b - 2}$

47. $\dfrac{2}{t^2 - t - 2} + \dfrac{1}{t^2 + 2t + 1}$

48. $\dfrac{3t}{t^2 + t - 2} - \dfrac{2}{t^2 - 1}$

49. $\dfrac{5}{w + 5} + \dfrac{w}{w - 4} - \dfrac{11w - 8}{w^2 + w - 20}$

50. $\dfrac{w}{w - 6} + \dfrac{1}{w + 1} - \dfrac{5w - 2}{w^2 - 5w - 6}$

51. $\dfrac{4a + 1}{a - 8} - \dfrac{3a + 2}{a + 4} - \dfrac{49a + 4}{a^2 - 4a - 32}$

52. $\dfrac{3a + 1}{a - 1} - \dfrac{a - 1}{a - 3} + \dfrac{a + 1}{a^2 - 4a + 3}$

[Example 6] **53.** $\dfrac{7}{m - 5} + \dfrac{2}{5 - m}$

54. $\dfrac{12}{8 - m} + \dfrac{5}{m - 8}$

55. $\dfrac{3u}{3u - v} + \dfrac{v}{v - 3u}$

56. $\dfrac{u}{u - 2v} + \dfrac{2v}{2v - u}$

57. $\dfrac{-8}{x^2 + 2x - 15} - \dfrac{1}{3 - x}$

58. $\dfrac{-8}{x^2 + 4x - 12} - \dfrac{1}{2 - x}$

59. $\dfrac{1}{4 - t^2} + \dfrac{1}{t^2 - 2t}$

60. $\dfrac{1}{3t - 12} + \dfrac{1}{16 - t^2}$

In exercises **61–68**, do the indicated operations.

[Example 7] **61.** $\dfrac{3}{2} + \dfrac{7x - 35}{2x^2 + 10x} \cdot \dfrac{x + 5}{3x - 15} - \dfrac{2}{3x}$

62. $\dfrac{9}{4x} - \dfrac{x + 2}{4x - 12} \cdot \dfrac{5x - 15}{3x^2 + 6x} + \dfrac{2}{3}$

63. $\dfrac{5}{6y} + \dfrac{2y - 14}{4y - 20} \cdot \dfrac{2y - 10}{6y^2 - 6y} - \dfrac{1}{y - 1}$

64. $\dfrac{2}{y - 2} + \dfrac{5y - 20}{3y^2 - 6y} \cdot \dfrac{2y - 12}{10y - 40} - \dfrac{5}{3y}$

65. $\dfrac{z - 2}{z - 6} + \dfrac{2z}{6 - z} + \dfrac{z^2 - 11z + 28}{z^2 - 13z + 42} \div \dfrac{z^2 + z - 20}{z^2 + 7z + 10}$

66. $\dfrac{6z + 3}{1 + 3z} - \dfrac{6z + 4}{3z + 1} + \dfrac{3z^2 + 14z + 8}{2z^2 + 7z - 4} \div \dfrac{3z^2 + 16z + 5}{2z^2 + 9z - 5}$

67. $\dfrac{1}{t + 3} + \dfrac{t - 3}{4t^3 - 12t^2 + 36t} \div \dfrac{t^2 - 9}{24t^2} - \dfrac{t}{t^2 - 3t + 9}$

68. $\dfrac{t^2 + 2u^2}{t^2 - tu - 2u^2} - \dfrac{t^2 - 2tu}{t^2 + 2tu + u^2} \div \dfrac{t^2 - 4tu + 4u^2}{3tu + 3u^2}$

SECTION 5-5.   Ten Review Exercises

In exercises **1–10**, solve each problem.

1. Multiply: $(10t)^2$

2. Multiply: $(10 + t)^2$

3. Divide: $\dfrac{4k^2}{2k}$

4. Divide: $\dfrac{4 - k^2}{2 - k}$

5. Simplify: $m + 5(m - 5)$

6. Multiply: $(m + 5)(m - 5)$

7. Divide: $\dfrac{27u^3v^3}{9u^2v^2}$

8. Divide: $\dfrac{27u^3 - v^3}{9u^2 - v^2}$

9. Evaluate for $x = 3$ and $y = -2$:
$5(x^2 + 7xy - 10y^2) - 4(x^2 + 9xy - 12y^2)$

10. **a.** Simplify: $5(x^2 + 7xy - 10y^2) - 4(x^2 + 9xy - 12y^2)$

　b. Evaluate the expression obtained in part **a** for $x = 3$ and $y = -2$.

SECTION 5-5.　Supplementary Exercises

In exercises **1–8**,

a. replace x by 3; then add or subtract as indicated.

b. add or subtract the given expressions and reduce, if possible.

c. evaluate the expression in part **b** for $x = 3$.

1. $\dfrac{1}{3x} + \dfrac{1}{2x}$

2. $\dfrac{1}{5x} + \dfrac{1}{4x}$

3. $\dfrac{1}{x^2} - \dfrac{1}{2x}$

4. $\dfrac{1}{x^2} - \dfrac{1}{4x}$

5. $\dfrac{2x}{3x + 9} + \dfrac{x}{4x + 12}$

6. $\dfrac{x}{4x - 2} - \dfrac{4x}{6x - 3}$

7. $\dfrac{3}{x^2 - 5x} + \dfrac{2}{x^2 + 5x} - \dfrac{4}{x^2 - 25}$

8. $\dfrac{x}{x + 1} + \dfrac{1}{x - 1} - \dfrac{2x}{x^2 - 1}$

In exercises **9–20**, simplify.

9. $\dfrac{2y - 3}{3 - 2y} + \dfrac{5y + 1}{1 + 5y}$

10. $\dfrac{6y - 5}{5 - 6y} + \dfrac{3y + 1}{1 + 3y}$

11. $\dfrac{3}{p - 2} + \dfrac{2}{2 - p}$

12. $\dfrac{4}{5 - p} + \dfrac{3}{p - 5}$

13. $\dfrac{18}{q^2 - 9} + \dfrac{3}{3 - q}$

14. $\dfrac{q - 1}{q^2 - 2q + 1} + \dfrac{q + 1}{1 - q}$

15. $\dfrac{4 + k}{k^2 - k - 42} + \dfrac{1}{7 - k}$

16. $\dfrac{3 + k}{k^2 - 3k - 10} + \dfrac{1}{5 - k}$

17. $\dfrac{w + 3}{w^3 - 27} - \dfrac{1}{w^2 - 9}$

18. $\dfrac{w + 5}{w^3 - 125} - \dfrac{1}{w^2 - 25}$

19. $1 + \dfrac{2}{x} - \dfrac{x-1}{x^2 - x}$

20. $\dfrac{2}{x-5} + \dfrac{3}{x+5} - 1$

In exercises **21** and **22**, answer each part.

21. a. Replace w by 3 and compute $\dfrac{1}{w-1} - \dfrac{1}{w}$.

    b. Compute $\dfrac{1}{w-1} - \dfrac{1}{w}$ and simplify the difference.

    c. Based on the answer to part **b**, would it be true that $\dfrac{1}{742} - \dfrac{1}{743} = \dfrac{1}{743(742)}$?

    d. Replace w by 3 and compute $\dfrac{1}{w-1} + \dfrac{1}{w}$ and simplify the sum.

    e. Compute $\dfrac{1}{w-1} + \dfrac{1}{w}$ and simplify the sum.

    f. Based on the answer to part **e**, would it be true that $\dfrac{1}{538} + \dfrac{1}{539} = \dfrac{1}{539(538)}$?

22. a. Replace w by 3 and compute $\dfrac{w}{1-w} + \dfrac{1}{w-1}$.

    b. Compute $\dfrac{w}{1-w} + \dfrac{1}{w-1}$ and simplify the sum.

    c. Based on the answer to part **b**, would it be true that $\dfrac{35}{-34} + \dfrac{1}{34} = -1$?

    d. Replace w by 3 and compute $\dfrac{w}{1-w} - \dfrac{1}{w-1}$.

    e. Compute $\dfrac{w}{1-w} - \dfrac{1}{w-1}$ and simplify the difference.

    f. Based on the answer to **e**, would it be true that $\dfrac{107}{-106} - \dfrac{1}{106} = -1$?

SECTION 5-6.   Complex Fractions

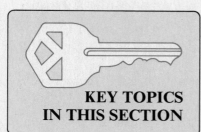

**KEY TOPICS
IN THIS SECTION**

1. A definition of a complex fraction

2. Changing a mixed expression to a rational expression

3. Mixed expressions with more than two terms

4. Plan A for simplifying complex fractions

5. Plan B for simplifying complex fractions

6. Complex fractions written with negative exponents

A rational expression can be written as a ratio of polynomials. However, in mathematics, science, and statistics, we frequently also encounter expressions that are written as ratios of rational expressions. In this section we will study ways in which such expressions can be simplified.

A Definition of a Complex Fraction

Definition 5.4 is a statement of the general form of a complex fraction.

Definition 5.4.   Complex fractions
   A **complex fraction** is an expression in which the numerator, denominator, or both contain a rational expression.

Examples **a** and **b** illustrate two complex fractions:

a. $\dfrac{t - \dfrac{4}{t}}{1 + \dfrac{2}{t}}$ **b.** $t + \dfrac{1 + t}{t - \dfrac{1}{t}}$

In example **a** the numerator has two terms, and one of them is the rational expression $\dfrac{4}{t}$. Likewise the denominator has two terms, and one of them is the rational expression $\dfrac{2}{t}$.

   A complex fraction is **simplified** when it is written as a rational expression; that is, a polynomial divided by a polynomial.

Changing a Mixed Expression to a Rational Expression

The first step in simplifying many complex fractions is changing **mixed expressions** to rational expressions. Examples **a** and **b** contain three such mixed expressions.

In **a**, $t - \dfrac{4}{t}$.

In **a**, $1 + \dfrac{2}{t}$.   These expressions consist of a term of a polynomial plus or minus a rational expression, and they can be changed to a single fraction.

In **b**, $t - \dfrac{1}{t}$.

A mixed expression is similar to a mixed number, such as $2\frac{3}{4}$. A mixed number is a sum of a whole number and a fraction. For example,

$2\frac{3}{4}$ means $2 + \frac{3}{4}$.   The fraction $\frac{3}{4}$ is added to the whole number 2.

Study the well-known steps that change $2\frac{3}{4}$ to $\frac{11}{4}$.

$$2 + \frac{3}{4} = 2 \cdot \frac{4}{4} + \frac{3}{4} = \frac{8}{4} + \frac{3}{4} = \frac{8 + 3}{4} = \frac{11}{4}$$

Example 1.   Write as a single rational expression $y + \dfrac{y}{y - 1}$.

Solution.

$$y + \frac{y}{y-1}$$ The given expression

$$= y \cdot \frac{y-1}{y-1} + \frac{y}{y-1}$$ Multiply y by $\frac{y-1}{y-1}$.

$$= \frac{y(y-1)+y}{y-1}$$ The LCM is $y-1$.

$$= \frac{y^2-y+y}{y-1}$$ $y(y-1)=y^2-y$

$$= \frac{y^2}{y-1}$$ Simplify the numerator.

Mixed Expressions with More than Two Terms

In simplifying a complex fraction it may be necessary to change mixed expressions of more than two terms to a single rational expression.

Example 2. Write as a single rational expression.

$$4 - \frac{4}{a} + \frac{1}{a^2}$$

Solution. **Discussion.** The LCM of the denominators is a^2.

$$4 - \frac{4}{a} + \frac{1}{a^2}$$ The given expression

$$= 4 \cdot \frac{a^2}{a^2} - \frac{4}{a} \cdot \frac{a}{a} + \frac{1}{a^2}$$ Write each term over a^2.

$$= \frac{4a^2-4a+1}{a^2}$$ Combine the numerators.

$$\text{or } \frac{(2a-1)^2}{a^2}$$ The numerator can be factored.

Plan A for Simplifying Complex Fractions

Examples **a** and **b** are complex fractions that can be simplified by a technique we simply call "Plan A."

a. $\dfrac{t - \dfrac{4}{t}}{1 + \dfrac{2}{t}}$ **b.** $\dfrac{\dfrac{2}{2x+1} - \dfrac{3}{3x-1}}{\dfrac{6}{3x-1} - \dfrac{4}{2x+1}}$

Plan A requires writing the numerator and denominator of a fraction as single rational expressions. We then use the definition of division of rational expressions to further simplify the fraction.

Example 3. Simplify.

$$\frac{t - \dfrac{4}{t}}{1 + \dfrac{2}{t}}$$

Solution.　　$\dfrac{t - \dfrac{4}{t}}{1 + \dfrac{2}{t}}$　　The given expression

$= \dfrac{\dfrac{t^2 - 4}{t}}{\dfrac{t + 2}{t}}$　　Write numerator and denominator as single rational expressions.

$= \dfrac{t^2 - 4}{t} \div \dfrac{t + 2}{t}$　　Write as a division.

$= \dfrac{t^2 - 4}{t} \cdot \dfrac{t}{t + 2}$　　$\dfrac{P}{Q} \div \dfrac{R}{S} = \dfrac{P}{Q} \cdot \dfrac{S}{R}$

$= \dfrac{t(t + 2)(t - 2)}{t(t + 2)}$　　$t^2 - 4 = (t + 2)(t - 2)$

$= (t - 2)$　　Reduce the fraction.

Solution.　　Simplify:

$$\dfrac{\dfrac{2}{2x + 1} - \dfrac{3}{3x - 1}}{\dfrac{6}{3x - 1} - \dfrac{4}{2x + 1}}$$

Example 4.　　$\dfrac{\dfrac{2}{2x + 1} - \dfrac{3}{3x - 1}}{\dfrac{6}{3x - 1} - \dfrac{4}{2x + 1}}$　　The given expression

$= \dfrac{\dfrac{2(3x - 1) - 3(2x + 1)}{(2x + 1)(3x - 1)}}{\dfrac{6(2x + 1) - 4(3x - 1)}{(2x + 1)(3x - 1)}}$　　The LCM of both numerator and denominator is $(2x + 1)(3x - 1)$.

$= \dfrac{\dfrac{6x - 2 - 6x - 3}{(2x + 1)(3x - 1)}}{\dfrac{12x + 6 - 12x + 4}{(2x + 1)(3x - 1)}}$　　Remove parentheses in numerator and denominator.

$= \dfrac{\dfrac{-5}{(2x + 1)(3x - 1)}}{\dfrac{10}{(2x + 1)(3x - 1)}}$　　Simplify the numerators.

$= \dfrac{-5}{(2x + 1)(3x - 1)} \div \dfrac{10}{(2x + 1)(3x - 1)}$　　Write as a division

$$= \frac{-5}{(2x + 1)(3x - 1)} \cdot \frac{(2x + 1)(3x - 1)}{10} \qquad \text{Change to a multiplication.}$$

$$= \frac{-5}{10} = \frac{-1}{2} \qquad \text{Multiply and reduce.}$$

Based on the results of Example 4, for any replacement of x in the given complex fraction $\left(\text{except the restricted values of } \dfrac{-1}{2} \text{ and } \dfrac{1}{3}\right)$, the fraction will always simplify to $\dfrac{-1}{2}$. You might experiment by replacing x by 2.

Plan B for Simplifying Complex Fractions

Example **a** is a complex fraction that can be simplified by a technique we call "Plan B."

a. $\dfrac{2 + \dfrac{y}{x} + \dfrac{x}{y}}{\dfrac{1}{y^2} - \dfrac{1}{x^2}}$
Notice that the denominators of the rational expressions on the top and bottom of this fraction are monomials, namely x, y, y^2 or x^2.

Plan B requires multiplying the top and bottom of the fraction by the LCM of the denominators of the rational expressions in the fraction.

Example 5. Simplify:

$$\frac{2 + \dfrac{y}{x} + \dfrac{x}{y}}{\dfrac{1}{y^2} - \dfrac{1}{x^2}}$$

Solution. **Discussion.** The LCM of the denominators of the fractions on the top and bottom is $x^2 y^2$. We therefore multiply by $\dfrac{x^2 y^2}{x^2 y^2}$ and distribute to each term.

$$\frac{2 + \dfrac{y}{x} + \dfrac{x}{y}}{\dfrac{1}{y^2} - \dfrac{1}{x^2}} \cdot \frac{x^2 y^2}{x^2 y^2}$$

$$= \frac{2x^2 y^2 + y(xy^2) + x(x^2 y)}{x^2 - y^2} \qquad \text{Distribute to each term and reduce the products.}$$

$$= \frac{xy(x^2 + 2xy + y^2)}{x^2 - y^2} \qquad \text{Factor an } xy \text{ and write in descending powers of } x.$$

$$= \frac{xy(x + y)(x + y)}{(x + y)(x - y)} \qquad \text{Completely factor.}$$

$$= \frac{xy(x + y)}{x - y} \qquad \text{Reduce the fraction.}$$

Complex Fractions Written with Negative Exponents

A rational expression may have a term or factor with a negative exponent. If the definition of a negative exponent is used, then the expression may take on the form of a complex fraction.

Example 6.   Simplify: $\dfrac{u^{-1} + 4v^{-1}}{v^2 - 16u^2}$

Solution.

$$\dfrac{u^{-1} + 4v^{-1}}{v^2 - 16u^2} \qquad \text{The given expression}$$

$$= \dfrac{\dfrac{1}{u} + \dfrac{4}{v}}{v^2 - 16u^2} \qquad u^{-1} = \dfrac{1}{u} \text{ and } 4v^{-1} = \dfrac{4}{v}$$

$$= \dfrac{\dfrac{v + 4u}{uv}}{v^2 - 16u^2} \qquad \begin{array}{l}\text{Write the numerator as a single}\\ \text{rational expression.}\end{array}$$

$$= \dfrac{v + 4u}{uv} \div (v^2 - 16u^2) \qquad \text{Write as a division.}$$

$$= \dfrac{v + 4u}{uv} \cdot \dfrac{1}{v^2 - 16u^2} \qquad \text{Change to a multiplication.}$$

$$= \dfrac{(v + 4u)}{uv(v - 4u)(v + 4u)} \qquad \text{Factor the denominator.}$$

$$= \dfrac{1}{uv(v - 4u)} \qquad \text{Reduce.}$$

SECTION 5-6.   Practice Exercises

In exercises **1–18**, write as a single rational expression.

[Example 1]   **1.** $a - \dfrac{2}{a}$ **2.** $a^2 - \dfrac{5}{a}$

3. $1 + \dfrac{1}{b - 1}$ **4.** $1 - \dfrac{b}{b - 1}$

5. $x - \dfrac{2x}{x + 2}$ **6.** $x - \dfrac{3x}{x + 3}$

7. $y - \dfrac{1}{1 - y}$ **8.** $y + \dfrac{1}{1 - y}$

[Example 2]   **9.** $6 - \dfrac{7}{z} - \dfrac{5}{z^2}$ **10.** $1 + \dfrac{9}{z} + \dfrac{14}{z^2}$

11. $3 - \dfrac{9}{t} + \dfrac{6}{t^2}$

12. $\dfrac{20}{t} + 10 - \dfrac{350}{t^2}$

13. $\dfrac{u}{4} - 1 + \dfrac{1}{u}$

14. $\dfrac{1}{u^2} + 4 - \dfrac{4}{u}$

15. $\dfrac{x}{y} + 1 - \dfrac{6y}{x}$

16. $\dfrac{4x}{y} - 4 - \dfrac{3y}{x}$

17. $\dfrac{3}{2a + 3} + \dfrac{3}{2a - 3}$

18. $\dfrac{2}{a + 1} + \dfrac{2}{a - 1}$

In exercises **19–30**, simplify using Plan A.

[Examples 3 and 4]

19. $\dfrac{b - \dfrac{4}{b}}{1 + \dfrac{2}{b}}$

20. $\dfrac{b - \dfrac{1}{b}}{1 - \dfrac{1}{b}}$

21. $\dfrac{x + \dfrac{1}{x}}{x - \dfrac{1}{x}}$

22. $\dfrac{2 - \dfrac{1}{x}}{2 + \dfrac{1}{x}}$

23. $\dfrac{1 + \dfrac{1}{y - 1}}{1 - \dfrac{1}{y + 1}}$

24. $\dfrac{1 - \dfrac{y}{y - 1}}{1 + \dfrac{y}{y - 1}}$

25. $\dfrac{2 + \dfrac{y}{x} + \dfrac{x}{y}}{\dfrac{1}{y^2} - \dfrac{1}{x^2}}$

26. $\dfrac{\dfrac{1}{x^2} + 4 - \dfrac{4}{x}}{x - \dfrac{1}{4x}}$

27. $\dfrac{\dfrac{3}{9b^2 - 1}}{\dfrac{2}{3b + 1} + \dfrac{2}{3b - 1}}$

28. $\dfrac{\dfrac{4}{4b^2 - 9}}{\dfrac{3}{2b + 3} + \dfrac{3}{2b - 3}}$

29. $\dfrac{\dfrac{1}{p + 1} - \dfrac{1}{p - 1}}{\dfrac{2}{p + 1} + \dfrac{2}{p - 1}}$

30. $\dfrac{\dfrac{3}{p - 5} + \dfrac{3}{p + 5}}{\dfrac{4}{p - 5} - \dfrac{4}{p + 5}}$

In exercises **31–40**, simplify using Plan B.

[Example 5]

31. $\dfrac{\dfrac{1}{u} - \dfrac{1}{v}}{\dfrac{2}{u} + \dfrac{2}{v}}$

32. $\dfrac{\dfrac{1}{u^2} - \dfrac{1}{v^2}}{\dfrac{3}{v} + \dfrac{3}{u}}$

33. $\dfrac{\dfrac{w}{4} - \dfrac{1}{w}}{\dfrac{w}{4} - 1 + \dfrac{1}{w}}$

34. $\dfrac{1 + \dfrac{1}{w} - \dfrac{2}{w^2}}{1 + \dfrac{3}{w} + \dfrac{2}{w^2}}$

35. $\dfrac{\dfrac{2m}{n} + 1 - \dfrac{6n}{m}}{\dfrac{4m}{n} - 4 - \dfrac{3n}{m}}$

36. $\dfrac{\dfrac{2m}{n} + 5 - \dfrac{3n}{m}}{\dfrac{m}{n} + 1 - \dfrac{6n}{m}}$

37. $\dfrac{\dfrac{1}{x+3} - \dfrac{1}{x-3}}{\dfrac{2}{x^2-9}}$

38. $\dfrac{\dfrac{1}{x-2} - \dfrac{1}{x+2}}{\dfrac{2}{x^2-4}}$

39. $\dfrac{\dfrac{2}{2y+1} - \dfrac{3}{3y-1}}{\dfrac{6}{3y-1} - \dfrac{4}{2y+1}}$

40. $\dfrac{\dfrac{3}{3y+2} - \dfrac{4}{4y-1}}{\dfrac{1}{4y-1} + \dfrac{2}{3y+2}}$

In exercises **41–50**, simplify.

[Example 6]   **41.** $\dfrac{1}{x^{-1} + y^{-1}}$

42. $\dfrac{1}{2x^{-1} + 3y^{-1}}$

43. $\dfrac{1 + t^{-1}}{1 - t^{-2}}$

44. $\dfrac{t + t^{-1}}{t^2 - t^{-2}}$

45. $\dfrac{a - b}{ab^{-1} - a^{-1}b}$

46. $\dfrac{a^2 - b^2}{b^{-1} - a^{-1}}$

47. $(k^{-1} - 2^{-1})^{-1}$

48. $(k^{-1} + 3^{-1})^{-1}$

49. $\dfrac{u - 15(u-2)^{-1}}{u - 20(u-1)^{-1}}$

50. $\dfrac{u + 35(u+12)^{-1}}{u - 63(u-2)^{-1}}$

SECTION 5-6.   Ten Review Exercises

In exercises **1–7**, do the indicated operations.

1. $(-3u^2)(2u)$

2. $(u^2 - 3)(u + 2)$

3. $\dfrac{-3u^2}{4v^3} \cdot \dfrac{-20v}{2u}$

4. $\dfrac{15t^4}{5t}$

5. $\dfrac{15t^4 + 80t^3 + 30t^2 + 20t}{5t}$

6. $\dfrac{15t^4 + 80t^3 + 30t^2 + 20t}{t + 5}$

7. $\dfrac{15t^4}{16u^2} \div \dfrac{5t}{12u^3}$

In exercises **8–10**, solve and check.

8. $9 + 6(16 - a) = 3(3a - 5)$ **9.** $7(b - 1) + 1 = 3(2 + b) + 4b$

10. $t^2 + 6t = 27$

SECTION 5-6. Supplementary Exercises

In exercises **1–6**, simplify.

1. $\dfrac{1 - \dfrac{y}{y - 1}}{1 + \dfrac{y}{1 - y}}$ **2.** $\dfrac{y - \dfrac{1}{y - 1}}{y + \dfrac{1}{1 - y}}$

3. $\left(1 + \dfrac{5}{w - 5}\right) \div \left(1 - \dfrac{5}{w + 5}\right)$ **4.** $\left(1 - \dfrac{1}{w + 1}\right) \div \left(1 + \dfrac{1}{w - 1}\right)$

5. $\dfrac{1}{1 - \dfrac{1}{1 - \dfrac{1}{m}}}$ **6.** $\dfrac{1}{m - \dfrac{1}{m + \dfrac{1}{m}}}$

In exercises **7–12**,

1. Replace x by 2 and y by 3 in expression **a** and simplify.

2. Replace x by 2 and y by 3 in expression **b** and simplify.

3. Based on the results obtained in **1** and **2**, are expressions **a** and **b** equivalent?

7. a. $\dfrac{x^{-1} + y^{-1}}{x^{-1}}$ **b.** $\dfrac{x}{x + y}$

8. a. $\dfrac{x^{-1} - y^{-1}}{y^{-1}}$ **b.** $\dfrac{y}{x - y}$

9. a. $\dfrac{x^{-1} + y^{-1}}{x^{-1} - y^{-1}}$ **b.** $\dfrac{x - y}{x + y}$

10. a. $\dfrac{y^{-1} - x^{-1}}{y^{-1} + x^{-1}}$ **b.** $\dfrac{y + x}{y - x}$

11. a. $2x^{-1} + 6y^{-1}$ **b.** $\dfrac{1}{2x} + \dfrac{1}{3y}$

12. a. $4x^{-1} + 3y^{-1}$ **b.** $\dfrac{1}{4x} + \dfrac{1}{3y}$

In exercises **13** and **14**, evaluate the formula from trigonometry using the given values.

$$\frac{(\tan a) + (\tan b)}{1 - (\tan a)(\tan b)}$$

13. $(\tan a) = \frac{1}{4}$ and $(\tan b) = \frac{1}{3}$ **14.** $(\tan a) = \frac{-4}{5}$ and $(\tan b) = \frac{-3}{5}$

In exercises **15** and **16**, evaluate the formula from statistics.

$$\frac{[d - (p + q)]^2}{\dfrac{p(1 - p)}{m} + \dfrac{q(1 - q)}{n}}$$

15. $d = \dfrac{1}{10}$, $p = \dfrac{1}{2}$, $q = \dfrac{1}{5}$, $m = 50$, and $n = 50$

16. $d = \dfrac{1}{20}$, $p = \dfrac{1}{4}$, $q = \dfrac{1}{2}$, $m = 100$, and $n = 100$

SECTION 5-7.   Equations with Rational Expressions

**KEY TOPICS
IN THIS SECTION**

1. A description of a fractional equation

2. A procedure for solving a fractional equation

3. A restricted value cannot be a solution.

In previous sections we have solved linear and quadratic equations in one variable, such as those in examples **a** and **b**.

a. $2t + 4(2t + 3) = 14 - 6(t - 1)$ A linear equation in t

b. $2t^2 - 5t = 3$ A quadratic equation in t

In mathematics we frequently must solve equations that have one or more rational expressions. Examples **c** and **d** illustrate equations that have more than one rational expression. These equations are frequently called **fractional.**

c. $\dfrac{1}{2} - \dfrac{2}{5t} = \dfrac{t + 2}{5t}$ A fractional equation in t

d. $\dfrac{x + 2}{4} - \dfrac{x}{2x + 4} = \dfrac{x + 3}{x + 2}$ A fractional equation in x

In this section we will study a technique for solving fractional equations.

A Description of a Fractional Equation

Definition 5.5 is a statement of the general form of a fractional equation.

Definition 5.5.   Fractional equation
  A **fractional equation** is one in which at least one term is a rational expression.

Examples **c** and **d** above are fractional equations in which the largest degree of any denominator is one. Examples **e** and **f** are fractional equations in which at least one denominator is degree two.

e. $\dfrac{1}{a^2 - 1} = \dfrac{6}{a^2 - 15a - 16}$ **f.** $\dfrac{x + 2}{x} = \dfrac{x + 2}{x + 6} + \dfrac{12 - x^2}{x^2 + 6}$

A Procedure for Solving a Fractional Equation

A technique that can be used to solve a fractional equation is to change it to one that can be solved by a method already known. This task can be accomplished by using the multiplication property of equality to write the given fractional equation as one without fractions.

Multiplication property of equality
  If $A = B$ and $C \neq 0$, then $A \cdot C = B \cdot C$.

For a given fractional equation, the C that the equation is multiplied by is the LCM of the rational expressions in the equation. The following six steps provide a recommended procedure for using the LCM and the multiplication property of equality.

A procedure for solving a fractional equation

Step 1. If necessary, factor the denominators.

Step 2. Determine the LCM of all the denominators.

Step 3. Determine all values of the variable that would make the LCM zero. (These values are the *restricted values* of the rational expressions in the equation.)

Step 4. Multiply both sides of the equation by the LCM.

Step 5. Solve the resulting equation obtained in Step 4.

Step 6. For a check, compare the solutions of Step 5 with the restricted values of Step 3. Reject any solutions that are restricted values.

Notice, the root (or roots) obtained in Step 5 can be considered **apparent roots** of the given fractional equation. An apparent solution is a number k that is a root of the equation $AC = BC$. However, k may or may not be a root of the equation $A = B$, depending on whether k is a restricted value. If k is a restricted value, then it is rejected. (It was only an apparent root of $A = B$.)

Example 1.   Solve and check:

$$\frac{1}{2} - \frac{2}{5t} = \frac{t + 2}{5t}$$

Solution.   **Step 1.** $\dfrac{1}{2} - \dfrac{2}{5t} = \dfrac{t + 2}{5t}$ The denominators are 2 and $5t$.

Step 2. The LCM is $10t$.

Step 3. If $10t = 0$, then $t = 0$. Thus, $t \neq 0$.

Step 4. $\left(\dfrac{1}{2} - \dfrac{2}{5t}\right) \cdot 10t = \left(\dfrac{t + 2}{5t}\right) \cdot 10t$ Multiply both sides by $10t$.

$\dfrac{1}{2} \cdot 10t - \dfrac{2}{5t} \cdot 10t = \dfrac{(t + 2)10t}{5t}.$ The distributive property

$5t - 4 = 2t + 4$ Simplify the products.

Step 5. $3t - 4 = 4$ Subtract $2t$ from both sides.

$3t = 8$ Add 4 to both sides.

$t = \dfrac{8}{3}$ Divide both sides by 3.

Step 6. Since the apparent root is not a restricted value, the root of the equation is $\frac{8}{3}$.

Example 2.   Solve and check:

$$\dfrac{1}{a^2 - 1} = \dfrac{6}{a^2 - 15a - 16}$$

Solution.   **Step 1.** $\dfrac{1}{(a + 1)(a - 1)} = \dfrac{6}{(a - 16)(a + 1)}$

Step 2. The LCM is $(a + 1)(a - 1)(a - 16)$

Step 3. From Step 2, $a \neq -1, 1,$ or 16.

Step 4. $\dfrac{1}{(a + 1)(a - 1)} \cdot (a + 1)(a - 1)(a - 16)$

$= \dfrac{6}{(a - 16)(a + 1)} \cdot (a + 1)(a - 1)(a - 16)$

$a - 16 = 6(a - 1)$

Step 5. $a - 16 = 6a - 6$

$-16 = 5a - 6$

$-10 = 5a$

$-2 = a$

Step 6. The root of the given equation is -2, because (-2) is not a restricted value.

A Restricted Value Cannot be a Solution

As previously stated, any solution of the equation obtained in Step 5 is only an *apparent root* of the given fractional equation. If an apparent solution makes the denominator of any rational expression zero, then the solution is rejected as a root of the fractional equation.

Example 3. Solve and check:

$$\frac{x+2}{4} - \frac{x}{2x+4} = \frac{x+3}{x+2}$$

Solution.

Step 1. $\dfrac{x+2}{4} - \dfrac{x}{2(x+2)} = \dfrac{x+3}{x+2}$

Step 2. The LCM is $4(x + 2)$.

Step 3. From Step 2, $x \neq -2$.

Step 4. $\left(\dfrac{x+2}{4} - \dfrac{x}{2(x+2)} \right) \cdot 4(x+2) = \dfrac{x+3}{x+2} \cdot 4(x+2)$

$$(x+2)(x+2) - 2x = 4(x+3)$$

$$x^2 + 4x + 4 - 2x = 4x + 12$$

$$x^2 - 2x - 8 = 0$$

Step 5. $(x-4)(x+2) = 0$

$$x - 4 = 0 \text{ or } x + 2 = 0$$

$$x = 4 \qquad x = -2$$

Step 6. Since (-2) is a restricted value, it is rejected. However, 4 is not a restricted value and is accepted as a root.

Example 4. Solve and check:

$$\frac{x+2}{x} = \frac{x+2}{x+6} + \frac{12-x^2}{x^2+6x}$$

Solution.

Step 1. $\dfrac{x+2}{x} = \dfrac{x+2}{x+6} + \dfrac{12-x^2}{x(x+6)}$

Step 2. The LCM is $x(x + 6)$.

Step 3. From Step 2, $x \neq 0$ or -6.

Step 4. $\dfrac{x+2}{x} \cdot x(x+6) = \left(\dfrac{x+2}{x+6} + \dfrac{12-x^2}{x(x+6)} \right) x(x+6)$

$$(x+2)(x+6) = x(x+2) + (12 - x^2)$$

$$x^2 + 8x + 12 = x^2 + 2x + 12 - x^2$$

$$x^2 + 6x = 0$$

Step 5. $x(x+6) = 0$

$$x = 0 \quad \text{or} \quad x + 6 = 0$$

$$x = -6$$

Step 6. Since 0 and -6 are both restricted values, both apparent solutions are rejected. Since there are no other solutions, the given equation has no roots.

SECTION 5-7.　Practice Exercises

In exercises **1–40**, solve and check.

[Examples 1 and 2]

1. $y = \dfrac{5y}{2} + 9$

2. $y = \dfrac{2y}{5} + 3$

3. $b = \dfrac{-3b}{2} - \dfrac{5}{2}$

4. $b = \dfrac{7b}{3} + \dfrac{16}{3}$

5. $\dfrac{4x + 3}{3} = \dfrac{3x - 2}{2}$

6. $\dfrac{6x - 5}{5} = \dfrac{3x - 2}{2}$

7. $\dfrac{3y - 1}{2} = 3 + \dfrac{4y}{5}$

8. $\dfrac{3y + 1}{2} = 1 + \dfrac{5y - 5}{4}$

9. $\dfrac{2}{7} + \dfrac{3}{b} = \dfrac{3}{7b} - \dfrac{1}{7}$

10. $\dfrac{2}{5} - \dfrac{14}{15b} = \dfrac{13}{15b} + \dfrac{1}{10}$

11. $\dfrac{2u - 1}{u} = \dfrac{2u + 4}{3u} + \dfrac{1}{3u}$

12. $\dfrac{2u + 15}{3u} - \dfrac{1}{6} = \dfrac{5 - u}{3u}$

13. $\dfrac{k + 3}{k + 1} = 3$

14. $4 = \dfrac{5k + 1}{k + 2}$

15. $\dfrac{6p + 11}{p + 3} - 5 = 0$

16. $0 = \dfrac{4p + 11}{p + 2} - 3$

17. $\dfrac{3}{v - 2} = \dfrac{2}{v - 3}$

18. $\dfrac{4}{v - 5} = \dfrac{3}{v - 3}$

19. $\dfrac{3}{2t} - \dfrac{2t}{t + 1} = -2$

20. $\dfrac{3t}{t - 2} - \dfrac{2}{3t} = 3$

21. $\dfrac{2}{a - 2} = \dfrac{5}{3a - 4}$

22. $\dfrac{5}{2a + 13} = \dfrac{2}{a + 4}$

23. $\dfrac{5}{y^2 + 3y + 2} = \dfrac{1}{y^2 - y - 2}$

24. $\dfrac{1}{y^2 - 1} = \dfrac{6}{y^2 - 15y - 16}$

[Examples 3 and 4]

25. $\dfrac{x + 2}{3} + \dfrac{1}{x - 3} = \dfrac{x - 2}{x - 3}$

26. $\dfrac{x - 3}{5} + \dfrac{1}{x + 5} = \dfrac{-x - 4}{x + 5}$

27. $\dfrac{2z + 1}{z} = \dfrac{z - 5}{z - 1} + \dfrac{4}{z(z - 1)}$

28. $\dfrac{3z - 1}{z} = \dfrac{z - 5}{z - 3} + \dfrac{6}{z(z - 3)}$

29. $\dfrac{2}{u - 2} - 4 = \dfrac{u}{u - 2}$

30. $\dfrac{4}{u - 3} = \dfrac{u + 1}{u - 3} - 6$

31. $\dfrac{w - 1}{2} + \dfrac{w}{2w + 4} = \dfrac{2w + 3}{w + 2}$

32. $\dfrac{w + 2}{4} + \dfrac{1 - w}{4w - 20} = \dfrac{4 - w}{w - 5}$

33. $\dfrac{3}{b - 3} = \dfrac{5}{2b - 6} + 1$

34. $\dfrac{35}{3b - 15} = \dfrac{5}{b - 5} - 2$

35. $\dfrac{2m}{m - 1} = 1 - \dfrac{m - 5}{m^2 - 1}$

36. $\dfrac{1 - m}{m - 3} = m - \dfrac{2}{m - 3}$

37. $\dfrac{y}{y-4} = \dfrac{12}{y^2 - 5y + 4} - \dfrac{3}{y-1}$ 38. $\dfrac{y}{y-2} = \dfrac{4}{y^2 - 8y + 12} - \dfrac{1}{y-6}$

39. $\dfrac{1}{x^2 + x - 2} - \dfrac{3}{x^2 - 2x - 8} = \dfrac{1}{x^2 - 5x + 4}$

40. $\dfrac{5}{7x + 7} = \dfrac{7x - 2}{2x^2 + x - 1} + \dfrac{3}{14x - 7}$

SECTION 5-7. Ten Review Exercises

In exercises **1–10**, do the indicated operations. Simplify answers.

1. $3(2t^3 + 2t^2 - 5t + 10) - 2(t^3 + 3t^2 - t + 15)$

2. $(3u + 4)(4u^2 + 3u - 2)$ 3. $(9v^2 - 1)(4v - 3)$

4. $(5k + 6)^2$ 5. $\dfrac{15a^5 b - 36a^3 b^2 + 3ab^3}{3ab}$

6. $(4x^4 + 6x^3 - 12x^2 + 30) \div (2x + 3)$

7. $\dfrac{3y^2 - y}{4y^2 - 9} \cdot \dfrac{2y^2 + 7y + 6}{3y^2 + 5y - 2}$ 8. $\dfrac{a^2}{a^2 - b^2} + \dfrac{3ab + 2b^2}{a^2 - b^2}$

9. $\dfrac{3}{14mn^2} - \dfrac{2}{21m^2 n}$ 10. $\dfrac{\dfrac{a}{b} - \dfrac{3}{2} - \dfrac{b}{a}}{\dfrac{2a}{b} - \dfrac{b}{2a}}$

SECTION 5-7. Supplementary Exercises

In exercises **1–10**, solve and check.

1. $\dfrac{6}{5t} + \dfrac{7}{10t^2} = \dfrac{12t + 7}{10t^2}$ 2. $\dfrac{5}{6t^2} + \dfrac{1}{9t} = \dfrac{2t + 15}{18t^2}$

3. $\dfrac{3}{u-2} - \dfrac{3}{3u} - \dfrac{18}{3u^2 - 6u} = \dfrac{7}{3u}$ 4. $\dfrac{2u}{(u-4)^2} + \dfrac{2}{u-4} = \dfrac{3u - 4}{(u-4)^2}$

5. $\dfrac{x}{x-3} + \dfrac{4}{3-x} = \dfrac{4}{5}$ 6. $\dfrac{x}{2x-1} - \dfrac{2}{1-2x} = \dfrac{1}{3}$

7. $\dfrac{y-8}{y-3} - \dfrac{y+8}{y+3} = \dfrac{y}{9 - y^2}$ 8. $\dfrac{y}{y+2} - \dfrac{3}{y-2} = \dfrac{8 - y^2}{4 - y^2}$

9. $\dfrac{-12}{z+1} = 0$ 10. $\dfrac{17}{z-3} = 0$

Many literal equations can be written in which at least one term is a rational expression. In the following example the given equation is solved for y in terms of x. The solution shows how the equation can be solved for x in terms of y.

Example.   Solve $y = \dfrac{x - 2}{3x + 1}$ for x.

Solution. $y = \dfrac{x - 2}{3x + 1}$ The LCM is $(3x + 1)$.

$y(3x + 1) = \dfrac{x - 2}{3x + 1}(3x + 1)$ Multiply both sides by $(3x + 1)$.

$3xy + y = x - 2$ Simplify both sides.

$y + 2 = x - 3xy$ Collect x-terms on right side.

$y + 2 = (1 - 3y)x$ Factor an x.

$\dfrac{y + 2}{1 - 3y} = x$ Divide both sides by $(1 - 3y)$.

The equation is solved for x in terms of y.

In exercises **11–20** solve each equation for the indicated variable.

11. $y = \dfrac{5 - 2x}{1 + x}$ for x **12.** $y = \dfrac{3x}{2x - 5}$ for x

13. $\dfrac{4}{y} - \dfrac{5}{3} = \dfrac{7}{x}$ for y **14.** $\dfrac{3}{x} + \dfrac{2}{y} = \dfrac{1}{5}$ for y

15. $\dfrac{3}{t + a} - \dfrac{2}{t - a} = \dfrac{1}{t}$ for t **16.** $\dfrac{a - 2}{t} + \dfrac{3}{2t} = \dfrac{2}{a}$ for t

17. $\dfrac{P}{N} = \dfrac{P + 1}{N + M}$ for N **18.** $\dfrac{P}{N} = \dfrac{P - 2}{N + M}$ for M

19. $\dfrac{1}{R} = \dfrac{1}{r_1} + \dfrac{1}{r_2}$ for R **20.** $\dfrac{1}{R} = \dfrac{1}{r_1} + \dfrac{1}{r_2}$ for r_1

Example.   Rewrite the following equation in the form $Ax + By = C$:

$$y + 3 = \dfrac{-2}{5}(x - 1)$$

Solution.   **Discussion.** To write the given equation in the $Ax + By = C$ form we want the x and y terms to be on the left side and the constant term to be on the right.

$5(y + 3) = 5\left(\dfrac{-2}{5}(x - 1)\right)$ The LCM is 5.

$5(y + 3) = -2(x - 1)$ $5\left(\dfrac{-2}{5}\right) = -2.$

$5y + 15 = -2x + 2$ Use the distributive property.

$2x + 5y = -13$ Add $2x$ and -15 to both sides.

Thus, $y + 3 = \dfrac{-2}{5}(x - 1)$ can be written as $2x + 5y = -13$.

In exercises **21–26**, write each equation in the form $Ax + By = C$.

21. $y = \dfrac{-3}{2}(x - 1)$

22. $y = \dfrac{-3}{5}(x - 4)$

23. $y - 4 = \dfrac{-2}{3}(x + 5)$

24. $y - 2 = \dfrac{-3}{4}(x + 2)$

25. $y + 5 = \dfrac{10}{3}(x - 3)$

26. $y + 9 = \dfrac{4}{7}(x + 15)$

SECTION 5-8. Proportions

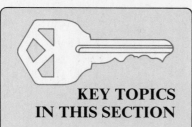

KEY TOPICS IN THIS SECTION

1. A definition of a proportion

2. The means-extremes product property of a proportion

3. Some applications of proportions

4. Using proportions to solve similar triangles

Equations **a** and **b** qualify as fractional because each one has at least one rational expression.

a. $\dfrac{4}{3t} = \dfrac{5}{2}$ **b.** $\dfrac{x}{x + 1} = \dfrac{6}{2x + 5}$

These equations contain exactly two rational expressions set equal to each other. Equations of this type are called **proportions.** In this section we will study the procedure commonly used for solving proportions.

A Definition of a Proportion

Definition 5.6 is a statement that identifies the general form of a proportion.

> **Definition 5.6.   Proportions**
>   A **proportion** is a fractional equation that states that one rational expression is equal to a second rational expression.

The letters a, b, c, and d are frequently used to represent the four terms of a proportion. Using these symbols,

$$\frac{a}{b} \text{ and } \frac{c}{d} \text{ are the rational expressions, and}$$

$$\frac{a}{b} = \frac{c}{d} \text{ is the corresponding proportion.}$$

Since b and d are denominators of rational expressions, we understand that $b \neq 0$ and $d \neq 0$.

The terms of a proportion are divided into two categories—**means** and **extremes**—based on their location in the proportion.

Extremes $\longrightarrow$ $\dfrac{a}{b} = \dfrac{c}{d}$ $\longleftarrow$ Means   **The mean terms are b and c.**
The extreme terms are a and d.

Example 1.   In the proportion $\dfrac{4}{3t} = \dfrac{5}{2}$,

  a. identify the mean terms.

  b. identify the extreme terms.

Solution.   **a.** The mean terms are $3t$ and 5.

  b. The extreme terms are 4 and 2.

The Means-Extremes Product Property of a Proportion

Consider again the proportions in examples **a** and **b**. We will solve them using the procedure given in the last section for solving any fractional equation.

a. Step 1. $\dfrac{4}{3t} = \dfrac{5}{2}$

Step 2. The LCM is $6t$.

Step 3. $t \neq 0$

Step 4. $\dfrac{4}{3t} \cdot 6t = \dfrac{5}{2} \cdot 6t$

$4 \cdot 2 = 5 \cdot 3t$

b. Step 1. $\dfrac{x}{x+1} = \dfrac{6}{2x+5}$

Step 2. The LCM is $(x+1)(2x+5)$.

Step 3. $x \neq -1$, or $\dfrac{-5}{2}$

Step 4. $\dfrac{x}{x+1} \cdot \text{LCM} = \dfrac{6}{2x+5} \cdot \text{LCM}$

$x(2x+5) = 6(x+1)$

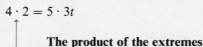

The product of the extremes

The product of the means

As shown, in the solving process we have obtained an equation in which the product of the extremes is on the left side and the product of the means is on the right side. Such an equation can always be obtained when solving a proportion.

The means-extremes product property of a proportion
  In a proportion having the form

$$\frac{a}{b} = \frac{c}{d}$$

$$a \cdot d = b \cdot c$$

the product of the extremes equals the product of the means.

Example 2.   Use the means-extremes product property to solve:

  a. $\dfrac{4}{3t} = \dfrac{5}{2}$

  b. $\dfrac{x}{x+1} = \dfrac{6}{2x+5}$

Solution.   **a.** $\dfrac{4}{3t} = \dfrac{5}{2}$ The given equation

$\qquad\qquad\quad 4 \cdot 2 = 3t \cdot 5$ Means-extremes product property

$\qquad\qquad\qquad 8 = 15t$ Simplify the products.

$\qquad\qquad\quad \dfrac{8}{15} = t$ Divide both sides by 15.

b. $\dfrac{x}{x + 1} = \dfrac{6}{2x + 5}$

$\quad x(2x + 5) = 6(x + 1)$

$\quad 2x^2 + 5x = 6x + 6$

$\quad 2x^2 - x - 6 = 0$

$\quad (2x + 3)(x - 2) = 0$

$\quad 2x + 3 = 0 \qquad \text{or} \qquad x - 2 = 0$

$\qquad\qquad x = \dfrac{-3}{2} \qquad\qquad\qquad x = 2$

For the given proportion the roots are $\dfrac{-3}{2}$ and 2.

Some Applications of Proportions

The proportion can be used to solve many types of applied problems. One such use of the proportion is illustrated by the way in which a proportion can be read:

$$\frac{a}{b} = \frac{c}{d} \text{ can be read } \text{``}a \text{ is to } b \text{ as } c \text{ is to } d.\text{''}$$

For example, suppose a car averages 28 miles for each gallon of gas. Thus, "*a* is to *b*" can be interpreted as "28 miles is to 1 gallon" and can be written as follows:

$$\frac{28 \text{ miles}}{1 \text{ gallon}} = \frac{n \text{ miles}}{g \text{ gallons}} \qquad \frac{\text{Miles is the unit on top.}}{\text{Gallons is the unit on bottom.}}$$

Now, for a specified number of miles, the gallons needed can be found. Or, for a specified number of gallons, the miles traveled can be calculated.

Find the number of miles (n) traveled on 13 gallons of gas.

$$\frac{28}{1} = \frac{n}{13}$$

$$(28)(13) = n(1)$$

$$364 = n$$

The car can travel 364 miles on 13 gallons of gas.

Find the number of gallons (g) needed to travel 980 miles.

$$\frac{28}{1} = \frac{980}{g}$$

$$28g = 980$$

$$g = 35$$

The car needs 35 gallons of gas to travel 980 miles.

Example 3.   If 2 gallons of paint cover 325 square feet of surface, approximately how many gallons are needed to paint 2,900 square feet.

Solution.   **Discussion.** The data "2 gallons of paint cover 325 square feet" can be written $\dfrac{2 \text{ gallons}}{325 \text{ square feet}}$.

Let $y =$ the number of gallons to cover 2,900 square feet.

$$\frac{2}{325} = \frac{y}{2,900} \qquad \frac{\text{gallons}}{\text{square feet}}$$

$$(2)(2,900) = 325y$$

$$\frac{5,800}{325} = y$$

$$y = 17.846\dots$$

Thus, approximately 18 gallons of paint are needed.

Example 4.   If 4 cans of tomatoes are on sale for 93¢, find the cost of a case of 24 cans.

Solution.   **Discussion.** The data "4 cans of tomatoes cost 93¢" can be written $\dfrac{4 \text{ cans}}{93¢}$.

Let $c =$ the cost of 24 cans.

$$\frac{4}{93} = \frac{24}{c} \qquad \frac{\text{cans}}{\text{cost}}$$

$$4c = (93)(24)$$

$$c = 558$$

Thus, the cost of 24 cans is $5.58.

Using Proportions to Solve Similar Triangles

In Figure 5-1, triangles ABC and XYZ have the same shape but are different in size. We say that such triangles are **similar,** and we write

$$\triangle ABC \sim \triangle XYZ.$$

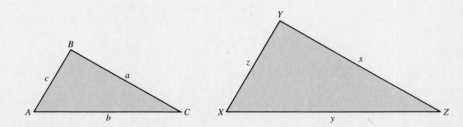

Figure 5-1. $\triangle ABC \sim \triangle XYZ$

For such triangles we may conclude the following:

$$A = X \qquad B = Y \qquad C = Z$$ Corresponding angles are equal.

$$\frac{a}{x} = \frac{b}{y} \qquad \frac{a}{x} = \frac{c}{z} \qquad \frac{b}{y} = \frac{c}{z}$$ Corresponding sides have lengths that are proportional.

Example 5. In $\triangle XYZ$, $x = 8$ centimeters, $y = 10$ centimeters, and $z = 16$ centimeters. In $\triangle ABC$, $a = 3$ centimeters, and the triangles are similar. Find b and c.

Solution. **Discussion.** The triangles are similar, therefore

$$\frac{a}{x} = \frac{b}{y} \qquad \text{and} \qquad \frac{a}{x} = \frac{c}{z}.$$

With $a = 3$, $x = 8$, and $y = 10$,   With $a = 3$, $x = 8$, and $z = 16$,

$$\frac{3}{8} = \frac{b}{10}. \qquad\qquad\qquad\qquad \frac{3}{8} = \frac{c}{16}.$$

$$3(10) = 8b \qquad\qquad\qquad\qquad 3(16) = 8c$$

$$b = \frac{30}{8} = 3.75 \text{ centimeters} \qquad c = \frac{48}{8} = 6 \text{ centimeters}$$

SECTION 5-8.   Practice Exercises

In exercises **1–6**, identify

a. the mean terms.

b. the extreme terms.

[Example 1]   **1.** $\dfrac{7}{5} = \dfrac{3x}{y}$ **2.** $\dfrac{6}{5x} = \dfrac{2y}{9}$

3. $\dfrac{x}{70 - x} = \dfrac{3}{4}$ **4.** $\dfrac{x}{24 - x} = \dfrac{1}{2}$

5. $\dfrac{x + 3}{x - 3} = \dfrac{4}{x}$ **6.** $\dfrac{x + 10}{x + 2} = \dfrac{x}{6}$

In exercises **7–20**, solve.

[Example 2]   **7.** $\dfrac{x}{105} = \dfrac{3}{35}$ **8.** $\dfrac{x}{12} = \dfrac{69}{92}$

9. $\dfrac{5}{6} = \dfrac{75}{y}$ **10.** $\dfrac{7}{4} = \dfrac{35}{y}$

11. $\dfrac{7}{w} = \dfrac{21}{2}$ **12.** $\dfrac{8}{w} = \dfrac{16}{3}$

13. $\dfrac{a}{2a + 10} = \dfrac{1}{3}$

14. $\dfrac{a}{2a - 3} = \dfrac{3}{5}$

15. $\dfrac{7}{8} = \dfrac{2b - 3}{2b}$

16. $\dfrac{1}{4} = \dfrac{2b - 1}{7b}$

17. $\dfrac{m - 1}{m} = \dfrac{m + 2}{2m}$

18. $\dfrac{m + 6}{2m} = \dfrac{m - 1}{m}$

19. $\dfrac{n}{n + 1} = \dfrac{n + 7}{2n + 2}$

20. $\dfrac{n}{n + 2} = \dfrac{n + 9}{2n + 4}$

In exercises **21–28**, answer parts **a** and **b**.

[Examples 3 and 4]

21. If 3 cans of soup sell for 89¢,
    a. find the cost of 15 cans.
    b. how many cans can be purchased for $3.56?

22. If 4 cans of tomatoes sell for 93¢,
    a. find the cost of 24 cans.
    b. how many cans can be purchased for $4.65?

23. If 5 centimeters on a map corresponds to 56 kilometers on the road,
    a. how many centimeters on the map corresponds to 392 kilometers on the road?
    b. how many kilometers on the road correspond to 7 centimeters on the map?

24. The scale on an old mining map is 2 inches to 25 miles.
    a. How many inches on the map correspond to 75 miles?
    b. How many miles are represented by 7 inches on the map?

25. Sixty mph corresponds to a speed of 88 feet per second (fps). If a baseball pitcher can throw the ball
    a. 90 mph, what is the speed in fps?
    b. 110 fps, what is the speed in mph?

26. A speed of 1,760 feet per minute (fpm) corresponds to 20 mph. If a bicyclist can pedal
    a. 22 mph, what is her speed in fpm?
    b. 1,584 fpm, what is her speed in mph?

27. In a mountain lake, 30 fish were caught, marked, and released. A month later
    a. 70 fish were caught, and 2 were marked. What is the estimated fish population in the lake?
    b. 90 fish were caught, and 3 were marked. What is the estimated fish population in the lake?

28. On a state game preserve 20 rabbits were marked and released. Several weeks later,
    a. of 50 rabbits observed, 2 were marked. What is the estimated rabbit population on this reserve?
    b. of 75 rabbits observed, 3 were marked. What is the estimated rabbit population in this reserve?

In exercises **29–36**, triangles *ABC* and *XYZ* are similar. The lengths of the corresponding sides of the triangles are labeled *a* and *x*, *b* and *y*, and *c* and *z*. In each exercise, find the lengths of the unknown sides.

[Example 5] **29.** Find y and z if $a = 3$, $b = 4$, $c = 6$, and $x = 36$.

30. Find b and c if $a = 5$, $x = 15$, $y = 12$, and $z = 21$.

31. Find x and y if $a = 2$, $b = 4$, $c = 5$, and $z = 20$.

32. Find a and b if $c = 7$, $x = 18$, $y = 15$, and $z = 21$.

33. Find x and z if $a = 3.0$, $b = 5.0$, $c = 7.0$, and $y = 8.0$.

34. Find a and c if $b = 6.0$, $x = 18.0$, $y = 15.0$, and $z = 27.0$.

35. Find x and b if $a = 15$, $c = 25$, $y = 54$, and $z = 75$.

36. Find a and y if $b = 24$, $c = 40$, $x = 132$, and $z = 160$.

In exercises **37–42**, solve each proportion problem.

37. One inch corresponds to 2.54 centimeters. How many centimeters are there in a piece of material that measures 120 inches?

38. One inch corresponds to 2.54 centimeters. How many inches are there in a piece of material that measures 215.9 centimeters?

39. If 22 calories of heat are needed to raise the temperature of 10 grams of aluminum from 20°C to 30°C, how many calories are needed to raise the temperature of 75 grams of aluminum from 20°C to 30°C?

40. If 60 calories of heat are needed to raise the temperature of 100 grams of copper from 20°C to 40°C, how many calories are needed to raise the temperature of 450 grams of copper from 20°C to 40°C?

41. One rod corresponds to 5.5 yards. How many yards are in a field that measures 72 rods?

42. One rod corresponds to 5.5 yards. How many rods are in a football field that is 100 yards long?

SECTION 5-8. Ten Review Exercises

In exercises **1–10**, solve each problem.

1. Simplify: $3(k - 1) - 2(4k + 3) + 4$

2. Evaluate for $k = \frac{1}{5}$: $3(k - 1) - 2(4k + 3) + 4$

3. Solve: $3(k - 1) = 2(4k + 3) + 4$

4. Factor: $2t^2 - t - 6$

5. Evaluate for $t = \frac{1}{2}$ the expression of Exercise **4**.

6. Solve: $2t^2 = t + 6$

7. Simplify: $\dfrac{5}{3} - \dfrac{3}{2a} - \dfrac{3}{2} - \dfrac{1}{3a}$

8. Evaluate for $a = \dfrac{1}{6}$: $\dfrac{5}{3} - \dfrac{3}{2a} - \dfrac{3}{2} - \dfrac{1}{3a}$

9. Solve: $\dfrac{5}{3} - \dfrac{3}{2a} = \dfrac{3}{2} + \dfrac{1}{3a}$

10. Solve: $\dfrac{k}{k+2} = \dfrac{5}{k-4}$

SECTION 5-8.   Supplementary Exercises

In exercises **1–4**, solve for x.

1. $\dfrac{x}{x+3} = \dfrac{2x}{2x+6}$

2. $\dfrac{2x+1}{x} = \dfrac{4x+2}{2x}$

3. $\dfrac{x-8}{x+1} = \dfrac{3x-24}{3+3x}$

4. $\dfrac{x+5}{x-2} = \dfrac{20+4x}{4x-8}$

In the following proportion, both means terms are b. For such a proportion, b is called **the mean proportional.**

$$\frac{a}{b} = \frac{b}{c}, \text{ where } a < b < c.$$

In exercises **5–10**, find the mean proportional for the given values of a and c.

5. $a = 4, c = 9$

6. $a = 3, c = 27$

7. $a = 6, c = 24$

8. $a = 5, c = 20$

9. $a = x, c = 4x$ and $x > 0$

10. $a = x, c = 9x$ and $x > 0$

In exercises **11–18**, find the missing value, given that b is the mean proportional to a and c.

11. $b = 4, c = 8$; find a

12. $b = 8, c = 32$; find a

13. $b = 4, c = 16$; find a

14. $b = 10, c = 100$; find a

15. $a = 7, b = 14$; find c

16. $a = 8, b = 16$; find c

17. $a = 3x, b = 12x$; find c if $x > 0$

18. $a = 2x, b = 6x$; find c if $x > 0$

The **golden section** (also called the **divine proportion**) is the division of a length into two parts such that the ratio of the shorter part to the longer part is the same as the ratio of the longer part to the original length. Both ratios are approximately equal to $\frac{5}{8}$.

For example, if a string 13 inches long is divided into pieces that are 5 inches and 8 inches, then

$$\left(\frac{\text{shorter part}}{\text{longer part}}\right) \quad \frac{5}{8} \text{ is approximately equal to } \frac{8}{13} \quad \left(\frac{\text{longer part}}{\text{original length}}\right).$$

0.625 is approximately equal to 0.615 . . .

In exercises **19–24**, divide a length into two parts that satisfy the golden section. Write answers to the indicated number of places. (Hint: Let x be the *longer part* and k be the *original length*. Write the proportion $\frac{x}{k} = \frac{5}{8}$. Solve for x and round the answer.)

19. The length is 26 centimeters. Solve to the nearest whole number.

20. The length is 39 centimeters. Solve to the nearest whole number.

21. The length is 6.5 inches. Solve to one decimal place.

22. The length is 19.5 inches. Solve to one decimal place.

23. The length is 3.25 meters. Solve to two decimal places.

24. The length is 22.75 meters. Solve to two decimal places.

SECTION 5-9. Applied Problems with Rational Expressions

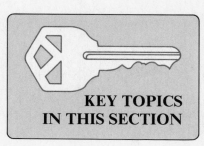

KEY TOPICS IN THIS SECTION

1. Number problems

2. Work problems

3. Uniform motion problems

Many applied problems require equations that include one or more rational expressions. In this section we will study three types of problems to illustrate such equations: **number problems, work problems,** and **uniform motion problems.** We will describe each kind of problem, and then we will give examples to illustrate any special characteristics of that type.

Number Problems

Recall that a rational number can be written as a divided by b, where a and b are integers and $b \neq 0$. Three terms will be used in connection with such a rational number:

Rational Number $\dfrac{a}{b}$

a is the **numerator** of the number.

b is the **denominator** of the number.

$\dfrac{b}{a}$ is the **reciprocal,** and $a \neq 0$.

Example 1.   A number is added to the numerator of $\frac{13}{20}$. Two times the same number is subtracted from the denominator. The resulting rational number is $\frac{9}{5}$. Find the number.

Solution.   **Discussion.** The numerator of $\frac{13}{20}$ is 13, and the denominator is 20.

Let n be the unknown number.
"A number is added to the numerator" is written $13 + n$.
"Two times the number is subtracted from the denominator" is written $20 - 2n$.
"The resulting rational number is $\frac{9}{5}$" yields the following equation:

$$\frac{13 + n}{20 - 2n} = \frac{9}{5} \qquad \text{The LCM of the denominators is } 5(20 - 2n).$$

$$\frac{13 + n}{20 - 2n} \cdot 5(20 - 2n) = \frac{9}{5} \cdot 5(20 - 2n)$$

$$5(13 + n) = 9(20 - 2n)$$

$$65 + 5n = 180 - 18n$$

$$65 + 23n = 180$$

$$23n = 115$$

$$n = 5$$

Check: $\dfrac{13 + 5}{20 - 2(5)} = \dfrac{13 + 5}{20 - 10} = \dfrac{18}{10} = \dfrac{9}{5}$

Thus, the number is 5.

Example 2.   The sum of a number and three times its reciprocal is $\frac{19}{4}$. Find the number.

Solution.   **Discussion.** If we let n stand for the unknown number, then $\frac{1}{n}$ is the reciprocal of the number.

"Three times its reciprocal" is written $3 \cdot \frac{1}{n}$, or $\frac{3}{n}$.
"The sum ... is $\frac{19}{4}$" yields the following equation:

$$n + \frac{3}{n} = \frac{19}{4} \qquad \text{The LCM of the denominators is } 4n.$$

$$\left(n + \frac{3}{n}\right) \cdot 4n = \frac{19}{4} \cdot 4n$$

$$4n^2 + 12 = 19n \qquad \text{A quadratic equation}$$

$$4n^2 - 19n + 12 = 0 \qquad \text{Set equal to zero.}$$

$$(4n - 3)(n - 4) = 0 \qquad \text{Factor the trinomial.}$$

$$4n - 3 = 0 \qquad \text{or} \qquad n - 4 = 0 \qquad \text{Zero-product property}$$

$$n = \frac{3}{4} \qquad\qquad n = 4 \qquad \text{Solve each equation.}$$

Check: If $\qquad n = \dfrac{3}{4}$ $\qquad\qquad$ If $\qquad n = 4$

$\qquad\qquad$ then $\qquad \dfrac{1}{n} = \dfrac{4}{3}$ $\qquad\qquad$ then $\qquad \dfrac{1}{n} = \dfrac{1}{4}$

$\qquad\qquad\qquad\qquad \dfrac{3}{n} = 4$ $\qquad\qquad\qquad\qquad \dfrac{3}{n} = \dfrac{3}{4}$

$\qquad n + \dfrac{3}{n} = \dfrac{3}{4} + 4 = \dfrac{19}{4}$ $\qquad\qquad n + \dfrac{3}{n} = 4 + \dfrac{3}{4} = \dfrac{19}{4}$

Both numbers check; thus the number is either $\frac{3}{4}$ or 4.

Work Problems

A work problem usually includes some statement about a job being done. The job may consist of such tasks as processing data, completing a building project, or filling or emptying a receptacle of fluid. The following three questions need to be answered by the given information in any work problem:

Question 1. How long does it take to complete the job?

Question 2. How many machines are used on the job?

Question 3. At what rate does each machine work?

The following concepts may also be helpful in setting up an equation for a work problem:

Concept 1. If a machine can complete a job in t hours, minutes, and so on working by itself, then the machine works at a rate of $\frac{1}{t}$. For example, if it takes five days to paint some house, then one-fifth of the house is painted each day.

Concept 2. If the product of time (t) worked on a job and the rate (r) at which work is done equals 1, then the job is completed.

$r \cdot t = 1$ means "one completed job."

Thus ($\frac{1}{5}$ house painted per day)(5 days) = one painted house.

Concept 3. If more than one rate and time is involved in completing a job, then the sum of the products of the times worked and the rates at which work is performed must equal 1.

Example 3. A model 1029W computer can process a certain quantity of data in 10 hours. A Model 475Q computer can process the same data in 15 hours. If both computers are used to process the data, how long would it take?

Solution. **Discussion.** The information can be displayed in a table as shown below. Let t stand for the time it would take the two computers working together to process the data.

Machine	Time needed working alone	Fractional part of job done in 1 hour	Fractional part of job done in t hours
Model 1029W	10 hours	$\frac{1}{10}th$ of job/hour	$\frac{1}{10} \cdot t = \frac{t}{10}$ of job
Model 475Q	15 hours	$\frac{1}{15}th$ of job/hour	$\frac{1}{15} \cdot t = \frac{t}{15}$ of job

An equation for the problem is

$$\frac{t}{10} + \frac{t}{15} = 1.$$ ⬚ Work done by 1029W ⬚ + ⬚ Work done by 475Q ⬚ = ⬚ One job completed ⬚

$$\left(\frac{t}{10} + \frac{t}{15}\right) \cdot 30 = 1 \cdot 30$$

$$3t + 2t = 30$$

$$5t = 30$$

$$t = 6$$

Check: If $t = 6$, then $\frac{6}{10} + \frac{6}{15} = \frac{18}{30} + \frac{12}{30} = \frac{30}{30} = 1$.

Thus, it would take about 6 hours for both computers to process the data.

Example 4.   A hose can completely fill a tank used to water horses in 18 minutes. An outlet valve used to clean and drain the tank can empty a full tank in 30 minutes. Linda leaves the drain valve open to flush out the tank while filling the tank at the same time. Using this technique, how long will it take to completely fill the tank before turning off the outlet valve?

Solution.   **Discussion.** Since the inlet and outlet valves are working in opposition to each other, the rates at which the valves work will be given opposite signs. A positive rate will be given to the inlet valve and a negative rate to the outlet valve. Let t represent the time needed to fill the tank.

Machine	Time needed working alone	Fractional part of job done in 1 minute	Fractional part of job done in t minutes
Inlet valve	18 minutes	$\frac{1}{18}th$ of job/minute	$\frac{1}{18} \cdot t = \frac{t}{18}$ of job
Outlet valve	30 minutes	$\frac{-1}{30}th$ of job/minute	$\frac{-1}{30} \cdot t = \frac{-t}{30}$ of job

An equation for the problem is

$$\frac{t}{18} - \frac{t}{30} = 1.$$ ⬚ Water in ⬚ − ⬚ Water out ⬚ = ⬚ One full tank ⬚

$$\left(\frac{t}{18} - \frac{t}{30}\right) \cdot 90 = 1 \cdot 90$$

$$5t - 3t = 90$$

$$2t = 90$$

$$t = 45$$

Check: $\dfrac{45}{18} - \dfrac{45}{30} = \dfrac{225}{90} - \dfrac{135}{90} = \dfrac{90}{90} = 1$

Thus, it would take about 45 minutes to fill the tank.

Uniform Motion Problems

The equation $d = r \cdot t$ was used to compute the distance (d) that an object moved in a specified time (t) when moving at a constant rate (r). If the distance is known in a distance-rate-time problem, then equation **1** or **2** can be used to find the unknown quantity.

1. $r = \dfrac{d}{t}$ **2.** $t = \dfrac{d}{r}$

Displaying the given information in a table usually helps determine an equation for solving the problem.

Example 5. Virginia Colbourn, an executive of a large cosmetic company, uses her private plane to make business trips. On a recent trip she flew 445 miles in $1\frac{1}{4}$ hours aided by a 30-mph tail wind. On the return trip, against the wind, she was 75 miles short of her starting point after $1\frac{1}{4}$ hours of flying time. Find the speed of Virginia's plane with no wind, using the flying conditions on this trip.

Solution. **Discussion.** The time out and the time back are the same; that is, $1\frac{1}{4}$ hours. As a consequence we can use equation **2** to set the ratios of distances over rates equal to each other. Let r stand for the unknown rate of the plane with no wind. Thus the rate *with the wind* is $(r + 30)$, and the rate *against the wind* is $(r - 30)$.

Virginia's plane	Rate	Distance	Time
With the wind	$(r + 30)$ mph	445 miles	$1\dfrac{1}{4} = \dfrac{445}{r + 30}$
Against the wind	$(r - 30)$ mph	370 miles	$1\dfrac{1}{4} = \dfrac{370}{r - 30}$

Using equation **2**,

$$\dfrac{445}{r + 30} = \dfrac{370}{r - 30} \qquad \dfrac{d_1}{r_1} = \dfrac{d_2}{r_2}$$

$$445(r - 30) = 370(r + 30) \qquad \text{Means-extremes product property}$$

$$445r - 13{,}350 = 370r + 11{,}100$$

$$75r = 24{,}450$$

$$r = 326$$

Thus, Virginia's plane on this trip could travel 326 mph with no wind.

Example 6.   The distance between cities S and B is 220 miles. The distance between cities B and LA is 300 miles. On a recent business trip Shelby White made the trips between S and B and between B and LA at the same rate. He drove from S to B in 1.6 hours less time than it took him to drive from B to LA. Find Shelby's rate on these trips.

Solution.   **Discussion.** The rates on the two trips are the same; therefore we can use equation **1**. Let r represent the unknown rate.

Shelby's car	Rate	Distance	Time
From S to B	r mph	220 miles	$\dfrac{220}{r}$ hours
From B to LA	r mph	300 miles	$\dfrac{300}{r}$ hours

Based on the given data, the difference between the times from S to B and from B to LA is 1.6 hours. Thus, an equation for the problem is

$$\frac{220}{r} = \frac{300}{r} - 1.6. \qquad \text{The LCM of the denominators is } r.$$

$$220 = 300 - 1.6r \qquad \text{Multiply both sides by } r.$$

$$1.6r = 80$$

$$r = 50$$

Thus, Shelby averaged 50 mph on these trips.

SECTION 5-9.   Practice Exercises

In exercises **1–12**, solve the number problems.

[Examples 1 and 2]   **1.** A number is added to the numerator and subtracted from the denominator of $\frac{2}{7}$. The resulting fraction is the reciprocal of $\frac{2}{7}$. Find the number.

2. A number is subtracted from the numerator and added to the denominator of $\frac{4}{9}$. The resulting fraction is $\frac{1}{12}$. Find the number.

3. A number is added to the numerator, and twice the number is subtracted from the denominator, of $\frac{5}{12}$. The resulting fraction is $\frac{4}{3}$. Find the number.

4. Twice a number is subtracted from the numerator and twice the same number is added to the denominator of $\frac{19}{11}$. The resulting fraction is the reciprocal of $\frac{19}{11}$. Find the number.

5. Three times a number is added to the numerator of $\frac{15}{28}$, and five times the same number is subtracted from the denominator. The resulting fraction is equivalent to $\frac{7}{6}$. Find the number.

ANS 21

6. Twice a number is subtracted from the numerator of $\frac{7}{18}$, and three times the same number is added to the denominator. The resulting fraction is equivalent to $\frac{-1}{11}$. Find the number.

7. The sum of a number and its reciprocal is $\frac{13}{6}$. Find the number.

8. The sum of a number and its reciprocal is $\frac{34}{15}$. Find the number.

9. The difference between a number and its reciprocal is $\frac{21}{10}$. Find the number.

10. The difference between a number and its reciprocal is $\frac{33}{28}$. Find the number.

11. The sum of a number and twice its reciprocal is $\frac{9}{2}$. Find the number.

12. The difference between twice a number and its reciprocal is $\frac{1}{6}$. Find the number.

In exercises **13–22**, solve the work problems.

[Example 3] 13. Machine A can process a quantity of data in 8 hours, while machine B can process the same data in 12 hours. How long would it take both machines, working together, to process the data?

14. Tim can plow a field in 8 hours using his own equipment. It takes Gene 10 hours to plow the same field using *his* own equipment. How long would it take both of them to plow the field working together?

15. Linda can feed and water 35 horses at the barn in 45 minutes. It takes Kimberley, her daughter, exactly one hour to do the same job. Working together, how long would it take both of them to do the job?

16. Three machines, A, B, and C, operate in a gravel pit. A certain quantity of gravel can be removed by A in 9 hours, or by B in 12 hours, or by C in 18 hours. Working together, in how many hours can A, B, and C remove the quantity of gravel?

17. Machine A can process a quantity of data in 12 hours, and machine B can process the same data in 15 hours. After machine A has been working 2 hours, B is also put into operation. How long does it take to complete the work?

18. A quantity of milk can be bottled by machine A in 6 hours or by machine B in 10 hours. Machine A is used alone for 3 hours, and then B is also put into operation. How long does it take to bottle the milk?

19. It takes Tom 9 hours, working alone, to stucco the walls of a home in a subdivision. It takes Todd 12 hours to do the same amount of work. Tom works on a house for 4 hours by himself, and then Todd helps finish the job. How long does it take to complete the job?

20. One well can irrigate a field in 6 hours, and a second well can irrigate the same field in 10 hours. The second well is put into operation 1.5 hours after the first well started the job. How long does it take to irrigate the field?

21. A pipe can fill a trough in 20 minutes. The trough is punctured with a hole that drains the trough in 90 minutes. If the hole is not patched, how long does it take to fill the trough?

22. An air mattress can be filled with a hand-operated air pump in 10 minutes. The mattress has a puncture that empties it in 25 minutes. With the puncture, how long does it take to fill the mattress with the pump?

In exercises **23–30**, solve the uniform motion problems.

[Examples 4 and 5]   **23.** A boat traveled 8 miles against a river current of 5 mph, in the same time it traveled 12 miles with the current. Find the speed of the boat without the river current.

24. Cycling up a grade, a bicyclist can travel 10 miles in the same time that she can travel 14 miles down the grade. If her rate on the grade differs from her rate on level ground by 2 miles per hour, find her rate on level ground.

25. With a wind of 20 miles per hour, a light airplane can travel 560 miles, which is 160 miles farther than it can travel against the wind in the same time. Find the speed of the plane with no wind.

26. Traveling at the same rate, a family went 495 miles the first day and 660 miles the second day of their vacation. If they traveled 3 hours longer the second day than the first, find the rate at which they traveled.

27. On Monday Rick jogged 6 miles. On Tuesday he covered 10 miles, but he jogged for 30 more minutes than on Monday. At what rate does he jog?

28. An airliner made the trip from city A to city B, a distance of 560 miles, in 1.5 hours less time that it made the trip from city B to city C, a distance of 1,130 miles. Assuming no wind assistance or resistance, find the rate of the plane.

29. An incoming tidal current is "flooding" at $1\frac{1}{2}$ mph. A sailboat can travel 5 miles with the current in the same time that it can travel 3 miles against the current. Find the speed of the sailboat without the current.

30. A small, ultra-light aircraft can fly 75 miles with a 10-mph tail wind in the same time that it can fly 63 miles against the same wind. Find the speed of the ultra-light without the wind.

SECTION 5-9.   Ten Review Exercises

In exercises **1** and **2**, simplify.

1. $\dfrac{2^4 - 1}{3^2 - 1} + \dfrac{3(5^2 - 7 \cdot 3 - 1)}{3 \cdot 5 - 7}$ **2.** $-2[3^2 - 5(2^3 - 7)] - (5^2 - 3 \cdot 11)$

In exercises **3–6**, solve each problem.

3. Solve $3x - 8y = 24$ for y.

4. Solve for y: $3x - 8y = 24$, given $x = 48$.

5. Solve $3x - 8y = 24$ for x.

6. Solve for x: $3x - 8y = 24$, given $y = -9$.

In exercises **7–10**, factor completely.

7. $5x^3y + 80xy$ **8.** $5x^4y + 135xy^4$

9. $100u^2 - 9v^2$ **10.** $100u^2 - 60uv + 9v^2$

SECTION 5-9.   Supplementary Exercises

In exercises **1–6**, do parts **a–g**.

1. Use the rational expression $\dfrac{3x + 1}{x + 1}$.

   In parts **a–e**, approximate the values to three decimal places.
   a. Approximate for $x = 1$.
   b. Approximate for $x = 5$.
   c. Approximate for $x = 10$.
   d. Approximate for $x = 100$.
   e. Approximate for $x = 1,000$.
   f. Based on the results of parts **a–e**, what value will the expression approach (but never equal) as x becomes larger and larger?
   g. What value would $\dfrac{7x + 1}{x + 1}$ approach (but never equal) as x becomes larger and larger?

2. Use the rational expression $\dfrac{5x - 1}{x + 1}$.

   In parts **a–e**, approximate the values to three decimal places.
   a. Approximate for $x = 1$.
   b. Approximate for $x = 5$.
   c. Approximate for $x = 10$.
   d. Approximate for $x = 100$.
   e. Approximate for $x = 1,000$.
   f. Based on the results of parts **a–e**, what value will the expression approach (but never equal) as x becomes larger and larger?
   g. What value would $\dfrac{4x - 1}{x + 1}$ approach (but never equal) as x becomes larger and larger?

3. Use the expression $\dfrac{x + 1}{x^2 + 1}$.

   In parts **a–e**, approximate the values to three decimal places.
   a. Approximate for $x = 1$.
   b. Approximate for $x = 5$.
   c. Approximate for $x = 10$.
   d. Approximate for $x = 50$.
   e. Approximate for $x = 100$.
   f. Based on the results of parts **a–e**, what value will the expression approach (but never equal) as x becomes larger and larger?
   g. What value would $\dfrac{x - 1}{x^2 + 1}$ approach (but never equal) as x becomes larger and larger?

4. Use the expression $\dfrac{x + 1}{x^2 + x}$.

   In parts **a–e**, approximate the values to three decimal places.
   a. Approximate for $x = 1$.
   b. Approximate for $x = 5$.
   c. Approximate for $x = 10$.

d. Approximate for $x = 50$.

e. Approximate for $x = 100$.

f. Based on the results of parts **a–e**, what value will the expression approach (but never equal) as x becomes larger and larger?

g. What value would $\dfrac{x + 1}{x^2 - x}$ approach (but never equal) as x becomes larger and larger?

5. Use the expression $\dfrac{x^2}{x + 1}$.

In parts **a–e**, approximate the values to three decimal places.

a. Approximate for $x = 1$.

b. Approximate for $x = 5$.

c. Approximate for $x = 10$.

d. Approximate for $x = 100$.

e. Approximate for $x = 1{,}000$.

f. Based on the results of parts **a–e**, what will happen to the values of the expression as x becomes larger and larger?

g. What will happen to the values of $\dfrac{x^2}{x - 1}$ as x becomes larger and larger?

6. Use the expression $\dfrac{x^2 + x}{x + 1}$.

In parts **a–e**, approximate the values to three decimal places.

a. Approximate for $x = 1$.

b. Approximate for $x = 5$.

c. Approximate for $x = 10$.

d. Approximate for $x = 100$.

e. Approximate for $x = 1{,}000$.

f. Based on the results of parts **a–e**, what will happen to the values of the expression as x becomes larger and larger?

g. What will happen to the values of $\dfrac{x^2 - 1}{x + 1}$ as x becomes larger and larger?

SECTION 5-1. Summary Exercises

Answer

In exercises **1–3**, evaluate for the stated value of the variable or variables.

1. $\dfrac{x^2 + 9}{x^2 + x - 3}$ and $x = -3$

1. _____

2. $\dfrac{2m - n}{2m^2 - 3mn + n^2}$ and $m = 3$, $n = 2$

2. _____

3. $\dfrac{a^2 - b^2 + c^2}{a^3 + c^3}$ and $a = -2$, $b = 6$, $c = 5$

3. _____

In exercises **4** and **5**, find any restricted values.

4. $\dfrac{p + 8}{p^2 - 5p}$

4. _____

5. $\dfrac{a^2 + 10}{a^2 - b^2}$

5. _____

In exercises **6–8**, reduce. Assume no denominator is zero.

6. $\dfrac{6c^2 + 16c}{9c^2 - 64}$

6. _____

7. $\dfrac{2m^2 - 8m + 8}{6m^2 - 24}$

7. _____

8. $\dfrac{27u^3 - 125}{6u^2 + 9uv - 10u - 15v}$

8. _____

SECTION 5-2.   Summary Exercises

In exercises **1–3**, write with only positive exponents.

1. $4^{-3}x^{-2}y^5$

1. _____

2. $\dfrac{10}{p^{-3}}$

2. _____

3. $\left(\dfrac{s}{3}\right)^{-2}$

3. _____

In exercises **4–7**, simplify and write answers with only positive exponents.

4. $w^{-11}w^7$

4. _____

5. $\dfrac{t^{-10}}{t^{-6}}$

5. _____

6. $(9m^{-3})^{-2}$

6. _____

7. $\left(\dfrac{b^{-5}}{5a^{-1}}\right)^{-2}$

7. _____

8. a. Write the following number in scientific notation:

27,500,000

8. a. _____

b. Write the following amount in ordinary notation:

6.3×10^{-4}

b. _____

SECTION 5-3.   Summary Exercises

In exercises **1–8**, do the indicated operations and reduce.

1. $\dfrac{9y^2}{10x^2} \cdot \dfrac{4x}{3y}$

1. _____

2. $\dfrac{30}{x^2y^2} \div \dfrac{18}{xy}$

2. _____

3. $\dfrac{2x^2 - 5x - 3}{x^2 - 6x + 9} \cdot \dfrac{2x^4}{2x^2 + x}$

3. _____

4. $\dfrac{a^2 + 2ab + b^2}{4a^2b + 2ab^2} \div \dfrac{a + b}{2a + b}$

4. _____

5. $\dfrac{2a^2 + 3ab + b^2}{a^3 - b^3} \cdot \dfrac{a^2 + ab + b^2}{2a^2 - ab - b^2}$

5. _____

6. $\dfrac{xy - y^3}{2x^3 + 2xy} \div \dfrac{2x^2 - 2xy^2 + 3xy - 3y^3}{2x^3 + 3x^2y + 2xy + 3y^2}$

6. _____

7. $\left(\dfrac{8x^2 + 18x - 5}{10x^2 - 9x + 2} \div \dfrac{8x^2 + 22x + 15}{10x^2 + 11x - 6} \right) \cdot \dfrac{4x + 5}{4x - 1}$

7. _____

8. $\left[\left(\dfrac{3n^2}{6m^2 - 3mn} \div \dfrac{m^2n + 2mn^2}{m^2 - 4n^2} \right) \cdot \dfrac{2m}{45m - 15n} \right] \div \dfrac{mn}{10m^2n - 5mn^2}$

8. _____

SECTION 5-4. Summary Exercises

In exercises **1–8**, simplify.

1. $\dfrac{3a^3}{4(a+3)} + \dfrac{9a^2}{4(a+3)}$

1. _____

2. $\dfrac{5y+2}{2y+3} - \dfrac{y-4}{2y+3}$

2. _____

3. $\dfrac{a+4}{a+2} - \dfrac{a}{a+2}$

3. _____

4. $\dfrac{x+y}{x^2+y^2} + \dfrac{3x-y}{x^2+y^2}$

4. _____

5. $\dfrac{x^2 + 1}{x^2 - 3x + 2} + \dfrac{x - 3}{x^2 - 3x + 2}$

5. _____

6. $\dfrac{5m^2 + 2mn}{2m^2 + 3mn - 2n^2} - \dfrac{m^2 + 2mn + n^2}{2m^2 + 3mn - 2n^2}$

6. _____

7. $\dfrac{y^2 - 4y}{3y^2 + 5y - 2} + \dfrac{3(y - 2)}{3y^2 + 5y - 2}$

7. _____

8. $\dfrac{3a^2}{a^2 + 5a + 6} + \dfrac{2a + 6}{a^2 - 4} \div \dfrac{a^2 + 6a + 9}{a^2 - 2a} + \dfrac{2 + 5a}{a^2 + 5a + 6}$

8. _____

Name _____

Date _____

Score _____

SECTION 5-5. Summary Exercises

Answer

In exercises **1** and **2**, find the LCM.

1. $6m^2n$ and $15mn$

1. _____

2. $x^2 - 7x$, $2x^2 - 98$, and $5x + 35$

2. _____

In exercises **3–8**, simplify.

3. $\dfrac{5}{12a^2b} - \dfrac{1}{8ab}$

3. _____

4. $\dfrac{5y}{6y - 18} + \dfrac{3y}{6 - 2y}$

4. _____

5. $\dfrac{2}{b^2 + 6b + 9} + \dfrac{2}{b^2 - 9}$

5. _____

6. $\dfrac{2p + 5}{p^2 + p - 2} + \dfrac{4p + 9}{3p + 6}$

6. _____

7. $\dfrac{z}{z + 7} + \dfrac{z}{z - 9} - \dfrac{z^2 + 4z + 27}{z^2 - 2z - 63}$

7. _____

8. $\dfrac{2}{t} - \dfrac{1}{t + 5} + \dfrac{t^2 - 2t - 8}{t^2 + 7t + 10} \div \dfrac{t^2 - 11t + 28}{t^2 - t - 42}$

8. _____

SECTION 5-6. Summary Exercises

In exercises **1–3**, simplify using Plan A.

1. $\dfrac{6 + \dfrac{7}{x} - \dfrac{3}{x^2}}{9 + \dfrac{3}{x} - \dfrac{2}{x^2}}$

1. _____

2. $\dfrac{3 - \dfrac{1}{y}}{9 - \dfrac{1}{y^2}}$

2. _____

3. $\dfrac{\dfrac{a}{a-b} - \dfrac{b}{a+b}}{\dfrac{a}{a+b} + \dfrac{b}{a-b}}$

3. _____

In exercises **4–7**, simplify using Plan B.

4. $\dfrac{t - \dfrac{9}{t}}{t - 7 + \dfrac{12}{t}}$

4. _____

5. $\dfrac{\dfrac{1}{p} - \dfrac{2}{p^2} - \dfrac{3}{p^3}}{1 - \dfrac{9}{p^2}}$

5. _____

6. $\dfrac{\dfrac{m}{n} + \dfrac{n}{5}}{\dfrac{mn}{10}}$

6. _____

7. $\dfrac{1 + \dfrac{1}{x + 1}}{\dfrac{1}{2} + \dfrac{1}{x}}$

7. _____

8. Simplify: $\dfrac{a + 3a^{-1}}{a^2 - 9a^{-2}}$

8. _____

SECTION 5-7. Summary Exercises

In exercises **1–8**, solve and check.

1. $\dfrac{2x + 1}{3} - 1 = \dfrac{x}{2}$

1. _____

2. $\dfrac{1}{3} + \dfrac{y + 1}{4y} = \dfrac{1}{3y} - \dfrac{y + 5}{6y}$

2. _____

3. $\dfrac{4}{2a^2 + 9a + 4} = \dfrac{1}{2a^2 - 3a - 2}$

3. _____

4. $\dfrac{4b}{b^2 + b} + \dfrac{3}{b^2 - b} = \dfrac{4}{b}$

4. _____

5. $\dfrac{u-1}{u-3} = \dfrac{2u}{u-2} + \dfrac{4}{u^2-5u+6}$

5. _____

6. $\dfrac{v+1}{v+2} + \dfrac{v+1}{3-v} = \dfrac{5}{v^2-v-6}$

6. _____

7. $\dfrac{5}{5-a} - \dfrac{a^2}{5-a} = -2$

7. _____

8. $\dfrac{x}{x-2} = 2 - \dfrac{1}{2-x}$

8. _____

Name _____

Date _____

Score _____

SECTION 5-8. Summary Exercises

Answer

In exercises **1–8**, solve each problem.

1. $\dfrac{12}{7} = \dfrac{132}{t}$

1. _____

2. $\dfrac{u+3}{12} = \dfrac{3}{4}$

2. _____

3. $\dfrac{x-1}{x} = \dfrac{2x+3}{3x}$

3. _____

4. $\dfrac{y}{2y-1} = \dfrac{15}{5y+2}$

4. _____

5. If 3 ounces of breakfast cereal contain 260 calories, how many calories are in   **5.** _____
12 ounces?

6. Avocados are on special at 5 for $1.98. Find the cost of 45 avocados at this   **6.** _____
price.

7. If two out of every five registered voters voted in the last election in Williamsburg,   **7.** _____
and 246 people voted, how many registered voters are there?

8. Triangle ABC is similar to triangle XYZ. Sides with lengths a, b, and c corre-   **8.** $c =$ _____
spond to sides with lengths x, y, and z. Find c and x if $a = 4$, $b = 6$, $y = 10$, $x =$ _____
and $z = 18$.

SECTION 5-9.   Summary Exercises

Answer

1. If a number is added to the numerator of $\frac{9}{13}$, and one less than the number is added to the denominator, the fraction has the value $\frac{4}{5}$. What is the number?

1. _____

2. If three times the reciprocal of a number is subtracted from the number, the difference is $\frac{23}{6}$. Find the number.

2. _____

3. A tub can be drained in 12 minutes. With the hot and cold water faucets opened, the tub can be filled in 8 minutes. How long does it take to fill the tub with the drain open and using both faucets?

3. _____

4. Charlotte Strausen is a strong swimmer. She lives on a river that regularly has a one-half mph current. Charlotte can swin one-fifth of a mile against the current in the same time she can swim one-third of a mile with the current. How fast can Charlotte swim with no current?

4. _____

CHAPTER 5   Review Exercises

In exercises **1–6**, use expressions R and S.

$$R: \frac{2t + 16}{t^2 + 8t} \qquad S: \frac{m^2 - n^2}{m^2 - mn - 2n^2}$$

1. Evaluate R for $t = -3$.

2. Evaluate S for $m = -2$ and $n = 3$.

3. Find any restricted values for R.

4. Find any restricted values for S.

5. Write R in reduced form.

6. Write S in reduced form.

In exercises **7–18**, simplify and write using only positive exponents. Assume variables are nonzero numbers.

7. $10^{-1}p^{-2}q$

8. $2^{-4}p^3q^{-1}$

9. $\dfrac{7x}{y^{-3}}$

10. $\dfrac{5^{-1}x}{y^{-4}}$

11. $\left(\dfrac{7}{m}\right)^{-1}$

12. $\left(\dfrac{m^2}{3}\right)^{-3}$

13. $(s^{-3}t^5)^{-4}$

14. $x^5 \cdot x^{-9}$

15. $\dfrac{z^{-9}}{z^{-4}}$

16. $-x^{-3}$

17. Write in scientific notation.

0.00000065

18. Write in ordinary notation.

7.41×10^8

In exercises **19** and **20**, reduce.

19. $\dfrac{6z^2 + 7z - 3}{4z^2 + 12z + 9}$

20. $\dfrac{125x^3 - y^3}{100x^3y + 20x^2y^2 + 4xy^3}$

In exercises **21–34**, do the indicated operations.

21. $\dfrac{2x}{x^2 - 9} \cdot \dfrac{3 - x}{x}$

22. $\dfrac{x^2 - 2x}{3x^2 - 8x - 3} \cdot \dfrac{2x - 6}{2x}$

23. $\dfrac{8a^3}{15b^3} \div \dfrac{2a^2}{3b^3}$

24. $\dfrac{a^2 - 2ab - 3b^2}{2a^2 - 5ab - 3b^2} \div \dfrac{2a^2 - 4ab}{2a^2 - 3ab - 2b^2}$

25. $\dfrac{13}{15u} + \dfrac{7}{15u}$

26. $\dfrac{u^2}{u^2 - v^2} + \dfrac{3uv + 2v^2}{u^2 - v^2}$

27. $\dfrac{8k + 5}{k + 3} - \dfrac{2k - 13}{k + 3}$

28. $\dfrac{3}{20k} + \dfrac{2}{45k}$

29. $\dfrac{5}{4a + 8} - \dfrac{5}{6a + 12}$

30. $\dfrac{3a + 4b}{a^2 - 16b^2} + \dfrac{1}{a + 4b}$

31. $\dfrac{\dfrac{12x}{5y}}{\dfrac{6x^2}{25y^2}}$

32. $\dfrac{\dfrac{3}{x^2 - 2xy + y^2}}{\dfrac{9}{x^2 - y^2}}$

33. $\dfrac{\dfrac{4}{3} + \dfrac{m}{n}}{\dfrac{9}{16} - \dfrac{n^2}{m^2}}$

34. $\dfrac{mn^{-1} - 1.5 - m^{-1}n}{2mn^{-1} - 2^{-1}m^{-1}n}$

In exercises **35–38**, solve and check.

35. $\dfrac{5}{4} - \dfrac{3}{2x} = \dfrac{2}{3} + \dfrac{5}{6x}$

36. $\dfrac{7}{x^2 - 5x - 50} = \dfrac{3}{x^2 - x - 30}$

37. $\dfrac{t + 12}{t^2 - 16} + \dfrac{1}{4 - t} = \dfrac{1}{4 + t}$

38. $\dfrac{5t^2}{t^2 - t - 20} = \dfrac{3t + 2}{t + 4} - \dfrac{2t + 3}{5 - t}$

In exercises **39** and **40**, find the value of n.

39. $\dfrac{20}{5} = \dfrac{2}{n}$

40. $\dfrac{n + 2}{n} = \dfrac{21}{3n}$

In exercises **41** and **42**, solve.

41. A car can travel 150 miles on 8 gallons of gasoline. How far can the car travel on 30 gallons of gasoline?

42. If 4 acres of a 620-acre field of wheat yields 170 bushels of grain, how many bushels will the whole field yield?

Triangles ABC and XYZ are similar. The lengths of the corresponding sides are a and x, b and y, and c and z. Find the lengths of the unknown sides with the information in exercises **43** and **44**.

43. $a = 4$, $x = 12$, $y = 18$, $z = 24$; find b and c.

44. $a = 3.0$, $b = 7.0$, $y = 24.5$, $z = 28.0$; find c and x.

In exercises **45–47**, solve.

45. The sum of a number and twice its reciprocal is $\frac{11}{3}$. Find the number.

46. Secretary A can type an article in 6 hours, and secretary B can type the same article in 8 hours. After A has been typing for 1 hour on the article, B helps finish the job. How long does it take to type the article?

47. With a wind of 30 mph pushing, an airplane can travel 650 miles. In the same time, the plane can travel 500 miles against the same wind. Find the speed of the plane with no wind.

6
Linear Equations and Inequalities in Two Variables

"The equations and inequalities we have worked with so far have basically involved one variable," Ms. Glaston said as she started the class. "But much work is done in mathematics that involves two or more variables at the same time. For example, suppose I tell you I have two sets of numbers in mind. For simplicity, let's refer to them as sets D and R."

She then turned to the board and wrote the following display:

SET D	**SET R**
$x_1, x_2, x_3, x_4, \ldots$	$y_1, y_2, y_3, y_4, \ldots$

"As you can see," Ms. Glaston continued, "I'm using the variable x with subscripts to represent the numbers in set D, and the variable y with subscripts to represent the numbers in set R. The symbol x_1 is read 'x sub-one,' and y_2 is read 'y sub-two.' Whenever a subscript is put on a variable, it represents a number and is treated as a constant. Now suppose that each y in R is related to some x in D. That is, once a value of x is known, the corresponding value of y can be determined. Furthermore, there is an equation, or formula, that we can use to find a value of y. Now I'll list some of the possible pairings of x and y."

She turned again to the board to construct the following table.

Value of x	0	1	3	6	10	12	19	x_3	x_4
Corresponding y	5	8	14	23	35	y_1	y_2	50	95

"Can anyone in the class find the values I have labeled y_1, y_2, x_3, and x_4?" asked Ms. Glaston.

"Could you tell us the formula for x and y?" Jain Simmons asked. "Then we could easily find the missing numbers."

"You're right, Jain," Ms. Glaston replied. "Why not let that be our first task, then. What formula could we write that would give the observed values of y for the given values of x?"

The class looked intently at the numbers in the table. Finally Carrie Mattaini raised her hand and said, "I think I see something that might help. The values of y seem to be increasing three times as fast as the values of x." She then walked to the board and added to the table as follows:

	+1	+2	+3	+4				
Value of x	0	1	3	6	10	12	19	x_3	x_4
Corresponding y	5	8	14	23	35	y_1	y_2	50	95
	+3	+6	+9	+12				

"Very good, Carrie," Ms. Glaston said. "Based on your observation we might consider this equation as a possible formula." She then wrote on the board

$$y = 3x.$$

"There's a problem with that equation, Ms. Glaston," Raoul Savoie said. "Based on the first cell in the table, y is 5 when x is 0. That equation won't give us 5 for y when x is replaced by 0."

"OK," Carrie returned, "then maybe we should just add 5 to $3x$ in the equation. That would give us 5 for y when x is 0."

"That's right, Carrie," Ms. Glaston replied, "and if that gives us the observed values of y in the table, then we may have our formula." She then revised the equation as suggested and added the following series of equations under the formula.

$$y = 3x + 5$$

Value of x	Corresponding Value of y	
0	$y = 3(0) + 5 = 0 + 5 = 5$	Check
1	$y = 3(1) + 5 = 3 + 5 = 8$	Check
3	$y = 3(3) + 5 = 9 + 5 = 14$	Check
6	$y = 3(6) + 5 = 18 + 5 = 23$	Check
10	$y = 3(10) + 5 = 30 + 5 = 35$	Check

"For all five pairings," Ms. Glaston continued, "the equation yields the correct value for y. So it seems that we have found a suitable formula. Now, as Jain suggested earlier, we can replace x by 12, and then by 19, to find the missing values y_1 and y_2. And we can replace y by 50, and then by 95, to find the missing values x_3 and x_4. The major obstacle was overcome when we found an equation that described the relationship between the numbers in set D and the numbers in set R.

"This exercise was designed to show you a few of the factors we encounter when we work with two or more variables. Today we begin the task of studying these factors in detail. We will learn how to write paired values of x and y and we will develop a geometric model on which the pairs of numbers can be displayed. We will also study a special kind of equation in x and y that has many applications in problems related to the world. If there are no questions, we can begin looking at Section 6-1."

SECTION 6-1.   Equations in Two Variables

KEY TOPICS IN THIS SECTION

1. Equations in two variables

2. The general form of a solution of an equation in two variables

3. Independent and dependent variables

4. Finding a solution when one number is given

In previous chapters we have solved equations in one variable. An equation in one variable is solved when number replacements for the variable are found that make the equation true. Each number that makes the equation true is called a **root,** or **solution,** of the equation. The set of all roots for a given equation is called the **solution set.**

Recall from the section on literal equations that equations can have more than one variable. Many types of equations involving two variables dominate the study of elementary mathematics. Examples **a** and **b** illustrate two such equations.

a. $3x + 5y = 15$ $\qquad$ An equation in x and y

b. $s = 2t^2 - t - 1$ $\qquad$ An equation in s and t

In this chapter we will begin the study of equations in two variables.

Equations in Two Variables

The equations given in examples **a** and **b** contain *two variables*. Such equations become true or false when *both variables* are replaced with numbers. The following replacements use the equations in examples **a** and **b**.

Equation	Replacements for variables	True or false
a. $3x + 5y = 15$	$x = 0$ and $y = 3$	$3(0) + 5(3) = 15$, true
	$x = -5$ and $y = 6$	$3(-5) + 5(6) = 15$, true
	$x = 3$ and $y = 1$	$3(3) + 5(1) = 15$, false

Equation	Replacements for variables	True or false
b. $s = 2t^2 - t - 1$	$s = 0$ and $t = 1$	$0 = 2(1)^2 - 1 - 1$, true
	$s = 14$ and $t = 3$	$14 = 2(3)^2 - 3 - 1$, true
	$s = 3$ and $t = 2$	$3 = 2(2)^2 - 2 - 1$, false

For equations in two variables there are generally an infinite number of replacements for the variables that will make the equation either true or false.

The General Form of a Solution of an Equation in Two Variables

As shown above, a replacement is needed for both variables to make an equation in two variables true. *Thus, each solution of an equation in two variables consists of two numbers.* The **ordered pair** notation is frequently used to list such solutions.

An ordered pair consists of two numbers written inside parentheses and separated by a comma.

To identify which number in the ordered pair is the replacement for which variable in the equation, the general form of the order may be given. Suppose, for example, the solutions in example **a** are written in the form (x, y). The first number in the pair would be the replacement for x and the second number would be the replacement for y. Writing the two solutions above in ordered pair notation yields the following results:

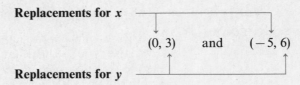

Replacements for x

$(0, 3)$ and $(-5, 6)$

Replacements for y

Definition 6.1.   A solution of an equation in x and y

Let (x, y) be the form of the ordered pair solutions for an equation in x and y. In general, (x_1, y_1) is a **solution** of the equation if and only if the equation is true when x is replaced by x_1 and y is replaced by y_1, where x_1 and y_1 are numbers.

Example 1.   Determine whether the following ordered pairs are solutions of $7x - 5y = 10$:

a. $(10, 12)$

b. $(-3, -6)$

Solution.   **Discussion.** Replace x by the first number in the ordered pair and y by the second number. If the resulting number statement is true, then the ordered pair is a solution. If the number statement is false, then it is not a solution.

a. $7(10) - 5(12) = 10$ Replace x by 10 and y by 12.

$$70 - 60 = 10$$

$$10 = 10, \text{ true}$$

Thus, $(10, 12)$ is a solution.

b. $7(-3) - 5(-6) = 10$ Replace x by -3 and y by -6.

$$-21 - (-30) = 10$$

$$9 = 10, \text{ false}$$

Thus, $(-3, -6)$ is not a solution.

Independent and Dependent Variables

An equation in two variables has an infinite number of solutions. A solution can be found by replacing one of the variables by a number, after which the resulting equation has only one variable. The root of this equation is the second number in the ordered pair.

> **Definition 6.2. Independent and dependent variables**
> In an equation in two variables, the variable that is replaced by arbitrarily selected values from a replacement set is called the **independent variable**. The other variable in the equation is called the **dependent variable**.

When the solution of an equation in two variables is stated, the first number in the ordered pair is the replacement for the independent variable and the second number is the replacement for the dependent variable.

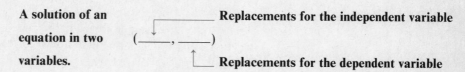

A solution of an equation in two variables.   **Replacements for the independent variable**

$(\underline{\quad}, \underline{\quad})$ **Replacements for the dependent variable**

Unless the independent and dependent variables in an equation are specifically identified, we shall assume the roles are based on *alphabetical order*. Examples **c** and **d** illustrate this assumption.

c. $2m^2 - n = 4$ m is independent n is dependent (m, n) are solutions

d. $4u^2 + 9v^2 = 36$ u is independent v is dependent (u, v) are solutions

Example 2.   Determine whether $(3, -4)$ is a solution of $5r - s = 19$.

Solution.   **Discussion.** Since r precedes s in the alphabet, the form of a solution is (r, s). Therefore, 3 is the replacement for r, and -4 is the replacement for s.

$$5(3) - (-4) = 19 \qquad \text{Replace } r \text{ by 3 and } s \text{ by } -4.$$

$$15 - (-4) = 19$$

$$19 = 19, \text{ true}$$

Thus, $(3, -4)$ is a solution of $5r - s = 19$.

Finding a Solution When One Number is Given

If (x_1, y_1) is a solution of an equation in x and y, and either x_1 or y_1 is known, then the value of the other element of the ordered pair can be found. The technique consists of two steps.

> If (x_1, y_1) is a solution of an equation in x and y, and x_1 or y_1 is given, then to find the unknown element do the following procedures:
>
> **Step 1.** Replace x by x_1, (or y by y_1).
>
> **Step 2.** Solve the resulting equation for y (or x), the remaining variable.

Example 3. If $x = \frac{4}{3}$, find the solution of $3x - 10y = 2$.

Solution.   **Step 1.** $3\left(\dfrac{4}{3}\right) - 10y = 2$ Replace x by $\frac{4}{3}$.

   Step 2. $4 - 10y = 2$ Solve for y.

   $-10y = -2$

   $y = \dfrac{-2}{-10} = \dfrac{1}{5}$

Thus, $\left(\frac{4}{3}, \frac{1}{5}\right)$ is a solution.

Sometimes a value is known for the dependent variable, the second element in the ordered pair. The value (or values) for the independent variable can then be found.

Example 4. If $n = 7$, find the solution(s) of $m^2 - n = 18$.

Solution.   **Step 1.** $m^2 - 7 = 18$ Replace n by 7.

   Step 2. $m^2 - 25 = 0$ Set equal to 0.

   $(m + 5)(m - 5) = 0$ Factor $m^2 - 25$.

   $m + 5 = 0$ or $m - 5 = 0$ Zero- product property

   $m = -5$ $m = 5$ Solve both equations.

Thus, $(-5, 7)$ and $(5, 7)$ are solutions of the equation.

SECTION 6-1.   Practice Exercises

In exercises **1–18**, determine whether the ordered pairs are solutions of the given equations.

[Example 1]   **1.** $5x - y = 9$
   a. $(4, 11)$ **b.** $(-3, -24)$

2. $4x - y = 11$
   a. $(6, 13)$ **b.** $(-6, -35)$

3. $8x + y = 8$
   a. $(-5, 40)$ **b.** $(-2, 24)$

4. $7x + y = 4$
   a. $(-3, 25)$ **b.** $(-5, 40)$

5. $4x + 3y = 12$
   a. $(3, 0)$ **b.** $(0, 4)$

6. $5x + 2y = 10$
   a. $(2, 0)$ **b.** $(0, 5)$

7. $x - 3y = -5$
   a. $(-8, 1)$ **b.** $(4, 3)$

8. $2x - 5y = 3$
   a. $(6, 2)$ **b.** $(-6, -3)$

[Example 2]   **9.** $8u - 2v = 2$
   a. $(-3, -11)$ **b.** $(3, 13)$

10. $6u - 5v = -5$
   a. $(-5, -7)$ **b.** $(5, 5)$

11. $2s + 7t = 3$
   a. $(5, -1)$ **b.** $(-4, 2)$

12. $3s + 4t = 5$
   a. $(-5, 5)$ **b.** $(7, -4)$

13. $7p - 3q = -1$
   a. $(-2, -4)$ **b.** $(5, 12)$

14. $7p - q = -2$
   a. $(-1, -2)$ **b.** $(2, 16)$

15. $2a = 5b - 3$

   a. $\left(-2, \dfrac{-1}{5}\right)$ **b.** $\left(3, \dfrac{9}{5}\right)$

16. $7a = -8b - 6$

   a. $\left(-1, \dfrac{1}{8}\right)$ **b.** $(-2, 1)$

17. $8m + 9n = 7$

   a. $\left(\dfrac{1}{2}, \dfrac{1}{3}\right)$ **b.** $\left(\dfrac{1}{4}, \dfrac{2}{3}\right)$

18. $8m - 6n = 5$

   a. $\left(\dfrac{3}{2}, \dfrac{1}{3}\right)$ **b.** $\left(\dfrac{5}{2}, \dfrac{5}{2}\right)$

In exercises **19–34**, find the solution for the given member of the ordered pair.

[Example 3]

19. $x + y = 5$

   a. $x = 2$ **b.** $y = -4$

20. $3x - y = 8$

   a. $x = -2$ **b.** $y = 7$

21. $5m + 2n = 10$

   a. $n = -5$ **b.** $m = 6$

22. $8m - 2n = 4$

   a. $n = 6$ **b.** $m = -3$

23. $v + 4u = -3$

   a. $v = 1$ **b.** $u = \dfrac{-1}{2}$

24. $8v - 7u = 10$

   a. $v = \dfrac{-1}{2}$ **b.** $u = -4$

25. $n = \dfrac{5}{2}m - 4$

   a. $m = \dfrac{4}{5}$ **b.** $n = 6$

26. $n = \dfrac{2}{3}m + 9$

   a. $m = \dfrac{-3}{2}$ **b.** $n = 13$

[Example 4]

27. $x^2 + y = 10$

   a. $x = 3$ **b.** $x = -3$

28. $x^2 + 2y = 10$

   a. $x = 2$ **b.** $x = -2$

29. $t = \dfrac{1}{2}s^2 - 5$

   a. $s = 4$ **b.** $s = -6$

30. $t = \dfrac{1}{4}s^2 + 1$

   a. $s = -2$ **b.** $s = 4$

31. $q = p^2 - 3$

   a. $p = 1$ **b.** $q = 1$

32. $q = p^2 - 15$

   a. $p = -1$ **b.** $q = 1$

33. $x^2 + y^2 = 4$

   a. $x = 2$ **b.** $y = -2$

34. $x^2 + y^2 = 25$

   a. $x = -5$ **b.** $y = 5$

SECTION 6-1. Ten Review Exercises

In exercises **1–8**, solve each problem.

1. Simplify: $15 + 5(2y + 1) + y - 4(2 - y)$

2. Solve and check: $15 + 5(2y + 1) + y = 4(2 - y)$

3. Factor: $a^2 - 3a - 54$

4. Solve and check: $a^2 = 3a + 54$

5. Simplify: $7(3 - t) - 5(t - 7) + 4$

6. Solve and graph: $7(3 - t) > 5(t - 7) - 4$

7. Simplify: $\dfrac{7}{b^2 - 5b - 50} - \dfrac{3}{b^2 - b - 30}$

8. Solve and check: $\dfrac{7}{b^2 - 5b - 50} = \dfrac{3}{b^2 - b - 30}$

In exercises **9** and **10**, simplify.

9. $\dfrac{\dfrac{3}{4} + \dfrac{m}{n}}{\dfrac{9}{16} - \dfrac{m^2}{n^2}}$

10. $\dfrac{\dfrac{3}{x^2 - 2xy + y^2}}{\dfrac{9}{x^2 - y^2}}$

SECTION 6-1. Supplementary Exercises

In exercises **1–20**, determine whether the given ordered pair is a solution of the equation.

1. $x - 2y = 1$, and $\left(\dfrac{1}{2}, \dfrac{-1}{4}\right)$

2. $8x + 9y = 6$ and $\left(\dfrac{1}{4}, \dfrac{4}{9}\right)$

3. $10m + 3n = 5$ and $\left(0, \dfrac{5}{3}\right)$

4. $2m - 7n = -3$ and $\left(0, \dfrac{3}{7}\right)$

5. $y = 2x - x^2$ and $(5, 15)$

6. $y = 3x + x^2$ and $(4, 24)$

7. $u = v^2 + 4v - 9$ and $(2, 2)$

8. $u = v^2 + 3v - 1$ and $(14, 3)$

9. $a^2 + 4b^2 = 16$ and $(0, -2)$

10. $5a^2 + b^2 = 5$ and $(-1, 0)$

11. $x^2 - 2y^2 = 0$ and $(0, 0)$

12. $5x^2 - y^2 = 0$ and $(0, 0)$

13. $a + 8b = -1$ and $\left(-4, \dfrac{3}{8}\right)$

14. $a - 6b = 7$ and $\left(-3, \dfrac{-5}{3}\right)$

15. $8p - 3q = -2$ and $\left(\dfrac{3}{2}, \dfrac{13}{3}\right)$

16. $8p + 4q = -8$ and $\left(\dfrac{5}{2}, \dfrac{-15}{2}\right)$

17. $6s + 5t = 2$ and $(-2.5, 3.4)$

18. $7s + 5t = -3$ and $(2.5, -4.1)$

19. $5x - 2y = -5$ and $(-4.4, -8.5)$

20. $3x + 7y = -7$ and $(-2.8, 0.2)$

In exercises **21–38**, find the solution for the given member of the ordered pair.

21. $2x + 3y = 0$ and $x = 0$

22. $4x - 5y = 0$ and $x = 0$

23. $x^2 - 2y = 25$ and $y = 0$

24. $x^2 - 6y = 49$ and $y = 0$

25. $2x^2 - 3y^2 = -48$ and $x = 0$

26. $5x^2 - 2y^2 = -32$ and $x = 0$

27. $b = a^2 - 2a + 4$ and $a = 2$

28. $b = a^2 - 3a + 1$ and $a = -2$

29. $b = a^2 - 2a + 4$ and $a = -3$

30. $b = a^2 - 3a + 1$ and $a = 3$

31. $6u - v + 1 = 0$ and $u = \dfrac{1}{3}$

32. $6u - v + 1 = 0$ and $u = \dfrac{1}{2}$

33. $y = \frac{2}{3}x$ and $x = \frac{9}{4}$

34. $y = \frac{4}{5}x$ and $x = \frac{15}{8}$

35. $y = x^2 + 5x + 6$ and $y = 0$

36. $y = x^2 + 6x + 5$ and $y = 0$

37. $x - 3 = 0$ and $y = 2$

38. $x - 4 = 0$ and $y = 1$

The equation $2x + y - 3z = 10$ is an example of an equation in three variables. A solution of such an equation is an **ordered triple** of the form (x, y, z).

Example. Determine whether $(4, -4, -2)$ is a solution of

$$2x + y - 3z = 10.$$

Solution. **Discussion.** Replace x by 4, y by -4, and z by -2. If the resulting number statement is true, the ordered triple is a solution. Otherwise it is not a solution.

$$2(4) + (-4) - 3(-2) = 10$$

$$8 + (-4) - (-6) = 10$$

$$4 - (-6) = 10, \text{ true}$$

Thus, $(4, -4, -2)$ is a solution of $2x + y - 3z = 10$.

In exercises **39–46**, determine whether the ordered triple is a solution of the equation.

39. $4x - 3y + 6z = 1$ $\qquad (-2, 1, 2)$

40. $3x - 4y + 7z = -16$ $\qquad (-5, 2, 1)$

41. $x + 5y - 3z = 9$ $\qquad (7, -2, -4)$

42. $x - 4y + 6z = 9$ $\qquad (9, -3, -2)$

43. $6x + 3y - 10z = -5$ $\qquad \left(\frac{1}{2}, \frac{-2}{3}, \frac{3}{5}\right)$

44. $8x - 9y - 15z = 15$ $\qquad \left(\frac{3}{2}, \frac{1}{3}, \frac{-2}{5}\right)$

45. $x^2 + 2y^2 - 5z^2 = 17$ $\qquad (-2, 3, -1)$

46. $8x^2 - 9y^2 + 25z^2 = 7$ $\qquad \left(\frac{-1}{2}, \frac{2}{3}, \frac{3}{5}\right)$

SECTION 6-2. Graphing Linear Equations

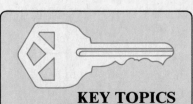

**KEY TOPICS
IN THIS SECTION**

1. The rectangular coordinate system

2. Plotting points in a coordinate system

3. The general form of a linear equation in two variables

4. Graphing a linear equation

5. Horizontal and vertical lines

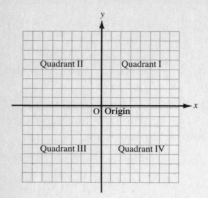

Figure 6-1. An xy-coordinate system.

A geometric line can be used as a visual model for the set of real numbers. Every real number can be graphed as a point on a **number line.** Furthermore every point on a line can be assigned a real number **coordinate.**

As we saw in Section 6-1, a solution of an equation in two variables is an **ordered pair of real numbers.** It is therefore helpful to have a geometric model for graphing pairs of numbers. Two number lines can be used to form a coordinate system for graphing pairs of numbers. We will study the details of such a system in this section. Furthermore, we will use the system to graph the solutions of one of the simplest equations in two variables, the **linear equation.**

The Rectangular Coordinate System

In Figure 6-1 a horizontal line and a vertical line intersect each other at a point labeled O. The horizontal line is used as a number line for the independent variable in the equation. Since many equations in two variables are written in terms of x and y, and since x is the independent variable, the horizontal line is labeled the **x axis.**

The vertical line is therefore used as a number line for the dependent variable, which is y in an equation in x and y. The vertical line is labeled the **y axis.**

We can use the points of the plane on which the axes are drawn as graphs of ordered pairs of real numbers. The plane with the lines forms an **xy-coordinate system.** Notice the following features of this coordinate system.

Feature 1. Point O, the intersection of the axes, is called the **origin** of the system.

Feature 2. The positive numbers are graphed to the right of the origin on the x axis and above the origin on the y axis.

Feature 3. The axes divide the plane on which they are drawn into four parts. Each part is called a **quadrant.**

Feature 4. The quadrants are labeled I, II, III, and IV, beginning in the upper right corner and moving counterclockwise, as shown.

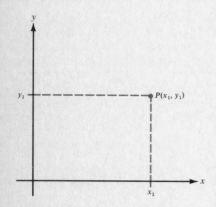

Figure 6-2. $P(x_1, y_1)$ in an xy-system.

Plotting Points in a Coordinate System

Every point in the plane can be paired with an ordered pair of real numbers. These numbers are called the **coordinates** of the point, and the point is called the **graph** of the ordered pair.

To illustrate how coordinates are assigned to points in the system, consider point P in Figure 6-2. A vertical line through P intersects the x axis at x_1. *The first number in the coordinates of P is x_1.* A horizontal line through P intersects the y axis at y_1. *Thus, the second number in the coordinates of P is y_1.* As shown in the figure, the coordinates of P are (x_1, y_1).

> If P is a point with coordinates (x_1, y_1), then x_1 specifies the *distance* and *direction* from the vertical axis to P. y_1 specifies the distance and direction from the horizontal axis to P.

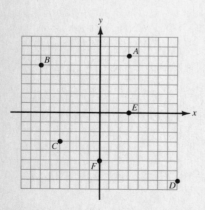

Figure 6-3. Points A–F.

Example 1.   State the coordinates of points labeled A–F in Figure 6-3.

Solution.   **Discussion.** Imagine that vertical and horizontal lines are drawn through each of the points. The coordinates of the intersections of these lines with the coordinate axes are the coordinates of the corresponding points.

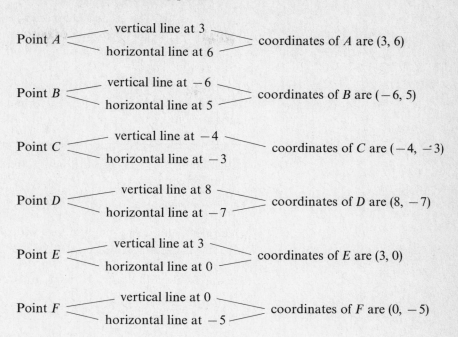

Point A — vertical line at 3 / horizontal line at 6 — coordinates of A are (3, 6)

Point B — vertical line at −6 / horizontal line at 5 — coordinates of B are (−6, 5)

Point C — vertical line at −4 / horizontal line at −3 — coordinates of C are (−4, −3)

Point D — vertical line at 8 / horizontal line at −7 — coordinates of D are (8, −7)

Point E — vertical line at 3 / horizontal line at 0 — coordinates of E are (3, 0)

Point F — vertical line at 0 / horizontal line at −5 — coordinates of F are (0, −5)

Notice that points E and F in Figure 6-3 are on the x- and y-axes respectively. As a consequence, one of the coordinates of these points is 0.

> Any point on the x axis has coordinates $(x, 0)$. Any point on the y axis has coordinates $(0, y)$.

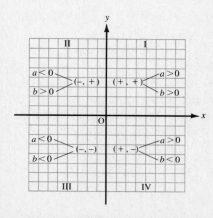

Figure 6-4. Graphs of G, H, and I.

Example 2.   Plot the following points: $G(−6, 7)$, $H(5, −4)$, and $I(0, 3)$

Solution.   G: From −6 on the x axis move seven units up.

H: From 5 on the x axis move four units down.

I: On the y axis move three units up from the origin. (See Figure 6-4.)

If (a, b) is an ordered pair in which $a \neq 0$ and $b \neq 0$, then the graph will be in one of the four quadrants. Precisely which quadrant can be determined by the signs of a and b. The details are shown in Figure 6-5.

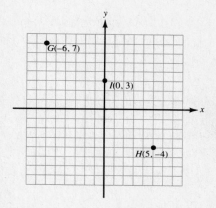

Figure 6-5. Signs of coordinates of points in Quadrants I–IV.

Example 3.   Identify the quadrant in which the following points lie. If a point is on an axis, identify which one.

a. $J(−18, −20)$

b. $K(25, −40)$

c. $L(−30, 0)$

Solution.   **a.** Since both coordinates are negative, the graph of J is in Quadrant III.

b. The first component is positive and the second component is negative. Therefore, the graph of K is in Quadrant IV.

c. The second component is zero; therefore, the graph of L is on the horizontal axis.

The General Form of a Linear Equation in Two Variables

Definition 6.3 identifies the general form of a linear equation in x and y. Such an equation is one of the simplest types of equations in two variables.

> **Definition 6.3.   The general form of a linear equation in x and y**
> A linear equation in x and y can be written in the form
> $$Ax + By = C$$
> where A, B, and C are real numbers with A and B not both zero.

Examples **a** and **b** illustrate linear equations in x and y:

a. $3x + 2y = 6$ **b.** $5x - 3y = 2$

Example 4.   Write the following linear equations in x and y in the general form.

a. $y = \dfrac{3}{4}x - 1$

b. $x = \dfrac{5}{2}$

Solution.   **a. Discussion.** As written, the equation is solved for y in terms of x.

$$y = \frac{3}{4}x - 1 \qquad \text{The given equation}$$

$$4y = 3x - 4 \qquad \text{Multiply both sides by 4.}$$

$$-3x + 4y = -4 \qquad \text{Subtract } 3x \text{ from both sides.}$$

In this form $A = -3$, $B = 4$, and $C = -4$. The **preferred form** would be to write the equation with A positive. Multiplying both sides of the equation by -1,

$$3x - 4y = 4 \qquad \text{The preferred form with } A > 0$$

b. Discussion. The equation appears to be a linear equation in one variable. However, we are told that it is a linear equation in x and y. The variable y is not written because the coefficient is 0. We can write this term and show the 0 coefficient.

$$x + 0 \cdot y = \frac{5}{2} \qquad \text{The given equation}$$

$$2x + 0 \cdot y = 5 \qquad \text{Multiply both sides by 2.}$$

The preferred form is with A, B, and C as integers.

Graphing a Linear Equation

As previously stated, examples **a** and **b** are linear equations in x and y:

a. $3x + 2y = 6$ **b.** $5x - 3y = 2$

A solution of either equation is an ordered pair of real numbers. For example, $(-2, 6)$, $(2, 0)$, and $(6, -6)$ are solutions of $3x + 2y = 6$.

$$(-2, 6): \quad 3(-2) + 2(6) = 6$$
$$-6 + 12 = 6, \text{ true}$$
$$(2, 0): \quad 3(2) + 2(0) = 6$$
$$6 + 0 = 6, \text{ true}$$
$$(6, -6): \quad 3(6) + 2(-6) = 6$$
$$18 + (-12) = 6, \text{ true}$$

In Figure 6-6 these three ordered pairs are plotted as points P, Q, and R. The number of solutions of $3x + 2y = 6$ is infinite. Therefore, a graph is one way in which all the solutions can be represented. The xy-coordinate system can be used to graph these solutions.

Notice that points P, Q, and R in Figure 6-6 appear to *line up*. That is, one can imagine drawing a straight line through these points. This appearance is not deceptive, since the solutions of $3x + 2y = 6$ are coordinates of points that do lie on a line. In fact, a graph of any equation having the form $Ax + By = C$ is a line.

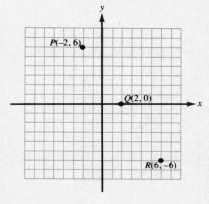

Figure 6-6. Three solutions of $3x + 2y = 6$.

To graph the solutions of $Ax + By = C$,

Step 1. find two solutions of $Ax + By = C$.

Step 2. plot these ordered pairs as points in an xy-coordinate system.

Step 3. draw the line through these points.

Step 4. check the accuracy of the line by plotting a third solution of the equation. If the third point falls on the line, the graph is probably correct. If the third point does not fall on the line, you have made an error and you must check your work.

Example 5. Graph: $2x + 5y = 20$.

Solution. If x is replaced by 0, then $5y = 20$ and $y = 4$.
If y is replaced by 0, then $2x = 20$ and $x = 10$.
If x is replaced by 5, then $10 + 5y = 20$ and $y = 2$.

The solutions can be displayed in a table.

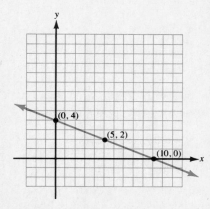

Figure 6-7. A graph of $2x + 5y = 20$.

x	y
0	4
10	0
5	2

These ordered pairs are plotted in Figure 6-7. The line through these points is a graph of $2x + 5y = 20$.

The statement that the line in Figure 6-7 is a graph of the equation $2x + 5y = 20$ has two important implications:

Implication 1. The plot of every solution of the equation is a point on the line.
   The line can be considered a geometric representation of all the solutions of the equation. The arrows on the ends of the line indicate the infinite extent of the line and the corresponding infinite number of solutions.

Implication 2. The coordinates of every point on the line are solutions of the equation.
   The line has points whose coordinates are rational and irrational numbers (the subsets of the real numbers). All of these ordered pairs are solutions of the equations, not just integers such as the coordinates of the points plotted in Figure 6-7.

Horizontal and Vertical Lines

Equations **a** and **b** can be written in the form of linear equations.

a. $x = 3$ **b.** $y = 5$

Writing both equations in the general form of a linear equation yields the following results:

a. $x + 0 \cdot y = 3$ **b.** $0 \cdot x + y = 5$

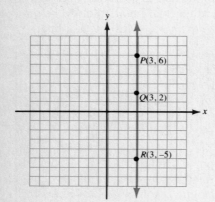

> If a and b are real numbers, then the graph of $x = a$ is a vertical line that crosses the x axis at $(a, 0)$. $y = b$ is a horizontal line that crosses the y axis at $(0, b)$.

Figure 6-8. Graph of $x = 3$.

Example 6.   Graph:

   a. $x = 3$

   b. $y = 5$

Solution.   **a. Discussion.** This equation can be written as follows:

$$x + 0 \cdot y = 3$$

Since the coefficient of y is zero, y can be replaced by any real number, but x must be replaced by 3. Therefore, the following ordered pairs are solutions: $(3, 6)$, $(3, 2)$, and $(3, -5)$. These three ordered pairs are plotted as points P, Q, and R in Figure 6-8. The vertical line through these points is the graph of $x = 3$.

b. Discussion. This equation can be written as follows:

$$0 \cdot x + y = 5$$

Since the coefficient of x is 0, x can be replaced by any real number, but y must be replaced by 5. Therefore, the following ordered pairs are solutions: $(-5, 5)$, $(2, 5)$, and $(8, 5)$. These three ordered pairs are plotted as points P, Q, and R in Figure 6-9. The horizontal line through these points is the graph of $y = 5$.

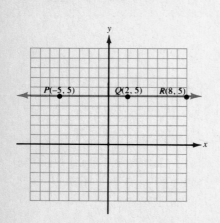

Figure 6-9. Graph of $y = 5$.

SECTION 6-2. Practice Exercises

In exercises **1–4**, state the coordinates of each point in Figures 6-10–6-13.

[Example 1]

1. A
B
C
D
E
F
G

2. A
B
C
D
E
F
G

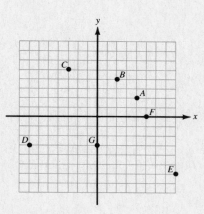

Figure 6-10. Points *A–G*.

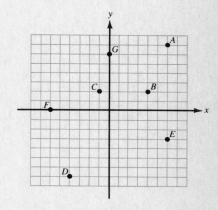

Figure 6-11. Points *A–G*.

3. A
B
C
D
E
F
G

4. A
B
C
D
E
F
G

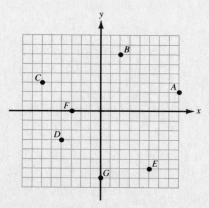

Figure 6-12. Points *A–G*.

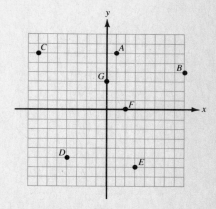

Figure 6-13. Points *A–G*.

In exercises **5–14**, plot the points on a rectangular coordinate system.

[Example 2]   **5.** $A(0, 0)$
$\quad\quad B(-4, 4)$
$\quad\quad C(6, -6)$

6. $D(2, -2)$
$\quad\quad E(0, 0)$
$\quad\quad F(-3, 3)$

7. $G(2, 7)$
$\quad\quad H(0, 3)$
$\quad\quad I(-2, -1)$

8. $J(1, -8)$
$\quad\quad K(0, -2)$
$\quad\quad L(-1, 4)$

9. $M(1, 4)$
$\quad\quad N(2, 1)$
$\quad\quad O(4, -5)$

10. $P(-5, 3)$
$\quad\quad Q(-1, -1)$
$\quad\quad R(4, -6)$

11. $S(-5, -2)$
$\quad\quad T(3, 6)$
$\quad\quad U(-1, 2)$

12. $V(-1, -8)$
$\quad\quad W(2, 4)$
$\quad\quad X(3, 8)$

13. $A(4, 5)$
$\quad\quad B(4, -2)$
$\quad\quad C(4, 3)$

14. $D(2, 3)$
$\quad\quad E(-1, 3)$
$\quad\quad F(-4, 3)$

In exercises **15–28**, identify the quadrant in which the following points lie. If a point is on an axis, identify which axis.

[Example 3]   **15.** $G(-50, -21)$

16. $H(-53, 72)$

17. $I(43, -57)$

18. $J(-73, -82)$

19. $K(-10, 0)$

20. $L(0, -15)$

21. $M(\sqrt{2}, -\sqrt{2})$

22. $N(-\sqrt{5}, \sqrt{5})$

23. $O(0, 1)$

24. $P(2, 0)$

25. $Q(-3.5, 4.7)$

26. $R(4.8, -9.3)$

27. $S\left(\dfrac{-5}{3}, \dfrac{-3}{8}\right)$

28. $T\left(\dfrac{9}{2}, \dfrac{7}{4}\right)$

In exercises **29–38**, write the linear equation in the general form.

[Example 4]   **29.** $y = -3x + 5$

30. $y = -6x + 1$

31. $y = \dfrac{-2x - 1}{5}$

32. $y = \dfrac{-4x - 3}{2}$

33. $y = \dfrac{-4x}{3} + 2$

34. $y = \dfrac{-3x}{5} + 3$

35. $y = \dfrac{6x}{5} - 3$

36. $y = \dfrac{4x}{3} - 2$

37. $y = 4x - \dfrac{3}{2}$

38. $y = 2x - \dfrac{2}{5}$

In exercises **39–60**, graph.

[Examples 5 and 6]

39. $2x - y = 0$

40. $3x + y = 0$

41. $x + 3y = 0$

42. $x - 2y = 0$

43. $2x - y = 1$

44. $2x + y = -2$

45. $3x + y = 1$

46. $3x - y = -2$

47. $4x - 5y = 20$

48. $4x + 5y = 20$

49. $2x + 3y = 6$

50. $2x - 3y = 12$

51. $5x + 2y = 10$

52. $5x + 3y = 30$

53. $x = 4$

54. $x = 6$

55. $y = -2$

56. $y = -4$

57. $x = -1$

58. $x = -5$

59. $y = 3$

60. $y = 2$

SECTION 6-2. Ten Review Exercises

In exercises **1–8**, solve each problem.

1. Simplify: $3t - (t + 12) - 2(3t - 4) - 4(t + 5)$

2. Evaluate the expression of exercise **1** for $t = -3$.

3. Based on your answer to exercise **2**, is -3 a root of

$$3t - (t + 12) - 2(3t - 4) - 4(t + 5) = 0?$$

4. Solve and check: $3t - (t + 12) = 2(3t - 4) + 4(t + 5)$

5. Factor: $2k^2 - 11k - 21$

6. Evaluate the factored form of exercise **5** for $k = 7$.

7. Based on your answer to exercise **6**, is 7 a root of

$$2k^2 - 11k - 21 = 0?$$

8. Solve and check: $11k + 21 = 2k^2$

In exercises **9** and **10**, do the indicated divisions.

9. $\dfrac{a^2 - 4b^2}{a^2 + 3ab - 10b^2} \div \dfrac{4a^2 - b^2}{2a^2 - 5ab + 2b^2}$

10. $(8y^4 + 3y^2 + 3) \div (y^2 + y + 1)$

SECTION 6-2.   Supplementary Exercises

In exercises **1** and **2**, state the coordinates of each point in Figures 6-14 and 6-15.

1. A
   B
   C
   D
   E

2. A
   B
   C
   D
   E

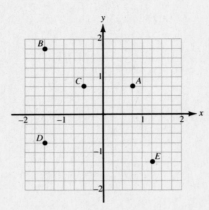

Figure 6-14. Points A–E.

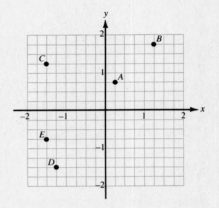

Figure 6-15. Points A–E.

In exercises **3–10**, determine in which quadrant $P(x, y)$ lies, given:

3. $x > 0$ and $y < 0$

4. $x > 0$ and $y > 0$

5. $x < 0$ and $y > 0$

6. $x < 0$ and $y < 0$

7. $x = 4.3$ and $y > 0$

8. $x = 9.8$ and $y < 0$

9. $x = \dfrac{-3}{4}$ and $y > 0$

10. $x = \dfrac{-9}{2}$ and $y > 0$

In exercises **11–30**, graph.

11. $x + y = 0$

12. $x - y = 0$

13. $2x = 9$

14. $2x + 3 = 0$

15. $x = y - 3$

16. $x = -y + 4$

17. $2y + 7 = 0$

18. $2y - 5 = 0$

19. $y = \dfrac{-1}{2}x + 3$

20. $y = \dfrac{-1}{3}x + 1$

21. $y = \dfrac{x}{4} - 2$

22. $y = \dfrac{x}{3} - 1$

23. $y = 0$

24. $x = 0$

25. $9x + 8y = 72$

26. $10x + 3y = 30$

27. $y = \dfrac{5}{3}x + \dfrac{7}{3}$

28. $y = \dfrac{11}{2}x - \dfrac{7}{2}$

29. $x = \dfrac{-3}{5}y - \dfrac{9}{5}$

30. $x = \dfrac{-9}{4}y + \dfrac{3}{4}$

For many linear equations, x and y are related so that the ratio of y to x is a nonzero constant k. For relationships such as these y is said to vary directly as x. The general form of direct variation equations is as follows:

$$\frac{y}{x} = k \qquad \text{or} \qquad y = kx$$

The equation $y = kx$ implies that

1. y increases directly as x increases, if $k > 0$.

2. y decreases directly as x increases, if $k < 0$.

In exercises **31–36**, use a large piece of graph paper to draw the graphs.

31. Joe Bennett owns a fishing boat that averages 5 mph in still water. The distance (d) that the boat travels varies directly with the time (t) that the boat travels. Written as an equation,

$$d = 5t.$$

a. Label the horizontal axis t and the vertical axis d to graph the equation. Plot points (t, d), where $t = 1$, $t = 3$, and $t = 4$.
b. Use the graph to estimate d for $t = 2.5$ hours.
c. Use the graph to estimate d for $t = 0.5$ hours.

32. Janet Peters is an avid hiker. She can average 3 mph on a section of a Sierra trail through Yosemite Park. The distance (d) that she can hike varies directly with the time (t) that she hikes. Written as an equation,

$$d = 3t.$$

a. Label the horizontal axis t and the vertical axis d to graph the equation. Plot points (t, d), where $t = 1$, $t = 2$, and $t = 5$.
b. Use the graph to estimate d for $t = 4.5$ hours.
c. Use the graph to estimate d for $t = 6.5$ hours.

33. The weight of an object in kilograms (k) varies directly with the weight of the object in pounds (p). Written as an equation,

$$k = 2.2p.$$

a. Label the horizontal axis p and the vertical axis k to graph the equation. Plot three points of the form (p, k) to locate the line.
b. Use the graph to estimate k for $p = 2.5$.
c. Use the graph to estimate p for $k = 9.5$.

34. The length of an object in meters (m) varies directly with the length of the object in yards (y). Written as an equation,

$$m = 1.1y.$$

a. Label the horizontal axis y and the vertical axis m to graph the equation. Plot three points of the form (y, m) to locate the line.

b. Use the graph to estimate m for $y = 5$.

c. Use the graph to estimate y for $m = 6.5$.

35. Monica Stutzman is standing at the edge of a very high cliff. She throws a rock in such a way that it initially goes up, then reverses direction and falls down past Monica into the valley below. The speed of the rock (s) varies directly with the time (t) that the rock is in motion. Written as an equation,

$$s = 64 - 32t \begin{cases} t \text{ is measured in seconds.} \\ s \text{ is measured in feet per second.} \end{cases}$$

a. Label the horizontal axis t and the vertical axis s to graph the equation. Use different unit lengths on the t and s axes. For example, let one unit on the t axis correspond to five units on the s axis.

b. For what value of t does $s = 0$?

c. For what values of t is $s > 0$?

d. For what values of t is $s < 0$?

e. For what value of t does s have the greatest positive value?

f. Does $t = -10$ have any practical meaning?

g. Does $t = 1,000$ have any practical meaning?

36. Ted Barnett is a professional baseball player. Recently he hit a baseball upward with an initial speed of 96 feet per second. The speed of the ball at a time t after being hit can be described by the following equation:

$$s = 96 - 32t$$

a. Label the horizontal axis t and the vertical axis s to graph the equation. Use different unit lengths on the t and s axes. For example, let one unit on the t axis correspond to ten units on the s axis.

b. For what value of t does $s = 0$?

c. For what values of t is $s > 0$?

d. For what values of t is $s < 0$?

e. For what value of t does s have the greatest positive value?

f. Does $t = -10$ have any practical meaning?

g. Does $t = 1,000$ have any practical meaning?

Example.   Graph: $5x - 3y = 2$

Solution.   **Discussion.** It is usually easier to find solutions for a linear equation by first solving the equation for y in terms of x. Make arbitrary replacements for x and find the corresponding values for y. If possible, select integer values for x that will yield integer values for y.

$5x - 3y = 2$	The given equation
$5x - 2 = 3y$	Isolate the y-term.
$y = \dfrac{5x - 2}{3}$	Divide both sides by 3.

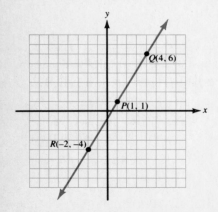

Figure 6-16. Graph of $5x - 3y = 2$.

Step 1. Replacing x by 1, $y = \dfrac{5(1) - 2}{3} = 1$. Thus, $(1, 1)$ is a solution.

Replacing x by 4, $y = \dfrac{5(4) - 2}{3} = 6$. Thus, $(4, 6)$ is a solution.

Step 2. $P(1, 1)$ and $Q(4, 6)$ are plotted in Figure 6-16.

Step 3. The line is drawn through P and Q.

Step 4. Replacing x by (-2), $y = \dfrac{5(-2) - 2}{3} = -4$.

The point $R(-2, -4)$ is plotted in Figure 6-16. Since this point falls on the line through P and Q, we accept the line as a graph of $5x - 3y = 2$.

In exercises **37–44**, graph.

37. $2x - 3y = 3$ **38.** $5x - 2y = 12$

39. $3x + 2y = 8$ **40.** $2x + 5y = 25$

41. $4x + 3y = 7$ **42.** $7x + 2y = 9$

43. $5x - 6y = 1$ **44.** $4x - 3y = -5$

SECTION 6-3.   Slope and Intercepts of a Line

KEY TOPICS IN THIS SECTION

1. A definition of the slope of a line

2. Slope in relation to vertical and horizontal lines

3. Using slope to graph a line

4. A definition of the x and y intercepts of a line

When a line is drawn in a rectangular coordinate system, two physical characteristics of the line can be used to describe it. One characteristic is the "inclination" of the line; that is, by how much and in what direction the line leans as it is traced from left to right. The other characteristic is **the points of intersection of the line with the coordinate axes.** These characteristics are studied in this section.

A Definition of the Slope of a Line

In Figure 6-17a, b, and c, three lines are drawn. On each line two points are labeled. Suppose we are instructed to move from points P_1 to P_2 on each line.

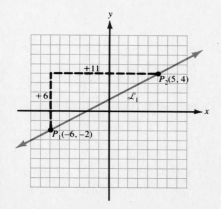

Figure 6-17a. $\mathscr{L}_1$.

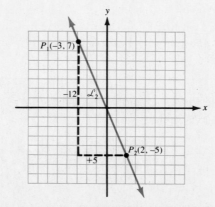

Figure 6-17b. $\mathscr{L}_2$.

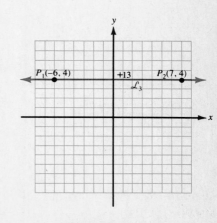

Figure 6-17c. $\mathscr{L}_3$.

However, in each case we can move only vertically (up or down) and horizontally (left or right).

Each of the required movements can be verified from the figures.

For $\mathscr{L}_1$: 6 units *up* (written $+6$), and then 11 units *to the right* (written $+11$).

For $\mathscr{L}_2$: 12 units *down* (written -12), and then 5 units *to the right* (written $+5$).

For $\mathscr{L}_3$: 0 units *up* or *down*, and then 13 units *to the right* (written $+13$).

The ratios of the vertical changes over the corresponding horizontal changes are called the **slopes of the lines.**

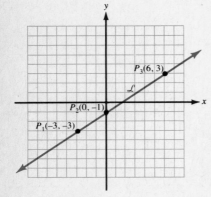

For $\mathscr{L}_1$: The slope is $\frac{6}{11}$.   Notice: $\dfrac{6}{11} = \dfrac{4-(-2)}{5-(-6)} = \dfrac{y_2 - y_1}{x_2 - x_1}$

For $\mathscr{L}_2$: The slope is $\frac{-12}{5}$.   Notice: $\dfrac{-12}{5} = \dfrac{-5-7}{2-(-3)} = \dfrac{y_2 - y_1}{x_2 - x_1}$

For $\mathscr{L}_3$: The slope is $\frac{0}{13} = 0$.   Notice: $0 = \dfrac{4-4}{7-(-6)} = \dfrac{y_2 - y_1}{x_2 - x_1}$

Definition 6.4.   The slope of a line
   The **slope of a line** is the ratio of the amount of vertical change in the line to the corresponding horizontal change.

Figure 6-18. P_1, P_2, and P_3 on $\mathscr{L}$.

Example 1.   Find the slope of $\mathscr{L}$ through P_1, P_2, and P_3 in Figure 6-18.

Solution.   **Discussion.** To show that the slope of a line is the same regardless of which points are used, we will calculate the slope using P_1 and P_2, P_1 and P_3, and P_2 and P_3.

For P_1 and P_2: Count 2 units up and 3 units to the right. Slope $= \frac{2}{3}$.

For P_1 and P_3: Count 6 units up and 9 units to the right. Slope $= \frac{6}{9} = \frac{2}{3}$.

For P_2 and P_3: Count 4 units up and 6 units to the right. Slope $= \frac{4}{6} = \frac{2}{3}$.

Thus, the slope of $\mathscr{L}$ is $\frac{2}{3}$, regardless of which pair of points is used to compute it.

The slope of a line can be computed when the coordinates of any two different points on the line are known. In Figure 6-19 points $P_1(x_1, y_1)$ and $P_2(x_2, y_2)$ are shown on line $\mathscr{L}$. The vertical *rise* in $\mathscr{L}$ in moving from P_1 to P_2 is $y_2 - y_1$. The corresponding *run* in $\mathscr{L}$ is $x_2 - x_1$. The ratio of these two subtractions is the slope of $\mathscr{L}$.

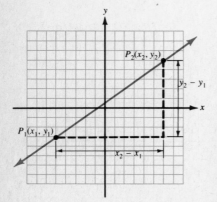

If $P_1(x_1, y_1)$ and $P_2(x_2, y_2)$ are two different points on $\mathscr{L}$, then the slope of $\mathscr{L}$ is m, and

$$m = \frac{y_1 - y_2}{x_1 - x_2} \qquad \text{or} \qquad m = \frac{y_2 - y_1}{x_2 - x_1} \qquad \frac{\text{rise}}{\text{run}}$$

provided $x_1 \neq x_2$.

Figure 6-19. Points $P_1(x_1, y_1)$ and $P_2(x_2, y_2)$ on $\mathscr{L}$.

Example 2. Find the slope of the lines through the following pairs of points:

    a. $P_1(8, 4)$ and $P_2(-4, -6)$

    b. $P_3(-6, 1)$ and $P_4(4, -5)$

Solution.   **a. Discussion.** The given points are plotted in Figure 6-20. Notice that the line is inclined upward to the right.

$$m = \frac{4 - (-6)}{8 - (-4)} = \frac{10}{12} = \frac{5}{6}$$

Thus, the line *rises* 5 units for every 6 units of *run*. A line inclined upward to the right has a **positive slope.**

b. Discussion. The given points are plotted in Figure 6-21. Notice that the line is inclined downward to the right.

$$m = \frac{1 - (-5)}{-6 - 4} = \frac{6}{-10} = \frac{-3}{5}$$

Thus the line *falls* 3 units for every 5 units of *run*. A line inclined downward to the right has a **negative slope.**

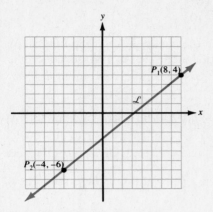

Figure 6-20. $P_1(8, 4)$ and $P_2(-4, -6)$ on $\mathscr{L}$.

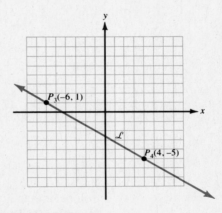

Figure 6-21. $P_3(-6, 1)$ and $P_4(4, -5)$ on $\mathscr{L}$.

Slope in Relation to Vertical and Horizontal Lines

In Figure 6-22 the vertical line defined by $x = 5$ is graphed. $P_1(5, 6)$ and $P_2(5, -4)$ are points on the line. When the equation for computing slope is used on these points, a zero is obtained in the denominator:

$$m = \frac{6 - (-4)}{5 - 5} = \frac{10}{0} \qquad \text{Division by 0 is undefined.}$$

A denominator of zero is obtained every time the equation for computing slope is applied to the coordinates of points on a vertical line. As a consequence, *slope is not defined for vertical lines.*

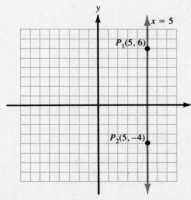

Figure 6-22. The vertical line $x = 5$.

> Slope is undefined for vertical lines, and any line for which the slope is undefined is vertical.

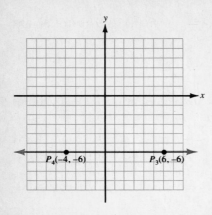

Figure 6-23. The horizontal line $y = -6$.

In Figure 6-23 the horizontal line defined by $y = -6$ is graphed. $P_3(6, -6)$ and $P_4(-4, -6)$ are points on the line. When the equation for computing slope is used on these points, a zero is obtained in the numerator:

$$m = \frac{-6 - (-6)}{6 - (-4)} = \frac{0}{10} = 0$$

A numerator of zero is obtained every time the equation for computing slope is applied to the coordinates of points on a horizontal line. As a consequence, *the slope of any horizontal line is zero.*

> The slope of any horizontal line is zero, and any line with zero slope is horizontal.

Example 3. Determine whether the following pairs of points lie on a vertical or horizontal line:

    a. $P_1(-10, 8)$ and $P_2(10, 8)$

    b. $P_3(-10, 8)$ and $P_4(-10, -8)$

Solution.   **a.** $m = \dfrac{8 - 8}{10 - (-10)} = \dfrac{0}{20} = 0$ $\dfrac{\text{rise}}{\text{run}}$

Since the slope is 0, P_1 and P_2 lie on a horizontal line.

    b. $m = \dfrac{-8 - 8}{-10 - (-10)} = \dfrac{-16}{0}$ $\dfrac{\text{rise}}{\text{run}}$

Since $\frac{-16}{0}$ is undefined, P_3 and P_4 lie on a vertical line.

Using Slope to Graph a Line

If the slope and any point on a line are known, then a graph of the line can be easily determined. The techniques uses the fact that slope describes the amount of vertical change in the graph for a given amount of horizontal change.

> $\text{slope} = \dfrac{\text{amount of vertical change}}{\text{amount of horizontal change}}$ $\dfrac{\text{rise}}{\text{run}}$

Example 4. Graph the following lines.

    a. $7x - 5y = 20$; the slope is $\frac{7}{5}$.

    b. $4x + 9y = 45$; the slope is $\frac{-4}{9}$.

Solution.   **a.** Replacing x by 0 in the given equation,

$$7(0) - 5y = 20$$

$$y = -4$$

In Figure 6-24, $P_1(0, -4)$ is plotted. The slope is given as $\frac{7}{5}$. Beginning at P_1, count up 7 units to P_2. Now count 5 units to

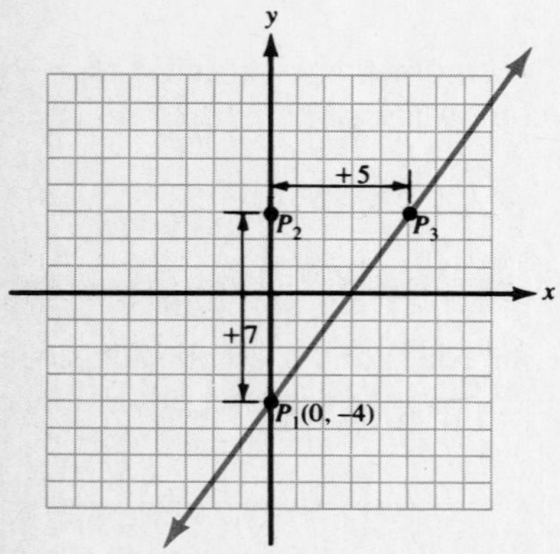

Figure 6-24. A graph of $7x - 5y = 20$.

the right of P_2 to P_3, and P_3 is on the line. Draw the line through P_1 and P_3 to obtain a graph of $7x - 5y = 20$.

b. Replacing x by 0 in the given equation,

$$4(0) + 9y = 45$$

$$y = 5$$

In Figure 6-25, $P_1(0, 5)$ is plotted. Beginning at P_1, count down 4 units to P_2. Now count 9 units to the right of P_2 to P_3, and P_3 is on the line. Draw the line through P_1 and P_3 to obtain a graph of $4x + 9y = 45$.

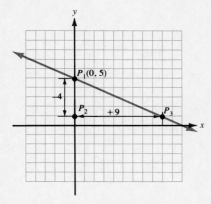

Figure 6-25. A graph of $4x + 9y = 45$.

A Definition of the x and y Intercepts of a Line

If $\mathscr{L}$ is the graph of a linear equation and $\mathscr{L}$ is neither a horizontal nor a vertical line, then $\mathscr{L}$ will intersect both the x-axis and the y-axis. The nonzero coordinates of these points are called the **x- and y-intercepts** of the line.

> **Definition 6.5. x-intercepts and y-intercepts**
> If $P_1(a, 0)$ and $P_2(0, b)$ are points of a nonvertical and nonhorizontal line $\mathscr{L}$, then
>
> **(i)** a is the **x-intercept** of $\mathscr{L}$.
>
> **(ii)** b is the **y-intercept** of $\mathscr{L}$.

Example 5.   For the line defined by $2x - 5y = 5$,

a. find the x-intercept.

b. find the y-intercept.

Solution.   **a. Discussion.** To find the x-intercept, replace y by 0 in the equation. The root of the resulting equation is the x-intercept.

$$2x - 5(0) = 5 \qquad \text{Replace } y \text{ by 0.}$$

$$2x = 5 \qquad \text{Solve for } x.$$

$$x = \frac{5}{2} \qquad \text{The } x\text{-intercept is } \tfrac{5}{2}.$$

Thus, $P_1(\tfrac{5}{2}, 0)$ is a point on the graph of the equation.

b. Discussion. To find the y-intercept, replace x by 0 in the equation. The root of the resulting equation is the y-intercept.

$$2(0) - 5y = 5 \qquad \text{Replace } x \text{ by 0.}$$

$$-5y = 5 \qquad \text{Solve for } y.$$

$$y = -1 \qquad \text{The } y\text{-intercept is } -1.$$

Thus, $P_2(0, -1)$ is a point on the graph of the equation.

SECTION 6-3. Practice Exercises

In exercises **1–8**, find the slopes of the lines in Figures 6-26–6-33.

[Example 1] **1.**

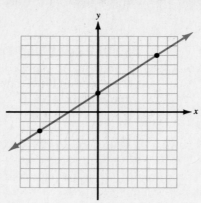

Figure 6-26.

2.

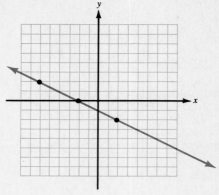

Figure 6-27.

3.

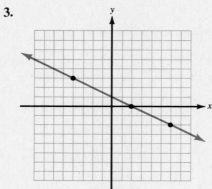

Figure 6-28.

4.

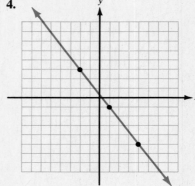

Figure 6-29.

5.

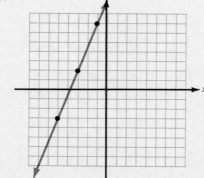

Figure 6-30.

6.

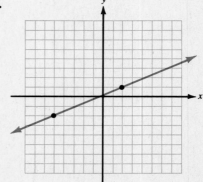

Figure 6-31.

7.

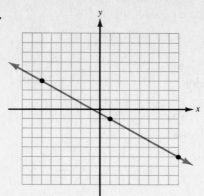

Figure 6-32.

8.

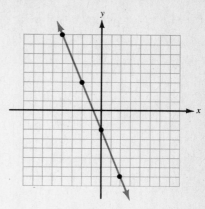

Figure 6-33.

In exercises **9–22**, find the slope of a line through P_1 and P_2.

[Example 2]

9. $P_1(3, 1)$, $P_2(5, 7)$

10. $P_1(2, 5)$, $P_2(4, 9)$

11. $P_1(2, 3)$, $P_2(4, 2)$

12. $P_1(8, 2)$, $P_2(5, 8)$

13. $P_1(6, -3)$, $P_2(5, 1)$

14. $P_1(8, -5)$, $P_2(7, 2)$

15. $P_1(0, 4)$, $P_2(-3, 0)$

16. $P_1(0, -5)$, $P_2(2, 0)$

17. $P_1(-10, -4)$, $P_2(-8, 6)$

18. $P_1(1, -8)$, $P_2(9, 8)$

19. $P_1(13, -8)$, $P_2(-12, 7)$

20. $P_1(-9, 7)$, $P_2(-12, 2)$

21. $P_1(0, 6)$, $P_2(10, -3)$

22. $P_1(0, -7)$, $P_2(8, 3)$

In exercises **23–30**, determine whether the points with the given coordinates lie on a vertical or a horizontal line.

[Example 3]

23. $P_1(4, -2)$, $P_2(4, 3)$

24. $P_1(6, 8)$, $P_2(6, -1)$

25. $P_1(0, 5)$, $P_2(-6, 5)$

26. $P_1(4, 3)$, $P_2(0, 3)$

27. $P_1(7, -3)$, $P_2(-5, -3)$

28. $P_1(4, -7)$, $P_2(-6, -7)$

29. $P_1(-2, 9)$, $P_2(-2, 12)$

30. $P_1(-5, 6)$, $P_2(-5, -11)$

In exercises **31–40**, graph.

[Example 4]

31. $2x - y = -3$; the slope is $\frac{2}{1}$.

32. $3x - y = -1$; the slope is $\frac{3}{1}$.

33. $2x + 3y = 9$; the slope is $\frac{-2}{3}$.

34. $2x + 4y = 8$; the slope is $\frac{-1}{2}$.

35. $5x - 2y = 6$; the slope is $\frac{5}{2}$.

36. $4x - 3y = 6$; the slope is $\frac{4}{3}$.

37. $x - 3y = 6$; the slope is $\frac{1}{3}$.

38. $2x - 3y = 3$; the slope is $\frac{2}{3}$.

39. $3x + 4y = 0$; the slope is $\frac{-3}{4}$.

40. $3x + 5y = 0$; the slope is $\frac{-3}{5}$.

In exercises **41–50**, state

a. the x-intercept.

b. the y-intercept.

[Example 5]

41. $4x + 2y = 8$

42. $5x + 3y = 15$

43. $2x + 5y = -10$

44. $6x + 5y = -30$

45. $4x - 3y = 12$

46. $2x - 7y = -14$

47. $5x + 6y = 15$

48. $4x - 7y = -14$

49. $15x + 7y = 21$

50. $4x - 7y = -7$

SECTION 6-3.   Ten Review Exercises

In exercises **1–4**, multiply.

1. $(2k^2)(3k)$

2. $(2 + k^2)(3 + k)$

3. $(5t^2)^2$

4. $(5 + t^2)^2$

In exercises **5** and **6**, divide.

5. $\dfrac{9x^2}{3x}$

6. $\dfrac{9 - x^2}{3 - x}$

In exercises **7** and **8**, do the indicated operations.

7. $(6a^2 - 3a - 5) + (2a - 3)$

8. $(6a^2 - 3a - 5)(2a - 3)$

In exercises **9** and **10**, solve each word problem.

9. The difference between three times an integer and two times the next consecutive integer is 11. Find both integers.

10. How many ounces of artificial flavoring costing $1.90 per ounce should be mixed with a natural extract costing $3.65 per ounce, to make 50 ounces of flavoring mixture costing $2.95 per ounce?

SECTION 6-3.   Supplementary Exercises

In exercises **1–12**, find the slopes of the lines that go through the points with the given coordinates. If a given pair lie on a vertical line, write "Slope is undefined."

1. $P_1(-3, 4)$, $P_2(3, -4)$

2. $P_1(7, -8)$, $P_2(-7, 8)$

3. $P_1\left(\dfrac{1}{2}, \dfrac{2}{3}\right)$, $P_2\left(\dfrac{3}{4}, \dfrac{-1}{6}\right)$

4. $P_1\left(\dfrac{3}{5}, \dfrac{-1}{2}\right)$, $P_2\left(\dfrac{-2}{5}, \dfrac{5}{2}\right)$

5. $P_1(0, 0)$, $P_2(-14, 7)$

6. $P_1(6, -2)$, $P_2(0, 0)$

7. $P_1\left(\dfrac{1}{2}, \dfrac{3}{4}\right)$, $P_2\left(\dfrac{1}{2}, \dfrac{-1}{4}\right)$

8. $P_1\left(\dfrac{2}{3}, \dfrac{-1}{5}\right)$, $P_2\left(\dfrac{2}{3}, \dfrac{4}{5}\right)$

9. $P_1(3.2, -1.4)$, $P_2(0.8, -1.4)$

10. $P_1(-1.6, 2.1)$, $P_2(0.4, 2.1)$

11. $P_1(2.1, 4.2)$, $P_2(3.1, -0.8)$

12. $P_1(4.6, 3.4)$, $P_2(1.6, 0.4)$

In exercises **13–16**, do parts **a–d**.

13. Given the equation $x - 2y = -2$,
   a. find x if $y = 0$.
   b. find y if $x = 4$.
   c. use the ordered pairs from parts **a** and **b** to find the slope of the line defined by the equation.
   d. use the point from part **a** and the slope from part **c** to graph the line.

14. Given the equation $x - 2y = 4$,
   a. find x if $y = 0$.
   b. find y if $x = -2$.
   c. use the ordered pairs from parts **a** and **b** to find the slope of the line defined by the equation.
   d. use the point from part **a** and the slope from part **c** to graph the line.

15. Given the equation $2x + y = 3$,
   a. find y if $x = 0$.
   b. find x if $y = -1$.
   c. use the ordered pairs from parts **a** and **b** to find the slope of the line defined by the equation.
   d. use the point from part **a** and the slope from part **c** to graph the line.

16. Given the equation $x + 3y = 6$,
   a. find y if $x = 0$.
   b. find x if $y = -3$.
   c. use the ordered pairs from parts **a** and **b** to find the slope of the line defined by the equation.
   d. use the point from part **a** and the slope from part **c** to graph the line.

One technique for graphing equations is called the **intercept method,** which uses $P_1(a, 0)$ and $P_2(0, b)$ to graph the line.

Example. Graph using the intercept method:

$$4x - 7y = 28$$

Solution. If $y = 0$, then $4x = 28$.

$$x = 7 \qquad P_1(7, 0) \text{ is on the graph.}$$

If $x = 0$, then $-7y = 28$.

$$y = -4 \qquad P_2(0, -4) \text{ is on the graph.}$$

In Figure 6-34, points P_1 and P_2 are plotted and a line is drawn through them.

In exercises **17–28**, use the intercept method to graph the line.

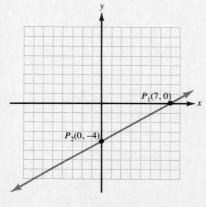

Figure 6-34.

17. $7x - 4y = -28$ **18.** $5x - 3y = -15$

19. $4x + 5y = 20$ **20.** $3x + 4y = 12$

21. $x - 6y = 6$ **22.** $x - 5y = 5$

23. $2x + 4y = -8$ **24.** $2x + 8y = 16$

25. $4x + 9y = 18$ **26.** $6x + 7y = 21$

27. $30x + y = 6$ **28.** $12x - y = 6$

In exercises **29–32**, if point P is on a line, identify any other point on the line if

a. the line is horizontal.

b. the line is vertical.

29. $P(-2, 3)$ **30.** $P(6, -1)$

31. $P(0, 5)$ **32.** $P(-7, 4)$

SECTION 6-4.   The Slope-Intercept Equation of a Line

**KEY TOPICS
IN THIS SECTION**

1. The slope-intercept equation of a line

2. Using the slope-intercept equation to graph

3. Slope and parallel lines

4. Slope and perpendicular lines

5. Horizontal and vertical lines in relation to parallel and perpendicular lines

The previous section covered the slope and intercepts of a line. In this section we will study a form of an equation of a line that reveals the slope and y-intercept. We will also use this form to determine whether two lines are parallel or perpendicular.

The Slope-Intercept Equation of a Line

Consider the line defined by $3x - 2y = 10$. Let us solve this equation for y and find two solutions for the equation.

$$3x - 2y = 10$$

$$-2y = -3x + 10$$

1. $y = \dfrac{3}{2}x - 5$ ⟨ If $x = 2$, then $y = \frac{3}{2}(2) - 5 = -2$.
          If $x = 8$, then $y = \frac{3}{2}(8) - 5 = 7$.

Thus, $(2, -2)$ and $(8, 7)$ are solutions of the equation, and $P_1(2, -2)$ and $P_2(8, 7)$ are points on the graph. Using the equation for computing slope on the coordinates of P_1 and P_2,

$$m = \frac{7 - (-2)}{8 - 2} = \frac{9}{6} = \frac{3}{2}.$$

The slope of the line defined by $3x - 2y = 10$ is $\frac{3}{2}$. Notice that $\frac{3}{2}$ is also the coefficient of x in equation **1**:

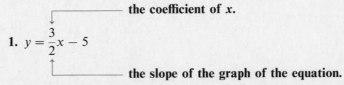

the coefficient of x.

1. $y = \dfrac{3}{2}x - 5$

the slope of the graph of the equation.

The results of this example are not unique. That is, when the linear equation $Ax + By = C$ is solved for y in terms of x (and $B \neq 0$), then the coefficient of x will be the slope of the line defined by that equation.

Furthermore, consider again equation **1**. If x is replaced by 0, then $y = \frac{3}{2}(0) - 5 = -5$. Thus $(0, -5)$ is a solution, and -5 is the y-intercept of the equation.

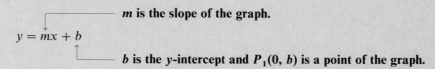

1. $y = \dfrac{3}{2}x - 5$

the constant term

the y-intercept of the equation

This result is true for any linear equation that does not define a vertical line. That is, when the linear equation $Ax + By = C$ is solved for y in terms of x (and $B \neq 0$), then the constant term will be the y-intercept.

The slope-intercept equation of a line

If $Ax + By = C$ and $B \neq 0$ is solved for y in terms of x, then the form of the equation can be written as follows:

m is the slope of the graph.

$y = mx + b$

b is the y-intercept and $P_1(0, b)$ is a point of the graph.

Example 1. For the line defined by $2x + 5y = 25$,

    a. state the slope.

    b. state the y-intercept.

Solution.   **Discussion.** Change the given equation to the slope-intercept form. The coefficient of x is the slope, and the constant term is the y-intercept.

$2x + 5y = 25$ The given equation

$5y = -2x + 25$ Subtract $2x$ from both sides.

$y = \dfrac{-2}{5}x + 5$ Divide each term by 5.

    a. The slope is $\frac{-2}{5}$.   The coefficient of x is the slope.

    b. The y-intercept is 5.   The constant term is the y-intercept.

Using the Slope-Intercept Equation to Graph

In the previous section the slope of a line was used to graph a linear equation. The slope-intercept form of an equation of a nonvertical line is a form of $Ax + By = C$ that uses $P_1(0, b)$ as the known point on the line. The slope of the line can then be used to find additional points on the line.

Example 2. Use the slope-intercept equation to graph $7x - 4y = 24$.

Solution. $7x - 4y = 24$ The given equation

         $-4y = -7x + 24$ Subtract $7x$ from both sides.

         $y = \dfrac{7}{4}x - 6$ Divide each term by -4.

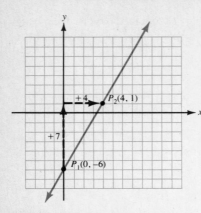

Figure 6-35. A graph of $7x - 4y = 24$.

In Figure 6-35 the point $P_1(0, -6)$ is plotted. The slope of the line is $\frac{7}{4}$. Counting up 7 units and to the right 4 units yields $P_2(4, 1)$ on the line. A line through P_1 and P_2 is a graph of the given equation.

Example 3. Use the slope-intercept equation to graph $\dfrac{y+3}{3} = \dfrac{5-x}{5}$.

Solution.

$\dfrac{y+3}{3} = \dfrac{5-x}{5}$	The given equation
$15\left(\dfrac{y+3}{3}\right) = 15\left(\dfrac{5-x}{5}\right)$	The LCD is 15.
$5(y+3) = 3(5-x)$	$15 \div 3 = 5$ and $15 \div 5 = 3$
$5y + 15 = 15 - 3x$	Remove parentheses.
$5y = -3x$	Subtract 15 from both sides.
$y = \dfrac{-3}{5}x$	The slope is $\frac{-3}{5}$ and the y-intercept is 0.

In Figure 6-36 the point $P_1(0, 0)$ is plotted. The slope of the line is $\frac{-3}{5}$. Counting down 3 units and to the right 5 units yields $P_2(5, -3)$ on the line. A line through P_1 and P_2 is a graph of the given equation.

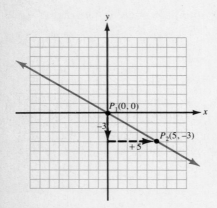

Figure 6-36. A graph of $\dfrac{y+3}{3} = \dfrac{5-x}{5}$.

Slope and Parallel Lines

In Figure 6-37 two lines are graphed. The lines can be defined by the following equations:

(1) $\quad y = \frac{1}{3}x + 3$ Same slope of $\frac{1}{3}$.

(2) $\quad y = \frac{1}{3}x - 5$ Different y-intercepts of 3 and -5

One can imagine that these lines will not intersect no matter how far they are extended. Such lines are called **parallel.** Parallel lines have the same slope but different y-intercepts.

> If $\mathscr{L}_1$ and $\mathscr{L}_2$ are nonhorizontal and nonvertical lines defined respectively by $y = m_1x + b_1$ and $y = m_2x + b_2$, then $\mathscr{L}_1$ and $\mathscr{L}_2$ are parallel if and only if
>
>   **(i)** $m_1 = m_2$ and   **(ii)** $b_1 \neq b_2$.

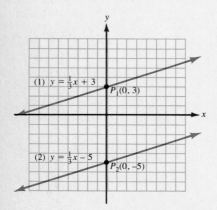

Figure 6-37. Graphs of two parallel lines.

Example 4. Determine whether the graphs of the following equations are parallel:

(1) $\quad y = \dfrac{-9}{5}x + 2$

(2) $\quad 9x - 5y = 10$

Solution. **Discussion.** Equation (1) is written in the slope-intercept form. Write equation (2) in this form and then compare the slopes and y-intercepts.

(1) $y = \dfrac{-9}{5}x + 2$ \qquad The slope is $\frac{-9}{5}$.

The y-intercept is 2.

(2) $9x - 5y = 10$

$-5y = -9x + 10$

$y = \dfrac{9}{5}x - 2$ \qquad The slope is $\frac{9}{5}$.

The y-intercept is -2.

The slopes of the two lines are not the same ($\frac{-9}{5} \neq \frac{9}{5}$); thus the lines are not parallel.

Slope and Perpendicular Lines

In Figure 6-38, two lines are graphed. The lines can be defined by the following equations:

(1) $y = \frac{1}{3}x + 3$ \qquad Slopes are $\frac{1}{3}$ and -3.

(2) $y = -3x - 7$ \qquad y-intercepts are 3 and -7.

As indicated by the square in a corner of the intersection of the lines, the angle of intersection is $90°$, a **right angle.** Such lines are called **perpendicular.** Perpendicular lines have slopes whose product is -1.

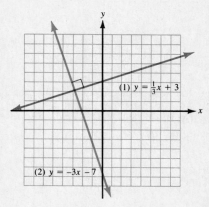

Figure 6-38. Graphs of two perpendicular lines.

(1) $y = \frac{1}{3}x + 3$

(2) $y = -3x - 7$

> If $\mathscr{L}_1$ and $\mathscr{L}_2$ are nonvertical and nonhorizontal lines defined respectively by $y = m_1x + b_1$ and $y = m_2x + b_2$, then $\mathscr{L}_1$ and $\mathscr{L}_2$ are perpendicular if and only if
>
> $$m_1 \cdot m_2 = -1 \qquad \text{or} \qquad m_2 = \frac{-1}{m_1}.$$

Example 5. Determine whether the graphs of the following equations are perpendicular:

(1) $y = \dfrac{3}{7}x + 9$

(2) $y + 5 = \dfrac{9 - 7x}{3}$

Solution. **Discussion.** Equation (1) is written in the slope-intercept form. Write equation (2) in this form and then compute the product of the slopes:

(1) $y = \dfrac{3}{7}x + 9$ \qquad The slope is $\frac{3}{7}$.

(2) $y + 5 = \dfrac{9 - 7x}{3}$ \qquad The given equation

$3y + 15 = 9 - 7x$ \qquad Multiply both sides by 3.

$3y = -7x - 6$ \qquad Subtract 15 from both sides.

$y = \dfrac{-7}{3}x - 2$ \qquad Divide both sides by 3.

The slopes of the two lines are $\frac{3}{7}$ and $\frac{-7}{3}$.

$$\frac{3}{7} \cdot \frac{-7}{3} = -1 \qquad \text{The lines are perpendicular.}$$

Horizontal and Vertical Lines in Relation to Parallel and Perpendicular Lines

In the statements regarding parallel and perpendicular lines, $\mathscr{L}_1$ and $\mathscr{L}_2$ were restricted from being horizontal and vertical. The restrictions were needed because slope is not defined for vertical lines. However, the following relationships exist for any horizontal or vertical lines.

> **(i)** Horizontal lines are parallel.
>
> **(ii)** Vertical lines are parallel.
>
> **(iii)** Horizontal and vertical lines are perpendicular.

Example 6. Determine whether lines defined by the following pairs of equations are parallel, perpendicular, or neither:

  a. (1) $x = 3$

  (2) $x = -5$

  b. (1) $x = 3$

  (2) $y = 9$

Solution. **a.** Equations (1) and (2) define vertical lines. Thus, the graphs of these equations are parallel.

  b. Equations (1) and (2) define a vertical and a horizontal line respectively. Thus, the graphs of these equations are perpendicular.

SECTION 6-4. Practice Exercises

In exercises **1–8**,

a. write equations in $y = mx + b$ form.

b. state the slope of a graph of the equation.

c. state the y-intercept of the equation.

[Example 1] **1.** $2x + y = 3$ **2.** $x + 2y = 4$

  3. $3x - y = 4$ **4.** $x - 3y = -6$

  5. $5x - 4y = 0$ **6.** $4x + 5y = 0$

  7. $3y - 2x = 9$ **8.** $3y + 4x = 12$

In exercises **9–18**,

a. write equations in $y = mx + b$ form.

b. use the slope-intercept form to graph each equation.

[Example 2]   **9.** $x + 3y = 6$ **10.** $x - 3y = -6$

11. $4x - y = 3$ **12.** $2x + y = -3$

13. $5x - 4y = 0$ **14.** $3x + 7y = 0$

15. $5x + y = 3$ **16.** $6x - y = 5$

17. $2x - 5y = -5$ **18.** $4x + 3y = -6$

In exercises **19–26**, use the slope-intercept form to graph each equation.

[Example 3]   **19.** $\dfrac{y - 2}{2} = \dfrac{x + 3}{3}$ **20.** $\dfrac{y + 7}{4} = \dfrac{5 - x}{5}$

21. $\dfrac{6 - 7x}{4} = \dfrac{y - 3}{2}$ **22.** $\dfrac{4x + 3}{3} = \dfrac{y + 4}{2}$

23. $\dfrac{x - 2}{2} = \dfrac{y + 3}{5}$ **24.** $\dfrac{x + 4}{4} = \dfrac{4 - y}{3}$

25. $\dfrac{y}{4} - \dfrac{x - 3}{2} = 0$ **26.** $\dfrac{y + 1}{9} - \dfrac{x + 1}{3} = 0$

In exercises **27–44**, determine whether the graphs of the pairs of equations are parallel, perpendicular, or neither.

[Examples 4, 5, and 6]   **27.** $3x + 2y = 0$ **28.** $4x - 5y = 0$
 $3x + 2y = -3$ $4x - 5y = -2$

29. $y = \dfrac{-2}{3}x + 4$ **30.** $y = \dfrac{7}{4}x - 5$
 $2x + 3y = 10$ $7x - 4y = 15$

31. $x = 5$ **32.** $y = 7$
 $x = -3$ $y = -2$

33. $3x - 5y = -5$ **34.** $2x - 3y = -6$
 $y = \dfrac{-5}{3}x + 1$ $y = \dfrac{-3}{2}x + 4$

35. $x - 2 = 0$ **36.** $y + 5 = 0$
 $2y + 3 = 0$ $3x - 1 = 0$

37. $x - 2y = 3$ **38.** $4x - y = 7$
 $2x - y = 3$ $x - 4y = 7$

39. $x = \dfrac{5}{3}y + 2$ **40.** $x = \dfrac{5}{3}y + 2$
 $3x + 5y = 6$ $3x - 5y = 8$

41. $\dfrac{y - 1}{2} = \dfrac{x - 1}{3}$ **42.** $\dfrac{y + 2}{3} = \dfrac{x - 3}{5}$
 $3x + 2y = 1$ $5x + 3y = 10$

43. $5x + 2y = 4$

$\quad y = 4 - \dfrac{5}{2}x$

44. $2x - 7y = 35$

$\quad y = 3 + \dfrac{2}{7}x$

SECTION 6-4.   Ten Review Exercises

In exercises **1–6**, factor completely.

1. $t^2 + 3t - 70$

2. $8u^2 - 18u - 35$

3. $16x^2 - 56xy + 49y^2$

4. $27z^3 + 125$

5. $2k^3 - 8k^2 + 3k - 12$

6. $25p^2 - 144q^2$

In exercises **7–10**, use the equation $3x - 8y = 24$.

7. Write the equation in the $y = mx + b$ form.

8. State the slope of a graph of the equation.

9. Identify the y-intercept.

10. Sketch a graph of the equation.

SECTION 6-4.   Supplementary Exercises

In exercises **1–16**,

a. if possible, write equations in the $y = mx + b$ form.

b. state the slope, if it exists.

c. state the y-intercept, if it exists.

d. graph.

1. $3(x + 4) = 2(y + 3)$

2. $4(x - 1) = -3(y + 2)$

3. $\dfrac{2x + 1}{3} = \dfrac{y - 2}{2}$

4. $\dfrac{x - 2}{3} = \dfrac{2y - 1}{4}$

5. $-y = -3x + 4$

6. $-y = -8x - 2$

7. $x = 3$

8. $x = -5$

9. $13 - 2y = 0$

10. $4 - 3y = 0$

11. $\dfrac{x}{7} - \dfrac{y}{3} = 1$

12. $\dfrac{x}{5} + \dfrac{y}{3} = 1$

13. $x = 2y + 3$

14. $x = -3y + 4$

15. $x = \dfrac{3}{4}y + \dfrac{5}{4}$

16. $x = \dfrac{-4}{3}y + \dfrac{7}{3}$

In exercises **17–24**, find m_2 so that lines $\mathscr{L}_1$ and $\mathscr{L}_2$ are

a. parallel.

b. perpendicular.

17. $\mathscr{L}_1: 2x - 3y = 5$
$\mathscr{L}_2: y = m_2 x + 1$

18. $\mathscr{L}_1: 4x - 3y = 7$
$\mathscr{L}_2: y = m_2 x - 2$

19. $\mathscr{L}_1: x + y = 0$
$\mathscr{L}_2: y = m_2 x + 11$

20. $\mathscr{L}_1: x - y = 0$
$\mathscr{L}_2: y = m_2 x$

21. $\mathscr{L}_1: 4x - 9y = 18$
$\mathscr{L}_2: y = m_2 x$

22. $\mathscr{L}_1: 5x - 6y = 12$
$\mathscr{L}_2: y = m_2 x$

23. $\mathscr{L}_1: \dfrac{x}{3} + \dfrac{y}{2} = 2$
$\mathscr{L}_2: y = m_2 x - \dfrac{2}{3}$

24. $\mathscr{L}_1: \dfrac{x}{6} + \dfrac{y}{5} = 1$
$\mathscr{L}_2: y = m_2 x - \dfrac{1}{3}$

In exercises **25–32**, graph both equations on the same axes and state the co-ordinates of the point of intersection.

25. $x + y = 3$
$x - y = 5$

26. $2x + y = 3$
$2x - 3y = 15$

27. $x - 2y = 4$
$2x - y = -4$

28. $2x - y = 3$
$x + y = 6$

29. $5x + 2y = 10$
$3x + 4y = -8$

30. $2x - 3y = 6$
$x + 2y = 10$

31. $x = -4$
$y = 3$

32. $x = 2$
$y = -5$

In exercises **33–40**, write in $y = mx + b$ form an equation for the lines in Figures 6-39–6-46.

33.

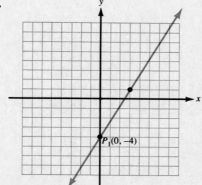

Figure 6-39.

34.

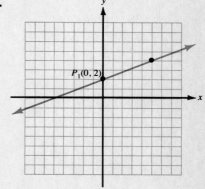

Figure 6-40.

35.

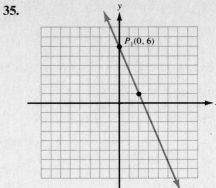

Figure 6-41.

36.

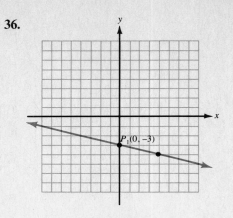

Figure 6-42.

37.

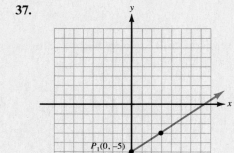

Figure 6-43.

38.

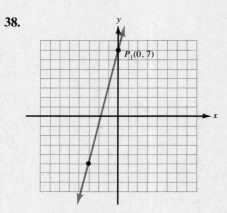

Figure 6-44.

39.

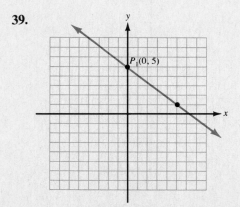

Figure 6-45.

40.

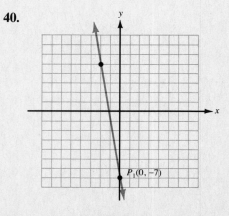

Figure 6-46.

SECTION 6-5. Writing an Equation of a Line

**KEY TOPICS
IN THIS SECTION**

1. The point-slope equation of a line

2. Writing equations of nonhorizontal and nonvertical lines

3. Writing equations of horizontal and vertical lines

Two forms of a linear equation in two variables have been given:

1. $Ax + By = C$ The general form of an equation

2. $y = mx + b$ The slope-intercept form of an equation

The third form of an equation of a line is called the **point-slope equation.** This form is especially useful for writing an equation of a line.

The Point-Slope Equation of a Line

In Figure 6-47, a line $\mathscr{L}$ with slope $\frac{3}{2}$ is drawn through $P_1(3, 5)$ on an xy-coordinate system. Thus, the graph of the line is given, but the equation of the line is not. Suppose we need to find numbers A, B and C such that

$$Ax + By = C$$

has as solutions the coordinates of every point of $\mathscr{L}$. The task may seem formidable, but the equation for computing slope makes it relatively simple.

In Figure 6-48, $\mathscr{L}$ is shown with another point $P(x, y)$. Imagine that this point is not fixed, but it represents every other point on $\mathscr{L}$ except $P_1(3, 5)$. That is, $P(x, y)$ may be found anywhere on $\mathscr{L}$ other than at $(3, 5)$.

For any line the slope is constant. Thus, the difference between the y-coordinates divided by the corresponding difference in the x-coordinates is the same for any pair of points on the line. Using the variable point $P(x, y)$, the fixed point $P_1(3, 5)$, and the given slope $\frac{3}{2}$,

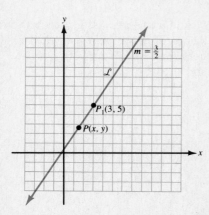

Figure 6-47. Line $\mathscr{L}$ through $P_1(3, 5)$ with slope $\frac{3}{2}$.

Figure 6-48. $\mathscr{L}$ with variable point $P(x, y)$ shown.

slope = slope

$$\frac{3}{2} = \frac{y - 5}{x - 3} \qquad\qquad m = \frac{y_1 - y_2}{x_1 - x_2}$$

$$\frac{3}{2}(x - 3) = y - 5 \qquad\qquad \text{Multiply both sides by } (x - 3).$$

$$y - 5 = \frac{3}{2}(x - 3) \qquad \textbf{1.} \qquad \text{If } a = b, \text{ then } b = a.$$

Equation **1** is an example of the **point-slope equation of a line.** The equation is so named because it requires knowing a point on the line and the slope of the line in order to write an equation for the line.

The point-slope equation of a nonhorizontal and nonvertical line
 If $P_1(x_1, y_1)$ is a known point on a line $\mathscr{L}$, and m is the slope, then

$$y - y_1 = m(x - x_1)$$

is the **point-slope equation** of $\mathscr{L}$.

Writing Equations of Nonhorizontal and Nonvertical Lines

To write an equation for a line, two items of information must be known:

1. a point on the line, namely $P_1(x_1, y_1)$

2. the slope of the line, m

Examples 1–4 illustrate ways of using the given information to determine P_1 and m.

Example 1. Write an equation of the line containing $P_1(-5, 4)$ and slope $\frac{-2}{3}$.

Solution. **Discussion.** In this example P_1 and m are given. The point-slope equation will be used to write an equation, but the answer will be stated in the $Ax + By = C$ form.

$$y - \boxed{y_1} = \boxed{m}(x - \boxed{x_1})$$ The point-slope equation

$$y - 4 = \frac{-2}{3}(x - (-5))$$ Replace y_1 with 4, m with $\frac{-2}{3}$, and x_1 with (-5).

$$3(y - 4) = -2(x + 5)$$ Multiply both sides by 3.

$$3y - 12 = -2x - 10$$ Use the distributive property.

$$2x + 3y = 2$$ Add $2x$ and 12 to both sides.

Thus, an equation of the described line is $2x + 3y = 2$.

A quick check of the equation can be made by verifying that the coordinates of P_1 are a solution of the equation.

$$2(-5) + 3(4) = 2$$ Replace x by -5 and y by 4.

$$-10 + 12 = 2$$ Find the indicated products.

$$2 = 2, \text{ true}$$ Thus $(-5, 4)$ is a solution.

Example 2. Write an equation of the line containing $P_1(7, 2)$ and $P_2(-5, -4)$.

Solution. **Discussion.** The coordinates of P_1 and P_2 can be used to compute the slope of the line. The coordinates of either point can then be used as the fixed point. Since the coordinates of P_1 are both positive, we will use these values for x_1 and y_1.

$$m = \frac{2 - (-4)}{7 - (-5)} \qquad m = \frac{y_1 - y_2}{x_1 - x_2}$$

$$m = \frac{6}{12} \text{ or } \frac{1}{2}$$ The slope is $\frac{1}{2}$.

$$y - 2 = \frac{1}{2}(x - 7)$$ The point-slope equation

$$2(y - 2) = (x - 7)$$ Multiply both sides by 2.

$$2y - 4 = x - 7$$ Use the distributive property.

$$3 = x - 2y$$ Subtract $2y$ and add 7 to both sides.

$$x - 2y = 3$$ If $a = b$, then $b = a$.

Recall that parallel lines have the same slope. We use this relationship to write an equation for the line in Example 3.

Example 3.   Write an equation for the line containing $P_1(-2, 7)$ and parallel to the line defined by $x + 3y = 9$.

Solution.   **Discussion.** First find the slope of the line defined by $x + 3y = 9$ by writing the equation in the $y = mx + b$ form.

$$x + 3y = 9 \qquad \text{The given equation.}$$

$$3y = -x + 9 \qquad \text{Subtract } x \text{ from both sides.}$$

$$y = \frac{-1}{3}x + 3 \qquad \text{Divide each term by 3.}$$

The slope of the line is $\frac{-1}{3}$; thus, the slope of the line whose equation we need to write is also $\frac{-1}{3}$.

$$y - 7 = \frac{-1}{3}(x - (-2)) \qquad \text{Use } P_1(-2, 7) \text{ and } m = \frac{-1}{3}.$$

$$3(y - 7) = -1(x + 2) \qquad \text{Multiply both sides by 3.}$$

$$3y - 21 = -x - 2 \qquad \text{Use the distributive property.}$$

$$x + 3y = 19 \qquad \text{The } Ax + By = C \text{ form}$$

Recall that perpendicular lines have slopes whose product is (-1). We use this relationship to write an equation for the line in Example 4.

Example 4.   Write an equation for the line containing $P_1(-2, 7)$ and perpendicular to the line defined by $x + 3y = 9$.

Solution.   **Discussion.** From Example 3 the slope of the line defined by $x + 3y = 9$ is $\frac{-1}{3}$. Thus, the slope of any line perpendicular to this line is 3. (Notice, $3(\frac{-1}{3}) = -1$.)

$$y - 7 = 3(x - (-2)) \qquad \text{Replace } y_1 \text{ by 7, } m \text{ by 3, and } x_1 \text{ by } (-2).$$

$$y - 7 = 3x + 6 \qquad \text{Remove parentheses.}$$

$$-3x + y = 13 \qquad \text{Subtract } 3x \text{ and add 7 to both sides}$$

$$3x - y = -13 \qquad \text{Multiply both sides by } (-1).$$

Writing Equations of Horizontal and Vertical Lines

In the statement of the point-slope equation, $\mathscr{L}$ is restricted from being a horizontal or vertical line. These restrictions were given because the equation can be simplified for horizontal lines and cannot be used for vertical lines.

In Figure 6-49 a horizontal line $\mathscr{L}$ is drawn through $P_1(-6, -3)$ and $P_2(4, -3)$. For any horizontal line the slope is zero. Using the coordinates of P_1 and the point-slope equation,

$$y - (-3) = 0(x - (-6))$$

$$y + 3 = 0$$

$$y = -3$$

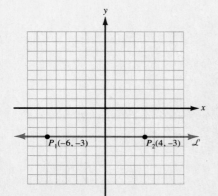

Figure 6-49. A horizontal line $\mathscr{L}$.

This equation states that the value of the y-coordinate of every point on $\mathscr{L}$ is -3. On any horizontal line the y-coordinate of every point is the same. The following is a general statement of this fact.

Equations for horizontal lines

If $\mathscr{L}$ is a horizontal line with constant y-coordinate y_1, then an equation for $\mathscr{L}$ is

$$y = y_1.$$

In Figure 6-50, a vertical line $\mathscr{L}$ is drawn through $P_1(9, 6)$ and $P_2(9, -4)$. *For any vertical line slope is undefined.* Therefore, we cannot replace m in the point-slope equation with the word "undefined."

However, for any point on $\mathscr{L}$ in Figure 6-50, the x-coordinate is 9. The following equation is an assertion of this fact:

$$x = 9 \qquad \text{An equation for } \mathscr{L} \text{ in Figure 6-50}$$

Equations for vertical lines

If $\mathscr{L}$ is a vertical line with constant x-coordinate x_1, then an equation for $\mathscr{L}$ is

$$x = x_1.$$

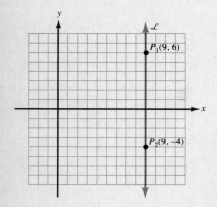

Figure 6-50. A vertical line $\mathscr{L}$.

Example 5. Write equations for lines through the following points:

a. $P_1(-5, 8)$ and $P_2(-5, -10)$

b. $P_3(10, 7)$ and $P_4(-3, 7)$

Solution.
a. **Discussion.** Observe that the x-coordinate of both P_1 and P_2 is -5. We may conclude that the line through these points is vertical with constant x-coordinate -5. The following equation is a statement of this fact:

$$x = -5 \qquad \text{Replace } x_1 \text{ by } -5 \text{ in } x = x_1.$$

b. **Discussion.** Observe that the y-coordinate of both P_3 and P_4 is 7. We may conclude that the line through these points is horizontal with constant y-coordinate 7. The following equation is a statement of this fact:

$$y = 7 \qquad \text{Replace } y_1 \text{ by } 7 \text{ in } y = y_1.$$

SECTION 6-5. Practice Exercises

In exercises **1–28**, write an equation of the line based on the information given.

a. Use the $y - y_1 = m(x - x_1)$ form.

b. Use the $Ax + By = C$ form with $A > 0$.

[Example 1] **1.** Containing $P(1, 2)$ and slope 3

2. Containing $P(7, -2)$ and slope 4

3. Containing $Q(-4, 6)$ and slope -1

4. Containing $Q(-3, -4)$ and slope -2

5. Containing $R(1, -1)$ and slope $\frac{1}{2}$

6. Containing $R(3, 3)$ and slope $\frac{10}{3}$

7. Containing $S(-2, 2)$ and slope $\frac{-3}{2}$

8. Containing $S(3, -5)$ and slope $\frac{-2}{5}$

[Example 2]　**9.** Containing $T_1(0, 4)$ and $T_2(-6, 0)$

10. Containing $T_1(0, -5)$ and $T_2(-5, 0)$

11. Containing $U_1(2, -2)$ and $U_2(5, 2)$

12. Containing $U_1(2, -2)$ and $U_2(-1, 5)$

13. Containing $V_1(-3, 1)$ and $V_2(-8, 4)$

14. Containing $V_1(-2, -3)$ and $V_2(1, -1)$

15. Containing $W_1(0, 0)$ and $W_2(-3, 4)$

16. Containing $W_1(0, 0)$ and $W_2(-5, -4)$

[Example 3]　**17.** Containing $X(3, 0)$ and parallel to the graph of $6x + 3y = 5$

18. Containing $X(0, -8)$ and parallel to the graph of $8x - 2y = 9$

19. Containing $Y(5, 1)$ and parallel to the graph of $2x - 5y = 10$

20. Containing $Y(-2, 8)$ and parallel to the graph of $7x + 4y = 20$

21. Containing $Z(-9, -9)$ and parallel to the graph of $8x - 5y = 15$

22. Containing $Z(10, -6)$ and parallel to the graph of $3x + 7y = 21$

[Example 4]　**23.** Containing $A(1, -2)$ and perpendicular to the graph of $3x - y = 0$

24. Containing $A(-3, 4)$ and perpendicular to the graph of $2x - y = 0$

25. Containing $B(0, 5)$ and perpendicular to the graph of $3x + 4y = 8$

26. Containing $B(0, -4)$ and perpendicular to the graph of $5x + 2y = 8$

27. Containing $C(-6, 8)$ and perpendicular to the graph of $5x + 8y = 48$

28. Containing $C(-9, 5)$ and perpendicular to the graph of $6x + 7y = 14$

In exercises **29–36**, write an equation of a line that contains P and that is

a. horizontal.

b. vertical.

[Example 5]　**29.** $P(5, -3)$　　　　　　　**30.** $P(6, -1)$

31. $P(-7, -8)$　　　　　　**32.** $P(-10, -9)$

33. $P(0, 4)$　　　　　　　　**34.** $P(0, 10)$

35. $P\left(\dfrac{-1}{2}, \dfrac{2}{3}\right)$　　　　　**36.** $P\left(\dfrac{-3}{4}, \dfrac{7}{8}\right)$

SECTION 6-5. Ten Review Exercises

In exercises **1–10** do the indicated operations.

1. $(7t^2 + t - 6) - (6t^2 + 5t - 1) + (t^2 + 4t - 3)$

2. $(-2x^2y)^4$

3. $-9ab^2(4a^2 + ab - 3b^2)$

4. $(9u - 4v)^2$

5. $\dfrac{60z^5 - 96z^3 + 12z}{12z}$

6. $\dfrac{m^2 - 16}{2m + 1} \cdot \dfrac{2m^2 + m}{m + 4}$

7. $\dfrac{12x^2}{35y^4} \div \dfrac{2x}{7y^3}$

8. $\dfrac{3t^2 + 8t}{3t^2 - t - 2} - \dfrac{3t - 2}{3t^2 - t - 2}$

9. $\dfrac{3}{20u} + \dfrac{2}{45u}$

10. $\dfrac{3a + 4}{a^2 - 16} + \dfrac{1}{a + 4}$

SECTION 6-5. Supplementary Exercises

In exercises **1–14**, write an equation of the line.

a. Use the $y - y_1 = m(x - x_1)$ form.

b. Use the $Ax + By = C$ form, with $A > 0$.

1. Containing $Q(0, 0)$ and with slope $\frac{-5}{3}$

2. Containing $Q(0, 0)$ and with slope $\frac{-4}{7}$

3. x intercept 4 and parallel to the graph of $x - y = 5$

4. y intercept -6 and parallel to the graph of $12x - 5y = 10$

5. Containing $R(\frac{1}{2}, \frac{1}{3})$ and parallel to the graph of $2x - 3y = 6$

6. Containing $R(\frac{5}{4}, \frac{-7}{5})$ and parallel to the graph of $4x + 5y = 10$

7. Containing $S(-1, 6)$ and perpendicular to the graph of $x = 3$

8. Containing $S(3, -2)$ and perpendicular to the graph of $x = -7$

9. x-intercept 3 and y-intercept -9 **10.** x-intercept 4 and y-intercept -3

11. x-intercept -8 and y-intercept -7 **12.** x-intercept -3 and y-intercept -5

13. x-intercept 2.5 and y-intercept -3 **14.** x-intercept -1.5 and y-intercept 2

If the slope and y-intercept of a line are given, then the slope-intercept equation can more readily be used to write an equation. In exercises **15–22**, write an equation of the line using the slope-intercept form.

15. $m = 6$ and $b = -3$

16. $m = 5$ and $b = -4$

17. $m = \frac{-3}{2}$ and $b = 7$

18. $m = \frac{-5}{4}$ and $b = 2$

19. $m = \frac{2}{3}$ and $b = 0$ **20.** $m = \frac{4}{7}$ and $b = 0$

21. $m = 0$ and $b = -3$ **22.** $m = 0$ and $b = -7$

In exercises **23–26** (Figures 6-51–6-54),

a. state the coordinates of P_1.

b. state the coordinates of P_2.

c. use the coordinates of P_1 and P_2 to compute the slope of $\mathscr{L}$.

d. use P_1 and the slope found in part **c** to write an equation of $\mathscr{L}$ in $Ax + By = C$ form.

e. use P_2 and the slope found in part **c** to write an equation of $\mathscr{L}$ in $Ax + By = C$ form and compare with the equation in part **d**.

f. state the x-intercept of the equation.

g. state the y-intercept of the equation.

23.

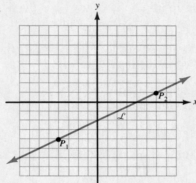

Figure 6-51.

24.

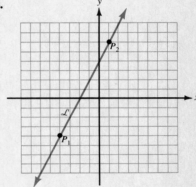

Figure 6-52.

25.

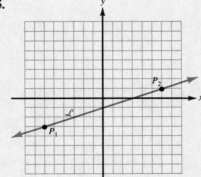

Figure 6-53.

26.

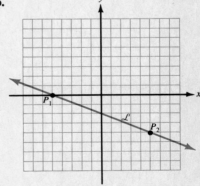

Figure 6-54.

A parallelogram is a four-sided figure whose opposite sides are parallel. In exercises **27–30**;

a. plot P_1, P_2, P_3 and P_4.

b. find the slope of the line containing P_1 and P_2.

c. find the slope of the line containing P_3 and P_4.

d. are the slopes in parts **b** and **c** equal?

e. find the slope of the line containing P_1 and P_4.

f. find the slope of the line containing P_2 and P_3.

g. are the slopes in parts **e** and **f** equal?

h. is the figure a parallelogram?

27. $P_1(-3, 2), P_2(3, 4), P_3(4, -1), P_4(-2, -3)$

28. $P_1(-3, 3)\ P_2(1, 2), P_3(-2, -4), P_4(-6, -3)$

29. $P_1(-6, 4), P_2(1, 7), P_3(5, -2), P_4(-2, -5)$

30. $P_1(-4, 3), P_2(4, 1), P_3(3, -1), P_4(-5, 1)$

In exercises **31–34**, answer each problem based on the following information: Johnny Lee's grandmother always has a dish filled with green and red candies. Each time Johnny took candy from the dish he took 5 green and 2 red ones. Suppose after his last trip to the candy dish it held 3 green candies and no red candies.

31. If initially 22 red candies were in the dish, how many green candies were initially in the dish?

32. If initially 34 red candies were in the dish, how many green candies were initially in the dish?

33. Could the dish initially have had 29 red candies in it? Justify your answer.

34. Could the dish initially have had 48 red candies in it? Justify your answer.

For exercises **35** and **36**, suppose that after Johnny's last trip to the candy dish it held 4 green and 1 red candies.

35. If the dish initially had 11 red candies, how many green candies did it initially have?

36. If the dish initially had 25 red candies, how many green candies did it initially have?

SECTION 6-6. Graphs of Linear Inequalities in Two Variables

KEY TOPICS IN THIS SECTION

1. A line divides a plane into three subsets of points

2. A definition of a linear inequality in x and y

3. Graphing the solution set of a linear inequality in x and y

4. Compound inequalities in two variables

5. Writing an inequality that defines a half-plane

Examples **a** and **b** illustrate linear inequalities in x that were solved in Chapter 2.

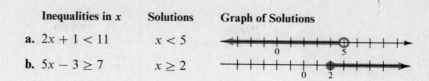

Inequalities in x	Solutions	Graph of Solutions
a. $2x + 1 < 11$	$x < 5$	
b. $5x - 3 \geq 7$	$x \geq 2$	

In this section we will graph the solutions of linear inequalities in two variables. The solution of an inequality in two variables is an ordered pair of real numbers that makes the inequality true. As will be seen, the solution set of such an inequality can be graphed as a region of points in a rectangular coordinate system.

A Line Divides a Plane into Three Subsets of Points

In Figure 6-55, a point P has been given a coordinate k, where k is a constant. Suppose x is a variable that represents any real number. There are exactly three ways of relating x to the number k.

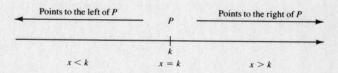

Figure 6-55. P divides the line into three subsets of points.

Relationship 1. $x < k$ x is the coordinate of a point to the left of P.

Relationship 2. $x = k$ x is the coordinate of P.

Relationship 3. $x > k$ x is the coordinate of a point to the right of P.

The relationships shown in Figure 6-55 may seem trivial. However, when the relationship is extended to two variables, we get a picture such as the one in Figure 6-56.

As shown in the figure, a line $\mathscr{L}$ divides the plane into three subsets of points. If $Ax + By = C$ is an equation that defines $\mathscr{L}$, then $Ax + By < C$ and $Ax + By > C$ are inequalities that can define the **half-planes** above and below $\mathscr{L}$. That is, depending on the values of A, B, and C we could have the following relationships between the coordinates of the points in the plane and the ordered pair solutions of the inequalities and equations:

Relationship 1. (x_1, y_1) is a solution of $Ax + By > C$.
   (x_1, y_1) is the coordinate of a point above $\mathscr{L}$.

Relationship 2. (x_1, y_1) is a solution of $Ax + By = C$.
   (x_1, y_1) is the coordinate of a point on $\mathscr{L}$.

Relationship 3. (x_1, y_1) is a solution of $Ax + By < C$.
   (x_1, y_1) is the coordinate of a point below $\mathscr{L}$.

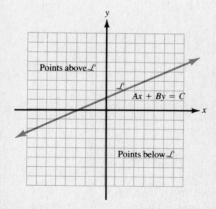

Figure 6-56. $\mathscr{L}$ divides the plane into three subsets of points.

Notice that $\mathscr{L}$ is a boundary to the two half-planes defined by the inequalities.

A Definition of a Linear Inequality in x and y

A linear inequality in two variables can be used to describe a half-plane in a rectangular coordinate system.

Definition 6.6.   A linear inequality in x and y
   A linear inequality in x and y can be written in one of the following forms:

   (i) $Ax + By < C$ or   **(ii)** $Ax + By > C$ or

   (iii) $Ax + By \leq C$ or   **(iv)** $Ax + By \geq C$

Inequalities of the form **(iii)** and **(iv)** are called **compound.**

Examples **a** and **b** illustrate linear inequalities **(i)** and **(ii)**.

a. $2x - y > 4$ **b.** $x + 4y > -4$

A *solution* of a linear inequality in x and y is an ordered pair of real numbers that makes the inequality a true number statement.

Example 1.   Determine whether

   a. $(6, 5)$ or   **b.** $(-9, -20)$

   is a solution of $2x - y > 4$.

Solution.   **Discussion.** Replace x and y in the given inequality with the numbers in the ordered pairs and simplify. If the resulting number statement is true, the ordered pair is a solution. If the number statement is false, the ordered pair is not a solution.

a. $2(6) - 5 > 4$	Replace x by 6 and y by 5.
$12 - 5 > 4$	Multiply first.
$7 > 4$, true	$(6, 5)$ is a solution.
b. $2(-9) - (-20) > 4$	Replace x by (-9) and y by (-20).
$-18 - (-20) > 4$	Multiply first.
$2 > 4$, false	$(-9, -20)$ is not a solution.

Graphing the Solution Set of a Linear Inequality in x and y

In Example 1 the ordered pair $(6, 5)$ was shown to be a solution of $2x - y > 4$. The number of solutions of any inequality in two variables is infinite. A graph is a suitable way of displaying all the solutions.

In Figure 6-57 the *shaded half-plane* is a graph of the solutions of $2x - y > 4$. The *boundary* between the two half-planes is the line defined by $2x - y = 4$. Notice the line is *dotted* because the coordinates of the points on the line are not solutions of the inequality.

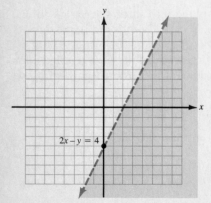

Figure 6-57. A graph of solutions of $2x - y > 4$.

To graph the solutions of $Ax + By < C$ or $Ax + By > C$,

Step 1. graph $Ax + By = C$ with a dotted line.

Step 2. use $P(x_1, y_1)$ as a **test point,** where P is any point in one of the half-planes. Replace x with x_1 and y with y_1 in the given inequality.

Step 3. if (x_1, y_1) is a solution of the given equality, shade the half-plane containing P. If (x_1, y_1) is not a solution of the given equality, shade the half-plane that does not contain P.

Example 2. Graph the solutions of $x + 4y > -4$.

Solution. **Step 1.** Graph the boundary line $x + 4y = -4$.

$$y = \frac{-1}{4}x - 1 \qquad \text{Write in } y = mx + b \text{ form.}$$

With $P_1(0, -1)$ and $m = \frac{-1}{4}$, the line is shown as a dotted line in Figure 6-58.

Step 2. Use $P(0, 0)$ as a test point.

$$0 + 4(0) > -4$$

$$0 > -4, \text{ true}$$

Step 3. Shade the half-plane containing P because $(0, 0)$ is a solution of the inequality.

The shaded half-plane in Figure 6-58 with the boundary line defined by $x + 4y = -4$ is a graph of the solutions.

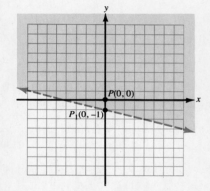

Figure 6-58. Graph of the solutions of $x + 4y > -4$.

Example 3. Graph the solutions of $3x + y > 6$.

Solution. **Step 1.** Graph the boundary line $3x + y = 6$.

$$y = -3x + 6 \qquad \text{Write in } y = mx + b \text{ form.}$$

With $P_1(0, 6)$ and $m = -3$, the line is shown as a dotted line in Figure 6-59.

Step 2. Use $P(0, 0)$ as a test point.

$$3(0) + 0 > 6$$

$$0 > 6, \text{ false}$$

Step 3. Shade the half-plane that does not contain P because $(0, 0)$ is not a solution.

The shaded half-plane in Figure 6-59 with the boundary line defined by $3x + y = 6$ is a graph of the solutions.

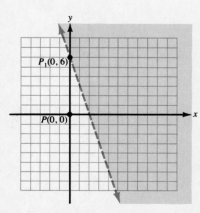

Figure 6-59. Graph of the solutions of $3x + y > 6$.

Compound Inequalities in Two Variables

If the linear inequality is written $\leq$ or $\geq$, then the boundary line to the half-plane containing the solutions is included in the graph. A solid line is used instead of a dotted line for the boundary.

Example 4.   Graph the solutions of $5x - 4y \leq 0$.

Solution.   **Step 1.** Graph the boundary line $5x - 4y = 0$.

$$y = \frac{5}{4}x + 0 \qquad \text{Write in } y = mx + b \text{ form.}$$

With $P_1(0, 0)$ and $m = \frac{5}{4}$, the line is shown as a solid line in Figure 6-60.

Step 2. The boundary passes through the origin. A point anywhere else on either axis can be used as a suitable test point. Arbitrarily selecting $P(5, 0)$,

$$5(5) - 4(0) \leq 0$$

$$25 \leq 0, \text{ false}$$

Step 3. Shade the half-plane that does not contain P because $(5, 0)$ is not a solution.

The shaded half-plane in Figure 6-60, including the boundary line defined by $5x - 4y = 0$, is a graph of the solutions.

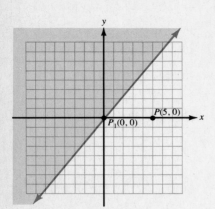

Figure 6-60. Graph of the solutions of $5x - 4y \leq 0$.

Writing an Inequality that Defines a Half-Plane

If the appropriate information is given regarding a line, then we can write an equation that defines the line. Similarly, if a graph of the solutions of an inequality is given, then we can write the corresponding inequality that defines the half-plane.

Example 5.   Write an inequality in x and y that defines the half-plane shaded in Figure 6-61.

Solution.   **Discussion.** Use the points on the boundary to write an equation for the line. Then use a test point within the shaded region to determine the order of the inequality ($<$ or $>$) that defines the region.

$$m = \frac{1 - (-2)}{-5 - 0} \qquad \text{Use } P_1(-5, 1) \text{ and } P_2(0, -2).$$

$$= \frac{-3}{5} \qquad \text{The slope of the boundary line is } \frac{-3}{5}.$$

$$y = \frac{-3}{5}x - 2 \qquad \text{The } y\text{-intercept is } -2.$$

$$5y = -3x - 10 \qquad \text{Multiply both sides by 5.}$$

$$3x + 5y = -10 \qquad \text{The } Ax + By = C \text{ form}$$

Notice that $P(0, 0)$ is within the shaded region. Thus, $(0, 0)$ is a solution of the inequality.

$$3(0) + 5(0) \ \square \ -10 \qquad \text{Replace } x \text{ and } y \text{ by 0.}$$

$$0 \ \square \ -10 \qquad \text{Simplify the left side.}$$

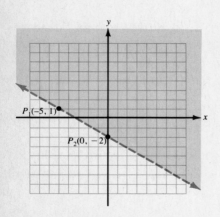

Figure 6-61. Graph of a solution set.

Since $0 > -10$ is a true statement, we place a $>$ symbol inside the box to get an inequality that defines the shaded region.

$$3x + 5y > -10$$

SECTION 6-6.   Practice Exercises

In exercises **1–8**, determine which, if any, of the stated ordered pairs are solutions of the given inequalities.

[Example 1]

1. $3x + y < 4$
  a. $(2, -3)$ **b.** $(-3, 2)$

2. $x + 2y < 10$
  a. $(5, -2)$ **b.** $(-1, 5)$

3. $5x - y > 12$
  a. $(4, 3)$ **b.** $(-2, -10)$

4. $2x + 4y > -8$
  a. $(-6, 1)$ **b.** $(-1, 3)$

5. $x + 5y \leq 10$
  a. $(-3, 2)$ **b.** $(0, 2)$

6. $4x - 9y \geq 18$
  a. $(0, -2)$ **b.** $(-5, -4)$

7. $x \leq 8$
  a. $(3, -10)$ **b.** $(6, 953)$

8. $y > -5$
  a. $(869, -6)$ **b.** $(0, 1)$

In exercises **9–28**, graph the solutions.

[Examples 2 and 3]

9. $x + y < 2$

10. $x + y < 5$

11. $2x - y > 3$

12. $x - 3y > -3$

13. $x + 3y > 0$

14. $2x + 3y > 0$

15. $3x + 4y < -12$

16. $2x + 5y < -10$

17. $x - y < 0$

18. $x + y < 0$

19. $2x - 4y > 12$

20. $2x + 6y > 12$

[Example 4]

21. $3x + 5y \leq 15$

22. $x - y \leq 3$

23. $4x - 9y \geq 18$

24. $2x + 3y \geq 6$

25. $x + 2y \geq 0$

26. $2x - y \geq 0$

27. $x + 3y \leq -6$

28. $x - 4y \leq 8$

In exercises **29–36** (Figures 6-62–6-69), write an inequality in x and y that defines the half-plane.

29.

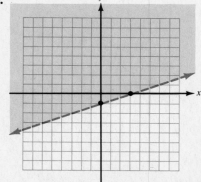

Figure 6-62.

30.

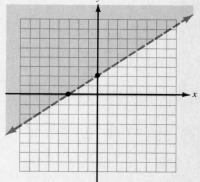

Figure 6-63.

31.

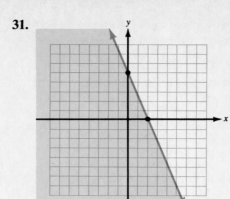

Figure 6-64.

32.

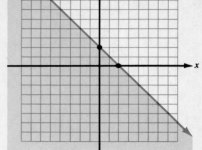

Figure 6-65.

33.

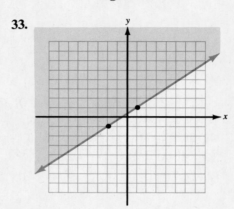

Figure 6-66.

34.

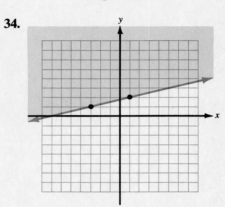

Figure 6-67.

35.

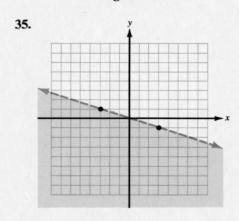

Figure 6-68.

36.

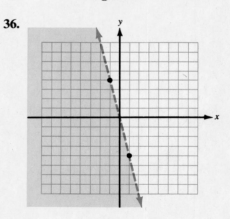

Figure 6-69.

SECTION 6-6.   Ten Review Exercises

In exercises **1–5**, solve and check.

1. $7t + 17 = -23 - 3t$

2. $11m + 3(3m - 2) = 2[5 - 2(3m - 2)]$

3. $2z^2 = z + 6$

4. $\dfrac{2k+1}{k} - \dfrac{k+2}{k-1} + \dfrac{1}{k^2-k} = 0$

5. $16u^2 = 25$

In exercises **6–10**, do the indicated operations and simplify.

6. $2(3xy - 10x^2 - 7y^2) - (8x^2 + y^2 + 6xy) - 3(xy - 9x^2 - 5y^2)$

7. $(-5u^2v)(-2uv^2)^2$

8. $(8c - 7)(9c + 10)$

9. $\dfrac{3p^2 - 8p - 3}{p^2 + 6p + 9} \div \dfrac{6p^2 - 7p - 3}{2p^2 + 3p - 9}$

10. $\dfrac{8q + 5}{q + 3} - \dfrac{2q - 13}{q + 3}$

SECTION 6-6.   Supplementary Exercises

In exercises **1–4**, determine which, if any, of the stated ordered pairs are solutions of the given inequalities.

1. $3x + y < 4$
   a. $\left(\dfrac{1}{3}, \dfrac{-5}{9}\right)$ **b.** $(-0.4, 3.8)$

2. $x + 2y < 1$
   a. $\left(\dfrac{4}{5}, \dfrac{1}{10}\right)$ **b.** $(8.3, -4.1)$

3. $5x - y > 12$
   a. $\left(\dfrac{11}{4}, \dfrac{-5}{2}\right)$ **b.** $(1.3, -6.2)$

4. $2x + 4y > -8$
   a. $\left(\dfrac{-7}{8}, \dfrac{-5}{8}\right)$ **b.** $(0.3, -2.3)$

In exercises **5–20**, graph the solution set.

5. $y > x + 2$

6. $y \le x - 3$

7. $y \le -x - 4$

8. $y > -x + 4$

9. $y \ge -3$

10. $y < -1$

11. $x < 4$

12. $x \ge 1$

13. $4 \ge -2y + x$

14. $2 \ge -y + 3x$

15. $y < \dfrac{-1}{2}x + 1$

16. $y \le \dfrac{-1}{2}x - 1$

17. $\dfrac{1}{2}x - \dfrac{3}{4}y > \dfrac{9}{4}$

18. $\dfrac{1}{2}x + \dfrac{2}{3}y > 2$

19. $\dfrac{3}{10}x + \dfrac{1}{2}y \le \dfrac{1}{2}$

20. $\dfrac{-1}{2}x + \dfrac{1}{5}y \ge \dfrac{2}{5}$

In exercises **21–26**, determine whether to insert $<$ or $>$ so that the given ordered pair is a solution.

21. $7x + 5y \,\square\, 10$ and $(-2, 5)$

22. $3x + 10y \,\square\, 21$ and $(12, -1)$

23. $2x - 9y \,\square\, 20$ and $(-4, -3)$

24. $8x - y \,\square\, 49$ and $(7, 8)$

25. $\frac{2}{3}x + \frac{3}{4}y \ \square \ 9$ and $(-6, 20)$

26. $\frac{2}{5}x - \frac{5}{2}y \ \square \ 51$ and $(30, -16)$

In exercises **27–32**, choose the inequality that defines each half-plane.

a. $y \leq \dfrac{-2}{3}x + 2$ b. $y \geq \dfrac{-2}{3}x + 2$

c. $y \leq \dfrac{2}{3}x - 2$ d. $y \geq \dfrac{2}{3}x - 2$

e. $y \leq \dfrac{2}{3}x + 2$ f. $y \geq \dfrac{2}{3}x + 2$

g. $y \leq \dfrac{-2}{3}x - 2$ h. $y \geq \dfrac{-2}{3}x - 2$

27.
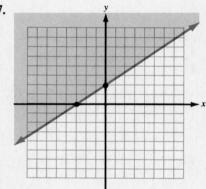

Figure 6-70.

28.
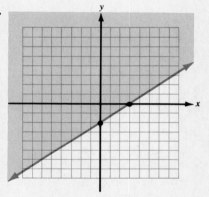

Figure 6-71.

29.
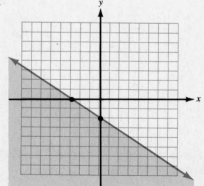

Figure 6-72.

30.
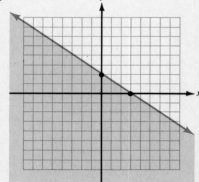

Figure 6-73.

31.

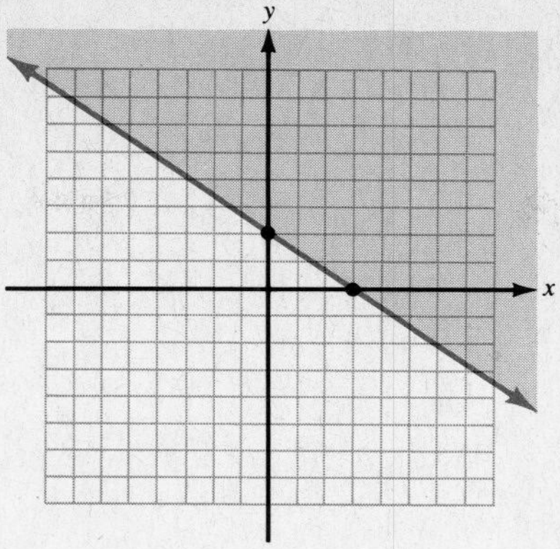

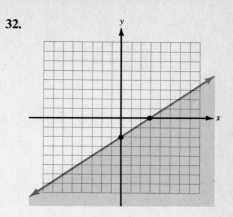

Figure 6-74.

32.

Figure 6-75.

SECTION 6-1.   Summary Exercises

In exercises **1–4**, determine whether the given ordered pair is a solution of the equation.

1. $3x - 4y = 9$ and $(5, 2)$

1. _____

2. $4x + 3y = 11$ and $(-4, 9)$

2. _____

3. $r - 7s = 21$ and $(0, -3)$

3. _____

4. $2a - 3b = 13$ and $\left(6, \dfrac{-1}{3}\right)$

4. _____

In exercises **5–8**, find the solution for the given member of the ordered pair.

5. $u + v = 9$ and $u = -2$

5. _____

6. $b = \dfrac{1}{2}a - 3$ and $a = 4$

6. _____

7. $n = \dfrac{5}{2}m - 1$ and $n = \dfrac{3}{2}$

7. _____

8. $r^2 + 2s = 33$ and $r = -5$

8. _____

SECTION 6-2. Summary Exercises

1. State the coordinates of each point.

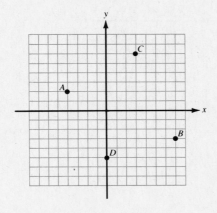

1. $A($, $)$

$B($, $)$

$C($, $)$

$D($, $)$

2. Plot the following points:

$A(0, 7)$ $B(4, -2)$ $C(-2, 1)$

2.

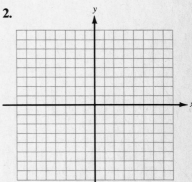

3. Identify the quadrant in which the following points lie.

a. $P(-6, -1)$

b. $Q(4, 10)$

c. $R(-19, 8)$

d. $S(16, -13)$

3. a. _____

b. _____

c. _____

d. _____

4. Write the following linear equation in the general form.

$y = \dfrac{2}{3}x + 5$

4. _____

427

In exercises **5–8**, graph.

5. $x + 4y = 0$

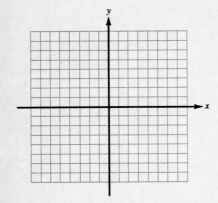

6. $2x + y = 3$

7. $5x - 4y = 20$

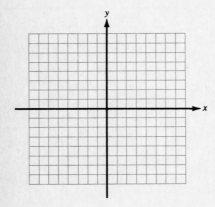

8. $y = -3$

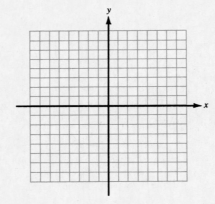

Name

Date

Score

SECTION 6-3. Summary Exercises

Answer

In exercises **1–4**, find the slope, if it exists, for the given pair of points.

1. $P(3, 7), Q(-2, 1)$

1. _____

2. $P(-6, 2), Q(6, -2)$

2. _____

3. $P(0, 0), Q(-4, 5)$

3. _____

4. $P(-4, 6), Q(-7, 6)$

4. _____

In exercises **5** and **6**, solve parts **a** and **b**.

5. Determine whether the following pairs of points lie on a vertical or a horizontal line.

    a. $P(0, 7), Q(-3, 7)$

    b. $R(4, -2), S(4, 10)$

5. a. _____

b. _____

6. For the line $3x - 6y = 18$, find:

    a. the x-intercept.

    b. the y-intercept.

6. a. _____

b. _____

In exercises **7** and **8**, graph using the given slope and a point on the line.

7. $5x - 2y = 6$

    The slope is $\frac{5}{2}$.

7.

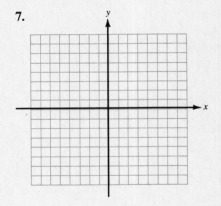

8. $2x - 3y = -1$

    The slope is $\frac{-2}{3}$.

8.

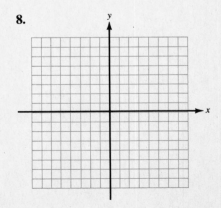

Name _____

Date _____

Score _____

SECTION 6-4. Summary Exercises

Answer

In exercises **1** and **2**,

a. write equations in $y = mx + b$ form.

b. state the slope of the graph of the equation.

c. state the y-intercept of the equation.

1. $x - 5y = -15$

1. **a.** _____

   b. _____

   c. _____

2. $6x + 5y = 35$

2. **a.** _____

   b. _____

   c. _____

In equations **3** and **4**, determine whether the graph of the pairs of equations are *parallel, perpendicular,* or *neither.*

3. $y = \dfrac{-3}{5}x$

   $3x + 5y = 15$

3. _____

4. $7x - 4y = 28$

   $4x + 7y = -14$

4. _____

431

In exercises **5–8**, graph.

5. $y = \dfrac{2}{3}x - 3$

5.
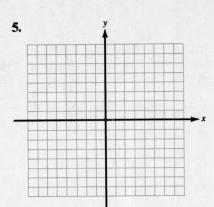

6. $4x + 3y = -12$

6.
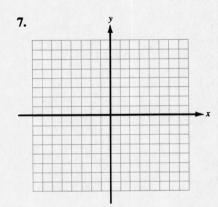

7. $\dfrac{3x + 2}{4} = \dfrac{y - 2}{2}$

7.
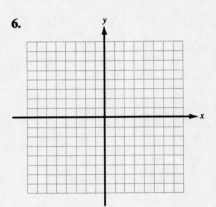

8. $x = -6$

8.
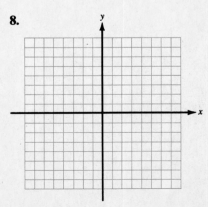

SECTION 6-5.   Summary Exercises

Answer

In exercises **1–8**, write an equation of the line based on the given conditions:

a. Use the point-slope form.

b. Use the $Ax + By = C$ form.

1. Containing $P(-3, 4)$ and slope $\frac{-6}{7}$

1. a. _____

    b. _____

2. Containing $Q(0, 0)$ and slope $\frac{1}{2}$

2. a. _____

    b. _____

3. Containing $R_1(-3, 7)$ and $R_2(2, -1)$

3. a. _____

    b. _____

4. Containing $S_1(-7, -7)$ and $S_2(-3, 0)$

4. a. _____

    b. _____

5. Containing $T(-4, -2)$ and parallel to the graph of $x + 3y = 6$

5. a. _____

b. _____

6. Containing $U(9, -4)$ and perpendicular to the graph of $6x - 5y = 0$

6. a. _____

b. _____

7. A horizontal line containing $V(11, -5)$

7. a. _____

b. _____

8. A vertical line containing $W(-13, -11)$

8. a. _____

b. _____

SECTION 6-6. Summary Exercises

In exercises **1–3**, determine if the stated ordered pair is a solution of the inequality.

1. $y > \dfrac{-1}{2}x - 1$ $\qquad$ $(-6, 1)$

1. _____

2. $2x + y \leq 3$ $\qquad$ $(3, -5)$

2. _____

3. $x \geq -4$ $\qquad$ $(-6, 2)$

3. _____

In exercises **4–7**, graph the solutions.

4. $4x - y \geq 4$

4.

5. $6x + 2y < 6$

5.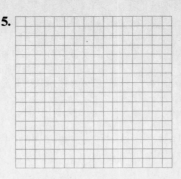

6. $x - y \leq 2$

6.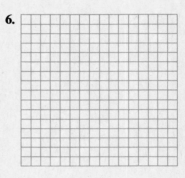

7. $2x - 5y > -1$

7.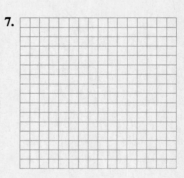

8. Write an inequality in x and y that defines the half-plane.

8.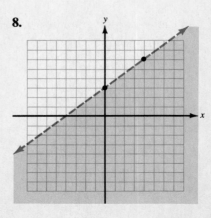

CHAPTER 6. Review Exercises

In exercises **1–10**, solve each problem.

1. Determine which of the following ordered pairs are solutions of $3x - 5y = 15$.
   a. $(-5, -6)$ **b.** $(5, 6)$ **c.** $(0, 3)$ **d.** $(5, 0)$

2. Find the solution of $4x + 5y = 20$ for the given member of the ordered pair.
   a. $y = \dfrac{4}{5}$ **b.** $x = -10$

3. State the coordinates of A–E.

   a. $A($, $)$

   b. $B($, $)$

   c. $C($, $)$

   d. $D($, $)$

   e. $E($, $)$

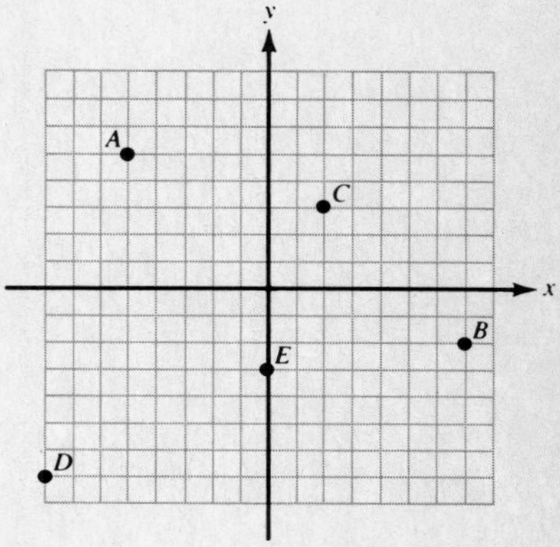

Figure 6-76.

4. Plot the following points:
   $F(-1, 2)$
   $G(3, 6)$
   $H(0, 5)$
   $I(-4, 8)$
   $J(2, -7)$
   $K(-6, -6)$
   $L(-8, 0)$

5. Identify the quadrant in which each point lies.

   a. $P(-7, -10)$ **b.** $Q(3.2, -0.5)$ **c.** $R\left(\dfrac{-3}{4}, \dfrac{7}{8}\right)$

6. Identify the axis (horizontal or vertical) on which each point lies.
   a. $S(-35, 0)$ **b.** $T(0, 101)$

In exercises **7–10**, graph.

7. $5x + 2y = 10$

8. $y = \dfrac{3}{4}x - 5$

9. $x - 3 = 0$

10. $y + 5 = 0$

In exercises **11 and 12**, state

a. the x-intercept.

b. the y-intercept.

11. $7x - 8y = 24$

12. $\dfrac{2x + 3y}{4} = \dfrac{9 - x}{3}$

In exercises **13–16**, find the slope of a line containing each pair of points. If slope is undefined for a given line, write "Undefined."

13. $P(2, -4)$, $Q(8, 0)$

14. $R(0, -5)$, $S(10, 5)$

15. $T(7, -3)$, $U(7, 5)$

16. $V(-8, 2)$, $W(6, 2)$

In exercises **17** and **18**, state

a. the slope.

b. the y-intercept.

17. $15x + 5y = 20$

18. $6x - 8y = 0$

In exercises **19** and **20**, for the lines defined by the given equations,

a. state the slope of a line parallel to the graph.

b. state the slope of a line perpendicular to the graph.

19. $3x + 5y = 10$

20. $x - 3y + 6 = 0$

In exercises **21–24**, find the equation of the line with the stated conditions. Express equations in $Ax + By = C$ form with $A > 0$.

21. Containing $P(-4, 2)$ and slope $\frac{-7}{3}$

22. Containing $P(0, 5)$ and $Q(-6, -3)$

23. Containing $P(-5, -6)$ and $Q(0, -6)$

24. Containing $P(4, 1)$ and parallel to the line with the equation $2x + y = 7$

In exercises **25** and **26**, graph.

25. $2x - 7y < 14$

26. $5x + 4y \geq 8$

27. Write an inequality in x and y that defines the half-plane.

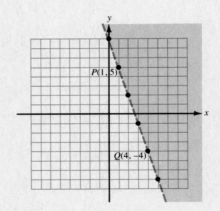

Figure 6-77.

7
Systems of Equations

Ms. Glaston was carrying two paper bags when she entered the classroom. She emptied their contents on the desk at the front of the room. The contents of one bag formed a pile of hard candies. The contents of the second bag formed a pile of chocolate candies individually wrapped in foil.

She then turned to the board and wrote the following information:

Hard candies at $1.00 per pound

Chocolate candies at $1.75 per pound

Wanted: 10 pounds of mixed candies costing $1.27 per pound.

She then turned to the class and said, "I have two piles of candy on the desk. The hard candies in the plastic wraps cost $1.00 per pound. The chocolate candies in the foil wraps cost $1.75 per pound. I want to take some candy from both piles and mix them so that I have exactly ten pounds of candy that would cost $1.27 per pound. What we need to figure out is the weight of candy to take from each pile."

After a period of silence, Brian Hickman said, "Ms. Glaston, to be honest, I don't even know where to begin. It seems we have too many conditions to satisfy at the same time. First of all, the weights of the candies from both piles must add up to 10 pounds. Secondly, the values of the candies must add up to ten times $1.27, or $12.70. How can we figure out the right combination for both conditions?"

"We could try some combinations," Debbie Bacus suggested. "Then we might get some idea of how to approach the problem."

"Good suggestion, Debbie," Ms. Glaston replied. "Let's use a table to organize our work." She then turned to the board and wrote the following table:

$1.00 candy		$1.75 candy		
Amount	Value	Amount	Value	Total value
1 pound	$1.00	9 pounds	$15.75	$16.75
2 pounds	2.00	8 pounds	14.00	16.00
3 pounds	3.00	7 pounds	12.25	15.25
4 pounds	4.00	6 pounds	10.50	14.50
5 pounds	5.00	5 pounds	8.75	13.75
6 pounds	6.00	4 pounds	7.00	13.00
7 pounds	7.00	3 pounds	5.25	12.25

"We can stop here," Ms. Glaston said. "As Brian noted, 10 pounds at $1.27 per pound has a value of $12.70. The last two entries ($13.00 and $12.25) bracket this amount. Therefore, the correct mixture requires between 6 and 7 pounds of the $1.00 candy and between 3 and 4 pounds of the $1.75 candy. The question is, what are the exact values?"

After another period of silence, Wendy Eledge said, "Ms. Glaston, it looks like we need two variables—one for the amount of $1.00 candy and another for the amount of $1.75 candy."

"That would make our work easier Wendy," Ms. Glaston replied. "If we use two variables, then maybe we can write two different equations involving both variables. One equation can describe the weight requirement and the other one can describe the value restriction." She then wrote the following sentences on the board:

Let x represent the amount of $1.00 candy needed.

Let y represent the amount of $1.75 candy needed.

(weight) + (weight) = (weight)

(1) $x \ + \ y \ = \ 10$ The sum of the weights is 10 pounds.

(value) + (value) = (value)

(2) $1.00x \ + \ 1.75y \ = \ 10(1.27)$ The value of the mixture is $12.70.

"Equations (1) and (2) form a **system of equations**," Ms. Glaston said. "As we will learn in this chapter, a solution of a system of equations must be a solution of every equation in the system. Furthermore, we will learn several ways in which a system of equations can be solved. After we learn some preliminary details about a system of equations, we will return to this candy problem to determine exactly how much of each kind of candy is needed to get the proper mixture. If there are no questions, let's begin our study with the first section."

SECTION 7-1.   Solutions of Linear Systems

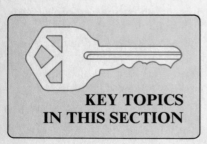

**KEY TOPICS
IN THIS SECTION**

1. A definition of a linear system in two variables

2. A solution of a linear system in two variables

3. Finding a solution, given one of the members

4. A solution of a linear system in three variables

In this section we will begin a study of systems of equations. A **system of equations** consists of two or more equations that are taken to be a single unit. A definition will be given for a basic system that contains two linear equations in two variables.

A Definition of a Linear System in Two Variables

Suppose x and y stand for two numbers. Furthermore, suppose the value of y paired with a given x can be described by an equation. As an example, equation (1) states that "y is always five less than two times x."

$$(1) \quad y = 2x - 5$$

Infinitely many pairs of numbers are solutions to equation (1). Suppose, however, that the value of y paired with a given x can also be described by equation (2). This equation states that "y is always nine more than negative one-third times x."

$$(2) \quad y = \frac{-1}{3}x + 9$$

Infinitely many pairs of numbers are solutions of equation (2). *However, when both (1) and (2) are taken at the same time, there may be only one pair of numbers that simultaneously satisfies both equations.* If two or more equations or inequalities are considered at the same time, then a **system of equations or inequalities** is formed. The simplest system is one composed of two linear equations in two variables.

Definition 7.1. A system of linear equations in two variables
  If $Ax + By = C$ and $Dx + Ey = F$ are linear equations, then

$$(1) \quad Ax + By = C$$

$$(2) \quad Dx + Ey = F$$

is a **system of linear equations** in x and y.

Example **a** illustrates a system of linear equations in x and y. Example **b** illustrates a system of linear equations in s and t. Variables other than x and y may be used for a system, but *both equations must have the same variables.*

a. (1) $x + 3y = 5$ **b.** (1) $2s - t = 14$
  (2) $3x - 2y = -7$ (2) $s + 2t = -3$

Example 1. Write a system of linear equations in x and y for the following:

(1) Two times a number x added to five times a number y is 0.

(2) The difference between six times x and five times y is 12.

Solution.   "Two times x" is written $2x$.
"Five times y" is written $5y$.
"[The sum of these products] is 0" yields the following equation:

(1) $2x + 5y = 0$

"Six times x" is written $6x$.
"Five times y" is written $5y$.
"[The difference in these products] is 12" yields the following equation:

(2) $6x - 5y = 12$

Thus, a system that describes the problem is

(1) $2x + 5y = 0$.

(2) $6x - 5y = 12$.

A Solution of a Linear System in Two Variables

Consider the numbers x and y of equations (1) and (2) stated on pages 440–41.

$$(1) \quad y = 2x - 5$$

$$(2) \quad y = \frac{-1}{3}x + 9$$

Graphs of these equations are shown in Figure 7-1. As seen in the figure, the lines cross at $P_3(6, 7)$. Since this point is on both lines, $(6, 7)$ must be a solution of equations (1) and (2).

Replacing x by 6 and y by 7

$$(1) \quad 7 = 2(6) - 5 \qquad\qquad (2) \quad 7 = \frac{-1}{3}(6) + 9$$

$$7 = 12 - 5, \text{ true} \qquad\qquad 7 = -2 + 9, \text{ true}$$

Since $(6, 7)$ is a solution of both equations, we say that $(6, 7)$ is a solution of the system composed of equations (1) and (2).

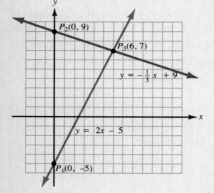

Figure 7-1. Graphs of

$$(1) \quad y = 2x - 5$$

$$(2) \quad y = \frac{-1}{3}x + 9$$

Definition 7.2.   A solution of a system of linear equations in x and y

The ordered pair (a, b) is a solution of

$$(1) \quad Ax + By = C$$

$$(2) \quad Dx + Ey = F$$

if (a, b) is a solution of both equations (1) and (2).

Example 2.   Determine whether $(-1, 2)$ is a solution of the following system:

$$(1) \quad x + 3y = 5$$

$$(2) \quad 3x - 2y = -7$$

Solution.   Replacing x with -1 and y with 2 in both equations,

$$(1) \quad -1 + 3(2) = 5 \qquad\qquad (2) \quad 3(-1) - 2(2) = -7$$

$$-1 + 6 = 5, \text{ true} \qquad\qquad -3 - 4 = -7, \text{ true}$$

Since $(-1, 2)$ is a solution of both equations, it is a solution of the system.

Example 3.   Determine whether $(4, -6)$ is a solution of the following system:

$$(1) \quad 2a - b = 14$$

$$(2) \quad a + 2b = -3$$

Solution.   Replacing a with 4 and b with -6 in both equations,

$$(1) \quad 2(4) - (-6) = 14 \qquad\qquad (2) \quad 4 + 2(-6) = -3$$

$$8 - (-6) = 14, \text{ true} \qquad\qquad 4 + (-12) = -3, \text{ false}$$

Since $(4, -6)$ is not a solution of $a + 2b = -3$, it is not a solution of the system.

Finding a Solution, Given One of the Members

A solution of a system of linear equations in two variables is an ordered pair of numbers. If one of those numbers is known, then the second number can be determined.

Example 4. Find the value of k so that $(k, 7)$ is a solution of the following system:

(1) $\quad 5m - 7n = -4$

(2) $\quad 7m - 2n = 49$

Solution. **Discussion.** The equations of this system are in the variables m and n. Therefore, the form of a solution of the system is (m, n). If the solution has the form $(k, 7)$, then 7 is the value of n. We need to find k, the missing value of m. Either equation can be used, so we arbitrarily select equation (1).

(1) $\quad 5m - 7(7) = -4$	Replace n by 7 in (1).
$5m - 49 = -4$	Do the multiplication on the left side.
$5m = 45$	Add 49 to both sides.
$m = 9$	Divide both sides by 5.

Checking $(9, 7)$ in equation (2),

(2) $\quad 7(9) - 2(7) = 49$

$\qquad 63 - 14 = 49$, true

Since $(9, 7)$ is a solution of both equation (1) and equation (2), it is a solution of the system, and $k = 9$.

A Solution of a Linear System in Three Variables

Definition 7.1 identifies the general form of a system of linear equations in two variables, and Definition 7.2 identifies the form of a solution of such a system. Example 5 illustrates a solution of a system of three equations in three variables.

Example 5. Verify that $(4, -1, 2)$ is a solution of the following system:

(1) $\quad 2x + 3y - z = 3$

(2) $\quad x - 2y + 3z = 12$

(3) $\quad 3x + y - 4z = 3$

Solution. **Discussion.** This system has three equations, and each equation has three variables. The **ordered triple** is a solution of this system if each equation is true when x is replaced by 4, y is replaced by -1, and z is replaced by 2. The form of the replacements in the ordered triple is alphabetical (x, y, z).

(1) $\quad 2(4) + 3(-1) - 2 = 3$,

$\qquad 8 + (-3) - 2 = 3$

$\qquad\qquad 5 - 2 = 3$, true

$$(2) \quad 4 - 2(-1) + 3(2) = 12,$$
$$4 - (-2) + 6 = 12$$
$$6 + 6 = 12, \text{ true}$$
$$(3) \quad 3(4) + (-1) - 4(2) = 3$$
$$12 + (-1) - 8 = 3$$
$$11 - 8 = 3, \text{ true}$$

Since $(4, -1, 2)$ is a solution of all three equations, it is a solution of the system.

SECTION 7-1.   Practice Exercises

In exercises **1–6**, write a system of linear equations in x and y for the given statements.

[Example 1]
1. If five times a number x is decreased by four times a number y, the result is 11. The difference between x and y is 2.

2. If three times a number x is decreased by five times a number y, the result is 11. The difference between x and y is 5.

3. The total number of red candies (x) and green candies (y) in a bag is 153. There are nine more red candies than green candies in the bag.

4. The total number of boys (x) and girls (y) in the cafeteria is 112. There are twelve more girls than boys in the cafeteria.

5. On Friday Ken purchased three fishing lures, each costing x dollars. He also purchased two jars of bait at y dollars each. The total cost of the purchase was $8.96. Sunday morning he purchased four more of the same lures and one jar of the bait for $7.48.

6. Rodney bought two rolls of slide film at x dollars each and three rolls of color print film at y dollars each. The total cost of the film was $18.15. His friend Mitzi bought three rolls of the same slide film and one roll of the same color print film for $17.60.

In exercises **7–22**, determine whether the given ordered pair is a solution of the stated system.

[Examples 2 and 3]
7. (1) $x + y = 7$
     (2) $x - y = 1,$ $(4, 3)$

8. (1) $2x + y = 8$
     (2) $x - 2y = 9,$ $(5, -2)$

9. (1) $a - 3b = 17$
     (2) $2a + b = -8,$ $(-1, -6)$

10. (1) $4a + 3b = 24$
      (2) $5a - b = 10,$ $(3, 4)$

11. (1) $2m - 3n = 8$
      (2) $7m + n = -40,$ $(-5, -6)$

12. (1) $4m + n = 27$
      (2) $m - 5n = -9,$ $(6, 3)$

13. (1) $2x + 3y = 10$
      (2) $3x - 4y = -53,$ $(-7, 8)$

14. (1) $5x + 2y = 26$
      (2) $9x - 8y = 128,$ $(8, -7)$

15. (1) $3x + 2y = 0$
    (2) $5x - 4y = -20$, $(-2, 3)$

16. (1) $4x + 10y = 0$
    (2) $3x - y = -17$, $(-5, 2)$

17. (1) $5a + 4b = -32$
    (2) $4a = 3b - 7$, $(-4, -3)$

18. (1) $\frac{2}{3}a - b = 3$

    (2) $3a + b = -25$, $(-6, -7)$

19. (1) $2y = 19 - 3x$
    (2) $y = \frac{-2}{3}x + 2$, $(9, -4)$

20. (1) $y = \frac{4x + 56}{5}$

    (2) $y = \frac{5x + 69}{6}$, $(-9, 4)$

21. (1) $a = \frac{-3b - 35}{2}$

    (2) $a = \frac{5b - 15}{4}$, $(-10, -5)$

22. (1) $b = \frac{55 - 3a}{4}$

    (2) $b = \frac{a + 13}{2}$, $(5, 10)$

In exercises **23–30**, find the value of the variable so that the ordered pair is a solution of the given system. Check your answer in each equation of the system.

[Example 4] **23.** (1) $4x + 3y = 11$
        (2) $2x - 5y = 25$, $(5, y)$

24. (1) $x + 3y = 0$
    (2) $x - 3y = -24$, $(x, 4)$

25. (1) $4a + 7b = -10$
    (2) $3a + 10b = 2$, $(-6, b)$

26. (1) $2a + 9b = 1$
    (2) $4a - 13b = 33$, $(a, -1)$

27. (1) $6m - 9n = 0$
    (2) $15n - 8m = 1$, $\left(\frac{1}{2}, n\right)$

28. (1) $6m - 4n = 1$
    (2) $12m + 12n = -13$, $\left(m, \frac{-3}{4}\right)$

29. (1) $9x + 4y = -5$
    (2) $18x + 48y = 0$, $\left(\frac{-2}{3}, y\right)$

30. (1) $15x + 10y = 12$
    (2) $15x - 10y = 0$, $\left(x, \frac{3}{5}\right)$

In exercises **31–36**, determine whether the ordered triple is a solution of the system.

[Example 5] **31.** (1) $x + y - z = 8$
        (2) $x - y + z = 2$
        (3) $3x - y - z = 4$, $(5, 7, 4)$

32. (1) $2x + y + 2z = 3$
    (2) $2x - y - 3z = 22$
    (3) $3x - 2y + 2z = -1$, $(5, 3, -5)$

33. (1) $x + 2y + z = 2$
    (2) $x + 2y - z = 8$
    (3) $2x - y + 2z = -6$, $(1, 2, -3)$

34. (1) $4x - y = 20 - 2z$
    (2) $3x + y = 3z$
    (3) $9x - y = 3z$, $(4, 12, 8)$

35. (1) $2x - y - z = 6$
    (2) $3x + y - 2z = 4$
    (3) $2x - 4y + z = 12$, $(6, -2, 8)$

36. (1) $3x + 2y + 4z = -10$
 (2) $4x - 3y + 2z = -14$
 (3) $2x + 4y + 3z = 21,$ $(-2, 4, -3)$

SECTION 7-1.   Ten Review Exercises

In exercises **1–6**, use the following points:

$$P(-4, 1) \text{ and } Q(-2, 4)$$

1. Find the slope of the line containing points P and Q.

2. State the slope of a line parallel to the line containing points P and Q.

3. Write in the *point-slope form* an equation for the line through points P and Q using the coordinates of point P.

4. Write the equation obtained in exercise **3** in the *slope-intercept* form.

5. Use the equation obtained in exercise **4** to identify the y-intercept.

6. Write an equation in $Ax + By = C$ form for the line perpendicular to the line through points P and Q at point Q.

In exercises **7–10**, use the expression $\dfrac{3b^2 + 7b + 2}{b^2 - b - 6}$

7. Find the value of the expression for $b = 4$.

8. Write the expression in factored form.

9. Reduce the expression obtained in exercise **8**.

10. Evaluate the reduced expression in exercise **9** for $b = 4$.

SECTION 7-1.   Supplementary Exercises

In exercises **1–4**, determine which, if any, of the ordered pairs **a**, **b**, or **c** are solutions of the system.

1. (1) $\dfrac{2}{3}a - \dfrac{3}{5}b = 12$

 (2) $10a - 9b = 180$
  a. $(9, -10)$
  b. $(18, 0)$
  c. $(0, -20)$

2. (1) $\dfrac{1}{2}a - \dfrac{1}{3}b = -1$

 (2) $\dfrac{3}{4}a + \dfrac{5}{3}b = 18$

  a. $(8, 15)$
  b. $(12, 21)$
  c. $(4, 9)$

3. (1) $\dfrac{4}{7}m + \dfrac{3}{7}n = 27$

    (2) $\dfrac{4}{7}m - \dfrac{3}{7}n = -3$

    a. (35, 21)
    b. (21, 35)
    c. (14, 7)

4. (1) $5m + 2n = 16$
    (2) $15m + 6n = 48$
    a. (2, 3)
    b. $(-2, 13)$
    c. $(4, -2)$

In exercises **5–12**, determine whether the given ordered pair is a solution of the stated system.

5. (1) $10a + 9b = 8$
    (2) $12a - 18b = 0,$ $\left(\dfrac{1}{2}, \dfrac{1}{3}\right)$

6. (1) $10a + 12b = -23$
    (2) $20a - 21b = -16,$ $\left(\dfrac{-3}{2}, \dfrac{-2}{3}\right)$

7. (1) $5m - 3n = 7$
    (2) $10m + 3n = 1,$ $\left(\dfrac{3}{5}, \dfrac{-4}{3}\right)$

8. (1) $6m + 6n = -14$
    (2) $10m - 12n = 27,$ $\left(\dfrac{-5}{2}, \dfrac{1}{6}\right)$

9. (1) $6x + 7y = 5$
    (2) $15x - 14y = 26,$ $\left(\dfrac{4}{3}, \dfrac{-3}{7}\right)$

10. (1) $2x - 3y = 3$
    (2) $4x + 9y = -44,$ $\left(\dfrac{-7}{2}, \dfrac{-10}{3}\right)$

11. (1) $\dfrac{3}{5}a + \dfrac{2}{5}b = \dfrac{2}{5}$

    (2) $\dfrac{15a}{8} + \dfrac{25b}{8} = 2,$ $\left(\dfrac{2}{5}, \dfrac{2}{5}\right)$

12. (1) $\dfrac{1}{2}a - \dfrac{3}{4}b = \dfrac{2}{3}$

    (2) $3a + \dfrac{1}{2}b = \dfrac{-28}{3},$ $\left(\dfrac{-8}{3}, \dfrac{-8}{3}\right)$

In exercises **13–20**,

a. Find the value of b so that the stated ordered pair is a solution of the given system.

b. State the solution of the system. Check the solution in both equations.

13. (1) $3x + 2y = 14$
    (2) $5x - y = 6,$ $(b, 2b)$

14. (1) $2x + y = 21$
    (2) $2x - y = 3,$ $(2b, 3b)$

15. (1) $4x - 3y = -20$
    (2) $x - 4y = -18,$ $(b, -2b)$

16. (1) $5x + y = -42$
    (2) $2x - 3y = -27,$ $(3b, -b)$

17. (1) $3x + 4y = 9$
    (2) $4x + 3y = -9,$ $(-3b, 3b)$

18. (1) $2x - 5y = 14$
    (2) $5x - 2y = 14,$ $(-2b, 2b)$

19. (1) $10x - 7y = -48$
    (2) $7x + 10y = 26,$ $\left(\dfrac{1}{2}b, -b\right)$

20. (1) $6x - 2y = -6$
    (2) $2x + 3y = 20,$ $\left(\dfrac{1}{2}b, 3b\right)$

Example. Verify that $(-4, 3)$ and $(3, -4)$ are both solutions of the following system:

    (1) $x^2 + y^2 = 25$

    (2) $x + y = -1$

Solution.   Checking $(-4, 3)$,

(1) $(-4)^2 + 3^2 = 25$ (2) $-4 + 3 = -1$

       $16 + 9 = 25$, true $-1 = -1$, true

Since $(-4, 3)$ is a solution of equations (1) and (2), it is a solution of the system.

Checking $(3, -4)$,

(1) $3^2 + (-4)^2 = 25$ (2) $3 + (-4) = -1$

       $9 + 16 = 25$, true $-1 = -1$, true

Since $(3, -4)$ is a solution of equations (1) and (2), it is a solution of the system.

In exercises **21–28**, determine which of the given ordered pairs are solutions of the system of equations.

21. (1) $4x^2 + y^2 = 100$ **22.** (1) $x^2 + y^2 = 25$
     (2) $2x = y + 2$; $(4, 6), (-3, -8)$ (2) $x + 3y = 5$; $(5, 0), (-4, 3)$

23. (1) $a^2 - b^2 = 9$ **24.** (1) $a^2 - b^2 = 16$
     (2) $2a - b = 6$; $(5, 4), (0, 3)$ (2) $a + b = 4$; $(3, -5), (4, 0)$

25. (1) $n = m^2 + 2m - 8$
     (2) $2m + n + 3 = 0$; $(1, 5), (-5, 7)$

26. (1) $m^2 + n = 2$
     (2) $4m^2 + n^2 = 5$; $(-1, -1), (1, 1)$

27. (1) $xy + 12$
     (2) $y = x^2 + 3x - 4$; $(2, 6), (-2, -6), (-3, -4)$

28. (1) $16x^2 + y^2 = 100$
     (2) $xy = 12$; $(2, 6), (-2, -6), \left(\dfrac{3}{2}, 8\right)$

SECTION 7-2.   The Graphical Method of Solving a System

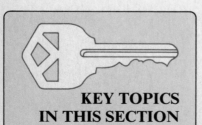

**KEY TOPICS
IN THIS SECTION**

1. The graphical method of solving a system

2. An inconsistent system of linear equations in x and y

3. A dependent system of linear equations in x and y

In this chapter we will study three methods of solving systems of linear equations in two variables. Two are algebraic and one is geometric. The geometric method, studied in this section, requires graphing the lines defined by the equations of the system. If the lines intersect, then the coordinates of the point of intersection are the solution of the system.

The Graphical Method for Solving a System of Linear Equations in Two Variables

The following three steps can be used to solve graphically a system of two linear equations in x and y.

The graphical method of solving a system of linear equations in 2 variables

Given: (1) $Ax + By = C$

       (2) $Dx + Ey = F$

Step 1. Graph the line defined by equation (1).

Step 2. Graph the line defined by equation (2).

Step 3. Determine, if possible, $P(a, b)$, the point of intersection of the lines.

If in Step 3 the lines intersect, then the coordinates of the point of intersection are a solution of both equations in the system. As a consequence, *the coordinates of the point of intersection are the solution of the system.* In this section, for any system that has exactly one solution, the lines will intersect in a point with integer coordinates.

Example 1. Solve and check:

(1) $3x + y = -2$

(2) $2x - y = -8$

Solution. **Step 1.** Graph the line defined by $3x + y = -2$.

$$y = -3x - 2 \begin{cases} \text{slope is } -3 \\ y\text{-intercept is } -2 \end{cases}$$

x	y
-2	4
0	-2
2	-8

A graph is drawn in Figure 7-2.

Step 2. Graph the line defined by $2x - y = -8$.

$$y = 2x + 8 \begin{cases} \text{slope is } 2 \\ y\text{-intercept is } 8 \end{cases}$$

x	y
-4	0
-2	4
0	8

A graph is drawn in Figure 7-2.

Step 3. $P(-2, 4)$ is the point of intersection.

Checking $(-2, 4)$,

(1) $3(-2) + 4 = -2$

      $-6 + 4 = -2$, true

(2) $2(-2) - 4 = -8$

      $-4 - 4 = -8$, true

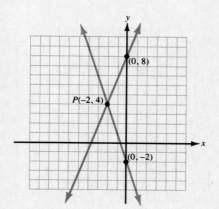

Figure 7-2. The lines intersect at $P(-2, 4)$.

Example 2.   Solve and check:

(1) $4x + 5y = 5$

(2) $2x - 5y = 25$

Solution.   **Step 1.** Graph the line defined by $4x + 5y = 5$.

$5y = -4x + 5$ Subtract $4x$ from both sides.

$y = \dfrac{-4x}{5} + \dfrac{5}{5}$ Divide each term by 5.

$y = \dfrac{-4}{5}x + 1$ ⟨ slope is $\frac{-4}{5}$
        ⟨ y-intercept is 1

x	y
-5	5
0	1
5	-3

Step 2. Graph the line defined by $2x - 5y = 25$.

$-5y = -2x + 25$ Subtract $2x$ from both sides.

$y = \dfrac{-2x}{-5} + \dfrac{25}{-5}$ Divide each term by -5.

$y = \dfrac{2}{5}x - 5$ ⟨ slope is $\frac{2}{5}$
        ⟨ y-intercept is -5

x	y
0	-5
5	-3
10	-1

Figure 7-3. The lines intersect at $P(5, -3)$.

Step 3. $P(5, -3)$ is the point of intersection. (See Figure 7-3.)

In both Example 1 and Example 2 the equations of the systems yielded *two different lines*. Furthermore, in both examples, the lines intersected in *exactly one point*. Systems containing equations yielding two different lines that intersect in exactly one point are classified **independent** (two different lines) and **consistent** (the lines intersect).

If the equations of the system

(1) $Ax + By = C$

(2) $Dx + Ey = F$

are **independent** and **consistent**, then

a. the graphs defined by equations (1) and (2) will be *two different lines*.

b. the lines will intersect in *exactly one point*.

An Inconsistent System of Linear Equations in x and y

Not all systems defined by equations (1) and (2) are consistent. To illustrate, suppose x and y are two numbers in (1) and (2):

(1) $x + y = 5$ "The sum of the numbers is 5."

(2) $x + y = 8$ "The sum of the numbers is 8."

Now the sum of two numbers cannot be 5 and also 8 at the same time. Therefore the system composed of these equations is classified *inconsistent*. The graphs of the lines defined by equations (1) and (2) are parallel.

Example 3.   Solve:

(1) $4x + 5y = 10$

(2) $y = \dfrac{-4}{5}x - 1$

Solution.   **Step 1.**   Graph the line defined by $4x + 5y = 10$.

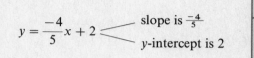

slope is $\dfrac{-4}{5}$

y-intercept is 2

x	y
−5	6
0	2
5	−2

Step 2.   Graph the line defined by $y = \dfrac{-4}{5}x - 1$.

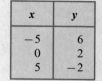

slope is $\dfrac{-4}{5}$

y-intercept is −1

x	y
−5	3
0	−1
5	−5

Step 3.   The graphs are parallel lines in Figure 7-4.
      Since the lines do not intersect, the system of equations is *inconsistent*. The system has no solution.

> If (1) $Ax + By = C$ and (2) $Dx + Ey = F$ have graphs that are **parallel lines** (same slopes but different y-intercepts), then the system composed of equations (1) and (2) is **inconsistent. The system has no solution.**

The graph at left:

(−5, 6)

(−5, 3)

(0, 2)

(0, −1)

(5, −2)

(5, −5)

Figure 7-4. The graphs are parallel lines.

A Dependent System of Linear Equations in x and y

Not all systems defined by equations (1) and (2) are independent. It may be that two apparently different equations are in fact the same. To illustrate, suppose x and y are two numbers in (1) and (2):

(1) $x + y = 5$ "The sum of the numbers is 5."

(2) $3x + 3y = 15$ "The sum of three times x and three times y is 15."

If the sum of two numbers is 5, then the sum of three times these numbers will also be 15. *That is, equations (1) and (2) are the same.* The graphs of (1) and (2) are the same line. Therefore the system is classified *dependent. Infinitely many ordered pairs are solutions of the system.*

Example 4.   Solve and check:

(1) $6x - 4y = -12$

(2) $y = \dfrac{3}{2}x + 3$

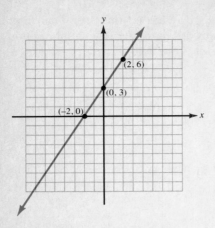

Figure 7-5. The graphs are the same line.

Solution.

Step 1. Graph the line defined by $6x - 4y = -12$.

$$y = \frac{6}{4}x + 3$$

slope is $\frac{6}{4}$, or $\frac{3}{2}$

y-intercept is 3

x	y
-2	0
0	3
2	6

Step 2. Graph the line defined by $y = \frac{3}{2}x + 3$.

$$y = \frac{3}{2}x + 3$$

slope is $\frac{3}{2}$

y-intercept is 3

x	y
-2	0
0	3
2	6

Step 3. The graphs are the same line in Figure 7-5.

The system of equations is *dependent*. The coordinates of every point on the line yield a solution of the system.

If (1) $Ax + By = C$ and (2) $Dx + Ey = F$ have graphs that are the *same line* (same slopes and same y-intercepts), then the system composed of equations (1) and (2) is *dependent*. *The system has infinitely many solutions.*

The following box summarizes the relationship between the graphs of the equations of a linear system in x and y, and the classification of the system.

If the graphs of two linear equations in x and y are not horizontal lines and not vertical lines, and the slope-intercept equations of the lines form the following linear system:

$$(1) \quad y = m_1 x + b_1$$

$$(2) \quad y = m_2 x + b_2$$

then the system is

a. independent and **consistent** if $m_1 \neq m_2$.

b. inconsistent if $m_1 = m_2$ and $b_1 \neq b_2$.

c. dependent if $m_1 = m_2$ and $b_1 = b_2$.

Example 5. Without graphing, determine whether the following system is independent and consistent, or inconsistent, or dependent.

$$(1) \quad 7x - 3y = 24$$

$$(2) \quad \frac{x + 2}{3} = \frac{y + 13}{7}$$

Solution.

Discussion. Write equations (1) and (2) in the $y = mx + b$ form. First compare the m's. If they are different, then the system is independent and consistent. If the m's are the same but the b's are different, then the system is inconsistent. If the m's and b's are the same, then the system is dependent.

(1) $7x - 3y = 24$

$7x - 24 = 3y$

$y = \dfrac{7}{3}x - 8$

(2) $\dfrac{x+2}{3} = \dfrac{y+13}{7}$

$7(x + 2) = 3(y + 13)$

$7x + 14 = 3y + 39$

$y = \dfrac{7}{3}x - \dfrac{25}{3}$

Since the slopes are equal ($\frac{7}{3} = \frac{7}{3}$), the system is not independent and consistent. The y-intercepts are different ($-8 \neq -\frac{25}{3}$). Therefore the system is *inconsistent*.

SECTION 7-2. Practice Exercises

In exercises **1–16**, use the graphical method to find the solution of each system.

[Examples 1 and 2]

1. (1) $x + y = 2$
(2) $2x - y = 1$

2. (1) $2x + y = 4$
(2) $2x - y = 0$

3. (1) $x - 2y = 0$
(2) $2x - y = 3$

4. (1) $x - 3y = -3$
(2) $5x - 3y = 9$

5. (1) $2x + y = -2$
(2) $x + 2y = 2$

6. (1) $x - y = 4$
(2) $5x + 2y = 6$

7. (1) $6x + y = -3$
(2) $y = 3$

8. (1) $5x + 3y = 12$
(2) $x = 3$

9. (1) $2x + y = -4$
(2) $2x - 3y = -12$

10. (1) $x - 3y = 9$
(2) $4x + 3y = 6$

11. (1) $x + 2y = 4$
(2) $3x - 4y = 12$

12. (1) $2x - y = -4$
(2) $4x + 3y = 12$

13. (1) $5x - 3y = -15$
(2) $4x + 3y = -12$

14. (1) $5x - 3y = 9$
(2) $7x + 2y = -6$

15. (1) $5x - 4y = 11$
(2) $7x + 2y = -15$

16. (1) $3x - 7y = -5$
(2) $2x - 3y = -5$

In exercises **17–30**, find the solution of each system with the graphical method. If the lines are parallel, write "inconsistent" for the answer. If the lines are the same, write "dependent" for the answer.

[Examples 3 and 4]

17. (1) $5x - 3y = 6$
(2) $5x - 3y = -6$

18. (1) $5x + 7y = 21$
(2) $5x + 7y = -14$

19. (1) $4x + 2y = -8$
(2) $2x + y = -4$

20. (1) $14x + 8y = -12$
(2) $7x + 4y = -6$

21. (1) $x - 3y = 7$
(2) $3x - 2y = 0$

22. (1) $x - 4y = 0$
(2) $4x - y = 0$

23. (1) $5x - 2y = -3$
(2) $10x - 4y = 8$

24. (1) $2x + 3y = 12$
(2) $4x + 6y = 0$

25. (1) $4x - 5y = 0$
    (2) $2x + 5y = -30$

26. (1) $4x - 3y = 0$
    (2) $4x + 3y = -24$

27. (1) $x - 3 = 0$
    (2) $y + 2 = 0$

28. (1) $x + 4 = 0$
    (2) $y - 1 = 0$

29. (1) $x - 4 = 0$
    (2) $x + 3 = 0$

30. (1) $y + 3 = 0$
    (2) $y + 1 = 0$

In exercises **31–40**, without graphing, determine whether each system is independent and consistent, or inconsistent, or dependent.

[Example 5]

31. (1) $5x - 10y = 70$
    (2) $11x + 13y = 39$

32. (1) $9x + 8y = 16$
    (2) $4x - 7y = 35$

33. (1) $3x + 7y = 35$
    (2) $3x + 7y = -35$

34. (1) $6x - 5y = 10$
    (2) $12x - 10y = 130$

35. (1) $x + 4y = 28$
    (2) $7x + 28y = 196$

36. (1) $10x - 7y = 56$
    (2) $40x - 28y = 224$

37. (1) $10x - 3y = 30$
    (2) $3x - 10y = 30$

38. (1) $12x + 7y = 7$
    (2) $7x - 12y = 12$

39. (1) $2x - 9y = 18$
    (2) $10x - 45y = 45$

40. (1) $2x - 9y = 18$
    (2) $10x - 45y = 90$

SECTION 7-2.   Ten Review Exercises

In exercises **1–10**, use the following points:

$$P_1(2, -1), \ P_2(6, 1), \ P_3(-6, 3), \text{ and } P_4(2, -9)$$

1. Find the slope of a line $\mathscr{L}_1$ through P_1 and P_2.

2. Write an equation of $\mathscr{L}_1$ in $Ax + By = C$ form.

3. Find the slope of a line $\mathscr{L}_2$ through P_3 and P_4.

4. Write an equation of $\mathscr{L}_2$ in $Ax + By = C$ form.

5. Write a system of linear equations in x and y using the equations of $\mathscr{L}_1$ and $\mathscr{L}_2$.

6. Graph $\mathscr{L}_1$ and $\mathscr{L}_2$ on the same set of coordinate axes.

7. Read the coordinates of the point of intersection as the solution of the system stated in exercise **5**.

8. Let x and y represent two different numbers and use them to write a sentence for the equation obtained in exercise **2**.

9. Let x and y represent two different numbers and use them to write a sentence for the equation obtained in exercise **4**.

10. To make the sentences in exercises **8** and **9** true, what numbers must x and y be?

SECTION 7-2. Supplementary Exercises

In Chapter 2 equations such as (1) were solved.

$$(1) \quad 3x + 3 = x - 3$$

The left and right sides of this equation can be used to form two equations in x and y. The solution of the system composed of these equations will have an x-value that is the root of equation (1).

Example. Solve as a system of linear equations in x and y.

$$(1) \quad 3x + 3 = x - 3$$

Solution. **Step 1.** Write the following system:

$$y = 3x + 3 \qquad \text{Set the left side of (1) equal to } y.$$

$$y = x - 3 \qquad \text{Set the right side of (1) equal to } y.$$

Step 2. Graph both equations of Step 1.
The lines are graphed in Figure 7-6.

Step 3. State the coordinates of the point of intersection.
The lines intersect at $P(-3, -6)$.

Step 4. Check the x-coordinate of P as a root of (1).
Replacing x by -3 in (1),

$$3(-3) + 3 = -3 - 3$$

$$-9 + 3 = -6, \text{ true}$$

Thus, the root of (1) is -3.

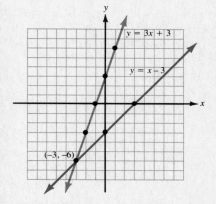

Figure 7-6. Graphs of $y = 3x + 3$ and $y = x - 3$.

In exercises **1–4**, solve by the method used in the above example.

1. $x + 1 = 2x - 2$

2. $2x + 4 = 4 - x$

3. $\dfrac{3}{2}x + 3 = \dfrac{-1}{2}x - 9$

4. $\dfrac{1}{2}x + 3 = \dfrac{3}{2}x + 5$

In exercises **5–12**,

a. write an equation for (1).

b. write an equation for (2).

c. graph the equations from **a** and **b** on the same set of axes.

d. use the point of intersection of the lines to answer the question.

5. (1) The sum of two times x and three times y is 1.
(2) The difference between x and two times y is -10.
Question: What are the numbers x and y?

6. (1) The sum of two times x and four times y is -2.
(2) The difference between two times x and three times y is -9.
Question: What are the numbers x and y?

7. (1) A number y is seven less than four times x.
   (2) A number y is one more than four-thirds x.
   Question: What are the numbers x and y?

8. (1) A number y is seven less than five times x.
   (2) A number y is seven minus two times x.
   Question: What are the numbers x and y?

9. (1) The length x of a rectangle is five meters more than the width y.
   (2) The perimeter of the rectangle of length x and width y is 26 meters.
   Question: What is the length of the rectangle?

10. (1) The length x of a rectangle is twice the width y.
    (2) The perimeter of the rectangle of length x and width y is 18 meters.
    Question: What is the length of the rectangle?

11. (1) x pounds of almonds at \$3.00 per pound are mixed with y pounds of peanuts at \$1.50 per pound. The total cost of the mix is \$7.50.
    (2) The number of pounds of almonds (x) is one more than the number of pounds of peanuts (y).
    Question: How many pounds are there of each type of nut?

12. (1) x pounds of candy at \$5.00 per pound is mixed with y pounds of mints at \$2.50 a pound. The total cost of the mix is \$20.00.
    (2) The number of pounds of mints (y) is two more than the number of pounds of candy (x).
    Question: How many pounds of each kind were mixed?

SECTION 7-3.   The Substitution Method of Solving a System of Linear Equations in Two Variables

**KEY TOPICS
IN THIS SECTION**

1. The substitution property of equality

2. The substitution method of solving an independent and consistent system of linear equations in two variables

3. Inconsistent systems and the substitution method

4. Dependent systems and the substitution method

A weakness in the graphical method of solving a system of linear equations in x and y becomes obvious when attempting to solve the following system by this method:

$$(1) \quad 3x + y = 5$$

$$(2) \quad 4x + 3y = 8$$

The solution of this system is $(\frac{7}{5}, \frac{4}{5})$. It would be difficult to read the coordinates of this point from an average-size grid. The algebraic method called **substitution** is a more practical way of obtaining the solution.

The Substitution Property of Equality

In previous chapters we have used the addition, multiplication, and symmetric properties of equality to solve equations. To solve a system of equations we may use the **substitution property of equality.**

The substitution property of equality
If $a = b$, then a can be replaced by b in any expression containing a.

Example 1. Solve:

(1) $y = 2x - 7$

(2) $y = 15$

Solution.   **Discussion.** If $y = 15$ in equation (2), then the substitution property of equality states that y can be replaced by 15 in any expression containing y.

(1) $15 = 2x - 7$ Replace y by 15 in (1).

$22 = 2x$ Add 7 to both sides.

$11 = x$ Divide both sides by 2.

Thus, (11, 15) is the solution of equations (1) and (2).

The Substitution Method of Solving an Independent and Consistent System of Linear Equations in Two Variables

The substitution method is one of the two algebriac techniques that we will study in this chapter. This method uses the substitution property of equality to replace one of the variables in one of the equations. The details are given in the following six-step procedure.

   To solve an independent and consistent system of linear equations in x and y by the substitution method,

Step 1. solve equation (1) or (2) for x (or y).

Step 2. replace x (or y) in the other equation of the system by the expression obtained for it in Step 1.

Step 3. solve the equation obtained in Step 2.

Step 4. replace the variable in equation (1) or (2) by the root and solve for the other variable.

Step 5. state the ordered pair as a solution of the system.

Step 6. check the solution in both equations.

Example 2.   Solve and check:

(1) $8x - 3y = -11$

(2) $2x + y = 6$

Solution.   **Discussion.** In the given system, y has a coefficient of 1 in equation (2). We therefore will solve this equation for y in terms of x.

Step 1. $y = 6 - 2x$ Solve (2) for y in terms of x.

Step 2. $8x - 3(6 - 2x) = -11$ Replace y by $6 - 2x$ in (1).

Step 3. $8x - 18 + 6x = -11$ Remove parentheses.

$$14x = 7 \qquad \text{Add 18 to both sides.}$$

$$x = \frac{7}{14} \qquad \text{Divide both sides by 14.}$$

$$x = \frac{1}{2} \qquad \text{Reduce.}$$

Step 4. Replace x by $\frac{1}{2}$ in equation (1) or (2).

$$2\left(\frac{1}{2}\right) + y = 6 \qquad \text{The coefficient of } y \text{ is 1 in (2).}$$

$$1 + y = 6 \qquad \text{Find the indicated product.}$$

$$y = 5 \qquad \text{Subtract 1 from both sides.}$$

Step 5. The solution of the system is $(\frac{1}{2}, 5)$.

Step 6. The check is omitted.

Example 3.   Solve and check:

(1) $10s - 3t = 8$

(2) $s + \dfrac{4}{5}t = \dfrac{1}{15}$

Solution.   **Discussion.** The coefficient of s in equation (2) is 1. We therefore solve this equation for s in terms of t.

Step 1. $s = \dfrac{1}{15} - \dfrac{4}{5}t$ Solve (2) for s in terms of t.

Step 2. $10\left(\dfrac{1}{15} - \dfrac{4}{5}t\right) - 3t = 8$ Replace s in (1) by $\dfrac{1}{15} - \dfrac{4}{5}t$.

Step 3. $\dfrac{10}{15} - 8t - 3t = 8$ Remove parentheses.

$$-11t = \frac{22}{3} \qquad 8 - \frac{2}{3} = \frac{24}{3} - \frac{2}{3} = \frac{22}{3}$$

$$t = \frac{-2}{3} \qquad \text{Divide both sides by } -11.$$

Step 4. Replace t by $\frac{-2}{3}$ in (1) or (2).

$$(1) \quad 10s - 3\left(\frac{-2}{3}\right) = 8$$

$$10s + 2 = 8$$

$$10s = 6$$

$$s = \frac{6}{10} = \frac{3}{5}$$

Step 5. The solution of the system is $(\frac{3}{5}, \frac{-2}{3})$.

Step 6. The check is omitted.

Inconsistent Systems and the Substitution Method

If the substitution method is used to solve a system of linear equations in x and y, then a root may not be obtained for the equations in Step 3. Specifically, both variables may be eliminated by the substitution, and the equation of Step 3 is a false number statement. When this happens, we may conclude that the system of equations is *inconsistent*. (Recall, the graphs of such equations are *parallel lines*.)

Example 4. Solve and check:

(1) $3y = x - 10$

(2) $7x = 21y + 50$

Solution. **Discussion.** In equation (1), x has a coefficient of 1. We therefore solve this equation for x in terms of y.

Step 1. (1) $3y + 10 = x$ Add 10 to both sides.

Step 2. (2) $7(3y + 10) = 21y + 50$ Replace x by $3y + 10$ in (2).

Step 3. $21y + 70 = 21y + 50$ Remove parentheses.

$70 = 50$, false   Subtract $21y$ from both sides.

The y terms combine to 0 and a *false number statement is the result*. Therefore the given system of equations is *inconsistent*.

If the graphs of (1) and (2) in Example 4 were drawn on the same set of axes, the result would be two parallel lines.

Dependent Systems and the Substitution Method

If the equation in Step 3 has all real numbers as solutions, then we may conclude that the system of equations is *dependent*. (Recall, the graphs of such equations are the *same line*.)

Example 5. Solve and check:

(1) $\frac{3}{10}a + b = 5$

(2) $3a + 10b = 50$

Solution.　**Discussion.** In equation (1), b has a coefficient of 1. We therefore solve this equation for b in terms of a.

Step 1. (1)　$b = 5 - \dfrac{3}{10}a$　　　　Subtract $\dfrac{3}{10}a$ from both sides.

Step 2. (2)　$3a + 10\left(5 - \dfrac{3}{10}a\right) = 50$　　Replace b by $5 - \dfrac{3}{10}a$ in (2).

Step 3. $3a + 50 - 3a = 50$　　　　Remove parentheses.

$50 = 50$, true　　　　$3a - 3a = 0$

The a terms combine to 0 and *a true number statement is obtained*. Therefore the given system of equations is *dependent*. If the graphs of equations (1) and (2) in Example 5 were drawn on the same set of axes, the result would be one line.

The following is a summary of the results of Examples 4 and 5.

If both variables are eliminated in Step 3,

a. the system is *inconsistent* if the resulting number statement is *false*.

b. The system is *dependent* if the resulting number statement is *true*.

SECTION 7-3.　Practice Exercises

In exercises **1–8**, find the solution of each system.

[Example 1]

1. (1) $3x + 5y = 15$
　　(2) $x = 10$

2. (1) $7x - 3y = 34$
　　(2) $x = 4$

3. (1) $4x - 3y = 22$
　　(2) $y = -10$

4. (1) $7x - 2y = 18$
　　(2) $y = 5$

5. (1) $y = 2x^2 + 11x + 15$
　　(2) $y = 3$

6. (1) $y = 4x^2 - 16x + 21$
　　(2) $y = 6$

7. (1) $x^2 + y^2 = 25$
　　(2) $y = 4$

8. (1) $x^2 - y^2 = 64$
　　(2) $x = 10$

In exercises **9–40**, solve using the substitution method.

[Examples 2 and 3]

9. (1) $5x - 3y = 7$
　　(2) $x = 2y$

10. (1) $10x - 4y = -78$
　　(2) $x = 3y$

11. (1) $5x - 3y = 18$
　　(2) $x = y + 2$

12. (1) $10x - 4y = -6$
　　(2) $y = x - 3$

13. (1) $6a - 8b = -12$
　　(2) $a = b - 5$

14. (1) $7a + 2b = -35$
　　(2) $a = b + 4$

15. (1) $9m - 2n = 22$
(2) $n = 3m - 2$

16. (1) $m - 3n = 31$
(2) $n = 6m + 1$

17. (1) $4s - t = -17$
(2) $3s + t = -11$

18. (1) $9s - 2t = 28$
(2) $5s + t = 5$

19. (1) $p - 2q = -10$
(2) $-p + 3q = 17$

20. (1) $p + 3q = -12$
(2) $-5p + 6q = -3$

21. (1) $b - a = 2$
(2) $7a - 3b = 6$

22. (1) $b - a = 3$
(2) $2a + 5b = 43$

23. (1) $2x + y = 16$
(2) $3x - 2y = 3$

24. (1) $x + 3y = 20$
(2) $5x - 3y = 28$

[Examples 4 and 5]   **25.** (1) $5m - n = 7$
(2) $-10m + 2n = 15$

26. (1) $-3m + 12n = 25$
(2) $m - 4n = 10$

27. (1) $3a + b = 10$
(2) $15a + 5b = 50$

28. (1) $28a - 7b = -70$
(2) $-4a + b = 10$

29. (1) $b = a - 10$
(2) $2a - 13b = 31$

30. (1) $a = 3 - b$
(2) $3a + 6b = 0$

31. (1) $x + 3y = 0$
(2) $y = 4x$

32. (1) $x = 3y$
(2) $y = \dfrac{x}{2}$

33. (1) $3x - 9y = 3$
(2) $3y + 1 = x$

34. (1) $2x - y = 4$
(2) $y = 2x + 3$

35. (1) $3s - 3t = 1$
(2) $t = s + 5$

36. (1) $2s = -4 - 6t$
(2) $-2 - 3t = s$

37. (1) $p + q = 1$
(2) $4p - 3q = 39$

38. (1) $p - q = 11$
(2) $8p + 7q = -2$

39. (1) $6a + b = 0$
(2) $18a - 7b = 50$

40. (1) $20a - b = -43$
(2) $36a + 7b = -7$

SECTION 7-3.   Ten Review Exercises

In exercises **1–3**, solve.

1. $5(8t + 6) - 33 = 3(12 - 4t)$

2. $2p^2 + p = 3$

3. $\dfrac{1}{x^2 + x} + \dfrac{4}{3x + 3} = \dfrac{2}{3x}$

In exercises **4** and **5**, solve each problem.

4. Solve and graph:

$3(y - 1) + 4 > 2(4y + 3) + 5$

5. Solve for a:

$S = at + v$

In exercises **6–10**, use $P_1(-4, 7)$ and $P_2(3, -7)$.

6. Compute the slope of $\mathcal{L}_1$ through P_1, and P_2.

7. Compute the slope of a line perpendicular to $\mathcal{L}_1$.

8. Write an equation of $\mathcal{L}_1$ in $Ax + By = C$ form.

9. Write in $Ax + By = C$ form an equation of a line perpendicular to $\mathcal{L}_1$ through P_1.

10. Graph the lines with equations in exercises **8** and **9** on the same axes.

SECTION 7-3. Supplementary Exercises

In exercises **1–12**, solve each system using the substitution method.

1. (1) $8x + y = 6$
 (2) $18x - 4y = 1$

2. (1) $15x - y = 7$
 (2) $18x + 5y = 27$

3. (1) $a + 12b = 6$
 (2) $4a + 20b = 3$

4. (1) $a + 5b = 3$
 (2) $7a - 40b = 51$

5. (1) $14m - n = -15$
 (2) $21m + 5n = 23$

6. (1) $6m + n = 9$
 (2) $60m - 7n = -46$

7. (1) $3x - y = 2$
 (2) $6x = 2y + 4$

8. (1) $4x + 4y = 5$
 (2) $8y = -8x + 3$

9. (1) $\dfrac{9}{2}x - y = 2$
 (2) $15x + 6y = 2$

10. (1) $\dfrac{15}{4}x + y = \dfrac{13}{4}$
 (2) $21x - 8y = 8$

11. (1) $\dfrac{14}{3}a + b = -3$
 (2) $20a + 9b = -16$

12. (1) $\dfrac{9}{2}a + b = \dfrac{-9}{2}$
 (2) $30a + 14b = -19$

In exercises **13–18**, on the same axis,

a. draw $\mathcal{L}_1$, the graph of (1).

b. draw $\mathcal{L}_2$, the graph of (2).

c. eliminate x in (1) and (2).

d. draw $\mathcal{L}_3$, the graph of the equation obtained in part **c**.

e. eliminate y in (1) and (2).

f. draw $\mathcal{L}_4$, the graph of the equation obtained in part **e**.

g. do $\mathcal{L}_1$, $\mathcal{L}_2$, $\mathcal{L}_3$, and $\mathcal{L}_4$ intersect in the same point?

13. (1) $x + y = 6$
 (2) $5x - 4y = 12$

14. (1) $x + y = 4$
 (2) $4x - 2y = -2$

15. (1) $x - y = -6$
 (2) $2x + 5y = -5$

16. (1) $x - y = -5$
 (2) $2x - 3y = -12$

17. (1) $x - y = 7$
(2) $4x + 4y = -12$

18. (1) $x - y = 5$
(2) $4x + 2y = 8$

In exercises **19–22**, determine whether the given ordered triple is a solution of the associated system. *The triple is a solution of the system if and only if it is a solution of every equation in the system.*

19. (1) $x + y + z = 9$
(2) $2x + 3y - z = 20$
(3) $3x - y + z = 5$, $(3, 5, 1)$

20. (1) $x + 2y - 2z = -6$
(2) $2x - 5y - 3z = 0$
(3) $5x + 3y - 5z = -8$, $(2, -1, 3)$

21. (1) $2x - 2y + 5z = 7$
(2) $3x + y - 2z = -25$
(3) $4x + y + 3z = -8$, $(-6, 3, 5)$

22. (1) $x - 2y + 3z = -14$
(2) $-2x - y + 5z = -15$
(3) $4x + 3y - 3z = 11$, $(-1, 2, -3)$

In exercises **23** and **24**, determine whether the given **ordered four-tuple** is a solution of the associated system. The four-tuple is a solution of the system if and only if it is a solution of every equation in the system. The form of the numbers is (a, b, c, d).

23. (1) $a - 2b + c - d = 9$
(2) $a + b - 2c + d = -7$
(3) $2a - b + 2c - d = 13$
(4) $3a + 2b - c + d = -1$, $(2, -1, 3, -2)$

24. (1) $2a + 3b - 6c - d = 7$
(2) $4a - 3b + 2c + d = 2$
(3) $6a + 9b - 4c - 2d = 2$
(4) $8a + 6b + 10c + 3d = -7$, $\left(\dfrac{1}{2}, \dfrac{-1}{3}, \dfrac{-3}{2}, 2\right)$

SECTION 7-4. The Addition Method of Solving a System of Linear Equations in Two Variables

KEY TOPICS IN THIS SECTION

1. The addition property of equality

2. The addition method of solving an independent and consistent system of linear equations in two variables

3. Inconsistent systems and the addition method

4. Dependent systems and the addition method

The substitution method of solving a system of linear equations in x and y is easier to use if at least one of the variables has a coefficient of one. If both variables have coefficients different from one in both equations, then the addition method is usually easier to use. This section covers this method of solving a system of linear equations.

The Addition Property of Equality

The addition property of equality has been used in previous chapters to solve equations.

> **Addition property of equality**
> If $A = B$, then $A + C = B + C$.

The substitution property of equality provides us with a slightly different form for the addition property. That is, suppose it is known that $C = D$. Then C can be added to the left side of the given equation and D can be added to the right side.

> **Alternate form of the addition property of equality**
> If $A = B$ and $C = D$, then $A + C = B + D$.

Example 1. If $5x - 3y = 8$ and $3y = 17$, find the value of x.

Solution.   **Discussion.** The given equations have the form

$$A = B \text{ and } C = D.$$

The two equations can therefore be added to form

$$A + C = B + D.$$

$(5x - 3y) + 3y = 8 + 17$	Add the two equations.
$5x = 25$	$-3y + 3y = 0$
$x = 5$	Divide both sides by 5.

Thus, x is 5 for the conditions specified by the given pair of equations.

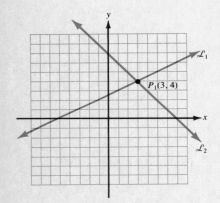

Figure 7-7. Graphs of (1) and (2).

Consider the following system of linear equations in x and y:

(1) $x - 2y = -5$

(2) $x + y = 7$

In Figure 7-7 lines $\mathscr{L}_1$ and $\mathscr{L}_2$ are graphs of (1) and (2) respectively. Since $P_1(3, 4)$ is the point of intersection of $\mathscr{L}_1$ and $\mathscr{L}_2$, the solution of the system is (3, 4).

The addition property of equations states that equations (1) and (2) can be added to form another equation, (3).

$$
\begin{array}{ll}
(1) & x - 2y = -5 \\
(2) & x + y = 7\,(+) \\
\hline
(3) & 2x - y = 2
\end{array}
$$

$x + x = 2x \longrightarrow (3)$

$-2y + y = -y$

$-5 + 7 = 2$

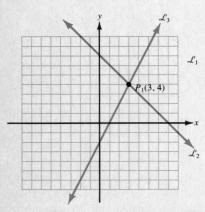

Figure 7-8 Graphs of (1), (2), and (3).

In Figure 7-8 the graph of (3) is $\mathscr{L}_3$. Notice that $\mathscr{L}_3$ also passes through $P_1(3, 4)$. Thus, (3, 4) is the solution of a system of equations formed by (1) and (3), or (2) and (3).

$$(1) \quad x - 2y = -5 \qquad (2) \quad x + y = 7$$
$$(3) \quad 2x - y = 2 \qquad (3) \quad 2x - y = 2$$

The solution for both systems is (3, 4).

This discussion shows that adding two equations of a system will yield an equation that is satisfied by any solutions of the given system. However, to use the addition property of equality to solve a system, we want to *eliminate* one of the variables when the equations are added. *This task will be accomplished whenever the coefficients of one of the variables in the system are equal in absolute value but opposite in sign.*

Consider again equations (1) and (2) graphed in Figures 7-7 and 7-8. Before adding the equations, let's multiply equation (2) by 2 to make the coefficients of y equal in absolute value ($|2| = |-2|$), but opposite in sign.

$$(2) \quad 2(x + y) = 2(7) \qquad (2^*)\ 2x + 2y = 14$$

Now adding (1) and (2*),

$$(1) \quad x - 2y = -5$$
$$(2^*) \quad \underline{2x + 2y = \ \ 14\ (+)}$$
$$(4) \quad 3x \qquad\ = \ \ \ 9 \qquad \text{An equation in only } x.$$
$$x = \ \ \ 3 \qquad \text{Divide both sides by 3.}$$

A graph of equation (4) would be a vertical line that also passes through $P_1(3, 4)$. Now that x is known, we can replace x by 3 in either equation (1) or equation (2) to find the value of y.

$$(1) \quad 3 - 2y = -5 \qquad \text{or} \qquad (2) \quad 3 + y = 7$$
$$y = \ \ \ 4 \qquad\qquad\qquad\qquad y = 4$$

Thus, the solution is (3, 4), as seen in the graphs of $\mathcal{L}_1$ and $\mathcal{L}_2$.

The Addition Method of Solving an Independent and Consistent System of Linear Equations in Two Variables

The following six steps provide a recommended procedure for solving a system of linear equations in two variables using the addition method.

To solve an independent and consistent system of linear equations in x and y by the addition method,

Step 1. If necessary, use the multiplication property of equality to form a system in which the coefficients of x (or y) are opposites.

Step 2. Add the two equations.

Step 3. Solve the equation from Step 2.

Step 4. Replace the variable in (1) or (2) by the root obtained in Step 3 and solve.

Step 5. State the solution of the system.

Step 6. Check the solution in (1) and (2).

Example 2. Solve and check:

(1) $4x + 3y = 18$

(2) $5x - 9y = 48$

Solution.

Discussion. The y-terms in equations (1) and (2) are added and subtracted respectively. Therefore, if the coefficients of these terms were the same, adding (1) and (2) would eliminate this variable.

Step 1. The least common multiple of 3 and 9 is 9. Therefore (1) is multiplied by 3.

Step 2. (1) $3(4x + 3y) = 3(18)$ Multiply both sides by 3.

(1*) $12x + 9y = 54$

Add (1*) and (2).

(1*) $12x + 9y = 54$

(2) $\underline{5x - 9y = 48\ (+)}$
$17x \qquad = 102$

Step 3. $x = 6$ Divide both sides by 6.

Step 4. (1) $4(6) + 3y = 18$ Replace x by 6 in (1).

$3y = -6$ Subtract 24 from both sides.

$y = -2$ Divide both sides by 3.

Step 5. The solution is $(6, -2)$.

Step 6. The check is omitted.

Example 3. Solve and check:

(1) $10a + 3b = 25$

(2) $15a - 4b = 12$

Solution.

Discussion. Both equations must be multiplied by some constant to get the coefficients of a or b to be opposites. Since the b-terms are added and subtracted, we arbitrarily eliminate these terms first.

Step 1. The least common multiple of 3 and 4 is 12. Therefore, we multiply equation (1) by 4 and equation (2) by 3.

(1) $4(10a + 3b) = 4(25)$ Multiply (1) by 4.

(1*) $40a + 12b = 100$

(2) $3(15a - 4b) = 3(12)$ Multiply (2) by 3.

(2*) $45a - 12b = 36$

Step 2. Add (1*) and (2*).

(1*) $40a + 12b = 100$

(2*) $\underline{45a - 12b = 36\ (+)}$
$85a \qquad = 136$

Step 3. $$a = \frac{136}{85}$$ Divide both sides by 85.

$$= \frac{8}{5}$$ Reduce the fraction.

Step 4. (1) $10 \cdot \dfrac{8}{5} + 3b = 25$ Replace a by $\frac{8}{5}$ in (1).

$$16 + 3b = 25$$

$$3b = 9$$

$$b = 3$$

Step 5. The solution is $(\frac{8}{5}, 3)$.

Step 6. The check is omitted.

Inconsistent Systems and the Addition Method

In Step 2 of the solving process, both variables in the system may be eliminated. If the resulting number statement is *false,* then the system is *inconsistent.*

Example 4.   Solve and check:

(1) $4s - 9t = 27$

(2) $-8s + 18t = 36$

Solution.   **Discussion.** To eliminate the s-terms, we can multiply equation (1) by 2.

Step 1. (1) $2(4s - 9t) = 2(27)$

(1*) $8s - 18t = 54$

Step 2. (1*) $8s - 18t = 54$

(2) $\underline{-8s + 18t = 36\,(+)}$

$0 = 90,$ false

Both variables have been eliminated and the resulting number statement is *false.* Thus the system is *inconsistent.* Graphs of equations (1) and (2) would yield *parallel lines* when drawn on the same axes.

Dependent Systems and the Addition Method

As previously stated, both variables in a system may be eliminated when the adjusted equations are added. If the resulting number statement is *true,* then the system is *dependent.*

Example 5.   Solve and check:

(1) $14x - 20y = 46$

(2) $\dfrac{x + 1}{10} = \dfrac{y + 3}{7}$

Solution. **Discussion.** Equation (2) must first be written in standard form before attempting to solve the system.

$$70\left(\frac{x+1}{10}\right) = 70\left(\frac{y+3}{7}\right) \qquad \text{The LCM is 70.}$$

$$7(x+1) = 10(y+3) \qquad 70 \div 10 = 7 \text{ and } 70 \div 7 = 10.$$

$$7x + 7 = 10y + 30 \qquad \text{Remove parentheses.}$$

$$(2) \quad 7x - 10y = 23 \qquad \text{Subtract 7 and } 10y \text{ from both sides.}$$

The system now appears as follows:

$$(1) \quad 14x - 20y = 46$$

$$(2) \quad 7x - 10y = 23$$

To eliminate the x-terms, multiply both sides of (2) by -2.

$$(2^*) \quad -14x + 20y = -46$$

$$(1) \quad \underline{14x - 20y = 46 \,(+)}$$

$$\quad 0 + 0 = 0, \text{ true}$$

Both variables have been eliminated and the resulting number statement is *true*. Thus the system is *dependent*. Graphs of (1) and (2) would yield only one line.

SECTION 7-4. Practice Exercises

In exercises **1–34**, solve using the addition method.

[Example 1] **1.** (1) $4x - 3y = -19$
        (2) $3y = 5$

2. (1) $8x - 5y = 6$
    (2) $5y = 8$

3. (1) $2u - 8v = -3$
    (2) $8v = 21$

4. (1) $4u - 9v = -11$
    (2) $9v = 31$

5. (1) $6m + 4n = -7$
    (2) $-4n = 11$

6. (1) $15m + 4n = 21$
    (2) $-4n = 19$

7. (1) $11x + 6y = 5$
    (2) $-11x = 13$

8. (1) $13x - 7y = -22$
    (2) $-13x = 8$

[Examples 2 and 3] **9.** (1) $2a + b = 12$
    (2) $3a - b = 13$ $(5, 7)$

10. (1) $10a + b = -21$
    (2) $2a - b = -15$

11. (1) $x + 6y = 7$
    (2) $-x + 4y = 13$

12. (1) $x - 8y = 33$
    (2) $-x + 3y = -18$

13. (1) $a + 3b = 9$
    (2) $4a - 3b = -24$

14. (1) $3a + 9b = 0$
    (2) $3a + 10b = 2$

15. (1) $s + t = 5$
    (2) $5s - 2t = 4$

16. (1) $s - t = 3$
    (2) $4s + 5t = 21$

17. (1) $m - 2n = 0$
(2) $6m - 13n = 5$

18. (1) $3m - 5n = -22$
(2) $5m + n = -18$

19. (1) $2a - 3b = 43$
(2) $3a + 2b = 6$

20. (1) $3a - 2b = 41$
(2) $2a + 3b = -3$

21. (1) $5x + 6y = -2$
(2) $8x - 9y = -28$

22. (1) $5x + 12y = 21$
(2) $3x - 8y = 24$

23. (1) $10u + 3v = -7$
(2) $-15u + 4v = -32$

24. (1) $4u + 9v = 0$
(2) $-8u + 5v = 23$

25. (1) $7a + 2b = -40$
(2) $-3a + 5b = 23$

26. (1) $9a + 2b = 92$
(2) $10a + 3b = 110$

[Examples 4 and 5] **27.** (1) $3a + 7b = 10$
(2) $15a + 35b = 49$

28. (1) $8a - 5b = 3$
(2) $32a - 20b = 10$

29. (1) $-6x + 5y = 9$
(2) $24x - 20y = -36$

30. (1) $7x - 10y = 6$
(2) $-70x + 100y = -60$

31. (1) $8m - 5n = 4$
(2) $4m + 15n = 9$

32. (1) $6m + 7n = 5$
(2) $9m + 28n = 45$

33. (1) $3a + 2b = 2$
(2) $12a + 8b = 8$

34. (1) $3a + 2b = 4$
(2) $15a + 10b = 10$

SECTION 7-4. Ten Review Exercises

In exercises **1–10**, solve each problem.

1. Simplify:

$$(8x^2 - xy + 9y^2) - 3(2xy + 4x^2 - y^2) + 2(3xy + 2x^2 - 6y^2)$$

2. Evaluate the expression in exercise **1** for $x = 5$ and $y = -2$.

3. Factor: $3t^2 - 5t - 8$.

4. Evaluate the expression of exercise **3** for $t = \frac{8}{3}$.

5. Simplify: $\dfrac{m^2 - 2m - 3}{2m^2 - 5m - 3} \div \dfrac{2m^2 - 4m}{2m^2 - 3m - 2}$

6. Evaluate the expression in exercise **5** for $m = 3$.

7. The sum of two numbers is 7. If 4 is subtracted from two times the larger, the result is the same as when 45 is added to five times the smaller. Find both numbers.

8. The sum of two integers is -3 and their product is -40. Find both numbers.

9. The area of a rectangle is 60 square feet. The length is 7 feet longer than the width. Find the length and width.

10. Solve and graph: $2(a - 4) + 6 \leq 5a - 2(6 - a)$.

SECTION 7-4.   Supplementary Exercises

In exercises **1–10**, solve using the addition method.

1. (1) $a + 0.3b = 2.5$
　　(2) $1.5a - 0.4b = 1.2$

2. (1) $0.4a + 1.1b = 0.72$
　　(2) $1.2a + 2.2b = 3.04$

3. (1) $\dfrac{1}{5}x - \dfrac{4}{5}y = -5$

　　(2) $\dfrac{1}{3}x + y = 8$

4. (1) $2x + \dfrac{3}{2}y = -6$

　　(2) $\dfrac{5}{2}x - \dfrac{9}{4}y = -24$

5. (1) $1.2x - 1.5 = 0$
　　(2) $0.8y + 3.5 = 0$

6. (1) $4.5y + 8.1 = 0$
　　(2) $3.2x - 8.0 = 0$

7. (1) $2a - \dfrac{1}{3}b = \dfrac{11}{9}$

　　(2) $\dfrac{9}{2}a + \dfrac{9}{2}b = 1$

8. (1) $\dfrac{4}{3}a + 5b = 1$

　　(2) $a - \dfrac{5}{6}b = -2$

9. (1) $2m + 2.1n = -1.32$
　　(2) $0.8m - 1.5n = 2.28$

10. (1) $0.8m - 2.1n = -7.32$
　　(2) $-1.2m + 3.5n = 12.80$

In exercises **11–16**,

a. Add equations (1) and (2). Label this equation (4).

b. Add equations (2) and (3). Label this equation (5).

c. Use the addition method to solve the system composed of equations (4) and (5).

d. Replace the variables in equation (1) with the values obtained in part **c** and solve for the remaining variable.

e. Check the ordered triple from parts **c** and **d** in equations (2) and (3).

f. State the solution of equations (1), (2), and (3) as an ordered triple (x, y, z).

11. (1) $x + 2y + z = 2$
　　(2) $-x - y + z = -6$
　　(3) $x - y + 2z = -7$

12. (1) $2x - 4y + 2z = 12$
　　(2) $-2x + y + z = 6$
　　(3) $2x + y - 2z = 7$

13. (1) $x - y + 3z = -1$
　　(2) $-2x + y - 4z = 2$
　　(3) $3x - y + 6z = -1$

14. (1) $2x + 2y - 4z = 2$
　　(2) $4x - 2y + z = 4$
　　(3) $4x + 2y - 2z = 16$

15. (1) $3x + y - z = 4$
　　(2) $5x - 2y + z = 6$
　　(3) $6x + 2y - z = 5$

16. (1) $-4x - 6y + 2z = 10$
　　(2) $2x + 4y - 2z = -8$
　　(3) $2x - 5y + 2z = 4$

In exercises **17–20**, solve.

a. Add equations (1) and (2).

b. Solve the equation in part **a** for x, obtaining two values: x_1 and x_2.

c. Replace x by x_1 and x_2 in equation (1), obtaining two values for y: y_1 and y_2.

d. Write the solutions of the system as (x_1, y_1) and (x_2, y_2).

17. (1) $y + 8 = x^2 + 2x$
   (2) $-y - 3 = 2x$

18. (1) $9 - y = x^2$
   (2) $-7 + y = x$

19. (1) $y = x^2 + 2x$
   (2) $-y + 12 = 2x$

20. (1) $y + 3 = x^2 + 4x$
   (2) $-y - 2 = -4x$

SECTION 7-5.   Applied Problems

**KEY TOPICS
IN THIS SECTION**

1. Solving number problems

2. Solving geometry problems

3. Solving mixture problems

4. Solving uniform motion problems

5. Solving age problems

In this section, we will use two variables to solve each problem in the examples and practice exercises. As a result, each problem will require two equations that will form a system of linear equations in two variables. The solution of the system should be the answer to the problem.

Solving Number Problems

In number problems, two different numbers are unknown. Each number is represented by a different variable. Sufficient information is given to write two equations involving both variables. These equations form a system of linear equations. The solution of the system is the unknown numbers.

Example 1.   The sum of two numbers is 7. When the smaller number is subtracted from the larger, the difference is 43. Find both numbers.

Solution.   Let a represent the larger number.

Let b represent the smaller number.

"The sum of the two numbers is 7" can be written as follows:

(1) $a + b = 7$

"When the smaller number is subtracted from the larger, the difference is 43" can be written as follows:

(2) $a - b = 43$

Equations (1) and (2) form the following system:

(1) $a + b = 7$

(2) $a - b = 43$

The addition method can be used as follows to solve the system:

(1) $a + b = 7$
(2) $a - b = 43 \,(+)$

$\overline{ 2a = 50}$

$ a = 25$ The larger number is 25.

(1) $25 + b = 7$ Replace a by 25 in (1).

$ b = -18$ The smaller number is -18.

Check: $25 + (-18) = 7$, true

$ 25 - (-18) = 43$, true

Thus, the numbers are 25 and -18.

Example 2.   The sum of three times the first of two numbers and two times the second is 84. When the second number is subtracted from two times the first, the difference is -7. Find both numbers.

Solution.   Let a represent the first number.

Let b represent the second number.

"The sum of three times the first ... and two times the second is 84" is written as follows:

(1) $3a + 2b = 84$

"... the second number is subtracted from two times the first, the difference is -7" is written

(2) $2a - b = -7.$

Equations (1) and (2) yield the following system:

(1) $3a + 2b = 84$

(2) $2a - b = -7$

Using the substitution method, solve equation (2) for b:

(2) $2a = b - 7$ Add b to both sides.

$ 2a + 7 = b$ Add 7 to both sides.

Replace b by $2a + 7$ in equation (1):

(1) $3a + 2(2a + 7) = 84$

$ 3a + 4a + 14 = 84$

$ 7a + 14 = 84$

$ 7a = 70$

$ a = 10$

Replace a by 10 in equation (2):

(2) $2(10) - b = -7$

$$-b = -27$$

$$b = 27$$

Checking,

(1) $3(10) + 2(27) = 84$

$$30 + 54 = 84, \text{ true}$$

(2) $2(10) - 27 = -7$

$$20 - 27 = -7, \text{ true}$$

The numbers are 10 and 27.

Solving Geometry Problems

A geometry problem usually gives information about some geometric figure. In this section the problems will require the use of two variables. It is recommended that a figure be drawn and the information be labeled on the figure.

Example 3. The difference between the length and width of a rectangular-shaped field is 15 meters. The perimeter of the field is 110 meters. Find the length and width of the field.

Solution. Let l represent the length.

Let w represent the width.

The field is drawn in Figure 7-9. The sides are labeled l and w.

"The difference between the length and width . . . is 15 meters" is written

(1) $l - w = 15.$

"The perimeter . . . is 110 meters" is written

(2) $2l + 2w = 110.$

To use the addition property, multiply equation (1) by 2.

(1) $2(l - w) = 2(15)$

(1*) $2l - 2w = 30$

(2) $2l + 2w = 110 \, (+)$

$$\overline{ 4l = 140}$$

$$l = 35$$

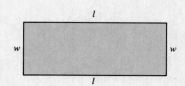

Figure 7-9. The perimeter is 110 meters.

Replace l by 35 in equation (1):

(1) $35 - w = 15$

$w = 20$

Checking,

(1) $35 - 20 = 15$, true

(1) $2(35) + 2(20) = 110$

$70 + 40 = 110$, true

The length is 35 meters and the width is 20 meters.

Solving Mixture Problems

In a **mixture problem,** two or more different kinds of objects are mixed. The given information usually states something about the *quantity* and *value* of items mixed. A table is usually a good way to organize the information.

Example 4.   Last Tuesday, when Bob Metheny closed Al's Supermarket Station, he noted that the cash box held $7.75 worth of dimes and quarters. He also noted that the number of dimes was one more than twice the number of quarters. How many dimes and quarters were in the cash box that night?

Solution.   Let d represent the number of dimes.

Let q represent the number of quarters.

Coins	Quantity	Value
Dimes	d dimes	$0.10d$
Quarters	q quarters	$0.25q$

Equation (1) is based on the *quantity* of coins:

(1) $d = 2q + 1$ "The number of dimes was one more than twice the number of quarters."

Equation (2) is based on the *value* of the coins:

(2) $0.10d + 0.25q = 7.75$ "The cash box held $7.75 worth of dimes and quarters."

Equations (1) and (2) yield the following system:

(1) $d = 2q + 1$

(2) $0.10d + 0.25q = 7.75$

Replace d by $2q + 1$ in (2):

(2) $0.10(2q + 1) + 0.25q = 7.75$

$0.20q + 0.10 + 0.25q = 7.75$

$0.45q = 7.65$

$$q = 17 \qquad \text{There were 17 quarters.}$$

$$d = 2(17) + 1$$

$$= 35 \qquad \text{There were 35 dimes.}$$

Check: $35(0.10) + 17(0.25) = 3.50 + 4.25 = 7.75$, true

Bob counted 35 dimes and 17 quarters in the cash box.

Example 5. Joyce Pierscinski works in a medical laboratory. To fill an order, she needed 140 liters of liquid that contained 44 liters of alcohol. The laboratory had two drums containing different percents of alcohol. Drum A had 25% alcohol and Drum B had 40% alcohol. How many liters of liquid from Drums A and B did Joyce mix to get the necessary percentage of alcohol for the order?

Solution. **Discussion.** The "value" of the items mixed is the *amount of alcohol* in the liquid taken from each drum.

Items	Quantity	Value
Liquid from Drum A	x liters	$x(25\%) = 0.25x$
Liquid from Drum B	y liters	$y(40\%) = 0.40y$

(1) $\quad x + y \qquad\quad = 140 \qquad$ liquid + liquid = liquid

(2) $\quad 0.25x + 0.40y = 44 \qquad$ alcohol + alcohol = alcohol

Solve (1) for x in terms of y.

(1) $\quad x + y = 140$

$$x = 140 - y$$

Replace x by $(140 - y)$ in equation (2).

(2) $\quad 0.25(140 - y) + 0.40y = 44$

$$35 - 0.25y + 0.40y = 44$$

$$0.15y = 9$$

$$y = 60$$

Replace y by 60 in equation (1).

(1) $\quad x + 60 = 140$

$$x = 80$$

Checking,

(1) $\quad 80 + 60 = 140$, true

(2) $\quad 0.25(80) + 0.40(60) = 44$

$$20 + 24 = 44, \text{ true}$$

Joyce mixed 80 liters from Drum A with 60 liters from Drum B.

Solving Uniform Motion Problems

The formula used in a **uniform motion problem** is

$$d = r \cdot t. \qquad \text{distance} = \text{rate} \cdot \text{time}$$

A typical problem will give information about two moving objects. The information will compare the distances, rates, or times traveled by the objects. Sufficient information will be given to form two equations using the formula $d = r \cdot t$.

Example 6. Lucinda Rodriquez left Davis in a Mack truck heading east on an interstate highway, averaging 60 mph. Tony Silvester left Davis one hour later on the same highway and in the same direction as Lucinda, averaging 68 mph. How many miles east of Davis were they when Tony overtook Lucinda?

Solution.   **Discussion.** A line diagram describing the conditions of motion for the two vehicles is useful in setting up the equation. If we let t represent the time Lucinda was on the road, then $t - 1$ represents the time Tony was on the road, since he left Davis one hour later.

Davis   **Car and truck meet**

Lucinda: $d = 60t$

Tony: $d = 60(t - 1)$

Vehicle	Time	Rate	Distance
Truck	t hours	60 mph	d miles
Car	$(t - 1)$ hours	68 mph	d miles

The following equations form a system in d and t:

(1) $d = 60t$

(2) $d = 68(t - 1)$

Replace d by $60t$ in equation (2):

(2) $60t = 68(t - 1)$

$$60t = 68t - 68$$

$$68 = 8t$$

$$t = 8.5 \text{ hours}$$

Thus, Lucinda was on the road 8.5 hours and Tony was on the road $(8.5 - 1)$ or 7.5 hours when they met. The distance each vehicle traveled when they met was

$d = 60(8.5) = 510$ miles   For the truck

$d = 68(7.5) = 510$ miles   For the car

Therefore, Lucinda and Tony were 510 miles east of Davis when they met.

Solving Age Problems

In an *age problem* the ages of two people or two objects are known. The age of each person or object is represented by a different variable. The given problem contains enough information to write two different equations involving both variables. The solution of the system formed by these equations is the unknown ages.

Example 7. Andy and Beth are brother and sister. The sum of their current ages is 35 years. In 5 years Andy's age will be two times Beth's age at that time. How old are Andy and Beth now?

Solution. **Discussion.** A table is also a useful way of organizing the information for an age problem.

Person	Current age	Age in 5 years
Andy	a years	$(a + 5)$ years
Beth	b years	$(b + 5)$ years

"The sum of their current ages is 35 years" is written

(1) $a + b = 35$.

"In five years Andy's age will be two times Beth's age at that time" is written

(2) $a + 5 = 2(b + 5)$.

Using the substitution method, solve equation (1) for a in terms of b:

(1) $a + b = 35$

$a = 35 - b$

Replace a by $(35 - b)$ in (2):

(2) $(35 - b) + 5 = 2(b + 5)$

$35 - b + 5 = 2b + 10$

$40 = 3b + 10$

$3b = 30$

$b = 10$

Replace b by 10 in equation (1):

(1) $a + 10 = 35$

$a = 25$

Thus, Andy is currently 25 years old and Beth is 10 years old.

SECTION 7-5.   Practice Exercises

In exercises **1–8**, find the two numbers described.

[Examples 1 and 2]

1. The sum of the first number and twice the second is 21. The difference between three times the first number and the second number is 7.

2. The first number is one more than twice the second. The sum of twice the first and three times the second is 51.

3. The sum of the numbers is −1. The difference between the numbers is 19.

4. The second number is three times the first. The difference between five times the first and eight times the second is 38.

5. The sum of three times the first number and four times the second is 2. The difference between nine times the first and eight times the second is 1.

6. Twice a first number added to five times a second is 0. Ten times the difference between the two numbers is 21.

7. Six times a first number added to three times a second is 0. Twelve times the difference between the two numbers is −42.

8. Five times the sum of two numbers is 4. The difference between the two numbers is 0.

In exercises **9–16**, find the dimensions of each rectangle described.

[Example 3]

9. The length of a picture frame is 2 inches more than the width. The perimeter of the frame is 36 inches.

10. The height of a door frame is 1 foot more than twice the width. The perimeter of the door frame is 20 feet.

11. Twice the width of a workbook added to its length is 28 inches. The perimeter of the book is 39 inches.

12. The length of the playing portion of an athletic field is 60 feet more than twice the width. The difference between the length and the width is 180 feet.

13. The length of a tapestry is 4 feet less than twice the width. The difference between the length and the width is 8 feet.

14. A pasture for horses is enclosed by 1,280 meters of fencing. Three times the width of the pasture is 120 meters more than twice the length.

15. The difference between the length and the width of a greeting card is 5 centimeters. The sum of three times the width and twice the length is 80 centimeters.

16. Twice the sum of the width and the length of a desk top is 102 inches. The difference between three times the width and twice the length is 8 inches.

In exercises **17–24**, find the unknown values.

[Examples 4 and 5]

17. A child's bank has a collection of 100 dimes and quarters. The value of the coins in the bank is $15.25. How many dimes and quarters are in the bank?

18. A coin purse contains 47 nickels and dimes. The coins are worth $3.75. How many of each kind of coin are in the purse?

19. A purchase of 45 stamps cost $9. The purchase consisted of 15¢ and 30¢ stamps. How many of each kind of stamp were bought?

20. The price of apples is 25¢ per pound, and the price of pears is 39¢ per pound. If the combined weight of apples and pears bought is 12 pounds and the total price is $3.70, how many pounds of apples and pears are purchased?

21. Linda bought $10\frac{1}{2}$ yards of two kinds of fabric. One fabric cost $2.25 per yard, and the other cost $2.90 per yard. The total cost was $26.55. How many yards of each fabric did she buy?

22. The attendance at a college football game was 5,400. Tickets cost $2 for students and $3 for general admission. Gate receipts were $12,600. How many of each kind of ticket were sold?

23. Bottle A contains liquid that is 20% alcohol, and bottle B contains liquid that is 60% alcohol. How much liquid from each bottle is needed to make 140 liters of liquid that is 35% alcohol?

24. The liquid in bottle A contains 80% alcohol, and the liquid in bottle B contains 15% alcohol. Some liquid from each bottle is mixed to yield 100 liters of liquid that contains 54 liters of alcohol. How many liters of liquid from each bottle are in the mixture?

For exercises **25–32**, find the values of the unknown quantities.

[Example 6]   **25.** A bus left Fresno headed in the same direction as a freight train but two hours later. The train averaged 30 mph, and the bus averaged 50 mph. How far from Fresno did the bus catch up with the train?

26. A hiker left the campground at Tuolumne Meadows walking along the road at 4 mph. A cyclist left the same campground two hours later on the same road at 12 mph. How far from the starting point did the two meet?

27. Two cars left Youngstown at the same time traveling in opposite directions. Car A averaged 43 mph, and car B averaged 52 mph. In how many hours were the two cars 380 miles apart?

28. Two ships left San Francisco at the same time. One headed north for Seattle and the other south for Los Angeles. Ship A moved at 18 knots (nautical miles per hour) and ship B at 24 knots. In how many hours were the ships 294 nautical miles apart?

29. A commercial airliner and a private plane left O'Hare International Airport at the same time headed for Atlanta. The speed of the airliner was 425 mph, and the private plane's speed was 250 mph. In how many hours was the airliner 350 miles ahead of the private plane?

30. Two cars left Denver headed for Kansas City. Car A averaged 44 mph, and car B averaged 58 mph. In how many hours was car B 112 miles ahead of car A?

31. A speed boat travels 60 miles up the Sacramento River against the current for three hours. The return trip with the current takes two hours. Find the speed of the boat in still water and the speed of the current in the river.

32. A private plane makes a trip of 1,500 miles in five hours with a tail wind, whereas a similar plane makes the same trip in six hours against the wind. Find the speed of the plane in still air and the speed of the wind.

In exercises **33–40**, find the ages of the individuals described.

[Example 7] **33.** Two sisters, Alice and Betty, have ages totaling 20 years. In 2 years, Betty will be twice as old as Alice. How old is each girl now?

34. Two sisters, Cindy and Darlene, have ages totaling 40 years. In 5 years, two times Cindy's age will equal three times Darlene's age. How old is each girl now?

35. Two brothers, Ed and Frank, have ages that total 76 years. Eight years ago, Ed was twice as old as Frank. How old is each now?

36. Two brothers, George and Harry, have ages that total 90 years. Ten years ago, three times George's age was 10 years more than Harry's age. How old is each now?

37. A wife and husband, Linda and Mark, have ages that total 80 years. Two years ago, two times Mark's age was 2 years more than three times Linda's age. How old is each now?

38. Nan is 20 years older than her son Oliver. In 2 years, her age will be six times the age of her son 3 years ago. How old are mother and son now?

39. Rusty is 37 years older than his daughter Sherry. In 2 years, he will be five times as old as Sherry was 1 year ago. How old are father and daughter now?

40. Terry and Vanessa are husband and wife. Three years ago, Terry was as old as Vanessa will be in 3 years. Two times Terry's age 2 years from now is the same as three times Vanessa's age 2 years ago. How old are Terry and Vanessa now?

SECTION 7-5. Ten Review Exercises

In exercises **1–5**, use points $P(10, -4)$ and $Q(-5, -1)$.

1. Find the slope of $\mathscr{L}_1$ that passes through P and Q.

2. Write an equation in $Ax + By = C$ form for $\mathscr{L}_1$.

3. Identify the x-intercept for $\mathscr{L}_1$.

4. Identify the y-intercept for $\mathscr{L}_1$.

5. Write an equation in $Ax + By = C$ form for $\mathscr{L}_2$ that passes through $R(3, 4)$ and is *perpendicular* to $\mathscr{L}_1$.

In exercises **6–10**, do the indicated operations.

6. $(7x + 3y)(7x - 3y)$

7. $(2t^2 - 5t + 1)(4t - 3)$

8. $(12m^3n + 28m^2n^2 - 56mn^3) \div 4mn$

9. $(13u^2 - v^2)^2$

10. $(10z^4 + 3z^3 - 3z - 2) \div (2z^2 + z + 1)$

SECTION 7-5. Supplementary Exercises

In exercises **1** and **2**, let

x represent the smallest of three numbers.

y represent the middle-sized number.

z represent the largest of the three numbers.

1. (1) The sum of the three numbers is 7.

(2) If the largest is subtracted from the sum of the smallest and two times the middle-sized number, the result is -8.

(3) If the largest is added to the sum of three times the smallest and the middle-sized number, the result is -3.

a. Write a system of three linear equations in x, y, and z for parts (1), (2), and and (3).
b. Form equation (4) by adding equations (1) and (2).
c. Form equation (5) by adding equations (2) and (3).
d. Solve equations (4) and (5) for x and y.
e. Use the values of x and y found in part **d** to find z.
f. What are the values of x, y, and z?

2. (1) The sum of the three numbers is 15.

(2) If the largest is subtracted from the sum of two times the smallest and three times the middle-sized number, the result is 1.

(3) If the largest is added to the difference between the smallest and the middle-sized number, the result is 1.

a. Write a system of three linear equations in x, y, and z for parts (1), (2), and (3).
b. Form equation (4) by adding equations (1) and (2).
c. Form equation (5) by adding equations (2) and (3).
d. Solve equations (4) and (5) for x and y.
e. Use the values of x and y found in part **d** to find z.
f. What are the values of x, y, and z?

In exercise **3**, let

x represent the cost of one pound of peaches.

y represent the cost of one pound of grapes.

z represent the unit cost of a watermelon.

3. (1) Mark Price bought two pounds of peaches, one pound of grapes, and a watermelon for $6.20.

(2) Larry Nance bought one pound of peaches, three pounds of grapes, and a watermelon for $7.60.

(3) Brad Daugherty bought two pounds of peaches, two pounds of grapes, and a watermelon for $7.30.

a. Write a system of three linear equations in x, y, and z for parts (1), (2), and (3).

b. Multiply equation (2) by -1 and add to equation (1) to form equation (4).

c. Multiply equation (2) by -1 and add to equation (3) to form equation (5).

d. Solve equations (4) and (5) for x and y.

e. Use the values of x and y found in part **d** to find z.

f. State the price of one pound of peaches, one pound of grapes, and one watermelon.

4. In exercise **4**, let

x represent the cost of one washcloth.

y represent the cost of one hand towel.

z represent the cost of one bath towel.

(1) Ron Harper bought one washcloth, two hand towels, and one bath towel for $20.00.

(2) Craig Ehlo bought two washcloths, three hand towels, and one bath towel for $26.75.

(3) Anne Williams bought three washcloths, two hand towels, and two bath towels for $33.25.

a. Write a system of three linear equations in x, y, and z for parts (1), (2), and (3).

b. Multiply equation (2) by -1 and add to equation (1) to form equation (4).

c. Multiply equation (2) by -2 and add to equation (3) to form equation (5).

d. Solve equations (4) and (5) for x and y.

e. Use the values of x and y found in part **d** to find z.

f. State the cost of one washcloth, one hand towel, and one bath towel.

SECTION 7-1.   Summary Exercises

1. Write a system of linear equations in x and y for the following: Four times a number x is three more than one-half a second number y. The sum of x and y is 12.

1. (1) _____

    (2) _____

In exercises **2–4**, determine whether the given ordered pair is a solution of the system.

2. (1) $3x + 3y = 9$

    (2) $3x - 6y = 0,$ $(2, 1)$

2. _____

3. (1) $3m - 2n = -6$

    (2) $9m + 20n = 216,$ $(4, 9)$

3. _____

4. (1) $10a + 9b = 8$

    (2) $12a - 18b = 0,$ $\left(\dfrac{1}{2}, \dfrac{1}{3}\right)$

4. _____

In exercises **5–7**, find the value of the variable so that the ordered pair is a solution of the system.

5. (1) $3x - 5y = 6$

    (2) $x - 2y = 1$, $(7, y)$

5. _____

6. (1) $24a - 15b = 35$

    (2) $5a - 3b = 7$, $\left(a, \dfrac{-7}{3}\right)$

6. _____

7. (1) $2p + 10q = 5$

    (2) $p - 4q = 7$, $(5, q)$

7. _____

In exercise **8**, determine whether the given ordered triple is a solution.

8. (1) $2x - 2y + z = 8$

    (2) $x + 3y - 3z = 10$

    (3) $3x - 5y + 4z = 12$, $(7, 5, 4)$

8. _____

SECTION 7-2. Summary Exercises

In exercises **1–7**, use the graphical method to find the solution of each system. If the lines are parallel, write *Inconsistent* for the answer. If the lines are the same, write *Dependent* for the answer.

1. (1) $x - y = 5$

(1) $x + 3y = 9$

2. (1) $2x + y = 0$

(2) $2x - y = 4$

1.

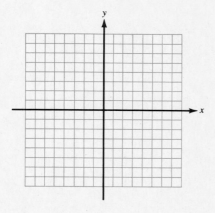

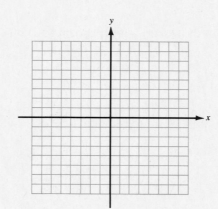

2. _____

3. (1) $x - 3y = -6$

(2) $2x - 6y = 0$

4. (1) $4x - 3y = 3$

(2) $x - 3y = -6$

3. _____

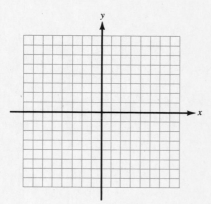

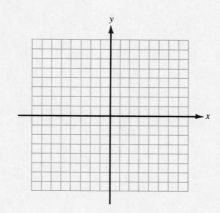

4. _____

485

5. (1) $4x + 5y = 10$

(2) $8x + 10y = 20$

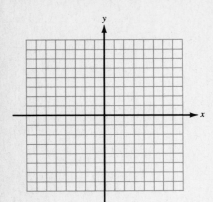

6. (1) $x = y - 5$

(2) $x + 4y - 20 = 0$

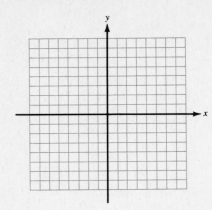

5. _____

6. _____

7. (1) $x + y = 1$

(2) $x = -3$

7. _____

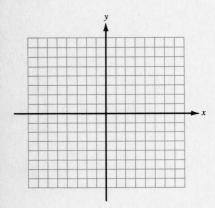

8. Without graphing, determine whether the system is independent and consistent, or inconsistent, or dependent.

(1) $2x - 5y = 70$

(2) $-4x + 10y = 20$

8. _____

Name _____

Date _____

Score _____

SECTION 7-3. Summary Exercises

Answer

In exercises **1–8**, use the substitution method to find the solution of each system.

1. (1) $7x - 9y = 64$

(2) $y = -4$

1. _____

2. (1) $r = 2s$

(2) $5r - 3s = 7$

2. _____

3. (1) $y = x - 1$

(2) $4x + 3y = 18$

3. _____

4. (1) $2m + n = -8$

(2) $3m - 5n = 14$

4. _____

5. (1) $x + y = 5$

   (2) $4x + 5y = 28$

5. _____

6. (1) $4a - b = 3$

   (2) $b = 4a$

6. _____

7. (1) $3s - 3 = 9t$

   (2) $s - 3t = 1$

7. _____

8. (1) $c - 4d = 10$

   (2) $5c + 8d = 1$

8. _____

SECTION 7-4. Summary Exercises

In exercises **1–8**, solve using the addition method.

1. (1) $4m - 2n = -11$

   (2) $2n = 19$

1. _____

2. (1) $x + y = 2$

   (2) $x - y = -6$

2. _____

3. (1) $s - 2t = 11$

   (2) $s - 3t = 18$

3. _____

4. (1) $4a - b = 3$

   (2) $5a + 2b = 20$

4. _____

5. (1) $2p - 3q = 18$

(2) $7p - 2q = -5$

5. _____

6. (1) $y - 2x = 6$

(2) $5y - 10x = 1$

6. _____

7. (1) $8a + 20b = 6$

(2) $4a + 10b = 3$

7. _____

8. (1) $6m - 12n = 7$

(2) $4m + 9n = -1$

8. _____

SECTION 7-5. Summary Exercises

Answer

In exercises **1–5**, solve.

1. The sum of two numbers is 9. When three times the smaller is subtracted from two times the larger, the difference is 93. Find both numbers.

   1. _____

2. The perimeter of a rectangular desk top is 24 feet. The difference between the length and width is 1 foot. Find the dimensions of the desk top.

   2. _____

3. A corner convenience store collected $18.55 in dimes and quarters for a local charity. It collected a total of 118 coins. How many dimes and how many quarters were collected?

   3. _____

491

4. An Amtrack train left Sacramento at 8:00 A.M. for Reno, Nevada, at an average speed of 45 mph. Fun Tours Bus #11 left Sacramento at 8:30 A.M. headed in the same direction as the train. How far from Sacramento did the bus overtake the train if the bus averaged 60 mph?

4. _____

5. Judy is nine years older than Jason, her brother. In ten years Judy's age will be nine years less than twice Jason's age. What are their current ages?

5. _____

CHAPTER 7.   Review Exercises

In exercises **1** and **2**, determine which ordered pairs are solutions.

1. (1) $3x + 4y = 14$

  (2) $\dfrac{4}{3}x + 3y = 5$

  a. $(2, 2)$

  b. $(6, -1)$

2. (1) $4t - 7u = 10$

  (2) $2t + 5u = 22$

  a. $(-6, 2)$

  b. $(6, 2)$

In exercises **3** and **4**, find the value of the variable so that the ordered pair is a solution of the system.

3. (1) $3x + 4y = 22$

  (2) $y = \dfrac{4x + 43}{5},$ $(-2, y)$

4. (1) $8a + 15b = 0$

  (2) $16a - 5b = -14,$ $\left(a, \dfrac{2}{5}\right)$

In exercises **5** and **6**, solve graphically.

5. (1) $2x + 5y = 25$

  (2) $6x - 5y = 15$

6. (1) $5x - 4y = -12$

  (2) $x - 4y = 4$

In exercises **7** and **8**, classify the equations as independent and consistent, or inconsistent, or dependent.

7. (1) $2x - 3y = 3$

  (2) $10x - 15y = 30$

8. (1) $8m + 9n = 90$

  (2) $-16m - 18n = -180$

In exercises **9** and **10**, use the substitution method to solve.

9. (1) $3s + 21 = 0$

  (2) $7s - 10t = -139$

10. (1) $y = 3x$

  (2) $4x + 3y = 39$

In exercises **11** and **12**, use the addition method to solve.

11. (1) $4x + 28y = 1$

  (2) $2x - 8y = -38$

12. (1) $10a - 15b = 3$

  (2) $3a + 2b = -3$

In exercises **13–17**, solve.

13. (1) $5p + 4q = 12$

  (2) $15p - 8q = -9$

14. (1) $2a - b = 2$

  (2) $9a + 18b = -11$

15. (1) $x = \dfrac{2}{3}y$

  (2) $x - 6y = -4$

16. (1) $5x - y = 3$

  (2) $-15x + 3y = -9$

*17. (1) $x + y = 10$

  (2) $1.00x + 1.75y = 10(1.27)$

In exercises **18–22**, solve the applied problems.

18. The sum of three times one number and two times a second is 15. The difference between twice the first and five times the second is 48. Find the two numbers.

* Ms. Glaston's candy problem.

19. The difference between three times the width and twice the length of the outside dimensions of a framed picture is 20 centimeters. The length is 50 centimeters less than twice the width. Find the length and width of the frame.

20. How many pounds of candy costing $1.20 per pound should be mixed with candy costing $1.65 per pound to make 150 pounds of candy costing $1.38 per pound?

21. Jill is eight years older than her brother Jack. In two years, two times Jill's age will be the same as five times Jack's age two years ago. How old is each now?

22. Two cars left Phoenix traveling in the same direction. Car *A* averaged 48 mph, and car *B* averaged 57 mph. How long after they left were the two cars 63 miles apart?

8

Radical Expressions

"*G*ood morning," Ms. Glaston said. "I'd like to begin today's class by having you all participate in a little exercise. First, I want everyone to pick a number between 1 and 20 inclusive. To help you remember the number, please write it down on a piece of paper. Is everyone ready?"

A mixture of nods and verbal *yes*'s signaled that everyone had selected a number.

"To keep a general record of what you will be doing," Ms. Glaston continued, "I'm going to let *n* stand for the possible numbers that you selected." She then wrote the following sentence on the board:

Let *n* stand for a number between 1 and 20 inclusive.

"O.K.," Ms. Glaston said, "now add 25 to your number and record the sum." She then wrote on the board

$$n + 25 \text{ is the sum.}$$

"Now," she continued, "subtract 25 from your sum and record the difference." Then she added to the expressions on the board

$$(n + 25) - 25 = n \text{ is the difference.}$$

"Based on the equation on the board," Ms. Glaston said, "I would guess that you all got your original number for the difference."

Kim Kimiecik raised her hand and said, "What's so surprising about that? Everytime we add and then subtract the same number from any number, we always get the original number."

"That's right, Kim," Ms. Glaston replied, "and for this reason we call addition and subtraction inverse operations. Recall that subtraction is defined in terms of addition." She then wrote on the board

$$a - b = a + (-b), \text{ where } -b \text{ is the opposite of } b.$$

"Many operations," Ms. Glaston continued, "in mathematics and also everyday life, have inverse operations. For example, we might claim that the inverse operation to 'putting on a pair of shoes' is the operation 'taking off the same pair of shoes.' In mathematics many operations such as addition and multiplication have inverse operations."

She then wrote on the board

$$n^2 = n \cdot n, \text{ or "the square of } n\text{."}$$

"In this chapter," she continued, "we will study the inverse operation to squaring a number, namely, taking the square root of a number. Unlike the simple process of squaring a number, we will find the operation of extracting a square root to be a little more complex. We will use one complete chapter of study time to explore the definition of the operation and several properties associated with it. Now let's begin our study of square roots."

SECTION 8-1.   Square Root of a Number

**KEY TOPICS
IN THIS SECTION**

1. The definition of a square root

2. The meaning of the square root symbol

3. Perfect squares and rational number square roots

4. Approximations of irrational number square roots

5. Square roots of negative numbers

   In this section "taking the square root of a number" is defined. As will be seen, this operation is defined in terms of "squaring a number."

The Definition of a Square Root

   The following is a definition of the square root of a number.

> **Definition 8.1.   Square root of a number**
> If $a^2 = b$, then a is a **square root** of b.

Example 1.   Verify the following:

   a. 7 is a square root of 49.

   b. -13 is a square root of 169.

   c. $\frac{3}{5}$ is a square root of $\frac{9}{25}$.

   d. -0.2 is a square root of 0.04.

Solution.   **Discussion.** Based on Definition 8.1, if $a^2 = b$, then a is a square root of b. Thus, to verify that the first numbers named are square roots of the second numbers, we simply square them.

   a. $7^2 = 49$; thus 7 is a square root of 49.

   b. $(-13)^2 = 169$; thus -13 is a square root of 169.

   c. $(\frac{3}{5})^2 = \frac{9}{25}$; thus $\frac{3}{5}$ is a square root of $\frac{9}{25}$.

   d. $(-0.2)^2 = 0.04$; thus -0.2 is a square root of 0.04.

The Meaning of the Square Root Symbol

   If b in Definition 8.1 is a positive number ($b > 0$), then there are two real numbers a such that $a^2 = b$.

$$3^2 = 9$$
$$(-3)^2 = 9$$

Therefore 3 and -3 are both square roots of 9.

The positive number square root of any positive number b is called the **principal square root.** Thus, 3 is the principal square root of 9.

> If $b > 0$, then
>
> **(i)** $\sqrt{b}$ is used to indicate the **principal square root** of b.
>
> **(ii)** the $\sqrt{}$ symbol is called the **radical sign.**
>
> **(iii)** b is called the **radicand.**
>
> **(iv)** $\sqrt{b}$ is the **positive number square root** of $b(\sqrt{b} > 0)$.
>
> **(v)** $-\sqrt{b}$ is the **negative number square root** of $b(-\sqrt{b} < 0)$.
>
> Note: $-\sqrt{b}$ can be read "the opposite of $\sqrt{b}$," and $-\sqrt{b}$ can be interpreted as $-1 \cdot \sqrt{b}$.

The expression $\sqrt{b}$ is read "the square root of b" or simply "radical b."

Example 2. Find the indicated square roots.

$$\textbf{a.}\ \sqrt{121} \qquad \textbf{b.}\ -\sqrt{225} \qquad \textbf{c.}\ \sqrt{\frac{81}{4}} \qquad \textbf{d.}\ -\sqrt{1.44}$$

Solution.   **Discussion.** To determine the square roots in parts **a–d**, look for positive numbers whose squares are the given radicands.

a. $11^2 = 121$; therefore $\sqrt{121} = 11$.

b. $15^2 = 225$; therefore $-\sqrt{225} = -15$.
  Note: $-\sqrt{225}$ is **the opposite** of $\sqrt{225}$.

c. $\left(\frac{9}{2}\right)^2 = \frac{81}{4}$; therefore $\sqrt{\frac{81}{4}} = \frac{9}{2}$.

d. $1.2^2 = 1.44$; therefore $-\sqrt{1.44} = -1.2$.

Perfect Squares and Rational Number Square Roots

Recall that real numbers can be divided into two large subsets: rational and irrational. The numbers in these subsets differ from each other in the way they are written using decimals. Furthermore, a rational number can always be written as an integer divided by a nonzero integer, but an irrational number cannot. If b is a positive number, then $\sqrt{b}$ and $-\sqrt{b}$ will either be rational numbers or irrational numbers.

Suppose in $\sqrt{b}$, the b is written as $\frac{m}{n}$, where m and n are positive integers. Furthermore, suppose $\frac{m}{n}$ is *reduced;* that is, m and n have no common factor other than one. Let us now replace b by $\frac{m}{n}$ as the radicand.

$$\sqrt{b} \text{ can be written as } \sqrt{\frac{m}{n}} \begin{array}{l} \rightarrow m \geq 0 \\ \rightarrow n > 0 \end{array}$$

In this form, we can now determine whether the square root is a rational number or an irrational number. Specifically, the square root will be rational if m and n are **perfect squares.** Otherwise the square root will be irrational. To study whether square roots of certain positive numbers are rational or irrational, we will first consider the definition of perfect square numbers.

Definition 8.2.   Perfect squares
If a is an integer greater than 0, then a^2 is a **perfect square.**

The following list contains the perfect squares from 1–625:

a	a^2	a	a^2	a	a^2	a	a^2	a	a^2
1	1	6	36	11	121	16	256	21	441
2	4	7	49	12	144	17	289	22	484
3	9	8	64	13	169	18	324	23	529
4	16	9	81	14	196	19	361	24	576
5	25	10	100	15	225	20	400	25	625

Example 3.   Determine whether each number is rational or irrational.

    a. $\sqrt{81}$ **b.** $\sqrt{50}$ **c.** $\sqrt{\dfrac{4}{9}}$

    d. $\sqrt{\dfrac{16}{31}}$ **e.** $\sqrt{\dfrac{3}{25}}$ **f.** $\sqrt{\dfrac{11}{17}}$

Solution.   **Discussion.** Check the number, or numbers, in each radicand with the list in the above table to determine whether they are perfect squares.

    a. $\sqrt{81}$ is a rational number because 81 is a perfect square. (Note: $\sqrt{81} = 9$, a rational number.)

    b. $\sqrt{50}$ is an irrational number because 50 is not a perfect square.

    c. $\sqrt{\frac{4}{9}}$ is a rational number because 4 and 9 are perfect squares. (Note: $\sqrt{\frac{4}{9}} = \frac{2}{3}$)

    d. $\sqrt{\frac{16}{31}}$ is an irrational number because 31 is not a perfect square. *Both numbers must be perfect squares.*

    e. $\sqrt{\frac{3}{25}}$ is an irrational number because 3 is not a perfect square. *Both numbers must be perfect squares.*

    f. $\sqrt{\frac{11}{17}}$ is an irrational number because 11 and 17 are not perfect squares.

Approximations of Irrational Number Square Roots

If $\sqrt{b}$ is an irrational number, then an **approximate value** for the number can be determined. Table 1 on the inside cover of the text contains the squares and square roots of the integers from 1–100. The decimal approximations of the irrational number square roots are rounded to three decimal places.

Example 4.   Use Table 1 to approximate the indicated square roots.

    a. $\sqrt{67}$

    b. $-\sqrt{35}$

Solution.   **Discussion.** Locate the radicands in the column headed **n**. The approximate square root is to the right in the column headed $\sqrt{n}$.

a. $\sqrt{67} \approx 8.185$

b. $-\sqrt{35} \approx -5.916$

A calculator with a square root key can also be used to approximate irrational number square roots. The approximations will be rounded to a specified number of decimal places.

Example 5.   Approximate to three decimal places.

a. $\sqrt{1,480}$

b. $-\sqrt{29.7}$

Solution.   **Discussion.** To use most calculators, simply enter the number under the radical sign, then press a key marked $\boxed{\sqrt{}}$. For some calculators it is necessary to press an $\boxed{\text{INV}}$ key, or $\boxed{\text{2nd}}$ key, and then the square root key. If you have any questions, consult the calculator manual or ask your instructor.

a. $\sqrt{1480} \approx 38.47076812\ldots$

≈ 38.471

b. $-\sqrt{29.7} \approx -5.449770637\ldots$

≈ -5.450

Example 6.   Use a calculator to approximate to three decimal places.

$$\frac{\sqrt{34} + \sqrt{10}}{5}$$

Solution.   **Discussion.** In the rule for order of operations, finding the square root has the same priority as raising to powers. Thus, the two square roots are approximated before the addition. *Furthermore, since a calculator is being used, only the final answer will be rounded.*

$$\frac{\sqrt{34} + \sqrt{10}}{5}$$

$$\approx \frac{5.8309\ldots + 3.1622\ldots}{5}$$

$$\approx \frac{8.9932\ldots}{5}$$

$$\approx 1.798645\ldots$$

$$\approx 1.799$$

Keystroke	Display	
Press $\boxed{(}$	[01	0.
Enter $\boxed{34}$		34.
Press $\boxed{\sqrt{}}$		5.83095189
Press $\boxed{+}$		5.83095189
Enter $\boxed{10}$		10.
Press $\boxed{\sqrt{}}$		3.1622776
Press $\boxed{)}$		8.9932295
Press $\boxed{\div}$		8.99322955
Enter $\boxed{5}$		5.
Press $\boxed{=}$		1.798645911

Square Roots of Negative Numbers

The squares of both positive numbers and negative numbers are greater than 0. Therefore, the square root of a number less than 0 is not a real number. In the Supplementary Exercises of this section the number i is defined that can be used to write square roots of negative numbers with positive radicands. However, for the present study of square roots, we will simply assert that

If $b < 0$, then $\sqrt{b}$ is not a real number.

To illustrate,

$$\sqrt{-4}, \sqrt{\frac{-3}{16}}, \text{ and } -\sqrt{-3.61} \text{ are not real numbers.}$$

Summary. If b is a real number and			
$b < 0$	$b = 0$	$b > 0$	
then $\sqrt{b}$ is not a real number. As examples, $\sqrt{-1}, \sqrt{-4}$, and $\sqrt{-9}$ are not real numbers.	then $\sqrt{b} = 0$. Notice that 0 has only one square root.	then $\sqrt{b}$ is a rational number if b can be expressed as the ratio of perfect squares.	then $\sqrt{b}$ is an irrational number if b cannot be expressed as the ratio of perfect squares.

SECTION 8-1.   Practice Exercises

In exercises **1–8**, determine which of the numbers **a**, **b**, or **c** is a square root of the given number.

[Example 1]

1. 64
  a. 4
  b. -4
  c. 8

2. 81
  a. -3
  b. 3
  c. 9

3. 256
  a. -16
  b. 4
  c. 12

4. 484
  a. -22
  b. 11
  c. 20

5. 9.61
  a. -3.1
  b. 3.1
  c. 5.5

6. 7.84
  a. 2.8
  b. -2.8
  c. 5.3

7. $\dfrac{16}{9}$
  a. $\dfrac{-4}{3}$
  b. 1.3
  c. $\dfrac{3}{4}$

8. $\dfrac{81}{16}$
  a. $\dfrac{-9}{4}$
  b. 2.2
  c. $\dfrac{4}{9}$

In exercises **9–22**, find the indicated square roots.

[Example 2]   9. $\sqrt{9}$ 10. $\sqrt{25}$ 11. $-\sqrt{144}$

12. $-\sqrt{100}$ 13. $\sqrt{225}$ 14. $\sqrt{441}$

15. $-\sqrt{\dfrac{81}{16}}$ 16. $-\sqrt{\dfrac{361}{49}}$ 17. $\sqrt{0.04}$

18. $\sqrt{0.01}$ 19. $-\sqrt{1.21}$ 20. $-\sqrt{1.44}$

21. $\sqrt{\dfrac{25}{196}}$ 22. $\sqrt{\dfrac{49}{36}}$

In exercises **23–30**, determine whether each number is rational or irrational.

[Example 3]   23. **a.** $\sqrt{121}$ 24. **a.** $\sqrt{169}$
               b. $\sqrt{120}$ **b.** $\sqrt{170}$

25. **a.** $\sqrt{17}$ 26. **a.** $\sqrt{11}$
  b. $\sqrt{16}$ **b.** $\sqrt{9}$

27. **a.** $\sqrt{\dfrac{100}{49}}$ 28. **a.** $\sqrt{\dfrac{25}{4}}$

  b. $\sqrt{\dfrac{3}{49}}$ **b.** $\sqrt{\dfrac{25}{7}}$

29. **a.** $\sqrt{\dfrac{64}{19}}$ 30. **a.** $\sqrt{\dfrac{23}{81}}$

  b. $\sqrt{\dfrac{63}{19}}$ **b.** $\sqrt{\dfrac{23}{82}}$

In exercises **31–38**, use Table 1 to approximate the indicated square roots to three decimal places.

[Example 4]   31. $\sqrt{71}$ 32. $\sqrt{33}$ 33. $-\sqrt{27}$

34. $-\sqrt{68}$ 35. $\sqrt{99}$ 36. $\sqrt{50}$

37. $-\sqrt{69}$ 38. $-\sqrt{96}$

In exercises **39–46**, use a calculator to approximate to three decimal places the indicated square roots.

[Example 5]   39. $\sqrt{5{,}221}$ 40. $\sqrt{4{,}858}$ 41. $-\sqrt{3{,}009}$

42. $-\sqrt{2{,}777}$ 43. $\sqrt{62.4}$ 44. $\sqrt{78.3}$

45. $-\sqrt{132.41}$ 46. $-\sqrt{218.64}$

In exercises **47–54**, use a calculator to approximate to three decimal places.

47. $\dfrac{\sqrt{10}+\sqrt{19}}{3}$ 48. $\dfrac{\sqrt{14}+\sqrt{7}}{8}$

49. $\dfrac{\sqrt{32}-\sqrt{46}}{\sqrt{51}}$ 50. $\dfrac{\sqrt{58}-\sqrt{27}}{\sqrt{63}}$

51. $\dfrac{\sqrt{75} + 8.1}{\sqrt{11} + 3.6}$ (Hint: Write expression as $(\sqrt{75} + 8.1) \div (\sqrt{11} + 3.6)$.)

52. $\dfrac{\sqrt{54} + 9.4}{\sqrt{72} + 10.8}$ (Hint: Write expression as $(\sqrt{54} + 9.4) \div (\sqrt{72} + 10.8)$.)

53. $\dfrac{\sqrt{52} - \sqrt{14}}{\sqrt{17} - \sqrt{30}}$ **54.** $\dfrac{\sqrt{28} - \sqrt{31}}{\sqrt{18} - \sqrt{21}}$

SECTION 8-1.   Ten Review Exercises

In exercises **1–5**, find the solution, or solutions, of each equation or system of equations.

1. $3(5k - 8) - 8 = 7k - 7(4k + 8)$ **2.** $\dfrac{3}{n - 1} - \dfrac{4}{n} = \dfrac{2}{n^2 - n}$

3. $5u^2 = 12 - 17u$ **4.** (1) $2x - 7y = 10$
                              (2) $2x - 3y = 2$

5. (1) $3x + 5y = 2$
       (2) $x = \dfrac{3 - 5y}{3}$

In exercises **6–10**, simplify. Assume variables are not zero.

6. $-3t^3(2t^2)(-5t)$ **7.** $(-6z^2)^2(-2z)$

8. $\dfrac{(5a^2b)^2}{15ab^2}$ **9.** $\left(\dfrac{-2m}{3n^2}\right)^3 \left(\dfrac{9n^4}{16m^3}\right)$

10. $\dfrac{2uv(-5v)^3}{(-uv^2)^2}$

SECTION 8-1.   Supplementary Exercises

If $b > 0$, then $\sqrt{b}$ can also be written as $b^{1/2}$.

$\sqrt{b}$ is called the **radical form.**

$b^{1/2}$ is called the **exponential form.**

In exercises **1–10**, change each radical form to exponential form.

1. $\sqrt{3}$ **2.** $\sqrt{10}$

3. $\sqrt{24}$ **4.** $\sqrt{31}$

5. $-\sqrt{6}$ **6.** $-\sqrt{19}$

7. $\sqrt{\dfrac{2}{3}}$

8. $\sqrt{\dfrac{5}{8}}$

9. $\sqrt{0.7}$

10. $\sqrt{6.3}$

In exercises **11–18**,

a. change each exponential form to radical form.

b. if necessary, use a table or calculator to approximate the square root to at most three decimal places.

11. $4^{1/2}$

12. $25^{1/2}$

13. $121^{1/2}$

14. $144^{1/2}$

15. $13^{1/2}$

16. $17^{1/2}$

17. $68^{1/2}$

18. $92^{1/2}$

Calculator Supplement A

Many scientific calculators allow the user to "fix" the number of decimal digits displayed. That is, the calculator can be set to display a specified number of decimal places, even if the digits are all zeros. This kind of display enables the user to view numbers close to zero without using scientific notation. For example, a typical key stroke sequence to fix the display to show five decimal digits would be:

Keystroke	Display	Comments
Press $\boxed{\text{MODE}}$	0.	
Press $\boxed{7}$	0.	The code for "fix"*
Press $\boxed{5}$	0.00000	5 decimal digits

Now all calculations will show five decimal digits, if possible. To change back to the standard number of decimal digits, simply turn the calculator off, and then on again.

Example.   Compare $\sqrt{15}$ to

   a. 3.87.

   b. 3.873.

Solution.   **Discussion.** The comparison consists of two parts:

   Step 1. Square the decimal approximation.

   Step 2. Subtract the number obtained in Step 1 from the radicand of the indicated square root.

---

* Code will vary for different brands of calculators.

Use a calculator and fix the number of decimal digits on the display at seven.

a. Step 1. $3.87^2 = 14.9769$

   Step 2. $15 - 14.9769 = 0.0231$
   The approximation is "under" by 0.0231.

b. Step 1. $3.873^2 = 15.000129$

   Step 2. $15 - 15.000129 = -0.000129$
   The approximation is "over" by 0.000129.

In exercises **19–24**, find the error in the approximations parts **a–c**.

19. $\sqrt{2}$
   a. 1.4
   b. 1.41
   c. 1.414

20. $\sqrt{3}$
   a. 1.7
   b. 1.73
   c. 1.732

21. $\sqrt{10}$
   a. 3.2
   b. 3.16
   c. 3.162

22. $\sqrt{12}$
   a. 3.5
   b. 3.46
   c. 3.464

23. $\sqrt{28}$
   a. 5.3
   b. 5.29
   c. 5.291

24. $\sqrt{32}$
   a. 5.7
   b. 5.66
   c. 5.657

A geometric interpretation of perfect squares can be given because a square array of dots can be formed using only a perfect square number of dots.

In exercises **25–28**, verify that a square array of dots can be formed with the specified number of dots.

25. 9 dots

26. 36 dots

27. 49 dots

28. 25 dots

In exercises **29–32**, verify that a square array of dots cannot be formed with the specified number of dots.

29. 10 dots

30. 32 dots

31. 50 dots

32. 24 dots

The square root of a negative number is not a real number. The following definition can be used to write the square root of a negative number with a positive radicand.

Definition 8.3. i is a number such that $i^2 = -1$. Thus,

$$i = \sqrt{-1}.$$

The indicated square root of a negative number can be written in the **i-form** using the number i and the square root of a positive number.

Example. Write in the *i*-form.

$$\textbf{a. } \sqrt{-100} \qquad \textbf{b. } \sqrt{\frac{-4}{9}}$$

Solution. **Discussion.** First use *i* to write the number with a positive radicand. Then simplify, if possible, the positive number radicand.

a. $\sqrt{-100} = i\sqrt{100} = i \cdot 10$ or $10i$, the preferred form

b. $\sqrt{\frac{-4}{9}} = i\sqrt{\frac{4}{9}} = i \cdot \frac{2}{3}$ or $\frac{2i}{3}$, the preferred form

In exercises **33–42**, write the indicated square roots in the *i*-form.

33. $\sqrt{-25}$ **34.** $\sqrt{-36}$

35. $\sqrt{-16}$ **36.** $\sqrt{-9}$

37. $\sqrt{-121}$ **38.** $\sqrt{-144}$

39. $\sqrt{\frac{-9}{25}}$ **40.** $\sqrt{\frac{-4}{49}}$

41. $\sqrt{-0.04}$ **42.** $\sqrt{-0.09}$

SECTION 8-2. Products and Quotients of Square Roots

KEY TOPICS IN THIS SECTION

1. The product property of square roots

2. Using the product property to change a radicand

3. The quotient property of square roots

4. Using the quotient property to change a radicand

In Chapter 3 we studied exponential expressions. At that time we used properties of exponents to change the forms of exponential expressions such as those in examples **a–e**.

a. $t^2 \cdot t^3 = t^5$ **b.** $(t^2)^3 = t^6$ **c.** $\dfrac{t^3}{t^2} = t^1$ or t

d. $\left(\dfrac{t}{3}\right)^2 = \dfrac{t^2}{9}$ **e.** $(2t)^3 = 8t^3$

In this section we will study some of the properties of radicals. These properties enable us to change the form of a radical expression without changing the value of the number the expression represents.

The Product Property of Square Roots

Consider expressions **f** and **g**:

f. $\sqrt{4} \cdot \sqrt{9}$ A product of the square roots of 4 and 9

g. $\sqrt{4 \cdot 9}$ A square root of the product of 4 and 9

Example **f** is simplified by first extracting the square roots and then multiplying.

f. $\sqrt{4} \cdot \sqrt{9} = 2 \cdot 3 = 6$ Extract roots, then multiply.

Example **g** is simplified by first multiplying under the radical sign and then extracting the square root.

g. $\sqrt{4 \cdot 9} = \sqrt{36} = 6$ Multiply, then extract the root.

Notice that both expressions simplified to 6. Thus, the two expressions are equivalent in that they both represent the same number. We can therefore write the following equality:

$$\sqrt{4} \cdot \sqrt{9} = \sqrt{4 \cdot 9}$$

Equation (1) is a general statement of this example.

The product property of square roots
   If $a \geq 0$ and $b \geq 0$, then

   (1) $\sqrt{a} \cdot \sqrt{b} = \sqrt{ab}$ and $\sqrt{ab} = \sqrt{a} \cdot \sqrt{b}$

Example 1.   Multiply and simplify, if possible.

a. $\sqrt{6} \cdot \sqrt{13}$ **b.** $(-\sqrt{3})(-\sqrt{7})$

c. $\sqrt{2} \cdot \sqrt{8}$ **d.** $-\sqrt{40} \cdot \sqrt{2.5}$

Solution.   **Discussion.** Use the first form of equation (1) to write the two square roots as one square root. If the product under the radical sign yields a perfect square, then take the square root and write an integer.

a. $\sqrt{6} \cdot \sqrt{13} = \sqrt{78}$ 78 is not a perfect square.

b. $(-\sqrt{3})(-\sqrt{7}) = \sqrt{21}$ 21 is not a perfect square.

c. $\sqrt{2} \cdot \sqrt{8} = \sqrt{16}$ $\sqrt{2} \cdot \sqrt{8} = \sqrt{2 \cdot 8}$

$\qquad\qquad = 4$ 16 is a perfect square.

d. $-\sqrt{40} \cdot \sqrt{2.5} = -\sqrt{100}$ 100 is a perfect square.

$\qquad\qquad\qquad = -10$ The negative square root of 100

Using the Product Property to Change a Radicand

The product property of radicals can be used to change the form of some radicands to what is considered a **simpler form.** To illustrate, consider the expressions in examples **a** and **b**:

a. $\sqrt{48}$ can be written as $\sqrt{16 \cdot 3}$.   16 is a perfect square.

b. $\sqrt{117}$ can be written as $\sqrt{9 \cdot 13}$.   9 is a perfect square.

In both examples the radicands can be changed to an indicated product of a perfect square and another integer. Equation (1) can now be used on these expressions to write the perfect square factor without a radical sign.

a. $\sqrt{16 \cdot 3} = \sqrt{16} \cdot \sqrt{3} = 4\sqrt{3}$

b. $\sqrt{9 \cdot 13} = \sqrt{9} \cdot \sqrt{13} = 3\sqrt{13}$

Read $4\sqrt{3}$ as "4 times the square root of 3" and $3\sqrt{13}$ as "3 times the square root of 13." That is, *a multiplication is indicated between the number in front of the radical and the square root, but no sign is used.*

Example 2. Simplify by extracting any perfect square factors.

   a. $\sqrt{242}$

   b. $3\sqrt{50}$

   c. $\sqrt{720}$

Solution.   **Discussion.** Check the radicands for factors that are perfect squares. Remove all perfect square factors from each radicand.

   a. $242 = 121 \cdot 2$ and 121 is a perfect square.

   $$\sqrt{242} = \sqrt{121 \cdot 2}$$
   $$= \sqrt{121} \cdot \sqrt{2} \qquad \sqrt{a \cdot b} = \sqrt{a} \cdot \sqrt{b}$$
   $$= 11\sqrt{2} \qquad \sqrt{121} = 11$$

   b. $50 = 25 \cdot 2$ and 25 is a perfect square.

   $$3\sqrt{50} = 3\sqrt{25 \cdot 2}$$
   $$= 3\sqrt{25}\sqrt{2}$$
   $$= 3(5)\sqrt{2} \qquad \sqrt{25} = 5$$
   $$= 15\sqrt{2} \qquad \text{Simplify the product of } 3 \cdot 5.$$

   c. 720 has more than one perfect square factor.

   $720 = 4 \cdot 180$ **4 is a perfect square.**

   $720 = 9 \cdot 80$ **9 is a perfect square.**

   $720 = 16 \cdot 45$ **16 is a perfect square.**

   $720 = 36 \cdot 20$ **36 is a perfect square.**

   $720 = 144 \cdot 5$ **144 is a perfect square.**

To write the simplest form, the **greatest perfect square** is removed from under the radical sign.

$$\sqrt{720} = \sqrt{144 \cdot 5}$$
$$= \sqrt{144} \cdot \sqrt{5}$$
$$= 12\sqrt{5}$$

A calculator can be used to verify that $\sqrt{720}$ and $12\sqrt{5}$ represent the same number.

$$\sqrt{720} \approx 26.8328 \ldots$$

$$12\sqrt{5} \approx 12(2.2360 \ldots) \approx 26.8328 \ldots$$

The Quotient Property of Square Roots

Consider expressions **a** and **b**:

a. $\dfrac{\sqrt{144}}{\sqrt{16}}$ A quotient of the square roots of 144 and 16

b. $\sqrt{\dfrac{144}{16}}$ A square root of the quotient of 144 and 16

Example **a** is simplified by first extracting the square roots and then dividing.

a. $\dfrac{\sqrt{144}}{\sqrt{16}} = \dfrac{12}{4} = 3$ Extract roots, then divide.

Example **b** is simplified by first dividing under the radical sign and then extracting the square root.

b. $\sqrt{\dfrac{144}{16}} = \sqrt{9} = 3$ Divide, then extract the root.

Notice that both expressions simplified to 3. Thus, the two expressions are equivalent in that they both represent the same number. We can therefore write the following equality:

$$\dfrac{\sqrt{144}}{\sqrt{16}} = \sqrt{\dfrac{144}{16}}$$

Equation (2) is a general statement of this equation.

The quotient property of square roots
 If $a \geq 0$ and $b > 0$, then

$$(2) \quad \dfrac{\sqrt{a}}{\sqrt{b}} = \sqrt{\dfrac{a}{b}} \quad \text{and} \quad \sqrt{\dfrac{a}{b}} = \dfrac{\sqrt{a}}{\sqrt{b}}.$$

Example 3.   Divide and simplify, if possible.

   a. $\dfrac{\sqrt{45}}{\sqrt{15}}$ **b.** $\dfrac{-\sqrt{78}}{\sqrt{13}}$ **c.** $\dfrac{\sqrt{150}}{\sqrt{6}}$ **d.** $\dfrac{-\sqrt{98}}{\sqrt{18}}$

Solution.   **Discussion.** Use the first form of equation (2) to write the two square roots as one square root. If the quotient under the radical sign yields a perfect square, or a ratio of perfect squares, then simplify the expression.

   a. $\dfrac{\sqrt{45}}{\sqrt{15}} = \sqrt{\dfrac{45}{15}} = \sqrt{3}$ 3 is not a perfect square.

   b. $\dfrac{-\sqrt{78}}{\sqrt{13}} = -\sqrt{\dfrac{78}{13}} = -\sqrt{6}$ 6 is not a perfect square.

   c. $\dfrac{\sqrt{150}}{\sqrt{6}} = \sqrt{25}$ $\dfrac{\sqrt{150}}{\sqrt{6}} = \sqrt{\dfrac{150}{6}}$

       $= 5$ 25 is a perfect square.

d. $\dfrac{-\sqrt{98}}{\sqrt{18}} = -\sqrt{\dfrac{98}{18}}$ Write under one radical.

$\qquad\qquad = -\sqrt{\dfrac{49}{9}}$ Reduce the fraction.

$\qquad\qquad = \dfrac{-7}{3}$ 49 and 9 are perfect squares.

Using the Quotient Property to Change a Radicand

The quotient property of radicals can be used to change the form of some radicands to what is considered a **simpler form.** Specifically, a **preferred form** of a radical expression is one in which the radicand is not a fraction. To illustrate, consider the expressions in examples **a** and **b**:

a. $\sqrt{\frac{3}{4}}$ can be written as $\frac{\sqrt{3}}{\sqrt{4}} = \frac{\sqrt{3}}{2}$.

b. $\sqrt{\frac{7}{100}}$ can be written as $\frac{\sqrt{7}}{\sqrt{100}} = \frac{\sqrt{7}}{10}$.

In both examples the denominators of the fractions were perfect squares. As a consequence, when equation (2) was used to write the expression as a ratio of square roots, the denominator could be written as an integer—the preferred form. A complete study of the topic of preferred forms for radical expressions is covered in Section 8-4.

Example 4. Simplify and write with a denominator that is an integer.

$\qquad$ **a.** $\sqrt{\dfrac{7}{3}}$ $\qquad$ **b.** $\sqrt{\dfrac{12}{5}}$

Solution. **Discussion.** In both example **a** and example **b**, the denominators are not perfect squares. The multiplication property of one will be used first to change the fractions to equivalent ones with perfect square denominators. Then equation (2) will be used to change the form to a ratio of square roots.

a. $\sqrt{\dfrac{7}{3}} = \sqrt{\dfrac{7}{3} \cdot \dfrac{3}{3}}$ Multiplication property of one

$\qquad = \sqrt{\dfrac{21}{9}}$ Do the indicated multiplication.

$\qquad = \dfrac{\sqrt{21}}{\sqrt{9}}$ $\sqrt{\dfrac{a}{b}} = \dfrac{\sqrt{a}}{\sqrt{b}}$

$\qquad = \dfrac{\sqrt{21}}{3}$ The preferred form

b. $\sqrt{\dfrac{12}{5}} = \sqrt{\dfrac{12}{5} \cdot \dfrac{5}{5}}$

$\qquad = \dfrac{\sqrt{60}}{\sqrt{25}}$ $\sqrt{60} = \sqrt{4 \cdot 15}$

$\qquad = \dfrac{2\sqrt{15}}{5}$ $\sqrt{4 \cdot 15} = 2\sqrt{15}$

A calculator can be used to verify that $\sqrt{\frac{7}{3}}$ and $\frac{\sqrt{21}}{3}$ represent the same number.

$$\sqrt{\frac{7}{3}} \approx \sqrt{2.3333\ldots} \approx 1.5275\ldots$$

$$\frac{\sqrt{21}}{3} \approx \frac{4.582575\ldots}{3} \approx 1.5275\ldots$$

SECTION 8-2.   Practice Exercises

In exercises **1–26**, write as one square root and simplify, if possible.

[Example 1]   **1.** $\sqrt{2}\sqrt{7}$ **2.** $\sqrt{3}\sqrt{5}$ **3.** $\sqrt{6}\sqrt{13}$

4. $\sqrt{13}\sqrt{10}$ **5.** $\sqrt{19}\sqrt{19}$ **6.** $\sqrt{37}\sqrt{37}$

7. $\sqrt{32}\sqrt{2}$ **8.** $\sqrt{50}\sqrt{2}$ **9.** $\sqrt{27}\sqrt{3}$

10. $\sqrt{24}\sqrt{6}$ **11.** $(-\sqrt{6})(-\sqrt{11})$ **12.** $(-\sqrt{3})(-\sqrt{10})$

13. $\sqrt{75}\sqrt{3}$ **14.** $\sqrt{18}\sqrt{8}$ **15.** $(-\sqrt{5})(-\sqrt{80})$

16. $(-\sqrt{20})(-\sqrt{45})$ **17.** $\sqrt{10}(-\sqrt{11})$ **18.** $(-\sqrt{15})(\sqrt{2})$

19. $\sqrt{2}\sqrt{3}\sqrt{5}$ **20.** $\sqrt{3}\sqrt{7}\sqrt{10}$ **21.** $\sqrt{2}\sqrt{6}\sqrt{3}$

22. $\sqrt{5}\sqrt{10}\sqrt{2}$ **23.** $\sqrt{98}\sqrt{\frac{1}{2}}$ **24.** $\sqrt{108}\sqrt{\frac{1}{3}}$

25. $(\sqrt{32})(-\sqrt{4.5})$ **26.** $(\sqrt{56})(-\sqrt{3.5})$

In exercises **27–44**, simplify by extracting any perfect square factors.

[Example 2]   **27.** $\sqrt{12}$ **28.** $\sqrt{45}$ **29.** $\sqrt{75}$

30. $\sqrt{20}$ **31.** $3\sqrt{24}$ **32.** $2\sqrt{50}$

33. $6\sqrt{63}$ **34.** $5\sqrt{8}$ **35.** $\sqrt{32}$

36. $\sqrt{72}$ **37.** $-\sqrt{180}$ **38.** $-\sqrt{192}$

39. $\sqrt{363}$ **40.** $\sqrt{243}$ **41.** $\sqrt{288}$

42. $\sqrt{343}$ **43.** $-5\sqrt{675}$ **44.** $-8\sqrt{2,000}$

In exercises **45–56**, divide and simplify, if possible.

[Example 3]   **45.** $\dfrac{\sqrt{10}}{\sqrt{2}}$ **46.** $\dfrac{\sqrt{18}}{\sqrt{3}}$ **47.** $\dfrac{-\sqrt{105}}{\sqrt{15}}$

48. $\dfrac{-\sqrt{84}}{\sqrt{12}}$ **49.** $\dfrac{\sqrt{48}}{\sqrt{3}}$ **50.** $\dfrac{\sqrt{20}}{\sqrt{5}}$

51. $\dfrac{\sqrt{50}}{\sqrt{18}}$ **52.** $\dfrac{\sqrt{72}}{\sqrt{98}}$ **53.** $\dfrac{\sqrt{486}}{\sqrt{6}}$

54. $\dfrac{\sqrt{275}}{\sqrt{11}}$ **55.** $\dfrac{12\sqrt{40}}{3\sqrt{10}}$ **56.** $\dfrac{15\sqrt{63}}{5\sqrt{7}}$

In exercises **57–70**, simplify and write with a denominator that is an integer.

[Example 4] **57.** $\sqrt{\dfrac{11}{9}}$ **58.** $\sqrt{\dfrac{15}{4}}$ **59.** $\sqrt{\dfrac{3}{49}}$

60. $\sqrt{\dfrac{5}{64}}$ **61.** $\sqrt{\dfrac{2}{5}}$ **62.** $\sqrt{\dfrac{3}{7}}$

63. $\sqrt{\dfrac{1}{3}}$ **64.** $\sqrt{\dfrac{1}{2}}$ **65.** $\sqrt{\dfrac{7}{10}}$

66. $\sqrt{\dfrac{10}{17}}$ **67.** $\sqrt{\dfrac{36}{7}}$ **68.** $\sqrt{\dfrac{25}{2}}$

69. $\sqrt{\dfrac{20}{27}}$ **70.** $\sqrt{\dfrac{28}{75}}$

SECTION 8-2.   Ten Review Exercises

In exercises **1–8**, use the equation $2x - 5y = 30$.

1. Write the equation in the $y = mx + b$ form.

2. State the slope of a graph of the equation.

3. State the y-intercept of the equation.

4. State the x-intercept of the equation.

5. Graph the solution set.

6. Write in $Ax + By = C$ form an equation of the line perpendicular to the graph of the equation that passes through $P(-4, 5)$.

7. Find the point of intersection of the graph of the given equation and the graph of $2x + 5y = -10$.

8. Are the graphs of the given equation and the line defined by $y = \dfrac{2x + 10}{5}$ parallel?

In exercises **9** and **10**, factor completely.

9. $24t^4 - 3t$ **10.** $10xy + 15y - 2x - 3$

SECTION 8-2.   Supplementary Exercises

In exercises **1–8**,

a. find an approximation to three decimal places for the given radicals.

b. simplify the given radicals by removing any perfect square factors.

c. approximate the simplified expression in part **b** to three decimal places and compare to part **a**.

1. $\sqrt{18}$ 2. $\sqrt{50}$ 3. $\sqrt{48}$

4. $\sqrt{27}$ 5. $\sqrt{125}$ 6. $\sqrt{180}$

7. $\sqrt{252}$ 8. $\sqrt{343}$

In exercises **9–16**, simplify.

9. $-\sqrt{\dfrac{256}{100}}$ 10. $-\sqrt{\dfrac{289}{100}}$ 11. $\sqrt{0.0081}$

12. $\sqrt{0.0049}$ 13. $\sqrt{\dfrac{0.09}{0.25}}$ 14. $\sqrt{\dfrac{0.49}{0.04}}$

15. $-\sqrt{0.0784}$ 16. $-\sqrt{0.0676}$

In exercises **17–24**, do not use a table or calculator.

a. State the positive integer whose square is closest to the given radicand.

b. Use the results of part **a** to approximate the indicated square root to one decimal place.

c. Square the approximation in part **b** and compare the product to the radicand.

d. Repeat parts **b** and **c** to obtain the best one-decimal digit approximation.

17. $\sqrt{18}$ 18. $\sqrt{13}$ 19. $\sqrt{27}$

20. $\sqrt{30}$ 21. $\sqrt{59}$ 22. $\sqrt{74}$

23. $\sqrt{92}$ 24. $\sqrt{109}$

In exercises **25–32**, use $<$, $>$, or $=$ to make a true statement.

25. $\sqrt{12}$ $3\sqrt{3}$ 26. $\sqrt{20}$ $3\sqrt{5}$ 27. $4\sqrt{2}$ $\sqrt{50}$

28. $4\sqrt{3}$ $\sqrt{75}$ 29. $-\sqrt{18}$ $-4\sqrt{2}$ 30. $-\sqrt{27}$ $-4\sqrt{3}$

31. $\sqrt{0.16}$ $\sqrt{0.49}$ 32. $\sqrt{0.81}$ $\sqrt{0.64}$

SECTION 8-3. Sums and Differences of Square Root Radicals

KEY TOPICS IN THIS SECTION

1. Sums and differences of like radicals

2. Sums and differences of unlike radicals

3. Further operations on radicals

The previous section covered two properties of square root radicals. These properties enabled us to write an indicated product or quotient of two square roots as one square root. In this section we will study a property that states the necessary conditions under which an indicated sum or difference of square roots can be written as one square root.

Sums and Differences of Like Radicals

Consider the expressions in examples **a**, **b**, and **c**:

a. $3\sqrt{25} + 2\sqrt{9}$

b. $7\sqrt{2} + 3\sqrt{2}$ $\qquad\left\{\begin{array}{l}\text{In each expression five operations are indicated:}\\ \text{two square roots, two multiplications, one addition}\end{array}\right.$

c. $4\sqrt{6} + 5\sqrt{3}$

Suppose we are instructed to "simplify" these expressions; that is, reduce, if possible, the number of symbols and operations. The rule for order of operations requires extracting square roots first, then multiplying and finally adding. In example **a** this causes no problem, since 25 and 9 are perfect squares.

a.	$3\sqrt{25} + 2\sqrt{9}$	The given expression
	$= 3(5) + 2(3)$	$\sqrt{25} = 5$ and $\sqrt{9} = 3$
	$= 15 + 6$	Do the multiplications.
	$= 21$	The value of the expression is 21.

In example **b**, the radicand 2 is not a perfect square. Therefore, $\sqrt{2}$ is an irrational number whose decimal representation can only be approximated. However, the distributive property of multiplication over addition can be used to "factor out" the square roots.

b.	$7\sqrt{2} + 3\sqrt{2}$	The given expression
	$= (7 + 3)\sqrt{2}$	$b \cdot a + c \cdot a = (b + c)a$
	$= 10\sqrt{2}$	The simplified form

This expression is "simpler" than the given one because it has fewer symbols. However, we cannot do the final operations of square root extraction and multiplication, unless we are willing to approximate $\sqrt{2}$ with decimals. The preferred form for this expression is $10\sqrt{2}$, as written.

In example **c**, the radicands 6 and 3 are not perfect squares. Therefore $\sqrt{6}$ and $\sqrt{3}$ are irrational numbers. Since these numbers are not common factors of both terms, the distributive property cannot be used to change the order of operations. Therefore the given expression cannot be simplified.

c. $4\sqrt{6} + 5\sqrt{3}$ $\qquad$ Cannot be simplified.

It is common practice to refer to square roots with the same radicands as **like radicals**. In example **b** the two terms contain like radicals, namely $\sqrt{2}$. In example **c** the two terms contain unlike radicals, namely $\sqrt{6}$ and $\sqrt{3}$. Equations (1) and (2) are statements that terms with like radicals can be **combined**, that is, added or subtracted.

The addition and subtraction properties of square roots
If $b \geq 0$, then

(3) $x\sqrt{b} + y\sqrt{b} = (x + y)\sqrt{b}.$ $\qquad$ Addition of like radicals

(4) $x\sqrt{b} - y\sqrt{b} = (x - y)\sqrt{b}.$ $\qquad$ Subtraction of like radicals

---

Example 1.   Simplify:

a. $4\sqrt{7} + 5\sqrt{7} - \sqrt{7}$

b. $6\sqrt{5} + 3\sqrt{3} - 2\sqrt{5} - 9\sqrt{3}$

Solution.

a. **Discussion.** The expression consists of three terms, each having a $\sqrt{7}$ factor. Thus they are like radicals and can be combined.

$$4\sqrt{7} + 5\sqrt{7} - \sqrt{7} \qquad \text{The given expression}$$

$$= (4 + 5 - 1)\sqrt{7} \qquad \text{Factor out } \sqrt{7}.$$

$$= 8\sqrt{7} \qquad \text{The simplified form}$$

b. **Discussion.** The expression consists of four terms. Two terms have a $\sqrt{5}$ factor and can be combined. Two terms have a $\sqrt{3}$ factor and can be combined.

$$6\sqrt{5} + 3\sqrt{3} - 2\sqrt{5} - 9\sqrt{3}$$

$$= 4\sqrt{5} - 6\sqrt{3}$$

Since $4\sqrt{5}$ and $6\sqrt{3}$ are unlike radicals, they cannot be combined.

Sums and Differences of Unlike Radicals

Equations (3) and (4) can be used to combine like radicals. Two **apparently unlike radicals** may also be combined if they both can be simplified to the same radicand. To illustrate, consider the expression in example **a**:

a. $8\sqrt{24} + \sqrt{150} - 2\sqrt{600}$ Unlike radicals

Each of the radicands contains a perfect square factor. Therefore, they can be written as equivalent expressions with reduced radicands.

$$8\sqrt{24} = 8\sqrt{4 \cdot 6} = 8\sqrt{4}\sqrt{6} = 8(2)\sqrt{6} = 16\sqrt{6}$$

$$\sqrt{150} = \sqrt{25 \cdot 6} = \sqrt{25}\sqrt{6} = 5\sqrt{6}$$

$$2\sqrt{600} = 2\sqrt{100 \cdot 6} = 2\sqrt{100}\sqrt{6} = 2(10)\sqrt{6} = 20\sqrt{6}$$

$$8\sqrt{24} + \sqrt{150} - 2\sqrt{600} \qquad \text{The given expression}$$

$$= 16\sqrt{6} + 5\sqrt{6} - 20\sqrt{6} \qquad \text{The simplified radicands}$$

$$= (16 + 5 - 20)\sqrt{6} \qquad \text{Factor out } \sqrt{6}.$$

$$= \sqrt{6} \qquad \text{The simplified expression}$$

Example 2.   Simplify:

a. $2\sqrt{108} + \sqrt{27} - 8\sqrt{3}$

b. $\sqrt{180} - 2\sqrt{20} + 6\sqrt{9} + 3\sqrt{5}$

Solution.

a. **Discussion.** The first two terms have radicands that can be simplified.

$$2\sqrt{108} = 2\sqrt{36 \cdot 3} = 2\sqrt{36}\sqrt{3} = 12\sqrt{3}$$

$$\sqrt{27} = \sqrt{9 \cdot 3} = \sqrt{9}\sqrt{3} = 3\sqrt{3}$$

$$2\sqrt{108} + \sqrt{27} - 8\sqrt{3} \qquad \text{The given expression}$$

$$= 12\sqrt{3} + 3\sqrt{3} - 8\sqrt{3} \qquad \text{The simplified radicands}$$

$$= (12 + 3 - 8)\sqrt{3} \qquad \text{Factor out } \sqrt{3}.$$

$$= 7\sqrt{3} \qquad \text{The simplified expression}$$

b. $\quad \sqrt{180} - 2\sqrt{20} + 6\sqrt{9} + 3\sqrt{5}$

$$= 6\sqrt{5} - 2(2)\sqrt{5} + 6(3) + 3\sqrt{5}$$

$$= (6 - 4 + 3)\sqrt{5} + 18$$

$$= 5\sqrt{5} + 18$$

Frequently a radicand may contain one or more variable factors. For simplicity, we will assume that any variables in a radicand represent only positive numbers.

Example 3. Simplify:

$$2\sqrt{90y} - 3\sqrt{40x} - 5\sqrt{10y} + \sqrt{250x}$$

Solution. **Discussion.** Two radicands have x as a factor and two have y as a factor. We may not assume x and y represent the same number. Therefore we must consider them to be unlike radicals.

$$2\sqrt{90y} - 3\sqrt{40x} - 5\sqrt{10y} + \sqrt{250x}$$

$$= 2(3)\sqrt{10y} - 3(2)\sqrt{10x} - 5\sqrt{10y} + 5\sqrt{10x}$$

$$= 6\sqrt{10y} - 6\sqrt{10x} - 5\sqrt{10y} + 5\sqrt{10x}$$

$$= \sqrt{10y} - \sqrt{10x} \qquad \text{The simplified expression}$$

Further Operations on Radicals

Examples 4–6 illustrate expressions that indicate operations involving square roots. The expressions are simplified by doing the operations.

Example 4. Simplify:

$$\frac{6 - \sqrt{18}}{15}$$

Solution. **Discussion.** The fraction bar is a grouping symbol. Therefore the fraction cannot be reduced until the numerator is written in factored form.

$$\frac{6 - \sqrt{18}}{15} \qquad \text{The given expression}$$

$$= \frac{6 - 3\sqrt{2}}{15} \qquad \sqrt{18} = \sqrt{9 \cdot 2} = 3\sqrt{2}$$

$$= \frac{3(2 - \sqrt{2})}{15} \qquad \text{Factor a 3 in the numerator.}$$

$$= \frac{2 - \sqrt{2}}{5} \qquad \text{Remove the common factor 3.}$$

Example 5.   Simplify:

a. $\sqrt{3}(2 + \sqrt{6})$

b. $(3 + 2\sqrt{5})(4 - \sqrt{5})$

Solution.   a. $\sqrt{3}(2 + \sqrt{6})$ The given expression

$= 2\sqrt{3} + \sqrt{6}\sqrt{3}$ $a(b + c) = b \cdot a + c \cdot a$

$= 2\sqrt{3} + \sqrt{18}$ $\sqrt{a} \cdot \sqrt{b} = \sqrt{a \cdot b}$

$= 2\sqrt{3} + 3\sqrt{2}$ Simplify $\sqrt{18}$.

b. $(3 + 2\sqrt{5})(4 - \sqrt{5})$ Use FOIL.

$= 3 \cdot 4 - 3 \cdot \sqrt{5} + 2 \cdot 4 \cdot \sqrt{5} - 2 \cdot \sqrt{5} \cdot \sqrt{5}$

$= 12 - 3\sqrt{5} + 8\sqrt{5} - 2\sqrt{25}$

$= 12 + 5\sqrt{5} - 10$ Combine the $\sqrt{5}$ terms.

$= 2 + 5\sqrt{5}$ The simplified product

Example 6.   Multiply: $(-3 + 2\sqrt{5})(-3 - 2\sqrt{5})$

Solution.   **Discussion.** The given binomials have the form $(x + y)(x - y)$, where x is -3 and y is $2\sqrt{5}$. Recall that such binomials have a multiplied form $x^2 - y^2$.

$(-3 + 2\sqrt{5})(-3 - 2\sqrt{5})$ The indicated product

$= (-3)^2 - (2\sqrt{5})^2$ $(x + y)(x - y) = x^2 - y^2$

$= 9 - 20$ $(2\sqrt{5})^2 = 2^2(\sqrt{5})^2 = 4 \cdot 5 = 20$

$= -11$

SECTION 8-3.   Practice Exercises

In exercises **1–26**, simplify any radical, if necessary, and then add or subtract, as indicated, any like radicals. Assume that any variable is positive.

[Example 1] **1.** $4\sqrt{2} + 3\sqrt{2}$ **2.** $2\sqrt{5} + 7\sqrt{5}$

3. $12\sqrt{6} - 8\sqrt{6}$ **4.** $9\sqrt{3} - 4\sqrt{3}$

5. $13\sqrt{11} + 3\sqrt{11} - \sqrt{11}$ **6.** $\sqrt{7} + 5\sqrt{7} - 3\sqrt{7}$

7. $4\sqrt{5} + 5\sqrt{2} - \sqrt{5} + 3\sqrt{2}$ **8.** $2\sqrt{3} + 3\sqrt{2} - \sqrt{3} - 2\sqrt{2}$

9. $8\sqrt{10} - \sqrt{3} - 2\sqrt{10} - 3\sqrt{3}$ **10.** $6\sqrt{6} - 2\sqrt{2} - 4\sqrt{6} - 2\sqrt{2}$

[Example 2] **11.** $\sqrt{8} + \sqrt{18}$ **12.** $\sqrt{50} + \sqrt{32}$

13. $3\sqrt{20} - \sqrt{45}$ **14.** $2\sqrt{125} - \sqrt{80}$

15. $\sqrt{200} + \sqrt{72} - \sqrt{2}$ **16.** $2\sqrt{128} - 3\sqrt{98} + 6\sqrt{2}$

17. $3\sqrt{54} - 4\sqrt{24} + \sqrt{6}$ **18.** $\sqrt{294} + \sqrt{384} - 10\sqrt{6}$

[Example 3]   **19.** $\sqrt{72x} + \sqrt{18x}$ **20.** $\sqrt{108x} + \sqrt{48x}$

21. $2\sqrt{500y} - 3\sqrt{125y}$ **22.** $\sqrt{600y} - 2\sqrt{24y}$

23. $4\sqrt{28a} + 5\sqrt{24b} - 2\sqrt{63a} + \sqrt{96b}$

24. $2\sqrt{40a} + 3\sqrt{27b} + 3\sqrt{90a} - 2\sqrt{75b}$

25. $6\sqrt{44s} - 2\sqrt{12t} - 2\sqrt{275s} - \sqrt{48t}$

26. $3\sqrt{52s} - 3\sqrt{150t} - \sqrt{117s} - \sqrt{216t}$

In exercises **27–34**, simplify.

[Example 4]   **27.** $\dfrac{3 + \sqrt{27}}{6}$ **28.** $\dfrac{2 + \sqrt{32}}{4}$

29. $\dfrac{4 - \sqrt{20}}{2}$ **30.** $\dfrac{6 - \sqrt{28}}{2}$

31. $\dfrac{5 + \sqrt{50}}{10}$ **32.** $\dfrac{10 + \sqrt{75}}{15}$

33. $\dfrac{4 - \sqrt{48}}{12}$ **34.** $\dfrac{8 - \sqrt{96}}{12}$

In exercises **35–50**, do the indicated operations. Simplify any radicals.

[Example 5]   **35.** $\sqrt{2}(7 + \sqrt{2})$ **36.** $\sqrt{5}(2 + \sqrt{5})$

37. $\sqrt{6}(4 - \sqrt{3})$ **38.** $\sqrt{6}(5 - \sqrt{2})$

39. $(4 + \sqrt{2})(3 - 2\sqrt{2})$ **40.** $(5 - 2\sqrt{3})(2 + \sqrt{3})$

41. $(7 - \sqrt{5})(2 + 3\sqrt{5})$ **42.** $(6 - \sqrt{7})(3 + 2\sqrt{7})$

[Example 6]   **43.** $(4 - \sqrt{2})(4 + \sqrt{2})$ **44.** $(1 - \sqrt{6})(1 + \sqrt{6})$

45. $(10 - 3\sqrt{5})(10 + 3\sqrt{5})$ **46.** $(6 + 2\sqrt{3})(6 - 2\sqrt{3})$

47. $(5 + 2\sqrt{6})(5 - 2\sqrt{6})$ **48.** $(2 + 3\sqrt{2})(2 - 3\sqrt{2})$

49. $(7 - 2\sqrt{6})(7 + 2\sqrt{6})$ **50.** $(9 - 2\sqrt{7})(9 + 2\sqrt{7})$

SECTION 8-3. Ten Review Exercises

In exercises **1–4**, solve each problem.

1. Do the indicated operations.

$(3a^2 - ab + 2b^2) - 2(a^2 - ab + b^2)$

2. Evaluate the expression in exercise **1** for $a = -3$ and $b = 5$.

3. Simplify:

$$-2[2^3 - 3(5^2 - 11 \cdot 2)] - \frac{7^2 - 3^2 \cdot 2}{7 \cdot 5 - 6^2}$$

4. Solve for t:

$$2t - a = 5 + bt$$

In exercises **5–8**, solve and check.

5. $3(5u + 7) - 3 = 3 + 5(1 - u)$ **6.** $16v^2 = 9$

7. $12 + 7x = 12x^2$ **8.** $\dfrac{2}{y^2 + y} + \dfrac{1}{3y + 3} = \dfrac{4}{3y}$

In exercises **9** and **10**, simplify.

9. $\dfrac{\dfrac{m}{3n} + \dfrac{7}{6} + \dfrac{n}{m}}{\dfrac{m}{3n} - \dfrac{n}{m} - \dfrac{1}{6}}$ **10.** $\dfrac{2 + \dfrac{1}{a}}{4 - \dfrac{1}{a^2}}$

SECTION 8-3.   Supplementary Exercises

In exercises **1–18**, simplify. Assume variables are positive.

1. $5\sqrt{49} - 3\sqrt{100} + (\sqrt{3})^2$ **2.** $4\sqrt{25} - 7\sqrt{64} + (\sqrt{2})^2$

3. $\dfrac{\sqrt{16} + 3\sqrt{49}}{5\sqrt{4}}$ **4.** $\dfrac{3\sqrt{121} + 2\sqrt{36}}{3\sqrt{25}}$

5. $\dfrac{\sqrt{81} - 2\sqrt{9}}{\sqrt{36}}$ **6.** $\dfrac{\sqrt{361} - \sqrt{169}}{\sqrt{144}}$

7. $3\sqrt{100} + \sqrt{8} - 5\sqrt{18} - 3\sqrt{64}$ **8.** $5\sqrt{49} + 2\sqrt{27} - 4\sqrt{9} + 2\sqrt{75}$

9. $\sqrt{256} - \sqrt{98} + \sqrt{121} + \sqrt{50}$ **10.** $\sqrt{324} + \sqrt{48} + \sqrt{108} - \sqrt{225}$

11. $-3\sqrt{4} - \sqrt{180} + 2\sqrt{9} - \sqrt{125}$ **12.** $-\sqrt{72} + 2\sqrt{441} + 2\sqrt{98} - 3\sqrt{196}$

13. $\sqrt{108t} + 5\sqrt{4u} - 3\sqrt{25u} - \sqrt{675t}$ **14.** $\sqrt{50t} - 10\sqrt{u} + \sqrt{200t} + 2\sqrt{16u}$

15. $\dfrac{\sqrt{9a}}{2} - \dfrac{\sqrt{8b}}{3} - \dfrac{\sqrt{25a}}{6} + \dfrac{\sqrt{242b}}{2}$ **16.** $\dfrac{-\sqrt{49a}}{3} + \dfrac{\sqrt{12b}}{3} + \dfrac{\sqrt{169a}}{2} - \dfrac{\sqrt{363b}}{6}$

17. $5\sqrt{0.16m} + \sqrt{6.25n} - \sqrt{4m} - \sqrt{2.25n}$

18. $\sqrt{0.09m} - \sqrt{1.44n} + 2\sqrt{0.01n} - 3\sqrt{0.04m}$

In exercises **19–32**, simplify.

19. $(\sqrt{3} + \sqrt{2})(\sqrt{3} - \sqrt{2})$ **20.** $(\sqrt{6} + \sqrt{5})(\sqrt{6} - \sqrt{5})$

21. $(\sqrt{3} + \sqrt{2})^2$ **22.** $(\sqrt{6} + \sqrt{5})^2$

23. $(\sqrt{3} - \sqrt{2})^2$ **24.** $(\sqrt{6} - \sqrt{5})^2$

25. $(2\sqrt{3} + \sqrt{2})(\sqrt{3} + 2\sqrt{2})$

26. $(2\sqrt{6} + \sqrt{5})(\sqrt{6} + 2\sqrt{5})$

27. $\sqrt{15}(\sqrt{3} - \sqrt{5})$

28. $\sqrt{6}(\sqrt{3} - \sqrt{2})$

29. $\left(\dfrac{1 + \sqrt{2}}{3}\right)\left(\dfrac{1 - \sqrt{2}}{3}\right)$

30. $\left(\dfrac{3 + \sqrt{7}}{2}\right)\left(\dfrac{3 - \sqrt{7}}{2}\right)$

31. $\left(\dfrac{1 + \sqrt{2}}{3}\right)^2$

32. $\left(\dfrac{3 + \sqrt{7}}{2}\right)^2$

In exercises **33–40**, simplify.

33. $\dfrac{3 + \sqrt{9 + 72}}{6}$

34. $\dfrac{3 - \sqrt{9 + 72}}{6}$

35. $\dfrac{8 - \sqrt{64 + 16}}{8}$

36. $\dfrac{8 + \sqrt{64 - 16}}{8}$

37. $\dfrac{-4 + \sqrt{16 - 4(-4)}}{8}$

38. $\dfrac{-4 - \sqrt{16 + 4(-4)}}{8}$

39. $\dfrac{-7 - \sqrt{49 - 4(6)(-3)}}{12}$

40. $\dfrac{-7 + \sqrt{49 - 4(6)(-3)}}{12}$

Calculator Supplement B

A calculator can simplify expressions such as $3\sqrt{5} - \sqrt{49}$ as follows:

Keystroke	Display	Comments
Enter $\boxed{3}$	3.	
Press $\boxed{\times}$	3.	
Enter $\boxed{5}$	5.	
Press $\boxed{\sqrt{\;}}$	2.236067977	$\sqrt{5}$
Press $\boxed{-}$	6.708203932	$3\sqrt{5}$
Enter $\boxed{49}$	49.	
Press $\boxed{\sqrt{\;}}$	7.	$\sqrt{49}$
Press $\boxed{=}$	-0.291796067	

The approximate value of the expression is -0.291796069.

In exercises **41–46**,

a. use a calculator to approximate to three decimal places the given expression.

b. simplify each expression.

c. use a calculator to approximate the simplified expression found in part **b** to three decimal places.

d. compare the approximations in parts **a** and **c**.

41. $2\sqrt{12} + \sqrt{27}$ **42.** $\sqrt{50} + 3\sqrt{32}$

43. $2\sqrt{125} + 3\sqrt{45}$ **44.** $\sqrt{180} - 2\sqrt{20}$

45. $2\sqrt{8} - 2\sqrt{18} + \sqrt{98}$ **46.** $\sqrt{27} - 2\sqrt{12} + 7\sqrt{3}$

SECTION 8-4. Preferred Form for a Square Root Radical

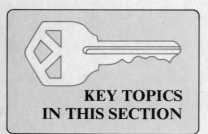

KEY TOPICS IN THIS SECTION

1. The conditions for the preferred form of an expression with square roots

2. Rationalizing the denominator

3. Conjugates

Throughout the study of mathematics we see evidence of what is frequently called the **preferred form** for expressions, or the **standard form** for equations and equalities. For example,

$\frac{1}{3}$ is the preferred form for $\frac{3}{9}$ and $\frac{5}{15}$.

$x^2 - 3x - 10 = 0$ is the standard form for $3x = x^2 - 10$.

In this section we will study the preferred form for writing expressions containing square roots of positive numbers. As is the case in most of mathematics, there is some reason for writing expressions in preferred form. We will give possible reasons why certain forms of square roots and expressions containing square roots are preferred over other forms.

The Conditions for the Preferred Form of an Expression with Square Roots

A square root of a number, or an expression containing a square root, must satisfy three conditions in order to qualify as preferred form. The term **simplified form** is also used for *preferred form*.

> An expression with a square root is in **preferred** (or **simplified**) **form** when it satisfies the following conditions:
>
> **Condition 1.** The radicand does not contain any perfect square factors.
>
> **Condition 2.** The radicand is not a fraction.
>
> **Condition 3.** Any denominator of the expression does not contain a radical.

When faced with the task of changing a single square root, or an expression containing square roots, to simplified form, ask the following question:
Why is this radical, or expression, not simplified?
First check for any observable deficiencies:

1. Is any radicand a fraction?

2. Is the expression a fraction with a radicand in the denominator?

Second, always check any radicand for perfect square factors. This rule applies also to the final form of the expression that you consider simplified.

Example 1. Simplify:

a. $\dfrac{2\sqrt{72}}{3}$

b. $\dfrac{\sqrt{320}}{4} - \dfrac{\sqrt{405}}{3}$

Solution. a. **Discussion.** The radicand is not a fraction, and there is no radical in the denominator. However, 72 contains perfect square factors 4, 9, and 36. The greatest perfect square factor is removed from the radicand.

$\dfrac{2\sqrt{72}}{3}$ The given expression

$= \dfrac{2(6)\sqrt{2}}{3}$ $\sqrt{72} = \sqrt{36 \cdot 2} = 6\sqrt{2}$

$= 4\sqrt{2}$ Reduce the fraction.

b. $\dfrac{\sqrt{320}}{4} - \dfrac{\sqrt{405}}{3}$ The given expression

$= \dfrac{8\sqrt{5}}{4} - \dfrac{9\sqrt{5}}{3}$ $\sqrt{320} = \sqrt{64 \cdot 5} = 8\sqrt{5}$
                      $\sqrt{405} = \sqrt{81 \cdot 5} = 9\sqrt{5}$

$= 2\sqrt{5} - 3\sqrt{5}$ Reduce the fractions.

$= -\sqrt{5}$ Do the indicated subtraction.

Before electronic calculators became readily available, it was difficult to obtain a decimal approximation of an irrational number square root. Consequently it was preferable to work with the smallest radicand possible. This motivated one to remove any perfect square factors. Using only paper and pencil, it would be much easier to approximate $-\sqrt{5}$ in part **b** of Example 1 than to approximate the given expression. However, we must keep in mind that both expressions name the same number. Using a calculator, we can obtain the same decimal approximation:

$$\frac{\sqrt{320}}{4} - \frac{\sqrt{405}}{3} \approx \frac{17.88\ldots}{4} - \frac{20.124\ldots}{3}$$

$$= 4.472\ldots - 6.708\ldots$$

$$= -2.236\ldots$$

$$-\sqrt{5} \approx -2.236\ldots$$

Example 2. Simplify: $\sqrt{\dfrac{2x}{3}} + \sqrt{\dfrac{3x}{2}} - \sqrt{\dfrac{x}{6}}$, and $x > 0$.

Solution. **Discussion.** Each radical has a fraction as radicand. We will use the multiplication property of one to change the denominator of each fraction to a perfect square. Then we will use the quotient property of radicals to write each radical as a quotient of square roots.

$$\sqrt{\frac{2x}{3}} + \sqrt{\frac{3x}{2}} - \sqrt{\frac{x}{6}}$$ The given expression

$$= \sqrt{\frac{2x}{3} \cdot \frac{3}{3}} + \sqrt{\frac{3x}{2} \cdot \frac{2}{2}} - \sqrt{\frac{x}{6} \cdot \frac{6}{6}}$$ Multiplication property of one

$$= \sqrt{\frac{6x}{9}} + \sqrt{\frac{6x}{4}} - \sqrt{\frac{6x}{36}}$$ Simplify the products.

$$= \frac{\sqrt{6x}}{3} + \frac{\sqrt{6x}}{2} - \frac{\sqrt{6x}}{6}$$ $\sqrt{\frac{a}{b}} = \frac{\sqrt{a}}{\sqrt{b}}$

$$= \frac{2\sqrt{6x}}{6} + \frac{3\sqrt{6x}}{6} - \frac{\sqrt{6x}}{6}$$ The LCM of denominators is 6.

$$= \frac{2\sqrt{6x} + 3\sqrt{6x} - \sqrt{6x}}{6}$$ Combine the numerators.

$$= \frac{4\sqrt{6x}}{6}$$ $x\sqrt{b} + y\sqrt{b} = (x + y)\sqrt{b}$

$$= \frac{2\sqrt{6x}}{3}$$ Reduce the fraction.

Rationalizing the Denominator

Condition 3 for the preferred form of an expression with radicals specifies that no term may contain a radical in the denominator. If a term does contain a radical in the denominator, and the number is irrational, then the multiplication property of one is used to change it to a rational number. The process of changing an irrational number in a denominator to a rational number is called **rationalizing the denominator.**

Example 3.   Simplify: $\dfrac{2}{\sqrt{45}}$

Solution.   **Discussion.** This expression is not in preferred form because the denominator is a radical. Before removing the radical from the denominator, we will simplify it by removing the perfect square factor 9.

$$\frac{2}{\sqrt{45}}$$ The given expression

$$= \frac{2}{3\sqrt{5}}$$ $\sqrt{45} = \sqrt{9 \cdot 5} = 3\sqrt{5}$

$$= \frac{2}{3\sqrt{5}} \cdot \frac{\sqrt{5}}{\sqrt{5}}$$ Multiplication property of one

$$= \frac{2\sqrt{5}}{15}$$ $\sqrt{5 \cdot 5} = 5$ and $3 \cdot 5 = 15$

One can imagine why $\frac{2\sqrt{5}}{15}$ is preferred to $\frac{2}{\sqrt{45}}$. Suppose you needed to approximate these expressions using only paper and pencil. Using a calculator for the following approximations,

$$\frac{2\sqrt{5}}{15} \approx \frac{2(2.23606\ldots)}{15} \approx \frac{4.47213\ldots}{15}$$ The irrational number is divided by the rational number 15.

$$\frac{2}{\sqrt{45}} \approx \frac{2}{6.70820\ldots}$$ The rational number 2 is divided by the irrational number $\sqrt{45}$.

both expressions simplify to $0.29814\ldots$, but if we are restricted to paper and pencil, the division by 15 appears more attractive, and thus traditionally has been the preferred form.

Conjugates

Consider the expressions in examples **a** and **b**.

a. $\dfrac{7}{4 + \sqrt{2}}$ **b.** $\dfrac{3 + \sqrt{6}}{5 - 2\sqrt{6}}$

These expressions are not simplified because they have radicals in the denominator. Notice also that the denominators are *binomials*. To rationalize these denominators requires more than simply multiplying top and bottom by the square root. To illustrate, consider what happens when we multiply the expression in example **a** by $\frac{\sqrt{2}}{\sqrt{2}}$:

$$\frac{7}{4 + \sqrt{2}} \cdot \frac{\sqrt{2}}{\sqrt{2}} = \frac{7\sqrt{2}}{4\sqrt{2} + 2}$$ The denominator still has a term with a square root factor.

The radical term can be removed from the denominator by multiplying the top and bottom by what is called the **conjugate** of the denominator. For $x + y$, the conjugate is $x - y$. For $x - y$, the conjugate is $x + y$. To illustrate, the conjugate of $4 + \sqrt{2}$ is $4 - \sqrt{2}$, and the conjugate of $5 - 2\sqrt{6}$ is $5 + 2\sqrt{6}$. The conjugates are used to rationalize the denominators of examples **a** and **b**.

a. $\dfrac{7}{4 + \sqrt{2}} \cdot \dfrac{4 - \sqrt{2}}{4 - \sqrt{2}}$ **b.** $\dfrac{3 + \sqrt{6}}{5 - 2\sqrt{6}} \cdot \dfrac{5 + 2\sqrt{6}}{5 + 2\sqrt{6}}$

$= \dfrac{7(4 - \sqrt{2})}{4^2 - (\sqrt{2})^2}$ $= \dfrac{15 + 6\sqrt{6} + 5\sqrt{6} + 2(6)}{5^2 - (2\sqrt{6})^2}$

$= \dfrac{7(4 - \sqrt{2})}{16 - 2}$ $= \dfrac{27 + 11\sqrt{6}}{25 - 24}$

$= \dfrac{4 - \sqrt{2}}{2}$ $= 27 + 11\sqrt{6}$

Example 4.   Rationalize the denominator: $\dfrac{6}{-2 - \sqrt{10}}$

Solution.   **Discussion.** The conjugate of $-2 - \sqrt{10}$ is $-2 + \sqrt{10}$.

$$\frac{6}{-2 - \sqrt{10}} \qquad \text{The given expression}$$

$$= \frac{6}{-2 - \sqrt{10}} \cdot \frac{-2 + \sqrt{10}}{-2 + \sqrt{10}} \qquad \text{Multiplication property of one}$$

$$= \frac{6(-2 + \sqrt{10})}{(-2)^2 - (\sqrt{10})^2} \qquad (a - b)(a + b) = a^2 - b^2$$

$$= \frac{6(-2 + \sqrt{10})}{4 - 10} \qquad (\sqrt{10})^2 = 10$$

$$= \frac{6(-2 + \sqrt{10})}{-6} \qquad \text{Simplify the denominator.}$$

$$= -1 \cdot (-2 + \sqrt{10}) \qquad \frac{6}{-6} = -1$$

$$= 2 - \sqrt{10}$$

SECTION 8-4.   Practice Exercises

In exercises **1–12**, write in simplified form.

[Example 1]   **1.** $3\sqrt{40}$ **2.** $4\sqrt{27}$ **3.** $-5\sqrt{72}$

4. $-8\sqrt{8}$ **5.** $2\sqrt{363}$ **6.** $5\sqrt{125}$

7. $\dfrac{10\sqrt{18}}{3}$ **8.** $\dfrac{8\sqrt{75}}{5}$ **9.** $-9\sqrt{90}$

10. $-7\sqrt{128}$ **11.** $\dfrac{3\sqrt{50}}{5} - \dfrac{7\sqrt{32}}{2}$ **12.** $\dfrac{5\sqrt{12}}{2} - \dfrac{4\sqrt{27}}{3}$

In exercises **13–24**, simplify. Assume variables are positive numbers.

[Example 2]   **13.** $\sqrt{\dfrac{2}{3}}$ **14.** $\sqrt{\dfrac{5}{7}}$ **15.** $\sqrt{\dfrac{t}{10}}$

16. $\sqrt{\dfrac{3t}{2}}$ **17.** $-\sqrt{\dfrac{18}{5}}$ **18.** $-\sqrt{\dfrac{20}{11}}$

19. $2\sqrt{\dfrac{u}{6}}$ **20.** $3\sqrt{\dfrac{u}{15}}$

21. $\sqrt{\dfrac{3x}{2}} + \sqrt{\dfrac{2x}{3}}$ **22.** $\sqrt{\dfrac{5x}{2}} - \sqrt{\dfrac{2x}{5}}$

23. $\sqrt{\dfrac{2a}{3}} - \sqrt{\dfrac{3b}{2}} + \sqrt{\dfrac{a}{6}} + \sqrt{\dfrac{b}{6}}$ **24.** $\sqrt{\dfrac{2a}{5}} - \sqrt{\dfrac{4b}{5}} + \sqrt{\dfrac{5a}{2}} - \sqrt{\dfrac{5b}{4}}$

In exercises **25–44**, simplify.

[Example 3]   **25.** $\dfrac{1}{\sqrt{10}}$ **26.** $\dfrac{1}{\sqrt{13}}$ **27.** $\dfrac{5}{\sqrt{15}}$

28. $\dfrac{3}{\sqrt{21}}$ **29.** $\dfrac{-2}{\sqrt{6}}$ **30.** $\dfrac{-6}{\sqrt{30}}$

31. $\dfrac{4}{\sqrt{24}}$ **32.** $\dfrac{3}{\sqrt{75}}$ **33.** $\dfrac{-2}{\sqrt{200}}$

34. $\dfrac{-5}{\sqrt{500}}$ **35.** $\dfrac{7x}{\sqrt{14}}$ **36.** $\dfrac{3x}{\sqrt{21}}$

[Example 4] **37.** $\dfrac{-2}{2+\sqrt{2}}$ **38.** $\dfrac{-6}{3+\sqrt{3}}$ **39.** $\dfrac{4}{3-\sqrt{5}}$

40. $\dfrac{14}{3-\sqrt{2}}$ **41.** $\dfrac{-8}{1-\sqrt{3}}$ **42.** $\dfrac{-20}{1-\sqrt{5}}$

43. $\dfrac{7}{3+2\sqrt{2}}$ **44.** $\dfrac{6}{5-3\sqrt{3}}$

SECTION 8-4. Ten Review Exercises

In exercises **1–10**, use the points $P(-3, 2)$ and $Q(1, -6)$.

1. Find the slope of $\mathscr{L}_1$ through P and Q.

2. Write in $Ax + By = C$ form an equation of $\mathscr{L}_1$.

3. Write the equation for $\mathscr{L}_1$ in $y = mx + b$ form.

4. Identify the y-intercept of the equation for $\mathscr{L}_1$.

5. Identify the x-intercept of the equation for $\mathscr{L}_1$.

6. Write in $Ax + By = C$ form an equation for $\mathscr{L}_2$ that passes through $R(0, 2)$ and is parallel to $\mathscr{L}_1$.

7. Write in $Ax + By = C$ form an equation for $\mathscr{L}_3$ that passes through $S(2, 2)$ and is perpendicular to $\mathscr{L}_1$.

8. Verify that $\mathscr{L}_3$ is also perpendicular to $\mathscr{L}_2$.

9. Use the equations for $\mathscr{L}_1$ and $\mathscr{L}_3$ to form a system of equations. Solve the system to find the point of intersection of the graphs of $\mathscr{L}_1$ and $\mathscr{L}_3$.

10. Graph $\mathscr{L}_1$, $\mathscr{L}_2$, and $\mathscr{L}_3$ on the same set of axes.

SECTION 8-4. Supplementary Exercises

In exercises **1–18**, simplify.

1. $\sqrt{\dfrac{24}{9}}$ **2.** $\sqrt{\dfrac{80}{49}}$

3. $\sqrt{\dfrac{2}{5}} + \sqrt{\dfrac{5}{2}}$

4. $\sqrt{\dfrac{5}{3}} + \sqrt{\dfrac{3}{5}}$

5. $\sqrt{8} + \dfrac{1}{\sqrt{2}}$

6. $\sqrt{27} + \dfrac{1}{\sqrt{3}}$

7. $\sqrt{12x} + \sqrt{\dfrac{x}{3}}$

8. $\sqrt{8y} + \sqrt{\dfrac{y}{2}}$

9. $3\sqrt{8} + \sqrt{12} - \sqrt{50} + \sqrt{3}$

10. $\sqrt{72} - \sqrt{75} + \sqrt{2} + \sqrt{27}$

11. $2\sqrt{3a} - 3\sqrt{2b} + \sqrt{108a} + \sqrt{200b}$

12. $3\sqrt{5a} + \sqrt{32b} - 4\sqrt{18b} + \sqrt{245a}$

13. $\sqrt{\dfrac{x}{2}} + \sqrt{\dfrac{y}{3}} + \sqrt{18x} - \sqrt{12y}$

14. $\sqrt{\dfrac{x}{2}} - \sqrt{\dfrac{y}{3}} + \dfrac{\sqrt{2x}}{2} - \dfrac{2\sqrt{3y}}{3}$

15. $\dfrac{1 + \sqrt{3}}{2} + \dfrac{1 - \sqrt{3}}{2}$

16. $\dfrac{3 + \sqrt{5}}{2} + \dfrac{3 - \sqrt{5}}{2}$

17. $\dfrac{\sqrt{2} + \sqrt{3}}{4} - \dfrac{\sqrt{2} - \sqrt{3}}{4}$

18. $\dfrac{\sqrt{5} + \sqrt{7}}{4} - \dfrac{\sqrt{5} - \sqrt{7}}{4}$

In advanced mathematics classes it is sometimes necessary to change an irrational number numerator to a rational number. The instructions are given to "rationalize the numerator."

In exercises **19–30**, rationalize the numerator.

19. $\dfrac{\sqrt{2}}{6}$

20. $\dfrac{\sqrt{3}}{12}$

21. $\dfrac{\sqrt{5}}{10}$

22. $\dfrac{\sqrt{7}}{14}$

23. $\dfrac{2 + \sqrt{2}}{4}$

24. $\dfrac{-3 - \sqrt{3}}{6}$

25. $\dfrac{-3 - \sqrt{5}}{2}$

26. $\dfrac{-6 - \sqrt{3}}{11}$

27. $\dfrac{3 - 2\sqrt{11}}{7}$

28. $\dfrac{5 + 3\sqrt{7}}{19}$

29. $\dfrac{\sqrt{3} + \sqrt{5}}{\sqrt{6} + \sqrt{5}}$

30. $\dfrac{\sqrt{3} - \sqrt{2}}{\sqrt{7} + \sqrt{2}}$

Definition 8.4. Cube root of a number
    If $a^3 = b$, then a is a **cube root** of b.

For any real number b there is exactly one cube root, which can be written $\sqrt[3]{b}$.

Example. Simplify:

  a. $\sqrt[3]{8}$ **b.** $\sqrt[3]{-27}$

Solution. **a.** Since $2^3 = 8$, it follows that $\sqrt[3]{8} = 2$.

  b. Since $(-3)^3 = -27$, it follows that $\sqrt[3]{-27} = -3$.

The cube roots of some numbers can be *simplified* by removing from under the radical sign any **perfect cube factors.** The following list contains the first ten perfect cubes.

$$1^3 = 1 \qquad 2^3 = 8 \qquad 3^3 = 27 \qquad 4^3 = 64 \qquad 5^3 = 125$$

$$6^3 = 216 \qquad 7^3 = 343 \qquad 8^3 = 512 \qquad 9^3 = 729 \qquad 10^3 = 1{,}000$$

In exercises **31–48**, simplify.

31. $\sqrt[3]{216}$ **32.** $\sqrt[3]{343}$ **33.** $\sqrt[3]{-1{,}000}$

34. $\sqrt[3]{-125}$ **35.** $\sqrt[3]{40}$ **36.** $\sqrt[3]{48}$

37. $\sqrt[3]{192}$ **38.** $\sqrt[3]{189}$ **39.** $\sqrt[3]{375}$

40. $\sqrt[3]{320}$ **41.** $\sqrt[3]{-250}$ **42.** $\sqrt[3]{-135}$

43. $\sqrt[3]{32}$ **44.** $\sqrt[3]{243}$ **45.** $\sqrt[3]{\dfrac{x}{8}}$

46. $\sqrt[3]{\dfrac{x^2}{729}}$ **47.** $\sqrt[3]{0.008}$ **48.** $\sqrt[3]{-0.064}$

SECTION 8-5. Equations with Square Root Radicals

KEY TOPICS IN THIS SECTION

1. A definition of a radical equation

2. A procedure for solving a radical equation

3. A definition of extraneous solutions

In previous chapters we have solved several different types of equations.

Previously studied equations	Example
a. Linear equation in one variable	$3t - 5 = 9 + 7t$
b. Quadratic equation in one variable	$6u^2 = 5 + 7u$
c. Fractional equation in one variable	$\dfrac{x}{2x-1} + \dfrac{1}{x+2} = \dfrac{5x+1}{2x^2+3x-2}$
d. Linear equation in two variables	$3x - 5y = 10$

In this section we will study equations in which at least one term has a square root factor.

A Definition of a Radical Equation

Examples **a** and **b** are equations that do not fit the form of any equation we have studied thus far in this text.

a. $\sqrt{2t - 3} - 5 = 2$ **b.** $\sqrt{u} + 4 = u - 2$

These equations are called **radical equations.**

> **Definition 8.5.   A radical equation**
>   A **radical equation** is one containing one or more terms with a variable in a radicand.

In example **a** the radical term is $\sqrt{2t - 3}$. In example **b** the radical term is $\sqrt{u}$.

A Procedure for Solving a Radical Equation

To solve equations that contain square root terms, we can use the following property of equality.

> **A property of equality**
>   If $X = Y$ has any real solutions, then $X^2 = Y^2$ contains the solutions.

This property of equality states that squaring both sides of an equation yields an equation containing the solutions of the given equation. Notice that this property does not guarantee that the squared equation will be equivalent to the given one. (Recall that equivalent equations have the same solutions.) Thus, squaring both sides of an equation will not *lose* any solutions. However, the squaring process may *pick up some apparent solutions* that are not solutions of the given equation. These **extra solutions** will be discussed later in this section.

> **A procedure for solving an equation with square-root terms:**
>
> **Step 1.** Isolate a square-root term on one side of the equation.
>
> **Step 2.** Square both sides of the equation.
>
> **Step 3.** If no square-root terms remain, then solve the equation obtained in Step 2. If a square-root term remains, then repeat Steps 1 and 2.
>
> **Step 4.** Check each solution in the given equation.

Example 1.   Solve: $\sqrt{2t - 3} - 5 = 2$

Solution.   **Discussion.** The given equation contains only one square-root term. This term can be isolated on one side of the equation.

Step 1. Isolate the square-root term on one side of the equation.

$$\sqrt{2t - 3} = 7$$

Step 2. Square both sides of the equation.

$$(\sqrt{2t - 3})^2 = 7^2$$

$$2t - 3 = 49 \qquad (\sqrt{b})^2 = b$$

Step 3. Solve the resulting equation.

$$2t = 52 \qquad \text{Add 3 to both sides.}$$

$$t = 26 \qquad \text{26 is an apparant solution.}$$

Step 4. Check the solution in the given equation.

$$\sqrt{2(26) - 3} - 5 = 2$$

$$\sqrt{49} - 5 = 2$$

$$7 - 5 = 2, \text{ true}$$

The solution of the given equation is 26.

Example 2. Solve: $a - \sqrt{5a - 6} = 0$

Solution. **Step 1.** Isolate the square-root term on one side of the equation.

$$a = \sqrt{5a - 6}$$

Step 2. Square both sides of the equation.

$$a^2 = (\sqrt{5a - 6})^2$$

$$a^2 = 5a - 6 \qquad (\sqrt{b})^2 = b$$

Step 3. Solve the resulting equation.

$$a^2 - 5a + 6 = 0 \qquad \text{A quadratic equation}$$

$$(a - 3)(a - 2) = 0 \qquad \text{Factor.}$$

$$a - 3 = 0 \quad \text{or} \quad a - 2 = 0 \qquad \begin{array}{l}\text{Zero-product}\\\text{property}\end{array}$$

$$a = 3 \qquad\qquad a = 2 \qquad \text{Solve for } a.$$

Step 4. Check each solution in the given equation.

$$\text{If } a = 3, 3 - \sqrt{5(3) - 6} = 0$$

$$3 - \sqrt{9} = 0, \text{ true}$$

$$\text{If } a = 2, 2 - \sqrt{5(2) - 6} = 0$$

$$2 - \sqrt{4} = 0, \text{ true}$$

Both apparent solutions check; thus the solutions of the given equation are 2 and 3.

A Definition of Extraneous Solutions

Squaring both sides of an equation does not necessarily yield an equivalent equation. That is, the squared equation may contain solutions that are not solutions of the given equation. This fact can be illustrated with the following equation:

$$t = 3 \qquad \text{The only solution is 3.}$$

$$t^2 = 9 \qquad \text{Now square both sides of the equation.}$$

$$t^2 - 9 = 0 \qquad \text{Set the equation equal to 0.}$$

$$(t + 3)(t - 3) = 0 \qquad \text{Factor } t^2 - 9.$$

$$t + 3 = 0 \quad \text{or} \quad t - 3 = 0 \qquad \text{Zero-product property}$$

$$t = -3 \qquad\qquad t = 3 \qquad \begin{array}{l}\text{The squared equation has } -3 \text{ and } 3\\\text{as roots.}\end{array}$$

Only the 3 is a solution of the given equation. The apparent solution -3 does not check in the given equation $(-3 \neq 3)$. The -3 is an **extraneous solution**.

Definition 8.6.   Extraneous solutions
   If c is a real number that is a solution of $X^2 = Y^2$ but not a solution of $X = Y$, then c is called an **extraneous solution** of $X = Y$.

Example 3.   Solve: $\sqrt{u} + 4 = u - 2$

Solution.   **Step 1.** $\sqrt{u} = u - 6$

   Step 2. $(\sqrt{u})^2 = (u - 6)^2$

                $u = u^2 - 12u + 36$

   Step 3. $0 = u^2 - 13u + 36$

                $0 = (u - 9)(u - 4)$

                $u - 9 = 0$ or $u - 4 = 0$

                $u = 9$ $u = 4$

   Step 4. If $u = 9$, $\sqrt{9} + 4 = 9 - 2$

                          $3 + 4 = 7$, true

                If $u = 4$, $\sqrt{4} + 4 = 4 - 2$

                          $2 + 4 = 2$, false

   Since 4 yields a false number statement, the apparent solution 4 is rejected as extraneous. Thus, the only solution of the given equation is 9.

   In example 4 the squaring process must be applied twice to eliminate all the square-root terms in the equation.

Example 4.   Solve: $\sqrt{k} + 2 = \sqrt{3k - 2}$

Solution.   **Discussion.** This equation contains three terms. Two are square-root terms. These terms cannot both be isolated on one side of the equation at the same time. Therefore, a square-root term will still remain after we square the equation the first time.

$$(\sqrt{k} + 2)^2 = (\sqrt{3k - 2})^2$$

$k + 4\sqrt{k} + 4 = 3k - 2$	$(x + y)^2 = x^2 + 2xy + y^2$
$4\sqrt{k} = 2k - 6$	Isolate the square-root term.
$2\sqrt{k} = k - 3$	Multiply both sides by $\frac{1}{2}$.
$(2\sqrt{k})^2 = (k - 3)^2$	Square both sides.
$4k = k^2 - 6k + 9$	$(2\sqrt{k})^2 = 2^2(\sqrt{k})^2$
$k^2 - 10k + 9 = 0$	Set equal to 0.
$(k - 9)(k - 1) = 0$	Factor the trinomial.

$$k - 9 = 0 \quad \text{or} \quad k - 1 = 0 \qquad \text{Zero-product property}$$

$$k = 9 \qquad\qquad\quad k = 1 \qquad \text{Solve for } k.$$

If $k = 9$, $\sqrt{9} + 2 = \sqrt{3(9) - 2}$

$$3 + 2 = \sqrt{25}, \text{ true}$$

If $k = 1$, $\sqrt{1} + 2 = \sqrt{3(1) - 2}$

$$1 + 2 = \sqrt{1}, \text{ false}$$

The apparent solution 1 is extraneous. The solution of the given equation is 9.

SECTION 8-5.   Practice Exercises

In exercises **1–20**, solve.

[Example 1]

1. $\sqrt{x} = 4$

2. $\sqrt{x} = 9$

3. $\sqrt{2y} = 6$

4. $\sqrt{5y} = 10$

5. $\sqrt{b + 2} = 3$

6. $\sqrt{b - 3} = 5$

7. $\sqrt{2y - 1} = 13$

8. $\sqrt{2y + 5} = 7$

9. $\sqrt{10t + 9} = 3$

10. $\sqrt{15t + 25} = 5$

[Example 2]

11. $x = \sqrt{7x - 10}$

12. $x = \sqrt{13x - 40}$

13. $\sqrt{4a - 3} - a = 0$

14. $a - \sqrt{5a - 6} = 0$

15. $b = \sqrt{6b - 5}$

16. $\sqrt{8b - 15} = b$

17. $\sqrt{5t + 1} - 1 = t$

18. $\sqrt{2t + 4} + 2 = t$

19. $2 - k = \sqrt{6 - 3k}$

20. $k + 3 = \sqrt{7k + 21}$

In exercises **21–44**, solve.

[Example 3]

21. $\sqrt{z} + 2 = z$

22. $2\sqrt{z} = z - 3$

23. $\sqrt{2q - 1} = q - 2$

24. $\sqrt{3q + 4} = q - 2$

25. $-p = \sqrt{-5p - 6}$

26. $-p = \sqrt{-6p - 5}$

27. $s = 7 + \sqrt{s - 5}$

28. $s = \sqrt{s + 3} - 1$

29. $\sqrt{x - 5} + 2 = 0$

30. $\sqrt{x + 8} + 3 = 0$

31. $\sqrt{t + 3} = \sqrt{2t + 3}$

32. $\sqrt{2t + 9} = \sqrt{3t - 5}$

33. $\sqrt{2a + 1} = \sqrt{3a - 4}$

34. $\sqrt{2a - 2} = \sqrt{3a - 4}$

[Example 4]

35. $\sqrt{p - 9} + \sqrt{p} = 1$

36. $\sqrt{p - 5} + \sqrt{p} = 5$

37. $\sqrt{m} = \sqrt{m - 5} + 1$

38. $\sqrt{m + 7} - 1 = \sqrt{m}$

39. $\sqrt{u} - \sqrt{u + 11} = -1$

40. $\sqrt{u + 24} - \sqrt{u} = 2$

41. $\sqrt{n + 3} + \sqrt{n - 3} = 3$ **42.** $\sqrt{n + 3} - \sqrt{n - 3} = 3$

43. $\sqrt{w + 5} - \sqrt{w - 5} = 5$ **44.** $\sqrt{w + 5} + \sqrt{w - 5} = 5$

SECTION 8-5.   Ten Review Exercises

In exercises **1–8**, do the indicated operations.

1. $(9ab + 11)(11ab - 9)$ **2.** $(7t - 3u)^2$

3. $(2x + 5)(3y - 7)$ **4.** $\dfrac{-27u^4v^2 + 108u^3v^3 + 9u^2v^4}{9u^2v^2}$

5. $\dfrac{3a^2 - 8a - 3}{a^2 - 2a} \cdot \dfrac{2a}{2a - 6}$ **6.** $\dfrac{2x^2 + 3x - 9}{6x^2 - 7x - 3} \div \dfrac{x^2 + 6x + 9}{3x^2 - 8x - 3}$

7. $\dfrac{13}{15y} + \dfrac{7}{15y}$ **8.** $\dfrac{8a + 5}{a + 3} - \dfrac{2a - 13}{a + 3}$

In exercises **9** and **10**, factor completely.

9. $a^3b^3 + 125$ **10.** $35yz - 63y + 10z - 18$

SECTION 8-5.   Supplementary Exercises

In exercises **1–14**, solve.

1. $2a + 1 = \sqrt{9 - 10a}$ **2.** $\sqrt{13 - 6a} = 3a + 1$

3. $\sqrt{18b - 5} + 1 = 6b$ **4.** $2b = \sqrt{6 + 5b} - 3$

5. $2c = 3 + \sqrt{29 - 10c}$ **6.** $\sqrt{79 - 18c} + 4 = 3c$

7. $\sqrt{5w + 1} + 3 = 8$ **8.** $\sqrt{3w + 1} - 7 = 1$

9. $\sqrt{y + 5} - 2 = \sqrt{y - 3}$ **10.** $\sqrt{y + 2} - 1 = \sqrt{y - 3}$

11. $\sqrt{3t + 1} - \sqrt{t - 7} = \sqrt{t}$ **12.** $\sqrt{5t + 4} - \sqrt{2t + 1} = \sqrt{t - 3}$

13. $\sqrt{3k + 18} = \sqrt{k} + \sqrt{2k + 6}$ **14.** $\sqrt{16k - 5} = \sqrt{k + 1} + \sqrt{k}$

The property of equality stated in this section can be extended to raising both sides of an equation to the power 3, or the power 4, and so on. Thus, the raising-to-powers property of equality can be used to solve cube root equations, and so on.

Example.   Solve the following cube-root equation.

$$\sqrt[3]{x + 12} = 5$$

Solution.   **Discussion.** The three in the radical sign indicates a cube-root. To solve the equation we cube both sides.

$$(\sqrt[3]{x + 12})^3 = 5^3 \qquad \text{Cube both sides.}$$

$$x + 12 = 125 \qquad (\sqrt[3]{x + 12})^3 = x + 12$$

$$x = 113 \qquad \text{Subtract 12 from both sides.}$$

$$\text{Check: } \sqrt[3]{113 + 12} = \sqrt[3]{125} = 5 \qquad \text{The solution is 113.}$$

In exercises **15–22**, solve.

15. $\sqrt[3]{x} = 2$

16. $\sqrt[3]{x} = 5$

17. $\sqrt[3]{s + 1} = 4$

18. $\sqrt[3]{s - 2} = 3$

19. $\sqrt[3]{2t - 3} = 1$

20. $\sqrt[3]{4t + 1} = 1$

21. $\sqrt[5]{p + 10} = 2$

22. $\sqrt[5]{2p - 8} = 2$

SECTION 8-1.   Summary Exercises

1. Determine whether **a**, **b**, **c**, or **d** is a square root of 289.

    a. -17

    b. -21

    c. 17

    d. 21

1. _____

In exercises **2–4**, find the indicated square roots.

2. $\sqrt{81}$

2. _____

3. $-\sqrt{\dfrac{25}{16}}$

3. _____

4. $\sqrt{0.49}$

4. _____

In exercises **5** and **6**, solve each problem.

5. Is the following number rational or irrational?

$$\sqrt{\frac{256}{101}}$$

5.

6. Use a table to approximate the indicated square root to 3 decimal places.

$$\sqrt{89}$$

6. _____

In exercises **7** and **8**, use a calculator to approximate to 3 decimal places.

7. $\sqrt{45.6}$

7. _____

8. $\dfrac{\sqrt{8} + \sqrt{12}}{\sqrt{20}}$

8. _____

SECTION 8-2. Summary Exericses

In exercises **1–8**, simplify. Write all fractions with a denominator that is an integer.

1. $\sqrt{45} \cdot \sqrt{5}$

1. _____

2. $\sqrt{7} \cdot \sqrt{14}$

2. _____

3. $\sqrt{529}$

3. _____

4. $\sqrt{363}$

4. _____

5. $\dfrac{-\sqrt{162}}{\sqrt{2}}$

5. _____

6. $\dfrac{\sqrt{350}}{\sqrt{14}}$

6. _____

7. $\sqrt{\dfrac{5}{36}}$

7. _____

8. $\sqrt{\dfrac{17}{3}}$

8. _____

SECTION 8-3. Summary Exercises

In exercises **1–8**, simplify. Assume variables are positive.

1. $\sqrt{11} - 3\sqrt{7} + 6\sqrt{11} + 14\sqrt{7}$

1. _____

2. $2\sqrt{108} + \sqrt{27}$

2. _____

3. $\sqrt{243t} - \sqrt{300t} + 3\sqrt{3t}$

3. _____

4. $10\sqrt{10x} - 12\sqrt{3y} - \sqrt{75y} + \sqrt{160x}$

4. _____

5. $\dfrac{9 + \sqrt{45}}{3}$

5. _____

6. $\sqrt{14}(3 - \sqrt{7})$

6. _____

7. $(8 + 5\sqrt{2})(3 - 2\sqrt{2})$

7. _____

8. $(9 - 3\sqrt{5})(9 + 3\sqrt{5})$

8. _____

SECTION 8-4. Summary Exercises

Answer

In exercises **1–8**, simplify.

1. $-9\sqrt{90}$

1. _____

2. $\dfrac{2\sqrt{147}}{21}$

2. _____

3. $\sqrt{\dfrac{8}{7}}$

3. _____

4. $\dfrac{\sqrt{45b}}{9} + \sqrt{20b} - \sqrt{\dfrac{b}{5}}$

4. _____

5. $\dfrac{7}{\sqrt{6}} + \sqrt{\dfrac{2}{3}} - \sqrt{\dfrac{3}{2}}$

5. _____

6. $\dfrac{-6}{\sqrt{60}} + \dfrac{3\sqrt{15}}{5}$

6. _____

7. $\dfrac{4}{2 + \sqrt{5}} - \dfrac{1 + \sqrt{5}}{2 - \sqrt{5}}$

7. _____

8. $\dfrac{-2}{1 - 3\sqrt{5}}$

8. _____

SECTION 8-5. Summary Exercises

Answer

In exercises **1–8**, solve.

1. $\sqrt{5a} = 10$

1. _____

2. $\sqrt{3u + 1} = 5$

2. _____

3. $3\sqrt{z} = \sqrt{7z + 2}$

3. _____

4. $x = \sqrt{10x - 24}$

4. _____

5. $\sqrt{2n - 1} + 2 = n$

5. _____

6. $p + 5 = \sqrt{2p + 13}$

6. _____

7. $\sqrt{3y + 9} = 2\sqrt{y}$

7. _____

8. $5 - \sqrt{u} = \sqrt{u + 5}$

8. _____

CHAPTER 8. Review Exercises

In exercises **1–6**, find the indicated square root. If the answer is not a real number, state "Not real."

1. Square roots of 144

2. $\sqrt{144}$

3. $\sqrt{\dfrac{100}{49}}$

4. $-\sqrt{64}$

5. $\sqrt{0.25}$

6. $\sqrt{-81}$

In exercises **7** and **8**, use a table or a calculator to approximate the indicated square roots to three decimal places.

7. $\sqrt{66}$

8. $-\sqrt{92}$

In exercises **9–28**, simplify. Give the answer in preferred form.

9. $\sqrt{12}\sqrt{3}$

10. $\sqrt{6}\sqrt{24}$

11. $\sqrt{448}$

12. $\sqrt{245}$

13. $\sqrt{\dfrac{3}{4}}$

14. $\sqrt{\dfrac{19}{100}}$

15. $\dfrac{\sqrt{252}}{\sqrt{7}}$

16. $\sqrt{\dfrac{13}{225}}$

17. $(-\sqrt{14})\sqrt{3}$

18. $\dfrac{-\sqrt{34}}{\sqrt{2}}$

19. $2\sqrt{20} + 3\sqrt{45} - \sqrt{125}$

20. $5\sqrt{216x} + 2\sqrt{54x},\ x > 0$

21. $\dfrac{3 + \sqrt{162}}{6}$

22. $\sqrt{10}(9 - \sqrt{2})$

23. $(2 + \sqrt{3})(5 - 2\sqrt{3})$

24. $(7 + 5\sqrt{2})(7 - 5\sqrt{2})$

25. $\dfrac{17\sqrt{125}}{10}$

26. $\sqrt{\dfrac{49}{5}}$

27. $\dfrac{4}{\sqrt{10}}$

28. $\dfrac{5}{2 - \sqrt{3}}$

In exercises **29–32**, solve.

29. $\sqrt{3x} = 6$

30. $2\sqrt{y} - 1 = y - 4$

31. $\sqrt{20 + a} = \sqrt{4 - 3a}$

32. $t + 2 = \sqrt{6 + 3t}$

9

Quadratic Equations

As Ms. Glaston was handing out the graded test on Chapter 8, Dan Coleman raised his hand and said, "Ms. Glaston, Doreen and I started studying the next chapter last night. We noticed the first section covered solving quadratic equations by factoring. Since we covered this in a previous chapter, it wasn't too tough. Then Doreen made up a couple of equations and we tried to solve them. But we couldn't factor them, so we couldn't solve them. Does that mean there are some quadratic equations that can't be solved?"

"Do you or Doreen remember what the equations were?" Ms. Glaston replied. "If you do, would you write them on the board for us?"

Doreen Clark found the equations in her notes and wrote them on the board.

$$(1) \quad 9x^2 + 9x + 1 = 0 \qquad (2) \quad 2y^2 - 2y + 5 = 0$$

"Well," Ms. Glaston asked, "can anyone in the class help Doreen and Dan factor the trinomials in these equations? Remember, we can only use integers to write expressions in factored form."

After studying the equations for some time, Lynne Pickens raised her hand and said, "The trinomial in the first equation can't be factored. The signs are both positive, so we need two integers whose product and sum are both 9. There just aren't two integers that will work."

"The same thing is true of the other trinomial," Dan added. "I tried every combination of integers that the factor-by-grouping technique permits, but none of them work."

"Since we can't factor these trinomials," Doreen said, "does that mean we can't solve the equations? In fact, does that mean the equations don't have any solutions? I just can't imagine math people having any equations that don't have solutions! They seem to have answers for everything!"

"Well, Doreen," Ms. Glaston said with a smile, "let me first assure you that 'math people' do not have answers for everything, although they are constantly working on the unsolved problems in an attempt to find answers. However, there *are* solutions to the equations you have written on the board. In this chapter we will study a formula that will enable us to solve any quadratic equation. Because the factoring method is the easiest way to solve such an equation, we will first review this technique. But before we begin Chapter 9, let's discuss the results of the Chapter 8 test that I just returned."

SECTION 9-1. Solving Quadratic Equations by Factoring

**KEY TOPICS
IN THIS SECTION**

1. The standard form of a quadratic equation

2. The zero-product property

3. Solving quadratic equations by factoring

4. Some applied problems

The study of quadratic equations will begin with a review of the topics studied in Section 4-6. In particular we will solve quadratic equations by factoring and using the zero product property.

The Standard Form of a Quadratic Equation

Quadratic equations in one variable can be written in many different forms. Definition 9.1 identifies the standard form for a quadratic equation in the variable x.

> **Definition 9.1. Standard form of a quadratic equation in x**
> In **standard form**, a quadratic equation in x is written as
>
> $$ax^2 + bx + c = 0 \qquad \text{or} \qquad 0 = ax^2 + bx + c$$
>
> where a, b, and c are real numbers and a is positive.

Many quadratic equations are not given in standard form. Such equations can be changed to standard form by making one side of the equation zero. The polynomial in the equation should then be written in descending powers of the variable.

Example 1. **a.** Write in standard form.

        b. Identify a, b, and c.

        $5t = 2t^2 - 12$

Solution. **a.** $5t = 2t^2 - 12$ The given equation

        $-2t^2 + 5t + 12 = 0$ Change the right side to 0.

        $2t^2 - 5t - 12 = 0$ Multiply both sides by (-1).

        b. Comparing $2t^2 - 5t - 12$ with $at^2 + bt + c$, $a = 2$, $b = -5$, and $c = -12$.

Example 2. **a.** Write in standard form.

        b. Identify a, b and c.

        $3x(x - 1) + 3 = (2x + 1)^2$

Solution. **a.** $3x(x - 1) + 3 = (2x + 1)^2$ The given equation

        $3x^2 - 3x + 3 = 4x^2 + 4x + 1$ $(a + b)^2 = a^2 + 2ab + b^2$

        $0 = x^2 + 7x - 2$ Collect terms on right side.

b. Comparing $x^2 + 7x - 2$ with $ax^2 + bx + c$, $a = 1$, $b = 7$, and $c = -2$.

The Zero-Product Property

The zero-product property is simply an assertion that a product can be zero only if at least one of the factors of the product is zero.

> **Zero-product property**
> If $x \cdot y = 0$, then $x = 0$, or $y = 0$, or both x and y are 0.

Example 3.　Solve and check:

$$(3m + 2)(5m - 8) = 0$$

Solution.　**Discussion.**　The given equation shows an indicated product of two factors that equals 0. According to the zero-product property, a product can only equal 0 if at least one of the factors is 0. This assertion permits us to replace the given equation with two linear equations.

$$3m + 2 = 0 \qquad \text{or} \qquad 5m - 8 = 0 \qquad \text{Zero-product property.}$$

$$3m = -2 \qquad\qquad 5m = 8$$

$$m = \frac{-2}{3} \qquad\qquad m = \frac{8}{5}$$

$$\text{If } m = \frac{-2}{3}, \left(3\left(\frac{-2}{3}\right) + 2\right)\left(5\left(\frac{-2}{3}\right) - 8\right) = 0$$

$$(-2 + 2)\left(\frac{-10}{3} - \frac{24}{3}\right) = 0$$

$$0\left(\frac{-34}{3}\right) = 0, \text{ true}$$

$$\text{If } m = \frac{8}{5}, \left(3\left(\frac{8}{5}\right) + 2\right)\left(5\left(\frac{8}{5}\right) - 8\right) = 0$$

$$\left(\frac{24}{5} + \frac{10}{5}\right)(8 - 8) = 0$$

$$\frac{34}{5}(0) = 0, \text{ true}$$

The solutions of the given equation are $\frac{-2}{3}$ and $\frac{8}{5}$.

Solving Quadratic Equations by Factoring

The following five steps provide a procedure for solving a quadratic equation by factoring.

To solve a quadratic equation using the zero-product property,

Step 1. if necessary, write the given equation in standard form.

Step 2. factor the nonzero polynomial.

Step 3. use the zero-product property to set both factors equal to zero.

Step 4. solve both equations obtained in Step 3.

Step 5. check the solutions obtained in Step 4 in the given equation.

Example 4. Solve and check:

$$(t + 2)^2 = 7 - 2(5t - 24)$$

Solution. **Step 1.** Change to standard form.

$$t^2 + 4t + 4 = 7 - 10t + 48 \qquad \text{Remove parentheses.}$$

$$t^2 + 14t - 51 = 0 \qquad \text{Collect terms on left side.}$$

Step 2. Factor the trinomial.

$$(t + 17)(t - 3) = 0$$

Step 3. Set both factors equal to zero.

$$t + 17 = 0 \qquad \text{or} \qquad t - 3 = 0$$

Step 4. Solve both equations.

$$t = -17 \qquad \text{or} \qquad t = 3$$

Step 5. Check -17 and 3 in the given equations.

$$\text{If } t = -17, (-17 + 2)^2 = 7 - 2(5(-17) - 24)$$

$$(-15)^2 = 7 - 2(-109)$$

$$225 = 7 - (-218), \text{ true}$$

$$\text{If } t = 3, (3 + 2)^2 = 7 - 2(5(3) - 24)$$

$$5^2 = 7 - 2(-9)$$

$$25 = 7 - (-18), \text{ true}$$

The solutions are -17 and 3.

Example 5. Solve and check:

$$\frac{y^2 + 5}{2y} = \frac{2y - 1}{3}$$

Solution. **Step 1.** $6y\left(\dfrac{y^2 + 5}{2y}\right) = 6y\left(\dfrac{2y - 1}{3}\right)$ The LCM of denominators is $6y$.

$$3(y^2 + 5) = 2y(2y - 1)$$

$$3y^2 + 15 = 4y^2 - 2y$$

$$0 = y^2 - 2y - 15$$

Step 2. $0 = (y + 3)(y - 5)$

Step 3. $y + 3 = 0$ or $y - 5 = 0$

Step 4. $y = -3$ $y = 5$

Step 5. The check is omitted.

The solutions are -3 and 5.

Some Applied Problems

Examples 6 and 7 are two applied problems that require quadratic equations to solve. The equations can be solved by factoring and using the zero-product property.

Example 6.   In right triangle ABC the length of the shorter leg is 2 meters less than the length of the longer leg. If the length of the hypotenuse is 10 meters, find the lengths of the legs.

Solution.   **Discussion.** A sketch of triangle ABC is shown in Figure 9-1. The Pythagorean Theorem will be used to solve the problem. Let x represent the length of the longer leg.

$x - 2$ represents the length of the shorter leg.

$$(x - 2)^2 + x^2 = 10^2$$

$$x^2 - 4x + 4 + x^2 = 100$$

$$2x^2 - 4x - 96 = 0$$

$$x^2 - 2x - 48 = 0$$

$$(x + 6)(x - 8) = 0$$

$x + 6 = 0$ or $x - 8 = 0$

$x = -6$ $x = 8$

Since x represents the length of a leg of a triangle, the -6 root is rejected.

The longer leg is 8 meters.
The shorter leg is $8 - 2 = 6$ meters.

Example 7.   Tim and Todd own and operate the TNT Lawn Service Company. One of their accounts is the Sunset Memorial Park. Working alone, Tim takes three hours longer to service the account than when Todd works alone. Last Friday Tim was working on the account by himself for four hours. Todd then arrived and together they completed the job in two more hours. How long would it take Tim working alone to do the job, and how long would it take Todd working alone to do the same job?

Solution.   Let t represent the time Todd takes to do the job alone. Then $t + 3$ represents the time Tim takes to do the job alone.

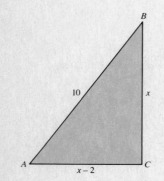

Figure 9-1. Right triangle ABC.

For the given conditions:
Todd worked a total of 2 hours.
Tim worked a total of $4 + 2 = 6$ hours.

	Time to do the Job Alone	Rate at which Work is Done	Amount of Work Done on Job
Todd	t hours	$\dfrac{1}{t}$	$2 \cdot \dfrac{1}{t} = \dfrac{2}{t}$
Tim	$t + 3$ hours	$\dfrac{1}{t + 3}$	$6 \cdot \dfrac{1}{t + 3} = \dfrac{6}{t + 3}$

$$\frac{2}{t} + \frac{6}{t + 3} = 1$$ The 1 represents a completed job.

$$t(t + 3)\left(\frac{2}{t} + \frac{6}{t + 3}\right) = t(t + 3)$$ The LCM of denominators is $t(t + 3)$.

$$2(t + 3) + 6t = t(t + 3)$$ The distributive property.

$$2t + 6 + 6t = t^2 + 3t$$ Remove parentheses.

$$t^2 - 5t - 6 = 0$$ In standard form

$$(t - 6)(t + 1) = 0$$ Factor the trinomial.

$$t = 6 \quad \text{or} \quad t = -1$$ The zero-product property

Since t represents time to complete the job, the -1 is rejected as a solution.

Todd takes 6 hours to work the account alone.
Tim takes $6 + 3 = 9$ hours to work the account alone.

SECTION 9-1. Practice Exercises

In exercises **1–10**,

a. write in standard form.

b. state the values of a, b and c.

[Examples 1 and 2]   **1.** $x^2 = 4x - 3$ **2.** $x^2 = -6x + 1$

3. $3t - 2t^2 = 7$ **4.** $9 = 5t - 6t^2$

5. $a^2 = 11a$ **6.** $a^2 = 14$

7. $5b(b + 3) = 2(b - 3)$ **8.** $3(b^2 - 3) = 2(b + 1)$

9. $x(5x - 3) = (x + 2)^2$ **10.** $(x - 4)^2 = 2x(x - 6)$

In exercises **11–18**, solve.

[Example 3]   **11.** $(x + 4)(3x - 5) = 0$ **12.** $(x - 2)(2x + 7) = 0$

13. $(3y - 1)(2y + 5) = 0$

14. $(4y + 3)(5y + 2) = 0$

15. $s(s + 9) = 0$

16. $s(s - 11) = 0$

17. $(3t + 8)(3t - 8) = 0$

18. $(5t + 6)(5t - 6) = 0$

In exercises **19–34**, solve.

[Examples 4 and 5]

19. $2(a^2 + 1) = 5a$

20. $3(a^2 + 1) = 10a$

21. $5b(b + 3) = 2(b - 3)$

22. $3(b^2 - 3) = b + 1$

23. $2(4y^2 - 1) = 2(y + 1) - 1$

24. $6(y^2 - y) = 1 - 6y^2 - 5y$

25. $7t(7t + 3) + 2 = 2(7t + 3) + 7t$

26. $8t(8t + 4) + 3 = 3(4 + 8t) + 8t$

27. $\dfrac{y^2 + 2y}{2} = \dfrac{y^2 + 4y}{3}$

28. $\dfrac{y(y + 4)}{4} = \dfrac{y(y + 2)}{3}$

29. $\dfrac{3x - 2}{4} = \dfrac{x^2 - 2}{x}$

30. $\dfrac{x^2 + 5}{2x} = \dfrac{2x - 1}{3}$

31. $\dfrac{b^2 - 1}{b} = \dfrac{b + 1}{2}$

32. $\dfrac{b - 1}{2} = \dfrac{b^2 - 1}{b}$

33. $\dfrac{(c - 2)^2}{20} = -3 - c$

34. $\dfrac{(c + 2)^2}{5} = -2c - 9$

In exercises **35–46**, solve.

[Example 6]

35. In a right triangle the length of the hypotenuse is 2 centimeters more than the shorter leg. The length of the longer leg is 1 centimeter more than the shorter leg. Find the lengths of the legs of the triangle.

36. In a right triangle the hypotenuse is 4 inches more than the shorter leg. The length of the longer leg is 2 inches more than the shorter leg. Find the lengths of the legs of the triangle.

37. The hypotenuse of a right triangle is one unit more than the length of the longer leg. The shorter leg is 7 units less than the longer leg. Find the lengths of the three sides.

38. The hypotenuse of a right triangle is two units more than the length of the longer leg. The shorter leg is 7 units less than the longer leg. Find the lengths of the three sides.

39. The longer leg of a right triangle is three inches more than three times the length of the shorter leg. The hypotenuse is 25 inches long. Find the lengths of the legs.

40. The longer leg of a right triangle is one centimeter longer than the shorter leg. The hypotenuse is 29 centimeters long. Find the lengths of the legs.

41. The length of rectangle *ABCD* is 4 meters more than the width. The length of the diagonal is 20 meters. Find the length and width of the rectangle.

42. The length of rectangle *ABCD* is 5 meters more than the width. The length of the diagonal is 25 meters. Find the length and width of the rectangle.

[Example 7] **43.** A painter's son working alone takes eight more hours to do a certain job than the painter. If the painter works alone for three hours and the son works alone for seven hours to complete the job, how long would it take each, working alone, to do the complete job?

44. George and Rapha from Custom Landscaping Company are doing a front yard. Rapha working alone would need 16 more hours to do the job than George. If Rapha works 14 hours and George works 6 hours to complete the project, how long would it take George to do the complete job alone?

45. Cindy and Paula service and maintain the city swimming pools. Cindy takes seven hours longer to do the job than Paula. The pools can be serviced if Paula works two hours and Cindy works ten hours. How long would it take Cindy working alone to service the pools?

46. The old John Deere tractor takes two hours longer to mow the field than the new Ford tractor. If the John Deere works five hours and the Ford works four hours to complete the project, how long would it take the Ford to mow the field alone?

SECTION 9-1.   Ten Review Exercises

In exercises **1–5**, simplify. Assume variables represent positive numbers.

1. $\sqrt{\dfrac{3t}{25}}$

2. $\sqrt{3a} \cdot \sqrt{15b}$

3. $\sqrt{\dfrac{2x}{7}} - \sqrt{\dfrac{7x}{2}} + \sqrt{\dfrac{x}{14}}$

4. $\sqrt{147}$

5. $\dfrac{\sqrt{3} + 2}{\sqrt{3} - 2}$

In exercises **6–10**, solve.

6. $5a - 24 = 10 - 7(4a + 8)$

7. $5a - 24 = -a^2$

8. $\sqrt{5a - 24} = 11$

9. $5a - 24 > 1$

10. (1) $5a - 3b = 27$
(2) $2a + b = 2$

SECTION 9-1.   Supplementary Exercises

In exercises **1–10**, solve.

1. $z^2 = 8z$

2. $z^2 + 5z = 0$

3. $3c^2 - 300 = 0$

4. $2c^2 = 288$

5. $(z - 2)(z + 1) = 4$

6. $(z - 3)(z + 4) = -6$

7. $(a - 2)(a - 3) = 6$ **8.** $(a + 1)(a + 2) = 2$

9. $(b + 3)(b - 3) = 6b + 7$ **10.** $(1 - b)(b + 1) = 10b + 26$

In exercises **11–14**,

a. write the trinomial in factored form.

b. find the solutions x_1 and x_2 of the factored equation.

c. plot the solutions on a number line.

d. pick a value of x between x_1 and x_2 and evaluate the corresponding inequality in the problem.

e. based on the results of part **d**, determine whether the solution set of the inequality is
   (i) $x_1 < x < x_2$
   (ii) $x < x_1$ or $x > x_2$

11. (1) $x^2 - x - 6 = 0$,
    (2) $x^2 - x - 6 < 0$

12. (1) $x^2 - 7x + 6 = 0$,
    (2) $x^2 - 7x + 6 < 0$

13. (1) $2x^2 - 7x - 4 = 0$,
    (2) $2x^2 - 7x - 4 > 0$

14. (1) $3x^2 + 16x + 5 = 0$,
    (2) $3x^2 + 16x + 5 > 0$

In exercises **15–22**, solve.

15. Find two consecutive odd integers, the sum of whose squares is 202.

16. Find two consecutive even integers, the sum of whose squares is 340.

17. One number is five more than another, and the sum of their squares is 193. Find both numbers.

18. Groups A and B are camping at Boulder Creek Campground. The two groups take day hikes and return to the campground every night. Group A can hike at a rate of one mph faster than B. On Tuesday, A headed east and B headed north. After three hours the groups were 15 miles apart. At what rate, in miles per hour, do both groups travel?

19. A boat heading due west travels at seven mph faster than a boat heading due south. If both boats leave port at the same time and are 26 miles apart after two hours, what are their speeds?

20. One number is three less than another. The sum of their squares is 89. Find both numbers.

21. One number is five more than another. The product of the two numbers is 24. Find the largest number.

22. A second number is four more than the first. The product of the two numbers is 77. Find the smaller number.

SECTION 9-2. Solving Equations of the Form $X^2 = k$ and $k > 0$

KEY TOPICS IN THIS SECTION

1. The square-root theorem

2. Equations having the form $X^2 = k$, where X is a monomial

3. Equations having the form $X^2 = k$, where X is a binomial

Many quadratic equations can be written in the form $X^2 = k$, where k is a number. The symbol X represents an expression that includes a variable. The expression can be a monomial or a binomial. The square-root theorem can be used to find the solutions of equations written in the form $X^2 = k$.

The Square–Root Theorem

Every positive number k has two square roots, one positive and one negative. The symbols $\sqrt{k}$ and $-\sqrt{k}$ can be used to indicate the positive and negative square roots respectively. Consider, therefore, an equation in which the solutions are the two square roots of a positive number. Examples **a** and **b** are two such equations.

a. $t^2 = 324$ **b.** $x^2 = 20$

The solutions of these equations are the positive and negative square roots of 324 and 20.

a. $t = \sqrt{324}$ or $t = -\sqrt{324}$ **b.** $x = \sqrt{20}$ or $x = -\sqrt{20}$

$= 18$ $= -18$ $= 2\sqrt{5}$ $= -2\sqrt{5}$

The solutions are 18 and -18.   The solutions are $2\sqrt{5}$ and $-2\sqrt{5}$.

The general statement of these examples is known as the square-root theorem.

> **Theorem 9.1.   The square-root theorem**
> If X is an algebraic expression and k is a real number and
> $$X^2 = k$$
> then $X = \sqrt{k}$ or $X = -\sqrt{k}$.

Example 1. Solve and check:

   a. $y^2 = 441$

   b. $z^2 = 192$

Solution. **Discussion.** Use the square-root theorem to rewrite the given equations as two equations involving square roots of the numbers in the equations. If possible, simplify the indicated square roots.

   a. If $y^2 = 441$, then $y = \sqrt{441}$ or $y = -\sqrt{441}$.

   $y = 21$ $y = -21$

   Check: $21^2 = (-21)^2 = 441$

   The solutions are 21 and -21.

b. If $z^2 = 192$, then $z = \sqrt{192}$ or $z = -\sqrt{192}$.

$$z = 8\sqrt{3} \qquad\qquad z = -8\sqrt{3}$$

Check: $(8\sqrt{3})^2 = (-8\sqrt{3})^2 = 64(3) = 192$

The solutions are $8\sqrt{3}$ and $-8\sqrt{3}$.

Equations Having the Form $X^2 = k$, where X is a Monomial

In the statement of the square-root theorem, X is identified as an algebraic expression. In Example 2, X is a monomial.

Example 2.   Solve and check:

$$(3t)^2 = 25$$

Solution.   **Discussion.** Use the square-root theorem to solve for $3t$. Then solve the resulting equation for t.

$(3t)^2 = 25$	The given equation
$3t = 5$ or $3t = -5$	The square-root theorem
$t = \dfrac{5}{3}$ $t = \dfrac{-5}{3}$	Divide both sides by 3.

The check is omitted and the solutions are $\frac{5}{3}$ and $\frac{-5}{3}$.

The symbol $\pm$, which is read "plus or minus" can be used to write these solutions in a more compact form. To illustrate, $\frac{5}{3}$ and $\frac{-5}{3}$ can be written $\frac{\pm 5}{3}$.

Equations Having the Form $X^2 = k$, where X is a Binomial

In Examples 3–6 the form of X is a binomial. As a consequence it requires additional steps to solve for the given variables.

Example 3.   Solve and check:

$$(a - 3)^2 = 49$$

Solution.

$(a - 3)^2 = 49$	The given equation
$a - 3 = \sqrt{49}$ or $a - 3 = -\sqrt{49}$	The square-root theorem
$a - 3 = 7$ or $a - 3 = -7$	Simplify radical.
$a = 10$ or $a = -4$	Add 3 to both sides.

The check is omitted and the solutions are 10 and -4.

Example 4.   Solve and check:

$$(k + 7)^2 = 72$$

Solution.

$(k + 7)^2 = 72$ or $k + 7 = -\sqrt{72}$	The given equation
$k + 7 = \sqrt{72}$	The square-root theorem

$$k + 7 = 6\sqrt{2} \qquad \text{or} \quad k + 7 = -6\sqrt{2} \qquad \text{Simplify radical.}$$

$$k = -7 + 6\sqrt{2} \quad \text{or} \qquad k = -7 - 6\sqrt{2} \qquad \text{Subtract 7 from both sides.}$$

The check is omitted and the solutions are $-7 + 6\sqrt{2}$ and $-7 - 6\sqrt{2}$, which can be written $-7 \pm 6\sqrt{2}$.

Example 5.   Solve and check:

$(2b + 1)^2 = 300$

Solution.

$(2b + 1)^2 = 300$	The given equation
$2b + 1 = \pm\sqrt{300}$	The square-root theorem
$2b + 1 = \pm 10\sqrt{3}$	$\sqrt{300} = \sqrt{100 \cdot 3} = 10\sqrt{3}$
$2b = -1 \pm 10\sqrt{3}$	Subtract 1 from both sides.
$b = \dfrac{-1 \pm 10\sqrt{3}}{2}$	Divide both sides by 2.

The check is omitted and the solutions are $\dfrac{-1 \pm 10\sqrt{3}}{2}$.

Example 6.   Solve and check:

$(16u^2 - 40u + 25) = 40$

Solution.

Discussion. The given equation is not written in the form required by the square-root theorem. Notice, however, that the trinomial on the left side is a perfect square that can be written in the form X^2.

$(16u^2 - 40u + 25) = 40$	The given equation
$(4u - 5)^2 = 40$	Factor the trinomial.

$$4u - 5 = \sqrt{40} \qquad \text{or} \qquad 4u - 5 = -\sqrt{40}$$

$$4u - 5 = 2\sqrt{10} \qquad \text{or} \qquad 4u - 5 = -2\sqrt{10}$$

$$4u = 5 + 2\sqrt{10} \qquad\qquad 4u = 5 - 2\sqrt{10}$$

$$u = \frac{5 + 2\sqrt{10}}{4} \qquad\qquad u = \frac{5 - 2\sqrt{10}}{4}$$

The check is omitted and the solutions can be written as $\dfrac{5 \pm 2\sqrt{10}}{4}$.

SECTION 9-2.   Practice Exercises

In exercises **1–20**, solve and check. Write irrational number solutions in simplified form.

[Examples 1 and 2]
1. $x^2 = 25$

2. $x^2 = 81$

3. $y^2 = 225$

4. $y^2 = 121$

5. $a^2 = \dfrac{49}{9}$ **6.** $a^2 = \dfrac{81}{100}$

7. $s^2 = 11$ **8.** $s^2 = 17$

9. $m^2 - 44 = 0$ **10.** $m^2 - 68 = 0$

11. $(2x)^2 = 25$ **12.** $(3x)^2 = 196$

13. $(5z)^2 - 2 = 0$ **14.** $(10z)^2 - 3 = 0$

15. $192 = (4t)^2$ **16.** $432 = (6t)^2$

17. $0 = 81 - 5u^2$ **18.** $0 = 121 - 7u^2$

19. $2x^2 - 1 = 0$ **20.** $3x^2 - 2 = 0$

In exercises **21–42**, solve and check. Write irrational number solutions in simplified form.

[Examples 3–5]   **21.** $(y + 1)^2 = 9$ **22.** $(y + 2)^2 = 25$

23. $(x - 2)^2 = 36$ **24.** $(x - 8)^2 = 49$

25. $(t - 9)^2 = 144$ **26.** $(t - 4)^2 = 121$

27. $(m + 7)^2 = 13$ **28.** $(m + 5)^2 = 19$

29. $(z - 4)^2 = 18$ **30.** $(z + 4)^2 = 12$

31. $147 = (w + 17)^2$ **32.** $180 = (w + 21)^2$

33. $(3b - 1)^2 = 25$ **34.** $(3b - 5)^2 = 49$

35. $(2m + 3)^2 = 169$ **36.** $(2m + 7)^2 = 225$

37. $(2p + 3)^2 = 6$ **38.** $(2p + 5)^2 = 10$

39. $(4c - 5)^2 = 63$ **40.** $(4c - 3)^2 = 54$

41. $(7x - 12)^2 - 124 = 0$ **42.** $(11x + 12)^2 - 117 = 0$

In exercises **43–50**, solve. Write irrational number solutions in simplified form.

[Example 6]   **43.** $x^2 + 6x + 9 = 25$ **44.** $x^2 + 10x + 25 = 49$

45. $z^2 + 8z + 16 = 81$ **46.** $z^2 - 8z + 16 = 1$

47. $25a^2 - 10a + 1 = 8$ **48.** $25a^2 + 10a + 1 = 27$

49. $9w^2 - 12w + 4 = 12$ **50.** $9w^2 + 12w + 4 = 28$

SECTION 9-2.   Ten Review Exercises

In exercises **1–10**, do the indicated operations.

1. $(7a^2 + 6a + 7) + (2a^2 - 7a - 4) - (8a^2 + a + 3)$

2. $(7\sqrt{3} + 6\sqrt{2} + 7) + (2\sqrt{3} - 7\sqrt{2} - 4) - (8\sqrt{3} + \sqrt{2} + 3)$

3. $(2x - 3y)(x + 4y)$

4. $(2\sqrt{3} - 3\sqrt{2})(\sqrt{3} + 4\sqrt{2})$

5. $(m + 3n)^2$

6. $(\sqrt{5} + 3)^2$

7. $\dfrac{4u^2 - 25v^2}{2u + 5v}$

8. $\dfrac{9}{2\sqrt{5} + \sqrt{2}}$

9. $(2b - 1)^2 - (b + 3)(4b - 5)$

10. $(2\sqrt{3} - 1)^2 - (\sqrt{3} + 3)(4\sqrt{3} - 5)$

SECTION 9-2.   Supplementary Exercises

In exercises **1–14**, solve and check. For any solutions that are not real numbers, write "No real solutions."

1. $18a^2 - 5 = 0$

2. $12a^2 - 1 = 0$

3. $4x^2 + 20x + 25 = 25$

4. $49x^2 + 28x + 4 = 4$

5. $(5y - 4)^2 = -81$

6. $(5y - 1)^2 = -100$

7. $\left(x + \dfrac{2}{3}\right)^2 = \dfrac{5}{9}$

8. $\left(x + \dfrac{4}{5}\right)^2 = \dfrac{3}{25}$

9. $\left(t - \dfrac{5}{6}\right)^2 = \dfrac{1}{36}$

10. $\left(t - \dfrac{7}{8}\right)^2 = \dfrac{1}{64}$

11. $\left(z + \dfrac{7}{2}\right)^2 - \dfrac{31}{4} = 0$

12. $\left(z + \dfrac{11}{9}\right)^2 - \dfrac{10}{81} = 0$

13. $(4w - 1)^2 = \dfrac{3}{4}$

14. $(2w - 5)^2 = \dfrac{5}{9}$

In exercises **15–20**, check the given numbers to determine whether they are solutions.

15. $a^2 + 4a = -1$; $-2 + \sqrt{3}$

16. $a^2 + 2a = 5$; $-1 + \sqrt{6}$

17. $y^2 = 10y - 20$; $5 - \sqrt{5}$

18. $y^2 - 5 = -6y$; $-3 - \sqrt{14}$

19. $p^2 - p = 1$; $\dfrac{1 + \sqrt{5}}{2}$

20. $p^2 = 5p - 3$; $\dfrac{5 - \sqrt{13}}{2}$

In exercises **21–28**, solve.

Example. $4t^4 - 13t^2 + 3 = 0$

Solution.   **Discussion.** Replace t^2 by x and t^4 by x^2. Factor and solve the equation for x. Then replace x by t^2 and solve for t using the square-root theorem.

$$4x^2 - 13x + 3 = 0 \qquad\qquad \text{Replace } t^2 \text{ by } x \text{ and } t^4$$
$$\text{by } x^2.$$

$$(4x - 1)(x - 3) = 0 \qquad\qquad \text{Factor.}$$

$$4x - 1 = 0 \qquad \text{or} \qquad x - 3 = 0 \qquad\qquad \text{The zero-product property.}$$

$$x = \frac{1}{4} \qquad\qquad x = 3 \qquad\qquad \text{Solve for } x.$$

$$t^2 = \frac{1}{4} \qquad\qquad t^2 = 3 \qquad\qquad \text{Replace } x \text{ by } t^2.$$

$$t = \pm\frac{1}{2} \qquad\qquad t = \pm\sqrt{3} \qquad\qquad \begin{array}{l}\text{If } X^2 = k, \text{ then}\\ X = \pm\sqrt{k}.\end{array}$$

The solutions are $\frac{-1}{2}, \frac{1}{2}, \sqrt{3}$, and $-\sqrt{3}$.

21. $t^4 - 34t^2 + 225 = 0$ **22.** $t^4 - 37t^2 + 36 = 0$

23. $4u^4 - 13u^2 + 9 = 0$ **24.** $9u^4 - 37u^2 + 4 = 0$

25. $v^4 - 66v^2 + 128 = 0$ **26.** $v^4 - 21v^2 + 108 = 0$

27. $2k^4 - 75k^2 + 108 = 0$ **28.** $3k^4 - 17k^2 + 20 = 0$

SECTION 9-3. Solving Quadratic Equations by Completing the Square

KEY TOPICS IN THIS SECTION

1. Changing binomials to perfect square trinomials

2. Solving equations of the form $x^2 + bx + c = 0$

3. Solving equations of the form $ax^2 + bx + c = 0, a \neq 1$ (optional)

Equations written in the form of examples **a** and **b** cannot be solved by using the square-root theorem.

a. $x^2 + 6x - 3 = 0$ **b.** $4y^2 - 4y - 17 = 0$

However, both these equations can be changed to the form $X^2 = k$. By changing the equations in **a*** and **b*** below to standard form, it can be shown that **a** and **a***, and also **b** and **b***, are the same equations. In the forms of **a*** and **b***, the equations can be solved using the square-root theorem.

a*. $(x + 3)^2 = 12$ **b*.** $(2y - 1)^2 = 18$

The process of changing equations such as **a** and **b** to the forms of **a*** and **b*** is called "completing the square." This process will be studied in this section.

Changing Binomials to Perfect Square Trinomials

Consider the binomials in examples **c** and **d**:

c. $t^2 + 10t$ **d.** $u^2 - 12u$

These binomials can be changed to perfect square trinomials by adding a third term. Precisely what term needs to be added depends on the multiplied forms of the squares of binomials. That is,

$$(1) \quad (x + y)^2 = x^2 + 2xy + y^2$$

$$(2) \quad (x - y)^2 = x^2 - 2xy + y^2$$

The binomial in example **c** fits the first two terms of equation (1). That is,

t^2 corresponds to x^2

$$t^2 + 10t \qquad\qquad x^2 + 2xy$$

$10t$ corresponds to $2xy$

We therefore need to add the equivalent of the missing y^2 term to transform $t^2 + 10t$ to a perfect square trinomial. Since 10 is two times y, we take one-half of 10 to get 5. Thus y is 5 and y^2 is 25. Adding 25 to $t^2 + 10t$,

$$t^2 + 10t + 25 \text{ can be written } (t + 5)^2.$$

The binomial in example **d** fits the first two terms of equation (2). That is,

u^2 corresponds to x^2

$$u^2 - 12u \qquad\qquad x^2 - 2xy$$

$12u$ corresponds to $2xy$

Since u^2 corresponds to x^2, it follows that u corresponds to x. We therefore take one-half of 12 (or 6) to compute y. Thus y is 6, and y^2 is 36. Adding 36 to $u^2 - 12u$,

$$u^2 - 12u + 36 = (u - 6)^2.$$

Example 1.   For the following binomials,

    1. add the missing term to make a perfect square trinomial.

    2. write the trinomial in factored form.

    a. $m^2 - 16m$

    b. $a^2 + 30a$

Solution.   **a. 1.** $\frac{1}{2} \cdot 16 = 8$ and $8^2 = 64$

      $m^2 - 16m + 64$ A perfect square trinomial.

    2. $m^2 - 16m + 64 = (m - 8)^2$

    b. 1. $\frac{1}{2} \cdot 30 = 15$ and $15^2 = 225$

      $a^2 + 30a + 225$ A perfect square trinomial.

    2. $a^2 + 30a + 225 = (a + 15)^2$

Solving Equations of the Form $x^2 + bx + c = 0$

The following five steps can be used to solve a quadratic equation by completing the square.

To solve $x^2 + bx + c = 0$ by completing the square,

Step 1. add $-c$ to both sides of the equation.

Step 2. add $(\frac{1}{2} \cdot b)^2$ to both sides of the equation.

Step 3. write the left side as the square of a binomial and simplify the right side.

Step 4. solve the equation of Step 3 by using the square-root theorem.

Step 5. state the solutions.

Example 2.   Solve by completing the square.

$$x^2 + 6x - 2 = 0$$

Solution.   **Step 1.** $x^2 + 6x = 2$ Add 2 to both sides.

Step 2. $\dfrac{1}{2} \cdot 6 = 3$ and $3^2 = 9$

$$x^2 + 6x + 9 = 2 + 9 \qquad \text{Add 9 to both sides.}$$

Step 3. $(x + 3)^2 = 11$

Step 4. $x + 3 = \sqrt{11}$ $\qquad$ or $\qquad$ $x + 3 = -\sqrt{11}$

$\qquad x = -3 + \sqrt{11}$ $\qquad$ or $\qquad$ $x = -3 - \sqrt{11}$

Step 5. $-3 + \sqrt{11}$ and $-3 - \sqrt{11}$, which can be written $-3 \pm \sqrt{11}$

Example 3.   Solve by completing the square.

$$y^2 - 10y + 7 = 0$$

Solution.   **Step 1.** $y^2 - 10y \qquad = -7$ Add -7 to both sides.

Step 2. $y^2 - 10y + 25 = -7 + 25$ Add 25 to both sides.

Step 3. $\qquad (y - 5)^2 = 18$ Factor the left side.

Step 4. $\qquad\quad y - 5 = \pm\sqrt{18}$ The square-root theorem

$\qquad\qquad\quad y = 5 \pm 3\sqrt{2}$ Add 5 and simplify $\sqrt{18}$.

Step 5. The solutions are $5 \pm 3\sqrt{2}$.

Solving Equations of the Form $ax^2 + bx + c = 0$, $a \neq 1$ (optional)

If the coefficient of the squared term in a quadratic equation is not one—as in Examples 2 and 3—then an additional step is needed to solve the equation.

Example 4.   Solve by completing the square.

$$4z^2 + 8z - 1 = 0$$

Solution.   **Step 1.** $4z^2 + 8z \quad = 1$ Add 1 to both sides.

Step 2. $z^2 + 2z \quad = \dfrac{1}{4}$ Divide each term by 4.

Step 3. $z^2 + 2z + 1 = \dfrac{1}{4} + 1$ Add 1 to both sides.

Step 4. $(z + 1)^2 = \dfrac{5}{4}$ $\dfrac{1}{4} + 1 = \dfrac{1}{4} + \dfrac{4}{4} = \dfrac{5}{4}$

Step 5. $z + 1 = \pm\sqrt{\dfrac{5}{4}}$ The square-root theorem

$z = -1 \pm \dfrac{\sqrt{5}}{2}$ $\sqrt{\dfrac{5}{4}} = \dfrac{\sqrt{5}}{\sqrt{4}} = \dfrac{\sqrt{5}}{2}$

$= \dfrac{-2}{2} \pm \dfrac{\sqrt{5}}{2}$ Write -1 as $\dfrac{-2}{2}$.

$= \dfrac{-2 \pm \sqrt{5}}{2}$ The common denominator is 2.

Step 6. The solutions are $\dfrac{-2 \pm \sqrt{5}}{2}$.

SECTION 9-3.   Practice Exercises

In exercises **1–10,**

a. add the missing term to make a perfect square trinomial.

b. write the trinomial in factored form.

[Example 1]
1. $x^2 + 8x$ **2.** $x^2 + 12x$

3. $y^2 - 18y$ **4.** $y^2 - 14y$

5. $m^2 - 20m$ **6.** $m^2 - 24m$

7. $p^2 + 28p$ **8.** $p^2 + 36p$

9. $z^2 - 60z$ **10.** $z^2 - 48z$

In exercises **11–34,** solve by completing the square.

[Examples 2 and 3]
11. $x^2 + 14x = -13$ **12.** $x^2 + 18x = -32$

13. $y^2 - 30y = -29$ **14.** $y^2 - 22y = -21$

15. $t^2 - 2t - 24 = 0$ **16.** $t^2 - 4t - 12 = 0$

17. $s^2 + 6s - 2 = 0$ **18.** $s^2 + 10s - 4 = 0$

19. $z^2 - 16z - 8 = 0$

20. $z^2 - 18z - 9 = 0$

21. $v^2 + 12v + 8 = 0$

22. $v^2 + 8v + 4 = 0$

23. $q^2 - 24q + 20 = 0$

24. $q^2 - 20q + 25 = 0$

25. $x^2 + 22x + 113 = 0$

26. $x^2 + 28x + 184 = 0$

[Example 4]   **27.** $5x^2 - 10x + 1 = 0$

28. $7x^2 - 28x + 10 = 0$

29. $3w^2 + 12w - 9 = 0$

30. $5w^2 + 20w - 30 = 0$

31. $5t^2 - 40t - 1 = 0$

32. $6t^2 - 36t + 5 = 0$

33. $4y^2 + 40y + 2 = 0$

34. $7y^2 + 56y - 16 = 0$

SECTION 9-3.   Ten Review Exercises

1. Evaluate $\dfrac{t^2 - t - 42}{t^2 - 49}$ for $t = 3$.

2. Factor and reduce $\dfrac{t^2 - t - 42}{t^2 - 49}$.

3. Evaluate the expression in exercise **2** for $t = 3$. Compare the answer to the one obtained in exercise **1**.

4. What numbers are not suitable replacements for t in $\dfrac{t^2 - t - 42}{t^2 - 49}$?

5. Solve: $\dfrac{t + 2}{t + 7} - \dfrac{t + 1}{t - 7} = \dfrac{5}{t^2 - 49}$

6. Solve $u^2 + 12u - 45 = 0$ by completing the square.

7. Solve $u^2 + 12u - 45 = 0$ by factoring and the zero-product property.

8. Solve $v^2 - 8v = 0$ by completing the square.

9. Solve $v^2 - 8v = 0$ by factoring and the zero-product property.

10. Based on the observations of exercises **6–9**, does it appear that any quadratic equation with integer coefficients that has rational number solutions can be solved by factoring and the zero-product property?

SECTION 9-3.   Supplementary Exercises

In exercises **1–22**, solve by any method.

1. $p^2 + 12p = 0$

2. $p^2 + 18p = 0$

3. $2q^2 - 40q = 0$

4. $3q^2 - 48q = 0$

5. $4s^2 + 20s = 0$

6. $5s^2 + 35s = 0$

7. $9t^2 - 72 = 0$ 8. $7t^2 - 28 = 0$

9. $3x^2 - 81 = 0$ 10. $4x^2 - 12 = 0$

11. $y^2 + 9y - 20 = 0$ 12. $y^2 + 7y - 10 = 0$

13. $x^2 - 5x + 2 = 0$ 14. $x^2 - 3x + 1 = 0$

15. $s^2 - 11s + 18 = 0$ 16. $s^2 - 13s + 42 = 0$

17. $2z^2 - 7z + 4 = 0$ 18. $2z^2 - 5z - 2 = 0$

19. $4t^2 - 3t - 6 = 0$ 20. $5t^2 - 4t - 5 = 0$

21. $3x^2 - 8x - 3 = 0$ 22. $7x^2 - 5x - 2 = 0$

In exercises **23–28**, solve by completing the square.

a. State the answer in exact form using square roots.

b. Approximate, to two decimal places, the solutions of part **a.**

23. $m^2 + 8m + 1 = 0$ 24. $m^2 + 12m + 2 = 0$

25. $p^2 - 2p - 10 = 0$ 26. $p^2 - 5p - 8 = 0$

27. $2y^2 + 10y + 3 = 0$ 28. $2y^2 + 12y + 5 = 0$

SECTION 9-4. Solving Equations with the Quadratic Formula

**KEY TOPICS
IN THIS SECTION**

1. The quadratic formula

2. Solving equations with the quadratic formula

3. Some applied problems

The standard form of a quadratic equation can be solved by completing the square. A guided solution of this process is given in the Supplementary Exercises of this section. The result of the process is known as the **quadratic formula.** This formula will be used to solve quadratic equations in this section.

The Quadratic Formula

Equations (1) and (2) express the two solutions x_1 and x_2 of $ax^2 + bx + c = 0$ in terms of a, b, and c.

> **The quadratic formula**
> If x_1 and x_2 are solutions of $ax^2 + bx + c = 0$ and $a \neq 0$, then
>
> $$(1) \quad x_1 = \frac{-b + \sqrt{b^2 - 4ac}}{2a} \quad \text{and} \quad (2) \quad x_2 = \frac{-b - \sqrt{b^2 - 4ac}}{2a}.$$

With equations (1) and (2), the solutions of any quadratic equation can be determined by simplifying the expressions using the values of a, b, and c in the equations.

Solving Equations with the Quadratic Formula

The following five steps provide a procedure for solving equations using the quadratic formula.

To solve a quadratic equation with the quadratic formula,

Step 1. if necessary, write the equation in the form $ax^2 + bx + c = 0$, with $a > 0$.

Step 2. list the values of a, b, and c.

Step 3. replace a, b, and c in equations (1) and (2) by the values in Step 2.

Step 4. simplify the numerical expressions by doing the indicated operations in the radicand and denominator. If possible, simplify the radical, factor, and reduce the expressions.

Step 5. state the solutions.

Example 1.   Solve: $5x^2 - 2 = x$

Solution.   **Step 1.** Write in standard form with $a > 0$.

$$5x^2 - x - 2 = 0 \qquad \text{Subtract } x \text{ from both sides.}$$

Step 2. Identify a, b, and c.

$$a = 5, b = -1, \text{ and } c = -2.$$

Step 3. Replace a, b, and c in equations (1) and (2) by 5, -1, and -2 respectively.

$$x_1 = \frac{-(-1) + \sqrt{(-1)^2 - 4(5)(-2)}}{2(5)} \qquad x_2 = \frac{-(-1) - \sqrt{(-1)^2 - 4(5)(-2)}}{2(5)}$$

Step 4. Simplify.

$$x_1 = \frac{1 + \sqrt{1 - (-40)}}{10} \qquad x_2 = \frac{1 - \sqrt{1 - (-40)}}{10}$$

$$x_1 = \frac{1 + \sqrt{41}}{10} \qquad x_2 = \frac{1 - \sqrt{41}}{10}$$

Step 5. State the solutions.

$$\frac{1 + \sqrt{41}}{10} \text{ and } \frac{1 - \sqrt{41}}{10}, \text{ which can be written as } \frac{1 \pm \sqrt{41}}{10}$$

Example 2.   Solve: $4t^2 = -7 - 12t$

Solution.   **Step 1.** Write in standard form.

$$4t^2 + 12t + 7 = 0 \qquad \text{Add 7 and } 12t \text{ to both sides.}$$

Step 2. Identify a, b, and c.

$$a = 4, b = 12, \text{ and } c = 7.$$

Step 3. Replace a, b, and c in equations (1) and (2) by 4, 12, and 7 respectively.

$$t_1 = \frac{-12 + \sqrt{12^2 - 4(4)(7)}}{2(4)} \qquad t_2 = \frac{-12 - \sqrt{12^2 - 4(4)(7)}}{2(4)}$$

Step 4. Simplify.

$$t_1 = \frac{-12 + \sqrt{32}}{8} \qquad t_2 = \frac{-12 - \sqrt{32}}{8}$$

$$= \frac{-12 + 4\sqrt{2}}{8} \qquad = \frac{-12 - 4\sqrt{2}}{8}$$

$$= \frac{4(-3 + \sqrt{2})}{8} \qquad = \frac{4(-3 - \sqrt{2})}{8}$$

$$= \frac{-3 + \sqrt{2}}{2} \qquad = \frac{-3 - \sqrt{2}}{2}$$

Step 5. State the solutions.

$$\frac{-3 + \sqrt{2}}{2} \text{ and } \frac{-3 - \sqrt{2}}{2}, \text{ or } \frac{-3 \pm \sqrt{2}}{2}$$

Example 3 illustrates solving a quadratic equation with the formula when the solutions are rational numbers.

Example 3. Solve: $x + 6 = 12x^2$

Solution. **Discussion.** To put the equation in standard form with $a > 0$, the three terms are written on the right side of the equation.

Step 1. $0 = 12x^2 - x - 6$ Subtract x and 6 from both sides.

$12x^2 - x - 6 = 0$ If $X = Y$, then $Y = X$.

Step 2. $a = 12, b = -1,$ Note: $12x^2 - x - 6$ can be written
and $c = -6.$ $12x^2 + (-1)x + (-6).$

Step 3. Using the $\pm$ symbol to combine equations (1) and (2),

$$x = \frac{-(-1) \pm \sqrt{(-1)^2 - 4(12)(-6)}}{2(12)}$$

Step 4. $x = \dfrac{1 \pm \sqrt{289}}{24}$

$$x = \frac{1 \pm 17}{24} \quad\begin{array}{l} x_1 = \dfrac{1 + 17}{24} = \dfrac{18}{24} = \dfrac{3}{4} \\[2mm] x_2 = \dfrac{1 - 17}{24} = \dfrac{-16}{24} = \dfrac{-2}{3} \end{array}$$

Step 5. The solutions are $\frac{3}{4}$ and $\frac{-2}{3}$.

In Example 3, once we determined that the radicand was a perfect square, we extracted the root and wrote two equations. By further simplifying these expressions, we identified the two solutions.

Some Applied Problems

The quadratic formula is used to solve the problems in Examples 4 and 5. In both examples the irrational number solutions are approximated to one decimal place.

Example 4. The sum of two numbers is three and the product is one. Find both numbers.

Solution. **Discussion.** This problem can be solved by using only one variable. However, it can also be solved by using two variables and a system of two equations. The two-variable-and-two-equation method will be demonstrated here.

Let x represent one of the numbers.
Let y represent the other number.

(1) $x + y = 3$ The sum of the numbers is 3.

(2) $x \cdot y = 1$ The product of the numbers is 1.

Solve equation (1) for y in terms of x.

(1) $y = 3 - x$

Replace y by $3 - x$ in equation (2).

(2) $x(3 - x) = 1$

$$3x - x^2 = 1$$ Remove parentheses.

$$0 = x^2 - 3x + 1$$ Write in standard form.

With $a = 1$, $b = -3$, and $c = 1$,

$$x = \frac{-(-3) \pm \sqrt{(-3)^2 - 4(1)(1)}}{2(1)}$$

$$x = \frac{3 \pm \sqrt{5}}{2}$$

If $x = \dfrac{3 + \sqrt{5}}{2}$, then $y = 3 - \dfrac{3 + \sqrt{5}}{2}$

$$= \frac{6 - 3 - \sqrt{5}}{2}$$

$$= \frac{3 - \sqrt{5}}{2}$$

Similarly, if $x = \dfrac{3 - \sqrt{5}}{2}$, then $y = \dfrac{3 + \sqrt{5}}{2}$.

Thus, the two numbers are $\dfrac{3 - \sqrt{5}}{2}$ and $\dfrac{3 + \sqrt{5}}{2}$.

To one decimal place,

$$\frac{3 - \sqrt{5}}{2} = \frac{3 - 2.236\ldots}{2} = 0.38\ldots \approx 0.4$$

$$\frac{3 + \sqrt{5}}{2} = \frac{3 + 2.236\ldots}{2} = 2.61\ldots \approx 2.6$$

The check is omitted.

Example 5. The length of a rectangle is three meters longer than the width. The area of the rectangle is 38 square meters. Approximate the length and width to one decimal place.

Solution.   Let l represent the length.
Let w represent the width.

(1) $l = w + 3$ The length is 3 meters longer.

(2) $l \cdot w = 38$ The area is 38 square meters.

Replace l in equation (2) by $w + 3$:

(2) $(w + 3)w = 38$

$$w^2 + 3w = 38$$

$$w^2 + 3w - 38 = 0$$

$a = 1$, $b = 3$, and $c = -38$:

$$w = \frac{-3 \pm \sqrt{3^2 - 4(1)(-38)}}{2(1)}$$

$$= \frac{-3 \pm \sqrt{161}}{2}$$

$$w_1 = \frac{-3 + \sqrt{161}}{2} \qquad\qquad w_2 = \frac{-3 - \sqrt{161}}{2}$$

≈ 4.8, to one place ≈ -7.8, to one place

Since w represents the width of a rectangle, the negative number solution is rejected.

If $w \approx 4.8$, then $l \approx 4.8 + 3 \approx 7.8$.

To one decimal place, the length is 7.8 meters and the width is 4.8 meters.

SECTION 9-4.   Practice Exercises

In exercises **1–16**,

a. list the values for a, b, and c.

b. solve with the quadratic formula.

[Examples 1–3]

1. $x^2 + x - 3 = 0$ **2.** $x^2 + 5x + 1 = 0$

3. $2y^2 + 3y = 2$ **4.** $3y^2 + 2y = 1$

5. $z^2 = 3z + 6$ **6.** $z^2 = z + 9$

7. $x^2 - 3 = 4x$ **8.** $x^2 + 1 = 4x$

9. $y^2 = 6y + 1$ **10.** $y^2 = 2y + 1$

11. $2z^2 = 7z + 9$ **12.** $20z^2 = 12 - z$

13. $4x^2 = 12x$ **14.** $6x = -9x^2$

15. $y^2 = 72$ **16.** $y^2 = 48$

In exercises **17–40**, solve by any method.

[Examples 1–3]   **17.** $4z + 1 = -z^2$ **18.** $z^2 + 4z = 1$

19. $4x^2 = 4x + 7$ **20.** $2x^2 = 4x - 1$

21. $2y^2 + 3 + 6y = 0$ **22.** $4y - 11 + 4y^2 = 0$

23. $2z = 15 - 8z^2$ **24.** $7z + 6 = 10z^2$

25. $6x(x + 2) = 14 - 5x$ **26.** $10x(x - 1) = 3x + 30$

27. $y(y - 5) = 5y - 17$ **28.** $y^2 + 2 = 10(y - 2)$

29. $9z^2 + 11 = 12(z + 1)$ **30.** $2z - 1 = z(z - 1)$

31. $a(a - 2) = 2a + 4$ **32.** $2a - 7 = a(a - 4)$

33. $(b - 1)^2 = 2b + 24$ **34.** $(b + 5)^2 = 21 + 4b$

35. $(3x + 1)^2 = 6x + 7$ **36.** $(3x + 2)^2 = 6(2x + 1)$

37. $(2y + 1)(y - 1) = -20y - 20$ **38.** $(3y + 1)^2 = 3(1 - 2y)$

39. $(2z - 3)^2 = 4(9 - 2z)$ **40.** $(3z + 4)^2 = 6(13 + 3z)$

In exercises **41–58**,

a. state answers in exact form using radicals.

b. approximate solutions to one decimal place.

[Example 4]   **41.** The sum of two numbers is 4 and the product is 2. Find both numbers.

42. The sum of two numbers is 12 and the product is 33. Find both numbers.

43. The sum of two numbers is -5 and the product is $\frac{11}{2}$. Find both numbers.

44. The sum of two numbers is -1 and the product is -11. Find both numbers.

45. If the smaller of two numbers is subtracted from the larger, the difference is $\frac{34}{15}$. The product of the numbers is -1. Find both numbers.

46. If the smaller of two numbers is subtracted from the larger the difference is $\frac{7}{6}$. The product of the numbers is $\frac{-1}{3}$. Find both numbers.

[Example 5]   **47.** The length of a rectangle is 5 centimeters longer than the width. The area is 40 square centimeters. Find the lengths of the sides of the rectangle.

48. The length of a rectangle is 8 feet longer than the width. The area is 92 square feet. Find the lengths of the sides of the rectangle.

49. The width of a rectangle is 4 meters shorter than the length. The area is 35 square meters. Find the lengths of the sides of the rectangle.

50. The width of a rectangle is 6 yards shorter than the length. The area is 75 square yards. Find the lengths of the sides of the rectangle.

51. In a right triangle, the length of the longer leg is 1 meter more than the shorter leg. Find the lengths of both legs if the length of the hypotenuse is 7 meters.

52. In a right triangle, the length of the shorter leg is 3 centimeters less than the longer leg. Find the lengths of both legs if the length of the hypotenuse is 9 centimeters.

53. In a right triangle, the shorter leg is 5 feet. The hypotenuse is 3 feet less than twice the length of the longer leg. Find the lengths of the longer leg and the hypotenuse.

54. In a right triangle, the longer leg is 12 inches. The hypotenuse is 3 inches more than twice the length of the shorter leg. Find the lengths of the shorter leg and the hypotenuse.

55. The area of a triangle is 35 square feet. The altitude to the base is 3 feet shorter than the length of the base. Find the lengths of the base and altitude.

56. The area of a triangle is 20 square meters. The base is 4 meters more than the length of the altitude to the base. Find the lengths of the base and altitude.

For exercises **57** and **58**, refer to the two triangles in the following figure. Notice that $\mathscr{L}_1$ and $\mathscr{L}_2$ are parallel and line segment AB is the base of both triangles ABC and ABD.

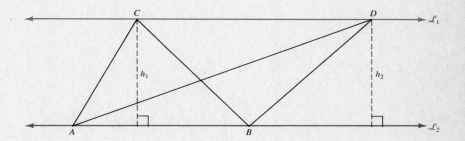

57. a. The base of triangle ABC is 4 centimeters less than the height (labeled h_1). If the area of the triangle is 47 square centimeters, find the lengths of the base and height.
   b. The base of triangle ABD is also 4 centimeters less than the height (labeled h_2). If the area of this triangle is also 47 square centimeters, find the lengths of the base and height.
   c. Compare the answers to part **a** and part **b**. Does the comparison agree with your visual comparisons of the areas?

58. a. The base of triangle ABC is 5 inches less than the height (labeled h_1). If the area of the triangle is 51 square inches, find the lengths of the base and height.
   b. The base of triangle ABD is 5 inches less than the height (labeled h_2). If the area of the triangle is 51 square inches, find the lengths of the base and height.
   c. Compare the answers to part **a** and part **b**. Does the comparison agree with your visual comparisons of the areas?

SECTION 9-4. Ten Review Exercises

In exercises **1–10**, use points $P(-3, 9)$, $Q(0, 5)$, and $R(6, -3)$.

1. Find the slope of the line $\mathscr{L}_1$ through P and R.

2. Using the slope from exercise **1** and the coordinates of P, write an equation of $\mathscr{L}_1$ in point-slope form.

3. Write the equation of exercise **2** in $Ax + By = C$ form with $A > 0$.

4. Are the coordinates of Q a solution of the equation obtained in exercise **3**?

5. Based on your answer to exercise **4**, what is the y-intercept of the equation of $\mathscr{L}_1$?

6. If $\mathscr{L}_2$ is perpendicular to $\mathscr{L}_1$, then what is the slope of $\mathscr{L}_2$?

7. If $\mathscr{L}_2$ passes through R, then what is the $Ax + By = C$ form of the equation of $\mathscr{L}_2$?

8. If $\mathscr{L}_3$ is parallel to $\mathscr{L}_1$, then what is the slope of $\mathscr{L}_3$?

9. If $\mathscr{L}_3$ is parallel to $\mathscr{L}_1$, then $\mathscr{L}_3$ is _____ to $\mathscr{L}_2$. (Fill in the blank with the correct word.)

10. If equations for $\mathscr{L}_1$ and $\mathscr{L}_3$ were used to form a system of linear equations in x and y, then would the system have a solution?

SECTION 9-4.   Supplementary Exercises

In exercises **1–12**, solve.

1. $x^2 + \sqrt{2}x - 3 = 0$

2. $x^2 + \sqrt{5}x - 1 = 0$

3. $2z^2 - \sqrt{3}z - 4 = 0$

4. $3z^2 - \sqrt{6}z - 2 = 0$

5. $6t^2 = \sqrt{7}t + 5$

6. $5t^2 = \sqrt{10}t + 7$

7. $\sqrt{2}p^2 + 3p - \sqrt{2} = 0$

8. $\sqrt{5}p^2 + 5p - \sqrt{5} = 0$

9. $m^2 - 0.8m + 0.15 = 0$

10. $m^2 + 1.1m - 0.42 = 0$

11. $0.1n^2 + 0.047n - 0.0161 = 0$

12. $0.1n^2 - 0.12n - 0.54 = 0$

The expression $b^2 - 4ac$ in the quadratic formula is called the **discriminant.** The value of this expression can be used to determine whether the solutions of a quadratic equation are rational numbers, irrational numbers, or complex numbers.

> If $ax^2 + bx + c = 0$ with a, b, and c *rational numbers,* then
>
> **(i)** the roots are *rational numbers* if $b^2 - 4ac$ is a *perfect square.*
>
> **(ii)** the roots are *irrational numbers* if $b^2 - 4ac$ is a *positive number that is not a perfect square.*
>
> **(iii)** the roots are *complex numbers* if $b^2 - 4ac$ is a *negative number.*

Example.   Determine whether the roots of $3t^2 - t - 10 = 0$ are rational, irrational, or complex numbers.

Solution.   With $a = 3$, $b = -1$, and $c = -10$, $b^2 - 4ac$ becomes

$$(-1)^2 - 4(3)(-10) = 1 - (-120)$$

$$= 121$$

Since 121 is a perfect square, the roots are rational numbers.

In exercises **13–20**, use the value of the discriminant to determine whether the solutions of the quadratic equations are

a. rational numbers.

b. irrational numbers.

c. complex numbers.

13. $x^2 + 14x + 49 = 0$ **14.** $x^2 - 16x + 64 = 0$

15. $t^2 + t + 1 = 0$ **16.** $t^2 - t + 1 = 0$

17. $2u^2 + 12u - 32 = 0$ **18.** $3u^2 - 12u - 36 = 0$

19. $5x^2 - 10x - 10 = 0$ **20.** $3x^2 + 18x + 12 = 0$

21. A guided solution to the development of the quadratic formula.
Solve $ax^2 + bx + c = 0$, with $a > 0$, by completing the square.

    a. Subtract c from both sides of the equation.
    b. Divide each term by a.
    c. Add $(\frac{b}{2a})^2$ to both sides of the equation.
    d. Write the left side as the square of a binomial.
    e. Simplify the right side by writing the two terms over the common denominator $4a^2$.
    f. Use the square-root theorem.
    g. Subtract $\frac{b}{2a}$ from both sides of the equation.
    h. Write the right side as one term over the common denominator $2a$.

SECTION 9-5.   Complex Solutions of Quadratic Equations (Optional)

KEY TOPICS IN THIS SECTION

1. A definition of the imaginary unit

2. The $a + bi$ form of a complex number

3. Solving quadratic equations with complex number solutions

The quadratic equations in examples **a** and **b** do not have real number solutions.

a. $t^2 + 25 = 0$ **b.** $4u^2 + 9 = 0$

For any real number t or u, t^2 and u^2 are greater than or equal to zero. Since the sums of t^2 and $4u^2$ with 25 and 9 respectively must therefore be positive, no real numbers exist that can make these equations true number statements. To find numbers that are solutions to equations such as **a** and **b**, we need to define numbers whose squares are negative. The **set of complex numbers** contains such numbers. We will now briefly study these numbers, and then we will solve quadratic equations whose solutions are complex numbers that are not real numbers.

A Definition of the Imaginary Unit

One of the first persons to work on defining a set of number whose squares are negative was the German mathematician Karl Friedrich Gauss (1777–1855). Gauss decided that a new set of numbers was needed, since the squares of all real numbers are not negative.

Definition 9.2.   The imaginary unit i
   The number i is called the imaginary unit, and i is the number such that $i^2 = -1$ and $i = \sqrt{-1}$.

The imaginary unit can be used to write the square roots of negative radicands in terms of square roots of positive radicands. Writing such numbers with the number i is called the **i-form.**

Example 1.   Write each number in the i-form.

   a. $\sqrt{-100}$

   b. $\sqrt{-8}$

Solution.   **Discussion.** The definition of i is used to write the given negative radicands as positive numbers. The resulting radicand is then simplified.

   a. $\sqrt{-100}$ The given radical

   $= \sqrt{-1 \cdot 100}$ Write -100 as $-1 \cdot 100$.

   $= \sqrt{-1} \cdot \sqrt{100}$ Write as a product of radicals.

   $= i \cdot 10$ $\sqrt{-1} = i$

   $= 10i$ The preferred form

   b. $\sqrt{-8}$ The given radical

   $= \sqrt{-1 \cdot 8}$ Write -8 as $-1 \cdot 8$.

   $= i\sqrt{8}$ $\sqrt{-1} = i$

   $= i2\sqrt{2}$ $\sqrt{8} = \sqrt{4 \cdot 2} = 2\sqrt{2}$

   $= 2i\sqrt{2}$ The preferred form

The $a + bi$ Form of a Complex Number

The imaginary unit is now used to define complex numbers.

Definition 9.3.   The set of complex numbers
   If a and b are real numbers and $i = \sqrt{-1}$, then a **complex number** is one that can be written in the form $a + bi$.

Example 2.   Write in the $a + bi$ form of a complex number.

   a. $-3 + \sqrt{-196}$

   b. $\sqrt{40} - \sqrt{-90}$

Solution.   **a.** $-3 + \sqrt{-196}$ The given number

$= -3 + i\sqrt{196}$ The i-form of the radical

$= -3 + 14i$ The $a + bi$ form

b. $\sqrt{40} - \sqrt{-90}$ The given number

$= \sqrt{40} - i\sqrt{90}$ The i-form of the radical

$= 2\sqrt{10} - 3i\sqrt{10}$ Simplify both radicals.

Solving Quadratic Equations with Complex Number Solutions

In Examples 3 and 4 the solutions of the quadratic equations are complex numbers.

Example 3.   Solve: $t^2 + 63 = 0$

Solution.   **Discussion.** The given quadratic equation contains a binomial expression. As a consequence, the equation can be solved using the square-root theorem.

$t^2 + 63 = 0$ The given equation

$t^2 = -63$ Solve for t^2.

$t = \pm\sqrt{-63}$ If $X^2 = k$, then $X = \pm\sqrt{k}$.

$t = \pm i\sqrt{63}$ Write in the i-form.

$t = \pm 3i\sqrt{7}$ Simplify the radical.

The solutions are $\pm 3i\sqrt{7}$.

Example 4.   Solve: $9u^2 = 30u - 29$

Solution.   **Discussion.** This quadratic equation contains a trinomial expression when written in standard form. Therefore, we will use the quadratic formula to solve the equation.

Step 1. $9u^2 - 30u + 29 = 0$

Step 2. $a = 9$, $b = -30$, and $c = 29$

Step 3. $u = \dfrac{-(-30) \pm \sqrt{(-30)^2 - 4(9)(29)}}{2(9)}$

Step 4. $u = \dfrac{30 \pm \sqrt{-144}}{18}$

$= \dfrac{30 \pm 12i}{18}$

$= \dfrac{6(5 \pm 2i)}{18}$

$= \dfrac{5 \pm 2i}{3}$

Step 5. The solutions are $\dfrac{5 \pm 2i}{3}$.

SECTION 9-5. Practice Exercises

In exercises **1–20**, write each number in the *i*-form.

[Example 1]

1. $\sqrt{-16}$ **2.** $\sqrt{-121}$ **3.** $-\sqrt{-4}$

4. $-\sqrt{-225}$ **5.** $\sqrt{-10}$ **6.** $\sqrt{-3}$

7. $-\sqrt{-23}$ **8.** $-\sqrt{-41}$ **9.** $\sqrt{\dfrac{-1}{9}}$

10. $\sqrt{\dfrac{-25}{49}}$ **11.** $\sqrt{\dfrac{-2}{81}}$ **12.** $\sqrt{\dfrac{-7}{400}}$

13. $-\sqrt{-27}$ **14.** $-\sqrt{-28}$ **15.** $\sqrt{-288}$

16. $\sqrt{-507}$ **17.** $\sqrt{\dfrac{-5}{12}}$ **18.** $\sqrt{\dfrac{-7}{45}}$

19. $\dfrac{\sqrt{-60}}{6}$ **20.** $\dfrac{\sqrt{-63}}{12}$

In exercises **21–30**, write each number in the $a + bi$ form.

[Example 2]

21. $7 + \sqrt{-64}$ **22.** $13 - \sqrt{-169}$

23. $-1 - \sqrt{-361}$ **24.** $-4 + \sqrt{-441}$

25. $10 + \sqrt{-150}$ **26.** $-9 + \sqrt{-112}$

27. $-2 - \sqrt{-44}$ **28.** $17 - \sqrt{-99}$

29. $\sqrt{\dfrac{4}{9}} - \sqrt{\dfrac{-1}{9}}$ **30.** $-\sqrt{\dfrac{25}{36}} + \sqrt{\dfrac{-49}{36}}$

In exercises **31–50**, solve.

[Examples 3 and 4]

31. $t^2 + 64 = 0$ **32.** $9t^2 + 100 = 0$

33. $25u^2 + 1 = 0$ **34.** $4u^2 + 3 = 0$

35. $v^2 + 50 = 0$ **36.** $v^2 + 288 = 0$

37. $24 + 5x^2 = 0$ **38.** $75 + 2x^2 = 0$

39. $y^2 - 2y + 10 = 0$ **40.** $10y - 26 = y^2$

41. $6z + z^2 + 45 = 0$ **42.** $z^2 + 29 = -10z$

43. $m^2 = 14m - 52$ **44.** $6m = 59 + m^2$

45. $-12k - 5 = 9k^2$ **46.** $-5 - 6k = 2k^2$

47. $3p^2 = 2p - 2$ **48.** $4p = 5 + 2p^2$

49. $0 = -25q^2 - 10q - 12$ **50.** $8 = -4q - 3q^2$

SECTION 9-5. Ten Review Exercises

1. Solve by completing the square.

$x^2 + 2x - 6 = 0$

2. Solve by using the quadratic formula.

$x + 3 = \dfrac{x + 6}{x}$

3. Solve by factoring and using the zero-product property.

$(3x - 1)(2x + 1) = 2x(x - 2) + 5$

In exercises **4** and **5**, solve.

4. $4t^2 - 81 = 0$ **5.** $4t^2 + 81 = 0$

In exercises **6–10**, do the indicated operations.

6. $3(2a + a^2 - 1) - (a^2 + 5a - 6) - 2(a^2 - a + 2)$

7. $(b^2 + 2b - 1)^2$ **8.** $\left(\dfrac{4c^2 - 9}{15}\right)\left(\dfrac{10}{4c^2 + 12c + 9}\right)$

9. $\left(\dfrac{p^2 - 7p}{p^2 - 16}\right) \div \left(\dfrac{p^2 - 4p - 21}{p^2 - 7p + 12}\right)$

10. $(8y^5 - 12y^3 + y^2 - 6y + 9) \div (2y^2 + y - 3)$

SECTION 9-5. Supplementary Exercises

In exercises **1–10**, write each number in the i-form, and then do the indicated operations.

1. $\sqrt{-36} \cdot \sqrt{-81}$ **2.** $\sqrt{-100} \cdot \sqrt{-9}$

3. $\dfrac{\sqrt{-64}}{\sqrt{-4}}$ **4.** $\dfrac{\sqrt{-256}}{\sqrt{-16}}$

5. $\sqrt{-3} \cdot \sqrt{-147}$ **6.** $\sqrt{-5} \cdot \sqrt{-405}$

7. $\dfrac{-\sqrt{-175}}{\sqrt{-7}}$ **8.** $\dfrac{-\sqrt{-968}}{\sqrt{-2}}$

9. $\dfrac{\sqrt{-1} \cdot \sqrt{-121}}{\sqrt{-75}}$ **10.** $\dfrac{\sqrt{-4} \cdot \sqrt{-2}}{\sqrt{-1} \cdot \sqrt{-20}}$

In exercises **11–20**, use the equation $i^2 = -1$ to simplify. Each number can be reduced to i, -1, $-i$, or 1.

11. i^3 **12.** i^4 **13.** i^5

14. i^6 **15.** i^7 **16.** i^8

17. i^9 **18.** i^{10} **19.** i^{25}

20. i^{99}

Two axes and a plane are required in order to graph complex numbers. In Figure 9-2, horizontal and vertical axes are drawn to form such a **complex plane.** The horizontal and vertical axes are labeled **real** and **i-axis** respectively.

Example. Graph:

   a. $4 + 7i$ **b.** $-6 - 5i$ **c.** 3   **d.** $4i$

Solution. **Discussion.** Locate the real part of the complex number on the real axis and the *i*-part on the vertical axis. The point of intersection of the vertical and horizontal lines through these points is a graph of the complex number.

   a. Locate 4 on the real axis.
   Locate 7*i* on the vertical axis.
   The vertical line through 4 and the horizontal line through 7*i* is the graph of $4 + 7i$. See Figure 9-3.

   b. The point in Figure 9-3 with coordinate $-6 - 5i$ is a graph of the complex number $-6 - 5i$.

   c. The point with coordinate 3 on the real axis is a graph of the complex number 3, which can also be written $3 + 0i$.

   d. The point with coordinate 4*i* on the *i*-axis is a graph of the complex number 4*i*, which can also be written $0 + 4i$.

Figure 9-2. A complex plane.

Figure 9-3. Graphs of four complex numbers.

In exercises **21–30**, graph the complex numbers.

21. $5 + 2i$ **22.** $1 + 8i$

23. $-3 + 7i$ **24.** $-8 + 3i$

25. $-6 - 4i$ **26.** $-2 - 7i$

27. $4 - 5i$ **28.** $7 - i$

29. $-2i$ **30.** $4i$

In exercises **31–40**, write the complex number associated with each point.

31. A   **32.** B

33. C   **34.** D

35. E   **36.** F

37. G   **38.** H

39. I   **40.** J

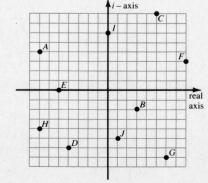

In exercises **41–50**, solve by any method.

41. $2t^2 = 7t + 15$　　　　　　　**42.** $14 - 19t = 3t^2$

43. $9x = x^2 - 41$　　　　　　　**44.** $x^2 = 8 - 4x$

45. $-61 = 12y + y^2$　　　　　　**46.** $y^2 = -109 - 20y$

47. $2z(z + 2) = 2z - 3$　　　　　**48.** $z(z + 12) = 7(z - 1)$

49. $(3u - 2)^2 = 18$　　　　　　　**50.** $(u + 3)^2 = 4u - 28 + 3u^2$

SECTION 9-6.　Solving Quadratic Inequalities (Optional)

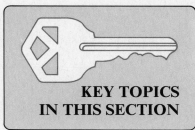

KEY TOPICS IN THIS SECTION

1. A definition of a quadratic inequality

2. Finding the boundaries to the solutions of a quadratic inequality

3. Determining the region or regions that contain the solutions

4. Some applied problems

　The trinomial $ax^2 + bx + c$ can also be used to write an inequality. As is the case with a linear inequality, the number of solutions is infinite. As a consequence, the solutions will be graphed on a number line. In fact, as will be seen, a number line will be used to help determine which numbers are solutions of a given inequality.

A Definition of a Quadratic Inequality

　The inequalities in Definition 9.4 identify the standard form of quadratic inequalities in one variable.

Definition 9.4.　A quadratic inequality in x
　A quadratic inequality in x is one that can be written in the form

$$ax^2 + bx + c < 0 \quad \text{or} \quad ax^2 + bx + c > 0 \quad \text{or}$$

$$ax^2 + bx + c \leq 0 \quad \text{or} \quad ax^2 + bx + c \geq 0$$

where a, b, and c are real numbers and $a \neq 0$.

　Examples **a** and **b** illustrate two inequalities that can be written in the standard form shown in one of the inequalities in Definition 9.4.

a. $t < 36 - 2t^2$　　　**b.** $3w^2 > 19w - 20$

Both of these inequalities can be changed to the standard form:

a. $t < 36 - 2t^2$　　　　　**b.** $3w^2 > 19w - 20$　　　　The given inequalities

　$2t^2 + t - 36 < 0$　　　　　$3w^2 - 19w + 20 > 0$　　　　In standard form

Finding the Boundaries to the Solutions of a Quadratic Inequality

Solving a quadratic inequality is similar to solving a linear inequality in one variable—we identify the **boundaries** between the real numbers that are solutions and those that are not solutions. The quadratic inequalities in this section have two boundaries for the solutions.

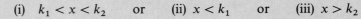

To solve a quadratic inequality in x means to find boundaries k_1 and k_2 such that

$$\text{(i)} \quad k_1 < x < k_2 \quad \text{or} \quad \text{(ii)} \ x < k_1 \quad \text{or} \quad \text{(iii)} \ x > k_2$$

The following three steps provide a procedure for finding k_1 and k_2.

To find the boundaries k_1 and k_2 for the solutions of a quadratic inequality,

Step 1. write a corresponding equation $ax^2 + bx + c = 0$.

Step 2. use the zero-product property to solve the equation of Step 1.

Step 3. state the solutions to the equation as the boundaries k_1 and k_2.

Example 1. Determine the boundaries for the solutions of

$$2t^2 + t - 36 < 0.$$

Solution.   **Step 1.** Write a corresponding equation.

$$2t^2 + t - 36 = 0$$

Step 2. Use the zero-product property to solve the equation.

$$(2t + 9)(t - 4) = 0$$

$$2t + 9 = 0 \qquad \text{or} \qquad t - 4 = 0$$

$$t = \frac{-9}{2} \qquad\qquad t = 4$$

Step 3. State the solutions as the boundaries k_1 and k_2.
The boundaries to the solutions of the given inequality are $k_1 = \frac{-9}{2}$ and $k_2 = 4$.

Determining the Region or Regions that Contain the Solutions

In Figure 9-4 the boundaries determined in Example 1 are graphed as hollow dots on a number line. The vertical dotted line segments above the dots separate the area above the line into three regions, labeled Region I, Region II, and Region III.

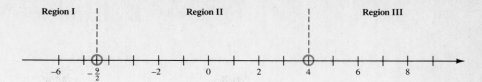

Figure 9-4. The boundaries separate the number line into three parts.

Consider now the inequality for which $\frac{-9}{2}$ and 4 are the boundaries of the solutions:

$$2t^2 + t - 36 < 0$$

To determine which region (or regions) contains the solutions, we evaluate the inequality with an arbitrarily selected test point in each region.

In Region I	In Region II	In Region III
Pick $t = -6$:	Pick $t = 0$:	Pick $t = 5$:
$2(-6)^2 + (-6) - 36 < 0$	$2(0)^2 + 0 - 36 < 0$	$2(5)^2 + 5 - 36 < 0$
$72 + (-6) - 36 < 0$	$0 + 0 - 36 < 0$	$50 + 5 - 36 < 0$
$30 < 0$, false	$-36 < 0$, true	$19 < 0$, false
Region I does not contain any solutions.	Region II contains solutions.	Region III does not contain any solutions.

Region II is the only one that contains solutions. Thus the solutions are any t, such that $\frac{-9}{2} < t < 4$. A graph of these solutions is given in Figure 9-5.

Figure 9-5. A graph of the solutions of $2t^2 + t - 36 < 0$.

Example 2.   Solve and graph $3w^2 - 19w + 20 > 0$.

Solution.   **Step 1.** Write a corresponding equation.

$$3w^2 - 19w + 20 = 0$$

Step 2. Use the zero-product property to solve the equation.

$$(3w - 4)(w - 5) = 0$$

$$3w - 4 = 0 \qquad \text{or} \qquad w - 5 = 0$$

$$w = \frac{4}{3} \qquad\qquad w = 5$$

Step 3. State the solutions as the boundaries k_1 and k_2.
The boundaries to the solutions of the given inequality are $k_1 = \frac{4}{3}$ and $k_2 = 5$. (See Figure 9-6.)

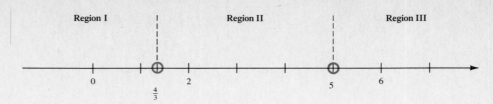

Figure 9-6. Number line with k_1 and k_2 plotted.

In Region I	**In Region II**	**In Region III**
Pick $w = 0$:	Pick $w = 2$:	Pick $w = 6$:
$3(0)^2 - 19(0) + 20 > 0$	$3(2)^2 - 19(2) + 20 > 0$	$3(6)^2 - 19(6) + 20 > 0$
$0 - 0 \quad + 20 > 0$	$12 - 38 + 20 > 0$	$108 - 114 \ + 20 > 0$
$20 > 0$, true	$-6 > 0$, false	$14 > 0$, true
Region I contains solutions.	Region II does not contain solutions.	Region III contains solutions.

Thus, the solutions for the given inequality are in Regions I or III; that is, $w < \frac{4}{3}$ or $w > 5$. (See Figure 9-7.)

Figure 9-7. A graph of the solutions of $3w^2 - 19w + 20 > 0$.

Example 3.   Solve and graph: $9 + 29m \le -6m^2$

Solution.   **Step 1.** $6m^2 + 29m + 9 = 0$

Step 2. $(3m + 1)(2m + 9) = 0$

$$3m + 1 = 0 \qquad \text{or} \qquad 2m + 9 = 0$$

$$m = \frac{-1}{3} \qquad\qquad m = \frac{-9}{2}$$

Step 3. The boundaries to the solutions are $k_1 = \frac{-1}{3}$ and $k_2 = \frac{-9}{2}$. (See Figure 9-8.)

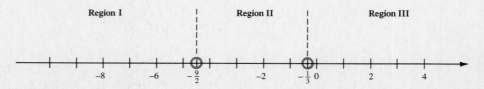

Figure 9-8. Number line with k_1 and k_2 plotted.

In Region I	In Region II	In Region III
Pick $m = -6$:	Pick $m = -2$:	Pick $m = 0$:
$9 + 29(-6) \leq -6(-6)^2$	$9 + 29(-2) \leq -6(-2)^2$	$9 + 29(0) \leq -6(0)^2$
$9 + (-174) \leq -216$	$9 + (-58) \leq -24$	$9 + \quad 0 \leq 0$
$-165 \leq -216$, false	$-49 \leq -24$, true	$9 \leq 0$, false
Region I does not contain solutions.	Region II does contain solutions.	Region III does not contain solutions.

The solutions are $\frac{-9}{2} \leq m \leq \frac{-1}{3}$. The boundaries are included in the solutions because the given inequality is compound. (See Figure 9-9.)

Figure 9-9. A graph of the solutions of $6m^2 + 29m + 9 \leq 0$.

Some Applied Problems

Example 4 is an illustration of a problem that involves a quadratic inequality.

Example 4. The Bries Picture Frame Manufactures produces a D3SP series of picture frames. Cindy Bries, the president of the company, has specified that the length of the frame for this series be 2 inches more than the width. The area enclosed by the frame should be less than or equal to 360 square inches, and the minimum width allowed should be 4 inches. Find the range of possible widths for the D3SP series.

Solution. **Discussion.** The area of a rectangle is the product of the length and width. If w represents the width, then for this series $w + 2$ represents the length, and $w(w + 2)$ represents the area. Based on the given information,

$$w(w + 2) \leq 360 \qquad and \qquad w \geq 4.$$

Step 1. $w(w + 2) = 360$

$$w^2 + 2w - 360 = 0$$

Step 2. $(w + 20)(w - 18) = 0$

$$w = -20 \qquad or \qquad w = 18$$

Step 3. The boundaries are $k_1 = -20$ and $k_2 = 18$.

Checking the three regions, we obtain as solutions

$$-20 \leq w \leq 18.$$

Based on these solutions, we might claim that the range of possible widths is from -20 inches through 18 inches. We are given, however, the restriction that w must be at least 4 inches. Therefore the range of possible widths is

$$4 \leq w \leq 18 \text{ inches.}$$

SECTION 9-6.   Practice Exercises

In exercises **1–8**, determine the boundaries for the solutions of the inequalities.

[Example 1]   **1.** $x^2 - x - 12 > 0$ **2.** $x^2 + 2x - 8 < 0$

3. $y^2 - 4y + 3 < 0$ **4.** $y^2 - 6y + 5 > 0$

5. $2a^2 - 9a - 5 < 0$ **6.** $2a^2 - 5a + 2 > 0$

7. $4b^2 + 27b + 35 > 0$ **8.** $3b^2 + 19b + 28 < 0$

In exercises **9–24**, solve and graph.

[Example 2]   **9.** $u^2 + 2u - 8 < 0$ **10.** $u^2 - 4u - 5 < 0$

11. $v^2 + 3v - 10 > 0$ **12.** $v^2 + 3v - 40 > 0$

13. $t^2 - 6t + 5 < 0$ **14.** $t^2 - 10t + 16 < 0$

15. $3w^2 + 29w + 18 > 0$ **16.** $2w^2 + 11w + 5 > 0$

[Example 3]   **17.** $8a \leq 20 - a^2$ **18.** $4a \leq 45 - a^2$

19. $b^2 \geq 2b + 63$ **20.** $b^2 \geq 24 - 5b$

21. $0 \geq 5 + 2w^2 + 11w$ **22.** $0 \geq 22w + 7 + 3w^2$

23. $18n \leq 5n^2 + 16$ **24.** $23n \leq 6n^2 + 15$

In exercises **25–30**, consider the picture frames in Example 4. Find the range of the possible values for the smaller sides.

[Example 4]   **25.** The length is 4 inches more than the width. The area enclosed is less than 480 square inches, and the minimum width is 10 inches.

26. The length is 6 centimeters more than the width. The area enclosed is less than 667 square centimeters and the minimum width is 6 centimeters.

27. The picture frame is a square with an area less than or equal to 144 square inches. The sides must be at least 10 inches.

28. The picture frame is a square with an area less than or equal to 225 square inches. The sides must be at least 8 inches.

29. The frame has a base that is 5 centimeters less than the height. The area is less than or equal to 456 square centimeters and the base must be at least 10 centimeters long.

30. The frame has a base that is 9 inches less than the height. The area is less than 630 square inches, and the base must be at least 15 inches long.

SECTION 9-6.   Ten Review Exercises

1. Divide $(3k^2 - 7k - 40)$ by $(3k + 8)$.

2. Divide $(3k^2 - 7k - 40)$ by $(k - 5)$.

3. Write $3k^2 - 7k - 40$ in factored form.

4. Evaluate $3k^2 - 7k - 40$ for $k = -3$.

5. Evaluate the factored form of exercise **3** for $k = -3$.

6. Evaluate $3k^2 - 7k - 40$ for $k = \frac{-8}{3}$.

7. Evaluate the factored form of exercise **3** for $k = \frac{-8}{3}$.

8. Solve and check: $3k^2 = 7k + 40$.

9. Solve and graph: $3k^2 - 40 < 7k$.

10. Solve and graph: $7k + 40 \leq 3k^2$.

SECTION 9-6.   Supplementary Exercises

In exercises **1–8**, determine the boundaries for the solutions.

1. $x^2 + 6x < 0$

2. $x^2 + 9x > 0$

3. $2y^2 - 8y > 0$

4. $3y^2 - 12y < 0$

5. $6x^2 + 7x - 20 \leq 0$

6. $20x^2 - 3x - 2 \geq 0$

7. $16a^2 - 49 \leq 0$

8. $9a^2 - 100 \geq 0$

In exercises **9–18**, solve and graph.

9. $b^2 - 5b \leq 0$

10. $3b^2 + 10b \geq 0$

11. $8k^2 + 2k < 0$

12. $15k^2 + 7k < 2$

13. $4t^2 \geq 28t - 45$

14. $9t^2 \geq -36t - 20$

15. $0 \geq 18 - 3w - w^2$

16. $0 \geq 22 + 9w - w^2$

17. $-8u > u^2 + 15$

18. $-11u > 28 + u^2$

Exercises **19** and **20** have *no solutions*. Give a reason why there are none.

19. $x^2 - 6x + 9 < 0$

20. $4x^2 + 20x + 25 < 0$

In exercises **21** and **22**, *all real numbers are solutions*. Give a reason why this is true.

21. $y^2 + 12y + 36 \geq 0$

22. $25y^2 + 20y + 4 \geq 0$

In exercises **23** and **24**, *only one number is a solution.*

a. Find the number.

b. Give a reason why this is the only number.

23. $z^2 - 14z + 49 \leq 0$

24. $4z^2 + 4z + 1 \leq 0$

In exercises **25** and **26**, *all real numbers are solutions except one number.*

a. Find the one exception.

b. Give a reason why this number is not a solution.

25. $t^2 + 16t + 64 > 0$ **26.** $36t^2 - 12t + 1 > 0$

Name _____

Date _____

Score _____

SECTION 9-1. Summary Exercises

Answer

1. Write in standard form.

$6 - 3(x + 2) = x(x - 5)$

1. _____

In exercises **2–8**, solve.

2. $(5x - 8)(9x + 1) = 0$

2. _____

3. $b^2 + 169 = 26b$

3. _____

4. $x(x + 4) = 21$

4. _____

5. $(x + 1)^2 = -2(x - 23)$

5. _____

6. $\dfrac{x + 4}{2} = \dfrac{8 - x}{x}$

6. _____

7. The length of the hypotenuse of a right triangle is 2 yards more than the length of the longer leg. The shorter leg is 7 yards shorter than the longer leg. Find the lengths of the three sides.

7. _____

8. To paint the Young's barn alone would take Scott nine hours longer than Juan. Working together, Scott and Juan could paint the barn in twenty hours. How long would it take each to paint the barn alone?

8. _____

Name _____

Date _____

Score _____

SECTION 9-2. Summary Exercises

Answer

In exercises **1–8**, solve and check. Write any irrational solutions in simplified form.

1. $s^2 = 175$

1. _____

2. $16y^2 = 45$

2. _____

3. $(5x)^2 = 100$

3. _____

4. $(3b)^2 = 4$

4. _____

5. $(p - 3)^2 = 220$

5. _____

6. $(3t + 7)^2 = 8$

6. _____

7. $9b^2 - 6b + 1 = 289$

7. _____

8. $36x^2 + 60x + 25 = 50$

8. _____

Name _____

Date _____

Score _____

SECTION 9-3. Summary Exercises

Answer

1. a. Add the term necessary to make a perfect square trinomial.
   b. Write the trinomial in factored form.

$$x^2 - 8x$$

1. **a.** _____

   b. _____

In exercises **2–8**, solve by completing the square.

2. $y^2 + 2y = 8$

2. _____

3. $z^2 - 6z = -2$

3. _____

4. $p^2 - 24p + 44 = 0$

4. _____

5. $s^2 - 20s + 50 = 0$

5. _____

6. $b^2 = -12b - 24$

6. _____

7. $8x^2 + 32x + 2 = 0$ (optional)

7. _____

8. $6y^2 - 48y + 3 = 0$ (optional)

8. _____

SECTION 9-4. Summary Exercises

In exercises **1** and **2**, state the values of a, b, and c.

1. $6x^2 = 3x + 4$

1. a is

b is

c is

2. $(x + 5)^2 = 2(3 + 5x)$

2. a is

b is

c is

In exercises **3–6**, solve using the quadratic formula.

3. $y^2 + 4y - 9 = 0$

3. _____

4. $z^2 = 3z + 11$

4. _____

5. $(2p + 1)(3p - 1) = -5p$

5. _____

6. $(2x - 3)^2 = 2(5 - 4x)$

6. _____

In exercises **7** and **8**, write the answers in exact form using square roots.

7. The product of two numbers is 20. If the smaller number is subtracted from the larger one, the difference is 10. Find both numbers.

7. The larger number is _____

The smaller number is _____

8. The height of a triangle is one meter longer than the length of the base. The area of the triangle is 8 square meters. Find the lengths of the base and height.

8. The height is _____

The base is _____

SECTION 9-5. Summary Exercises

In exercises **1** and **2**, write each number in the *i*-form.

1. $\sqrt{-72}$

2. $-\sqrt{\dfrac{-3}{25}}$

1. _____

2. _____

In exercises **3** and **4**, write each number in the $a + bi$ form.

3. $-13 + \sqrt{-144}$

3. _____

4. $\sqrt{20} - \sqrt{-125}$

4. _____

In exercises **5–8**, solve. Write complex solutions in simplified $a + bi$ form.

5. $9t^2 + 100 = 0$

5. _____

595

6. $k^2 = 6k - 58$

6. _____

7. $0 = 2a + 19 + a^2$

7. _____

8. $9b^2 - 12b + 11 = 0$

8. _____

SECTION 9-6. Summary Exercises

In exercises **1–8**, solve and graph.

1. $(z - 3)(z + 4) < 0$

1. _____

———————————————→

2. $(2x + 5)(x - 7) \leq 0$

2. _____

———————————————→

3. $3x(x + 8) \leq 0$

3. _____

———————————————→

4. $a^2 - 12a + 20 > 0$

4. _____

———————————————→

5. $b^2 + 99 < 20b$

5. _____

————————————————→

6. $26t \leq 9 - 3t^2$

6. _____

————————————————→

7. $18y + 8 > 5y^2$

7. _____

————————————————→

8. $25m^2 \geq 144$

8. _____

————————————————→

CHAPTER 9.   Review Exercises

In exercises **1** and **2**,

a. Write in standard form.

b. State the values of a, b, and c.

1. $x(2x - 5) = 4(x - 1)^2$ **2.** $y(2 - y) = 5y^2 + 8$

In exercises **3–6**, solve by factoring.

3. $(2p + 7)^2 = 56p$ **4.** $t^2 = 3t + 88$

5. $6w^2 = w$ **6.** $x(x + 4) - x = 5(x + 3)$

In exercises **7–10**, solve using the square-root theorem.

7. $x^2 = 121$ **8.** $(s + 5)^2 = 16$

9. $(3s + 5)^2 = 60$ **10.** $(25s^2 - 20s + 4) = 10$

In exercises **11–14**, solve by completing the square.

11. $m^2 - 12m - 13 = 0$ **12.** $m^2 - 12m + 13 = 0$

13. $z^2 + 20z = -20$ **14.** $y^2 = 6y + 10$

In exercises **15–18**, solve using the quadratic formula.

15. $z^2 - 5z - 1 = 0$ **16.** $4x^2 = 4x + 3$

17. $y^2 + 2 = 20(y - 2)$ **18.** $6x^2 - 14 = 0$

In exercises **19–24**, solve by any method.

19. $x^2 + x - 1 = 0$ **20.** $(2x + 3)(4x - 5) = 0$

21. $y^2 - 5y - 6 = 0$ **22.** $(2x + 3)^2 = 9$

23. $y^2 = 24$ **24.** $2x + 3x^2 = 1$

In exercises **25–28**, write the answers in exact form.

25. The length of the hypotenuse of a right triangle is 9 centimeters more than the length of the short leg. The length of the long leg is 1 centimeter more than the length of the short leg. Find the lengths of the sides.

26. Chris and Fred's lawn maintenance service maintains the Alexanders' summer mansion. Chris, new to the business, takes six hours longer than Fred to service the grounds. One weekend Chris worked alone for six hours and then Fred worked alone for three hours to complete the service. How long would it take each person, working alone, to complete the job?

27. Find two numbers such that their product is 25 and their difference is 5.

28. In right triangle ABC the length of the long leg is 2 inches more than the length of the short leg. If the hypotenuse is 10 inches long, how long are the legs?

Optional

In exercises **29–36**, solve.

29. $z^2 = -25$

30. $p^2 + 16 = 0$

31. $(3x + 2)^2 = -16$

32. $(2x - 5)^2 = -10$

33. $p^2 + p + 1 = 0$

34. $p^2 + 2p + 3 = 0$

35. $2y^2 + 5y + 9 = 0$

36. $y^2 + 5y + 7 = 0$

In exercises **37–40**, solve and graph.

37. $(2t - 3)(t + 5) < 0$

38. $(3t + 1)(t - 4) < 0$

39. $x^2 + 7x + 6 \geq 0$

40. $x^2 + x - 12 \geq 0$

10
Relations and Functions

W hen the students arrived for class today, the following equations were written on the board:

$$Ax + By = C \qquad \textbf{Standard form}$$

$$y = mx + b \qquad \textbf{Slope-intercept form}$$

$$y - y_1 = m(x - x_1) \qquad \textbf{Point-slope form}$$

"I trust you all recognize the equations on the board," Ms. Glaston said as she entered the room. "In Chapter 6 we studied these equations as part of the topic of linear equations in two variables. This study introduced us to the part of mathematics that deals with relationships between two or more quantities. Any relationship that can be described by one of these equations is called **linear.**"

"Ms. Glaston," Craig Behrndt asked, "are there equations other than linear that are used to describe relationships?"

"Yes, Craig," Ms. Glaston replied, "there are many other types of equations. To demonstrate a need for equations other than linear, let me show you a very simple experiment."

She then removed a small steel ball from a desk drawer and extended her arm upward as far as she could reach. When she was sure that everyone in the class was watching, she released the ball and it fell to the floor.

"OK," she continued, "you all saw the ball fall to the floor. What I want to know now is what two different quantities might we find related here? That is, what two measurable quantities were present in this experiment?"

"I know one thing," Netta Gilboa said. "We can measure the distance the ball fell."

"So distance is one measurable quantity of this activity," Ms. Glaston said.

"How about the weight of the ball?" Jeff Seeger asked.

"Good suggestion, Jeff," Ms. Glaston replied, "but did the weight of the ball change during the activity?"

"No," Jeff replied, "I'm sure it didn't. So we really want to identify quantities that changed during the time the ball fell."

"Time!" Shannon Schanze said. "Isn't time a quantity that changed during the experiment? It took a little while for the ball to hit the floor after you let it go."

"OK, so time is another variable quantity," Ms. Glaston replied. "Does it seem reasonable for us to assume that if an object is dropped and falls to the ground, then the distance it falls and the time it falls are somehow related?"

"Well," Jeff Seeger replied, "if you dropped that ball from a higher point, then it certainly would take longer to hit the floor."

"I think we would all agree to that fact," Ms. Glaston said. "The higher above the ground an object is at the time it is released, the longer it will take for the object to reach the ground. If certain assumptions are made regarding air resistance, this equation can be used to describe the relationship." She then turned to the board and wrote

$$d = 4.9t^2$$

d is the *distance* an object falls in meters.

t is the *time* the object falls in seconds.

"In this chapter," Ms. Glaston continued, "we will study several details about relationships between such quantities as distance and time. We will also study some of the more commonly encountered equations that describe relations."

SECTION 10-1.   A General Discussion of Relations and Functions

KEY TOPICS IN THIS SECTION

1. A definition of a relation

2. The definitions of domain and range of a relation

3. The definition of a function

4. The $f(x)$ notation

In this section we will study some of the terminology associated with mathematical relations. In addition, we will identify what is meant by the domain and range of a relation. Finally, the special relation called *function* will be defined.

A Definition of a Relation

The word **relation** implies a connection or pairing of objects from two different groups. In mathematics, the groups of objects are sets of numbers. For our study, the numbers in the two sets will be some or all of the set of real numbers. The way in which the pairs will be shown is by listing the numbers as ordered pairs.

> **Definition 10.1.   A mathematical relation**
> A **relation** is a set of ordered pairs of real numbers.

We can write a relation in three ways:

1. List the ordered pairs in the relation.

2. If possible, write an equation or inequality that describes the relationship between the ordered pairs.

3. Graph the ordered pairs on a rectangular coordinate system.

Example 1.   A relation is described by the equation $y = 3x - 7$.

    a. Find the ordered pairs in the relation for $x = 0$, $x = 2$, and $x = 4$.

b. Use the ordered pairs from part **a** to graph all the ordered pairs in the relation.

Solution.   **Discussion.** We obtain the ordered pairs in the relation for the three given values of x by replacing x in the equation by the given values to find the corresponding values of y.

a. If $x = 0$, then $y = 3(0) - 7$

$$= -7 \qquad (0, -7) \text{ is in the relation.}$$

If $x = 2$, then $y = 3(2) - 7$

$$= -1 \qquad (2, -1) \text{ is in the relation.}$$

If $x = 4$, then $y = 3(4) - 7$

$$= 5 \qquad (4, 5) \text{ is in the relation.}$$

b. In Figure 10-1 the three pairs are plotted as points in a coordinate system. The line through these points is a graph of the relation defined by $y = 3x - 7$.

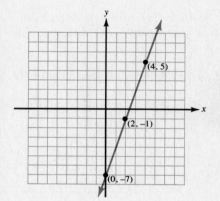

Figure 10-1. A graph of $y = 3x - 7$.

> If a relation is specified by ordered pairs of the form (x, y), then x is called the **independent variable** and y is called the **dependent variable.**

Example 2.   A relation is described by the equation $y = \dfrac{x + 3}{2x - 1}$. Find the values of the dependent variable for the following values of the independent variable: 0 and 4.

Solution.   **Discussion.** For the given relation, x is the independent variable. Thus 0 and 4 are replacements for x.

If $x = 0$, $y = \dfrac{0 + 3}{2(0) - 1}$

$$= -3 \qquad (0, -3) \text{ is in the relation.}$$

If $x = 4$, $y = \dfrac{4 + 3}{2(4) - 1}$

$$= 1 \qquad (4, 1) \text{ is in the relation.}$$

The Definition of Domain and Range of a Relation

The first and second components of the ordered pairs of a relation are separated into two different sets.

> **Definition 10.2.   The domain and range of a relation**
> The **domain** is the set of all first components of a relation.
> The **range** is the set of all second components of a relation.

Example 3.   List the elements in

    a. the domain

    b. the range

of the relation with the following ordered pairs:

$(-3, 32), (-2, 13), (-1, 6), (0, 5), (1, 4), (2, -3),$ and $(3, -22)$

Solution.

    a. Discussion. The domain consists of the first components in the ordered pairs.

        The domain: $-3, -2, -1, 0, 1, 2,$ and 3

    b. Discussion. The range consists of the second components in the ordered pairs.

        The range: $32, 13, 6, 5, 4, -3,$ and -22

The domain and range of a relation can be "read" from a graph of the relation. The domain is indicated by the **horizontal extent** in the graph. The range is indicated by the **vertical extent** in the graph.

Example 4.   Identify the domain and range of the relations graphed in **a** and **b**.

a.

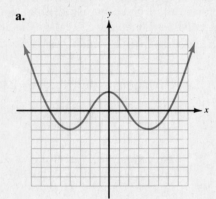

b.

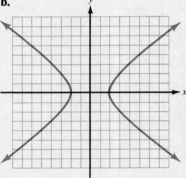

Solution.

    a. Discussion. The curve in the figure moves uninterrupted from left to right. Therefore, x can take on any real number. The domain contains all real numbers, which can be written "all x."

        The curve does not extend any further downward than -2 on the y-axis. However, y can take on any real number greater than -2, since the graph extends uninterrupted upward. The range contains all real numbers greater than -2, which can be written "all y, $y \geq -2$."

    b. Discussion. The graph of this relation consists of two separated curves. Since there is no graph between -2 and 2 on the x-axis, the domain does not contain any numbers between -2 and 2. However, the curves extend infinitely far to the left and to the right. Thus the domain contains all numbers less than or equal to -2 and greater than or equal to 2, which can be written "all x, $x \leq -2$ or $x \geq 2$."

        The curves extend infinitely far up and down and have no breaks. Thus the range contains all real numbers, written as "all y."

The Definition of a Function

The following definition identifies a special type of relation.

> **Definition 10.3.   Function**
>   A **function** is a relation in which each domain element is paired with exactly one range element.

If a function is defined by an equation in x and y, then for every x in the domain, there is one and only one y in the range.

Example 5.   Determine whether the following relations are functions:

  a. $\{(-2, 13), (-1, 7), (0, 5), (1, 5), (2, 7)\}$

  b. $\{(5, 0), (4, 3), (4, -3), (3, 4), (3, -4)\}$

Solution.   **a.** **Discussion.** Check each number in the domain. If no number in the domain is paired with two different numbers in the range, then the relation is a function. The domain contains -2, -1, 0, 1, and 2. Each number is paired with only one number; thus the relation is a function.

  b. The domain contains 5, 4, and 3. Since 4 is paired with 3 and -3, the relation is not a function. Note that 3 in the domain is paired with 4 and -4, which also violates the definition of a function.

The vertical line test can be used on a graph of a relation to determine whether the relation is a function.

> **Vertical line test**
>   If any vertical line intersects a graph of a relation in more than one point, then at least two different numbers in the range are paired with a number in the domain, and the relation is not a function.

Example 6.   Determine whether the relations in the following graphs are functions.

a.

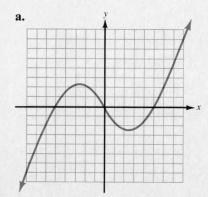

b.

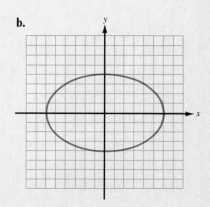

Solution.   **Discussion.** Imagine that vertical lines are drawn at several places. If any of these lines can intersect the graph in more than one point, then the corresponding relation is not a function. If, however, no vertical line can intersect the graph in more than one point, then the corresponding relation is a function.

a. The shape of the graph suggests that no vertical line can intersect it in more than one point. Thus, the graph is that of a function.

b. Any vertical line drawn between $x = -6$ and $x = 6$ will intersect the graph in two points. Thus, the graph is not a graph of a function.

The $f(x)$ Notation

If a relation is a function and x is the independent variable, then we frequently use $f(x)$ (read "f of x") to specify the value of the dependent variable paired with each x in the domain. In other words, $f(x)$ is used in place of y in the equation that defines the function.

Example 7.   If $f(x) = 3x^2 + x - 5$, find

a. $f(-2)$ **b.** $f(0)$

Solution.   **Discussion.** To find $f(-2)$, replace x everywhere in $3x^2 + x - 5$ by -2 and simplify. Similarly, replace x by 0 to find $f(0)$.

a. $f(-2) = 3(-2)^2 + (-2) - 5$ Replace x by -2.

$\qquad\quad = 12 + (-2) - 5$ $3(-2)^2 = 3(4) = 12$

$\qquad\quad = 5$ $f(-2) = 5$

b. $f(0) = 3(0)^2 + 0 - 5$ Replace x by 0.

$\qquad\quad = 0 + 0 - 5$

$\qquad\quad = -5$ $f(0) = -5$

SECTION 10-1.   Practice Exercises

In exercises **1–8**,

a. find the ordered pairs in the relation.

b. graph the ordered pairs.

[Example 1]   **1.** $y = 2x - 5$, and $x = 0$, $x = 4$, $x = 6$

2. $y = 5 - 3x$, and $x = -1$, $x = 2$, and $x = 4$

3. $y = \dfrac{3}{4}x - 3$, and $x = -4$, $x = 0$, and $x = 8$

4. $y = \dfrac{5}{2}x + 4$, and $x = -4$, $x = 0$, and $x = -2$

5. $y = 5$, and $x = -3$, $x = 0$, and $x = 6$

6. $y = -3$, and $x = -7$, $x = 0$, and $x = 7$

7. $2x = 5$ and $y = -5$, $y = 0$, and $y = 5$

8. $4x + 8 = 0$, and $y = -3$, $y = 0$, and $y = 3$

In exercises **9–20**, find the values of the dependent variable for the given values of the independent variable.

[Example 2]   **9.** $y = \dfrac{x}{x + 3}$; $-2, 0, 6$ **10.** $y = \dfrac{x - 5}{x}$; $-1, 5, 10$

11. $y = x^2 - 4$; $-3, 0, 3, 5$ **12.** $y = 8 - 2x^2$; $-2, 0, 2, 4$

13. $y = \sqrt{2x - 5}$; $3, 7, 15, 27$ **14.** $y = \sqrt{x + 1}$; $-1, 0, 3, 35$

15. $y = x^3 + 2$; $-2, 0, 2, 3$ **16.** $y = 2x^3 - 3x$; $-2, 0, 2, 4$

17. $y = \dfrac{x^2 - 1}{x^2 + 1}$; $-3, -1, 0, 2$ **18.** $y = \dfrac{4 - x^2}{4 + x^2}$; $-2, 0, 2, 4$

19. $y = x^4 - 2x^2$; $-2, 0, 2, 3$ **20.** $y = 4x^3 - 3x^4$; $-1, 0, 1, 2$

In exercises **21–26**, for each relation, list the elements in

a. the domain.

b. the range.

[Example 3]   **21.** $\{(-2, 3), (-1, 4), (0, 5), (1, 6), (2, 7)\}$

22. $\{(-2, -7), (-1, -5), (0, -3), (1, -1), (2, 1)\}$

23. $\left\{\left(-4, \dfrac{1}{7}\right), \left(-2, \dfrac{-1}{5}\right), (0, -1), (2, -5), (4, 1)\right\}$

24. $\left\{\left(-4, \dfrac{-8}{3}\right), \left(-1, \dfrac{-2}{3}\right), (0, 0), \left(1, \dfrac{2}{3}\right), \left(4, \dfrac{8}{3}\right)\right\}$

25. $\{(5, 0), (6, 1), (8, \sqrt{3}), (9, 2), (12, \sqrt{7})\}$

26. $\{(0, 1), (1, \sqrt{3}), (4, 3), (5, \sqrt{11}), (12, 5)\}$

In exercises **27–36**, identify the domain and range of the relations whose graphs are given.

[Example 4]   **27.**

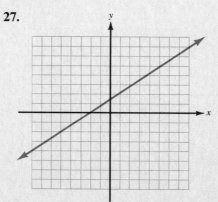

28.

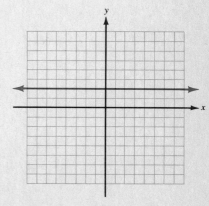

29.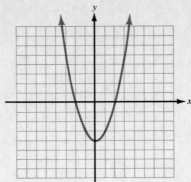

30.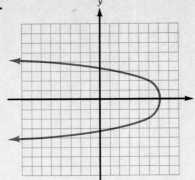

31.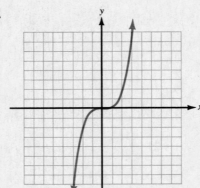

32.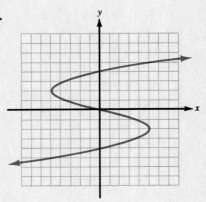

33.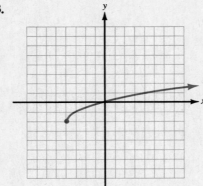

34.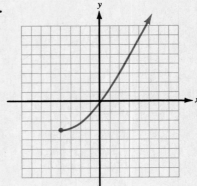

35.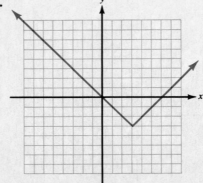

36.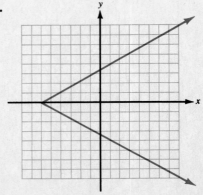

In exercises **37–42**, determine whether the relations are functions.

[Example 5]   **37.** $\{(-3, 80), (-2, 35), (-1, 8), (0, -1), (1, 8), (2, 35), (3, 80)\}$

38. $\{(-5, -14), (-3, -10), (-1, -6), (0, -4), (1, -2), (3, 2)\}$

39. $\{(-3, 1), (-3, -1), (-1, 2), (-1, -2), (0, 3), (0, -3)\}$

40. $\{(0, 4), (0, -4), (1, \sqrt{10}), (1, -\sqrt{2}), (2, \sqrt{3}), (2, -\sqrt{3})\}$

41. $\{(-3, 6), (-2, 6), (-1, 6), (0, 6), (1, 6), (2, 6), (3, 6)\}$

42. $\{(-1, 5), (-1, 3), (-1, 1), (-1, -1), (-1, -3), (-1, -5)\}$

In exercises **43–48**, determine whether the relations in the following graphs are functions.

[Example 6]   **43.**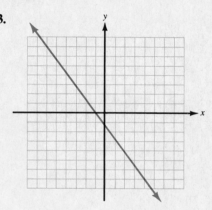

44.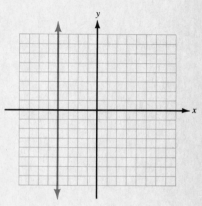

45.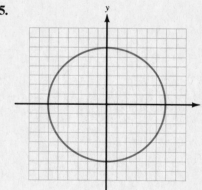

46.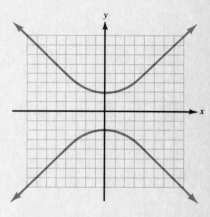

47.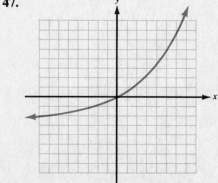

48.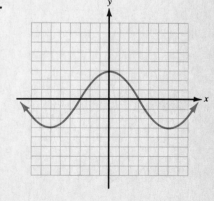

610 10 Relations and Functions

In exercises **49–56**, for each function find

a. $f(-3)$ **b.** $f(0)$ **c.** $f(2)$

49. $f(x) = 4x - 5$ **50.** $f(x) = x^2 + 4$

51. $f(x) = x^3 - x + 2$ **52.** $f(x) = \dfrac{x + 3}{x - 3}$

53. $f(x) = 9 - x^2$ **54.** $f(x) = \sqrt{x + 4}$

55. $f(x) = \sqrt{6 - x}$ **56.** $f(x) = \dfrac{-1}{3}x + 10$

SECTION 10-1. Ten Review Exercises

In exercises **1–5**, solve and check.

1. $9a = 9 + 2a^2$ **2.** $3b - 8 = 8 - 2(b - 2)$

3. $\dfrac{5}{4c} = \dfrac{1}{3} + \dfrac{5 + 2c}{4c}$ **4.** $\sqrt{d - 7} = \sqrt{d} - 1$

5. (1) $4x + y = 10$
    (2) $7x - 3y = 27$

In exercises **6–10**, factor completely.

6. $x^2 - 3x - 40$ **7.** $3t^2 - 12t$

8. $25u^2 + 10u + 1$ **9.** $8v^3 + 125$

10. $6mn - 3n + 4m - 2$

SECTION 10-1. Supplementary Exercises

In exercises **1–6**,

a. list all ordered pairs of the form (parent, child).

b. does each set of ordered pairs in part **a** constitute a relation?

c. if the answer to part **b** is yes, is the relation a function?

1. Judy and Jim Marta have two sons, Jason and Jacob.

2. Sam and Rebecca Johnson have two children, Carla and Woody.

3. Glenn and Jerry Katz have one daughter, Patricia.

4. Eric is the only son of Larry Hammond and Samantha Phillips-Hammond.

5. Fred Rockwell is a widower with three children: Ed, Mark, and Tammie.

6. Bunty Robinson is a single parent with two children, Martha and Clyde.

In exercises **7–10**,

a. list all ordered pairs of the following form: (restaurant, menu item).

b. list the domain elements.

c. list the range elements.

d. do the ordered pairs in part **a** constitute a function?

7. Ted's Diner serves roast beef, hamburgers, and meat loaf. Melba's Place offers chicken, ribs, and hamburgers.

8. Jill and Jack's Deli sells salami, pastrami, and ham. Clair's Corner serves tuna, ham, and bologna.

9. Smokie's drinks include cola and orange soda. The drinks at Tom's Place are coffee and tea.

10. The Red Top Eatery sells cherry pie and apple turnovers. The Fast Stop Beanery serves chocolate cake and donuts.

Exercises **11–14** list several ordered pairs of numbers. Use the given values of x as the domain elements of a relation. Select from the given ordered pairs the ones that belong to the relation.

11. $(7, 3)$, $(0, 2)$, $(6, 5)$, $(8, 10)$, $(3, 6)$, $(4, 12)$, $(2, 4)$

For the relation $x = 0$, 4, and 6

12. $(-1, 12)$, $(3, -1)$, $(14, -6)$, $(4, 3)$, $(1, 1)$, $(2, 5)$, $(7, 2)$

For the relation $x = -1$, 2, and 4.

13. $(5, 0)$, $(-4, 2)$, $(3, 8)$, $(0, 7)$, $(8, 0)$, $(4, 3)$, $(-4, 5)$, $(8, 1)$, $(0, -1)$, $(1, 7)$

For the relation $x = -4$, 0, and 8

14. $(3, 7)$, $(-5, 5)$, $(5, 2)$, $(9, 2)$, $(0, 9)$, $(5, 10)$, $(7, 10)$, $(4, 6)$, $(1, 5)$, $(9, 10)$

For the relation $x = 5$, 7, and 9

In exercises **15–20**, for each relation whose graph is shown

a. identify the domain.

b. identify the range.

c. determine whether the relation is a function.

15.

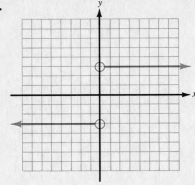

16.

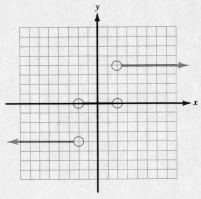

17.

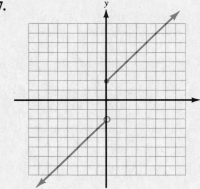

18.

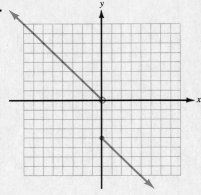

19.

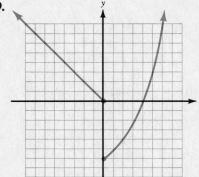

20.

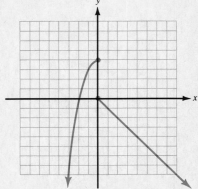

In exercises **21–26**, use the graphs of the relations to find the ordered pair or pairs for the given values of the variable.

21.

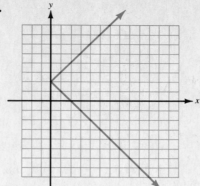

a. $x = 2$
b. $x = 4$
c. $y = 1$

22.

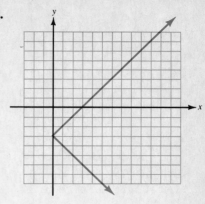

a. $x = 2$
b. $x = 4$
c. $y = 1$

23.

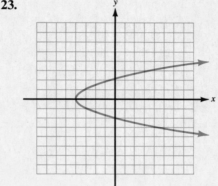

a. $x = -3$
b. $x = 0$
c. $y = -3$

24.

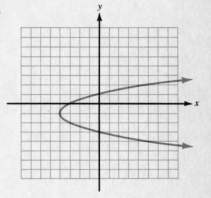

a. $x = -3$
b. $x = 0$
c. $y = -1$

25.

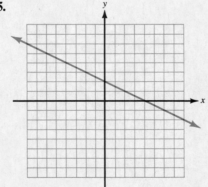

a. $x = -2$
b. $x = 0$
c. $y = 5$

26.

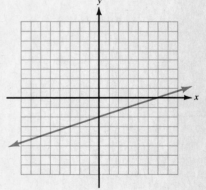

a. $x = -6$
b. $x = 0$
c. $y = -1$

SECTION 10-2. Linear and Quadratic Functions

KEY TOPICS IN THIS SECTION

1. Equations of the form $f(x) = mx + b$

2. A definition of a quadratic function

3. Some characteristics of a graph of a quadratic function

4. A recommended procedure for graphing a quadratic function

In this section we will again study the linear equation in two variables. This time we will look at these equations as functions. We will also study the quadratic functions.

Equations of the Form $f(x) = mx + b$

One of the most important functions is the linear function.

Definition 10.4.   Linear functions in x
 A linear function in x can be defined by an equation written in the form

$$f(x) = mx + b$$

where m and b are constants. The domain of any linear function in x is the set of real numbers, written "all x."

Example 1. Verify that the following equation is a linear function:

$$9x - 5y = -160$$

Solution. **Discussion.** To verify that the given equation is a linear function, we must write the equation in the form of Definition 10.4.

$9x - 5y = -160$	The given equation
$-5y = -9x - 160$	Subtract $9x$ from both sides.
$y = \dfrac{-9x}{-5} - \dfrac{160}{-5}$	Divide each term by -5.
$y = \dfrac{9}{5}x + 32$	Simplify the quotients.
$f(x) = \dfrac{9}{5}x + 32$	Replace y with $f(x)$.

In this form, m is $\frac{9}{5}$ and b is 32.

Example 2. Graph: $f(x) = \dfrac{3}{5}x - 2$

Solution. **Discussion.** The y-intercept is -2. Plot $P(0, -2)$. The slope is $\frac{3}{5}$. Beginning at P, count up 3 units, then 5 units to the right to $Q(1, 5)$. Draw the line through P and Q, as shown in Figure 10-2.

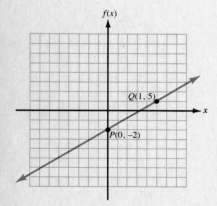

Figure 10-2. A graph of $f(x) = \frac{3}{5}x - 2$.

A Definition of a Quadratic Function

The standard form of a quadratic function is given in Definition 10.5.

> **Definition 10.5.   The standard form of a quadratic function in x**
>   A quadratic function in x is defined by an equation that can be written in the form
>
> $$f(x) = ax^2 + bx + c$$
>
> where a, b, and c are constants ($a \neq 0$). The domain of any quadratic function in x is the set of real numbers, written "all x."

Some Characteristics of a Graph of a Quadratic Function

To study the graphs of a quadratic function we begin with one defined by

$$f(x) = x^2.$$

In $f(x) = ax^2 + bx + c$, replace a by 1, b by 0, and c by 0. The following table of values yields the points plotted in Figure 10-3.

x	-3	-2	-1	0	1	2	3
$f(x)$	9	4	1	0	1	4	9

The smooth curve in Figure 10-3 is typical of a graph of any quadratic function. Such a curve is called a **parabola**. A parabola has some special characteristics that are aids in graphing a quadratic function.

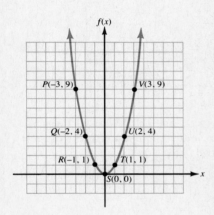

Figure 10-3. A graph of $f(x) = x^2$.

> **Characteristic 1.**
>   A graph of
>
> $$f(x) = ax^2 + bx + c$$
>
> is a parabola that opens
>
> **(i)** upward, if a is positive ($a > 0$).
>
> **(ii)** downward, if a is negative ($a < 0$).

In Figure 10-3 the parabola opens upward because in the equation $f(x) = x^2$, a is 1, a positive number. In Figure 10-4, a graph of

$$f(x) = \frac{-1}{2}x^2$$

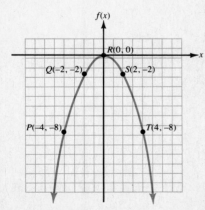

Figure 10-4. A graph of $f(x) = \frac{-1}{2}x^2$.

is shown. Since $a = \frac{-1}{2}$, a negative number, the parabola opens downward.

Characteristic 2.
  A parabola defined by

$$f(x) = ax^2 + bx + c$$

has a **turning point,** called the **vertex.**

  (i) The x-coordinate of the vertex is $\dfrac{-b}{2a}$.

  (ii) The y-coordinate of the vertex is $f\left(\dfrac{-b}{2a}\right)$.

The vertex is the **minimum point** on the parabola if it opens upward. The vertex is the **maximum point** on the parabola if it opens downward.

Characteristic 3.
  A parabola defined by

$$f(x) = ax^2 + bx + c$$

has an **axis of symmetry,** the vertical line $x = \dfrac{-b}{2a}$.

The axis of symmetry of a parabola is a vertical line that contains the vertex. If the plane on which the parabola is drawn were folded along the axis of symmetry, the left and right branches of the parabola would coincide.

Example 3.   For the parabola defined by $f(x) = x^2 - 2x - 3$,

  a. find the x-coordinate of the vertex.

  b. find the y-coordinate of the vertex.

  c. write an equation for the axis of symmetry.

Solution.   **a. Discussion.** For the given function, $a = 1$ *and* $b = -2$.
            Thus, $\frac{-b}{2a}$ becomes $\frac{-(-2)}{2(1)} = \frac{2}{2} = 1$.
            The x-coordinate of the vertex is 1.

          b. Discussion. To find the y-coordinate, compute $f(1)$.

            $f(1) = 1^2 - 2(1) - 3$

              $= -4$

            Thus, the y-coordinate of the vertex is -4. $V(1, -4)$ is the
            minimum point on the parabola.

          c. Since $\frac{-b}{2a} = 1$, an equation for the axis of symmetry is $x = 1$.

A Recommended Procedure for Graphing a Quadratic Function

The key to graphing the parabola for a given quadratic function is to exploit the three characteristics of the graph. The following procedure is recommended for locating the parabola.

A recommended procedure for graphing $f(x) = ax^2 + bx + c$

Step 1. Form $\frac{-b}{2a}$, the x-coordinate of the vertex.

Step 2. Evaluate $f(\frac{-b}{2a})$, the y-coordinate of the vertex.

Step 3. Plot the vertex and draw the axis of symmetry with a dotted line.

Step 4. Pick x_1 and x_2, two numbers "close to" but greater than the x-coordinate of the vertex.

Step 5. Evaluate $f(x_1)$ and $f(x_2)$, obtaining $P(x_1, y_1)$ and $Q(x_2, y_2)$, two points of the parabola to the right of the axis of symmetry.

Step 6. Using symmetry, locate points of the parabola to the left of the axis of symmetry that correspond to the points in Step 5.

Step 7. Draw a smooth curve through the five points plotted and extend to obtain a sketch of the parabola.

Example 4.   Sketch a graph of $f(x) = x^2 + 4x - 5$.

Solution.   **Step 1.** With $a = 1$ and $b = 4$, $\frac{-b}{2a}$ becomes $\frac{-4}{2(1)} = -2$.

Step 2. With $x = -2$,

$$f(-2) = (-2)^2 + 4(-2) - 5$$

$$= -9$$

Step 3. In Figure 10-5, $V(-2, -9)$ is plotted, and the axis of symmetry $x = -2$ is drawn.

Step 4. Arbitrarily select $x_1 = -1$ and $x_2 = 0$.

Step 5. $f(-1) = (-1)^2 + 4(-1) - 5$ and $f(0) = 0^2 + 4(0) - 5$

$\qquad\qquad = -8 \qquad\qquad\qquad\qquad\qquad = -5$

The points $P(-1, -8)$ and $Q(0, -5)$ are plotted in Figure 10-5.

Step 6. $R(-3, -8)$ is on a horizontal line with $P(-1, -8)$ but on *the opposite side of the axis of symmetry.* Thus P and R are called **symmetric points.** Similarly, $S(-4, -5)$ and $Q(0, -5)$ are symmetric points.

Step 7. A smooth curve through P, Q, R, S, and V is extended upward to obtain the parabola in Figure 10-5.

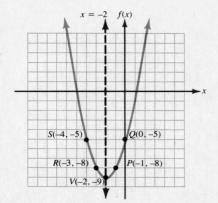

Figure 10-5. A graph of $f(x) = x^2 + 4x - 5$.

SECTION 10-2.   Practice Exercises

In exercises **1–8**, verify that each equation is a linear function by writing it in the form $f(x) = mx + b$.

[Example 1]   **1.** $2x - 3y = 16$ **2.** $5x - 2y + 2 = 0$

3. $4(y - 2) = 4 - 7x$ **4.** $3(y + 3) = 8 - (x + 14)$

5. $\dfrac{x - 2y}{3} = \dfrac{2x + 3}{2}$ **6.** $\dfrac{x + 2}{2} = \dfrac{y - 5}{4}$

7. $\dfrac{y + 3x}{3} = x + 2$ **8.** $\dfrac{6x + y}{6} = x - 2$

In exercises **9–22**, graph.

[Example 2]   **9.** $f(x) = \dfrac{2}{3}x - 8$ **10.** $f(x) = \dfrac{5}{4}x - 1$

11. $f(x) = \dfrac{-1}{4}x + 3$ **12.** $f(x) = \dfrac{-5}{2}x + 2$

13. $f(x) = 5x$ **14.** $f(x) = 2x + 1$

15. $f(x) = -3x - 2$ **16.** $f(x) = -6x$

17. $f(x) = 4$ **18.** $f(x) = -3$

19. $f(x) = \dfrac{2}{5}x + 3$ **20.** $f(x) = \dfrac{-3}{8}x - 2$

21. $f(x) = 0.8x - 1$ **22.** $f(x) = -1.2x + 4$

In exercises **23–30**,

a. find the x-coordinate of the vertex.

b. find the y-coordinate of the vertex.

c. write an equation for the axis of symmetry.

[Example 3]   **23.** $f(x) = 2x^2 + 4x$ **24.** $f(x) = -3x^2 + 12x$

25. $f(x) = x^2 - 4x + 4$ **26.** $f(x) = x^2 + 6x + 9$

27. $f(x) = \dfrac{-1}{3}x^2 + 4$ **28.** $f(x) = 3x^2 - 6$

29. $f(x) = 3x^2 - 12x + 19$ **30.** $f(x) = \dfrac{2}{3}x^2 + 12x + 2$

In exercises **31–44**, graph.

[Example 4]   **31.** $f(x) = 2x^2$ **32.** $f(x) = \dfrac{1}{2}x^2$

33. $f(x) = 2x - x^2$ **34.** $f(x) = x^2 + 6x$

35. $f(x) = \dfrac{-1}{2}x^2 + 3$

36. $f(x) = \dfrac{2}{3}x^2 - 5$

37. $f(x) = x^2 - 6x + 9$

38. $f(x) = x^2 + 8x + 16$

39. $f(x) = \dfrac{-3}{2}x^2 - 6x - 6$

40. $f(x) = \dfrac{1}{2}x^2 - 2x + 2$

41. $f(x) = 2x^2 - 12x + 9$

42. $f(x) = 3x^2 + 30x + 81$

43. $f(x) = -x^2 - 2x + 5$

44. $f(x) = -x^2 + 8x - 12$

SECTION 10-2. Ten Review Exercises

In exercises **1–5**, simplify.

1. $\dfrac{-12t^4u^3}{(2tu^2)^2}$

2. $(-a^3b)(3ab^2)$

3. $(-2xy^2z^3)^3$

4. $\dfrac{(-k^3)(6k)}{(3k)^2}$

5. $\left(\dfrac{-1}{2}m\right)^3\left(\dfrac{4}{5}m^2\right)(-5m)$

In exercises **6–10**, write in simplified form. Assume variables are positive.

6. $\sqrt{120t}$

7. $\sqrt{\dfrac{u}{18}}$

8. $\dfrac{3}{\sqrt{2}+1}$

9. $\dfrac{-10+\sqrt{40}}{6}$

10. $5\sqrt{24x} - 2\sqrt{54x}$

SECTION 10-2. Supplementary Exercises

In exercises **1–8**, graph the three linear functions on the same set of axes.

1. $f(x) = 2x + b$; where $b = -3$, $b = 0$, and $b = 2$

2. $f(x) = x + b$; where $b = -2$, $b = 0$, and $b = 3$

3. $f(x) = b$; where $b = 4$, $b = 0$, and $b = -2$

4. $f(x) = -b$; where $b = 4$, $b = 0$, and $b = -2$

5. $f(x) = mx + 1$; where $m = 1$, $m = 2$, and $m = 3$

6. $f(x) = mx + 1$; where $m = -1$, $m = -2$, and $m = -3$

7. $f(x) = mx + 6$; where $m = \dfrac{1}{2}$, $m = \dfrac{-1}{2}$, and $m = 2$

8. $f(x) = mx + 6$; where $m = \dfrac{1}{3}$, $m = \dfrac{-1}{3}$, and $m = 3$

In exercises **9–12**, refer to the figures in each problem.

a. Identify the coordinates of V, the vertex of the parabola.

b. Write an equation for the axis of symmetry.

c. State whether the a in the equation $f(x) = ax^2 + bx + c$ that defines the graph is positive or negative.

9.

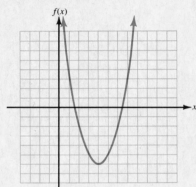

10.

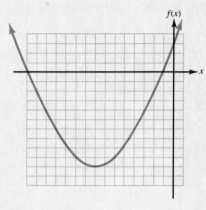

11.

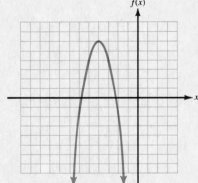

12.

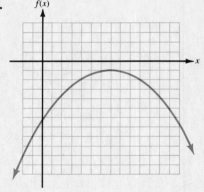

In exercises **13–18**,

a. graph each function on graph paper.

b. using the graph, locate the two positive integers (one below and one above the intersection of the right branch of the parabola and the x-axis).

c. using the graph, locate the two negative integers (one below and one above the intersection of the left branch of the parabola and the x-axis).

d. replace $f(x)$ by 0 and solve the equation with the quadratic formula.

e. use a calculator to approximate the positive root of the equation to one decimal place and verify that the number is between the integers from part **b**.

f. use a calculator to approximate the negative root of the equation to one decimal place and verify that the number is between the integers from part **c**.

13. $f(x) = x^2 - 6$ **14.** $f(x) = 12 - x^2$

15. $f(x) = x^2 + 2x - 6$ **16.** $f(x) = x^2 - 4x - 6$

17. $f(x) = x^2 - 4x - 9$ **18.** $f(x) = x^2 + 6x - 11$

SECTION 10-3.   Other Functions

**KEY TOPICS
IN THIS SECTION**

1. Functions defined by $f(x) = ax^n$, with n an odd integer

2. Functions defined by $f(x) = ax^n$, with n an even integer

3. Functions defined by $f(x) = a|x - h|$

In this section we will study the general shapes of the graphs of three more types of functions.

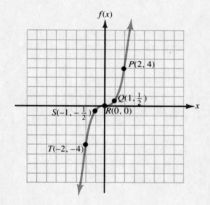

Figure 10-6. A graph of $f(x) = \frac{1}{2}x^3$.

Functions Defined by $f(x) = ax^n$, with n an Odd Integer

Examples **a** and **b** are functions in x defined by a monomial in which the exponent on x is an odd integer.

a. $f(x) = \frac{1}{2}x^3$ **b.** $f(x) = \frac{-3}{16}x^5$

Functions defined by equations such as these have graphs with special characteristics that help in sketching the graphs. To identify these characteristics, a graph of $f(x) = \frac{1}{2}x^3$ is shown in Figure 10-6. The coordinates of points P, Q, R, S, and T are solutions of the equation and are used to sketch the graph.

Characteristics of a graph defined by $f(x) = ax^n$, and n an odd integer

  (i) The graph passes through the origin.

 (ii) The graph extends infinitely far up and down.

(iii) As one traces the curve from left to right,
    a. it *rises* if $a > 0$.
    b. it *falls* if $a < 0$.

 (iv) The curve has a bend (or "kink") as it approaches $O(0, 0)$ from either direction.

Usually three points on either side of the origin are enough to make an accurate sketch of the graph.

Example 1.   Graph: $f(x) = \dfrac{-3}{16}x^5$

Solution.   **Discussion.** For the given equation, $a = \frac{-3}{16}$. Therefore, the curve will fall as it is traced from left to right. The following ordered pairs are used to plot the points in Figure 10-7.

x	-2	-1	0	1	2
$f(x)$	6	$\frac{3}{16}$	0	$\frac{-3}{16}$	-6

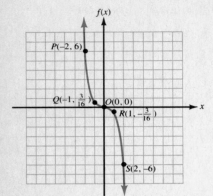

Figure 10-7. A graph of $f(x) = \frac{-3}{16}x^5$.

Functions Defined by $f(x) = ax^n$, with n an Even Integer

Examples **a** and **b** are functions in x defined by a monomial in which the exponent on x is an even integer.

a. $f(x) = \dfrac{3}{4}x^4$ **b.** $f(x) = \dfrac{-1}{16}x^6$

Functions defined by equations such as these have graphs that closely resemble those of quadratic functions. To illustrate, graphs of $f(x) = \frac{3}{4}x^2$ and $f(x) = \frac{3}{4}x^4$ are shown in Figure 10-8. Notice that both graphs pass through the origin. The graphs also intersect at the points for which $x = -1$ and $x = 1$. For $x < -1$ or $x > 1$, the graph of $f(x) = \frac{3}{4}x^4$ is closer to the y-axis than is the graph of $f(x) = \frac{3}{4}x^2$. The "steeper" sides are the result of the greater exponent; that is, x^4 has a greater degree than x^2.

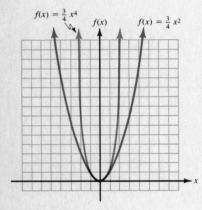

Figure 10-8. Graphs of $f(x) = \frac{3}{4}x^2$ and $f(x) = \frac{3}{4}x^4$.

Characteristics of a graph defined by $f(x) = ax^n$, and n an even integer

(i) The graph passes through the origin.

(ii) The graph is a "cup" that opens
   a. upward, if $a > 0$.
   b. downward, if $a < 0$.

(iii) The bottom of the cup becomes more square-shaped as n gets larger.

Example 2.   Graph: $f(x) = \dfrac{-1}{16}x^6$

Solution.   **Discussion.** For the given equation, $a = \frac{-1}{16}$. Therefore, the cup opens downward. The following ordered pairs are used to plot the points in Figure 10-9.

x	-2	-1	0	1	2
$f(x)$	-4	$\frac{-1}{16}$	0	$\frac{-1}{16}$	-4

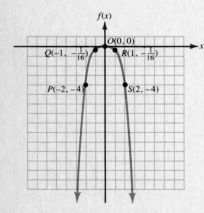

Figure 10-9. A graph of $f(x) = \frac{-1}{16}x^6$.

Functions Defined by $f(x) = a|x - h|$

If absolute value bars are applied to any real number, then the expression always names a number greater than or equal to zero.

$$|10| = 10 \qquad |0| = 0 \qquad |-13| = 13$$

If a variable is put inside absolute value bars, then an equation can be written that defines an **absolute value function.**

Definition 10.6.

An **absolute value function** is one that can be written in the form

$$f(x) = a|x - h|$$

where a and h are real numbers and $a \neq 0$.

To show the characteristics of a graph defined by an absolute value function, a graph of

$$f(x) = |x|$$

is pictured in Figure 10-10. The following ordered pairs are the coordinates of points P, Q, O, R, and S respectively.

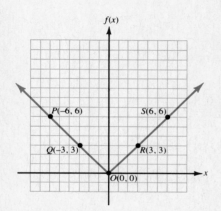

x	-6	-3	0	3	6
$f(x) = |x|$	6	3	0	3	6

The "V-shaped" graph is typical of a graph of an absolute value function.

Figure 10-10. A graph of $f(x) = |x|$.

Characteristics of a graph defined by $f(x) = a|x - h|$ where a and h are real numbers, and $a \neq 0$.

1. A graph is V-shaped,
   a. the V opens upward if $a > 0$.
   b. the V opens downward if $a < 0$.

2. The point of the V is called the **vertex.** The vertex is the point $V(h, 0)$.

3. The slopes of the sides are a and $-a$.

Example 3. Graph: $f(x) = \dfrac{-1}{2}|x + 2|$

Solution. **Discussion.** To compare the form of the given function with the general form in Definition 10-6, write

$$\frac{-1}{2}|x + 2| \qquad \text{as} \qquad \frac{-1}{2}|x - (-2)|$$

In this form, we see that $a = \frac{-1}{2}$ and $h = -2$. As a consequence, $V(-2, 0)$ is the vertex. The slopes of the sides are $\frac{-1}{2}$ and $\frac{1}{2}$, and the V opens downward. A graph of the function is shown in Figure 10-11. First locate the vertex. Then use the slopes to find points to the left and to the right of the vertex on the sides of the V.

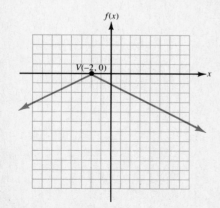

Figure 10-11. A graph of $y = \frac{-1}{2}|x + 2|$

SECTION 10-3.   Practice Exercises

In exercises **1–24**, graph.

[Example 1]   **1.** $f(x) = x^3$ **2.** $f(x) = x^3$

3. $f(x) = \dfrac{-1}{4}x^3$ **4.** $f(x) = \dfrac{3}{8}x^3$

5. $f(x) = \dfrac{1}{32}x^5$ **6.** $f(x) = \dfrac{-5}{16}x^5$

7. $f(x) = \dfrac{-1}{8}x^5$ **8.** $f(x) = \dfrac{5}{32}x^5$

[Example 2]   **9.** $f(x) = x^4$ **10.** $f(x) = \dfrac{-1}{2}x^4$

11. $f(x) = \dfrac{-1}{8}x^4$ **12.** $f(x) = \dfrac{3}{4}x^4$

13. $f(x) = \dfrac{1}{32}x^6$ **14.** $f(x) = \dfrac{-5}{8}x^6$

15. $f(x) = \dfrac{-5}{64}x^6$ **16.** $f(x) = \dfrac{5}{32}x^6$

[Example 3]   **17.** $f(x) = 2|x|$ **18.** $f(x) = -3|x|$

19. $f(x) = -|x - 1|$ **20.** $f(x) = |x + 3|$

21. $f(x) = \dfrac{1}{3}|x + 4|$ **22.** $f(x) = \dfrac{-3}{4}|x - 2|$

23. $f(x) = \dfrac{-3}{2}|x - 3|$ **24.** $f(x) = \dfrac{5}{3}|x + 1|$

SECTION 10-3.   Ten Review Exercises

In exercises **1–10**, do the indicated operations.

1. $4t - 3(4 - 2t) - 2(t - 3) - 6t - 2$

2. $5u - (8 - u) - 2[-4 - (3 + 5u) - 13]$

3. $(4x + 11y)(x - 13y)$

4. $(2p + q - 3)^2$

5. $(3m + 2)(m^3 + 2m^2 - m + 2)$

6. $(10n^4 + n^3 - 17n^2 + 41n - 14) \div (5n - 2)$

7. $\dfrac{t^2 - 9}{t^2 + t - 20} \cdot \dfrac{t^2 + 2t - 15}{t^2 - 4t + 3}$

8. $\dfrac{16u^2 - 9}{16u^2 - 24u + 9} \div \dfrac{16u^2 + 24u + 9}{16u^2 - 16u + 3}$

9. $\dfrac{a + b}{3a^2 + 2ab - b^2} - \dfrac{b - a}{6a^2 - 5ab + b^2}$

10. $\dfrac{\dfrac{x}{x + 1} + 1}{\dfrac{2x + 1}{x - 1}}$

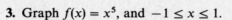

SECTION 10-3. Supplementary Exercises

In exercises **1–4**, use the grid on which 8 squares = 1 unit.

1. Graph $f(x) = x$, and $-1 \leq x \leq 1$.

2. Graph $f(x) = x^3$, and $-1 \leq x \leq 1$.

3. Graph $f(x) = x^5$, and $-1 \leq x \leq 1$.

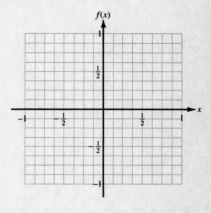

4. Based on the graphs in exercises 1, 2, and 3, describe in words the effect on the "bend" in the graphs of

$$f(x) = x^n, \text{ and } n \text{ an odd integer as } n \text{ gets larger.}$$

In exercises **5–8**, use the grid on which 8 squares = 1 unit.

5. Graph $f(x) = x^2$, and $-1 \leq x \leq 1$.

6. Graph $f(x) = x^4$, and $-1 \leq x \leq 1$.

7. Graph $f(x) = x^6$, and $-1 \leq x \leq 1$.

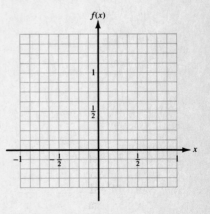

8. Based on the graphs in exercises 5, 6, and 7, describe in words the effect on the "cup" in the graphs of

$$f(x) = x^n, \text{ and } n \text{ an even integer as } n \text{ gets larger}$$

There are some striking similarities between the graphs of the functions defined by

$$\text{a. } f(x) = a(x - h)^2 \qquad \text{and} \qquad \text{b. } f(x) = a|x - h|$$

A graph of part **a** is a parabola with vertex $V(h, 0)$, and a graph of part **b** is a V with vertex $V(h, 0)$. For both graphs the axis of symmetry is a vertical line defined by $x = h$.

In exercises **9–14**, graph both functions on the same set of axes.

9. **a.** $f(x) = x^2$ and **b.** $f(x) = |x|$

10. **a.** $f(x) = (x - 2)^2$ and **b.** $f(x) = |x - 2|$

11. **a.** $f(x) = \dfrac{-1}{2}x^2$ and **b.** $f(x) = \dfrac{-1}{2}|x|$

12. **a.** $f(x) = -2x^2$ and **b.** $f(x) = -2|x|$

13. **a.** $f(x) = \dfrac{3}{4}(x + 1)^2$ and **b.** $f(x) = \dfrac{3}{4}|x + 1|$

14. **a.** $f(x) = \dfrac{-2}{3}(x - 3)^2$ and **b.** $f(x) = \dfrac{-2}{3}|x - 3|$

SECTION 10-1. Summary Exercises

Answer

1. a. find the ordered pairs in the relation defined by $y = -2x + 5$ for $x = -1$,   **1.** _____
    $x = 0$ and $x = 6$.

  b. graph the ordered pairs.

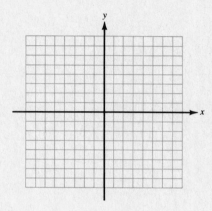

2. Find the values of the dependent variable y for the given values of the indepen-   **2.** _____
dent variable x.

$$y = \frac{x^2}{x^2 + 4}; \; -2, 0, 4$$

3. List the elements of   **3.** _____

  a. the domain

    _____

  b. the range

of the relation $(-6, 24)$, $(6, 19)$, $(-12, 0)$, $(15, 8)$, $(0, 0)$.

4. Identify the domain and range of the relation whose graph is given below.   **4. a.** Domain: _____

  b. Range: _____

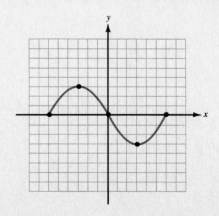

In exercises **5** and **6**, determine whether the relations are functions.

5. $\left\{\left(\dfrac{1}{2}, 2\right), (3, -4), (6, 17), (5, 17)\right\}$

5. _____

6.

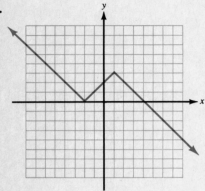

6. _____

In exercises **7** and **8**, find the indicated function values.

7. If $f(x) = 6x^2 + \dfrac{1}{x}$, find

  a. $f(2)$

  b. $f(-1)$

  c. $f\left(\dfrac{1}{2}\right)$

7. a. _____

  b. _____

  c. _____

8. If $f(x) = \dfrac{\sqrt{x - 4}}{x^2 - 1}$, find

  a. $f(4)$

  b. $f(5)$

  c. $f(8)$

8. a. _____

  b. _____

  c. _____

SECTION 10-2. Summary Exercises

Answer

In exercises **1** and **2**, verify that each equation is a linear function by writing it in the form $f(x) = mx + b$.

1. $5(x + y) = 3(x + 1)$

1. _____

2. $\dfrac{6(x + 4)}{5} = \dfrac{3 - y}{2}$

2. _____

In exercises **3** and **4**, graph.

3. $f(x) = \dfrac{-2}{3}x + 4$

3.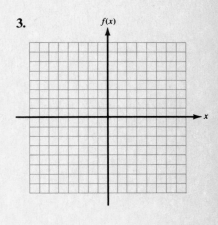

4. $f(x) = -2$

4.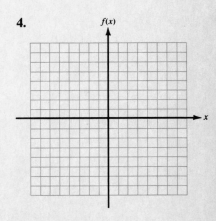

In exercises **5** and **6**.

a. find the x-coordinate of the vertex.

b. find the y-coordinate of the vertex.

c. write an equation for the axis of symmetry.

5. $f(x) = 10x - x^2 - 15$

6. $f(x) = 2x^2 - 12x + 9$

5. a. _____

b. _____

c. _____

6. a. _____

b. _____

c. _____

In exercises **7** and **8**, graph.

7. $f(x) = x^2 - 2x$

7.

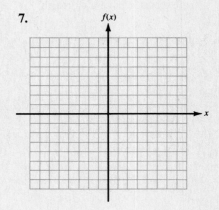

8. $f(x) = 3x^2 - 6x + 2$

8.

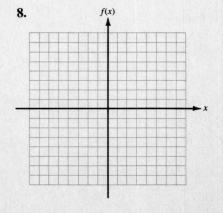

SECTION 10-3.   Summary Exercises

In exercises **1–6**, graph.

1. $f(x) = \dfrac{1}{64}x^5$

1.

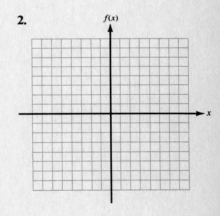

2. $f(x) = \dfrac{-5}{8}x^3$

2.

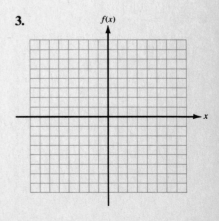

3. $f(x) = \dfrac{9}{64}x^6$

3.

4. $f(x) = \dfrac{-1}{32}x^4$

4.

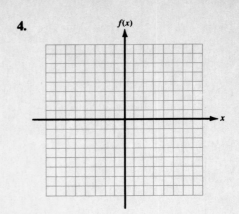

5. $f(x) = 2|x - 3|$

5.

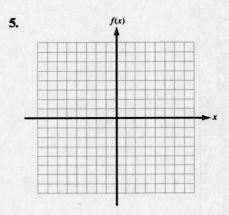

6. $f(x) = \dfrac{-1}{4}|x + 2|$

6.

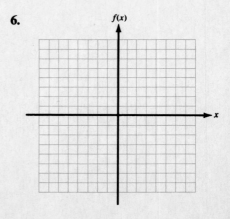

CHAPTER 10.   Review Exercises

In exercises **1–4**, find the ordered pair or pairs in the relation for the given values of the variables. In these exercises, the form of the ordered pairs is (x, y).

1. $y = -2x + 10$

   a. $x = -5$ **b.** $x = \dfrac{1}{2}$ **c.** $y = 6$

2. $y = \dfrac{x^2 - 4}{x}$

   a. $x = 1$ **b.** $x = -4$ **c.** $x = 2$

3. $(-5, 9), (8, 13), (-3, 0), (4, -11), \left(0, \dfrac{1}{2}\right), (4, 13)$

   a. $x = -3$ **b.** $x = 4$ **c.** $y = 13$

4. $(-5, 0), (-4, 1), (-1, 2), (0, \sqrt{5}), (4, 3), (7, 2\sqrt{3})$

   a. $x = -4$ **b.** $x = 4$ **c.** $y = 2$

In exercises **5** and **6**, identify

a. the domain.   **b.** the range.

5. $(6, -1), (4, 5), (6, 0), (3, 5), (4, -1), (-2, 0)$

6.

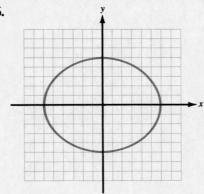

In exercises **7** and **8**, state whether the given relations are functions.

7. a. $(13, 7), (2, 9), (-4, 10), (4, 17)$
   b. $(18, 16), (14, -19), (3, 16), (-4, -19)$
   c. $(6, -3), (5, 11), (6, 4), (8, 2)$

8. a.   **b.**

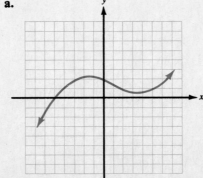 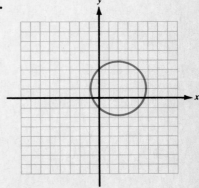

9. If $f(x) = 9 - x^2$, find
    a. $f(-3)$ **b.** $f(10)$

10. If $f(x) = \sqrt{x - 8}$, find
    a. $f(17)$ **b.** $f(8)$

In exercises **11** and **12**, verify that the given equation is a linear function by writing it in the form $f(x) = mx + b$.

11. $-2(x + 5) = 3(4 - y)$

12. $\dfrac{2y + x}{4} = x - 2$

In exercises **13** and **14**, graph.

13. $f(x) = \dfrac{4}{3}x - 7$

14. $f(x) = -5x + 6$

In exercises **15** and **16**

a. write the quadratic function using $f(x)$ notation.

b. compute $f(-2)$.

15. $(x + 6)(2x - 5) = y + 3$

16. $\dfrac{(x + 3)^2}{2} = \dfrac{y - 1}{4}$

In exercises **17–20**,

a. find the x-coordinate of the vertex.

b. find the y-coordinate of the vertex.

c. find the equation of the axis of symmetry.

d. graph.

17. $f(x) = 2x^2 - 8$ **18.** $f(x) = 4x^2 - 12x + 10$

19. $f(x) = -x^2 + 6x - 9$ **20.** $f(x) = -4x^2 + 8x$

In exercises **21–24**, graph.

21. $f(x) = \dfrac{-5}{8}x^3$ **22.** $f(x) = \dfrac{1}{4}x^4$

23. $f(x) = \dfrac{1}{2}|x - 3|$ **24.** $f(x) = -3|x + 1|$

Sets

The notion of "set" is found in almost all of mathematics. This short introduction to the subject includes most of the basic concepts and vocabulary.

**KEY TOPICS
IN THIS SECTION**

1. What is meant by a set

2. Notations used in sets

3. Special sets

4. Venn diagrams as representations of sets

5. Operations on sets

What Is Meant by a Set

A set is a collection of things. The collection must be **well-defined** so that there is no question as to whether or not a given thing belongs to the collection.
Examples **a** through **c** illustrate well-defined sets.

a. The set of people registered to vote in New York

b. The set of books that are printed in English

c. The set of names of African nations

Examples **d** through **f** illustrate sets that are not well defined.

d. The set of "elderly people" that are students at Purdue University

e. The set of "warm days" last June in Cleveland, Ohio

f. The set of "good books" in a public library

The sets that are most frequently used in mathematics are sets of points (graphs) and sets of numbers (integers, reals, complex).

Notations Used in Sets

Uppercase letters, such as *A*, *B*, and *C*, are used to represent sets. The **members,** or **elements,** of a given set may be listed between a pair of braces. Examples **g** through **i** illustrate this method of writing a set.

g. $A = \{1, 2, 3, 4, 5\}$
The set of counting numbers 1 through 5, inclusive

h. $B = \{$Alaska, California, Hawaii, Oregon, Washington$\}$
The set of states that border on the Pacific Ocean

i. $C = \{$red, orange, yellow, green, blue, indigo, violet$\}$
The set of primary colors

Another method of writing the elements of a set is with **set builder notation.** With this method a rule is stated that describes the property, or properties, that an element x must exhibit in order for x to be in that set. Examples **g*** through **i*** illustrate this method of writing sets A, B, and C above.

g.* $A = \{x \mid x$ is a counting number from 1 through 5, inclusive$\}$

h.* $B = \{x \mid x$ is a state of the United States with a border on the Pacific Ocean$\}$

i.* $C = \{x \mid x$ is a primary color$\}$

In this notation the vertical bar is read "such that." Thus, **g*** is read as follows: "Set A is the set of x's, such that x is a counting number from 1 through 5, inclusive." The following symbols from sets are also used in mathematics.

Symbol	How read	Example
$\in$	"is an element of"	$3 \in \{1, 2, 3, 4, 5\}$
$\notin$	"is not an element of"	$8 \notin \{1, 2, 3, 4, 5\}$
$=$	"equals"	$\{a, b, c, d\} = \{b, d, c, a\}$

(Two sets A and B are equal, $A = B$, if and only if both sets contain exactly the same elements.)

$\subseteq$	"is a subset of"	$\{a, b, c\} \subseteq \{b, d, c, a\}$

(If every element of A is an element of B, then A is a subset of B.)

$\subset$	"is a proper subset of"	$\{a, b\} \subset \{b, d, c, a\}$

(If $A \subseteq B$ but $A \neq B$, then A is a proper subset of B.)

Special Sets

There are two sets that are called special:

The **universal set,** written U, contains all of the elements that might be of interest in a given discussion.
The **null set,** or **empty set,** written $\varnothing$, contains no elements.

In plane geometry, the universal set is frequently the set of points on some plane. In algebra, the universal set is often the set of real numbers. The null set is a set that has no elements. Examples **j** through **l** illustrate three null sets.

j. $\{x \mid x$ is a triangle with four equal sides$\}$

k. $\{y \mid y$ is a living person who fought in the American Revolutionary War$\}$

l. $\{z \mid z$ is a counting number less than 0$\}$

Another special set is the complement of a given set A:

> If A is a set, then the set of all elements of the universal set that are not in A is called the **complement of A,** written A'. The symbol A' can be read "A-prime," or "complement of A."

Figure A-1. A Venn diagram of U and a set A.

Examples **m** through **o** illustrate the complements of three sets when the universal set is $U = \{1, 2, 3, 4, 5\}$.

m. $A = \{1, 3, 5\}$ $\qquad$ $A' = \{2, 4\}$

n. $B = \{1, 4, 3, 2, 5\}$ $\qquad$ $B' = \varnothing$

o. $C = \{\ \ \}$ $\qquad$ $C' = \{1, 2, 3, 4, 5\}$

Venn Diagrams as Representations of Sets

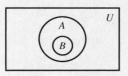

Figure A-2. $B \subseteq A$.

A geometric representation of sets can be made through the use of **Venn diagrams.** In a Venn diagram the universal set is represented as the region inside a rectangle. The subsets of the universal set are represented as regions inside closed curves that are drawn inside this rectangle.

In Figure A-1 a universal set, U, is shown with a set A that is a proper subset of U. Figure A-2 shows a Venn diagram representation of sets A and B, where $B \subseteq A$. In Figure A-3, the shaded region shows the complement of a set A.

Operations on Sets

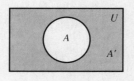

Figure A-3. The shaded region is A'.

There are two operations that are performed on two sets at the same time. These **binary operations** are called **intersection** and **union:**

> If A and B are two sets, then
>
> (1) the **intersection of A and B,** written $A \cap B$, is a set that contains elements that are in both A and B.
>
> $$A \cap B = \{x \mid x \in A \quad \text{and} \quad x \in B\}$$
>
> (2) the **union of A and B,** written $A \cup B$, is a set that contains elements that are in A or B, or both A and B.
>
> $$A \cup B = \{x \mid x \in A \quad \text{or} \quad x \in B\}$$

Example 1. If $A = \{a, b, c, d, e\}$ and $B = \{a, e, i, o, u\}$, find

   a. $A \cap B$ **b.** $A \cup B$

Solution.   **a.** $A \cap B = \{a, e\}$ *a* and *e* are in both *A* and *B*.

   b. $A \cup B = \{a, b, c, d, e, i, o, u\}$ *a, b, c, d, e, i, o,* and *u* are in *A* or *B*, or both *A* and *B*.

The shaded region in Figure A-4 shows the intersection of sets A and B. The shaded region in Figure A-5 shows the union of sets A and B.

When more than two operations are indicated on two or more sets, the following rules govern the order in which the operations are to be performed:

1. First do all operations within parentheses.

2. Find any indicated complements before intersections or unions.

3. When there are no parentheses, do any indicated intersections before any indicated unions.

Example 2. List the elements in $A \cap (B \cup C')$, where

$$U = \{1, 2, 3, 4, 5, 6, 7, 8, 9\}$$

$$A = \{1, 2, 4, 6, 9\}$$

$$B = \{2, 3, 4, 5\}$$

$$C = \{1, 4, 7, 9\}$$

Solution.   First, find C'. With $U = \{1, 2, 3, 4, 5, 6, 7, 8, 9\}$ and $C = \{1, 4, 7, 9\}$, then $C' = \{2, 3, 5, 6, 8\}$.

Second, form the union indicated within the parentheses.

$$B \cup C' = \{2, 3, 4, 5\} \cup \{2, 3, 5, 6, 8\}$$

$$= \{2, 3, 4, 5, 6, 8\}$$

Third, find the indicated intersection.

$$A \cap (B \cup C') = \{1, 2, 4, 6, 9\} \cap \{2, 3, 4, 5, 6, 8\}$$

$$= \{2, 4, 6\}$$

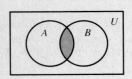

Figure A-4. Shaded region is $A \cap B$.

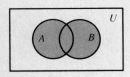

Figure A-5. Shaded region is $A \cup B$.

B
Fractions

**KEY TOPICS
IN THIS SECTION**

1. A definition of a fraction

2. Reduced form of a fraction

3. Multiplying fractions

4. Dividing fractions

5. Adding and subtracting fractions, like denominators

6. Writing a fraction with a different denominator

7. Adding and subtracting fractions, unlike denominators

A Definition of a Fraction

The **whole numbers** are the numbers $0, 1, 2, 3, \ldots$. When a whole number is divided by a whole number that is not 0, the ratio can be called a **fraction,** also called a **common fraction.**

> If a and b are whole numbers and $b \neq 0$, then $\dfrac{a}{b}$ can be called a **fraction.**

Following are examples of fractions:

a. $\dfrac{3}{4}$ **b.** $\dfrac{10}{7}$ **c.** $\dfrac{9}{1}$

In the form $\dfrac{a}{b}$, a is called the **numerator** and b is called the **denominator.** In examples **a**, **b**, and **c**, 3, 10, and 9 are the numerators, and 4, 7, and 1 are the denominators.

Reduced Form of a Fraction

The fractions $\frac{3}{4}$ and $\frac{15}{20}$ each name the same fraction. That is, $\frac{3}{4}$ and $\frac{15}{20}$ are **equal fractions,** written $\frac{3}{4} = \frac{15}{20}$.

The fraction $\frac{3}{4}$ is called the **reduced form** of $\frac{15}{20}$, since the numerator ($15 = 3 \cdot 5$) and the denominator ($20 = 4 \cdot 5$) have a common factor 5. To reduce $\frac{15}{20}$, the common factor is cancelled.

$$\frac{15}{20} = \frac{3 \cdot 5}{4 \cdot 5} = \frac{3 \cdot \cancel{5}}{4 \cdot \cancel{5}} \qquad \text{common factor 5 is cancelled}$$

$$= \frac{3}{4} \qquad \text{reduced form}$$

> To write a fraction in reduced form, any common factors of the numerator and denominator other than 1 are cancelled. If a, b, and c are whole numbers, $b \neq 0$, $c \neq 0$, and $c \neq 1$, then
>
> $$\frac{a \cdot c}{b \cdot c} = \frac{a \cdot \cancel{c}}{b \cdot \cancel{c}} = \frac{a}{b} \qquad\qquad \textbf{(1)}$$

Example 1. Reduce $\dfrac{315}{840}$.

Solution. $\dfrac{315}{840} = \dfrac{3 \cdot 105}{3 \cdot 280} = \dfrac{\cancel{3} \cdot 105}{\cancel{3} \cdot 280} = \dfrac{105}{280}$ 3 is a common factor.

$\dfrac{105}{280} = \dfrac{5 \cdot 21}{5 \cdot 56} = \dfrac{\cancel{5} \cdot 21}{\cancel{5} \cdot 56} = \dfrac{21}{56}$ 5 is a common factor.

$\dfrac{21}{56} = \dfrac{7 \cdot 3}{7 \cdot 8} = \dfrac{\cancel{7} \cdot 3}{\cancel{7} \cdot 8} = \dfrac{3}{8}$ 7 is a common factor.

Multiplying Fractions

The product of fractions can be found by writing the product of the numerators over the product of the denominators.

> If a, b, c, and d are whole numbers, $b \neq 0$, and $d \neq 0$, then
>
> $$\frac{a}{b} \cdot \frac{c}{d} = \frac{a \cdot c}{b \cdot d} \qquad\qquad \textbf{(2)}$$

It is common practice to write any fraction in reduced form. As a consequence, the product of two or more fractions should be reduced whenever possible. It is usually easier to find the reduced form of a product by cancelling any common factors in the numerator and denominator before finding the indicated products.

Example 2. Multiply $\dfrac{4}{15} \cdot \dfrac{25}{18}$.

Solution.   Using equation (2),

$$\frac{4}{15} \cdot \frac{25}{18} = \frac{4 \cdot 25}{15 \cdot 18}$$

$$= \frac{2 \cdot 2 \cdot 5 \cdot 5}{3 \cdot 5 \cdot 2 \cdot 9} \qquad \text{2 is a factor of 4 and 18.}$$
$$\text{5 is a factor of 25 and 15.}$$

$$= \frac{\cancel{2} \cdot 2 \cdot \cancel{5} \cdot 5}{3 \cdot \cancel{5} \cdot \cancel{2} \cdot 9}$$

$$= \frac{10}{27}$$

Dividing Fractions

The quotient of two fractions can be found by multiplying the first fraction by the **reciprocal** of the second fraction. The reciprocal of a fraction is obtained by interchanging the numerator and denominator.

If a, b, c, and d are whole numbers, $b \neq 0$, $c \neq 0$, and $d \neq 0$, then

$$\frac{a}{b} \div \frac{c}{d} = \frac{a}{b} \cdot \frac{d}{c} = \frac{a \cdot d}{b \cdot c} \qquad \textbf{(3)}$$

where $\dfrac{d}{c}$ is the reciprocal of $\dfrac{c}{d}$.

Example 3.   Divide $\dfrac{3}{8} \div \dfrac{9}{10}$.

Solution.   Using equation (3),

$$\frac{3}{8} \div \frac{9}{10} = \frac{3}{8} \cdot \frac{10}{9} = \frac{3 \cdot 10}{8 \cdot 9}$$

$$\frac{3 \cdot 10}{8 \cdot 9} = \frac{\cancel{3} \cdot \cancel{2} \cdot 5}{\cancel{2} \cdot 4 \cdot \cancel{3} \cdot 3} \qquad \text{Dividing out 2 and 3}$$

$$= \frac{5}{12} \qquad \text{The reduced quotient}$$

Adding and Subtracting Fractions, Like Denominators

The sum of fractions with the same denominator can be found by adding the numerators and writing the sum over the common denominator.

If a, b, and c are whole numbers and $c \neq 0$, then

$$\frac{a}{c} + \frac{b}{c} = \frac{a+b}{c} \qquad \textbf{(4)}$$

The difference between two fractions with the same denominator can be found by subtracting the numerators and writing the difference over the common denominator.

If a, b, and c are whole numbers and $c \neq 0$, then

$$\frac{a}{c} - \frac{b}{c} = \frac{a-b}{c} \qquad \textbf{(5)}$$

Example 4.   Find the indicated values.

   a. $\dfrac{2}{15} + \dfrac{7}{15}$ **b.** $\dfrac{17}{18} - \dfrac{5}{18}$

Solution.   **a.** Using equation (4) and finding the indicated sum in the numerator,

$$\frac{2}{15} + \frac{7}{15} = \frac{2+7}{15} = \frac{9}{15}$$

Now reducing the fraction,

$$\frac{9}{15} = \frac{3 \cdot \cancel{3}}{5 \cdot \cancel{3}} = \frac{3}{5} \qquad \text{The reduced sum is } \tfrac{3}{5}.$$

b. Using equation (5) and finding the indicated difference in the numerator,

$$\frac{17}{18} - \frac{5}{18} = \frac{17-5}{18} = \frac{12}{18}$$

Now reducing the fraction,

$$\frac{12}{18} = \frac{2 \cdot \cancel{6}}{3 \cdot \cancel{6}} = \frac{2}{3} \qquad \text{The reduced difference is } \tfrac{2}{3}.$$

Writing a Fraction with a Different Denominator

To use equations (4) and (5) to add or subtract fractions, the denominators of the fractions must be the same. Therefore, to find the sum or difference of fractions with different, or *unlike denominators,* two tasks must first be performed:

1. Change the denominators of the fractions being added or subtracted to the same number.

2. Keep the values of the fractions the same.

Task 1 is accomplished by finding a number that the denominators of the fractions being added or subtracted can divide evenly. Such a number is called a **common multiple.** Since the common multiple is a denominator, the number is frequently called a **common denominator.** To illustrate,

d. 24 is a common denominator for $\frac{5}{8}$ and $\frac{7}{12}$ because 8 and 12 evenly divide 24. That is, $24 \div 8 = 3$ and $24 \div 12 = 2$.

e. 30 is a common denominator of $\frac{1}{6}$, $\frac{7}{10}$, and $\frac{8}{15}$ because 6, 10, and 15 evenly divide 30. That is, $30 \div 6 = 5$, $30 \div 10 = 3$, and $30 \div 15 = 2$.

For any group of fractions there are many common denominators. To keep the numbers small, it is advisable to use the least or **smallest common denominator.** This number is usually called the **least common denominator,** written LCD. In examples **d** and **e**, the LCDs are 24 and 30, respectively.

Unless the fractions being added or subtracted have unusually large denominators, the least common denominator can be found experimentally. The skill in picking the LCD improves with practice.

Example 5. For the fractions $\dfrac{7}{15}$ and $\dfrac{1}{20}$,

    a. find the LCD.

    b. write each fraction as an equal fraction with the LCD as denominator.

Solution.   **a.** 60 is the smallest number that 15 and 20 can divide evenly ($60 \div 15 = 4$ and $60 \div 20 = 3$). Thus, 60 is the LCD.

    b. $\dfrac{7}{15} = \dfrac{7 \cdot 4}{15 \cdot 4} = \dfrac{28}{60}$

       $\dfrac{1}{20} = \dfrac{1 \cdot 3}{20 \cdot 3} = \dfrac{3}{60}$

Adding and Subtracting Fractions, Unlike Denominators

The following steps are recommended for adding or subtracting fractions with unlike denominators.

> **Step 1.** Find the least common denominator (written LCD) of the fractions being added or subtracted.
>
> **Step 2.** Change each fraction to an equal fraction with the LCD as denominator.
>
> **Step 3.** Use equation (4) or (5) to add or subtract the fractions with the same denominator.
>
> **Step 4.** Reduce, if possible, the resulting sum or difference.

Example 6. Add $\dfrac{3}{4} + \dfrac{7}{10}$.

Solution.   **Step 1.** By inspection, 20 is the LCD ($20 \div 4 = 5$ and $20 \div 10 = 2$).

Step 2. $\dfrac{3}{4} = \dfrac{3 \cdot 5}{4 \cdot 5} = \dfrac{15}{20}$ equal fractions with

$\dfrac{7}{10} = \dfrac{7 \cdot 2}{10 \cdot 2} = \dfrac{14}{20}$ LCD as denominators

Step 3.   Using equation (4),

$$\dfrac{3}{4} + \dfrac{7}{10} = \dfrac{15}{20} + \dfrac{14}{20}$$

$$= \dfrac{15 + 14}{20}$$

$$= \dfrac{29}{20}$$

Step 4.   Since 29 and 20 have no common factors, the fraction cannot be reduced.

Subtract $\dfrac{11}{12} - \dfrac{4}{15}$.

Example 7.   **Step 1.**   By inspection, 60 is the LCD ($60 \div 12 = 5$ and $60 \div 15 = 4$).

Solution.   **Step 2.** $\dfrac{11}{12} = \dfrac{11 \cdot 5}{12 \cdot 5} = \dfrac{55}{60}$ equal fractions with

$\dfrac{4}{15} = \dfrac{4 \cdot 4}{15 \cdot 4} = \dfrac{16}{60}$ LCD as denominators

Step 3.   Using equation (5),

$$\dfrac{11}{12} - \dfrac{4}{15} = \dfrac{55}{60} - \dfrac{16}{60}$$

$$= \dfrac{55 - 16}{60}$$

$$= \dfrac{39}{60}$$

Step 4.   39 and 60 have a common factor 3.

$$\dfrac{39}{60} = \dfrac{13 \cdot \cancel{3}}{20 \cdot \cancel{3}} = \dfrac{13}{20}$$

Thus, $\dfrac{11}{12} - \dfrac{4}{15} = \dfrac{13}{20}$.

APPENDIX B.   Exercises

In exercises **1–20**, write each fraction in reduced form. Use equation (1).

[Example 1]   **1.** $\dfrac{6}{12}$ **2.** $\dfrac{14}{21}$ **3.** $\dfrac{12}{20}$ **4.** $\dfrac{10}{35}$

5. $\dfrac{42}{7}$ 6. $\dfrac{90}{9}$ 7. $\dfrac{36}{12}$ 8. $\dfrac{180}{15}$

9. $\dfrac{30}{24}$ 10. $\dfrac{63}{18}$ 11. $\dfrac{168}{40}$ 12. $\dfrac{195}{90}$

13. $\dfrac{55}{88}$ 14. $\dfrac{49}{63}$ 15. $\dfrac{99}{143}$ 16. $\dfrac{165}{210}$

17. $\dfrac{1{,}260}{1{,}120}$ 18. $\dfrac{630}{441}$ 19. $\dfrac{275}{1{,}100}$ 20. $\dfrac{231}{2{,}079}$

In exercises **21–40**, find the indicated products. Use equation (2). Write answers in reduced form.

[Example 2] 21. $\dfrac{2}{3} \cdot \dfrac{1}{3}$ 22. $\dfrac{3}{2} \cdot \dfrac{1}{2}$ 23. $\dfrac{4}{5} \cdot \dfrac{3}{7}$ 24. $\dfrac{1}{8} \cdot \dfrac{3}{4}$

25. $\dfrac{12}{5} \cdot \dfrac{4}{1}$ 26. $\dfrac{5}{1} \cdot \dfrac{9}{4}$ 27. $\dfrac{13}{12} \cdot \dfrac{7}{8}$ 28. $\dfrac{15}{7} \cdot \dfrac{9}{17}$

29. $\dfrac{3}{10} \cdot \dfrac{15}{16}$ 30. $\dfrac{8}{9} \cdot \dfrac{5}{12}$ 31. $\dfrac{7}{12} \cdot \dfrac{1}{21}$ 32. $\dfrac{6}{11} \cdot \dfrac{33}{35}$

33. $\dfrac{6}{5} \cdot \dfrac{20}{21}$ 34. $\dfrac{18}{25} \cdot \dfrac{10}{39}$ 35. $\dfrac{28}{15} \cdot \dfrac{24}{49}$ 36. $\dfrac{35}{32} \cdot \dfrac{56}{45}$

37. $\dfrac{3}{8} \cdot \dfrac{10}{9} \cdot \dfrac{12}{13}$ 38. $\dfrac{1}{10} \cdot \dfrac{13}{3} \cdot \dfrac{25}{26}$ 39. $\dfrac{7}{12} \cdot \dfrac{10}{13} \cdot \dfrac{6}{35}$ 40. $\dfrac{24}{25} \cdot \dfrac{36}{11} \cdot \dfrac{15}{32}$

In exercises **41–50**, find the indicated quotients. Use equation (3). Write answers in reduced form.

[Example 3] 41. $\dfrac{3}{4} \div \dfrac{2}{3}$ 42. $\dfrac{5}{6} \div \dfrac{2}{5}$ 43. $\dfrac{7}{8} \div \dfrac{10}{3}$ 44. $\dfrac{4}{9} \div \dfrac{3}{2}$

45. $\dfrac{9}{10} \div \dfrac{1}{5}$ 46. $\dfrac{13}{12} \div \dfrac{1}{6}$ 47. $\dfrac{8}{15} \div \dfrac{12}{35}$ 48. $\dfrac{10}{27} \div \dfrac{5}{24}$

49. $\dfrac{7}{16} \div \dfrac{21}{20}$ 50. $\dfrac{18}{55} \div \dfrac{3}{44}$

In exercises **51–70**, add or subtract as indicated. Use equation (4) or (5). Write answers in reduced form.

[Example 4] 51. $\dfrac{5}{8} + \dfrac{3}{8}$ 52. $\dfrac{7}{6} + \dfrac{5}{6}$ 53. $\dfrac{13}{12} - \dfrac{1}{12}$ 54. $\dfrac{23}{9} - \dfrac{5}{9}$

55. $\dfrac{1}{5} + \dfrac{3}{5}$ 56. $\dfrac{10}{7} + \dfrac{3}{7}$ 57. $\dfrac{8}{9} - \dfrac{4}{9}$ 58. $\dfrac{18}{13} - \dfrac{10}{13}$

59. $\dfrac{7}{10} + \dfrac{1}{10}$ 60. $\dfrac{4}{15} + \dfrac{8}{15}$ 61. $\dfrac{11}{14} - \dfrac{3}{14}$ 62. $\dfrac{13}{20} - \dfrac{3}{20}$

63. $\dfrac{17}{30} + \dfrac{31}{30}$ 64. $\dfrac{41}{36} + \dfrac{7}{36}$ 65. $\dfrac{17}{40} - \dfrac{3}{40}$ 66. $\dfrac{16}{45} - \dfrac{4}{45}$

67. $\dfrac{35}{56} + \dfrac{53}{56}$ 68. $\dfrac{43}{100} + \dfrac{47}{100}$ 69. $\dfrac{75}{63} - \dfrac{5}{63}$ 70. $\dfrac{101}{80} - \dfrac{29}{80}$

In exercises **71–80**, for each pair of fractions, **a.** find the LCD (lowest common denominator), and **b.** write each of the pair as an equal fraction with the LCD as denominator.

[Example 5] **71.** $\dfrac{3}{4}, \dfrac{5}{8}$ **72.** $\dfrac{2}{3}, \dfrac{1}{6}$ **73.** $\dfrac{7}{10}, \dfrac{6}{5}$ **74.** $\dfrac{5}{12}, \dfrac{4}{9}$

75. $\dfrac{8}{21}, \dfrac{9}{14}$ **76.** $\dfrac{7}{15}, \dfrac{4}{3}$ **77.** $\dfrac{11}{18}, \dfrac{11}{12}$ **78.** $\dfrac{9}{16}, \dfrac{3}{20}$

79. $\dfrac{29}{24}, \dfrac{7}{30}$ **80.** $\dfrac{13}{36}, \dfrac{13}{30}$

In exercises **81–100**, add or subtract as indicated. Write answers in reduced form.

[Examples 6 & 7] **81.** $\dfrac{5}{6} + \dfrac{1}{4}$ **82.** $\dfrac{7}{8} + \dfrac{5}{12}$ **83.** $\dfrac{7}{10} - \dfrac{2}{5}$ **84.** $\dfrac{11}{15} - \dfrac{2}{3}$

85. $\dfrac{8}{9} + \dfrac{7}{6}$ **86.** $\dfrac{2}{9} + \dfrac{1}{12}$ **87.** $\dfrac{3}{4} - \dfrac{1}{16}$ **88.** $\dfrac{5}{8} - \dfrac{5}{24}$

89. $\dfrac{3}{20} + \dfrac{4}{15}$ **90.** $\dfrac{3}{8} + \dfrac{1}{5}$ **91.** $\dfrac{17}{18} - \dfrac{11}{12}$ **92.** $\dfrac{13}{24} - \dfrac{4}{9}$

93. $\dfrac{3}{20} + \dfrac{1}{16}$ **94.** $\dfrac{5}{13} + \dfrac{3}{5}$ **95.** $\dfrac{3}{8} - \dfrac{1}{4} + \dfrac{5}{6}$ **96.** $\dfrac{5}{12} + \dfrac{2}{9} - \dfrac{1}{18}$

97. $\dfrac{7}{20} + \dfrac{5}{8} - \dfrac{2}{5}$ **98.** $\dfrac{4}{7} - \dfrac{1}{5} + \dfrac{8}{35}$ **99.** $\dfrac{2}{3} - \dfrac{1}{10} + \dfrac{4}{15}$

100. $\dfrac{9}{16} + \dfrac{5}{12} - \dfrac{1}{24}$

Answers to Appendix B Exercises

1. $\dfrac{1}{2}$ **2.** $\dfrac{2}{3}$ **3.** $\dfrac{3}{5}$ **4.** $\dfrac{2}{7}$ **5.** 6 **6.** 10 **7.** 3 **8.** 12 **9.** $\dfrac{5}{4}$ **10.** $\dfrac{7}{2}$ **11.** $\dfrac{8}{5}$ **12.** $\dfrac{13}{6}$ **13.** $\dfrac{5}{8}$ **14.** $\dfrac{7}{9}$

15. $\dfrac{9}{13}$ **16.** $\dfrac{11}{14}$ **17.** $\dfrac{9}{8}$ **18.** $\dfrac{10}{7}$ **19.** $\dfrac{1}{4}$ **20.** $\dfrac{1}{9}$ **21.** $\dfrac{2}{9}$ **22.** $\dfrac{3}{4}$ **23.** $\dfrac{12}{35}$ **24.** $\dfrac{3}{32}$ **25.** $\dfrac{48}{5}$ **26.** $\dfrac{45}{4}$

27. $\dfrac{91}{96}$ **28.** $\dfrac{135}{119}$ **29.** $\dfrac{9}{32}$ **30.** $\dfrac{10}{27}$ **31.** $\dfrac{1}{36}$ **32.** $\dfrac{18}{35}$ **33.** $\dfrac{8}{7}$ **34.** $\dfrac{12}{65}$ **35.** $\dfrac{32}{35}$ **36.** $\dfrac{49}{36}$ **37.** $\dfrac{5}{13}$ **38.** $\dfrac{5}{12}$

39. $\dfrac{1}{13}$ **40.** $\dfrac{81}{55}$ **41.** $\dfrac{9}{8}$ **42.** $\dfrac{25}{12}$ **43.** $\dfrac{21}{80}$ **44.** $\dfrac{8}{27}$ **45.** $\dfrac{9}{2}$ **46.** $\dfrac{13}{2}$ **47.** $\dfrac{14}{9}$ **48.** $\dfrac{16}{9}$ **49.** $\dfrac{5}{12}$ **50.** $\dfrac{24}{5}$

51. 1 **52.** 2 **53.** 1 **54.** 2 **55.** $\dfrac{4}{5}$ **56.** $\dfrac{13}{7}$ **57.** $\dfrac{4}{9}$ **58.** $\dfrac{8}{13}$ **59.** $\dfrac{4}{5}$ **60.** $\dfrac{4}{5}$ **61.** $\dfrac{4}{7}$ **62.** $\dfrac{1}{2}$ **63.** $\dfrac{8}{5}$

64. $\dfrac{4}{3}$ **65.** $\dfrac{7}{20}$ **66.** $\dfrac{4}{15}$ **67.** $\dfrac{11}{7}$ **68.** $\dfrac{9}{10}$ **69.** $\dfrac{10}{9}$ **70.** $\dfrac{9}{10}$ **71. a.** 8 **b.** $\dfrac{6}{8}, \dfrac{5}{8}$ **72. a.** 6 **b.** $\dfrac{4}{6}, \dfrac{1}{6}$

73. a. 10 **b.** $\dfrac{7}{10}, \dfrac{12}{10}$ **74. a.** 36 **b.** $\dfrac{15}{36}, \dfrac{16}{36}$ **75. a.** 42 **b.** $\dfrac{16}{42}, \dfrac{27}{42}$ **76. a.** 15 **b.** $\dfrac{7}{15}, \dfrac{20}{15}$ **77. a.** 36

b. $\dfrac{22}{36}, \dfrac{33}{36}$ **78. a.** 80 **b.** $\dfrac{45}{80}, \dfrac{12}{80}$ **79. a.** 120 **b.** $\dfrac{145}{120}, \dfrac{28}{120}$ **80. a.** 180 **b.** $\dfrac{65}{180}, \dfrac{78}{180}$ **81.** $\dfrac{13}{12}$ **82.** $\dfrac{31}{24}$

83. $\dfrac{3}{10}$ **84.** $\dfrac{1}{15}$ **85.** $\dfrac{37}{18}$ **86.** $\dfrac{11}{36}$ **87.** $\dfrac{11}{16}$ **88.** $\dfrac{5}{12}$ **89.** $\dfrac{5}{12}$ **90.** $\dfrac{23}{40}$ **91.** $\dfrac{1}{36}$ **92.** $\dfrac{7}{72}$ **93.** $\dfrac{17}{80}$ **94.** $\dfrac{64}{65}$

95. $\dfrac{23}{24}$ **96.** $\dfrac{7}{12}$ **97.** $\dfrac{23}{40}$ **98.** $\dfrac{3}{5}$ **99.** $\dfrac{5}{6}$ **100.** $\dfrac{15}{16}$

C

Rounding a Number

The following procedure will be used to round off a number to a specified place value.

Step 1. Locate the digit (call it d) in the place value to which the number is to be rounded.

Step 2. Find the digit immediately to the right of d.
   a. If this digit is 0, 1, 2, 3, or 4, change this digit and all digits to the right to zeroes.
   b. If this digit is 5, 6, 7, 8, or 9, add one to d, then change all the digits to the right of $d + 1$ to zeroes.

Example 1. Round 3.1415926 to the nearest ten thousandth.

Solution.   **Step 1.** The 5 is in the ten thousandths position. Thus, d is 5.

        Step 2. The digit to the right of d is 9. Using part (b), one is added to 5 (and $5 + 1 = 6$). Now changing the digits to the right of 6 to zeroes, $3.1415926 = 3.1416000$, or simply 3.1416.

Example 2. Round 3.1415926 to the nearest hundredth.

Solution.   **Step 1.** The 4 is in the hundredths position. Thus, d is 4.

        Step 2. The digit to the right of d is 1. Using part (a), the 1 and all digits to the right are changed to zeroes. Therefore, $3.1415926 = 3.1400000$, or simply 3.14.

APPENDIX C. Exercises

In exercises **1–8**, round each whole number to the indicated place.

1. 5,280 to the nearest hundred

2. 17,065 to the nearest hundred

3. 1,324 to the nearest ten

4. 591 to the nearest ten

5. 86,525 to the nearest thousand

6. 189,702 to the nearest thousand

7. 3,099,200 to the nearest ten thousand

8. 10,123,675 to the nearest ten thousand

In exercises **9–16**, round each decimal to the indicated number of decimal places.

9. 0.72389 to two places

10. 0.90295 to two places

11. 0.05958 to three places

12. 0.00385 to three places

13. 0.604499 to three places

14. 0.000731 to three places

15. 3.141592654 to four places

16. 2.718281828 to four places

In exercises **17–24**, round each square root approximation to three decimal places.

17. $\sqrt{2} \approx 1.414213562$

18. $\sqrt{3} \approx 1.732050808$

19. $\sqrt{10} \approx 3.162277660$

20. $\sqrt{15} \approx 3.872983346$

21. $\sqrt{29} \approx 5.385164807$

22. $\sqrt{37} \approx 6.082762530$

23. $\sqrt{109} \approx 10.44030651$

24. $\sqrt{109} \approx 10.90871211$

In exercises **25–32**, use a calculator to do the indicated operations and round answers to two decimal places.

25. $\dfrac{156}{47}$

26. $\dfrac{391}{86}$

27. $\dfrac{97}{254}$

28. $\dfrac{683}{3,671}$

29. (1.05)(0.463)

30. (23.8)(0.081)

31. (467.2)(0.00379)

32. (5,438)(0.00108)

Answers to Appendix C Exercises

1. 5,300   **2.** 17,100   **3.** 1,320   **4.** 590   **5.** 87,000   **6.** 190,000   **7.** 3,100,000   **8.** 10,120,000   **9.** 0.72
10. 0.90   **11.** 0.060   **12.** 0.004   **13.** 0.604   **14.** 0.001   **15.** 3.1416   **16.** 2.7183   **17.** 1.414   **18.** 1.732
19. 3.162   **20.** 3.873   **21.** 5.385   **22.** 6.083   **23.** 10.440   **24.** 10.909   **25.** 3.32   **26.** 4.55   **27.** 0.38
28. 0.19   **29.** 0.49   **30.** 1.93   **31.** 1.77   **32.** 5.87

D

Table: Powers of Periodic Interest Rates

nt	$(1.01)^{nt}$	$(1.02)^{nt}$	$(1.03)^{nt}$	$(1.04)^{nt}$	$(1.05)^{nt}$
1	1.01000000	1.02000000	1.03000000	1.04000000	1.05000000
2	1.02010000	1.04040000	1.06090000	1.08160000	1.10250000
3	1.03030100	1.06120800	1.09272700	1.12486400	1.15762500
4	1.04060401	1.08243216	1.12550881	1.16985856	1.21550625
5	1.05101005	1.10408080	1.15927407	1.21665290	1.27628156
6	1.06152015	1.12616241	1.19405229	1.26531901	1.34009563
7	1.07213535	1.14868566	1.22987386	1.31593177	1.40710042
8	1.08285670	1.17165937	1.26677007	1.36856904	1.47745544
9	1.09368527	1.19509256	1.30477318	1.42331180	1.55132821
10	1.10462212	1.21899441	1.34391637	1.48024427	1.62889461
11	1.11566834	1.24337430	1.38423386	1.53945404	1.71033934
12	1.12682502	1.26824178	1.42576087	1.60103220	1.79585631
13	1.13809327	1.29360662	1.46853370	1.66507349	1.88564912
14	1.14947420	1.31947875	1.51258971	1.73167643	1.97993158
15	1.16096894	1.34586832	1.55796740	1.80094349	2.07892816
16	1.17257863	1.37278569	1.60470642	1.87298122	2.18287456
17	1.18430442	1.40024140	1.65284761	1.94790047	2.29201829
18	1.19614746	1.42824623	1.70243304	2.02581649	2.40661920
19	1.20810893	1.45681115	1.75350603	2.10684915	2.52695016
20	1.22019002	1.48594737	1.80611121	2.19112311	2.65329767
21	1.23239192	1.51566632	1.86029454	2.27876803	2.78596255
22	1.24471584	1.54597965	1.91610338	2.36991875	2.92526067
23	1.25716300	1.57689924	1.97358648	2.46471550	3.07152370
24	1.26973463	1.60843722	2.03279407	2.56330412	3.22509989
30	1.34784889	1.81136154	2.42726242	3.24339744	4.32194228
36	1.43076875	2.03988729	2.89827825	4.10393244	5.79181598
42	1.51878985	2.29724440	3.46069579	5.19278375	7.76158732
48	1.61222602	2.58707030	4.13225173	6.57052801	10.40126928
54	1.71141040	2.91346133	4.93412465	8.31381401	13.93869556
60	1.81669662	3.28103064	5.89160284	10.51962694	18.67918507

Answers for Exercises

Chapter 1

Section 1-1. Practice Exercises

1. 2, 4, 6, 8, 10, 12, 14 **2.** 4, 8, 12, 16, 20, 24, 28 **3.** 3, 6, 9, 12, 15, 18, 21 **4.** 5, 10, 15, 20, 25, 30, 35
5. 9, 18, 27, 36, 45, 54, 63 **6.** 8, 16, 24, 32, 40, 48, 56 **7.** 15, 30, 45, 60, 75, 90, 105 **8.** 20, 40, 60, 80, 100, 120, 140
9. 1, 2, 4, 5, 10, 20 **10.** 1, 2, 3, 4, 6, 8, 12, 24 **11.** 1, 2, 3, 5, 6, 10, 15, 30 **12.** 1, 2, 4, 8, 16, 32
13. 1, 2, 3, 6, 11, 22, 33, 66 **14.** 1, 5, 11, 55 **15.** 1, 2, 3, 6, 7, 14, 21, 42 **16.** 1, 2, 5, 7, 10, 14, 35, 70

17. $\frac{12}{1}$ **18.** $\frac{19}{1}$ **19.** $\frac{-3}{1}$ **20.** $\frac{-6}{1}$ **21.** $\frac{16}{3}$ **22.** $\frac{20}{3}$ **23.** $\frac{0}{1}$ **24.** $\frac{2}{2}$ **25.** $\frac{713}{100}$ **26.** $\frac{451}{500}$ **27.** >

28. > **29.** > **30.** < **31.** > **32.** > **33.** < **34.** < **35.** = **36.** = **37.** > **38.** >
39. 17 **40.** 42 **41.** 3.2 **42.** 7.8 **43.** 0 **44.** 0 **45.** −3 **46.** −7 **47.** −3 **48.** −7 **49.** 3

50. 7 **51.** −3 **52.** −7 **53.** 11 **54.** 24 **55.** 0 **56.** 0 **57.** $\frac{1}{4}$ **58.** $\frac{11}{20}$ **59.** 6.39 **60.** 1.54

61. $\frac{9}{5}$ **62.** $\frac{3}{5}$ **63.** 86.346 **64.** 374.591 **65.** $\frac{37}{4}$ **66.** $\frac{17}{2}$ **67.** 2.1 **68.** 3.3 **69.** 11 **70.** 30 **71.** 7

72. 66 **73.** 17 **74.** 3 **75.** 14 **76.** 21

Section 1-1. Supplementary Exercises

1. a. 7 **b.** 0, 7 **c.** −3, 0, 7 **d.** $-3, \frac{1}{2}, 0, 7$ **e.** None **2. a.** 6 **b.** 6, 0 **c.** 6, −11, 0

d. $6, -\frac{3}{4}, -11, 0$ **e.** None **3. a.** None **b.** None **c.** −3 **d.** $1.4, -3, \frac{17}{2}$ **e.** $\sqrt{2}$ **4. a.** 101 **b.** 101

c. 101 **d.** $\frac{4}{3}, -1.05, 101$ **e.** $\sqrt{5}$ **5. a.** 0 **b.** 0 **6. a.** 24 **b.** 0 **7. a.** 0 **b.** 30 **8. a.** 18 **b.** 0
9. 46 **10.** 16 **11.** 25 **12.** 76 **13.** 1 **14.** 1 **15.** < **16.** > **17.** < **18.** > **19.** < **20.** >
21. Cannot be determined **22.** Cannot be determined **23. a.** 12, 24, 36, 48, 60, 72, 84, 96
b. 18, 36, 54, 72, 90, 108, 126, 144 **c.** 36, 72 **24. a.** 21, 42, 63, 84, 105, 126, 147, 168
b. 28, 56, 84, 112, 140, 168, 196, 224 **c.** 84, 168

Section 1-2. Practice Exercises

1. −12 **2.** −13 **3.** −13 **4.** −16 **5.** 45 **6.** 56 **7.** −113 **8.** −59 **9.** 5 **10.** −17 **11.** 14
12. 50 **13.** −7 **14.** −11 **15.** −55 **16.** −31 **17.** −59 **18.** 43 **19.** 9 **20.** 47 **21.** 56 **22.** 47
23. −51 **24.** −45 **25.** −16 **26.** −12 **27.** −117 **28.** −187 **29.** −4 **30.** 4 **31.** −8 **32.** 7
33. 15 **34.** 27 **35.** 114 **36.** 109 **37.** 29 **38.** 6 **39.** 50 **40.** 72 **41.** 1,141 ft **42.** 1,410 ft

43. 6,863 ft **44.** 6,612 ft **45.** $36 **46.** 41\frac{1}{2}$ **47.** $48 **48.** 53\frac{1}{2}$ **49.** 23-yd line **50.** 26-yd line

51. 23-yd line **52.** 46-yd line

Section 1-2. Supplementary Exercises

1. 9 **2.** 18 **3.** −20 **4.** 12 **5.** 40 **6.** 20 **7.** −2 **8.** −4 **9.** 10 **10.** 36 **11.** $\frac{-11}{15}$ **12.** $\frac{-11}{15}$

13. $\frac{-1}{23}$ **14.** $\frac{-25}{23}$ **15.** $\frac{8}{11}$ **16.** $\frac{10}{17}$ **17.** $\frac{1}{2}$ **18.** $\frac{-8}{3}$ **19.** $\frac{-9}{20}$ **20.** $\frac{-2}{15}$ **21.** $\frac{-4}{5}$ **22.** $\frac{1}{6}$ **23.** 32.1

24. 33.6 **25.** 13.57 **26.** 23.33 **27.** 39.99 **28.** 49.1 **29.** −16.2 **30.** −50.28 **31. a.** 4 **b.** −4 **c.** 4
d. 4 **e.** No **f.** Yes **g.** Yes **32. a.** −2 **b.** 2 **c.** −2 **d.** −2 **e.** No **f.** Yes **g.** Yes **33. a.** 9
b. 17 **c.** 9 **d.** 9 **e.** No **f.** Yes **g.** Yes **34. a.** 5 **b.** 29 **c.** 5 **d.** 5 **e.** No **f.** Yes **g.** Yes
35. a. −8 **b.** 8 **c.** 3 **d.** −3 **e.** No **f.** No **36. a.** −12 **b.** 12 **c.** 12 **d.** −12 **e.** No **f.** No

Section 1-3. Practice Exercises

1. -21 **2.** -10 **3.** -95 **4.** -208 **5.** -9 **6.** -7 **7.** -6 **8.** -4 **9.** -3 **10.** -4 **11.** -170
12. -230 **13.** -77 **14.** -60 **15.** -240 **16.** -267 **17.** -13 **18.** -19 **19.** 9 **20.** -1 **21.** 9
22. 10 **23.** -1 **24.** -2 **25.** 28 **26.** 27 **27.** 132 **28.** 130 **29.** 24 **30.** 24 **31.** 6 **32.** 4
33. 816 **34.** 572 **35.** 0 **36.** 0 **37.** -1 **38.** -13 **39.** -1 **40.** -8 **41.** 56 **42.** 72 **43.** -2

44. 4 **45.** $\dfrac{36}{48}$ **46.** $\dfrac{27}{45}$ **47.** $\dfrac{-12}{28}$ **48.** $\dfrac{-28}{63}$ **49.** $\dfrac{-120}{36}$ **50.** $\dfrac{-143}{52}$ **51.** $\dfrac{-51}{39}$ **52.** $\dfrac{-45}{51}$ **53.** $\dfrac{-1}{3}$

54. $\dfrac{-1}{7}$ **55.** $\dfrac{-2}{3}$ **56.** $\dfrac{-1}{3}$ **57.** $\dfrac{-1}{10}$ **58.** $\dfrac{-1}{10}$ **59.** $\dfrac{-2}{5}$ **60.** $\dfrac{-1}{4}$ **61.** 7 **62.** 10 **63.** 0 **64.** 0

65. -23 **66.** -43 **67.** Undefined **68.** Undefined

Section 1-3. Supplementary Exercises

1. -37 **2.** -31 **3.** -22 **4.** -21 **5.** 8 **6.** 8 **7.** 1 **8.** 1 **9.** -6 **10.** -24 **11.** $\dfrac{-4}{7}$ **12.** $\dfrac{-4}{11}$

13. $\dfrac{-3}{2}$ **14.** $\dfrac{-2}{3}$ **15.** $\dfrac{7}{5}$ **16.** $\dfrac{5}{4}$ **17.** $\dfrac{21}{8}$ **18.** 7 **19.** $\dfrac{8}{5}$ **20.** 5 **21.** $\dfrac{1}{3}$ **22.** 0 **23.** -17.9

24. -41.7 **25.** -15.98 **26.** -10.62 **27.** -0.96 **28.** 17.68 **29.** $\dfrac{19}{3}$ **30.** -6 **31.** $\dfrac{-17}{5}$ **32.** $\dfrac{25}{3}$

33. a. 20 **b.** 20 **c.** 5 **d.** 0.2 **e.** Yes **f.** No **34. a.** 63 **b.** 63 **c.** 7 **d.** $\dfrac{1}{7}$ **e.** Yes **f.** No

35. a. 2 **b.** 18 **c.** No **36. a.** 5 **b.** 20 **c.** No **37. a.** 0 **b.** 0 **c.** 0 **d.** 1 **e.** 1 **f.** 1 **g.** Yes
h. Yes **38. a.** 0 **b.** 0 **c.** 0 **d.** 1 **e.** 1 **f.** 1 **g.** Yes **h.** Yes **39. a.** 0 **b.** 1

Section 1-4. Practice Exercises

1. 32 **2.** 128 **3.** 16 **4.** 16 **5.** -27 **6.** -125 **7.** $\dfrac{8}{125}$ **8.** $\dfrac{27}{64}$ **9.** -9 **10.** -16 **11.** Negative

12. Negative **13.** Positive **14.** Positive **15.** Negative **16.** Negative **17.** Negative **18.** Negative
19. Negative **20.** Negative **21.** 23 **22.** 31 **23.** -542 **24.** -181 **25.** -41 **26.** -5 **27.** 50
28. 50 **29.** -2 **30.** 2 **31.** -1 **32.** -1 **33.** -4 **34.** -2 **35.** 11 **36.** 9 **37.** -10 **38.** -17
39. 113 **40.** 65 **41.** -77 **42.** 15 **43.** 17 **44.** 3

Section 1-4. Supplementary Exercises

1. 2 **2.** 15 **3.** -1 **4.** 12 **5.** 42 **6.** 116 **7.** 17 **8.** -13 **9.** 45 **10.** 18 **11.** 83 **12.** -28
13. 81 **14.** 121 **15.** 0 **16.** 0 **17.** $\dfrac{3}{10}$ **18.** $\dfrac{1}{3}$ **19.** $\dfrac{1}{8}$ **20.** $\dfrac{7}{25}$ **21.** 7 **22.** 10 **23.** $\dfrac{-5}{4}$ **24.** $\dfrac{-171}{16}$
25. -16.42 **26.** -50.46 **27.** 10.165 **28.** -3.975 **29.** -160 **30.** -512 **31.** 45 **32.** 9 **33.** 23
34. 25 **35.** $(5 + (-3))(2) + 4$ **36.** $(5 + (-3))(2 + 4)$ **37.** $(-3)[6 + 4(-2)]$ **38.** $(-3)(6 + 4)(-2)$
39. $(-5 + 4)(-9) + 6 \div (-2)$ **40.** $-5 + [4(-9) + 6] \div (-2)$

Section 1-5. Practice Exercises

1. $-5 + 32$ **2.** $-9 + 19$ **3.** $(-3)(21)$ **4.** $(-8)(42)$ **5.** $\dfrac{3}{4} + \dfrac{1}{2}$ **6.** $\dfrac{7}{8} + \dfrac{1}{9}$ **7.** $\dfrac{2}{5} \cdot \dfrac{-3}{7}$ **8.** $\dfrac{5}{6} \cdot \dfrac{-4}{9}$

9. $0 + \square$ **10.** $* + \triangle$ **11.** $(0)(\square)$ **12.** $(*)(\triangle)$ **13. a.** 21 **b.** 21 **14. a.** 20 **b.** 20 **15. a.** 90 **b.** 90
16. a. 56 **b.** 56 **17. a.** -8 **b.** -8 **18. a.** 97 **b.** 97 **19. a.** 1 **b.** 1 **20. a.** 7 **b.** 7 **21. a.** 8
b. $\dfrac{-1}{8}$ **22. a.** 5 **b.** $\dfrac{-1}{5}$ **23. a.** $\dfrac{-2}{3}$ **b.** $\dfrac{3}{2}$ **24. a.** $\dfrac{-4}{5}$ **b.** $\dfrac{5}{4}$ **25. a.** $\dfrac{-7}{2}$ **b.** $\dfrac{2}{7}$ **26. a.** $\dfrac{-7}{3}$

b. $\dfrac{3}{7}$ **27. a.** 4.2 **b.** $\dfrac{-5}{21}$ **28. a.** 3.1 **b.** $\dfrac{-10}{31}$ **29.** 49 **30.** 105 **31.** 10 **32.** 13 **33.** -37

34. -17 **35.** 23.36 **36.** 24.36 **37.** 240 **38.** 240 **39.** 280 **40.** 180 **41.** -500 **42.** -400
43. 7,310 **44.** 5,850

Section 1-5. Supplementary Exercises

1. 0   **2.** 0   **3.** 1   **4.** 1   **5.** 1   **6.** 1   **7.** 0   **8.** 0   **9.** Yes   **10.** Yes   **11.** No   **12.** No   **13.** Yes
14. Yes   **15.** Yes   **16.** Yes   **17.** Yes   **18.** Yes   **19.** No   **20.** No   **21.** 7   **22.** 6   **23.** -6 **24.** -4
25. -3 **26.** 10   **27.** 17   **28.** 11   **29.** 5.3   **30.** 2.5   **31.** Commutative property of addition
32. Commutative property of addition   **33.** Identity element for addition   **34.** Identity element for addition
35. Identity element for multiplication   **36.** Identity element for multiplication   **37.** Associative property of multiplication
38. Associative property of multiplication   **39.** Distributive property   **40.** Distributive property

Section 1-6. Practice Exercises

1. One   **2.** One   **3.** Three   **4.** Three   **5.** Two   **6.** Two   **7.** Two   **8.** Two   **9.** Four   **10.** Four
11. Two   **12.** Two   **13. a.** 23   **b.** x^2 **14. a.** 45   **b.** y^3 **15. a.** 7   **b.** x^2y **16. a.** 4   **b.** a^2b

17. a. $\dfrac{-3}{2}$ **b.** p^2q^2 **18. a.** $\dfrac{-7}{10}$ **b.** xy **19. a.** 1   **b.** a^2b **20. a.** 1   **b.** x^2y **21. a.** 0.3   **b.** mn^2

22. a. 0.8   **b.** t^2w **23.** $5x^2z$ **24.** $8p^2q$ **25.** $-6ab^2c$ **26.** $-9pq^2r$ **27.** $-10(x+y)^4$ **28.** $-3(a-b)^5$
29. $-23uvw^2$ **30.** $-13q^2rs$ **31.** $19x$ **32.** $58y$ **33.** $8a^2$ **34.** $-3b^2$ **35.** $10xyz$ **36.** $6ab$ **37.** $21a^2b$
38. $42xy^2$ **39.** $3a^2$ **40.** $2k^3$ **41.** 0   **42.** 0   **43.** $4a+11b$ **44.** $22c-13d$ **45.** $34h^2-9k^2+3$
46. $10x^2+32y^2-7$ **47.** $-4+7y$ **48.** $39-29x$ **49.** a^3+2a^2+2a **50.** $13y^3+4y^2-2y$ **51.** $10x+14$
52. $12y+18$ **53.** $13b-19$ **54.** $37-12c$ **55.** $-11a^2-13a-19$ **56.** $-19m^2-26m+4$ **57.** $-v^2+64v+62$
58. $-19w^2-42w+40$

Section 1-6. Supplementary Exercises

1. a. Three   **b.** Two   **c.** Two   **d.** One   **2. a.** Three   **b.** Two   **c.** Two   **d.** One   **3. a.** Four   **b.** Two
c. Three   **d.** Two   **4. a.** Four   **b.** Two   **c.** Two   **d.** Three   **5. a.** 17   **b.** 6   **6. a.** 12   **b.** -5
7. a. -1 **b.** -6 **8. a.** -1 **b.** -4 **9. a.** 49   **b.** 0   **10. a.** 25   **b.** 0   **11. a.** -4 **b.** 3   **12. a.** -9
b. -2 **13.** $10xy+50$ **14.** $-10xy+10$ **15.** $8y^3+1$ **16.** $15x^2-6$ **17.** $4b^2-5$ **18.** $4b^2-6b$
19. $-20y$ **20.** $5x-13y$ **21.** $-10x+16y$ **22.** $12a-13b$ **23.** $2m+12$ **24.** $-12n-20$ **25.** $50a+60b$
26. $24c$ **27.** $-28m+72n$ **28.** $-6m$ **29.** 9   **30.** 2   **31.** 11   **32.** 3   **33.** 7   **34.** -7 **35.** 8   **36.** -8
37. $h=-8$ $i=-6$ $j=0$ **38.** $h=16$ $i=0$ $j=12$

Section 1-7. Practice Exercises

1. 23   **2.** 18   **3.** 83   **4.** 138   **5.** -11 **6.** -19 **7.** 7   **8.** 39   **9.** 33   **10.** 105   **11.** 5   **12.** -27
13. -87 **14.** -10 **15.** -16 **16.** -52 **17.** -280 **18.** -32 **19.** 30   **20.** 161   **21.** 8   **22.** 12

23. -14 **24.** -65 **25.** 79   **26.** 14   **27.** -4 **28.** 39   **29.** 60   **30.** 13   **31.** $\dfrac{-11}{20}$ **32.** $\dfrac{1}{64}$ **33.** 30 in^2

34. 60 in^2 **35.** 69 cm^2 **36.** 135 cm^2 **37.** 2,700 mm^2 **38.** 1,898 mm^2 **39.** \$4,079.77   **40.** \$1,343.92
41. a. \$4,433.70   **b.** \$933.70   **42. a.** \$31,743.37   **b.** \$6,743.37   **43.** 176 m   **44.** 78 m   **45.** 99 m   **46.** 31 m

Section 1-7. Supplementary Exercises

1. 149   **2.** -24 **3.** -17 **4.** -8.1 **5.** 0   **6.** -29 **7.** 22   **8.** 21   **9.** -1 **10.** -1 **11. a.** 24
b. 24   **c.** Same   **12. a.** 6   **b.** 6   **c.** Same   **13. a.** 36   **b.** 0   **c.** II   **14. a.** 0   **b.** 0   **c.** II   **15. a.** 30
b. 120   **c.** Four   **16. a.** 35   **b.** 315   **c.** Nine   **17. a.** \$1,790.85   **b.** \$1,816.70   **c.** \$25.85   **18. a.** \$36,018.87
b. \$36,333.92   **c.** \$315.06   **19. a.** 4.9 m   **b.** 19.6 m   **c.** Four times   **20. a.** 44.1 m   **b.** 176.4 m   **c.** Four times

Section 1-8. Practice Exercises

1. $(p+5)+3$ **2.** $(12+z)+9$ **3.** $25+(x+y)$ **4.** $2+(p+q)$ **5.** $(5+a)+(6+b)$ **6.** $(\mu+1)+(w+6)$
7. $m-7$ **8.** $13-p$ **9.** $(t+12)-5$ **10.** $(9+y)-10$ **11.** $(m-n)-(p+q)$ **12.** $(x+y)-(s-t)$
13. $9y-7w$ **14.** $23m+14p$ **15.** $(b+3)(b-7)$ **16.** $(x-5)(x+4)$ **17.** $5y+4x$ **18.** $11m-4n$ **19.** $4x-x^2$
20. $p^3+(y+3)$ **21.** y^5z **22.** a^6+3 **23.** z^3+5z **24.** x^2-10x **25.** b^3-a^3 **26.** x^2-y^2 **27.** $\dfrac{xy}{10}$

28. $\dfrac{p+q}{5}$ **29.** $\dfrac{(y-8)^2}{3y}$ **30.** $\dfrac{(7a)^2}{2-a}$ **31.** $\dfrac{\frac{1}{2}t}{\frac{1}{3}s}$ **32.** $\dfrac{\frac{1}{4}x}{\frac{1}{5}y}$ **33.** $\dfrac{a^2}{a^2-b^2}$ **34.** $\dfrac{p^3-q^3}{p^2}$

Section 1-8. Supplementary Exercises

1. $10z - 2$ **2.** $ab + 14$ **3.** $\dfrac{ab}{a+b}$ **4.** $\dfrac{x-y}{x+y}$ **5.** $\dfrac{m^2}{a+1}$ **6.** $\dfrac{x^3}{3y}$ **7.** $(z-4)^2 - 5$ **8.** $(9x)^2 + 7$ **9.** $\left(\dfrac{y}{8}\right)^3$

10. $\left(\dfrac{5}{b}\right)^2$ **11.** $x^6 - x^5$ **12.** $y^7 + y^5$ **13.** $\dfrac{z+5}{(z+2)^2}$ **14.** $\dfrac{x-3}{(6y)^2}$ **15.** lwh **16.** $a+b+c$ **17.** $\dfrac{1}{2}bh$

18. $\dfrac{a+b}{2}$ **19.** $\dfrac{pr}{s}$ **20.** $\dfrac{SI}{D}$ **21.** $\dfrac{wt^2 c}{l}$ **22.** $\dfrac{1}{12}bd^3$

Chapter 1 Review Exercises

1. $\dfrac{16}{5}$ **2.** 1, 2, 3, 4, 6, 12 **3. a.** 19 **b.** -19 **c.** -19 **4.** 10 **5. a.** $<$ **b.** $<$ **6.** -8 **7.** 18

8. -50 **9.** $-3xy$ **10.** 32 **11.** 8 **12.** -3 **13.** 0 **14. a.** $\dfrac{-20}{35}$ **b.** $\dfrac{35}{-56}$ **15.** 24 **16.** 60

17. -6 **18.** 3 **19.** 24 **20.** Commutative property of multiplication **21.** Distributive property
22. Commutative property of addition **23.** Associative property of addition **24.** Distributive property

25. a. 12 **b.** $\dfrac{-1}{12}$ **26.** $-6x^2 + 7x + 2$ **27.** $-4ab + 6a + 2b$ **28.** $2m^2 n + 12mn^2$ **29. a.** Three **b.** Two

c. One **30.** -40 **31.** 19 **32.** 3 **33.** 1 **34.** $25x + x^3$ **35.** $y^4 - \dfrac{x}{4}$ **36.** $(10-a)^2(c+d)$

Chapter 2 Linear Equations and Inequalities in One Variable

Section 2-1. Practice Exercises

1. True **2.** True **3.** Not true **4.** Not true **5.** True **6.** True **7.** True **8.** True **9.** True **10.** True
11. False **12.** True **13.** True **14.** False **15.** True **16.** False **17.** 2 **18.** 2 **19.** 1 **20.** 2

21. -2 **22.** -2 **23.** $\dfrac{1}{2}$ **24.** $\dfrac{1}{2}$ **25.** 6 **26.** 2 **27.** 1 and 4 **28.** 2 and 4 **29.** -2 and 1 **30.** -2 and 4
31. Yes **32.** Yes **33.** No **34.** No **35.** Yes **36.** Yes **37.** No **38.** No **39. a.** $-x - 3$ **b.** Yes
40. a. $22x + 10$ **b.** Yes **41. a.** $-4a + 7$ **b.** No **42. a.** $36 - 12b$ **b.** Yes **43. a.** $20c - 17$ **b.** Yes
44. a. $18m + 8$ **b.** No

Section 2-1. Ten Review Exercises

1. 0 **2.** 0 **3.** Yes **4.** Yes **5.** -6 **6.** -6 **7.** $k - 4$ **8.** -6 **9.** No **10.** No

Section 2-1. Supplementary Exercises

1. Not true **2.** Not true **3.** True **4.** True **5.** True **6.** True **7.** $-\dfrac{1}{4}$ **8.** $-\dfrac{1}{4}$ **9.** -3 and 6

10. -3 and 2 **11.** $-2, -1, 0,$ and 2 **12.** $-2, -1, 0,$ and 1 **13.** Yes **14.** Yes **15.** Yes **16.** Yes **17.** No
18. No **19. a.** -2 and 2 **b.** -2 and 2 **c.** Yes **20. a.** -3 and 3 **b.** -3 and 3 **c.** Yes **21. a.** -10 and 10
b. 10 **c.** No **22. a.** -4 and 4 **b.** 4 **c.** No **23. a.** \$2.00, \$1.75, \$1.50, \$1.05 **b.** \$3.00, \$2.50, \$2.00, \$1.10
c. \$0.25 and \$1.25 **24. a.** 110 lbs, 125 lbs, 145 lbs, 170 lbs **b.** 230 lbs, 260 lbs, 300 lbs, 350 lbs **c.** 150 lbs and 140 lbs

Section 2-2. Practice Exercises

1. 10 **2.** 8 **3.** 5 **4.** -16 **5.** 3 **6.** 7 **7.** 8 **8.** 9 **9.** -8 **10.** -19 **11.** -11 **12.** -2 **13.** 4
14. 6 **15.** -9 **16.** -13 **17.** 2 **18.** 1 **19.** 13 **20.** -5 **21.** 9 **22.** 4 **23.** 5 **24.** 3 **25.** -12
26. -16 **27.** 75 **28.** 90 **29.** -35 **30.** -36 **31.** 168 **32.** 144 **33.** 30 **34.** 20 **35.** -128
36. -162 **37.** 5 **38.** 6 **39.** 2 **40.** 9 **41.** -20 **42.** -8 **43.** -1 **44.** -4 **45.** 1 **46.** 0

Section 2-2. Ten Review Exercises

1. 0 **2.** $3x - 12$ **3.** 4 **4.** Yes **5.** Yes **6.** 0 **7.** $7y + 35$ **8.** -5 **9.** No **10.** No

Section 2-2. Supplementary Exercises

1. 0 **2.** 0 **3.** 0 **4.** 0 **5.** 1 **6.** $\dfrac{1}{2}$ **7.** $\dfrac{5}{8}$ **8.** $\dfrac{1}{10}$ **9.** -8 **10.** 8 **11.** 0 **12.** 0 **13.** -65

14. -32 **15.** -1.5 **16.** -1.2 **17.** -6 **18.** -5 **19.** 40 **20.** 39 **21.** 0 **22.** 0 **23.** -1 **24.** -2

25. 1.9   **26.** −3.1   **27.** 30   **28.** 6   **29.** −3   **30.** −6   **31.** 13.2   **32.** 18.2   **33.** −7   **34.** 8   **35.** 6
36. −8   **37. a.** 8 in   **b.** 5 cm   **38. a.** 8 hours   **b.** 48 mph   **39. a.** $2,050   **b.** 4 years   **40. a.** 7 in   **b.** 8 ft

Section 2-3. Practice Exercises

1. The number is 15.   **2.** The number is 23.   **3.** The number is −18.   **4.** The number is −16.
5. The number is 21.   **6.** The number is 15.   **7.** The number is 8.   **8.** The number is −12.
9. One hamburger cost $1.59.   **10.** One sandwich cost $2.55.   **11.** The shoes cost $48.
12. One large bar costs 70¢.   **13.** Common stamps cost 39¢; foreign stamps cost $1.17; rare stamp cost $2.73.
14. The first type cost 25¢; the second type cost 50¢; the third type cost 75¢.   **15.** The width is 48 mm.
16. The width is 11 in.   **17.** The width is 4 ft; the length is 8 ft.   **18.** The width is 5 in; the length is 15 in.
19. 17 cm, 21 cm, and 18 cm   **20.** 12 in, 12 in, and 15 in.   **21.** Marty's car averaged 40 mph.
22. Judy's bike averaged 11 mph.   **23.** Dave's horse is traveling 3 mph.   **24.** The Loftan's are traveling 49 mph.
25. The truck's rate is 44 mph.   **26.** The train's rate is 56 mph.   **27.** Tammy is 24; Gary is 28.
28. The car is 23 years old; the truck is 14 years old.   **29.** The cat is 6 years old.   **30.** Julia is 14 years old.
31. Gert is $3\frac{1}{2}$ months old.   **32.** The gander is 9 months old.

Section 2-3. Ten Review Exercises

1. 0   **2.** 0   **3.** 0   **4.** 0   **5.** 0   **6.** Yes   **7.** Yes   **8.** Yes   **9.** Yes   **10.** No

Section 2-3. Supplementary Exercises

1. a. 50 years old   **b.** 50 years old   **c.** 0 years   **2. a.** 8 years old   **b.** 10 years old   **c.** 2 years   **3. a.** 20 mi
b. 16 mi   **c.** 0 mi   **4. a.** 430 mi   **b.** 380 mi   **c.** 40 mi   **5. a.** 4-yd gain   **b.** 13-yd gain
c. On Mustang's 39-yard line.   **6. a.** 550-ft increase   **b.** 525-ft increase   **c.** 25-ft difference   **7. a.** 15   **b.** 24
c. 15   **d.** Eleven 40¢ and twenty-two 25¢   **e.** Fourteen 40¢ and sixteen 25¢   **8. a.** 30 ft, 35 ft, and 35 ft
b. 25 ft, 35 ft, and 40 ft   **c.** 30 ft, 40 ft, and 30 ft   **d.** 33 ft each

Section 2-4. Practice Exercises

1. 5   **2.** 6   **3.** −6   **4.** −3   **5.** −1   **6.** −4   **7.** $\frac{-3}{5}$ **8.** $\frac{1}{2}$ **9.** 3   **10.** −2   **11.** $\frac{10}{3}$ **12.** $\frac{8}{3}$
13. $\frac{-1}{2}$ **14.** $\frac{4}{3}$ **15.** −6   **16.** $\frac{1}{2}$ **17.** 4   **18.** −5   **19.** $\frac{-1}{2}$ **20.** $\frac{-3}{2}$ **21.** −2   **22.** 1   **23.** $\frac{-1}{4}$
24. $\frac{4}{3}$ **25.** $\frac{-2}{5}$ **26.** $\frac{9}{2}$ **27.** No roots   **28.** No roots   **29.** R **30.** R **31.** No roots   **32.** No roots
33. R **34.** R **35.** 11, 12, and 13   **36.** 21, 22, and 23   **37.** 71, 73, and 75   **38.** 37, 39, and 41   **39.** 10 and 11
40. 7 and 8   **41.** 9, 11, and 13   **42.** 26, 28, and 30   **43.** 30°, 70°, and 80°   **44.** 45°, 45°, and 90°
45. 25°, 45°, and 110°   **46.** 33°, 44°, and 103°   **47.** 118°, 28°, and 34°   **48.** 42°, 30° and 108°

Section 2-4. Ten Review Exercises

1. $3n + 8$ **2.** $5n − 12$ **3.** 38   **4.** 38   **5.** The number is 10.   **6.** $(2n + 4) − 10$ **7.** 0   **8.** $2n − 6$
9. The number is 3.   **10.** Yes

Section 2-4. Supplementary Exercises

1. $\frac{-1}{2}$ **2.** $\frac{-2}{3}$ **3.** 5   **4.** −6   **5.** No roots   **6.** No roots   **7.** R **8.** R **9.** $\frac{2}{3}$ **10.** $\frac{-5}{4}$ **11.** 0
12. 1   **13.** 3   **14.** 5   **15.** −2   **16.** 4   **17.** −2   **18.** 6   **19.** −2, 0, and 2   **20.** −3, −1, and 1
21. Any three consecutive even integers.   **22.** Any three consecutive odd integers.   **23.** 10 ft by 12 ft
24. 40 in by 120 in   **25.** 15 dimes and 25 nickels   **26.** 70 nickels and 250 pennies
27. Twenty 25¢ stamps, twenty-five 1¢ stamps, and eighty-five 15¢ stamps
28. 22 of kind A, 12 of kind B, and 5 of kind C **29.** 56 in, 56 in, and 72 in   **30.** 25 ft, 30 ft, and 40 ft
31. a. 360°   **b.** 50°, 50°, 55°, 60°, 70°, and 75°

Section 2-5. Practice Exercises

1. a. 6 hours   **b.** 8:00 A.M.   **2. a.** 7 hours   **b.** 12:30 P.M.   **3. a.** 155 mi   **b.** 17 minutes   **4. a.** 43.5 mi
b. 72.5 mph   **5.** 45.5 mph   **6.** 48 mph   **7.** 9:45 A.M.   **8.** 7:00 A.M.   **9.** 10 ft   **10.** 24 ft   **11.** 78.5 ft²
12. 200.96 ft²   **13.** 10 m   **14.** 36 in   **15.** $a = P − b − c$ **16.** $b = P − a − c$ **17.** $n = \frac{P}{a}$ **18.** $a = \frac{P}{n}$

19. $r = \dfrac{C}{2\pi}$ **20.** $\pi = \dfrac{C}{2r}$ **21.** $P = \dfrac{2(A - B)}{S}$ **22.** $S = \dfrac{2(A - B)}{P}$ **23.** $h = \dfrac{A - 2B}{P}$ **24.** $B = \dfrac{A - Ph}{2}$

25. $l = \dfrac{P - 2w}{2}$ **26.** $w = \dfrac{P - 2l}{2}$ **27.** $a = \dfrac{P - 2b}{2}$ **28.** $b = \dfrac{P - 2a}{2}$ **29.** $c = 2S - a - b$ **30.** $b = 2S - a - c$

31. $R = \dfrac{A - \pi rs}{\pi s}$ **32.** $r = \dfrac{A - \pi Rs}{\pi s}$ **33.** $s = \dfrac{A - \pi R^2}{\pi R}$ **34.** $\pi = \dfrac{A}{R(R + s)}$ **35.** $y = \dfrac{36 - 9x}{4}$ **36.** $x = \dfrac{36 - 4y}{9}$

37. $q = \dfrac{7 - 5p + 2r}{9}$ **38.** $r = \dfrac{5p + 9q - 7}{2}$ **39.** $a = \dfrac{10 - bx - c}{x^2}$ **40.** $b = \dfrac{10 - ax^2 - c}{x}$ **41.** $v = \dfrac{d - at^2}{t}$

42. $a = \dfrac{d - vt}{t^2}$ **43.** $x = \dfrac{5}{c - 4}$ **44.** $x = \dfrac{9}{c - 9}$ **45.** $x = \dfrac{25}{a + 5}$ **46.** $x = \dfrac{20}{d + 2}$ **47.** $x = \dfrac{11}{b - 1}$ **48.** $x = \dfrac{2}{1 - b}$

49. a. $y = 6 - 3x$ **b.** -9 **50. a.** $y = 15 - 5x$ **b.** -5 **51. a.** $y = \dfrac{16 - 2x - 4z}{3}$ **b.** 0 **52. a.** $x = \dfrac{12 - 2y - z}{5}$

b. 1 **53. a.** $b = \dfrac{P - a - c}{2}$ **b.** 8 **54. a.** $h = \dfrac{P - 2l - 2w}{2}$ **b.** 6

Section 2-5. Ten Review Exercises

1. 8   **2.** 8   **3.** 8   **4.** 8   **5.** -17 **6.** $\dfrac{8}{5}$ **7.** 33   **8.** -3 **9.** -7 **10.** -5.64

Section 2-5. Supplementary Exercises

1. $D = 360° - A - B - C$ **2.** $B = 360° - A - C - D$ **3.** $R = \dfrac{A}{2\pi h}$ **4.** $h = \dfrac{A}{2\pi R}$ **5.** $n = \dfrac{t - a + d}{d}$ **6.** $d = \dfrac{t - a}{n - 1}$

7. a. 104 mi   **b.** 260 mi   **c.** 156 mi   **8. a.** $t = \dfrac{I}{PR}$ **b.** 2 years   **9.** $x = \dfrac{12}{18 - a}$ **10.** $y = \dfrac{4}{p + 2}$ **11.** $a = \dfrac{bc}{d}$

12. $c = \dfrac{ad}{b}$ **13.** $a = \dfrac{b^2 - D}{4c}$ **14.** $c = \dfrac{b^2 - D}{4a}$ **15. a.** $y = 4x - 2$ **b.** 0 **16. a.** $y = \dfrac{x - 3}{3}$ **b.** -2

17. $A = ab + \dfrac{\pi b^2}{8}$ **18.** $A = a(b - d) + d(a - c)$ **19.** $A = \dfrac{\pi s^2}{2} + s^2$ **20.** $A = \dfrac{1}{2}bh + bw$

Section 2-6. Practice Exercises

1.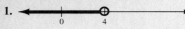

2.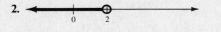

3.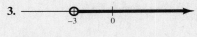

4.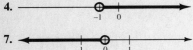

5.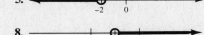

6.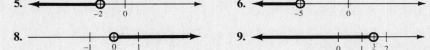

7.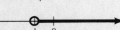

8.

9.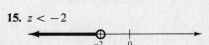

10.

11. $x < 5$

12. $x < 8$

13. $y > 3$

14. $z < 2$

15. $z < -2$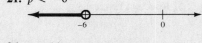

16. $p > -4$

17. $a > -5$

18. $t > -6$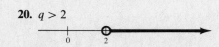

19. $x > 6$

20. $q > 2$

21. $p < -6$

22. $x < -5$

23. $x > -21$

24. $y < 11$

25. $w > -15$

26. $t > -7$

27. $b > -10$

28. $d < -9$

29. $x > -54$

30. $y < -100$

31. $z < 7$

32. $z > 2$

33. $t > -2$

34. $t < 6$

35. $k > -2$

36. $k < 3$

37. $b < -2$

38. $b < 8$

39. $x > -4$

40. $x < 5$

41. $t > -2$

42. $t < 4$

43. $z \le 6$

44. $a \ge 1$

45. $z \le 4$

46. $a \le -4$

47. $c \le 2$

48. $x \ge 3$

49. $x \ge -22$

50. $b \le 6$

51. All numbers less than or equal to -1 **52.** All numbers greater than 2 **53.** All numbers greater than -5
54. All numbers greater than or equal to 6 **55.** All numbers less than 1 **56.** Between 0 and 6 in
57. Between 0 and 8 in, including 8 **58.** Between 0 and 60 in **59.** More than 51 points **60.** More than 23 yds

Section 2-6. Ten Review Exercises

1. $18k - 54$ **2.** 3 **3.** $k < 3$

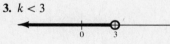

4. $45 - 15y$ **5.** 3

6. $y \ge 3$

7. 65 **8.** 53 **9.** 81 **10.** -24

Section 2-6. Supplementary Exercises

1. $z \ge 3$

2. $y \ge -3$

3. $z > \dfrac{-7}{10}$

4. $t < \dfrac{-6}{5}$

5. $n < -4$

6. $z > -1$

7. $z \ge 0$

8. $p \le 0$

9. $p > -24$

10. $p < -6$

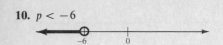

11. $x > \dfrac{-1}{2}$

12. $x \le 2$

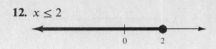

13. $y < 2$

14. $y < 4$

15. $x \le -5$

16. $x \le 9$

17. $x < 0.8$

18. $y > 3$

19.

20.

21.

22.

23.

24.

25.

26.

27.

28.

29.

30.

31. $-6 < x < 3$

32. $-4 < x < 4$

33. $-1 < x < 2$

34. $1 < x < 2$

Chapter 2 Review Exercises

1. -1 **2.** -5 **3.** 1 and 5 **4.** 3 and 4 **5.** Yes **6.** No **7.** 18 **8.** -49 **9.** $\dfrac{-9}{5}$ **10.** -11

11. 120 **12.** 51 **13.** -12 **14.** 3 **15.** -2 **16.** -3 **17.** The number is 6. **18.** The number is 7.

19. 6 dimes, 18 nickels **20.** Ten 45¢ stamps, twenty 25¢ stamps **21.** 17 cm **22.** 15 cm **23.** $\dfrac{-1}{3}$ **24.** 8

25. $\dfrac{3}{4}$ **26.** -1 **27.** 21, 22, and 23 **28.** 36, 37, and 38 **29.** $x = \dfrac{-3}{2}y$ **30.** $a = \dfrac{S - V}{t}$ **31.** $v = \dfrac{Fl}{kA}$

32. $y = \dfrac{1 - 3x^2}{4}$ **33. a.** $y = \dfrac{x - 10}{5}$ **b.** -5 **34. a.** $y = 4 - 2x$ **b.** 34

35. $x < 4$

36. $y \ge -7$

37. $y \ge -2$

38. $t < -4$

39. $y \le -3$

40. $x < 5$

Chapter 3 Polynomials

Section 3-1. Practice Exercises

1. a. 5, 3, 0 **b.** 5 **c.** Trinomial **2. a.** 0, 2, 4 **b.** 4 **c.** Trinomial **3. a.** 4, 1 **b.** 4 **c.** Binomial
4. a. 5, 1 **b.** 5 **c.** Binomial **5. a.** 4, 8, 2, 0 **b.** 8 **c.** Polynomial of 4 terms **6. a.** 7, 9, 1, 0 **b.** 9
c. Polynomial of 4 terms **7. a.** 2, 0, 7, 5, 3 **b.** 7 **c.** Polynomial of 5 terms **8. a.** 0, 3, 9, 10, 2 **b.** 10
c. Polynomial of 5 terms **9. a.** 4, 2, 0 **b.** 4 **10. a.** 6, 2, 0 **b.** 6 **11. a.** 5, 3 **b.** 5 **12. a.** 3, 6
b. 6 **13. a.** 6, 5, 9 **b.** 9 **14. a.** 6, 7, 5 **b.** 7 **15. a.** 6, 5, 4, 3 **b.** 6 **16. a.** 0, 5, 5, 6 **b.** 6

17. a. 6, 6, 6   **b.** 6   **18. a.** 5, 5, 5   **b.** 5   **19.** $4 + 3x + x^2$ **20.** $1 + 7x + 6x^2$ **21.** $10 + 2a - 4a^2 + 8a^3$
22. $27 - 3a + a^2 - 9a^3$ **23.** $b + 4b^3 + 7b^5 - 2b^{10}$ **24.** $3b + 3b^3 - 7b^5 - b^8$ **25.** $y^2 - 16y + 64$
26. $121y^2 + 22y + 1$ **27.** $-4m^5 + 8m^3 - m + 1$ **28.** $-6n^5 + 3n^2 + n - 5$ **29.** $8z^5 + 11z^3 + 6z^2 - 3z$
30. $z^7 + 14z^4 + 3z^3 - 8z$ **31. a.** $x^2 - 7xy + 10y^2$ **b.** $10y^2 - 7xy + x^2$ **32. a.** $4x^2 - 12xy + 9y^2$
b. $9y^2 - 12xy + 4x^2$ **33. a.** $-a^2 - 3ab + 5b^2 + 5$ **b.** $5b^2 - 3ab - a^2 + 5$ **34. a.** $-m^2 - mn + n^2 + 12$
b. $n^2 - mn - m^2 + 12$ **35. a.** $2r^5t^3 - r^4st^2 + 5r^3s^2 - 8s^4t^5$ **b.** $-8s^4t^5 + 5r^3s^2 - r^4st^2 + 2r^5t^3$
c. $-8s^4t^5 + 2r^5t^3 - r^4st^2 + 5r^3s^2$ **36. a.** $-4a^3bc + 6a^2b^2c^3 - 4ab^3 + b^4c^2$ **b.** $b^4c^2 - 4ab^3 + 6a^2b^2c^3 - 4a^3bc$
c. $6a^2b^2c^3 + b^4c^2 - 4a^3bc - 4ab^3$

Section 3-1. Ten Review Exercises

1. $5n + 2$ **2.** -23 **3.** $3n - 8$ **4.** -23 **5.** The number is -5.   **6.** $\dfrac{21}{2}$ **7.** $\dfrac{21}{2}$ **8.** $\dfrac{-1}{2}$

9. $z < \dfrac{-1}{2}$ **10.** $z \geq \dfrac{-1}{2}$

Section 3-1. Supplementary Exercises

1. a. 3   **b.** 1   **2. a.** 2   **b.** 1   **3. a.** 2   **b.** 4   **4. a.** 2   **b.** 6   **5. a.** $x^2 - 3xy + 5y^2 + 8$ **b.** $5y^2$ and 8
c. $5y^2$ **6. a.** $5y^2 - 3xy + x^2 + 8$ **b.** x^2 and 8   **c.** x^2 **7. a.** $a^3b^3 + 3a^2b + 6a^2 + 2ab^2 + 5b^2 - 9$
b. $3a^2b$ and $6a^2$ **c.** $3a^2b$ **8. a.** $a^3b^3 + 2ab^2 + 5b^2 - 3a^2b + 6a^2 - 9$ **b.** $2ab^2$ and $5b^2$ **c.** $2ab^2$ **9. a.** 3
b. 3   **c.** 1   **d.** 1   **10. a.** 5   **b.** 5   **c.** 1   **d.** 1   **11. a.** 52   **b.** -38 **c.** -8 **12. a.** -45 **b.** 115
c. 1   **13. a.** 68   **b.** 225   **c.** 147   **14. a.** 36   **b.** 160   **c.** 200

Section 3-2. Practice Exercises

1. $7x^2 + 2$ **2.** $13x^2 - 2$ **3.** $10z^2 - z - 7$ **4.** $3z^2 + 4z - 3$ **5.** $-y^3 + 15$ **6.** $2y^2 + 4y$ **7.** $a^4 + 2a^3 - 3a$
8. $5a^4 - 7a^3 - 2a^2$ **9.** $2a^2 + 3ab$ **10.** $5a^2 - 4ab$ **11.** $8m^3 - 2m - 1$ **12.** $m^3 + 4m^2 - 2$ **13.** $3x^3 + 2x^2 - 2$
14. $-2x^2 + 4x + 3$ **15.** $3xy + xz + 3yz$ **16.** $8xz + 2yz + 4$ **17.** $6ab - 2ac + bc - 3$ **18.** $ab + 7ac - 3bc + 3$
19. $-x^2 + 3x - 4$ **20.** $-x^2 + 5x - 10$ **21.** $a^2 - 5a - 7$ **22.** $a^2 - a - 6$ **23.** $-10a^2 - ab + 2b^2 + 13$
24. $-5a^2 + 3ab - b^2 - 4$ **25.** $-a^3 + b^3$ **26.** $-a^5 + b^4$ **27.** $-2y - 2$ **28.** $-5y + 5$ **29.** $2z^2 + 3z - 3$
30. $-2z^2 - 3z - 3$ **31.** $3a^4 + 4a^2 + 12a - 4$ **32.** $8a^4 + 4a^3 + 18a - 1$ **33.** $-8xy$ **34.** $-6xy + 2yz$
35. $-8x + 3$ **36.** $4x - 1$ **37.** $-3a^3 + 11a^2$ **38.** $-a^4 - 4a + 13$ **39.** $2a^2 + 2ab + 2b^2$ **40.** $2a^2 - 4ab + 12b^2$
41. $6x^2 - 20x + 34$ **42.** $-5x^2 + 27x - 7$ **43.** $-7t^3 - 27t^2 + 16t$ **44.** $26t^2 + 4t$ **45.** $-4a^5 - 18a^4 + 2a^3 + 2a$
46. $-6a^4 - a^3 + 24a^2 - 7a$ **47.** $-11k^2 + k$ **48.** $2k^2 + 12k + 1$ **49.** $11m^2 - 2mn - 16n^2$ **50.** $4m^2 + mn - 14n^2$

Section 3-2. Ten Review Exercises

1. 80   **2.** 80   **3.** $3y^2 - y + 10$ **4.** 80   **5.** 50   **6.** 50   **7.** $13x^2 + 3xy + 4y^2$ **8.** 50
9. $k^6 - 7k^4 + 6k^3 - 9k^2 + 3k + 1$ **10.** 6

Section 3-2. Supplementary Exercises

1. $2x^3 - 4x^2$ **2.** $7x^2 + 6$ **3.** -1 **4.** $-7x$ **5.** $-x^2 - 11x + 14$ **6.** 0   **7.** $3y - 11$ **8.** $-3t - 13$
9. $a^2 - 8a + 11$ **10.** $-c^2 - 7c + 22$ **11.** $-x^3 + 15x^2 - 10x + 7$ **12.** $3x^3 - 2x^2 - 9x + 9$ **13.** $3a^2 + 5b^2$
14. $12u^2 - 7v^2$ **15.** $2.1b^2 + 2.1b + 1.25$ **16.** $-0.9b^2 - 1.9b - 3.87$ **17.** $-2z$ **18.** $-2z^2$ **19.** -8 **20.** $24x^2$
21. $13p^2$ **22.** $19p^2 + 14pq$ **23.** $28m^2n - 7n^2$ **24.** $14m^2n$ **25. a.** 103   **b.** -1 **c.** $3t^2 - 5t + 2$ **d.** 102
e. $t^2 - 13t + 14$ **f.** 104   **26. a.** -6 **b.** 11   **c.** $t^2 - t - 1$ **d.** 5   **e.** $-7t^2 + 9t + 19$ **f.** -17

Section 3-3. Practice Exercises

1. x^{12} **2.** x^{10} **3.** t^{10} **4.** t^9 **5.** a^6b^4 **6.** a^7b^4 **7.** $w^{13}y^8z^6$ **8.** $w^{10}y^9z^8$ **9.** x^5y^5 **10.** x^6y^5
11. $-14a^3b^2$ **12.** $-15a^4b^3$ **13.** $12m^5n^7$ **14.** $30m^3n^5$ **15.** $160a^7b^9$ **16.** $-70a^8b^6$ **17.** $-60a^5b^8$
18. $-39a^{13}b^{10}$ **19.** $a^{12}b^6$ **20.** a^8b^{15} **21.** x^3y^{15} **22.** $x^{20}y^4$ **23.** $p^{12}q^{14}r^{24}$ **24.** $p^{15}q^{21}r^{16}$ **25.** $-5a^{14}b^{11}$
26. $-2a^7b^{23}$ **27.** $125p^3$ **28.** $49p^2$ **29.** $-32a^5b^5$ **30.** $10{,}000a^4b^4$ **31.** $9x^8$ **32.** $125x^9$ **33.** $64a^6b^{15}c^3$
34. $81a^4b^{16}c^8$ **35.** $-32x^5y^{10}$ **36.** $-27x^3y^{12}$ **37.** $72x^{13}$ **38.** $-128x^{13}$ **39.** $-1{,}125a^8b^7c^6$ **40.** $-972a^{14}b^7c^4$
41. $\dfrac{k^5}{32}$ **42.** $\dfrac{k^3}{1{,}000}$ **43.** $\dfrac{s^4t^4}{625}$ **44.** $\dfrac{-s^5t^5}{243}$ **45.** $\dfrac{27p^3}{8}$ **46.** $\dfrac{4p^2}{25}$ **47.** $\dfrac{x^8y^4}{z^{12}}$ **48.** $\dfrac{x^3y^9}{z^6}$ **49.** $\dfrac{4p^{13}}{q^8}$ **50.** $\dfrac{9p^{17}}{q^{11}}$
51. $\dfrac{a^{16}b^8c^{15}}{80{,}000}$ **52.** $\dfrac{a^7b^{17}c^6}{576}$ **53.** $6u^2v + 15uv^2$ **54.** $-42u^2v + 63uv^2$ **55.** $12x^4y - 6x^3y^2 + 10x^3y^3$
56. $6x^3y^2 + 21x^2y^3 - 27x^2y^4$ **57.** $15s^3t^4 - 30s^2t^2 + 20st^4$ **58.** $24s^3t^2 + 42s^2t^2 - 30st^3$
59. $21m^8n - 12m^6n^2 + 3m^4n^3 - 18m^3n^4$ **60.** $10m^6n^3 + 20m^4n^5 - 30m^2n^7 - 35m^2n^9$ **61.** $3p^6 - 5p^5 + 14p^4 - 2p^3$

62. $2p^9 + p^7 - 7p^5 + 10p^4$ **63.** $-6a^5b^5 + 4a^4b^3 + 7a^4b^2 - 9a^3b^2$ **64.** $-a^5b^7 - 5a^4b^5 + 10a^3b^3 + 7a^2b^4$

65. $-8x^6y^3 + \dfrac{29}{2}x^4y^6 - 12x^3y^7$ **66.** $-5x^7y^2 - \dfrac{20}{3}x^3y^4 + 12x^2y^6$

Section 3-3. Ten Review Exercises

1. $5t^3 - 10t^2$ **2.** $-3t - 10t^2 + 5t^3$ **3.** $10t^3 - 20t^2 - 3t$ **4.** $3t$ **5.** $2a^5 - a^3b + 3a^2b^2 - 5b^3$
6. $-5b^3 + 3a^2b^2 - a^3b + 2a^5$ **7.** 5   **8.** 4   **9.** $-2a^5 + a^3b - 3a^2b^2 + 5b^3$ **10.** Polynomial of four terms

Section 3-3. Supplementary Exercises

1. a. 32   **b.** 12   **c.** 36   **2. a.** 243   **b.** 36   **c.** 216   **3. a.** -128 **b.** 8   **c.** 24   **4. a.** -243 **b.** -18
c. 36   **5. a.** $13x^2$ **b.** $89x^2$ **c.** $1{,}600x^4$ **6. a.** $5s^2$ **b.** $13s^2$ **c.** $36s^4$ **7. a.** $-132p^2$ **b.** $-p$ **8. a.** $15q^8$
b. $8q^4$ **9.** 2^{4x} **10.** 5^{3p} **11.** a^{2n} **12.** b^{6n} **13.** $r^{2t}s^{3t}$ **14.** $a^{6t}b^{4t}$ **15.** $\dfrac{p^{2xz}}{q^{yz}}$ **16.** $\dfrac{m^{3xz}}{n^{6yz}}$

Section 3-4. Practice Exercises

1. $6a^2 + 7a + 2$ **2.** $5a^2 + 13a + 6$ **3.** $30s^2 + 7s - 15$ **4.** $6t^2 + 23t - 18$ **5.** $8m^2 - 2mn - 15n^2$
6. $14x^2 + xy - 3y^2$ **7.** $2a^2 - 11ab + 5b^2$ **8.** $18c^2 - 33cd + 14d^2$ **9.** $2x^3 + 6x^2 + x + 3$ **10.** $4x^3 + x^2 + 4x + 1$
11. $2y^3 + 9y^2 + 15y + 9$ **12.** $18x^3 + 17x^2 + 49x + 20$ **13.** $x^3 + x^2y - xy^2 + 2y^3$ **14.** $3a^3 - 2a^2b + 2ab^2 + b^3$
15. $4a^3 - 8a^2b + 5ab^2 - b^3$ **16.** $a^3 - 3a^2b - 27ab^2 - 27b^3$ **17.** $x^4 + 2x^3 + 3x^2 + 2x + 1$ **18.** $a^4 - 2a^3 + 3a^2 - 2a + 1$
19. $3y^4 - 7y^3 - 2y^2 - y + 1$ **20.** $10y^4 + 3y^3 - 5y^2 + 5y - 6$ **21.** $2a^5 + a^4 + 4a^2 - 3a$
22. $6b^5 + 10b^4 - 5b^3 + b^2 + 11b - 2$ **23.** $-9x^4 - 12x^3 + 14x^2 + 12x - 5$ **24.** $4y^4 - 9y^2 - 2y + 3$
25. $35y^2 - y - 12$ **26.** $6w^2 + 5w - 21$ **27.** $2z^3 + z^2 - 9z - 15$ **28.** $18y^3 - 7y^2 + 37y - 40$
29. $12p^4 + 5p^3 - 30p^2 - 15p + 8$ **30.** $14m^4 + m^3 + 18m^2 + 19m - 12$ **31.** $8c^4 + 26c^3 - 45c^2 + 6c + 27$
32. $15q^4 + 31q^3 + 10q^2 - 10q - 4$ **33.** $a^5 - a^4 + 2a^3 + 5a^2 - 35a$ **34.** $m^5 + 4m^4 + m^3 + 16m^2 - 12m$
35. $n^5 + 2n^4 + 5n^3 + 22n^2 - 12n - 18$ **36.** $a^5 - 5a^4 - a^3 + 16a^2 - 11a + 2$ **37. a.** $x + 3$ **b.** $x^2 + 3x$
c. 180 in^2 **38. a.** $s + 10$ **b.** $s^2 + 10s$ **c.** 651 bulbs   **39. a.** $n + 2$ **b.** $n^2 + 2n$ **c.** 80 people
40. a. $t + 18$ **b.** $t^2 + 18t$ **c.** 1,855 tiles   **41. a.** $30 - 2x$ **b.** $40 - 2x$ **c.** $1{,}200x - 140x^2 + 4x^3$
d. 1,872 cm^3 **42. a.** $10 - 2x$ **b.** $15 - 2x$ **c.** $150x - 50x^2 + 4x^3$ **d.** 132 cm^3

Section 3-4. Ten Review Exercises

1. 3   **2.** 2   **3.** $2t^3 - 2t^2 + 7$ **4.** 3   **5.** $6t^5 - 15t^4 + 14t^3 - 35t^2$ **6.** 5   **7.** $\dfrac{1}{4}$ **8.** $\dfrac{-1}{2}$ **9.** No roots   **10.** $\dfrac{2}{3}$

Section 3-4. Supplementary Exercises

1. $49x^2 + 42xy + 9y^2$ **2.** $25x^2 - 60xy + 36y^2$ **3.** $a^2 - 4ab + 6a + 4b^2 - 12b + 9$ **4.** $x^2 + 2xy - 6x + y^2 - 6y + 9$
5. a. $10t^2$ **b.** $10t + 2t^2$ **c.** $10 + 7t + t^2$ **6. a.** $18x^2$ **b.** $18x + 6x^2$ **c.** $18 + 9x + x^2$ **7. a.** $21y^2$
b. $21y - 3y^2$ **c.** $21 - 10y + y^2$ **8. a.** $40y^2$ **b.** $40y - 10y^2$ **c.** $40 - 14y + y^2$ **9.** $\dfrac{1}{8}t^3 + \dfrac{1}{48}t^2 + \dfrac{1}{6}t + \dfrac{5}{8}$
10. $\dfrac{1}{9}x^3 + \dfrac{53}{108}x^2 + \dfrac{11}{18}x + \dfrac{2}{9}$ **11.** $1.61u^3 + 2.41u^2 - 7.73u + 1.55$ **12.** $3.9u^3 + 8.16u^2 + 5.37u + 19.8$ **13.** $a^3 - b^3$
14. $a^3 + b^3$ **15.** $a^4 - 16$ **16.** $a^4 - 81$ **17.** $3x^2 + 26x + 35$ **18.** $2x^2 + 11x + 12$

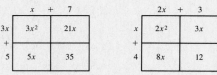

19. $x^2 + 2xy + 7x + 8y + 12$ **20.** $2x^2 + xy + 19x + 7y + 35$ **21.** $-2x^2$ **22.** $6x^2$ **23.** $2x^2 - 20$

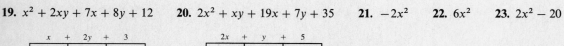

24. $2x^2 + 10$ **25.** $15x - 5$ **26.** $36x^2 + 45$

Section 3-5. Practice Exercises

1. $x^2 + 8x + 15$ **2.** $x^2 + 6x + 5$ **3.** $3z^2 + 14z + 8$ **4.** $2z^2 + 15z + 25$ **5.** $8a^2 + 34a + 21$ **6.** $15a^2 + 28a + 12$
7. $y^2 - 4y - 21$ **8.** $y^2 - 4y - 45$ **9.** $8a^2 + 22a - 21$ **10.** $18a^2 + 23a - 6$ **11.** $15y^2 - 46y + 16$
12. $24y^2 - 46y + 21$ **13.** $18x^3 - 6x^2 + 24x - 8$ **14.** $28x^3 + 4x^2 + 35x + 5$ **15.** $a^3b^3 - 4a^2b^2 + 3ab - 12$
16. $2a^3b^3 + a^2b^2 - 10ab - 5$ **17.** $-4a^2 + 11a - 6$ **18.** $-12a^2 + 20a - 3$ **19.** $-x^3 + 4x^2 - 3x + 12$
20. $9n^3 - 15n^2 + 3n - 5$ **21.** $-a^3 - 2a^2b + 3ab^2 + 6b^3$ **22.** $-6a^3 - 3a^2b + 4ab^2 + 2b^3$ **23.** $-2z^4 - 3z^2 + 20$
24. $4w^4 - 15w^2 - 54$ **25.** $y^2 - 16$ **26.** $x^2 - 36$ **27.** $9x^2 - 1$ **28.** $16y^2 - 1$ **29.** $25z^2 - 4$ **30.** $81b^2 - 16$
31. $49m^2 - 100$ **32.** $16x^2 - 9$ **33.** $64c^2 - 121d^2$ **34.** $4c^2 - 9d^2$ **35.** $16x^6 - 9$ **36.** $49m^6 - 100$
37. $9y^4 - 25z^4$ **38.** $4p^4 - 9q^4$ **39.** $64 - 9a^2b^2$ **40.** $36 - 49x^2y^2$ **41.** $121t^{10} - 49$ **42.** $169s^{12} - 25$
43. $x^2 + 10x + 25$ **44.** $m^2 + 6m + 9$ **45.** $4a^2 + 12a + 9$ **46.** $100c^2 + 180c + 81$ **47.** $36y^2 - 12yz + z^2$
48. $25w^2 - 40wx + 16x^2$ **49.** $16y^6 - 24y^3z + 9z^2$ **50.** $36m^6 - 12m^3n + n^2$ **51.** $a^4 + 6a^2b^2 + 9b^4$
52. $4c^4 + 20c^2d^2 + 25d^4$ **53.** $49m^2n^4 - 140mn^2 + 100$ **54.** $144x^4y^6 - 168x^2y^3 + 49$

Section 3-5. Ten Review Exercises

1. 240 **2.** $6m^2 + 7m - 5$ **3.** 240 **4.** 240 **5.** $6m^2 + 8m + 2$ **6.** 240 **7.** -7 **8.** $10u^6$ **9.** $\dfrac{a^8b^{12}}{81}$
10. $-20r^4s^5$

Section 3-5. Supplementary Exercises

1. $\dfrac{1}{4}x^2 + \dfrac{1}{3}x + \dfrac{1}{9}$ **2.** $\dfrac{1}{9}y^2 + \dfrac{2}{15}y + \dfrac{1}{25}$ **3.** $18z^2 - z - \dfrac{1}{9}$ **4.** $32a^2 + a - \dfrac{1}{16}$ **5.** $0.08b^2 - 0.16b - 0.1$
6. $0.18w^2 - 0.03w - 0.03$ **7.** $1.44m^2 - 5.04m + 4.41$ **8.** $1.69x^2 - 6.5x + 6.25$ **9.** $x^4 - 81$ **10.** $b^4 - 16$
11. $36x^2y^2 + 60xy + 25$ **12.** $81x^2y^2 + 36xy + 4$ **13.** $64y^2 - 80y + 25$ **14.** $49p^2 - 42p + 9$ **15.** $3a^2 + 3a - 18$
16. $2x^2 + 3x - 20$ **17.** $30t^3 - 14t^2 - 8t$ **18.** $14t^4 + 63t^3 - 126t^2$ **19.** $-4k^3 - 36k^2 - 81k$ **20.** $-75k^3 + 30k^2 - 3k$
21. $8b^3 + 12b^2 + 6b + 1$ **22.** $27b^3 - 54b^2 + 36b - 8$ **23.** $a^4 - 200a^2 + 10,000$ **24.** $81a^4 - 450a^2 + 625$ **25. a.** 121
b. 65 **c.** 121 **d.** 121 **e.** Parts **a**, **c**, and **d** **26. a.** 64 **b.** 34 **c.** 64 **d.** 64 **e.** Parts **a**, **c**, and **d**
27. a. 396 **b.** 18 **c.** 22 **d.** $x^2 - 4$ **e.** 396 **f.** Same **g.** $x^2 - 1$ **28. a.** 891 **b.** 27 **c.** 33
d. $x^2 - 9$ **e.** 891 **f.** Same **g.** $x^2 - 4$ **29. a.** 529 **b.** 23 **c.** $x^2 + 6x + 9$ **d.** 529 **e.** Same
f. $x^2 + 4x + 4$ **30. a.** 1024 **b.** 32 **c.** $x^2 + 4x + 4$ **d.** 1024 **e.** Same **f.** $x^2 + 10x + 25$

Section 3-6. Practice Exercises

1. x^4 **2.** $3x^6$ **3.** $5p^3$ **4.** $7p$ **5.** a^3b **6.** a^2b^2 **7.** $x^7y^8z^3$ **8.** $x^6y^4z^7$ **9.** $3p^2q^2$ **10.** $-3p^6r^2$
11. $27t$ **12.** $121s^6t$ **13.** $4x^6$ **14.** $9x^3y^5$ **15.** $36m^4n^7$ **16.** $2m^8n$ **17.** $-7a^2b$ **18.** $-6b^2c$ **19.** $6x^2z$
20. $15yz^3$ **21.** $\dfrac{-22}{3}m^2$ **22.** $\dfrac{-16}{3}n^2$ **23.** $\dfrac{5}{2}p^2r^2$ **24.** $\dfrac{14}{9}q^4r^5$ **25.** $3x^3 + 2$ **26.** $4x^2 + 1$ **27.** $9 - 25y^2$
28. $3y^4 - 2y^2$ **29.** $a^2 + 2a + 1$ **30.** $6a^2 + 4a + 1$ **31.** $b^4 + 5b^2 - 3$ **32.** $b^4 - 3b^2 + 2$ **33.** $x + 3$ **34.** $x + 2$
35. $4a + 3$ **36.** $5a - 3$ **37.** $2y^2 - 3y + 1$ **38.** $3y^2 + y - 2$ **39.** $m^2 - 2mn + 3n^2$ **40.** $10m^2 + 7mn - n^2$
41. $3a - 2$ **42.** $4a + 1$ **43.** $2s^2 - s + 2 + \dfrac{-6}{2s - 1}$ **44.** $3s^2 + 2s - 2 + \dfrac{6}{s - 2}$ **45.** $3y^2 + 2y - 3 + \dfrac{-1}{y + 2}$
46. $3y^2 - 2y - 5 + \dfrac{-9}{2y - 3}$ **47.** $2x^2 - 3x + 4 + \dfrac{-2}{2x - 3}$ **48.** $3x^2 - 4x + 5 + \dfrac{-5}{3x + 5}$ **49.** $u^3 + 4u^2 - 4u - 16$
50. $27u^3 - 18u^2 + 36u - 24$ **51.** $32x^5 - 16x^4 + 8x^3 + 4x^2 - 2x + 1$ **52.** $32x^5 + 16x^4 + 8x^3 - 4x^2 - 2x - 1$
53. $a^3 - 3a^2 + 9a - 27$ **54.** $a^3 + 5a^2 + 25a + 125$

Section 3-6. Ten Review Exercises

1. $2x^3 - x^2 - 13x - 6$ **2.** $2x^3 - x^2 - 13x - 6$ **3.** $x^2 - x - 6$ **4.** $x + 2$ **5.** $(2x + 1)(x - 3)(x + 2)$
6. $21u^2 - 40uv - 21v^2$ **7.** $49u^2 - 9v^2$ **8.** $49u^2 + 42uv + 9v^2$ **9.** $9u^2 - 42uv + 49v^2$ **10.** $81u^4 - 16v^4$

Section 3-6. Supplementary Exercises

1. a. $\dfrac{1}{4}x^2 + \dfrac{5}{4}x + 1$ **b.** $x^2 + x$ **2. a.** $\dfrac{1}{2}x^2 + \dfrac{5}{2}x + 3$ **b.** $x^2 + 3x$ **3. a.** $\dfrac{-1}{3}x^3 - 2x^2 + 9x$ **b.** $x^3 + 9x^2$

4. a. $\dfrac{-1}{2}x^3 - \dfrac{3}{2}x^2 + 5x$ **b.** $x^3 + 5x^2$ **5.** $y^2 + y - 3$ **6.** $2y^2 - y + 2$ **7.** $2b^3 - 2b^2 - b + 1$ **8.** $b^2 + 2b + 5$

9. $3a^2 + 5a - 1 + \dfrac{-2}{2a^2 - a + 3}$ **10.** $5a^2 - a - 2 + \dfrac{-1}{2a^2 + a + 1}$ **11. a.** 36 **b.** 6 **c.** 6 **d.** 3x **e.** 6 **f.** Same

12. a. 135 **b.** 15 **c.** 9 **d.** $3a$ **e.** 9 **f.** Same **13. a.** 12 **b.** $x + 2$ **c.** 12 **d.** Same **14. a.** 12
b. $x + 2$ **c.** 12 **d.** Same **15.** $10x$ **16.** $20x$ **17.** 0 **18.** $-a - c$ **19.** $2y$ **20.** $2y$

Chapter 3 Review Exercises

1. 5 **2.** 3 **3.** $5x^5 + 3x^2 - 8x + 7$ **4.** $m^3 + 3m^2n + 3mn^2 - n^3$ **5.** 4 **6.** 1 **7.** $3x^2 + x - 2$ **8.** $6x^2 - 8xy$

9. $t^5 + t^4 - t^3 + 5t - 7$ **10.** $8t^4 - t^2 - 1$ **11.** $-30u^6$ **12.** $-8u^6v^3$ **13.** $\dfrac{a^8b^{12}}{81}$ **14.** $6a^4b + 3a^3b^2 - 15a^2b^3$

15. $15z^3 + 20z^2 - 3z - 4$ **16.** $12z^3 + 25z^2 + 6z - 8$ **17.** $3m^2 + 10mn + 8n^2$ **18.** $8m^2 - 33mn - 35n^2$ **19.** $9k^2 - 4$
20. $k^4 - 9j^2$ **21.** $169t^2 - 104t + 16$ **22.** $100t^4 + 140t^2 + 49$ **23.** $-8a^2$ **24.** $-3a^9b^5$ **25.** $4y^2 - 3$

26. $\dfrac{4}{5}y^2 - \dfrac{5}{2}y + \dfrac{1}{2}$ **27.** $5b^3 - 4b^2 + b - 3$ **28.** $6b^3 + b^2 + 4b + 2$ **29.** -2 **30.** $-3x + 3$

Chapter 4 Factoring Polynomials

Section 4-1. Practice Exercises

1. $12(2 + 3x)$ **2.** $4(11x^2 - 3)$ **3.** $4(2m - 3n)$ **4.** $5(8m + 7n)$ **5.** $2(27a + 15b - 1)$ **6.** $3(12a - 14b + 1)$
7. $8(x^2 + 15x + 3)$ **8.** $10(x^2 + 25x - 9)$ **9.** $9(z^2 + 4z - 11)$ **10.** $16(5z^2 - z + 9)$ **11.** $11(a^2 - 6b^2)$
12. $7(7x^2 - 2y^2)$ **13.** $10(10a^2 - 20ab + b^2)$ **14.** $6(4a^3 - 2a^2b + ab^2 - 16b^3)$ **15.** $x(5x + 3)$ **16.** $x^2(4x + 9)$
17. $y(7y^2 - 2y + 13)$ **18.** $y^2(y^2 - 10y + 3)$ **19.** $b^3(5 + b^2 - 4b^4 - 2b^6)$ **20.** $b^4(2 - 7b - 3b^2 - b^3)$
21. $a^2(a^6 - a^4 + a^2 - 3)$ **22.** $a^3(5a^6 + 2a^4 - a^2 - 4)$ **23.** $a^2b(5a^2 + 7ab + b^2)$ **24.** $ab^2(3a^2 + 5ab + 3b^2)$
25. $x^2y^2(10x^2 - xy - y^2)$ **26.** $x^3y^3(2x^2 + 3xy - y^2)$ **27.** $m^2n^2(m^6 - 5m^4n^2 + 6m^2n^4 - 2n^6)$
28. $m^3n^3(m^6 + m^4n^2 - m^2n^4 + 3n^6)$ **29.** $uvw(u - v + w)$ **30.** $u^2v^2w^2(u + 2v - 3w)$ **31.** $2x(2x + 3)$ **32.** $3x^2(3x + 4)$
33. $5y^2(3y^2 - 4y + 7)$ **34.** $5y^3(5y^2 - 7y + 12)$ **35.** $3xy^2(25x^2 + 9xy - y^2)$ **36.** $4x^2y^3(x^2 - 36xy + 9y^2)$
37. $4tu(t^2 - 7tu - 10u^2 + 25)$ **38.** $2t^2u^2(3t^2 - 2tu - u^2 + 15)$ **39.** $4x(2x^4 - 10x^3 + 4x^2 + 9x - 3)$
40. $5x(2x^3 - 7x^2 + 5x - 3)$ **41.** $6z(-11z^6 + 6z^4 + 4z^2 - 7)$ **42.** $3z^2(-9z^6 + 4z^4 + 3z^2 - 6)$
43. $10t^3(t^4 + t^3 - 5t^2 + 3t - 7)$ **44.** $9t^3(-2t^4 + 8t^3 + 4t^2 + 6t - 9)$

Section 4-1. Ten Review Exercises

1. $21t^5 - 35t^4 + 63t^2$ **2.** $-15a^3b + 12a^2b^2 - 3ab^3$ **3.** $7t^2(3t^3 - 5t^2 + 9)$ **4.** $3ab(-5a^2 + 4ab - b^2)$
5. $14k^2 - 41k + 15$ **6.** $2x^3 - 5x^2 + 7x + 5$ **7.** $(7k - 3)(2k - 5)$ **8.** $(2x + 1)(x^2 - 3x + 5)$ **9.** 1 **10.** 1

Section 4-1. Supplementary Exercises

1. $-x^2y(9x^2 - 35x - 5)$ **2.** $-x^2y(6y^2 + 7y - 21)$ **3.** $-13a^2(7a^2 - 5a - 1)$ **4.** $17a^3(4a^2 - 5a + 2)$

5. $2x(3a^2 - 10b^2 + 4c^2)$ **6.** $3t(3a + 5b - 7c)$ **7.** $\dfrac{x}{4}(2x - 1)$ **8.** $\dfrac{y}{9}(y - 3)$ **9.** $\dfrac{2}{27}z^2(12z^2 - 5z + 3)$

10. $\dfrac{-1}{100}z^2(3z^2 - 2z - 4)$ **11.** $0.2(3a^2 - 4ab + 7b^2)$ **12.** $0.4ab(2a^2 - ab - 7b^2)$ **13.** $0.05x(x^4 + 5x^2 - 7)$
14. $0.06x^2(x^4 + 7x^2 - 12)$ **15.** $-2x(x^2 + 4x + 2)$ **16.** $-8x^2(2x^2 + 6x + 3)$ **17.** $-1 \cdot (y^2 + 4y + 2)$
18. $-y^3(y^2 + 3y + 5)$ **19. a.** Yes **b.** $x(x - 5)$ **c.** $x(x - 5) = 0$ **d.** Yes **e.** 0 **20. a.** Yes **b.** $y(y + 7)$

c. $y(y + 7) = 0$ **d.** Yes **e.** 0 **21. a.** Yes **b.** $k(5k - 3) = 0$ **c.** Yes **d.** $\dfrac{3}{5}$ **22. a.** Yes **b.** $s(10s - 3) = 0$

c. Yes **d.** $\dfrac{3}{10}$

Section 4-2. Practice Exercises

1. $(y + 7)(y + 11)$ **2.** $(y + 3)(y + 5)$ **3.** $(x - 2)(x - 5)$ **4.** $(x - 6)(x - 7)$ **5.** $(a + 3)(a + 18)$ **6.** $(a + 10)(a + 13)$
7. $(b - 3)(b + 11)$ **8.** $(b - 7)(b + 8)$ **9.** $(t + 2)(t - 19)$ **10.** $(t + 3)(t - 13)$ **11.** $(z - 4)(z - 8)$ **12.** $(z - 3)(z - 4)$
13. $(k + 6)(k - 12)$ **14.** $(k + 5)(k - 13)$ **15.** $(u - 4)(u + 14)$ **16.** $(u - 7)(u + 11)$ **17.** $(m - 2)(m + 25)$
18. $(m - 4)(m + 25)$ **19.** $(y + 3z)(y - 10z)$ **20.** $(y - 3z)(y - 4z)$ **21.** $(a + 6b)(a + 11b)$ **22.** $(a - 3b)(a + 20b)$
23. $(s - 2t)(s - 14t)$ **24.** $(s - 6t)(s + 8t)$ **25.** $(p + 5q)(p + 10q)$ **26.** $(p + 4q)(p + 16q)$ **27.** $3(x + 2)(x + 6)$
28. $4(x + 1)(x + 10)$ **29.** $2(z + 3)(z - 13)$ **30.** $4(z - 2)(z - 8)$ **31.** $5(a - 4b)(a + 12b)$ **32.** $3(a - 4b)(a - 12b)$
33. $10(m + 4)(m - 16)$ **34.** $12(m - 3)(m - 7)$ **35.** $2x(x + 7)(x - 8)$ **36.** $3x(x - 9)(x + 10)$ **37.** $-3y^2(y - 1)(y - 4)$
38. $-2x^4(x + 2)(x + 4)$ **39.** $2uv(u - v)(u + 2v)$ **40.** $3uv(u - 3v)(u + 5v)$ **41.** $-4x^2y^2(x + 2y)(x - 12y)$
42. $-5x^3y^3(x - 2y)(x - 5y)$ **43. a.** $(x + 3)(x + 4)$ **b.** $(x + 2)(x + 6)$ **c.** Cannot be factored. **44. a.** $(x + 3)(x + 5)$
b. Cannot be factored. **c.** $(x + 1)(x + 15)$ **45. a.** $(t + 4)(t - 5)$ **b.** Cannot be factored. **c.** $(t + 1)(t - 20)$
46. a. $(t + 3)(t - 10)$ **b.** $(t + 2)(t - 15)$ **c.** Cannot be factored. **47. a.** $(a - 2)(a + 21)$ **b.** $(a + 1)(a - 42)$

c. Cannot be factored. **48. a.** $(a - 4)(a + 18)$ **b.** $(a + 1)(a - 72)$ **c.** Cannot be factored. **49. a.** $(b + 9)(b + 11)$
b. $(b - 9)(b - 11)$ **c.** Cannot be factored. **50. a.** $(b + 8)(b + 8)$ **b.** $(b - 8)(b - 8)$ **c.** Cannot be factored.

Section 4-2. Ten Review Exercises

1. -11 **2.** 46 **3.** 3 **4.** -6 **5. a.** $112 \cdot \dfrac{5}{8} - 112 \cdot \dfrac{2}{7}$ **b.** 38 **6. a.** $\dfrac{5}{2} \cdot 24 - \dfrac{5}{2} \cdot 16$ **b.** 20 **7.** $8u^9v^3w^6$

8. $\dfrac{-u^7v^5}{108}$ **9.** $-12a^4b^5$ **10.** $6a^9b^7$

Section 4-2. Supplementary Exercises

1. $(x - 5)(x + 8)$ **2.** $(y - 7)(y - 9)$ **3.** Cannot be factored. **4.** Cannot be factored. **5.** $5(x + 2y)(x - 15y)$
6. $4(a - 2b)(a - 8b)$ **7.** $6z(z - 1)(z - 6)$ **8.** $5z(z + 4)(z + 5)$ **9.** $4b(b + 4)(b - 18)$ **10.** $6b(b + 11)(b + 12)$
11. $(r + s)(r - 14s)$ **12.** $(p + 5q)(p + 13q)$ **13. a.** $k = 13, (x + 1)(x + 12)$ **b.** $k = -13, (x - 1)(x - 12)$
c. $k = 8, (x + 2)(x + 6)$ **d.** $k = -8, (x - 2)(x - 6)$ **e.** $k = 7, (x + 3)(x + 4)$ **f.** $k = -7, (x - 3)(x - 4)$
14. a. $k = 11, (x - 1)(x + 12)$ **b.** $k = -11, (x + 1)(x - 12)$ **c.** $k = 4, (x - 2)(x + 6)$ **d.** $k = -4, (x + 2)(x - 6)$
e. $k = 1, (x - 3)(x + 4)$ **f.** $k = -1, (x + 3)(x - 4)$ **15. a.** $k = 8, (y - 1)(y + 9)$ **b.** $k = -8, (y + 1)(y - 9)$
c. $k = 0, (y + 3)(y - 3)$ **16. a.** $k = 24, (y - 1)(y + 25)$ **b.** $k = -24, (y + 1)(y - 25)$ **c.** $k = 0, (y + 5)(y - 5)$
17. a. $k = 32, (z - 1)(z + 33)$ **b.** $k = -32, (z + 1)(z - 33)$ **c.** $k = 8, (z - 3)(z + 11)$ **d.** $k = -8, (z + 3)(z - 11)$
18. a. $k = 54, (z - 1)(z + 55)$ **b.** $k = -54, (z + 1)(z - 55)$ **c.** $k = 6, (z - 5)(z + 11)$ **d.** $k = -6, (z + 5)(z - 11)$
19. a. $k = 5, (a + 1)(a + 4)$ **b.** $k = -5, (a - 1)(a - 4)$ **c.** $k = 4, (a + 2)(a + 2)$ **d.** $k = -4, (a - 2)(a - 2)$
20. a. $k = 10, (a + 1)(a + 9)$ **b.** $k = -10, (a - 1)(a - 9)$ **c.** $k = 6, (a + 3)(a + 3)$ **d.** $k = -6, (a - 3)(a - 3)$
21. $(t^2 + 7)(t^2 + 11)$ **22.** $(t^2 + 3)(t^2 + 5)$ **23.** $(z^3 - 2)(z^3 - 5)$ **24.** $(z^3 - 6)(z^3 - 7)$

Section 4-3. Practice Exercises

1. $(6y + 1)(y^2 + 3)$ **2.** $(y + 9)(3y^2 + 5)$ **3.** $(4b + 7)(6a + 5)$ **4.** $(3b + 5)(6a + 7)$ **5.** $(5x + 1)(xy + 4z)$
6. $(x + 7)(9xy + 2z)$ **7.** $(5t^2 + 8)(6t^3 + 1)$ **8.** $(2t + 3)(8t^4 + 1)$ **9.** $(7u + 8v)(2u + 3)$ **10.** $(u + 1)(18u + 5v)$
11. $(3c + 8d)(3a + 5b)$ **12.** $(c + 7d)(8a + 3b)$ **13.** $(7m + 4)(2n + 7)$ **14.** $(2n + 7)(m + 4)$ **15.** $(4y + 7)(4x + 3)$
16. $(y + 3)(2x + 9)$ **17.** $(3z - 1)(4z^2 - 3)$ **18.** $(z - 2)(10z^2 - 7)$ **19.** $(2s + 1)(8s - 11t)$ **20.** $(s + 2)(3s - 13t)$
21. $(4c - d)(8a - 3b)$ **22.** $(3c - 2d)(8a - b)$ **23.** $(r + 7t)(3r - 5)$ **24.** $(6r + 5t)(5r - 6)$ **25.** $6x(15x + 1)(2x^2 + 1)$
26. $2x^2(x + 2)(9x^2 + 5)$ **27.** $8xy(y + 3)(2x - 1)$ **28.** $24xy(y + 1)(x - 2)$ **29.** $16abc^2(3b + 4)(2a + 5)$
30. $2a(3b + d)(12a + c)$ **31.** $2y(x + 3)(x - 7y)$ **32.** $2z(x + 1)(16x - y)$

Section 4-3. Ten Review Exercises

1. $40 - 8k$ **2.** 0 **3.** 5 **4.** $9a^2 - 2$ **5.** 79 **6.** $10x^2 - 19x - 15$ **7.** $25x^2 - 9$ **8.** $25x^2 + 30x + 9$
9. $-12x^3 + 60x^2 - 75x$ **10.** $2x^3 - 2x^2 - x + 1$

Section 4-3. Supplementary Exercises

1. $(x + 3)(8x + 1)$ **2.** $(4x + 3)(5x + 7)$ **3.** $(3y + 7)(8y + 1)$ **4.** $(5y + 6)(4y + 5)$ **5.** $(6z + 7)(z - 1)$
6. $(2z + 5)(7z - 1)$ **7.** $(a - 4)(5a - 7)$ **8.** $(3a - 4)(2a - 1)$ **9.** $(a - 4b)(14a - 3b)$ **10.** $(a - 4b)(6a - b)$
11. $(x + 2z + 5)(x + y)$ **12.** $(x + y + 3)(x + a)$ **13.** $(x + 3y + 1)(x + 3z)$ **14.** $(x + 4y + 5)(x + z)$
15. $(m + 2n + 6p)(m - 4)$ **16.** $(m + 4n + 5p)(m - 5)$ **17.** $(s + 4t + 6)(4t - u)$ **18.** $(s + 3t + 2)(4t - 5u)$ **19. a.** 1
b. $(2t + 7)(t^2 - 8)$ **c.** 1 **20. a.** -4 **b.** $(5t + 9)(3t^2 - 8)$ **c.** -4 **21. a.** -3 **b.** $(3c + 2d)(4a - 3b)$ **c.** -3
22. a. -10 **b.** $(2c - 5d)(5a + 2b)$ **c.** -10

Section 4-4. Practice Exercises

1. $(4x + 1)(3x + 7)$ **2.** $(5x + 1)(x + 7)$ **3.** $(4y - 3)(2y - 5)$ **4.** $(3y - 5)(4y - 3)$ **5.** $(4r - 9)(5r + 1)$
6. $(2r - 9)(4r + 7)$ **7.** $(5s - 9)(3s + 7)$ **8.** $(5s + 1)(3s - 7)$ **9.** $(a - 2)(3a - 5)$ **10.** $(a - 2)(2a - 5)$
11. $(3b - 1)(b - 2)$ **12.** $(4b - 3)(b - 6)$ **13.** $(5t - 1)(t - 7)$ **14.** $(2t - 1)(t - 7)$ **15.** $(3u - 1)(2u - 3)$
16. $(3u - 1)(3u - 5)$ **17.** $(3m - 7)(2m + 5)$ **18.** $(m - 2)(5m + 3)$ **19.** $(3n + 10)(n - 1)$ **20.** $(n + 6)(5n - 2)$
21. $(5w - 4)(6w + 1)$ **22.** $(4w - 15)(2w + 1)$ **23.** $(2y - 5)(2y + 3)$ **24.** $(y - 2)(5y + 4)$ **25.** $2(2a + 5)(3a - 2)$
26. $3(a - 6)(4a - 1)$ **27.** $3b(3b - 5)(2b + 1)$ **28.** $2b(2b + 5)(4b + 3)$ **29.** $3y^2(y - 4)(2y + 3)$ **30.** $5y(y - 2)(3y - 4)$
31. $6x^3(2x - 3)(x + 1)$ **32.** $3x^2(x - 1)(2x - 3)$ **33.** $(3u + 2v)(2u - v)$ **34.** $(u - 4v)(5u - v)$ **35.** $(3a + 2b)(4a - b)$
36. $(5a - 3b)(2a - b)$ **37.** $(5r + s)(4r + s)$ **38.** $(3r - s)(6r - s)$ **39.** $(x + 6y)(2x - 9y)$ **40.** $(3x - 8y)(x + 5y)$
41. $(2t + 1)(t - 6)$ **42.** $(3t - 1)(t + 4)$ **43.** $(5u + 2)(2u - 1)$ **44.** $(6u + 1)(2u - 3)$ **45.** $(3v + 2)(2v + 5)$
46. $(4v + 1)(2v + 3)$ **47.** $(5y - 2)(3y - 1)$ **48.** $(7y - 2)(2y - 3)$ **49.** $ab(3a + 10b)(a - b)$ **50.** $ab(5a - 2b)(a + 6b)$

Section 4-4. Ten Review Exercises

1. 88   **2.** $8t + 48$ **3.** 88   **4.** -6 **5.** $t < -6$ **6.** $24k^2 - 31k - 15$

7. $24k^2 - 32k$ **8.** $k - 15$ **9.** -5 **10.** 15

Section 4-4. Supplementary Exercises

1. $(y + 8)(4y - 7)$ **2.** $(2y + 3)(3y - 4)$ **3.** $(2a - 9)(5a - 3)$ **4.** $(a - 6)(5a - 6)$ **5.** $(b + 10)(5b - 6)$
6. $(b + 6)(5b - 7)$ **7.** $(2t + 9)(2t - 5)$ **8.** $(t - 8)(2t + 3)$ **9.** $(m + 4n)(4m - 3n)$ **10.** $(m - 2n)(6m + 5n)$ **11. a.** 24
b. $(\pm 1)(\pm 24), (\pm 2)(\pm 12), (\pm 3)(\pm 8), (\pm 4)(\pm 6)$ **c.** $\pm 25, \pm 14, \pm 11, \pm 10$ **d.** Cannot be factored.   **12. a.** 36
b. $(\pm 1)(\pm 36), (\pm 2)(\pm 18), (\pm 3)(\pm 12), (\pm 4)(\pm 9), (\pm 6)(\pm 6)$ **c.** $\pm 37, \pm 20, \pm 15, \pm 13, \pm 12$ **d.** Cannot be factored.
13. a. -14 **b.** $(1)(-14), (-1)(14), (2)(-7), (-2)(7)$ **c.** $-13, 13, -5, 5$ **d.** Cannot be factored.   **14. a.** -18
b. $(1)(-18), (-1)(18), (2)(-9), (-2)(9), (3)(-6), (-3)(6)$ **c.** $-17, 17, -7, 7, -3, 3$ **d.** Cannot be factored.
15. a. $2x, 3, x, 6$ **b.** $(2x + 3)(x + 6)$ **16. a.** $x, 4, 3x, 1$ **b.** $(x + 4)(3x + 1)$ **17. a.** $x, 2, 3x, 10$
b. $(x + 2)(3x + 10)$ **18. a.** $3x, 8, x, 7$ **b.** $(3x + 8)(x + 7)$ **19. a.** $2x, 7, x, 4$ **b.** $(2x + 7)(x + 4)$
20. a. $x, 4, 3x, 8$ **b.** $(x + 4)(3x + 8)$

Section 4-5. Practice Exercises

1. $(4x + 3)(4x - 3)$ **2.** $(2x + 7)(2x - 7)$ **3.** $(6y + 1)(6y - 1)$ **4.** $(7y + 1)(7y - 1)$ **5.** $(9u + 10)(9u - 10)$
6. $(5u + 11)(5u - 11)$ **7.** $(2p + 9)(2p - 9)$ **8.** $(8p + 5)(8p - 5)$ **9.** $(3a + 4b)(3a - 4b)$ **10.** $(2a + 5b)(2a - 5b)$
11. $2(5m + 7n)(5m - 7n)$ **12.** $2(6m + n)(6m - n)$ **13.** $y(11x + 14y)(11x - 14y)$ **14.** $y^2(13x + 8y)(13x - 8y)$
15. $18pq(p + 3q)(p - 3q)$ **16.** $2pq(5p + 19q)(5p - 19q)$ **17.** $(x + 4)^2$ **18.** $(x + 10)^2$ **19.** $(y - 7)^2$ **20.** $(y - 6)^2$
21. $(3m + 7)^2$ **22.** $(9m + 2)^2$ **23.** $(7a - 3)^2$ **24.** $(8a - 5)^2$ **25.** $(2m - 3n)^2$ **26.** $(5m + n)^2$ **27.** $4(k - 3)^2$
28. $5(2k - 1)^2$ **29.** $6(2x - 5y)^2$ **30.** $4(3x + y)^2$ **31.** $-10(6a - 1)^2$ **32.** $-9(5a - 2)^2$ **33.** $-uv(6u + 11v)^2$
34. $-uv(2u - 7v)^2$ **35.** $(x - 2)(x^2 + 2x + 4)$ **36.** $(5x - 3)(25x^2 + 15x + 9)$ **37.** $(3y + 2)(9y^2 - 6y + 4)$
38. $(y + 7)(y^2 - 7y + 49)$ **39.** $(2t - 5)(4t^2 + 10t + 25)$ **40.** $(3t - 4)(9t^2 + 12t + 16)$ **41.** $(6w + 1)(36w^2 - 6w + 1)$
42. $(9w + 10)(81w^2 - 90w + 100)$ **43.** $4(x - 5)(x^2 + 5x + 25)$ **44.** $5(x - 4)(x^2 + 4x + 16)$
45. $x(x + 3y)(x^2 - 3xy + 9y^2)$ **46.** $2y^2(y + 6z)(y^2 - 6yz + 36z^2)$ **47.** $2st(5s + 2t)(25s^2 - 10st + 4t^2)$
48. $2s^2t(4s + 3t)(16s^2 - 12st + 9t^2)$ **49.** $5(2x^2 - y^2)(4x^4 + 2x^2y^2 + y^4)$ **50.** $2(5x^3 - y)(25x^6 + 5x^3y + y^2)$

Section 4-5. Ten Review Exercises

1. $20k^3$ **2.** $5k^3 + 20k^2$ **3.** $k^3 + 4k^2 + 5k + 20$ **4.** $49t^2$ **5.** $t^2 + 14t + 49$ **6.** $-9x^4$ **7.** $x^4 - 9$
8. $(8y + 1)(8y - 1)$ **9.** $(4y - 1)(16y^2 + 4y + 1)$ **10.** $(8y - 1)^2$

Section 4-5. Supplementary Exercises

1. $(4x^2 + 3)(4x^2 - 3)$ **2.** $(2x^2 + 7)(2x^2 - 7)$ **3.** $(17u - 1)^2$ **4.** $(15u + 1)^2$ **5.** $(9 + 10b)(9 - 10b)$
6. $(15 + 4b)(15 - 4b)$ **7.** $(6ab + 1)(6ab - 1)$ **8.** $(7ab + 10)(7ab - 10)$ **9.** $(a + 5b)^2$ **10.** $(2a - b)^2$
11. $(9u^3 + 10)(9u^3 - 10)$ **12.** $(5u^3 + 11)(5u^3 - 11)$ **13.** $v^3(4v - 5)^2$ **14.** $v^3(8v + 1)^2$ **15.** $3x^2y(8x + 1)(8x - 1)$
16. $5xy^2(7x + 1)(7x - 1)$ **17.** $(3m - 5n)(9m^2 + 15mn + 25n^2)$ **18.** $(2m - 7n)(4m^2 + 14mn + 49n^2)$
19. $3a(4a + 3)(16a^2 - 12a + 9)$ **20.** $2a(7 + 8a)(49 - 56a + 64a^2)$ **21.** $\left(\dfrac{t}{2} + \dfrac{1}{3}\right)\left(\dfrac{t}{2} - \dfrac{1}{3}\right)$ **22.** $\left(\dfrac{3t}{5} + \dfrac{1}{6}\right)\left(\dfrac{3t}{5} - \dfrac{1}{6}\right)$
23. $\left(\dfrac{u}{4} + \dfrac{2v}{3}\right)\left(\dfrac{u}{4} - \dfrac{2v}{3}\right)$ **24.** $\left(\dfrac{2u}{7} + \dfrac{v}{10}\right)\left(\dfrac{2u}{7} - \dfrac{v}{10}\right)$ **25.** $(0.3w + 0.5)(0.3w - 0.5)$ **26.** $(0.2w + 0.9)(0.2w - 0.9)$
27. $(k^2 + 9)(k + 3)(k - 3)$ **28.** $(k^2 + 4)(k + 2)(k - 2)$ **29.** $(p^2 + 25)(p + 5)(p - 5)$ **30.** $(p^2 + 16)(p + 4)(p - 4)$
31. $(9q^2 + 4)(3q + 2)(3q - 2)$ **32.** $(25q^2 + 16)(5q + 4)(5q - 4)$ **33.** $(x^2 + y^2)(x + y)(x - y)$
34. $(x^2y^2 + z^2)(xy + z)(xy - z)$ **35.** -6 or 6 **36.** -60 or 60 **37.** 25   **38.** 1   **39.** 100   **40.** 169
41. a. The subtraction of 25   **b.** Change to addition   **c.** $(4m - 5)^2$ **42. a.** The subtraction of 4
b. Change to addition   **c.** $(3m + 2)^2$ **43. a.** The factor 7   **b.** Change 7 to 49.   **c.** $(7n - 10)^2$
44. a. The term 5   **b.** Change 5 to 25.   **c.** $(6n + 5)^2$ **45. a.** The factor 33   **b.** Change 33 to 66.   **c.** $(11a + 3b)^2$
46. a. The factor 15   **b.** Change 15 to 30.   **c.** $(15a - b)^2$

Review Exercises for Sections 4-1 through 4-5

1. $3x(x^2 + 5)$ **2.** $2x^2(3x - 5)$ **3.** $2ab(a - b + 3)$ **4.** $3a^2b(2a + b - 1)$ **5.** $(x + 6)(x - 6)$ **6.** $(3x + 5)(3x - 5)$
7. $(5y - 1)(25y^2 + 5y + 1)$ **8.** $(2y - 3z)(4y^2 + 6yz + 9z^2)$ **9.** $(3a + 4)^2$ **10.** $(5a - 2)^2$ **11.** $(x - 4y)^2$
12. $(2x - 5y)^2$ **13.** $(m + 3n)(m + 5n)$ **14.** $(m + n)(m - 2n)$ **15.** $(k + 7)(k^2 - 8)$ **16.** $(k - 4)(k^3 - 6)$
17. $(x - 1)(2x - 3)$ **18.** $(x - 5)(3x + 2)$ **19.** $(y - 1)(3y + 5)$ **20.** $(y - 8)(2y - 1)$ **21.** $6(x^2 + xy + 5y^2)$

22. $xy(x^2 - xy - 7y^2)$ **23.** $(10z + 3)(10z - 3)$ **24.** $(4 + 5z)(4 - 5z)$ **25.** $(2a - 7)^2$ **26.** $(a + 10)^2$
27. $(b - 2)(b + 9)$ **28.** $(b - 4)(b - 7)$ **29.** $(c + 2)(5c + 3)$ **30.** $(3c - 1)(2c - 1)$ **31.** $6(m + 5)(m^2 - 5m + 25)$
32. $12n(m + 2n)(m^2 - 2mn + 4n^2)$ **33.** $(2b - 9)(3a + 5)$ **34.** $(7a - 4)(2a + t)$ **35.** $(2x + 7y)^2$ **36.** $(5x - 3y)^2$
37. $(a - 4)(a + 10)$ **38.** $(a - 5)(a - 6)$ **39.** $(m - 8)(6m + 1)$ **40.** $(4m + 3)(m + 2)$ **41.** $(11a + 5b)(11a - 5b)$
42. $(13a + 2b)(13a - 2b)$ **43.** $8(x^2 + x + 3)$ **44.** $6x(2x^2 + x + 1)$ **45.** $(5y - 6)^2$ **46.** $(2y + 9)^2$
47. $(z - 4)(3z + 2)$ **48.** $(2z + 3)(3z + 2)$ **49.** $(x - 5y)(x + 3y)$ **50.** $(x + 2y)(x + 7y)$ **51.** $4a(a - 4)(a^2 + 4a + 16)$
52. $20a^2(a + 2)(a^2 - 2a + 4)$ **53.** $(2a + 3)^2$ **54.** $(2a - 3)^2$ **55.** $(4a - 9)(a + 1)$ **56.** $(4a - 9)(a - 1)$
57. $(2a - 9)(2a + 1)$ **58.** $(a - 3)(4a + 3)$ **59.** $(a + 9)(4a + 1)$ **60.** $(a - 9)(4a + 1)$ **61.** $(2y - 3)(9y^2 + 4)$
62. $(y + 5)(y^2 - 8)$ **63.** $(3y + 4)(3y + 1)$ **64.** $(3y - 4)(3y + 1)$ **65.** $(y + 2)(9y + 2)$ **66.** $(y - 2)(9y + 2)$
67. $(3y - 2)^2$ **68.** $(3y + 2)^2$ **69.** $(9y - 4)(y + 1)$ **70.** $(y - 4)(9y - 1)$ **71.** $2(x + 6yz)(x - 6yz)$
72. $3(2x + 5y^2)(2x - 5y^2)$ **73.** $5(a + 10bc)(a^2 - 10abc + 100b^2c^2)$ **74.** $2a(a + 3b^2)(a^2 - 3ab^2 + 9b^4)$ **75.** $3(x - 5y)^2$
76. $5(2x + 3y)^2$ **77.** $2mn(m + 2n)^2$ **78.** $3m^2(2m - 3n)^2$ **79.** $5k(k + 1)(k + 2)$ **80.** $7k^3(k + 5)(k - 3)$ **81.** $a(3a + b)^2$
82. $ab(2a - 3b)^2$ **83.** $x(6x + 5)(6x - 5)$ **84.** $xy(5x + 7y)(5x - 7y)$ **85.** $3y^2(y^2 + y + 3)$ **86.** $5xy(x^2 + xy + y^2)$
87. $z(z - 10)(z + 7)$ **88.** $z^2(2z - 1)(z + 3)$ **89.** $2m(m - 2)(m - 5)$ **90.** $3m(m + 6)(m - 3)$ **91.** $3(3k + 4)(2k^2 + 1)$
92. $5(5k - 8)(3k^2 + 1)$ **93.** $5x(4x + 3)(4x - 3)$ **94.** $3x^2(2x + 7)(2x - 7)$ **95.** $4y(y + 7)(y - 1)$ **96.** $4y(3y + 7)(y - 2)$
97. $2z^2(2z + 1)(z - 6)$ **98.** $3z^2(3z + 1)(z - 4)$ **99.** $10xy(2x - 3y)^2$ **100.** $8xy(2x + 3y)(2x - 3y)$

Section 4-6. Practice Exercises

1. 0 **2.** 0 **3.** 0, 3 **4.** 0, 2 **5.** $-1, 5$ **6.** $-7, 3$ **7.** $\frac{2}{5}, -6$ **8.** $\frac{5}{3}, 4$ **9.** $\frac{-3}{2}, \frac{1}{2}$ **10.** $\frac{2}{3}, \frac{-1}{4}$

11. 2, 4 **12.** $-3, -6$ **13.** 6, -9 **14.** 10, -5 **15.** 8, -7 **16.** $-1, -36$ **17.** 10, -10 **18.** 7, -7

19. 0, 7 **20.** 0, -2 **21.** 2, 25 **22.** 4, 16 **23.** $\frac{-3}{2}, 2$ **24.** $\frac{2}{3}, 2$ **25.** $\frac{3}{4}, 1$ **26.** $\frac{-1}{3}, 3$ **27.** $\frac{5}{3}, 5$

28. $\frac{-4}{3}, -1$ **29.** $\frac{5}{4}, 2$ **30.** $\frac{5}{2}, 4$ **31.** $0, \frac{9}{5}$ **32.** $0, \frac{-7}{3}$ **33.** 7 **34.** 8 **35.** -5 **36.** -6 **37.** $\frac{-3}{2}$

38. $\frac{-5}{3}$ **39.** $\frac{1}{6}$ **40.** $\frac{7}{4}$ **41.** $\frac{-5}{2}, \frac{5}{2}$ **42.** $\frac{-10}{3}, \frac{10}{3}$ **43.** 0, 3 **44.** 0, 11 **45.** $\frac{-1}{4}, \frac{1}{4}$ **46.** $\frac{-1}{7}, \frac{1}{7}$

Section 4-6. Ten Review Exercises

1. $2t^2 + 11t + 5$ **2.** $t + 5$ **3.** $6t^2 - t - 2$ **4.** $6t^3 + 29t^2 - 7t - 10$ **5.** $(t + 5)(3t - 2)(2t + 1)$ **6.** 140 **7.** 140
8. 0 **9.** 0 **10.** $-5, \frac{2}{3}, \frac{-1}{2}$

Section 4-6. Supplementary Exercises

1. $\frac{3}{7}, \frac{8}{3}$ **2.** $\frac{-9}{5}, \frac{5}{9}$ **3.** No rational number solutions. **4.** No rational number solutions. **5.** 0, 5 **6.** $0, \frac{-13}{2}$

7. $\frac{-1}{10}, \frac{1}{10}$ **8.** $\frac{-3}{8}, \frac{3}{8}$ **9.** No rational number solutions. **10.** No rational number solutions. **11.** $\frac{4}{5}$ **12.** $\frac{-2}{3}$

13. $\frac{-4}{3}, \frac{4}{3}$ **14.** $\frac{-11}{2}, \frac{11}{2}$ **15.** $-7, 13$ **16.** 12, 15 **17.** $-3, -2, 1$ **18.** $-4, 3, 5$ **19.** $-6, 0, 7$

20. $-8, 0, 10$ **21.** $-3, \frac{-4}{3}, 3$ **22.** $\frac{-5}{2}, \frac{3}{2}, \frac{5}{2}$ **23.** -4 **24.** -3 **25. a.** $(x - 2)(x^2 + 2x + 4) = 0$

b. $x - 2 = 0$ or $x^2 + 2x + 4 = 0$ **c.** 2 **26. a.** $(x + 3)(x^2 - 3x + 9) = 0$ **b.** $x + 3 = 0$ or $x^2 - 3x + 9 = 0$

c. -3 **27. a.** $(2y + 5)(4y^2 - 10y + 25) = 0$ **b.** $2y + 5 = 0$ or $4y^2 - 10y + 25 = 0$ **c.** $\frac{-5}{2}$

28. a. $(3y - 10)(9y^2 + 30y + 100) = 0$ **b.** $3y - 10 = 0$ or $9y^2 + 30y + 100 = 0$ **c.** $\frac{10}{3}$ **29.** $-2, 3$ **30.** $-5, 4$

31. $0, \frac{2}{3}$ **32.** $\frac{1}{2}, 2$

Section 4-7. Practice Exercises

1. 5 and 12 **2.** -5 and -17 **3.** -7 and 10 **4.** -4 and -9, or 4 and 9 **5.** -8 and 7, or -7 and 8
6. 3 and 18, or -3 and -18 **7.** 6 and 8 feet **8.** 10 and 24 inches **9.** 8 and 15 meters **10.** 16 and 30 inches
11. 7 and 24 feet **12.** 14 and 48 feet **13.** 8 and 12 inches **14.** 9 and 18 meters **15.** 12 and 25 feet

16. 10 and 15 centimeters   **17.** 5 and 17 centimeters   **18.** 13 and 20 yards   **19.** 7 and 10 centimeters
20. 4 and 14 feet   **21.** 12 and 15 meters   **22.** 16 and 25 yards   **23.** 8 and 24 inches   **24.** 3 and 8 miles

Section 4-7. Ten Review Exercises

1. -4 **2.** 5   **3.** -2 **4.** -2 and 5   **5.** No roots   **6.** All real numbers   **7.** $y = \dfrac{12 - 15x}{7}$

8. $a < \dfrac{3}{5}$,   **9.** $2.65 for hand towels, $5.89 for bath towels

10. Truck traveled 4 hours; car traveled 3 hours

Section 4-7. Supplementary Exercises

1. 7 and 8, or -7 and -8 **2.** 11 and 12, or -11 and -12 **3.** 9 and 11, or -9 and -11
4. 10 and 12, or -10 and -12 **5.** 11, 12, and 13; or -11, -12, and -13 **6.** 14, 15, and 16; or -14, -15, and -16
7. a. 5, 6, and 7; or -3, -2, and -1 **b.** Odd   **8. a.** 8, 9, and 10; or -10, -9, and -8 **b.** Even
9. 7, 9, and 11; or -5, -3, and -1 **10.** 6, 8, and 10; or -4, -2, and 0   **11.** 11 feet and 16 feet
12. 6 centimeters and 11 centimeters or 14 centimeters and 19 centimeters

Chapter 4 Review Exercises

1. $8m(m^2 - 2m + 4)$ **2.** $9mn^2(4m^2 + mn - 7n^2)$ **3.** $(x - 12)(x + 2)$ **4.** $(x - 18)(x + 7)$ **5.** $a^2(a - 9)(a + 16)$
6. $-5a(a + 7)(a + 8)$ **7.** $(2t + 1)(3t + u)$ **8.** $(2u + 3)(5t - 1)$ **9.** $(y - 2)(5y + 3)$ **10.** $(4y - 3)(3y - 5)$
11. $(3k + 5)(3k - 7)$ **12.** $(3k - 5)(3k - 10)$ **13.** $2(3x + 5y)(3x - y)$ **14.** $(7x + y)(7x - y)$ **15.** $(7m - 3n)^2$
16. $(7m + 5n)(49m^2 - 35mn + 25n^2)$ **17.** $3pq(8p + q)(8p - q)$ **18.** $(2p + 15q)^2$ **19.** $3(uv + 5)^2$
20. $-5uv(3uv + 2)(uv - 5)$ **21.** $\dfrac{-5}{3}, \dfrac{5}{2}$ **22.** 3, -9 **23.** $\dfrac{-5}{12}, \dfrac{5}{12}$ **24.** $\dfrac{-3}{5}, \dfrac{1}{2}$ **25.** $\dfrac{-7}{5}$, 4   **26.** 0, 5
27. The integers are -8 and 5.   **28.** 15 and 20 meters   **29.** 16 by 25 feet   **30.** 7 and 20 centimeters

Chapter 5   Rational Expressions

Section 5-1. Practice Exercises

1. $\dfrac{4}{5}$ **2.** $\dfrac{7}{2}$ **3.** 4   **4.** -1 **5.** $\dfrac{-1}{2}$ **6.** $\dfrac{-2}{3}$ **7.** 5   **8.** $\dfrac{-1}{2}$ **9.** 7   **10.** -35 **11.** $\dfrac{11}{9}$ **12.** -4 **13.** $\dfrac{9}{7}$

14. 25   **15.** -1 **16.** -1 **17.** 3   **18.** 4   **19.** $\dfrac{3}{2}$ **20.** -2 **21.** -11 **22.** -5 **23.** -1 **24.** 1

25. 0   **26.** 0   **27.** 6   **28.** -5 **29.** $\dfrac{-5}{3}$ **30.** $\dfrac{9}{2}$ **31.** 0 and 3   **32.** 0 and -8 **33.** -4 and -5

34. -4 and 5   **35.** $\dfrac{-9}{2}$ and $\dfrac{9}{2}$ **36.** $\dfrac{-5}{3}$ and $\dfrac{5}{3}$ **37.** No restricted values   **38.** No restricted values

39. $n \neq -3m$ and $3n \neq 2m$ **40.** $n \neq 5m$ and $3n \neq -2m$ **41.** $\dfrac{2y + 1}{3}$ **42.** $\dfrac{y + 2}{3}$ **43.** $a - 2$ **44.** $a + 5$

45. $\dfrac{x + 3}{x - 4}$ **46.** $\dfrac{x - 2}{x + 1}$ **47.** -1 **48.** -1 **49.** $\dfrac{3y + 1}{y + 3}$ **50.** $\dfrac{y + 5}{5y + 1}$ **51.** $\dfrac{3a + 4}{4a + 3}$ **52.** $\dfrac{2a - 1}{a - 2}$ **53.** $\dfrac{a + 2}{3a}$
54. $\dfrac{3a^3}{a - 5}$ **55.** $\dfrac{t^2 + 3t + 9}{t + 3}$ **56.** $\dfrac{4t^2 - 10t + 25}{2t + 5}$ **57.** -1 **58.** -1 **59.** $\dfrac{-1 \cdot (z + 3)}{2z + 5}$ **60.** $\dfrac{z - 5}{3z + 4}$

Section 5-1. Ten Review Exercises

1. $3b + 1$ **2.** $16t + 1$ **3.** $(2x - 7)(3x^2 + 5)$ **4.** $(2x - 7)(4x^2 + 14x + 49)$ **5.** $\dfrac{4x^2 + 14x + 49}{3x^2 + 5}$ **6.** $18y^3 - 17y + 5$

7. $6y^2 + 2y - 5$ **8.** $6y^2 + 2y - 5$ **9.** $-150k^7$ **10.** $\dfrac{-x^2}{4}$

Section 5-1. Supplementary Exercises

1. No restricted values   **2.** No restricted values   **3.** No restricted values   **4.** No restricted values   **5.** -1 **6.** -3
7. 0, -3, and $\dfrac{1}{2}$ **8.** 0, $\dfrac{-1}{3}$, and 4   **9.** 0 and -10 **10.** 0 and 2   **11.** -2 **12.** $\dfrac{1}{3}$ **13.** $x \neq y$ **14.** $y \neq -2x$

15. $\dfrac{3-y}{3+y}$ **16.** $\dfrac{5-y}{5+y}$ **17.** $\dfrac{a^2+2a+4}{a+2}$ **18.** $a-3$ **19.** $\dfrac{a+5b}{a-5b}$ **20.** $\dfrac{4a-5b}{4a+5b}$ **21.** $\dfrac{2(b-1)}{b+2}$ **22.** $\dfrac{2(b-2)}{b+3}$

23. t^2-9 **24.** t^2-4 **25. 1.** $\dfrac{23}{13}$ **2.** $\dfrac{13}{5}$ **3.** No **26. 1.** $\dfrac{5}{8}$ **2.** 4 **3.** No **27. 1.** $\dfrac{21}{13}$ **2.** $\dfrac{12}{7}$ **3.** No

28. 1. $\dfrac{21}{4}$ **2.** 6 **3.** No **29. a.** 1 **b.** $\dfrac{a}{2}$ **c.** 1 **d.** Yes **30. a.** 3 **b.** $\dfrac{6}{a}$ **c.** 3 **d.** Yes **31. a.** $\dfrac{-7}{3}$

b. $\dfrac{a+5}{a-5}$ **c.** $\dfrac{-7}{3}$ **d.** Yes **32. a.** 9 **b.** $\dfrac{a+7}{a-1}$ **c.** 9 **d.** Yes **33. a.** $\dfrac{7}{10}$ **b.** $\dfrac{5}{8}$ **c.** $\dfrac{x+1}{x+4}$ **d.** $\dfrac{121}{124}$

34. a. $\dfrac{10}{13}$ **b.** $\dfrac{7}{10}$ **c.** $\dfrac{x+1}{x+4}$ **d.** $\dfrac{193}{196}$

Section 5-2. Practice Exercises

1. 1 **2.** 1 **3.** 1 **4.** 1 **5.** -1 **6.** -1 **7.** 1 **8.** $\dfrac{1}{4}$ **9.** $\dfrac{1}{16}$ **10.** $\dfrac{1}{125}$ **11.** $\dfrac{1}{64}$ **12.** $\dfrac{1}{16}$ **13.** $\dfrac{-1}{81}$

14. $\dfrac{-1}{36}$ **15.** $\dfrac{1}{a^8}$ **16.** $\dfrac{1}{a^{11}}$ **17.** $\dfrac{-1}{p^6}$ **18.** $\dfrac{-1}{p^4}$ **19.** $\dfrac{-1}{7}$ **20.** $\dfrac{-1}{6}$ **21.** $\dfrac{10}{p^3}$ **22.** $\dfrac{17}{p^5}$ **23.** $\dfrac{4m^2}{n^2}$ **24.** $\dfrac{3m^4}{n^3}$

25. $\dfrac{1}{a^5b^2}$ **26.** $\dfrac{1}{a^3b^7}$ **27.** $\dfrac{x}{5y^3}$ **28.** $\dfrac{y^3}{2x^2}$ **29.** $\dfrac{49s^5}{t^8}$ **30.** $\dfrac{25t^3}{s^4}$ **31.** 49 **32.** 121 **33.** q^7 **34.** q^3 **35.** $\dfrac{4y^3}{x^2}$

36. $\dfrac{5y^4}{x}$ **37.** $\dfrac{5}{2}$ **38.** $\dfrac{3}{8}$ **39.** $\dfrac{x^2}{9}$ **40.** $\dfrac{36}{x^2}$ **41.** $\dfrac{8b^6}{125a^3}$ **42.** $\dfrac{16b^2}{9a^6}$ **43.** x^2 **44.** $\dfrac{1}{x^2}$ **45.** $\dfrac{1}{y^6}$ **46.** y^3

47. $\dfrac{1}{z^2}$ **48.** $\dfrac{1}{z^6}$ **49.** $\dfrac{a^3}{b^6}$ **50.** $\dfrac{a^6}{b^4}$ **51.** $\dfrac{1}{p^{12}q^4}$ **52.** $\dfrac{p^{15}}{q^6}$ **53.** $\dfrac{1}{25}$ **54.** 5 **55.** $\dfrac{t^6}{8}$ **56.** $\dfrac{t^8}{4}$ **57.** n^2 **58.** m^5

59. $25s^4t^6$ **60.** $\dfrac{8}{s^3t^{12}}$ **61.** $\dfrac{-5a}{4}$ **62.** $\dfrac{1}{18t}$ **63.** $\dfrac{2y}{3}$ **64.** $\dfrac{-x}{24}$ **65.** 5.4×10^5 **66.** 6.5×10^4 **67.** 9.1×10^6

68. 1.6×10^7 **69.** 8.0×10^{-4} **70.** 6.0×10^{-5} **71.** 6.4×10^{-5} **72.** 5.2×10^{-6} **73.** 8.61×10^8 **74.** 3.65×10^6

75. 8.07×10^{-4} **76.** 4.21×10^{-6} **77.** 2.85×10^4 **78.** 7.25×10^5 **79.** 74 **80.** 380 **81.** 4,300,000

82. 58,000 **83.** 0.00017 **84.** 0.0091 **85.** 0.0000087 **86.** 0.000057 **87.** 115,000,000 **88.** 79,600,000

89. 0.00000000553 **90.** 0.00000681

Section 5-2. Ten Review Exercises

1. 1 **2.** -4 **3.** -11 **4.** 62 **5.** -22 **6.** 3700 **7.** $(k-4)(3k+2)$ **8.** 0 **9.** 4 and $\dfrac{-2}{3}$ **10.** $\dfrac{2}{3k+2}$

Section 5-2. Supplementary Exercises

1. 6.93×10^8 **2.** 6.08×10^8 **3.** 1.64×10^4 **4.** 4.34×10^5 **5.** 1.776×10^{-9} **6.** 6.192×10^{-10} **7.** 8.822×10^{18}

8. 6.048×10^{15} **9.** 1.1×10^5 **10.** 2.2×10^6 **11.** 5.1×10^4 **12.** 3.0×10^9 **13.** 2.5×10^{-3} **14.** 3.1×10^{-6}

15. 3.0×10^8 **16.** 2.2×10^{13} **17.** 6.51×10^{-5} **18.** 2.99×10^5 **19.** 4.42×10^6 **20.** 3.04×10^3 **21.** $x\neq0$

22. $y\neq0$ **23.** $p\neq0$ and $q\neq0$ **24.** $p\neq0$ and $q\neq0$ **25.** No restricted values **26.** No restricted values

27. $x\neq-2$ **28.** $x\neq5$ **29.** $y\neq6$ **30.** $y\neq-8$ **31.** $z\neq0$ or 6 **32.** $z\neq0$ or -10

Section 5-3. Practice Exercises

1. $\dfrac{3}{t}$ **2.** $\dfrac{3t^2}{2}$ **3.** $12x^2y$ **4.** $2xy^2$ **5.** $\dfrac{-5p^2}{p+3}$ **6.** $\dfrac{-2p}{5(p+7)}$ **7.** $\dfrac{4a}{3(a-1)}$ **8.** $\dfrac{a-2}{a^2}$ **9.** $\dfrac{a-b}{2ab}$ **10.** $\dfrac{a-b}{a+b}$

11. $\dfrac{2x+1}{5x}$ **12.** $\dfrac{x^2}{x+1}$ **13.** $\dfrac{x-1}{x+2}$ **14.** $\dfrac{x+5}{x-1}$ **15.** $3x+2y$ **16.** $2x-3y$ **17.** $\dfrac{t^2+5t+25}{2t(t-5)}$

18. $\dfrac{25t^2-10t+4}{7t(5t-2)}$ **19.** $\dfrac{ab}{3a-2b}$ **20.** $\dfrac{a+b}{3a-b}$ **21.** $\dfrac{15}{4z}$ **22.** $\dfrac{3z^2}{2}$ **23.** $\dfrac{7x^2}{y^3}$ **24.** $\dfrac{3y}{5x^2}$ **25.** $\dfrac{2a}{a-1}$ **26.** $\dfrac{a+2}{3a}$

27. $\dfrac{-(u+3)}{6u}$ **28.** $\dfrac{-5u}{u-5}$ **29.** $\dfrac{a+b}{2ab}$ **30.** $\dfrac{a+b}{a-b}$ **31.** $\dfrac{x(x+2)}{2(x-1)}$ **32.** $\dfrac{6x^4}{(x+1)(2x+1)}$ **33.** 1 **34.** $\dfrac{(x+1)(y-1)}{(y+1)(x-1)}$

35. $\dfrac{u^2+2uv+4v^2}{2u^2v^2}$ **36.** $3uv(u+3v)$ **37.** $\dfrac{-(t+3)}{5t}$ **38.** $\dfrac{2t-1}{2t}$ **39.** $\dfrac{3m^2-1}{2}$ **40.** $\dfrac{1}{(m+n)(m-n)}$ **41.** $\dfrac{4v^2}{tu}$

42. $\dfrac{9u^3}{2v^3}$ **43.** $\dfrac{x-3}{3x}$ **44.** $\dfrac{x+5}{5x}$ **45.** $\dfrac{3y+1}{y-3}$ **46.** $\dfrac{y-2}{2y+1}$ **47.** $\dfrac{a+b}{3a}$ **48.** $\dfrac{ab}{a-b}$ **49.** -1 **50.** -1

Section 5-3. Ten Review Exercises

1. -7 **2.** $\dfrac{3}{5}$ **3.** $\dfrac{-2}{3}$ and $\dfrac{2}{3}$ **4.** $\dfrac{-7}{5}$ and 4 **5.** $z \geq 2$ **6.** $x - 2$

7. $9t^2 - 30tu + 25u^2$ **8.** $10m^3 - 16m^2 - 3m + 2$ **9.** $3u^2 - 7uv + 4v^2$ **10.** $2k^2 - 5k + 6$

Section 5-3. Supplementary Exercises

1. a. $\dfrac{3}{5}$ **b.** $\dfrac{3}{5}$ **c.** $\dfrac{3}{5}$ **2. a.** $\dfrac{1}{14}$ **b.** $\dfrac{1}{14}$ **c.** $\dfrac{1}{14}$ **3. a.** $\dfrac{3}{2}$ **b.** $\dfrac{x}{x - 1}$ **c.** $\dfrac{3}{2}$ **4. a.** $\dfrac{4}{5}$ **b.** $\dfrac{4x}{3(x + 2)}$ **c.** $\dfrac{4}{5}$

5. a. $\dfrac{1}{8}$ **b.** $\dfrac{1}{2(x + 1)}$ **c.** $\dfrac{1}{8}$ **6. a.** $\dfrac{1}{2}$ **b.** $\dfrac{2}{x + 1}$ **c.** $\dfrac{1}{2}$ **7. a.** 6 **b.** $2x$ **c.** 6 **8. a.** 6 **b.** $2x$ **c.** 6

9. a. 2 **b.** $\dfrac{x + 1}{2}$ **c.** 2 **10. a.** 4 **b.** $\dfrac{2(x + 3)}{3}$ **c.** 4 **11.** -3 **12.** $\dfrac{-1}{3}$ **13.** $\dfrac{-s^3(s + 7)}{t(s + 4)}$ **14.** $\dfrac{-t(s + 10)}{s - 3}$

15. $\dfrac{-(a + 8)}{a - 4}$ **16.** $\dfrac{-(a + 5)}{a + 6}$ **17.** $-xy^2$ **18.** $\dfrac{-y}{x}$ **19.** k is 7 **20.** k is 8 **21.** k is 10 **22.** k is 6 **23.** k is 12

24. k is 15

Section 5-4. Practice Exercises

1. $\dfrac{1}{2y}$ **2.** $\dfrac{5}{y}$ **3.** $\dfrac{7}{3a^2}$ **4.** $\dfrac{8}{7a^2}$ **5.** $\dfrac{5x}{y}$ **6.** $\dfrac{4x}{y}$ **7.** $\dfrac{3}{ab}$ **8.** $\dfrac{8}{ab}$ **9.** $\dfrac{-7}{x^2 y}$ **10.** $\dfrac{-14}{xy^2}$ **11.** $\dfrac{10u + 5}{11w}$

12. $\dfrac{8u + 4}{3w}$ **13.** $\dfrac{3x - 12}{8y^2}$ **14.** $\dfrac{4x - 20}{7y^2}$ **15.** $\dfrac{7p + 1}{3p}$ **16.** $\dfrac{6p - 5}{2p}$ **17.** $\dfrac{5m - n}{7mn}$ **18.** $\dfrac{6m + 4n}{5mn}$ **19.** 3 **20.** 5

21. 10 **22.** 2 **23.** $\dfrac{5a}{a + 1}$ **24.** $\dfrac{4a}{a - 1}$ **25.** $\dfrac{2}{2a - 1}$ **26.** $\dfrac{3}{a + 2}$ **27.** $\dfrac{b + 2}{b - 2}$ **28.** $\dfrac{b - 1}{b + 1}$ **29.** $\dfrac{z + 2}{2z - 3}$

30. $\dfrac{2z - 1}{4z + 1}$ **31.** $\dfrac{a + 1}{a - 1}$ **32.** $\dfrac{a + 1}{a - 6}$ **33.** $\dfrac{z - 2}{z - 9}$ **34.** $\dfrac{1}{z - 5}$ **35.** $\dfrac{2}{x + 2}$ **36.** $\dfrac{2}{2x - 1}$ **37.** $\dfrac{1}{3}$ **38.** $\dfrac{-1}{4}$

39. $\dfrac{2}{5xy}$ **40.** $\dfrac{1}{8xy}$

Section 5-4. Ten Review Exercises

1. $4t^2 + 21t - 18$ **2.** $t + 6$ **3.** $t + 6$ **4.** -23 **5.** $\dfrac{-13}{2}$ **6.** $\dfrac{3}{4}$ and -6 **7.** $t + 6$ **8.** Not a root

9. $-3mn(m - 8n)^2$ **10.** $5uv(6u + v)(6u - v)$

Section 5-4. Supplementary Exercises

1. a. $\dfrac{2}{x}$ **b.** $3x + 7$ **2. a.** $\dfrac{2}{x}$ **b.** $12x - 2$ **3. a.** $\dfrac{1}{y}$ **b.** $7y^2 - 4y$ **4. a.** $\dfrac{3}{y}$ **b.** $7y^3 + 5y$ **5. a.** $\dfrac{x + 2}{x}$

b. $4x + 1$ **6. a.** $\dfrac{x - 3}{x}$ **b.** $5x - 15$ **7. a.** 2 **b.** 2 **c.** 2 **8. a.** 3 **b.** 3 **c.** 3 **9. a.** $\dfrac{4}{3}$ **b.** $\dfrac{x + 1}{x}$

c. $\dfrac{4}{3}$ **10. a.** $\dfrac{-1}{9}$ **b.** $\dfrac{2 - x}{3x}$ **c.** $\dfrac{-1}{9}$ **11. a.** $\dfrac{4}{5}$ **b.** $\dfrac{x + 1}{x + 2}$ **c.** $\dfrac{4}{5}$ **12. a.** $\dfrac{1}{2}$ **b.** $\dfrac{2x - 1}{2(x + 2)}$ **c.** $\dfrac{1}{2}$

13. $a^2 - ab + b^2$ **14.** $a^2 + ab + b^2$ **15.** $\dfrac{1}{x + 2}$ **16.** $\dfrac{1}{x - 3}$ **17.** $\dfrac{1}{5 + x}$ **18.** $\dfrac{1}{4 - x}$ **19.** k is 2 **20.** k is (-3)

21. k is 4 **22.** k is 3 **23.** k is 3 **24.** k is 4 **25.** k is 0 **26.** k is 0

Section 5-5. Practice Exercises

1. $12x$ **2.** $35x$ **3.** $24y^2$ **4.** $12y^3$ **5.** $18ab$ **6.** $40ab$ **7.** $10a^2 b^3$ **8.** $15a^3 b^2$ **9.** $9y(y - 1)$
10. $10y(y + 2)$ **11.** $6a^3(a - 1)$ **12.** $9a^2(a + 2)$ **13.** $(a + b)(a - b)^2$ **14.** $(2a - 3b)(2a + 3b)^2$ **15.** $2(x - 3y)(x + 2y)^2$
16. $6(x - y)(x + 3y)^2$ **17.** $9(m - 3)(m - 2)^2$ **18.** $10(2m - 1)(2m + 1)^2$ **19.** $3(x + 2)(x - 2)(x + 3)$
20. $2(x + 3)(x - 3)(x + 5)$ **21.** $6(x + 4)(x + 1)^2$ **22.** $15(x - 3)(x + 2)^2$ **23.** $2(2y + 1)(2y - 1)(4y^2 + 2y + 1)$

24. $2(y - z)(y + z)(y^2 - yz + z^2)$ **25. a.** $4x^2 y^2$ **b.** $\dfrac{7y}{4x^2 y^2}, \dfrac{10x}{4x^2 y^2}$ **26. a.** $4x^2 y$ **b.** $\dfrac{3x}{4x^2 y}, \dfrac{22}{4x^2 y}$ **27. a.** $6(a + 5)$

b. $\dfrac{15}{6(a + 5)}, \dfrac{8}{6(a + 5)}$ **28. a.** $10(a - 1)$ **b.** $\dfrac{16}{10(a - 1)}, \dfrac{35}{10(a - 1)}$ **29. a.** $3(a - 2)(a + 2)$

b. $\dfrac{a^2 + 2a}{3(a-2)(a+2)}$, $\dfrac{6a}{3(a-2)(a+2)}$ **30. a.** $2(a-5)(a-5)$ **b.** $\dfrac{3a-15}{2(a-5)(a-5)}$, $\dfrac{2a}{2(a-5)(a-5)}$

31. a. $(x-3)(x+3)(6x+1)$ **b.** $\dfrac{12x^2 + 2x}{(x-3)(x+3)(6x+1)}$, $\dfrac{x^2 + 5x + 6}{(x-3)(x+3)(6x+1)}$ **32. a.** $2(x-1)(x-4)^2$

b. $\dfrac{x^2 - 7x + 12}{2(x-1)(x-4)^2}$, $\dfrac{2x^2 - 2x}{2(x-1)(x-4)^2}$ **33. a.** $(m+2n)(m^2 - 2mn + 4n^2)$

b. $\dfrac{mn}{(m+2n)(m^2 - 2mn + 4n^2)}$, $\dfrac{m+2n}{(m+2n)(m^2 - 2mn + 4n^2)}$ **34. a.** $(4m-n)(16m^2 + 4mn + n^2)$

b. $\dfrac{3mn}{(4m-n)(16m^2 + 4mn + n^2)}$, $\dfrac{32m^2 + 8mn + 2n^2}{(4m-n)(16m^2 + 4mn + n^2)}$ **35. a.** $24u^2v^2(u+2v)$

b. $\dfrac{4uv}{24u^2v^2(u+2v)}$, $\dfrac{2v^2}{24u^2v^2(u+2v)}$, $\dfrac{3u^2}{24u^2v^2(u+2v)}$ **36. a.** $30u^3(u-3v)$ **b.** $\dfrac{3u^2}{30u^3(u-3v)}$, $\dfrac{2u^3}{30u^3(u-3v)}$, $\dfrac{5}{30u^3(u-3v)}$

37. $\dfrac{3b+2a}{2a^2b^2}$ **38.** $\dfrac{5a-6b}{15a^2b^2}$ **39.** $\dfrac{12x+7}{10x^2}$ **40.** $\dfrac{35-12x}{10x^2}$ **41.** $\dfrac{1}{6(m-2)}$ **42.** $\dfrac{19}{24(m+1)}$ **43.** $\dfrac{n+3}{4n(n+1)}$

44. $\dfrac{9n-5}{6n(n-1)}$ **45.** $\dfrac{-4}{(b+2)(b-2)^2}$ **46.** $\dfrac{2b+3}{(b+2)(b+1)}$ **47.** $\dfrac{3t}{(t-2)(t+1)^2}$ **48.** $\dfrac{3t+4}{(t+2)(t+1)}$ **49.** $\dfrac{w+3}{w+5}$

50. $\dfrac{w-4}{w-6}$ **51.** $\dfrac{a-2}{a+4}$ **52.** $\dfrac{2a+1}{a-1}$ **53.** $\dfrac{5}{m-5}$ **54.** $\dfrac{-7}{m-8}$ **55.** 1 **56.** 1 **57.** $\dfrac{1}{x+5}$ **58.** $\dfrac{1}{x+6}$

59. $\dfrac{2}{t(t+2)(t-2)}$ **60.** $\dfrac{t+1}{3(t-4)(t+4)}$ **61.** $\dfrac{3x+1}{2x}$ **62.** $\dfrac{11+4x}{6x}$ **63.** $\dfrac{-2}{y(y-1)}$ **64.** $\dfrac{2y+4}{3y(y-2)}$ **65.** 0 **66.** 1

67. $\dfrac{9}{(t+3)(t^2 - 3t + 9)}$ **68.** $\dfrac{t-u}{t+u}$

Section 5-5. Ten Review Exercises

1. $100t^2$ **2.** $100 + 20t + t^2$ **3.** $2k$ **4.** $2+k$ **5.** $6m-25$ **6.** $m^2 - 25$ **7.** $3uv$ **8.** $\dfrac{9u^2 + 3uv + v^2}{3u+v}$ **9.** 7

10. a. $x^2 - xy - 2y^2$ **b.** 7

Section 5-5. Supplementary Exercises

1. a. $\dfrac{5}{18}$ **b.** $\dfrac{5}{6x}$ **c.** $\dfrac{5}{18}$ **2. a.** $\dfrac{3}{20}$ **b.** $\dfrac{9}{20x}$ **c.** $\dfrac{3}{20}$ **3. a.** $\dfrac{-1}{18}$ **b.** $\dfrac{2-x}{2x^2}$ **c.** $\dfrac{-1}{18}$ **4. a.** $\dfrac{1}{36}$ **b.** $\dfrac{4-x}{4x^2}$

c. $\dfrac{1}{36}$ **5. a.** $\dfrac{11}{24}$ **b.** $\dfrac{11x}{12(x+3)}$ **c.** $\dfrac{11}{24}$ **6. a.** $\dfrac{-1}{2}$ **b.** $\dfrac{-5x}{6(2x-1)}$ **c.** $\dfrac{-1}{2}$ **7. a.** $\dfrac{-1}{6}$ **b.** $\dfrac{1}{x(x-5)}$ **c.** $\dfrac{-1}{6}$

8. a. $\dfrac{1}{2}$ **b.** $\dfrac{x-1}{x+1}$ **c.** $\dfrac{1}{2}$ **9.** 0 **10.** 0 **11.** $\dfrac{1}{p-2}$ **12.** $\dfrac{-1}{p-5}$ or $\dfrac{1}{5-p}$ **13.** $\dfrac{-3}{(q+3)}$ **14.** $\dfrac{-q}{q-1}$ or $\dfrac{q}{1-q}$

15. $\dfrac{-2}{(k+6)(k-7)}$ **16.** $\dfrac{1}{(k+2)(k-5)}$ **17.** $\dfrac{3w}{(w+3)(w-3)(w^2 + 3w + 9)}$ **18.** $\dfrac{5w}{(w+5)(w-5)(w^2 + 5w + 25)}$

19. $\dfrac{x+1}{x}$ **20.** $\dfrac{20 + 5x - x^2}{(x-5)(x+5)}$ **21. a.** $\dfrac{1}{6}$ **b.** $\dfrac{1}{w(w-1)}$ **c.** Yes **d.** $\dfrac{5}{6}$ **e.** $\dfrac{2w-1}{w(w-1)}$ **f.** No **22. a.** -1

b. -1 **c.** Yes **d.** -2 **e.** $\dfrac{-w-1}{w-1}$ or $\dfrac{1+w}{1-w}$ **f.** No

Section 5-6. Practice Exercises

1. $\dfrac{a^2 - 2}{a}$ **2.** $\dfrac{a^3 - 5}{a}$ **3.** $\dfrac{b}{b-1}$ **4.** $\dfrac{-1}{b-1}$ or $\dfrac{1}{1-b}$ **5.** $\dfrac{x^2}{x+2}$ **6.** $\dfrac{x^2}{x+3}$ **7.** $\dfrac{y^2 - y + 1}{y-1}$ **8.** $\dfrac{y^2 - y - 1}{y-1}$

9. $\dfrac{(3z-5)(2z+1)}{z^2}$ **10.** $\dfrac{(z+2)(z+7)}{z^2}$ **11.** $\dfrac{3(t-1)(t-2)}{t^2}$ **12.** $\dfrac{10(t-5)(t+7)}{t^2}$ **13.** $\dfrac{(u-2)^2}{4u}$ **14.** $\dfrac{(2u-1)^2}{u^2}$

15. $\dfrac{(x+3y)(x-2y)}{xy}$ **16.** $\dfrac{(2x+y)(2x-3y)}{xy}$ **17.** $\dfrac{12a}{(2a+3)(2a-3)}$ **18.** $\dfrac{4a}{(a+1)(a-1)}$ **19.** $b-2$ **20.** $b+1$

21. $\dfrac{x^2 + 1}{x^2 - 1}$ **22.** $\dfrac{2x-1}{2x+1}$ **23.** $\dfrac{y+1}{y-1}$ **24.** $\dfrac{1}{1-2y}$ **25.** $\dfrac{xy(x+y)}{x-y}$ **26.** $\dfrac{4(2x-1)}{x(2x+1)}$ **27.** $\dfrac{1}{4b}$ **28.** $\dfrac{1}{3b}$ **29.** $\dfrac{-1}{2p}$

30. $\dfrac{3p}{20}$ **31.** $\dfrac{v-u}{2(v+u)}$ **32.** $\dfrac{v-u}{3uv}$ **33.** $\dfrac{w+2}{w-2}$ **34.** $\dfrac{w-1}{w+1}$ **35.** $\dfrac{m+2n}{2m+n}$ **36.** $\dfrac{2m-n}{m-2n}$ **37.** -3 **38.** 2

39. $\dfrac{-1}{2}$ **40.** $\dfrac{-1}{y}$ **41.** $\dfrac{xy}{y+x}$ **42.** $\dfrac{xy}{2y+3x}$ **43.** $\dfrac{t}{t-1}$ **44.** $\dfrac{t}{t^2-1}$ **45.** $\dfrac{ab}{a+b}$ **46.** $ab(a+b)$ **47.** $\dfrac{2k}{2-k}$

48. $\dfrac{3k}{3+k}$ **49.** $\dfrac{(u+3)(u-1)}{(u-2)(u+4)}$ **50.** $\dfrac{(u+5)(u-2)}{(u+12)(u-9)}$

Section 5-6. Ten Review Exercises

1. $-6u^3$ **2.** u^3+2u^2-3u-6 **3.** $\dfrac{15u}{2v^2}$ **4.** $3t^3$ **5.** $3t^3+16t^2+6t+4$ **6.** $15t^3+5t^2+5t-5+\dfrac{25}{t+5}$ **7.** $\dfrac{9t^3u}{4}$

8. 8 **9.** No roots   **10.** -9 and 3

Section 5-6. Supplementary Exercises

1. 1 **2.** 1 **3.** $\dfrac{w+5}{w-5}$ **4.** $\dfrac{w-1}{w+1}$ **5.** $1-m$ **6.** $\dfrac{m^2+1}{m^3}$ **7. 1.** $\dfrac{5}{3}$ **2.** $\dfrac{2}{5}$ **3.** No   **8. 1.** $\dfrac{1}{2}$ **2.** -3

3. No   **9. 1.** 5 **2.** $\dfrac{-1}{5}$ **3.** No   **10. 1.** $\dfrac{-1}{5}$ **2.** 5 **3.** No   **11. 1.** 3 **2.** $\dfrac{13}{36}$ **3.** No   **12. 1.** 3

2. $\dfrac{17}{72}$ **3.** No   **13.** $\dfrac{7}{11}$ **14.** $\dfrac{-35}{13}$ **15.** $\dfrac{1,800}{41}$ **16.** 112

Section 5-7. Practice Exercises

1. -6 **2.** 5 **3.** -1 **4.** -4 **5.** 12 **6.** 0 **7.** 5 **8.** -3 **9.** -6 **10.** 6 **11.** 2 **12.** -4 **13.** 0

14. 7 **15.** 4 **16.** -5 **17.** 5 **18.** -3 **19.** $\dfrac{-3}{7}$ **20.** $\dfrac{-1}{4}$ **21.** -2 **22.** 6 **23.** 3 **24.** -2 **25.** 1

26. -2 **27.** -5 **28.** $\dfrac{-1}{2}$ **29.** No roots   **30.** No roots   **31.** 4 **32.** -5 **33.** $\dfrac{7}{2}$ **34.** $\dfrac{5}{3}$ **35.** -4

36. -1 **37.** -6 **38.** -1 **39.** -1 **40.** $\dfrac{1}{7}$

Section 5-7. Ten Review Exercises

1. $4t^3-13t$ **2.** $12u^3+25u^2+6u-8$ **3.** $36v^3-27u^2-4v+3$ **4.** $25k^2+60k+36$ **5.** $5a^4-12a^2b+b^2$

6. $2x^3-6x+9+\dfrac{3}{2x+3}$ **7.** $\dfrac{y}{2y-3}$ **8.** $\dfrac{a+2b}{a-b}$ **9.** $\dfrac{9m-4n}{42m^2n^2}$ **10.** $\dfrac{a-2b}{2a-b}$

Section 5-7. Supplementary Exercises

1. All real numbers except 0 **2.** All real numbers except 0 **3.** No solutions   **4.** No solutions   **5.** 8 **6.** -7

7. 0 **8.** $\dfrac{2}{5}$ **9.** No solutions   **10.** No solutions   **11.** $x=\dfrac{5-y}{2+y}$ **12.** $x=\dfrac{5y}{2y-3}$ **13.** $y=\dfrac{12x}{5x+21}$

14. $y=\dfrac{10x}{x-15}$ **15.** $t=\dfrac{a}{5}$ **16.** $t=\dfrac{2a^2-a}{4}$ **17.** $N=MP$ **18.** $M=\dfrac{-2N}{P}$ **19.** $R=\dfrac{r_1r_2}{r_1+r_2}$ **20.** $r_1=\dfrac{Rr_2}{r_2-R}$

21. $3x+2y=3$ **22.** $3x+5y=12$ **23.** $2x+3y=2$ **24.** $3x+4y=2$ **25.** $10x-3y=45$ **26.** $4x-7y=3$

Section 5-8. Practice Exercises

1. a. 5 and $3x$ **b.** 7 and y **2. a.** $5x$ and $2y$ **b.** 6 and 9 **3. a.** 3 and $(70-x)$ **b.** x and 4 **4. a.** 1 and $(24-x)$
b. x and 2 **5. a.** $(x-3)$ and 4 **b.** $(x+3)$ and x **6. a.** $(x+2)$ and x **b.** $(x+10)$ and 6 **7.** 9 **8.** 9 **9.** 90

10. 20 **11.** $\dfrac{2}{3}$ **12.** $\dfrac{3}{2}$ **13.** 10 **14.** 9 **15.** 12 **16.** 4 **17.** 4 **18.** 8 **19.** 7 **20.** 9 **21. a.** $\$4.45$

b. 12 cans   **22. a.** $\$5.58$ **b.** 20 cans   **23. a.** 35 cm   **b.** 78.4 km   **24. a.** 6 in   **b.** 87.5 mi   **25. a.** 132 fps
b. 75 mph   **26. a.** $1,936$ fpm   **b.** 18 mph   **27. a.** $1,050$ fish   **b.** 900 fish   **28. a.** 500 rabbits   **b.** 500 rabbits
29. $y=48,\ z=72$ **30.** $b=4,\ c=7$ **31.** $x=8,\ y=16$ **32.** $a=6,\ b=5$ **33.** $x=4.8,\ z=11.2$
34. $a=7.2,\ c=10.8$ **35.** $x=45,\ b=18$ **36.** $a=33,\ y=96$ **37.** 304.8 cm   **38.** 85 in   **39.** 165 cal

40. 270 cal   **41.** 396 yd   **42.** $\dfrac{200}{11}$ rods

Section 5-8. Ten Review Exercises

1. $-5k - 5$ 2. -6 3. $\dfrac{-13}{5}$ 4. $(2t + 3)(t - 2)$ 5. -6 6. $\dfrac{-3}{2}, 2$ 7. $\dfrac{a - 11}{6a}$ 8. $\dfrac{-65}{6}$ 9. 11

10. $-1, 10$

Section 5-8. Supplementary Exercises

1. All real numbers except -3 2. All real numbers except 0   3. All real numbers except -1
4. All real numbers except 2   5. 6   6. 9   7. 12   8. 10   9. $2x$ 10. $3x$ 11. 2   12. 2   13. 1   14. 1
15. 28   16. 32   17. $48x$ 18. $18x$ 19. 10 cm and 16 cm   20. 15 cm and 24 cm   21. 2.4 in and 4.1 in
22. 7.3 in and 12.2 in   23. 1.22 m and 2.03 m   24. 8.53 m and 14.22 m

Section 5-9. Practice Exercises

1. The number is 5.   2. The number is 3.   3. The number is 3.   4. The number is 4.   5. The number is 2.
6. The number is 5.   7. The number is $\frac{2}{3}$ or $\frac{3}{2}$.   8. The number is $\frac{3}{5}$ or $\frac{5}{3}$.   9. The number is $\frac{-2}{5}$ or $\frac{5}{2}$.

10. The number is $\frac{-4}{7}$ or $\frac{7}{4}$.   11. The number is $\frac{1}{2}$ or 4.   12. The number is $\frac{3}{4}$ or $\frac{-2}{3}$.   13. 4.8 hr   14. $4\frac{4}{9}$ hr

15. $25\frac{5}{7}$ min   16. 4 hr   17. $7\frac{5}{9}$ hr   18. $4\frac{7}{8}$ hr   19. $6\frac{6}{7}$ hr   20. $4\frac{5}{16}$ hr   21. $25\frac{5}{7}$ min   22. $16\frac{2}{3}$ min

23. 25 mph   24. 12 mph   25. 120 mph   26. 55 mph   27. 8 mph   28. 380 mph   29. 6 mph   30. 115 mph

Section 5-9. Ten Review Exercises

1. 3   2. 0   3. $y = \dfrac{3x - 24}{8}$ 4. 15   5. $x = \dfrac{8y + 24}{3}$ 6. -16 7. $5xy(x^2 + 16)$ 8. $5xy(x + 3y)(x^2 - 3xy + 9y^2)$

9. $(10u + 3v)(10u - 3v)$ 10. $(10u - 3v)^2$

Section 5-9. Supplementary Exercises

1. a. 2.000   b. 2.667   c. 2.818   d. 2.980   e. 2.998   f. 3   g. 7   2. a. 2.000   b. 4.000   c. 4.455
d. 4.941   e. 4.994   f. 5   g. 4   3. a. 1.000   b. 0.231   c. 0.109   d. 0.020   e. 0.010   f. 0   g. 0
4. a. 1.000   b. 0.200   c. 0.100   d. 0.020   e. 0.010   f. 0   g. 0   5. a. 0.500   b. 4.167   c. 9.091
d. 99.010   e. 999.001   f. Get larger and larger   g. Get larger and larger   6. a. 1.000   b. 5.000   c. 10.000
d. 100.000   e. 1,000.000   f. Get larger and larger   g. Get larger and larger

Chapter 5 Review Exercises

1. $\dfrac{-2}{3}$ 2. $\dfrac{5}{8}$ 3. 0 and (-8) 4. $m \neq 2n$ and $m \neq -n$ 5. $\dfrac{2}{t}$ 6. $\dfrac{m - n}{m - 2n}$ 7. $\dfrac{q}{10p^2}$ 8. $\dfrac{p^3}{16q}$ 9. $7xy^3$

10. $\dfrac{xy^4}{5}$ 11. $\dfrac{m}{7}$ 12. $\dfrac{27}{m^6}$ 13. $\dfrac{s^{12}}{t^{20}}$ 14. $\dfrac{1}{x^4}$ 15. $\dfrac{1}{z^5}$ 16. $\dfrac{-1}{x^3}$ 17. 6.5×10^{-7} 18. 741,000,000

19. $\dfrac{3z - 1}{2z + 3}$ 20. $\dfrac{5x - y}{4xy}$ 21. $\dfrac{-2}{x + 3}$ 22. $\dfrac{x - 2}{3x + 1}$ 23. $\dfrac{4a}{5}$ 24. $\dfrac{a + b}{2a}$ 25. $\dfrac{4}{3u}$ 26. $\dfrac{u + 2v}{u - v}$ 27. 6

28. $\dfrac{7}{36k}$ 29. $\dfrac{5}{12(a + 2)}$ 30. $\dfrac{4a}{(a + 4b)(a - 4b)}$ 31. $\dfrac{10y}{x}$ 32. $\dfrac{x + y}{3(x - y)}$ 33. $\dfrac{16m^2}{3n(3m - 4n)}$ 34. $\dfrac{m - 2n}{2m - n}$ 35. 4

36. 3   37. 12   38. 1   39. $\dfrac{1}{2}$ 40. 5   41. 562.5 mi   42. 26,350 bu   43. b is 6; c is 8.   44. x is 10.5; c is 8.0

45. The number is $\frac{2}{3}$ or 3.   46. $3\frac{6}{7}$ hr   47. 230 mph

CHAPTER 6 Linear Equations and Inequalities in Two Variables

Section 6-1. Practice Exercises

1. a. Yes   b. Yes   2. a. Yes   b. Yes   3. a. No   b. Yes   4. a. Yes   b. No   5. a. Yes   b. Yes
6. a. Yes   b. Yes   7. a. No   b. Yes   8. a. No   b. Yes   9. a. No   b. No   10. a. No   b. No
11. a. Yes   b. No   12. a. Yes   b. Yes   13. a. No   b. Yes   14. a. No   b. Yes   15. a. Yes   b. Yes

16. a. Yes **b.** Yes **17. a.** Yes **b.** No **18. a.** No **b.** Yes **19. a.** (2, 3) **b.** (9, −4) **20. a.** (−2, −14)
b. (5, 7) **21. a.** (4, −5) **b.** (6, −10) **22. a.** (2, 6) **b.** (−3, −14) **23. a.** (−1, 1) **b.** $\left(\dfrac{-1}{2}, -1\right)$
24. a. $\left(-2, \dfrac{-1}{2}\right)$ **b.** $\left(-4, \dfrac{-9}{4}\right)$ **25. a.** $\left(\dfrac{4}{5}, -2\right)$ **b.** (4, 6) **26. a.** $\left(\dfrac{-3}{2}, 8\right)$ **b.** (6, 13) **27. a.** (3, 1)
b. (−3, 1) **28. a.** (2, 3) **b.** (−2, 3) **29. a.** (4, 3) **b.** (−6, 13) **30. a.** (−2, 2) **b.** (4, 5) **31. a.** (1, −2)
b. (−2, 1) or (2, 1) **32. a.** (−1, −14) **b.** (−4, 1) or (4, 1) **33. a.** (2, 0) **b.** (0, −2) **34. a.** (−5, 0) **b.** (0, 5)

Section 6-1. Ten Review Exercises

1. $15y + 12$ **2.** $\dfrac{-4}{5}$ **3.** $(a - 9)(a + 6)$ **4.** 9 and −6 **5.** $60 - 12t$ **6.** $t < 5$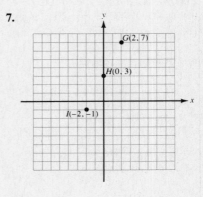

7. $\dfrac{4b - 12}{(b - 10)(b + 5)(b - 6)}$ **8.** 3 **9.** $\dfrac{4n}{3n - 4m}$ **10.** $\dfrac{x + y}{3(x - y)}$

Section 6-1. Supplementary Exercises

1. Yes **2.** Yes **3.** Yes **4.** Yes **5.** No **6.** No **7.** No **8.** No **9.** Yes **10.** Yes **11.** Yes
12. Yes **13.** Yes **14.** Yes **15.** No **16.** No **17.** Yes **18.** Yes **19.** Yes **20.** Yes **21.** (0, 0)
22. (0, 0) **23.** (−5, 0), (5, 0) **24.** (−7, 0), (7, 0) **25.** (0, −4), (0, 4) **26.** (0, −4), (0, 4) **27.** (2, 4) **28.** (−2, 11)
29. (−3, 19) **30.** (3, 1) **31.** $\left(\dfrac{1}{3}, 3\right)$ **32.** $\left(\dfrac{1}{2}, 4\right)$ **33.** $\left(\dfrac{9}{4}, \dfrac{3}{2}\right)$ **34.** $\left(\dfrac{15}{8}, \dfrac{3}{2}\right)$ **35.** (−2, 0), (−3, 0)
36. (−1, 0), (−5, 0) **37.** (3, 2) **38.** (4, 1) **39.** Yes **40.** Yes **41.** Yes **42.** Yes **43.** Yes **44.** Yes
45. Yes **46.** Yes

Section 6-2. Practice Exercises

1. $A(4, 2)$, $B(2, 4)$, $C(-3, 5)$, $D(-7, -3)$, $E(8, -6)$, $F(5, 0)$, $G(0, -3)$
2. $A(6, 7)$, $B(4, 2)$, $C(-2, 2)$, $D(-4, -7)$, $E(6, -3)$, $F(-6, 0)$, $G(0, 6)$
3. $A(8, 2)$, $B(2, 6)$, $C(-6, 3)$, $D(-4, -3)$, $E(5, -6)$, $F(-3, 0)$, $G(0, -7)$
4. $A(1, 6)$, $B(8, 4)$, $C(-7, 6)$, $D(-4, -5)$, $E(3, -6)$, $F(2, 0)$, $G(0, 3)$

5.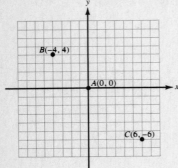

6.

7.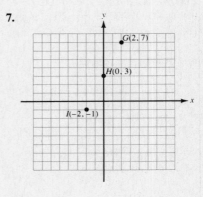

8.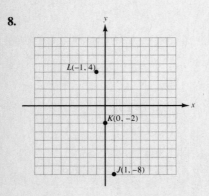

9.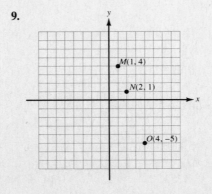

10.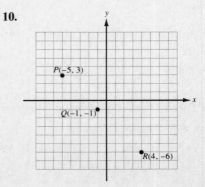

11.
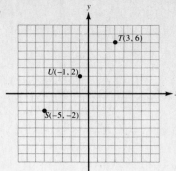

12.
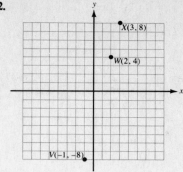

13.
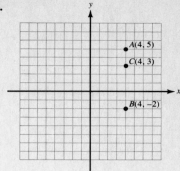

14.
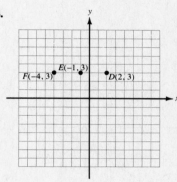

15. Q III   **16.** Q II   **17.** Q IV   **18.** Q III   **19.** Horizontal axis

20. Vertical axis   **21.** Q IV   **22.** Q II   **23.** Vertical axis   **24.** Horizontal axis   **25.** Q II   **26.** Q IV

27. Q III   **28.** Q I   **29.** $3x + y = 5$ **30.** $6x + y = 1$ **31.** $2x + 5y = -1$ **32.** $4x + 2y = -3$ **33.** $4x + 3y = 6$

34. $3x + 5y = 15$ **35.** $6x - 5y = 15$ **36.** $4x - 3y = 6$ **37.** $8x - 2y = 3$ **38.** $10x - 5y = 2$

39.
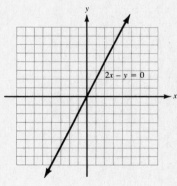

40.
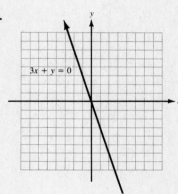

41.
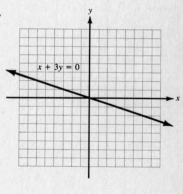

42.
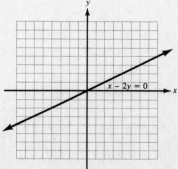

43.
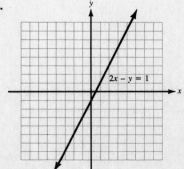

44.
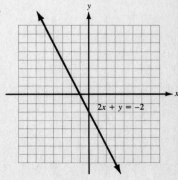

45.

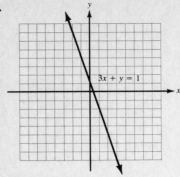

$3x + y = 1$

46.

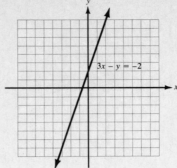

$3x - y = -2$

47.

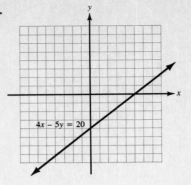

$4x - 5y = 20$

48.

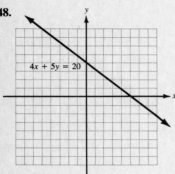

$4x + 5y = 20$

49.

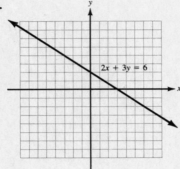

$2x + 3y = 6$

50.

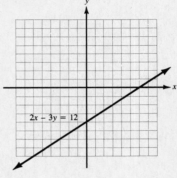

$2x - 3y = 12$

51.

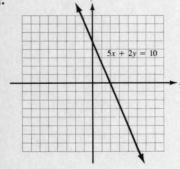

$5x + 2y = 10$

52.

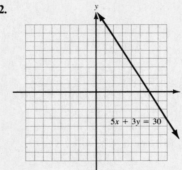

$5x + 3y = 30$

53.

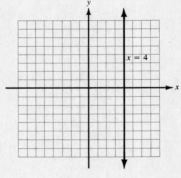

$x = 4$

54.

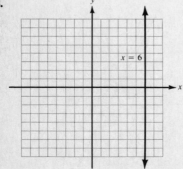

$x = 6$

55.

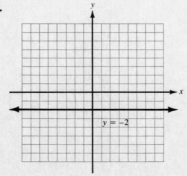

$y = -2$

56.

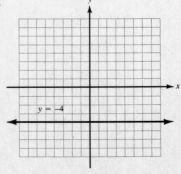

$y = -4$

57.

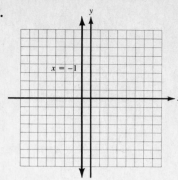

58.

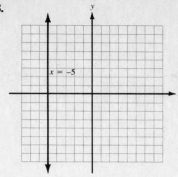

59.

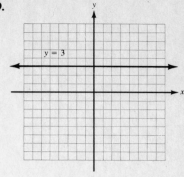

60.

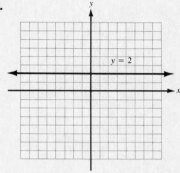

Section 6-2. Ten Review Exercises

1. $-8t - 24$ **2.** 0 **3.** Yes **4.** -3 **5.** $(2k + 3)(k - 7)$ **6.** 0 **7.** Yes **8.** $\dfrac{-3}{2}$ and 7 **9.** $\dfrac{(a + 2b)(a - 2b)}{(a + 5b)(2a + b)}$

10. $8y^2 - 8y + 3\ R5y$

Section 6-2. Supplementary Exercises

1. $A\left(\dfrac{3}{4}, \dfrac{3}{4}\right)$, $B\left(\dfrac{-3}{2}, \dfrac{7}{4}\right)$, $C\left(\dfrac{-1}{2}, \dfrac{3}{4}\right)$, $D\left(\dfrac{-3}{2}, \dfrac{-3}{4}\right)$, $E\left(\dfrac{5}{4}, \dfrac{-5}{4}\right)$ **2.** $A\left(\dfrac{1}{4}, \dfrac{3}{4}\right)$, $B\left(\dfrac{5}{4}, \dfrac{7}{4}\right)$, $C\left(\dfrac{-3}{2}, \dfrac{5}{4}\right)$, $D\left(\dfrac{-5}{4}, \dfrac{-3}{2}\right)$, $E\left(\dfrac{-3}{2}, \dfrac{-3}{4}\right)$

3. Q IV **4.** Q I **5.** Q II **6.** Q III **7.** Q I **8.** Q IV **9.** Q II **10.** Q II

11.

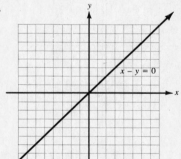

12.

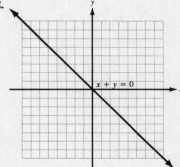

13.

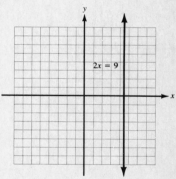

14.

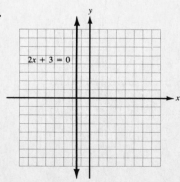

15.

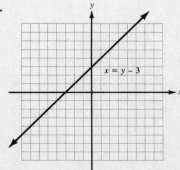

16.

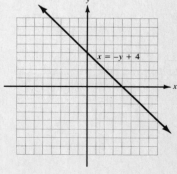

17.

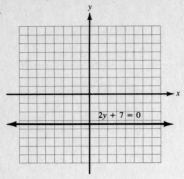

$2y + 7 = 0$

18.

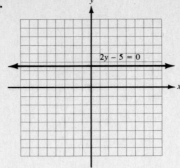

$2y - 5 = 0$

19.

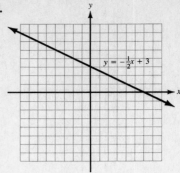

$y = -\frac{1}{2}x + 3$

20.

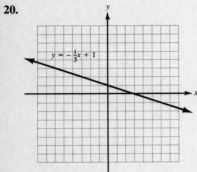

$y = -\frac{1}{3}x + 1$

21.

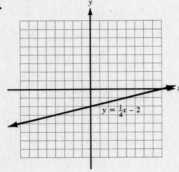

$y = \frac{1}{4}x - 2$

22.

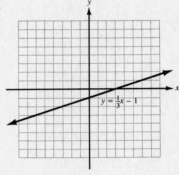

$y = \frac{1}{3}x - 1$

23.

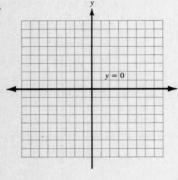

$y = 0$

24.

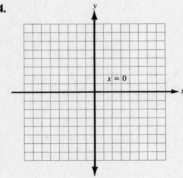

$x = 0$

25.

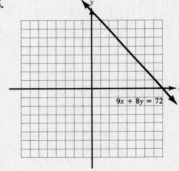

$9x + 8y = 72$

26.

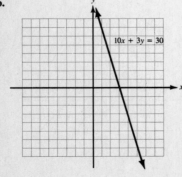

$10x + 3y = 30$

27.

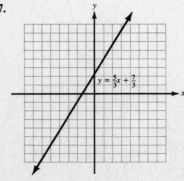

$y = \frac{5}{3}x + \frac{7}{3}$

28.

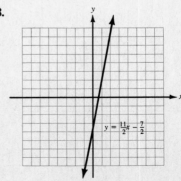

$y = \frac{11}{2}x - \frac{7}{2}$

29.

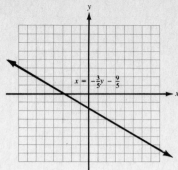

$x = -\frac{3}{5}y - \frac{9}{5}$

30.

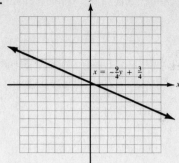

$x = -\frac{9}{4}y + \frac{3}{4}$

31. a.

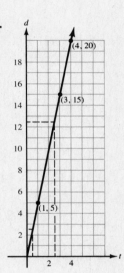

b. About 12.5 mi. **c.** About 2.5 mi.

32. a.

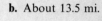

b. About 13.5 mi.

c. About 19.5 mi. **33. a.**

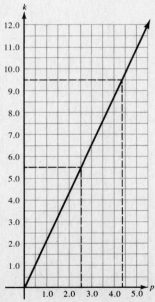

b. About 5.5k **c.** About 4.3p

34. a. 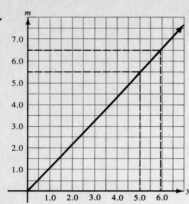   **b.** About 5.5m   **c.** About 5.8y

35. a. 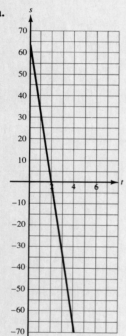   **b.** 3   **c.** $0 \le t < 3$ **d.** $t > 3$ **e.** $t = 0$ **f.** No   **g.** No

36. a. 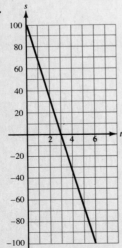   **b.** 3   **c.** $0 \le t < 3$ **d.** $t > 3$ **e.** $t = 0$ **f.** No   **g.** No

37.

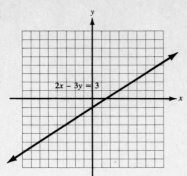

38.

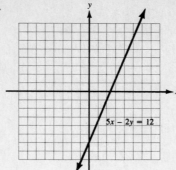

39.

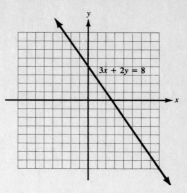

40.

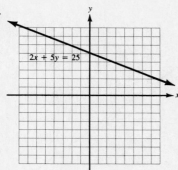

41.

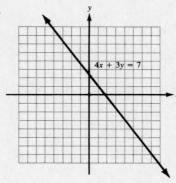

42.

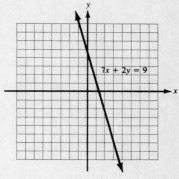

43.

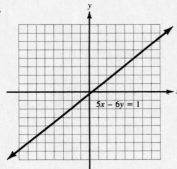

44.

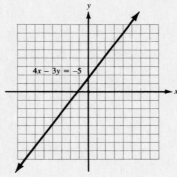

Section 6-3. Practice Exercises

1. $\dfrac{2}{3}$ **2.** $\dfrac{-1}{2}$ **3.** $\dfrac{-1}{2}$ **4.** $\dfrac{-4}{3}$ **5.** $\dfrac{5}{2}$ **6.** $\dfrac{3}{7}$ **7.** $\dfrac{-4}{7}$ **8.** $\dfrac{-5}{2}$ **9.** 3   **10.** 2   **11.** $\dfrac{-1}{2}$ **12.** -2

13. -4 **14.** -7 **15.** $\dfrac{4}{3}$ **16.** $\dfrac{5}{2}$ **17.** 5   **18.** 2   **19.** $\dfrac{-3}{5}$ **20.** $\dfrac{5}{3}$ **21.** $\dfrac{-9}{10}$ **22.** $\dfrac{5}{4}$ **23.** Vertical line

24. Vertical line   **25.** Horizontal line   **26.** Horizontal line   **27.** Horizontal line   **28.** Horizontal line

29. Vertical line   **30.** Vertical line

31.

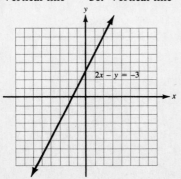

32.

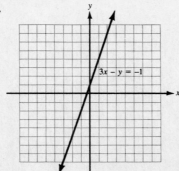

33.

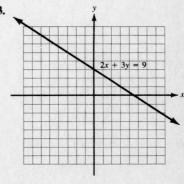

34.

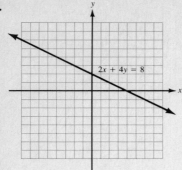

35.

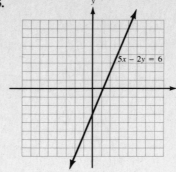

36.

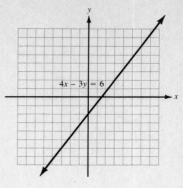

37.

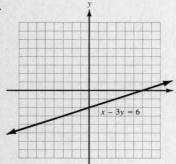

38.

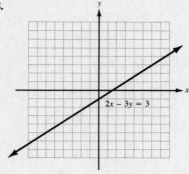

39.

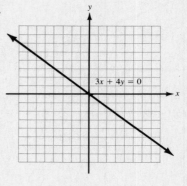

40.

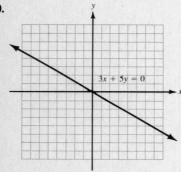

41. a. 2   **b.** 4   **42. a.** 3   **b.** 5   **43. a.** -5 **b.** -2 **44. a.** -5

b. -6 **45. a.** 3   **b.** -4 **46. a.** -7 **b.** 2   **47. a.** 3   **b.** $\dfrac{5}{2}$ **48. a.** $\dfrac{-7}{2}$ **b.** 2   **49. a.** $\dfrac{7}{5}$ **b.** 3

50. a. $\dfrac{-7}{4}$ **b.** 1

Section 6-3. Ten Review Exercises

1. $6k^3$ **2.** $6 + 2k + 3k^2 + k^3$ **3.** $25t^4$ **4.** $25 + 10t^2 + t^4$ **5.** $3x$ **6.** $3 + x$ **7.** $6a^2 - a - 8$
8. $12a^3 - 24a^2 - a + 15$ **9.** 13 and 14   **10.** 20 oz

Section 6-3. Supplementary Exercises

1. $\dfrac{-4}{3}$ **2.** $\dfrac{-8}{7}$ **3.** $\dfrac{-10}{3}$ **4.** -3 **5.** $\dfrac{-1}{2}$ **6.** $\dfrac{-1}{3}$ **7.** Slope is undefined   **8.** Slope is undefined   **9.** Zero

10. Zero   **11.** -5 **12.** 1   **13. a.** -2 **b.** 3   **c.** $\frac{1}{2}$ **d.**

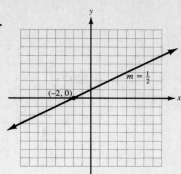

14. a. 4   **b.** -3

c. $\frac{1}{2}$ **d.**

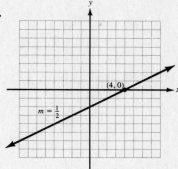

15. a. 3   **b.** 2   **c.** -2 **d.**

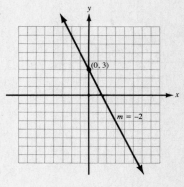

16. a. 2   **b.** 15   **c.** $\frac{-1}{3}$ **d.**

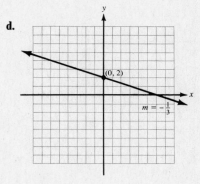

17.

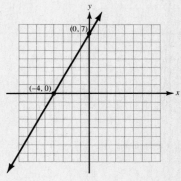

18.

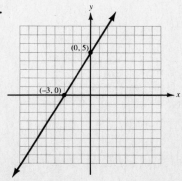

19.

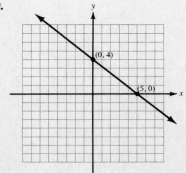

20.

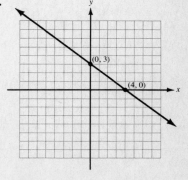

21.

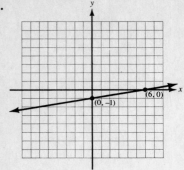

22.

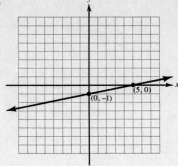

23.

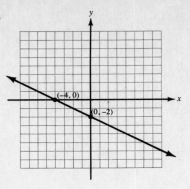

24.

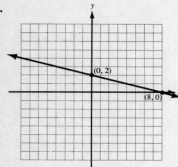

25.

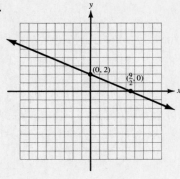

26.

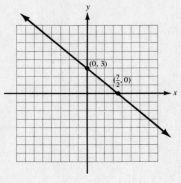

27.

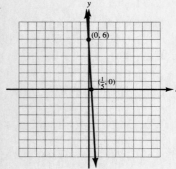

28.

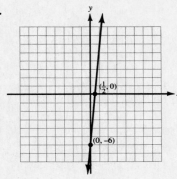

29–32. Answers may vary.

Section 6-4. Practice Exercises

1. a. $y = -2x + 3$ **b.** -2 **c.** 3   **2. a.** $y = \dfrac{-1}{2}x + 2$ **b.** $\dfrac{-1}{2}$ **c.** 2   **3. a.** $y = 3x - 4$ **b.** 3   **c.** -4

4. a. $y = \dfrac{1}{3}x + 2$ **b.** $\dfrac{1}{3}$ **c.** 2   **5. a.** $y = \dfrac{5}{4}x$ **b.** $\dfrac{5}{4}$ **c.** 0   **6. a.** $y = \dfrac{-4}{5}x$ **b.** $\dfrac{-4}{5}$ **c.** 0   **7. a.** $y = \dfrac{2}{3}x + 3$

b. $\dfrac{2}{3}$ **c.** 3   **8. a.** $y = \dfrac{-4}{3}x + 4$ **b.** $\dfrac{-4}{3}$ **c.** 4   **9. a.** $y = \dfrac{-1}{3}x + 2$ **b.**

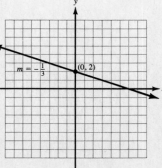

10. a. $y = \frac{1}{3}x + 2$ **b.**

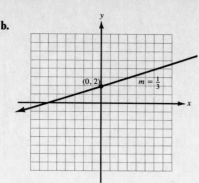

11. a. $y = 4x - 3$ **b.**

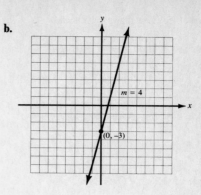

12. a. $y = -2x - 3$ **b.**

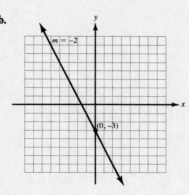

13. a. $y = \frac{5}{4}x$ **b.**

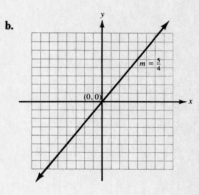

14. a. $y = \frac{-3}{7}x$ **b.**

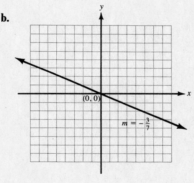

15. a. $y = -5x + 3$ **b.**

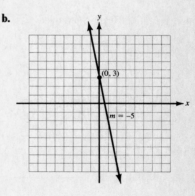

16. a. $y = 6x - 5$ **b.**

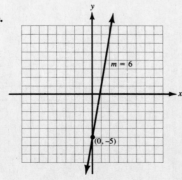

17. a. $y = \frac{2}{5}x + 1$ **b.**

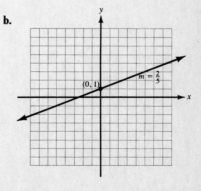

Section 8-5. Ten Review Exercises

1. $99a^2b^2 + 40ab - 99$ **2.** $49t^2 - 42tu + 9u^2$ **3.** $6xy - 14x + 15y - 35$ **4.** $-3u^2 + 12uv + v^2$ **5.** $\dfrac{3a+1}{a-2}$

6. $\dfrac{x-3}{x+3}$ **7.** $\dfrac{4}{3y}$ **8.** 6   **9.** $(ab+5)(a^2b^2 - 5ab + 25)$ **10.** $(5z-9)(7y+2)$

Section 8-5. Supplementary Exercises

1. $\dfrac{1}{2}$ **3.** $\dfrac{1}{3}$ and $\dfrac{1}{2}$ **5.** $\dfrac{5}{2}$ **7.** $\dfrac{24}{5}$ **9.** 4   **11.** 16   **13.** 3   **15.** 8   **17.** 63   **19.** 2   **21.** 22

Chapter 8 Review Exercises

1. -12 and 12   **3.** $\dfrac{10}{7}$ **5.** 0.5   **7.** 8.124   **9.** 6   **11.** $8\sqrt{7}$ **13.** $\dfrac{\sqrt{3}}{2}$ **15.** 6   **17.** $-\sqrt{42}$ **19.** $8\sqrt{5}$

21. $\dfrac{1+3\sqrt{2}}{2}$ **23.** $4+\sqrt{3}$ **25.** $\dfrac{17\sqrt{5}}{2}$ **27.** $\dfrac{2\sqrt{10}}{5}$ **29.** 12   **31.** -4

Chapter 9 Quadratic Equations

Section 9-1. Practice Exercises

1. a. $x^2 - 4x + 3 = 0$ **b.** 1, -4, and 3   **3. a.** $0 = 2t^2 - 3t + 7$ **b.** 2, -3, and 7   **5. a.** $a^2 - 11a = 0$
b. 1, -11, and 0   **7. a.** $5b^2 + 13b + 6 = 0$ **b.** 5, 13, and 6   **9. a.** $4x^2 - 7x - 4 = 0$ **b.** 4, -7, and -4

11. -4 and $\dfrac{5}{3}$ **13.** $\dfrac{1}{3}$ and $\dfrac{-5}{2}$ **15.** 0 and -9 **17.** $\dfrac{-8}{3}$ and $\dfrac{8}{3}$ **19.** $\dfrac{1}{2}$ and 2   **21.** $\dfrac{-3}{5}$ and -2

23. $\dfrac{3}{4}$ and $\dfrac{-1}{2}$ **25.** $\dfrac{-2}{7}$ and $\dfrac{2}{7}$ **27.** 0 and 2   **29.** -4 and 2   **31.** 2 and -1 **33.** -8

35. 3, 4, and 5 cm   **37.** 5, 12, and 13 units   **39.** 7 and 24 in   **41.** 12 m by 16 m   **43.** 6 hr and 14 hr
45. 14 hr

Section 9-1. Ten Review Exercises

1. $\dfrac{\sqrt{3t}}{5}$ **2.** $3\sqrt{5ab}$ **3.** $\dfrac{-2\sqrt{14x}}{7}$ **4.** $7\sqrt{3}$ **5.** $-(7+4\sqrt{3})$ **6.** $\dfrac{-2}{3}$ **7.** -8 and 3   **8.** 29   **9.** $a > 5$
10. $(3, -4)$

Section 9-1. Supplementary Exercises

1. 0 and 8   **3.** -10 and 10   **5.** -2 and 3   **7.** 0 and 5   **9.** -2 and 8   **11. a.** $(x+2)(x-3) = 0$
b. -2 and 3   **c.** ●——————|——————● (marked -2, 0, 3)   **d.** Pick $x = 0$ **e.** $-2 < x < 3$ **13. a.** $(2x+1)(x-4) = 0$

b. $\dfrac{-1}{2}$ and 4   **c.** ●—|——————————● (marked $-\frac{1}{2}$, 0, 4)   **d.** Pick $x = 0$ **e.** $x < \dfrac{-1}{2}$ or $x > 4$

15. -11 and -9, or 9 and 11   **17.** -7 and -12, or 7 and 12   **19.** South boat: 5 mph, west boat: 12 mph
21. -8 and -3, or 3 and 8

Section 9-2. Practice Exercises

1. -5 and 5   **3.** ± 15 **5.** $\pm\dfrac{7}{3}$ **7.** $\pm\sqrt{11}$ **9.** $\pm 2\sqrt{11}$ **11.** $\pm\dfrac{5}{2}$ **13.** $\pm\dfrac{\sqrt{2}}{5}$ **15.** $\pm 2\sqrt{3}$ **17.** $\pm\dfrac{9\sqrt{5}}{5}$

19. $\pm\dfrac{\sqrt{2}}{2}$ **21.** 2 and -4 **23.** 8 and -4 **25.** 21 and -3 **27.** $-7 \pm \sqrt{13}$ **29.** $4 \pm 3\sqrt{2}$

31. $-17 \pm 7\sqrt{3}$ **33.** 2 and $\dfrac{-4}{3}$ **35.** 5 and -8 **37.** $\dfrac{-3 \pm \sqrt{6}}{2}$ **39.** $\dfrac{5 \pm 3\sqrt{7}}{4}$ **41.** $\dfrac{12 \pm 2\sqrt{31}}{7}$

43. 2 and -8 **45.** 5 and -13 **47.** $\dfrac{1 \pm 2\sqrt{2}}{5}$ **49.** $\dfrac{2 \pm 2\sqrt{3}}{3}$

Section 6-4. Ten Review Exercises

1. $(t + 10)(t - 7)$ **2.** $(2u - 7)(4u + 5)$ **3.** $(4x - 7y)^2$ **4.** $(3z + 5)(9z^2 - 15z + 25)$ **5.** $(k - 4)(2k^2 + 3)$

6. $(5p + 12q)(5p - 12q)$ **7.** $y = \dfrac{3}{8}x - 3$ **8.** $\dfrac{3}{8}$ **9.** -3 **10.**

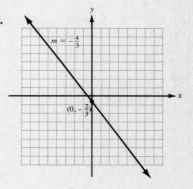

Section 6-4. Supplementary Exercises

1. a. $y = \dfrac{3}{2}x + 3$ **b.** $\dfrac{3}{2}$ **c.** 3 **d.**

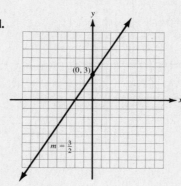

2. a. $y = \dfrac{-4}{3}x - \dfrac{2}{3}$ **b.** $\dfrac{-4}{3}$ **c.** $\dfrac{-2}{3}$

d.

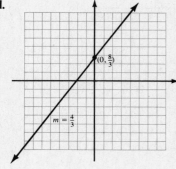

3. a. $y = \dfrac{4}{3}x + \dfrac{8}{3}$ **b.** $\dfrac{4}{3}$ **c.** $\dfrac{8}{3}$ **d.**

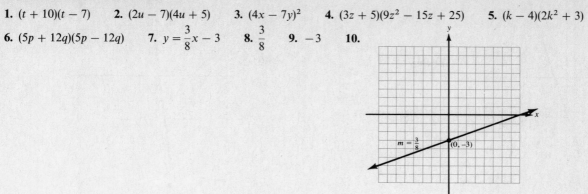

4. a. $y = \dfrac{2}{3}x - \dfrac{5}{6}$ **b.** $\dfrac{2}{3}$ **c.** $\dfrac{-5}{6}$ **d.**

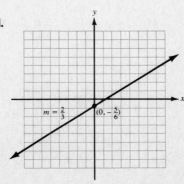

5. a. $y = 3x - 4$ **b.** 3 **c.** -4

d.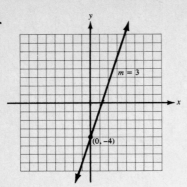

6. a. $y = 8x + 2$ **b.** 8   **c.** 2   **d.**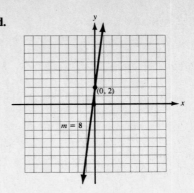

7. a. $x = 3$ **b.** Undefined   **c.** None   **d.**

8. a. $x = -5$ **b.** Undefined   **c.** None

d.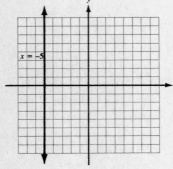

9. a. $y = \dfrac{13}{2}$ **b.** 0   **c.** $\dfrac{13}{2}$ **d.**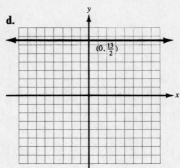

10. a. $y = \dfrac{4}{3}$

b. 0   **c.** $\dfrac{4}{3}$ **d.**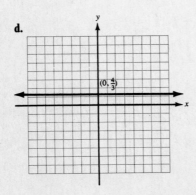

11. a. $y = \dfrac{3}{7}x - 3$ **b.** $\dfrac{3}{7}$ **c.** -3 **d.**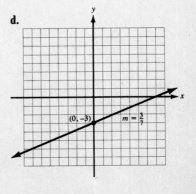

12. a. $y = \dfrac{-3}{5}x + 3$ **b.** $\dfrac{-3}{5}$ **c.** 3 **d.**

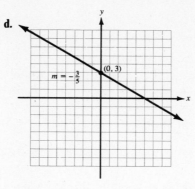

13. a. $y = \dfrac{1}{2}x - \dfrac{3}{2}$ **b.** $\dfrac{1}{2}$ **c.** $\dfrac{-3}{2}$

d.

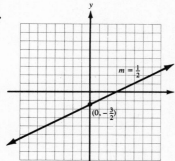

14. a. $y = \dfrac{-1}{3}x + \dfrac{4}{3}$ **b.** $\dfrac{-1}{3}$ **c.** $\dfrac{4}{3}$ **d.**

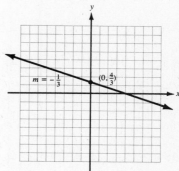

15. a. $y = \dfrac{4}{3}x - \dfrac{5}{3}$ **b.** $\dfrac{4}{3}$ **c.** $\dfrac{-5}{3}$ **d.**

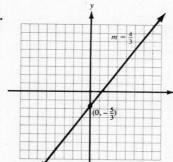

16. a. $y = \dfrac{-3}{4}x + \dfrac{7}{4}$ **b.** $\dfrac{-3}{4}$ **c.** $\dfrac{7}{4}$

d.

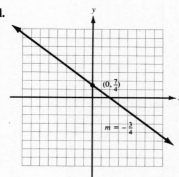

17. a. $\dfrac{2}{3}$ **b.** $\dfrac{-3}{2}$ **18. a.** $\dfrac{4}{3}$ **b.** $\dfrac{-3}{4}$ **19. a.** -1

b. 1 **20. a.** 1 **b.** -1 **21. a.** $\dfrac{4}{9}$ **b.** $\dfrac{-9}{4}$ **22. a.** $\dfrac{5}{6}$ **b.** $\dfrac{-6}{5}$ **23. a.** $\dfrac{-2}{3}$ **b.** $\dfrac{3}{2}$ **24. a.** $\dfrac{-5}{6}$ **b.** $\dfrac{6}{5}$

25. $(4, -1)$ **26.** $(3, -3)$ **27.** $(-4, -4)$ **28.** $(3, 3)$ **29.** $(4, -5)$ **30.** $(6, 2)$ **31.** $(-4, 3)$ **32.** $(2, -5)$

33. $y = \dfrac{5}{3}x - 4$ **34.** $y = \dfrac{2}{5}x + 2$ **35.** $y = \dfrac{-5}{2}x + 6$ **36.** $y = \dfrac{-1}{4}x - 3$ **37.** $y = \dfrac{2}{3}x - 5$ **38.** $y = 4x + 7$

39. $y = \dfrac{-4}{5}x + 5$ **40.** $y = -6x - 7$

Section 6-5. Practice Exercises

1. a. $y - 2 = 3(x - 1)$ **b.** $3x - y = 1$ **2. a.** $y + 2 = 4(x - 7)$ **b.** $4x - y = 30$ **3. a.** $y - 6 = -(x + 4)$

b. $x + y = 2$ **4. a.** $y + 4 = -2(x + 3)$ **b.** $2x + y = -10$ **5. a.** $y + 1 = \frac{1}{2}(x - 1)$ **b.** $x - 2y = 3$

6. a. $y - 3 = \frac{10}{3}(x - 3)$ **b.** $10x - 3y = 21$ **7. a.** $y - 2 = \frac{-3}{2}(x + 2)$ **b.** $3x + 2y = -2$ **8. a.** $y + 5 = \frac{-2}{5}(x - 3)$

b. $2x + 5y = -19$ **9. a.** $y - 4 = \frac{2}{3}(x - 0)$ **b.** $2x - 3y = -12$ **10. a.** $y + 5 = -1(x - 0)$ **b.** $x + y = -5$

11. a. $y - 2 = \frac{4}{3}(x - 5)$ **b.** $4x - 3y = 14$ **12. a.** $y - 5 = \frac{-7}{3}(x + 1)$ **b.** $7x + 3y = 8$ **13. a.** $y - 1 = \frac{-3}{5}(x + 3)$

b. $3x + 5y = -4$ **14. a.** $y + 3 = \frac{2}{3}(x + 2)$ **b.** $2x - 3y = 5$ **15. a.** $y - 0 = \frac{-4}{3}(x - 0)$ **b.** $4x + 3y = 0$

16. a. $y - 0 = \frac{4}{5}(x - 0)$ **b.** $4x - 5y = 0$ **17. a.** $y - 0 = -2(x - 3)$ **b.** $2x + y = 6$ **18. a.** $y + 8 = 4(x - 0)$

b. $4x - y = 8$ **19. a.** $y - 1 = \frac{2}{5}(x - 5)$ **b.** $2x - 5y = 5$ **20. a.** $y - 8 = \frac{-7}{4}(x + 2)$ **b.** $7x + 4y = 18$

21. a. $y + 9 = \frac{8}{5}(x + 9)$ **b.** $8x - 5y = -27$ **22. a.** $y + 6 = \frac{-3}{7}(x - 10)$ **b.** $3x + 7y = -12$

23. a. $y + 2 = \frac{-1}{3}(x - 1)$ **b.** $x + 3y = -5$ **24. a.** $y - 4 = \frac{-1}{2}(x + 3)$ **b.** $x + 2y = 5$ **25. a.** $y - 5 = \frac{4}{3}(x - 0)$

b. $4x - 3y = -15$ **26. a.** $y + 4 = \frac{2}{5}(x - 0)$ **b.** $2x - 5y = 20$ **27. a.** $y - 8 = \frac{8}{5}(x + 6)$ **b.** $8x - 5y = -88$

28. a. $y - 5 = \frac{7}{6}(x + 9)$ **b.** $7x - 6y = -93$ **29. a.** $y = -3$ **b.** $x = 5$ **30. a.** $y = -1$ **b.** $x = 6$

31. a. $y = -8$ **b.** $x = -7$ **32. a.** $y = -9$ **b.** $x = -10$ **33. a.** $y = 4$ **b.** $x = 0$ **34. a.** $y = 10$ **b.** $x = 0$
35. a. $3y = 2$ **b.** $2x = -1$ **36. a.** $8y = 7$ **b.** $4x = -3$

Section 6-5. Ten Review Exercises

1. $2t^2 - 8$ **2.** $16x^8y^4$ **3.** $-36a^3b^2 - 9a^2b^3 + 27ab^4$ **4.** $81u^2 - 72uv + 16v^2$ **5.** $5z^4 - 8z^2 + 1$ **6.** $m(m - 4)$

7. $\frac{6x}{5y}$ **8.** $\frac{t + 1}{t - 1}$ **9.** $\frac{7}{36u}$ **10.** $\frac{4a}{(a + 4)(a - 4)}$

Section 6-5. Supplementary Exercises

1. a. $y - 0 = \frac{-5}{3}(x - 0)$ **b.** $5x + 3y = 0$ **2. a.** $y - 0 = \frac{-4}{7}(x - 0)$ **b.** $4x + 7y = 0$ **3. a.** $y - 0 = 1(x - 4)$

b. $x - y = 4$ **4. a.** $y + 6 = \frac{12}{5}(x - 0)$ **b.** $12x - 5y = 30$ **5. a.** $y - \frac{1}{3} = \frac{2}{3}\left(x - \frac{1}{2}\right)$ **b.** $2x - 3y = 0$

6. a. $y + \frac{7}{5} = \frac{-4}{5}\left(x - \frac{5}{4}\right)$ **b.** $4x + 5y = -2$ **7. a.** $y - 6 = 0(x + 1)$ **b.** $y = 6$ **8. a.** $y + 2 = 0(x - 3)$ **b.** $y = -2$

9. a. $y - 0 = 3(x - 3)$ **b.** $3x - y = 9$ **10. a.** $y - 0 = \frac{3}{4}(x - 4)$ **b.** $3x - 4y = 12$ **11. a.** $y - 0 = \frac{-7}{8}(x + 8)$

b. $7x + 8y = -56$ **12. a.** $y - 0 = \frac{-5}{3}(x + 3)$ **b.** $5x + 3y = -15$ **13. a.** $y - 0 = 1.2(x - 2.5)$ **b.** $1.2x - y = 3$

14. a. $y - 0 = \frac{4}{3}(x + 1.5)$ **b.** $4x - 3y = -6$ **15.** $y = 6x - 3$ **16.** $y = 5x - 4$ **17.** $y = \frac{-3}{2}x + 7$

18. $y = \frac{-5}{4}x + 2$ **19.** $y = \frac{2}{3}x$ **20.** $y = \frac{4}{7}x$ **21.** $y = -3$ **22.** $y = -7$ **23. a.** $P_1(-4, -4)$ **b.** $P_2(6, 1)$ **c.** $\frac{1}{2}$

d. $x - 2y = 4$ **e.** $x - 2y = 4$ **f.** 4   **g.** -2 **24. a.** $P_1(-4, -4)$ **b.** $P_2(1, 6)$ **c.** 2   **d.** $2x - y = -4$

e. $2x - y = -4$ **f.** -2 **g.** 4   **25. a.** $P_1(-6, -3)$ **b.** $P_2(6, 1)$ **c.** $\frac{1}{3}$ **d.** $x - 3y = 3$ **e.** $x - 3y = 3$ **f.** 3

g. -1 **26. a.** $P_1(-5, 0)$ **b.** $P_2(5, -4)$ **c.** $\frac{-2}{5}$ **d.** $2x + 5y = -10$ **e.** $2x + 5y = -10$ **f.** -5 **g.** -2

27. a.

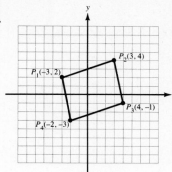

b. $\dfrac{1}{3}$ **c.** $\dfrac{1}{3}$ **d.** Yes **e.** -5 **f.** -5 **g.** Yes **h.** Yes

28. a.

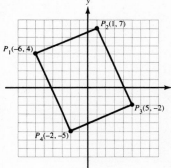

b. $\dfrac{-1}{4}$ **c.** $\dfrac{-1}{4}$ **d.** Yes **e.** 2 **f.** 2 **g.** Yes **h.** Yes

29. a.

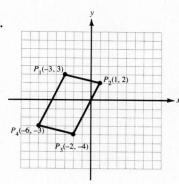

b. $\dfrac{3}{7}$ **c.** $\dfrac{3}{7}$ **d.** Yes **e.** $\dfrac{-9}{4}$ **f.** $\dfrac{-9}{4}$ **g.** Yes **h.** Yes

30. a.

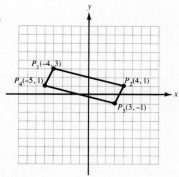

b. $\dfrac{-1}{4}$ **c.** $\dfrac{-1}{4}$ **d.** Yes **e.** 2 **f.** 2 **g.** Yes **h.** Yes **31.** 58

32. 88 **33.** No. It would require initially 75.5 green candies. **34.** Yes. It would require 123 green candies. **35.** 29
36. 64

Section 6-6. Practice Exercises

1. **a.** Yes   **b.** Yes   2. **a.** Yes   **b.** Yes   3. **a.** Yes   **b.** No   4. **a.** No   **b.** Yes   5. **a.** Yes   **b.** Yes
6. **a.** Yes   **b.** No   7. **a.** Yes   **b.** Yes   8. **a.** No   **b.** Yes

9.
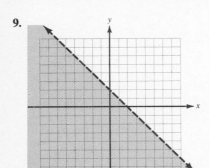

10.
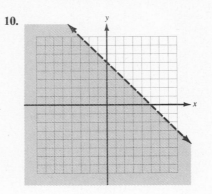

11.
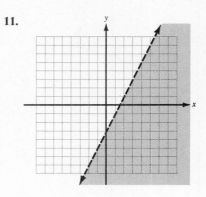

12.
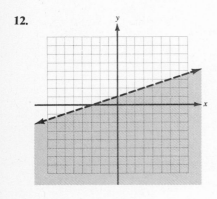

13.
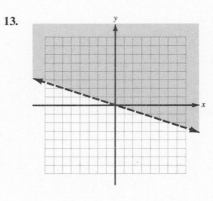

14.
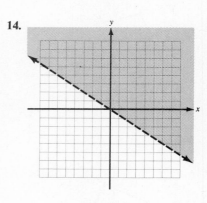

15.
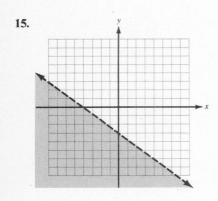

16.
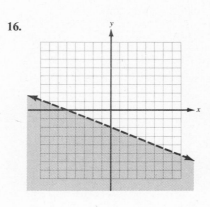

17.
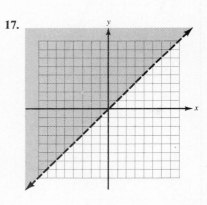

18.
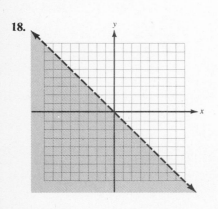

19.
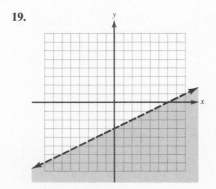

20.
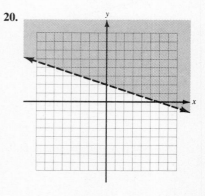

21.

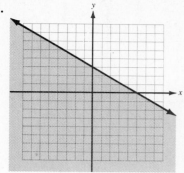

22.

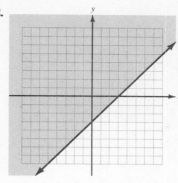

23.

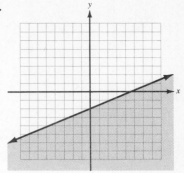

24.

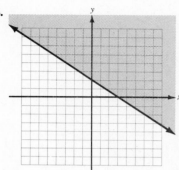

25.

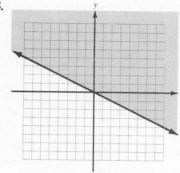

26.

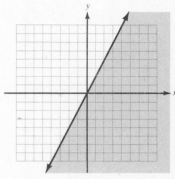

27.

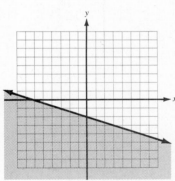

28.

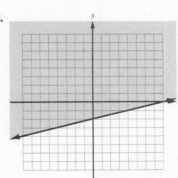

29. $x - 3y < 3$ **30.** $2x - 3y > -6$

31. $5x + 2y \le 10$ **32.** $x + y \le 2$ **33.** $2x - 3y \le -1$ **34.** $x - 4y \le -7$ **35.** $x + 3y < 0$ **36.** $4x + y < 0$

Section 6-6. Ten Review Exercises

1. -4 **2.** $\dfrac{3}{4}$ **3.** $\dfrac{-3}{2}, 2$ **4.** 3 **5.** $\dfrac{-5}{4}, \dfrac{5}{4}$ **6.** $-x^2 - 3xy$ **7.** $-20u^4v^5$ **8.** $72c^2 + 17c - 70$ **9.** $\dfrac{p-3}{p+3}$

10. 6

Section 6-6. Supplementary Exercises

1. a. Yes   **b.** Yes   **2. a.** No   **b.** Yes   **3. a.** Yes   **b.** Yes   **4. a.** Yes   **b.** No

5.

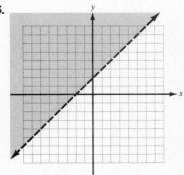

6.

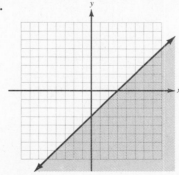

7.

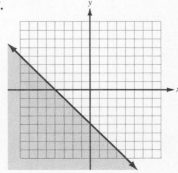

8.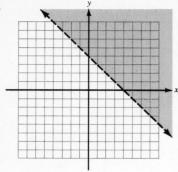

9.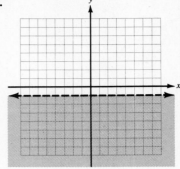

10.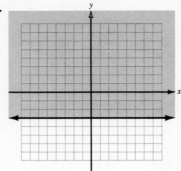

11.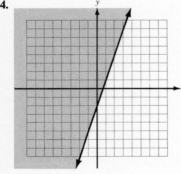

12.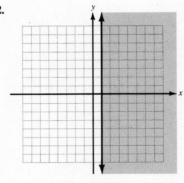

13.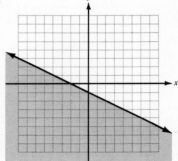

14.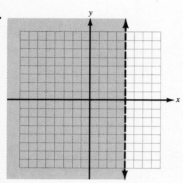

15.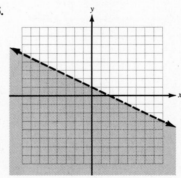

16.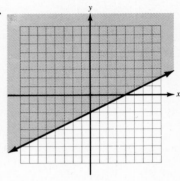

17.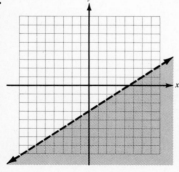

18.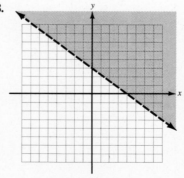

19.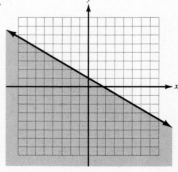

20.

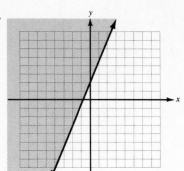

21. > **22.** > **23.** < **24.** < **25.** > **26.** > **27.** *f* **28.** *d*

29. *g* **30.** *a* **31.** *b* **32.** *c*

Chapter 6 Review Exercises

1. a. Yes **b.** No **c.** No **d.** Yes **2. a.** $\left(4, \frac{4}{5}\right)$ **b.** $(-10, 12)$ **3. a.** $A(-5, 5)$ **b.** $B(7, -2)$ **c.** $C(2, 3)$
d. $D(-8, -7)$ **e.** $E(0, -3)$ **4.** **5. a.** Q III **b.** Q IV **c.** Q II

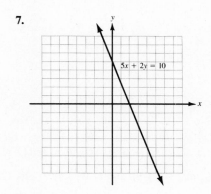

6. a. Horizontal axis **b.** Vertical axis **7.** **8.**

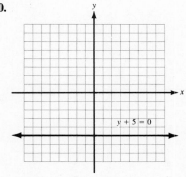

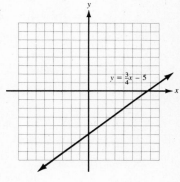

9. **10.** **11. a.** $\frac{24}{7}$ **b.** -3 **12. a.** $\frac{18}{5}$ **b.** 4

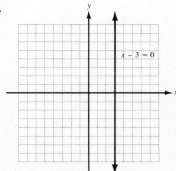

13. $\dfrac{2}{3}$ **14.** 1   **15.** Undefined   **16.** 0   **17. a.** -3 **b.** 4   **18. a.** $\dfrac{3}{4}$ **b.** 0   **19. a.** $\dfrac{-3}{5}$ **b.** $\dfrac{5}{3}$

20. a. $\dfrac{1}{3}$ **b.** -3 **21.** $7x + 3y = -22$ **22.** $4x - 3y = -15$ **23.** $y = -6$ **24.** $2x + y = 9$

25.   **26.**   **27.** $3x + y > 8$

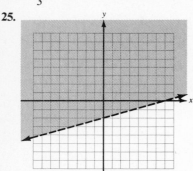   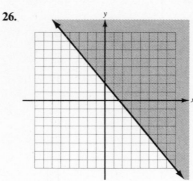

Chapter 7 Systems of Equations

Section 7-1. Practice Exercises

1. (1) $5x - 4y = 11$ (2) $x - y = 2$ **2.** (1) $3x - 5y = 11$ (2) $x - y = 5$ **3.** (1) $x + y = 153$ (2) $x = y + 9$
4. (1) $x + y = 112$ (2) $y = x + 12$ **5.** (1) $3x + 2y = 8.96$ (2) $4x + y = 7.48$ **6.** (1) $2x + 3y = 18.15$
(2) $3x + y = 17.60$ **7.** Yes   **8.** Yes   **9.** Yes   **10.** No   **11.** No   **12.** Yes   **13.** Yes   **14.** Yes   **15.** No
16. Yes   **17.** Yes   **18.** Yes   **19.** Yes   **20.** Yes   **21.** Yes   **22.** No   **23.** $(5, -3)$ **24.** $(-12, 4)$ **25.** $(-6, 2)$
26. $(5, -1)$ **27.** $\left(\dfrac{1}{2}, \dfrac{1}{3}\right)$ **28.** $\left(\dfrac{-1}{3}, \dfrac{-3}{4}\right)$ **29.** $\left(\dfrac{-2}{3}, \dfrac{1}{4}\right)$ **30.** $\left(\dfrac{2}{5}, \dfrac{3}{5}\right)$ **31.** Yes   **32.** Yes   **33.** Yes   **34.** Yes
35. No   **36.** No

Section 7-1. Ten Review Exercises

1. $\dfrac{3}{2}$ **2.** $\dfrac{3}{2}$ **3.** $y - 1 = \dfrac{3}{2}(x + 4)$ **4.** $y = \dfrac{3}{2}x + 7$ **5.** 7   **6.** $2x + 3y = 8$ **7.** 13   **8.** $\dfrac{(3b + 1)(b + 2)}{(b - 3)(b + 2)}$ **9.** $\dfrac{3b + 1}{b - 3}$
10. 13

Section 7-1. Supplementary Exercises

1. a. Yes   **b.** Yes   **c.** Yes   **2. a.** No   **b.** No   **c.** Yes   **3. a.** No   **b.** Yes   **c.** No   **4. a.** Yes   **b.** Yes
c. Yes   **5.** Yes   **6.** Yes   **7.** No   **8.** No   **9.** Yes   **10.** Yes   **11.** Yes   **12.** Yes   **13. a.** 2   **b.** $(2, 4)$
14. a. 3   **b.** $(6, 9)$ **15. a.** -2 **b.** $(-2, 4)$ **16. a.** -3 **b.** $(-9, 3)$ **17. a.** 3   **b.** $(-9, 9)$ **18. a.** -1
b. $(2, -2)$ **19. a.** -4 **b.** $(-2, 4)$ **20. a.** 2   **b.** $(1, 6)$ **21.** $(4, 6)$ and $(-3, -8)$ **22.** $(5, 0)$ and $(-4, 3)$ **23.** $(5, 4)$
24. $(4, 0)$ **25.** $(-5, 7)$ **26.** $(1, 1)$ **27.** $(2, 6), (-2, -6)$, and $(-3, -4)$ **28.** $(2, 6), (-2, -6)$, and $\left(\dfrac{3}{2}, 8\right)$

Section 7-2. Practice Exercises

1.   **2.**   **3.**

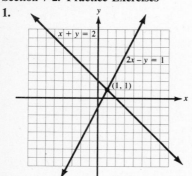   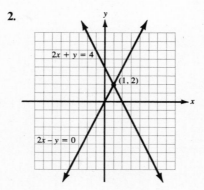   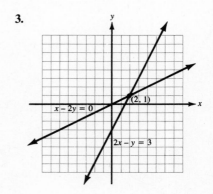

4.
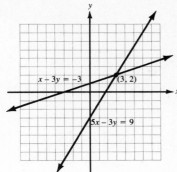
$x - 3y = -3$ (3, 2) $5x - 3y = 9$

5.
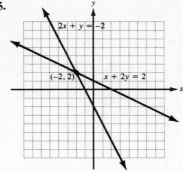
$2x + y = -2$ (−2, 2) $x + 2y = 2$

6.
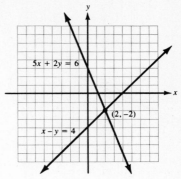
$5x + 2y = 6$ (2, −2) $x - y = 4$

7.
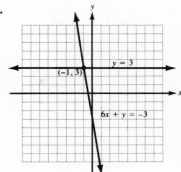
$y = 3$ (−1, 3) $6x + y = -3$

8.
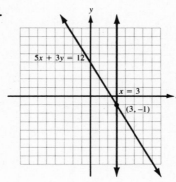
$5x + 3y = 12$ $x = 3$ (3, −1)

9.
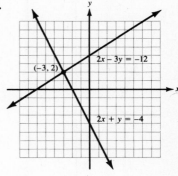
(−3, 2) $2x - 3y = -12$ $2x + y = -4$

10.
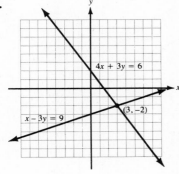
$4x + 3y = 6$ $x - 3y = 9$ (3, −2)

11.
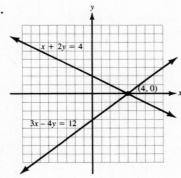
$x + 2y = 4$ (4, 0) $3x - 4y = 12$

12.
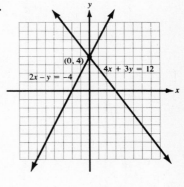
(0, 4) $4x + 3y = 12$ $2x - y = -4$

13.
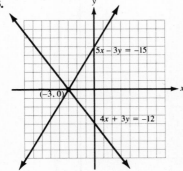
$5x - 3y = -15$ (−3, 0) $4x + 3y = -12$

14.
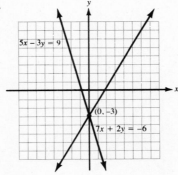
$5x - 3y = 9$ (0, −3) $7x + 2y = -6$

15.
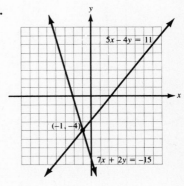
$5x - 4y = 11$ (−1, −4) $7x + 2y = -15$

16.

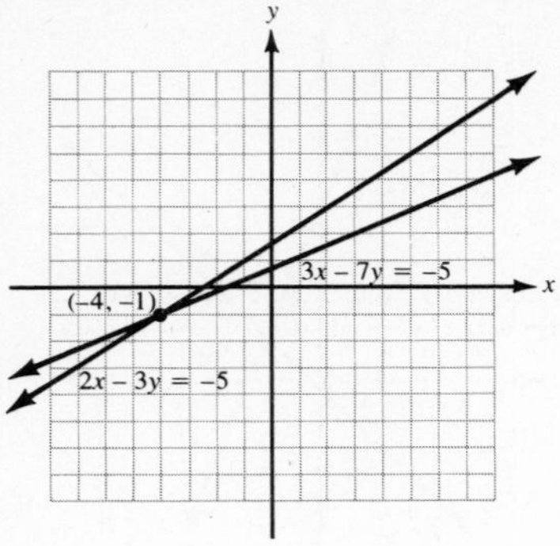

17.

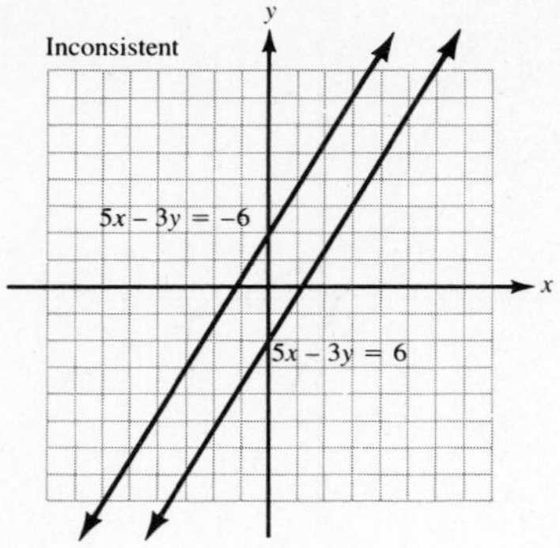

18.

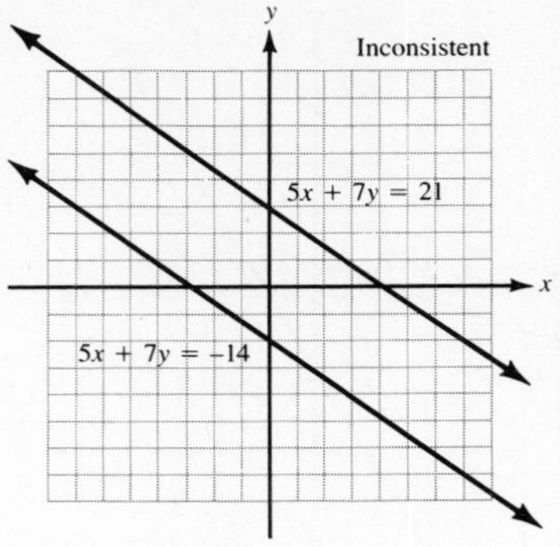

19.

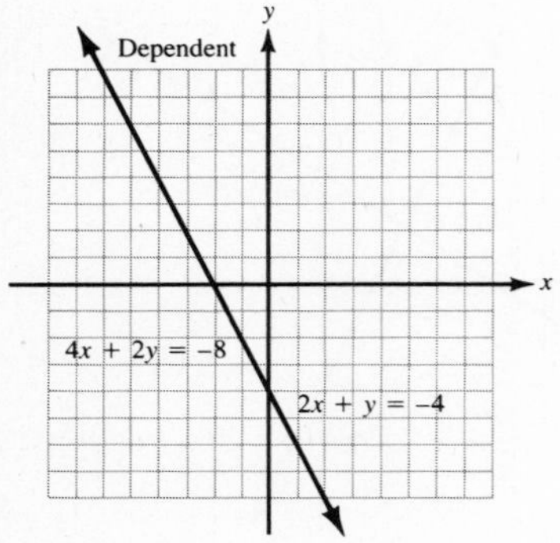

20.

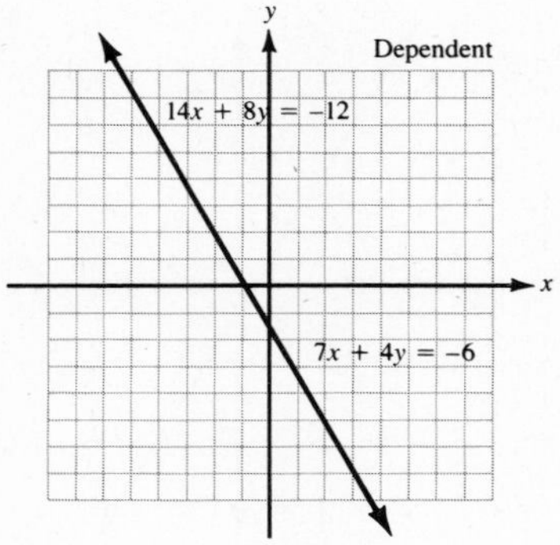

21.

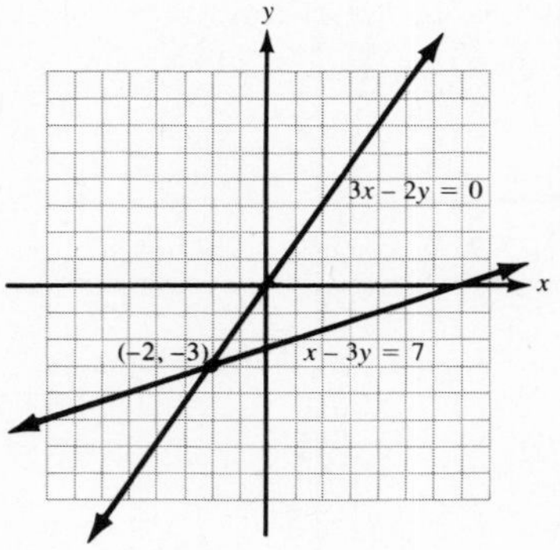

22.

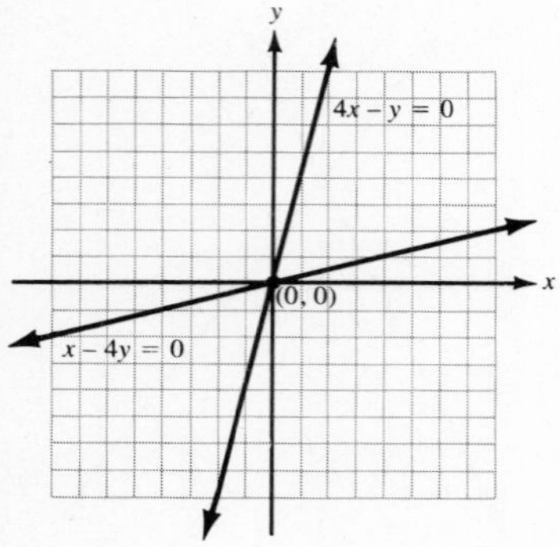

23.

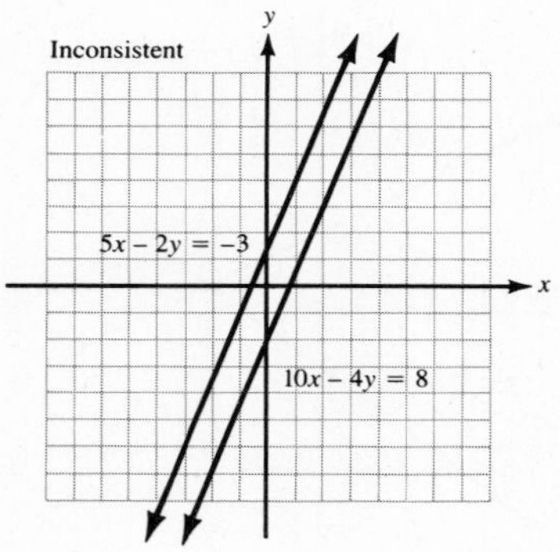

24.

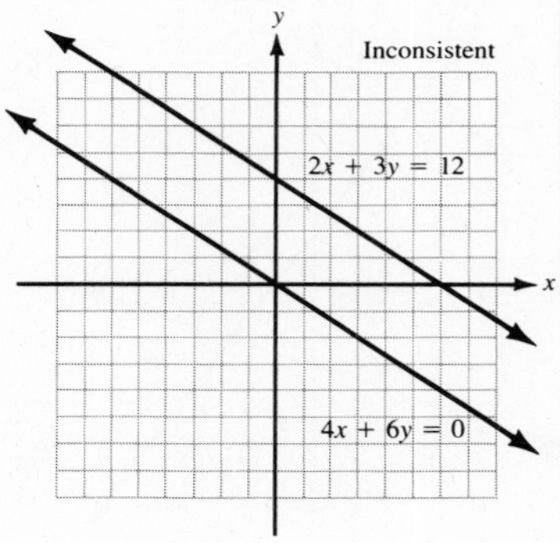

25.

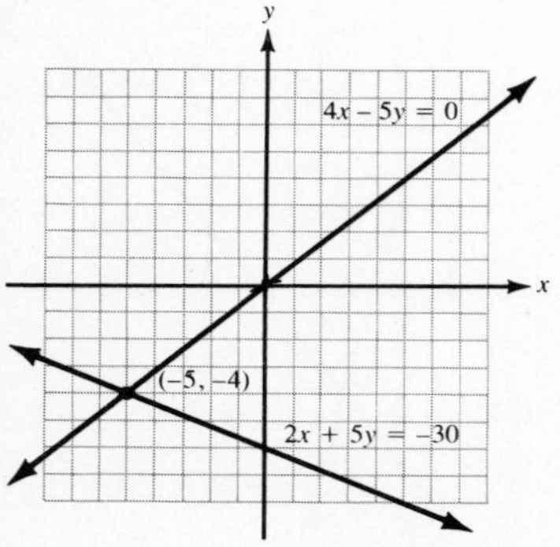

26.

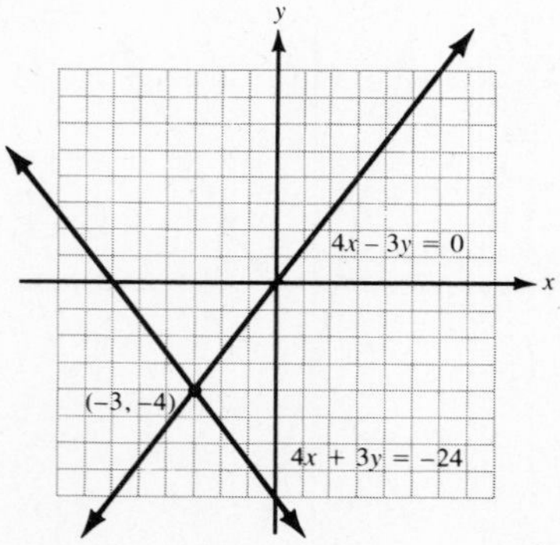

27.

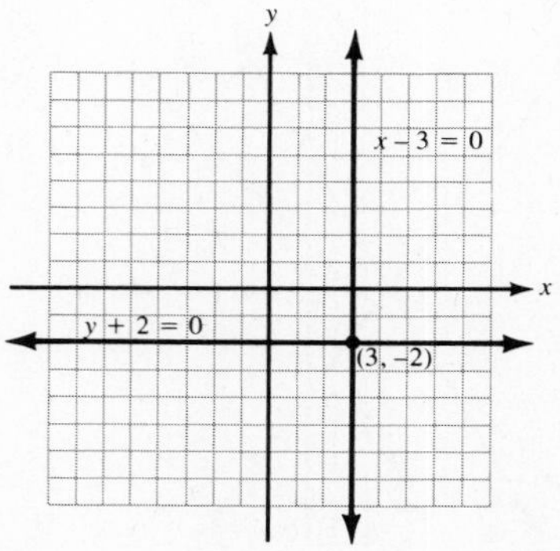

28.

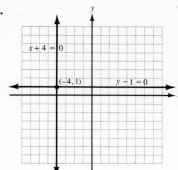

29.

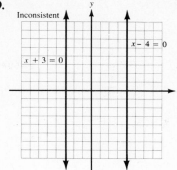

30.

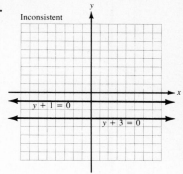

31. Independent and consistent   **32.** Independent and consistent   **33.** Inconsistent   **34.** Inconsistent   **35.** Dependent
36. Dependent   **37.** Independent and consistent   **38.** Independent and consistent   **39.** Inconsistent   **40.** Dependent

Section 7-2. Ten Review Exercises

1. $\dfrac{1}{2}$ **2.** $x - 2y = 4$ **3.** $\dfrac{-3}{2}$ **4.** $3x + 2y = -12$ **5.** (1) $x - 2y = 4$ (2) $3x + 2y = -12$

6.

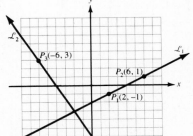

7. $(-2, -3)$ **8.** The difference between a number and 2 times a second number is 4.

9. The sum of 3 times the first number and 2 times the second number is -12.
10. The first number is -2 and the second number is -3.

Section 7-2. Supplementary Exercises

1. 3

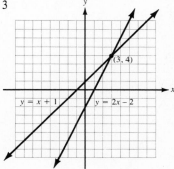

2. 0

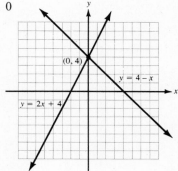

3. -6

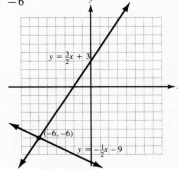

4. -2

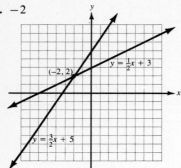

5. a. (1) $2x + 3y = 1$ **b.** (2) $x - 2y = -10$ **c.**

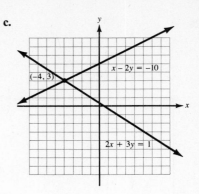

d. x is -4 and y is 3.   **6. a.** (1) $2x + 4y = -2$ **b.** (2) $2x - 3y = -9$ **c.**

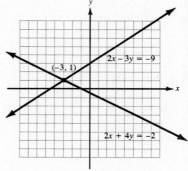

d. x is -3 and y is 1.   **7. a.** (1) $y = 4x - 7$ **b.** (2) $y = \dfrac{4}{3}x + 1$ **c.**

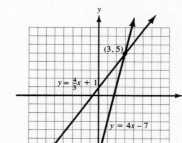

d. x is 3 and y is 5.   **8. a.** (1) $y = 5x - 7$ **b.** (2) $y = 7 - 2x$ **c.**

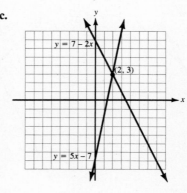

d. x is 2 and y is 3.

9. a. (1) $x = y + 5$ **b.** (2) $2x + 2y = 26$ **c.**

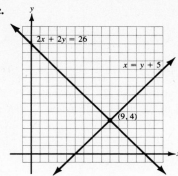

d. The length is 9 m.

10. a. (1) $x = 2y$ **b.** (2) $2x + 2y = 18$ **c.**

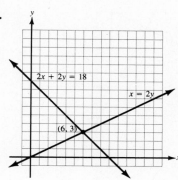

d. The length is 6 m.

11. a. (1) $3.0x + 1.5y = 7.5$ **b.** (2) $x = y + 1$ **c.**

d. 2 lb of almonds and 1 lb of peanuts

12. a. (1) $5.0x + 2.5y = 20.0$ **b.** (2) $y = x + 2$ **c.**

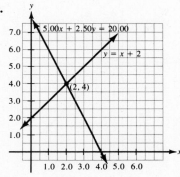

d. 2 lb of candy and 4 lb of mints

Section 7-3. Practice Exercises

1. $(10, -3)$ **2.** $(4, -2)$ **3.** $(-2, -10)$ **4.** $(4, 5)$ **5.** $\left(\dfrac{-3}{2}, 3\right)$ and $(-4, 3)$ **6.** $\left(\dfrac{5}{2}, 6\right)$ and $\left(\dfrac{3}{2}, 6\right)$

7. $(3, 4)$ and $(-3, 4)$ **8.** $(10, 6)$ and $(10, -6)$ **9.** $(2, 1)$ **10.** $(-9, -3)$ **11.** $(6, 4)$ **12.** $(-3, -6)$ **13.** $(-14, -9)$

14. $(-3, -7)$ **15.** $(6, 16)$ **16.** $(-2, -11)$ **17.** $(-4, 1)$ **18.** $(2, -5)$ **19.** $(4, 7)$ **20.** $(-3, -3)$ **21.** $(3, 5)$

22. $(4, 7)$ **23.** $(5, 6)$ **24.** $(8, 4)$ **25.** Inconsistent   **26.** Inconsistent   **27.** Dependent   **28.** Dependent

29. $(9, -1)$ **30.** $(-3, 6)$ **31.** $(0, 0)$ **32.** $(0, 0)$ **33.** Dependent   **34.** Inconsistent   **35.** Inconsistent

36. Dependent   **37.** $(6, -5)$ **38.** $(5, -6)$ **39.** $\left(\dfrac{5}{6}, -5\right)$ **40.** $\left(\dfrac{-7}{4}, 8\right)$

Section 7-3. Ten Review Exercises

1. $\dfrac{3}{4}$ **2.** $\dfrac{-3}{2}$ and 1 **3.** $\dfrac{-1}{2}$ **4.** $y < -2$ **5.** $a = \dfrac{S - v}{t}$ **6.** -2 **7.** $\dfrac{1}{2}$

8. $2x + y = -1$ **9.** $x - 2y = -18$ **10.**

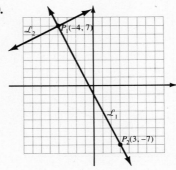

Section 7-3. Supplementary Exercises

1. $\left(\dfrac{1}{2}, 2\right)$ **2.** $\left(\dfrac{2}{3}, 3\right)$ **3.** $\left(-3, \dfrac{3}{4}\right)$ **4.** $\left(5, \dfrac{-2}{5}\right)$ **5.** $\left(\dfrac{-4}{7}, 7\right)$ **6.** $\left(\dfrac{1}{6}, 8\right)$ **7.** Dependent   **8.** Inconsistent

9. $\left(\dfrac{1}{3}, \dfrac{-1}{2}\right)$ **10.** $\left(\dfrac{2}{3}, \dfrac{3}{4}\right)$ **11.** $\left(\dfrac{-1}{2}, \dfrac{-2}{3}\right)$ **12.** $\left(\dfrac{-4}{3}, \dfrac{3}{2}\right)$ **13. a, b, d, f**   **c.** $y = 2$

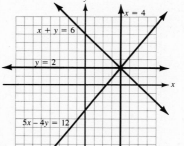

e. $x = 4$ **g.** Yes   **14. a, b, d, f**   **c.** $y = 3$ **e.** $x = 1$ **g.** Yes

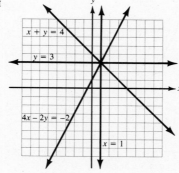

15. a, b, d, f   **c.** $y = 1$ **e.** $x = -5$ **g.** Yes   **16. a, b, d, f**

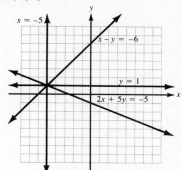

c. $y = 2$ **e.** $x = -3$ **g.** Yes   **17. a, b, d, f**

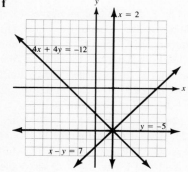

c. $y = -5$ **e.** $x = 2$ **g.** Yes

18. a, b, d, f   **c.** $y = -2$ **e.** $x = 3$ **g.** Yes

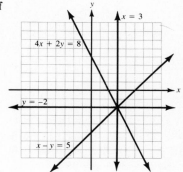

19. Yes, a solution   **20.** Yes, a solution   **21.** No, not a solution of (3)   **22.** No, not a solution of (3)
23. Yes, a solution   **24.** Yes, a solution

Section 7-4. Practice Exercises

1. $\left(\dfrac{-7}{2}, \dfrac{5}{3}\right)$ **2.** $\left(\dfrac{7}{4}, \dfrac{8}{5}\right)$ **3.** $\left(9, \dfrac{21}{8}\right)$ **4.** $\left(5, \dfrac{31}{9}\right)$ **5.** $\left(\dfrac{2}{3}, \dfrac{-11}{4}\right)$ **6.** $\left(\dfrac{8}{3}, \dfrac{-19}{4}\right)$ **7.** $\left(\dfrac{-13}{11}, 3\right)$ **8.** $\left(\dfrac{-8}{13}, 2\right)$

9. $(5, 2)$ **10.** $(-3, 9)$ **11.** $(-5, 2)$ **12.** $(9, -3)$ **13.** $(-3, 4)$ **14.** $(-6, 2)$ **15.** $(2, 3)$ **16.** $(4, 1)$

17. $(-10, -5)$ **18.** $(-4, 2)$ **19.** $(8, -9)$ **20.** $(9, -7)$ **21.** $\left(-2, \dfrac{4}{3}\right)$ **22.** $\left(6, \dfrac{-3}{4}\right)$ **23.** $\left(\dfrac{4}{5}, -5\right)$

24. $\left(\dfrac{-9}{4}, 1\right)$ **25.** $(-6, 1)$ **26.** $(8, 10)$ **27.** Inconsistent   **28.** Inconsistent   **29.** Dependent   **30.** Dependent

31. $\left(\dfrac{3}{4}, \dfrac{2}{5}\right)$ **32.** $\left(\dfrac{-5}{3}, \dfrac{15}{7}\right)$ **33.** Dependent   **34.** Inconsistent

Section 7-4. Ten Review Exercises

1. $-xy$ **2.** 10   **3.** $(3t - 8)(t + 1)$ **4.** 0   **5.** $\dfrac{m + 1}{2m}$ **6.** $\dfrac{2}{3}$ **7.** The numbers are -5 and 12

8. The numbers are -8 and 5.　　**9.** The width is 5 ft; the length is 12 ft.　　**10.** $a \geq 2$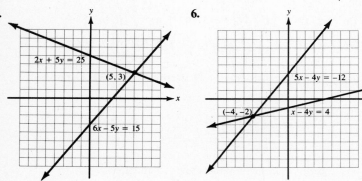

Section 7-4. Supplementary Exercises

1. $(1.6, 3.0)$　　**2.** $(4.0, -0.8)$　　**3.** $(3, 7)$　　**4.** $(-6, 4)$　　**5.** $(1.25, -4.375)$　　**6.** $(2.5, -1.8)$　　**7.** $\left(\dfrac{5}{9}, \dfrac{-1}{3}\right)$　　**8.** $\left(\dfrac{-3}{2}, \dfrac{3}{5}\right)$

9. $(0.6, -1.2)$　　**10.** $(4.5, 5.2)$　　**11. a.** $y + 2z = -4$　　**b.** $-2y + 3y = -13$　　**c.** $(2, -3)$　　**d.** 1　　**f.** $(1, 2, -3)$

12. a. $-3y + 3z = 18$　　**b.** $2y - z = 13$　　**c.** $(19, 25)$　　**d.** 19　　**f.** $(19, 19, 25)$　　**13. a.** $-x - z = 1$　　**b.** $x + 2z = 1$

c. $(-3, 2)$　　**d.** 4　　**f.** $(-3, 4, 2)$　　**14. a.** $6x - 3z = 6$　　**b.** $8x - z = 20$　　**c.** $(3, 4)$　　**d.** 6　　**f.** $(3, 6, 4)$

15. a. $8x - y = 10$　　**b.** $11x = 11$　　**c.** $(1, -2)$　　**d.** -3　　**f.** $(1, -2, -3)$　　**16. a.** $-2x - 2y = 2$　　**b.** $4x - y = -4$

c. $(-1, 0)$　　**d.** 3　　**f.** $(-1, 0, 3)$　　**17. a.** $5 = x^2 + 4x$　　**b.** -5 and 1　　**c.** 7 and -5　　**d.** $(-5, 7)$ and $(1, -5)$

18. a. $2 = x^2 + x$　　**b.** -2 and 1　　**c.** 5 and 8　　**d.** $(-2, 5)$ and $(1, 8)$　　**19. a.** $12 = x^2 + 4x$　　**b.** -6 and 2

c. 24 and 8　　**d.** $(-6, 24)$ and $(2, 8)$　　**20. a.** $1 = x^2$　　**b.** -1 and 1　　**c.** -6 and 2　　**d.** $(-1, -6)$ and $(1, 2)$

Section 7-5. Practice Exercises

1. The first number is 5; the second number is 8.　　**2.** The first number is 15; the second number is 7.

3. The numbers are 9 and -10.　　**4.** The numbers are -2 and -6.　　**5.** The numbers are $\frac{1}{3}$ and $\frac{1}{4}$.

6. The numbers are $\frac{3}{2}$ and $\frac{-3}{5}$.　　**7.** The numbers are $\frac{-7}{6}$ and $\frac{7}{3}$.　　**8.** The numbers are $\frac{2}{5}$ and $\frac{2}{5}$.　　**9.** 8 in by 10 in

10. 3 ft by 7 ft　　**11.** 8.5 in by 11 in　　**12.** 120 ft by 300 ft　　**13.** 12 ft by 20 ft　　**14.** 280 m by 360 m　　**15.** 14 cm by 19 cm

16. 22 in by 29 in　　**17.** 65 dimes and 35 quarters　　**18.** 19 nickels and 28 dimes　　**19.** Thirty 15¢ stamps and fifteen 30¢ stamps

20. 7 lb of apples and 5 lb of pears　　**21.** 6 yd at \$2.25 and 4.5 yd at \$2.90　　**22.** 3,600 students and 1,800 general admission

23. 87.5 l of 20% and 52.5 l of 60% alcohol　　**24.** 60 l of 80% with 40 l of 15% alcohol　　**25.** 150 mi from Fresno

26. 12 mi from starting point　　**27.** 4 hr　　**28.** 7 hr　　**29.** 2 hr　　**30.** 8 hr　　**31.** 25 mph for boat, 5 mph for current

32. 275 mph for plane, 25 mph for wind　　**33.** Alice is 6 and Betty is 14.　　**34.** Cindy is 25 and Darlene is 15.

35. Ed is 48 and Frank is 28.　　**36.** George is 30 and Harry is 60.　　**37.** Mark is 48 and Linda is 32.

38. Oliver is 8 and Nan is 28.　　**39.** Sherry is 11 and Rusty is 48.　　**40.** Vanessa is 22 and Terry is 28.

Section 7-5. Ten Review Exercises

1. $\dfrac{-1}{5}$　　**2.** $x + 5y = -10$　　**3.** -10　　**4.** -2　　**5.** $5x - y = 11$　　**6.** $49x^2 - 7xy - 12y^2$　　**7.** $8t^3 - 26t^2 + 19t - 3$

8. $3m^2 + 7mn - 14n^2$　　**9.** $169u^4 - 26u^2v^2 + v^4$　　**10.** $5z^2 - z - 2$

Section 7-5. Supplementary Exercises

1. a. (1) $x + y + z = 7$　　(2) $x + 2y - z = -8$　　(3) $3x + y + z = -3$　　**b.** (4) $2x + 3y = -1$　　**c.** (5) $4x + 3y = -11$

d. $(-5, 3)$　　**e.** 9　　**f.** x is -5; y is 3; z is 9　　**2. a.** (1) $x + y + z = 15$　　(2) $2x + 3y - z = 1$　　(3) $x - y + z = 1$

b. (4) $3x + 4y = 16$　　**c.** (5) $3x + 2y = 2$　　**d.** $(-4, 7)$　　**e.** 12　　**f.** x is -4; y is 7; z is 12　　**3. a.** (1) $2x + y + z = 6.20$

(2) $x + 3y + z = 7.60$　　(3) $2x + 2y + z = 7.30$　　**b.** (4) $x - 2y = -1.40$　　**c.** (5) $x - y = -0.30$　　**d.** $(0.80, 1.10)$　　**e.** 3.50

f. Peaches \$0.80, grapes \$1.10, watermelon \$3.50　　**4. a.** (1) $x + 2y + z = 20.00$　　(2) $2x + 3y + z = 26.75$

(3) $3x + 2y + 2z = 33.25$　　**b.** (4) $-x - y = -6.75$　　**c.** (5) $-x - 4y = -20.25$　　**d.** $(2.25, 4.50)$　　**e.** 8.75

f. Wash cloths \$2.25, hand towels \$4.50, bath towels \$8.75

Chapter 7 Review Exercises

1. a. No　　**b.** Yes　　**2. a.** No　　**b.** Yes　　**3.** $(-2, 7)$　　**4.** $\left(\dfrac{-3}{4}, \dfrac{2}{5}\right)$

5.　　　　　　　　　　　　　　　**6.**　　　　　　　　　　　　**7.** Inconsistent　　**8.** Dependent

9. $(-7, 9)$ **10.** $(3, 9)$ **11.** $\left(-12, \dfrac{7}{4}\right)$ **12.** $\left(\dfrac{-3}{5}, \dfrac{-3}{5}\right)$ **13.** $\left(\dfrac{3}{5}, \dfrac{9}{4}\right)$ **14.** $\left(\dfrac{5}{9}, \dfrac{-8}{9}\right)$ **15.** $\left(\dfrac{1}{2}, \dfrac{3}{4}\right)$ **16.** Dependent

17. 6.4 lb of \$1.00 and 3.6 lb of \$1.75   **18.** The numbers are 9 and -6.   **19.** 80 cm by 110 cm

20. 90 lb of \$1.20 and 60 lb of \$1.65   **21.** Jack's age is 10; Jill's age is 18.   **22.** 7 hr

Chapter 8 Radical Expressions

Section 8-1. Practice Exercises

1. a. No   **b.** No   **c.** Yes   **2. a.** No   **b.** No   **c.** Yes   **3. a.** Yes   **b.** No   **c.** No   **4. a.** Yes   **b.** No
c. No   **5. a.** Yes   **b.** Yes   **c.** No   **6. a.** Yes   **b.** Yes   **c.** No   **7. a.** Yes   **b.** No   **c.** No   **8. a.** Yes

b. No   **c.** No   **9.** 3   **10.** 5   **11.** -12 **12.** -10 **13.** 15   **14.** 21   **15.** $\dfrac{-9}{4}$ **16.** $\dfrac{-19}{7}$ **17.** 0.2

18. 0.1   **19.** -1.1 **20.** -1.2 **21.** $\dfrac{5}{14}$ **22.** $\dfrac{7}{6}$ **23. a.** Rational   **b.** Irrational   **24. a.** Rational   **b.** Irrational

25. a. Irrational   **b.** Rational   **26. a.** Irrational   **b.** Rational   **27. a.** Rational   **b.** Irrational   **28. a.** Rational
b. Irrational   **29. a.** Irrational   **b.** Irrational   **30. a.** Irrational   **b.** Irrational   **31.** 8.426   **32.** 5.745
33. -5.196 **34.** -8.246 **35.** 9.950   **36.** 7.071   **37.** -8.307 **38.** -9.798 **39.** 72.256   **40.** 69.699
41. -54.854 **42.** -52.697 **43.** 7.899   **44.** 8.849   **45.** -11.507 **46.** -14.786 **47.** 2.507   **48.** 0.798
49. -0.158 **50.** 0.305   **51.** 2.423   **52.** 0.868   **53.** -2.562 **54.** 0.813

Section 8-1. Ten Review Exercises

1. $\dfrac{-2}{3}$ **2.** 2   **3.** -4 and $\dfrac{3}{5}$ **4.** $(-2, -2)$ **5.** Inconsistent   **6.** $30t^6$ **7.** $-72z^5$ **8.** $\dfrac{5a^3}{3}$ **9.** $\dfrac{-1}{6n^2}$

10. $\dfrac{-250}{u}$

Section 8-1. Supplementary Exercises

1. $3^{1/2}$ **2.** $10^{1/2}$ **3.** $24^{1/2}$ **4.** $31^{1/2}$ **5.** $-6^{1/2}$ **6.** $-19^{1/2}$ **7.** $\left(\dfrac{2}{3}\right)^{1/2}$ **8.** $\left(\dfrac{5}{8}\right)^{1/2}$ **9.** $0.7^{1/2}$ **10.** $6.3^{1/2}$

11. a. $\sqrt{4}$ **b.** 2   **12. a.** $\sqrt{25}$ **b.** 5   **13. a.** $\sqrt{121}$ **b.** 11   **14. a.** $\sqrt{144}$ **b.** 12   **15. a.** $\sqrt{13}$ **b.** 3.606
16. a. $\sqrt{17}$ **b.** 4.123   **17. a.** $\sqrt{68}$ **b.** 8.246   **18. a.** $\sqrt{92}$ **b.** 9.592   **19. a.** 0.04 under   **b.** 0.0119 under
c. 0.000604 under   **20. a.** 0.11 under   **b.** 0.0071 under   **c.** 0.000176 under   **21. a.** 0.24 over   **b.** 0.0144 under
c. 0.001756 under   **22. a.** 0.25 over   **b.** 0.0284 under   **c.** 0.000704 under   **23. a.** 0.09 over   **b.** 0.0159 under
c. 0.005319 under   **24. a.** 0.49 over   **b.** 0.0356 over   **c.** 0.001649 over

25.   **26.**   **27.**   **28.**   **29.**   **30.**   **31.**

32.   **33.** $5i$ **34.** $6i$ **35.** $4i$ **36.** $3i$ **37.** $11i$ **38.** $12i$ **39.** $\dfrac{3}{5}i$ **40.** $\dfrac{2}{7}i$ **41.** $0.2i$ **42.** $0.3i$

Section 8-2. Practice Exercises

1. $\sqrt{14}$ **2.** $\sqrt{15}$ **3.** $\sqrt{78}$ **4.** $\sqrt{130}$ **5.** 19   **6.** 37   **7.** 8   **8.** 10   **9.** 9   **10.** 12   **11.** $\sqrt{66}$ **12.** $\sqrt{30}$
13. 15   **14.** 12   **15.** 20   **16.** 30   **17.** $-\sqrt{110}$ **18.** $-\sqrt{30}$ **19.** $\sqrt{30}$ **20.** $\sqrt{210}$ **21.** 6   **22.** 10   **23.** 7
24. 6   **25.** -12 **26.** -14 **27.** $2\sqrt{3}$ **28.** $3\sqrt{5}$ **29.** $5\sqrt{3}$ **30.** $2\sqrt{5}$ **31.** $6\sqrt{6}$ **32.** $10\sqrt{2}$ **33.** $18\sqrt{7}$
34. $10\sqrt{2}$ **35.** $4\sqrt{2}$ **36.** $6\sqrt{2}$ **37.** $-6\sqrt{5}$ **38.** $-8\sqrt{3}$ **39.** $11\sqrt{3}$ **40.** $9\sqrt{3}$ **41.** $12\sqrt{2}$ **42.** $7\sqrt{7}$

43. $-75\sqrt{3}$ **44.** $-160\sqrt{5}$ **45.** $\sqrt{5}$ **46.** $\sqrt{6}$ **47.** $-\sqrt{7}$ **48.** $-\sqrt{7}$ **49.** 4   **50.** 2   **51.** $\dfrac{5}{3}$ **52.** $\dfrac{6}{7}$

53. 9   **54.** 5   **55.** 8   **56.** 9   **57.** $\dfrac{\sqrt{11}}{3}$ **58.** $\dfrac{\sqrt{15}}{2}$ **59.** $\dfrac{\sqrt{3}}{7}$ **60.** $\dfrac{\sqrt{5}}{8}$ **61.** $\dfrac{\sqrt{10}}{5}$ **62.** $\dfrac{\sqrt{21}}{7}$ **63.** $\dfrac{\sqrt{3}}{3}$

64. $\dfrac{\sqrt{2}}{2}$ **65.** $\dfrac{\sqrt{70}}{10}$ **66.** $\dfrac{\sqrt{170}}{17}$ **67.** $\dfrac{6\sqrt{7}}{7}$ **68.** $\dfrac{5\sqrt{2}}{2}$ **69.** $\dfrac{2\sqrt{15}}{9}$ **70.** $\dfrac{2\sqrt{21}}{15}$

Section 8-2. Ten Review Exercises

1. $y = \dfrac{2}{5}x - 6$ **2.** $\dfrac{2}{5}$ **3.** -6 **4.** 15 **5.** **6.** $5x + 2y = -10$ **7.** $(5, -4)$

8. Yes **9.** $3t(2t - 1)(4t^2 + 2t + 1)$ **10.** $(2x + 3)(5y - 1)$

Section 8-2. Supplementary Exercises

1. a. 4.243 **b.** $3\sqrt{2}$ **c.** 4.243 **2. a.** 7.071 **b.** $5\sqrt{2}$ **c.** 7.071 **3. a.** 6.928 **b.** $4\sqrt{3}$ **c.** 6.928
4. a. 5.196 **b.** $3\sqrt{3}$ **c.** 5.196 **5. a.** 11.180 **b.** $5\sqrt{5}$ **c.** 11.180 **6. a.** 13.416 **b.** $6\sqrt{5}$ **c.** 13.416
7. a. 15.875 **b.** $6\sqrt{7}$ **c.** 15.875 **8. a.** 18.520 **b.** $7\sqrt{7}$ **c.** 18.520 **9.** $\dfrac{-8}{5}$ **10.** $\dfrac{-17}{10}$ **11.** 0.09 **12.** 0.07
13. $\dfrac{3}{5}$ **14.** $\dfrac{7}{2}$ **15.** -0.28 **16.** -0.26 **17. a.** 4 **b.** 4.2 **c.** 17.64 **d.** 4.2 **18. a.** 4 **b.** 3.6 **c.** 12.96
d. 3.6 **19. a.** 5 **b.** 5.2 **c.** 27.04 **d.** 5.2 **20. a.** 5 **b.** 5.5 **c.** 30.25 **d.** 5.5 **21. a.** 8 **b.** 7.7
c. 59.29 **d.** 7.7 **22. a.** 9 **b.** 8.6 **c.** 73.96 **d.** 8.6 **23. a.** 10 **b.** 9.6 **c.** 92.16 **d.** 9.6 **24. a.** 10
b. 10.4 **c.** 108.16 **d.** 10.4 **25.** $<$ **26.** $<$ **27.** $<$ **28.** $<$ **29.** $>$ **30.** $>$ **31.** $<$ **32.** $>$

Section 8-3. Practice Exercises

1. $7\sqrt{2}$ **2.** $9\sqrt{5}$ **3.** $4\sqrt{6}$ **4.** $5\sqrt{3}$ **5.** $15\sqrt{11}$ **6.** $3\sqrt{7}$ **7.** $3\sqrt{5} + 8\sqrt{2}$ **8.** $\sqrt{3} + \sqrt{2}$ **9.** $6\sqrt{10} - 4\sqrt{3}$
10. $2\sqrt{6} - 4\sqrt{2}$ **11.** $5\sqrt{2}$ **12.** $9\sqrt{2}$ **13.** $3\sqrt{5}$ **14.** $6\sqrt{5}$ **15.** $15\sqrt{2}$ **16.** $\sqrt{2}$ **17.** $2\sqrt{6}$ **18.** $5\sqrt{6}$
19. $9\sqrt{2x}$ **20.** $10\sqrt{3x}$ **21.** $5\sqrt{5y}$ **22.** $6\sqrt{6y}$ **23.** $2\sqrt{7a} + 14\sqrt{6b}$ **24.** $13\sqrt{10a} - \sqrt{3b}$ **25.** $2\sqrt{11s} - 8\sqrt{3t}$
26. $3\sqrt{13s} - 21\sqrt{6t}$ **27.** $\dfrac{1 + \sqrt{3}}{2}$ **28.** $\dfrac{1 + 2\sqrt{2}}{2}$ **29.** $2 - \sqrt{5}$ **30.** $3 - \sqrt{7}$ **31.** $\dfrac{1 + \sqrt{2}}{2}$ **32.** $\dfrac{2 + \sqrt{3}}{3}$
33. $\dfrac{1 - \sqrt{3}}{3}$ **34.** $\dfrac{2 - \sqrt{6}}{3}$ **35.** $7\sqrt{2} + 2$ **36.** $2\sqrt{5} + 5$ **37.** $4\sqrt{6} - 3\sqrt{2}$ **38.** $5\sqrt{6} - 2\sqrt{3}$ **39.** $8 - 5\sqrt{2}$
40. $4 + \sqrt{3}$ **41.** $-1 + 19\sqrt{5}$ **42.** $4 + 9\sqrt{7}$ **43.** 14 **44.** -5 **45.** 55 **46.** 24 **47.** 1 **48.** -14 **49.** 25
50. 53

Section 8-3. Ten Review Exercises

1. $a^2 + ab$ **2.** -6 **3.** 33 **4.** $t = \dfrac{5 + a}{2 - b}$ **5.** $\dfrac{-1}{2}$ **6.** $\dfrac{-3}{4}, \dfrac{3}{4}$ **7.** $\dfrac{-3}{4}, \dfrac{4}{3}$ **8.** $\dfrac{2}{3}$ **9.** $\dfrac{m + 2n}{m - 2n}$ **10.** $\dfrac{a}{2a - 1}$

Section 8-3. Supplementary Exercises

1. 8 **2.** -34 **3.** $\dfrac{5}{2}$ **4.** 3 **5.** $\dfrac{1}{2}$ **6.** $\dfrac{1}{2}$ **7.** $6 - 13\sqrt{2}$ **8.** $23 + 16\sqrt{3}$ **9.** $27 - 2\sqrt{2}$ **10.** $3 + 10\sqrt{3}$
11. $-11\sqrt{5}$ **12.** $8\sqrt{2}$ **13.** $-9\sqrt{3t} - 5\sqrt{u}$ **14.** $15\sqrt{2t} - 2\sqrt{u}$ **15.** $\dfrac{2\sqrt{a}}{3} + \dfrac{29\sqrt{2b}}{6}$ **16.** $\dfrac{25\sqrt{a}}{6} - \dfrac{7\sqrt{3b}}{6}$ **17.** $\sqrt{n}$
18. $-0.3\sqrt{m} - \sqrt{n}$ **19.** 1 **20.** 1 **21.** $5 + 2\sqrt{6}$ **22.** $11 + 2\sqrt{30}$ **23.** $5 - 2\sqrt{6}$ **24.** $11 - 2\sqrt{30}$
25. $10 + 5\sqrt{6}$ **26.** $22 + 5\sqrt{30}$ **27.** $3\sqrt{5} - 5\sqrt{3}$ **28.** $3\sqrt{2} - 2\sqrt{3}$ **29.** $\dfrac{-1}{9}$ **30.** $\dfrac{1}{2}$ **31.** $\dfrac{3 + 2\sqrt{2}}{9}$
32. $\dfrac{8 + 3\sqrt{7}}{2}$ **33.** 2 **34.** -1 **35.** $\dfrac{2 - \sqrt{5}}{2}$ **36.** $\dfrac{2 + \sqrt{3}}{2}$ **37.** $\dfrac{-1 + \sqrt{2}}{2}$ **38.** $\dfrac{-1}{2}$ **39.** $\dfrac{-3}{2}$ **40.** $\dfrac{1}{3}$

41. a. 12.124 **b.** $7\sqrt{3}$ **c.** 12.124 **d.** Same **42. a.** 24.042 **b.** $17\sqrt{2}$ **c.** 24.042 **d.** Same **43. a.** 42.485
b. $19\sqrt{5}$ **c.** 42.485 **d.** Same **44. a.** 4.472 **b.** $2\sqrt{5}$ **c.** 4.472 **d.** Same **45. a.** 7.071 **b.** $5\sqrt{2}$ **c.** 7.071
d. Same **46. a.** 10.392 **b.** $6\sqrt{3}$ **c.** 10.392 **d.** Same

Section 8-4. Practice Exercises

1. $6\sqrt{10}$ **2.** $12\sqrt{3}$ **3.** $-30\sqrt{2}$ **4.** $-16\sqrt{2}$ **5.** $22\sqrt{3}$ **6.** $25\sqrt{5}$ **7.** $10\sqrt{3}$ **8.** $8\sqrt{5}$ **9.** $-27\sqrt{10}$
10. $-56\sqrt{2}$ **11.** $-11\sqrt{2}$ **12.** $\sqrt{3}$ **13.** $\dfrac{\sqrt{6}}{3}$ **14.** $\dfrac{\sqrt{35}}{7}$ **15.** $\dfrac{\sqrt{10t}}{10t}$ **16.** $\dfrac{\sqrt{6t}}{2}$ **17.** $\dfrac{-3\sqrt{10}}{5}$ **18.** $\dfrac{-2\sqrt{55}}{11}$
19. $\dfrac{\sqrt{6u}}{3}$ **20.** $\dfrac{\sqrt{15u}}{5}$ **21.** $\dfrac{5\sqrt{6x}}{6}$ **22.** $\dfrac{3\sqrt{10x}}{5}$ **23.** $\dfrac{\sqrt{6a}}{2}-\dfrac{\sqrt{6b}}{3}$ **24.** $\dfrac{7\sqrt{10a}}{10}-\dfrac{9\sqrt{5b}}{10}$ **25.** $\dfrac{\sqrt{10}}{10}$ **26.** $\dfrac{\sqrt{13}}{13}$
27. $\dfrac{\sqrt{15}}{3}$ **28.** $\dfrac{\sqrt{21}}{7}$ **29.** $\dfrac{-\sqrt{6}}{3}$ **30.** $\dfrac{-\sqrt{30}}{5}$ **31.** $\dfrac{\sqrt{6}}{3}$ **32.** $\dfrac{\sqrt{3}}{5}$ **33.** $\dfrac{-\sqrt{2}}{10}$ **34.** $\dfrac{-\sqrt{5}}{10}$ **35.** $\dfrac{x\sqrt{14}}{2}$
36. $\dfrac{x\sqrt{21}}{7}$ **37.** $-2+\sqrt{2}$ **38.** $-3+\sqrt{3}$ **39.** $3+\sqrt{5}$ **40.** $2(3+\sqrt{2})$ **41.** $4(1+\sqrt{3})$ **42.** $5(1+\sqrt{5})$
43. $7(3-2\sqrt{2})$ **44.** $-3(5+3\sqrt{3})$

Section 8-4. Ten Review Exercises

1. -2 **2.** $2x+y=-4$ **3.** $y=-2x-4$ **4.** -4 **5.** -2 **6.** $2x+y=2$ **7.** $x-2y=-2$ **8.** $-2\cdot\dfrac{1}{2}=-1$
9. $(-2,0)$ **10.**

Section 8-4. Supplementary Exercises

1. $\dfrac{2\sqrt{6}}{3}$ **2.** $\dfrac{4\sqrt{5}}{7}$ **3.** $\dfrac{7\sqrt{10}}{10}$ **4.** $\dfrac{8\sqrt{15}}{15}$ **5.** $\dfrac{5\sqrt{2}}{2}$ **6.** $\dfrac{10\sqrt{3}}{3}$ **7.** $\dfrac{7\sqrt{3x}}{3}$ **8.** $\dfrac{5\sqrt{2y}}{2}$ **9.** $\sqrt{2}+3\sqrt{3}$
10. $7\sqrt{2}-2\sqrt{3}$ **11.** $8\sqrt{3a}+7\sqrt{2b}$ **12.** $10\sqrt{5a}-8\sqrt{2b}$ **13.** $\dfrac{7\sqrt{2x}}{2}-\dfrac{5\sqrt{3y}}{3}$ **14.** $\sqrt{2x}-\sqrt{3y}$ **15.** 1 **16.** 3
17. $\dfrac{\sqrt{3}}{2}$ **18.** $\dfrac{\sqrt{7}}{2}$ **19.** $\dfrac{1}{3\sqrt{2}}$ **20.** $\dfrac{1}{4\sqrt{3}}$ **21.** $\dfrac{1}{2\sqrt{5}}$ **22.** $\dfrac{1}{2\sqrt{7}}$ **23.** $\dfrac{1}{2(2-\sqrt{2})}$ **24.** $\dfrac{1}{-3+\sqrt{3}}$ **25.** $\dfrac{2}{-3+\sqrt{5}}$
26. $\dfrac{3}{-6+\sqrt{3}}$ **27.** $\dfrac{-5}{3+2\sqrt{11}}$ **28.** $\dfrac{-2}{5-3\sqrt{7}}$ **29.** $\dfrac{-2}{3\sqrt{2}-\sqrt{30}+\sqrt{15}-5}$ **30.** $\dfrac{1}{\sqrt{21}+\sqrt{14}+\sqrt{6}+2}$ **31.** 6
32. 7 **33.** -10 **34.** -5 **35.** $2\sqrt[3]{5}$ **36.** $2\sqrt[3]{6}$ **37.** $4\sqrt[3]{3}$ **38.** $3\sqrt[3]{7}$ **39.** $5\sqrt[3]{3}$ **40.** $4\sqrt[3]{5}$ **41.** $-5\sqrt[3]{2}$
42. $-3\sqrt[3]{5}$ **43.** $2\sqrt[3]{4}$ **44.** $3\sqrt[3]{9}$ **45.** $\dfrac{\sqrt[3]{x}}{2}$ **46.** $\dfrac{\sqrt[3]{x^2}}{9}$ **47.** 0.2 **48.** -0.4

Section 8-5. Practice Exercises

1. 16 **2.** 81 **3.** 18 **4.** 20 **5.** 7 **6.** 28 **7.** 85 **8.** 22 **9.** 0 **10.** 0 **11.** 2 and 5 **12.** 5 and 8
13. 1 and 3 **14.** 2 and 3 **15.** 1 and 5 **16.** 3 and 5 **17.** 0 and 3 **18.** 6 **19.** -1 and 2 **20.** -3 and 4
21. 4 **22.** 9 **23.** 5 **24.** 7 **25.** -2 and -3 **26.** -1 and -5 **27.** 9 **28.** 1 **29.** No solution
30. No solution **31.** 0 **32.** 14 **33.** 5 **34.** 2 **35.** No solution **36.** 9 **37.** 9 **38.** 9 **39.** 25 **40.** 25
41. $\dfrac{13}{4}$ **42.** No solution **43.** No solution **44.** $\dfrac{29}{4}$

Section 8-5. Ten Review Exercises

1. $99a^2b^2 + 40ab - 99$ **2.** $49t^2 - 42tu + 9u^2$ **3.** $6xy - 14x + 15y - 35$ **4.** $-3u^2 + 12uv + v^2$ **5.** $\dfrac{3a+1}{a-2}$

6. $\dfrac{x-3}{x+3}$ **7.** $\dfrac{4}{3y}$ **8.** 6 **9.** $(ab+5)(a^2b^2 - 5ab + 25)$ **10.** $(5z-9)(7y+2)$

Section 8-5. Supplementary Exercises

1. $\dfrac{1}{2}$ **2.** $\dfrac{2}{3}$ **3.** $\dfrac{1}{3}$ and $\dfrac{1}{2}$ **4.** $\dfrac{-3}{4}$ and -1 **5.** $\dfrac{5}{2}$ **6.** 3 **7.** $\dfrac{24}{5}$ **8.** 21 **9.** 4 **10.** 7 **11.** 16 **12.** 12

13. 3 **14.** $\dfrac{9}{16}$ **15.** 8 **16.** 125 **17.** 63 **18.** 29 **19.** 2 **20.** 0 **21.** 22 **22.** 20

Chapter 8 Review Exercises

1. -12 and 12 **2.** 12 **3.** $\dfrac{10}{7}$ **4.** -8 **5.** 0.5 **6.** Not real **7.** 8.124 **8.** 9.592 **9.** 6 **10.** 12

11. $8\sqrt{7}$ **12.** $7\sqrt{5}$ **13.** $\dfrac{\sqrt{3}}{2}$ **14.** $\dfrac{\sqrt{19}}{10}$ **15.** 6 **16.** $\dfrac{\sqrt{13}}{15}$ **17.** $-\sqrt{42}$ **18.** $-\sqrt{17}$ **19.** $8\sqrt{5}$ **20.** $36\sqrt{6x}$

21. $\dfrac{1+3\sqrt{2}}{2}$ **22.** $9\sqrt{10} - 2\sqrt{5}$ **23.** $4 + \sqrt{3}$ **24.** -1 **25.** $\dfrac{17\sqrt{5}}{2}$ **26.** $\dfrac{7\sqrt{5}}{5}$ **27.** $\dfrac{2\sqrt{10}}{5}$ **28.** $5(2 + \sqrt{3})$

29. 12 **30.** 9 **31.** -4 **32.** -2 and 1

Chapter 9 Quadratic Equations

Section 9-1. Practice Exercises

1. a. $x^2 - 4x + 3 = 0$ **b.** 1, -4, and 3 **2. a.** $x^2 + 6x - 1 = 0$ **b.** 1, 6, and -1 **3. a.** $0 = 2t^2 - 3t + 7$
b. 2, -3, and 7 **4. a.** $6t^2 - 5t + 9 = 0$ **b.** 6, -5, and 9 **5. a.** $a^2 - 11a = 0$ **b.** 1, -11, and 0 **6. a.** $a^2 - 14 = 0$
b. 1, 0, and -14 **7. a.** $5b^2 + 13b + 6 = 0$ **b.** 5, 13, and 6 **8. a.** $3b^2 - 2b - 11 = 0$ **b.** 3, -2, and -11

9. a. $4x^2 - 7x - 4 = 0$ **b.** 4, -7, and -4 **10. a.** $0 = x^2 - 4x - 16$ **b.** 1, -4 and -16 **11.** -4 and $\dfrac{5}{3}$

12. 2 and $\dfrac{-7}{2}$ **13.** $\dfrac{1}{3}$ and $\dfrac{-5}{2}$ **14.** $\dfrac{-3}{4}$ and $\dfrac{-2}{5}$ **15.** 0 and -9 **16.** 0 and 11 **17.** $\dfrac{-8}{3}$ and $\dfrac{8}{3}$

18. $\dfrac{-6}{5}$ and $\dfrac{6}{5}$ **19.** $\dfrac{1}{2}$ and 2 **20.** $\dfrac{1}{3}$ and 3 **21.** $\dfrac{-3}{5}$ and -2 **22.** $\dfrac{-5}{3}$ and 2 **23.** $\dfrac{3}{4}$ and $\dfrac{-1}{2}$

24. $\dfrac{1}{3}$ and $\dfrac{-1}{4}$ **25.** $\dfrac{-2}{7}$ and $\dfrac{2}{7}$ **26.** $\dfrac{-3}{8}$ and $\dfrac{3}{8}$ **27.** 0 and 2 **28.** 0 and 4 **29.** -4 and 2

30. 5 and -3 **31.** 2 and -1 **32.** -2 and 1 **33.** -8 **34.** -7 **35.** 3, 4, and 5 cm **36.** 6, 8, and 10 in
37. 5, 12, and 13 units **38.** 8, 15, and 17 units **39.** 7 and 24 in **40.** 20 and 21 cm **41.** 12 m by 16 m
42. 15 m by 20 m **43.** 6 hr and 14 hr **44.** 12 hr **45.** 14 hr **46.** 8 hr

Section 9-1. Ten Review Exercises

1. $\dfrac{\sqrt{3t}}{5}$ **2.** $3\sqrt{5ab}$ **3.** $\dfrac{-2\sqrt{14x}}{7}$ **4.** $7\sqrt{3}$ **5.** $-(7 + 4\sqrt{3})$ **6.** $\dfrac{-2}{3}$ **7.** -8 and 3 **8.** 29 **9.** $a > 5$
10. $(3, -4)$

Section 9-1. Supplementary Exercises

1. 0 and 8 **2.** 0 and -5 **3.** -10 and 10 **4.** -12 and 12 **5.** -2 and 3 **6.** -3 and 2
7. 0 and 5 **8.** 0 and -3 **9.** -2 and 8 **10.** -5 **11. a.** $(x+2)(x-3) = 0$ **b.** -2 and 3
c. **d.** Pick $x = 0$ **e.** $-2 < x < 3$ **12. a.** $(x-1)(x-6) = 0$ **b.** 1 and 6

c. **d.** Pick $x = 2$ **e.** $1 < x < 6$ **13. a.** $(2x+1)(x-4) = 0$ **b.** $\dfrac{-1}{2}$ and 4

c. **d.** Pick $x = 0$ **e.** $x < \dfrac{-1}{2}$ or $x > 4$ **14. a.** $(3x+1)(x+5) = 0$

b. $\dfrac{-1}{3}$ and -5 **c.**   **d.** Pick $x = -1$ **e.** $x < -5$ or $x > \dfrac{-1}{3}$

15. -11 and -9, or 9 and 11 **16.** -14 and -12, or 12 and 14 **17.** -7 and -12, or 7 and 12
18. 3 mph for Group B; 4 mph for Group A   **19.** South boat: 5 mph; west boat: 12 mph   **20.** -8 and -5, or 5 and 8
21. -8 and -3, or 3 and 8 **22.** -11 and -7, or 7 and 11

Section 9-2. Practice Exercises

1. -5 and 5 **2.** -9 and 9 **3.** ± 15 **4.** ± 11 **5.** $\pm\dfrac{7}{3}$ **6.** $\pm\dfrac{9}{10}$ **7.** $\pm\sqrt{11}$ **8.** $\pm\sqrt{17}$ **9.** $\pm 2\sqrt{11}$

10. $\pm 2\sqrt{17}$ **11.** $\pm\dfrac{5}{2}$ **12.** $\pm\dfrac{14}{3}$ **13.** $\pm\dfrac{\sqrt{2}}{5}$ **14.** $\pm\dfrac{\sqrt{3}}{10}$ **15.** $\pm 2\sqrt{3}$ **16.** $\pm 2\sqrt{3}$ **17.** $\pm\dfrac{9\sqrt{5}}{5}$ **18.** $\pm\dfrac{11\sqrt{7}}{7}$

19. $\pm\dfrac{\sqrt{2}}{2}$ **20.** $\pm\dfrac{\sqrt{6}}{3}$ **21.** 2 and -4 **22.** 3 and -7 **23.** 8 and -4 **24.** 15 and 1

25. 21 and -3 **26.** 15 and -7 **27.** $-7 \pm \sqrt{13}$ **28.** $-5 \pm \sqrt{19}$ **29.** $4 \pm 3\sqrt{2}$ **30.** $-4 \pm 2\sqrt{3}$

31. $-17 \pm 7\sqrt{3}$ **32.** $-21 \pm 6\sqrt{5}$ **33.** 2 and $\dfrac{-4}{3}$ **34.** 4 and $\dfrac{-2}{3}$ **35.** 5 and -8 **36.** 4 and -11

37. $\dfrac{-3 \pm \sqrt{6}}{2}$ **38.** $\dfrac{-5 \pm \sqrt{10}}{2}$ **39.** $\dfrac{5 \pm 3\sqrt{7}}{4}$ **40.** $\dfrac{3 \pm 3\sqrt{6}}{4}$ **41.** $\dfrac{12 \pm 2\sqrt{31}}{7}$ **42.** $\dfrac{-12 \pm 3\sqrt{13}}{11}$

43. 2 and -8 **44.** 2 and -12 **45.** 5 and -13 **46.** 5 and 3   **47.** $\dfrac{1 \pm 2\sqrt{2}}{5}$ **48.** $\dfrac{-1 \pm 3\sqrt{3}}{5}$

49. $\dfrac{2 \pm 2\sqrt{3}}{3}$ **50.** $\dfrac{-2 \pm 2\sqrt{7}}{3}$

Section 9-2. Ten Review Exercises

1. $a^2 - 2a$ **2.** $\sqrt{3} - 2\sqrt{2}$ **3.** $2x^2 + 5xy - 12y^2$ **4.** $5\sqrt{6} - 18$ **5.** $m^2 + 6mn + 9n^2$ **6.** $6\sqrt{5} + 14$ **7.** $2u - 5v$
8. $\dfrac{2\sqrt{5} - \sqrt{2}}{2}$ **9.** $-11b + 16$ **10.** $-11\sqrt{3} + 16$

Section 9-2. Supplementary Exercises

1. $\pm\dfrac{\sqrt{10}}{6}$ **2.** $\pm\dfrac{\sqrt{3}}{6}$ **3.** 0 and -5 **4.** 0 and $\dfrac{-4}{7}$ **5.** No real solutions   **6.** No real solutions.   **7.** $\dfrac{-2 \pm \sqrt{5}}{3}$

8. $\dfrac{-4 \pm \sqrt{3}}{5}$ **9.** 1 and $\dfrac{2}{3}$ **10.** 1 and $\dfrac{3}{4}$ **11.** $\dfrac{-7 \pm \sqrt{31}}{2}$ **12.** $\dfrac{-11 \pm \sqrt{10}}{9}$ **13.** $\dfrac{2 \pm \sqrt{3}}{8}$ **14.** $\dfrac{15 \pm \sqrt{5}}{6}$

15. Yes   **16.** Yes   **17.** Yes   **18.** Yes   **19.** Yes   **20.** Yes   **21.** ± 5 and ± 3 **22.** ± 1 and ± 6

23. $\pm\dfrac{3}{2}$ and ± 1 **24.** $\pm\dfrac{1}{3}$ and ± 2 **25.** ± 8 and $\pm\sqrt{2}$ **26.** ± 3 and $\pm 2\sqrt{3}$ **27.** $\pm\dfrac{\sqrt{6}}{2}$ and ± 6

28. $\pm\dfrac{\sqrt{15}}{3}$ and ± 2

Section 9-3. Practice Exercises

1. a. $x^2 + 8x + 16$ **b.** $(x + 4)^2$ **2. a.** $x^2 + 12x + 36$ **b.** $(x + 6)^2$ **3. a.** $y^2 - 18y + 81$ **b.** $(y - 9)^2$
4. a. $y^2 - 14y + 49$ **b.** $(y - 7)^2$ **5. a.** $m^2 - 20m + 100$ **b.** $(m - 10)^2$ **6. a.** $m^2 - 24m + 144$ **b.** $(m - 12)^2$
7. a. $p^2 + 28p + 196$ **b.** $(m + 14)^2$ **8. a.** $p^2 + 36p + 324$ **b.** $(p + 18)^2$ **9. a.** $z^2 - 60z + 900$ **b.** $(z - 30)^2$
10. a. $z^2 - 48z + 576$ **b.** $(z - 24)^2$ **11.** -1 and -13 **12.** -2 and -16 **13.** 1 and 29   **14.** 1 and 21
15. -4 and 6   **16.** -2 and 6   **17.** $-3 \pm \sqrt{11}$ **18.** $-5 \pm \sqrt{29}$ **19.** $8 \pm 6\sqrt{2}$ **20.** $9 \pm 3\sqrt{10}$
21. $-6 \pm 2\sqrt{7}$ **22.** $-4 \pm 2\sqrt{3}$ **23.** $12 \pm 2\sqrt{31}$ **24.** $10 \pm 5\sqrt{3}$ **25.** $-11 \pm 2\sqrt{2}$ **26.** $-14 \pm 2\sqrt{3}$
27. $\dfrac{5 \pm 2\sqrt{5}}{5}$ **28.** $\dfrac{14 \pm 3\sqrt{14}}{7}$ **29.** $-2 \pm \sqrt{7}$ **30.** $-2 \pm \sqrt{10}$ **31.** $\dfrac{20 \pm 9\sqrt{5}}{5}$ **32.** $\dfrac{18 \pm 7\sqrt{6}}{6}$ **33.** $\dfrac{-10 \pm 7\sqrt{2}}{2}$

34. $\dfrac{-28 \pm 8\sqrt{14}}{7}$

Section 9-3. Ten Review Exercises

1. $\dfrac{9}{10}$ **2.** $\dfrac{t+6}{t+7}$ **3.** $\dfrac{9}{10}$ **4.** $t \neq -7, 7$ **5.** -2 **6.** 3 and -15 **7.** 3 and -15 **8.** 0 and 8
9. 0 and 8   **10.** Yes

Section 9-3. Supplementary Exercises

1. 0 and -12 **2.** 0 and -18 **3.** 0 and 20   **4.** 0 and 16   **5.** 0 and -5 **6.** 0 and -7
7. $\pm 2\sqrt{2}$ **8.** ± 2 **9.** $\pm 3\sqrt{3}$ **10.** $\pm\sqrt{3}$ **11.** $\dfrac{-9 \pm \sqrt{161}}{2}$ **12.** $\dfrac{-7 \pm \sqrt{89}}{2}$ **13.** $\dfrac{5 \pm \sqrt{17}}{2}$ **14.** $\dfrac{3 \pm \sqrt{5}}{2}$

15. 2 and 9   **16.** 6 and 7   **17.** $\dfrac{7 \pm \sqrt{17}}{4}$ **18.** $\dfrac{5 \pm \sqrt{41}}{4}$ **19.** $\dfrac{3 \pm \sqrt{105}}{8}$ **20.** $\dfrac{2 \pm \sqrt{29}}{5}$ **21.** $\dfrac{-1}{3}$ and 3

22. $\dfrac{-2}{7}$ and 1   **23. a.** $-4 \pm \sqrt{15}$ **b.** $-0.13, -7.87$ **24. a.** $-6 \pm \sqrt{34}$ **b.** $-0.17, -11.83$ **25. a.** $1 \pm \sqrt{11}$

b. $4.32, -2.32$ **26. a.** $\dfrac{5 \pm \sqrt{57}}{2}$ **b.** $6.27, -1.27$ **27. a.** $\dfrac{-5 \pm \sqrt{19}}{2}$ **b.** $-0.32, -4.68$ **28. a.** $\dfrac{-6 \pm \sqrt{26}}{2}$
b. $-0.45, -5.55$

Section 9-4. Practice Exercises

1. a. 1, 1, and -3 **b.** $\dfrac{-1 \pm \sqrt{13}}{2}$ **2. a.** 1, 5, and 1   **b.** $\dfrac{-5 \pm \sqrt{21}}{2}$ **3. a.** 2, 3, and -2 **b.** 1 and -4

4. a. 3, 2, and -1 **b.** $\dfrac{1}{3}$ and -1 **5. a.** 1, -3, and -6 **b.** $\dfrac{3 \pm \sqrt{33}}{2}$ **6. a.** 1, -1, and -9 **b.** $\dfrac{1 \pm \sqrt{37}}{2}$
7. a. 1, -4, and -3 **b.** $2 \pm \sqrt{7}$ **8. a.** 1, -4, and 1   **b.** $2 \pm \sqrt{3}$ **9. a.** 1, -6, and -1 **b.** $3 \pm \sqrt{10}$
10. a. 1, -2, and -1 **b.** $1 \pm \sqrt{2}$ **11. a.** 2, -7, and -9 **b.** $\dfrac{9}{2}$ and -1 **12. a.** 20, 1, and -12 **b.** $\dfrac{3}{4}$ and $\dfrac{-4}{5}$

13. a. 4, -12, and 0   **b.** 3 and 0   **14. a.** 9, 6, and 0   **b.** 0 and $\dfrac{-2}{3}$ **15. a.** 1, 0, and -72 **b.** $\pm 6\sqrt{2}$

16. a. 1, 0, and -48 **b.** $\pm 4\sqrt{3}$ **17.** $-2 \pm \sqrt{3}$ **18.** $-2 \pm \sqrt{5}$ **19.** $\dfrac{1 \pm 2\sqrt{2}}{2}$ **20.** $\dfrac{2 \pm \sqrt{2}}{2}$ **21.** $\dfrac{-3 \pm \sqrt{3}}{2}$

22. $\dfrac{-1 \pm 2\sqrt{3}}{2}$ **23.** $\dfrac{5}{4}$ and $\dfrac{-3}{2}$ **24.** $\dfrac{6}{5}$ and $\dfrac{-1}{2}$ **25.** $\dfrac{2}{3}$ and $\dfrac{-7}{2}$ **26.** $\dfrac{-6}{5}$ and $\dfrac{5}{2}$ **27.** $5 \pm 2\sqrt{2}$

28. $5 \pm \sqrt{3}$ **29.** $\dfrac{2 \pm \sqrt{5}}{3}$ **30.** $\dfrac{3 \pm \sqrt{5}}{2}$ **31.** $2 \pm 2\sqrt{2}$ **32.** $3 \pm \sqrt{2}$ **33.** $2 \pm 3\sqrt{3}$ **34.** $-3 \pm \sqrt{5}$

35. $\pm \dfrac{\sqrt{6}}{3}$ **36.** $\pm \dfrac{\sqrt{2}}{3}$ **37.** $\dfrac{-19 \pm \sqrt{209}}{4}$ **38.** $\dfrac{-2 \pm \sqrt{6}}{3}$ **39.** $\dfrac{1 \pm 2\sqrt{7}}{2}$ **40.** $\dfrac{-1 \pm 3\sqrt{7}}{3}$

41. a. $2 + \sqrt{2}$ and $2 - \sqrt{2}$ **b.** 3.4 and 0.6   **42. a.** $6 + \sqrt{3}$ and $6 - \sqrt{3}$ **b.** 7.7 and 4.3
43. a. $\dfrac{-5 + \sqrt{3}}{2}$ and $\dfrac{-5 - \sqrt{3}}{2}$ **b.** -1.6 and -3.4 **44. a.** $\dfrac{-1 + 3\sqrt{5}}{2}$ and $\dfrac{-1 - 3\sqrt{5}}{2}$ **b.** 2.9 and -3.9

45. a. $\dfrac{3}{5}$ and $\dfrac{-5}{3}$, or $\dfrac{5}{3}$ and $\dfrac{-3}{5}$ **b.** 0.6 and -1.7, or 1.7 and -0.6 **46. a.** $\dfrac{2}{3}$ and $\dfrac{-1}{2}$, or $\dfrac{1}{2}$ and $\dfrac{-2}{3}$

b. 0.7 and -0.5, or 0.5 and -0.7 **47. a.** $\dfrac{-5 + \sqrt{185}}{2}$ and $\dfrac{5 + \sqrt{185}}{2}$ **b.** 4.3 cm and 9.3 cm

48. a. $-4 + 6\sqrt{3}$ and $4 + 6\sqrt{3}$ **b.** 6.4 ft and 14.4 ft   **49. a.** $-2 + \sqrt{39}$ and $2 + \sqrt{39}$ **b.** 4.2 m and 8.2 m
50. a. $-3 + 2\sqrt{21}$ and $3 + 2\sqrt{21}$ **b.** 6.2 yd and 12.2 yd   **51. a.** $\dfrac{-1 + \sqrt{97}}{2}$ and $\dfrac{1 + \sqrt{97}}{2}$

b. 4.4 m and 5.4 m   **52. a.** $\dfrac{-3 + 3\sqrt{17}}{2}$ and $\dfrac{3 + 3\sqrt{17}}{2}$ **b.** 4.7 cm and 7.7 cm

53. a. $\dfrac{6 + 2\sqrt{21}}{3}$ and $\dfrac{3 + 4\sqrt{21}}{3}$ **b.** 5.1 ft and 7.1 ft   **54. a.** 5 and 13   **b.** 5.0 in and 13.0 in
55. a. 7 and 10   **b.** 7.0 ft and 10.0 ft   **56. a.** $-2 + 2\sqrt{11}$ and $2 + 2\sqrt{11}$ **b.** 4.6 m and 8.6 m

57. a. $2 + 7\sqrt{2}$ cm and $-2 + 7\sqrt{2}$ cm   **b.** Same as **a**   **c.** Yes   **58. a.** $\dfrac{5 + \sqrt{433}}{2}$ in and $\dfrac{-5 + \sqrt{433}}{2}$ in
b. Same as **a**   **c.** Yes

Section 9-4. Ten Review Exercises

1. $\dfrac{-4}{3}$ **2.** $y - 9 = \dfrac{-4}{3}(x - (-3))$ **3.** $4x + 3y = 15$ **4.** Yes   **5.** 5   **6.** $\dfrac{3}{4}$ **7.** $3x - 4y = 30$ **8.** $\dfrac{-4}{3}$
9. Perpendicular   **10.** No solution

Section 9-4. Supplementary Exercises

1. $\dfrac{-\sqrt{2} \pm \sqrt{14}}{2}$ **2.** $\dfrac{-\sqrt{5} \pm 3}{2}$ **3.** $\dfrac{\sqrt{3} \pm \sqrt{35}}{4}$ **4.** $\dfrac{\sqrt{6} \pm \sqrt{30}}{6}$ **5.** $\dfrac{\sqrt{7} \pm \sqrt{127}}{12}$ **6.** $\dfrac{\sqrt{10} \pm 5\sqrt{6}}{10}$ **7.** $\dfrac{-3\sqrt{2} \pm \sqrt{34}}{4}$

8. $\dfrac{-\sqrt{5} \pm 3}{2}$ **9.** 0.5 and 0.3   **10.** 0.3 and -1.4 **11.** 0.23 and -0.70 **12.** 3 and -1.8 **13.** Rational

14. Rational   **15.** Complex   **16.** Complex   **17.** Rational   **18.** Rational   **19.** Irrational   **20.** Irrational

21. a. $ax^2 + bx = -c$ **b.** $x^2 + \dfrac{b}{a}x = \dfrac{-c}{a}$ **c.** $x^2 + \dfrac{b}{a}x + \dfrac{b^2}{4a^2} = \dfrac{b^2}{4a^2} - \dfrac{c}{a}$ **d.** $\left(x + \dfrac{b}{2a}\right)^2$ **e.** $\dfrac{b^2 - 4ac}{4a^2}$

f. $x + \dfrac{b}{2a} = \pm\sqrt{\dfrac{b^2 - 4ac}{4a^2}}$ **g.** $x = \dfrac{-b}{2a} \pm \dfrac{\sqrt{b^2 - 4ac}}{2a}$ **h.** $x = \dfrac{-b \pm \sqrt{b^2 - 4ac}}{2a}$

Section 9-5. Practice Exercises

1. $4i$ **2.** $11i$ **3.** $-2i$ **4.** $-15i$ **5.** $i\sqrt{10}$ **6.** $i\sqrt{3}$ **7.** $-i\sqrt{23}$ **8.** $-i\sqrt{41}$ **9.** $\dfrac{1}{3}i$ **10.** $\dfrac{5}{7}i$ **11.** $\dfrac{i\sqrt{2}}{9}$

12. $\dfrac{i\sqrt{7}}{20}$ **13.** $-3i\sqrt{3}$ **14.** $-2i\sqrt{7}$ **15.** $12i\sqrt{2}$ **16.** $13i\sqrt{3}$ **17.** $\dfrac{i\sqrt{15}}{6}$ **18.** $\dfrac{i\sqrt{35}}{15}$ **19.** $\dfrac{i\sqrt{15}}{3}$ **20.** $\dfrac{i\sqrt{7}}{4}$

21. $7 + 8i$ **22.** $13 - 13i$ **23.** $-1 - 19i$ **24.** $-4 - 21i$ **25.** $10 + 5i\sqrt{6}$ **26.** $-9 + 4i\sqrt{7}$ **27.** $-2 - 2i\sqrt{11}$

28. $17 - 3i\sqrt{11}$ **29.** $\dfrac{2}{3} - \dfrac{1}{3}i$ **30.** $-\dfrac{5}{6} + \dfrac{7}{6}i$ **31.** $\pm 8i$ **32.** $\pm\dfrac{10}{3}i$ **33.** $\pm\dfrac{1}{5}i$ **34.** $\pm\dfrac{i\sqrt{3}}{2}$ **35.** $\pm 5i\sqrt{2}$

36. $\pm 12i\sqrt{2}$ **37.** $\pm\dfrac{2i\sqrt{30}}{5}$ **38.** $\pm\dfrac{5i\sqrt{6}}{2}$ **39.** $1 \pm 3i$ **40.** $5 \pm i$ **41.** $-3 \pm 6i$ **42.** $-5 \pm 2i$ **43.** $7 \pm i\sqrt{3}$

44. $3 \pm 5i\sqrt{2}$ **45.** $\dfrac{-2 \pm i}{3}$ **46.** $\dfrac{-3 \pm i}{2}$ **47.** $\dfrac{1 \pm i\sqrt{5}}{3}$ **48.** $\dfrac{2 \pm i\sqrt{6}}{2}$ **49.** $\dfrac{-1 \pm i\sqrt{11}}{5}$ **50.** $\dfrac{-2 \pm 2i\sqrt{5}}{3}$

Section 9-5. Ten Review Exercises

1. $-1 \pm \sqrt{7}$ **2.** $-1 \pm \sqrt{7}$ **3.** $\dfrac{3}{4}$ and -2 **4.** $\pm\dfrac{9}{2}$ **5.** $\pm\dfrac{9}{2}i$ **6.** $3a - 1$ **7.** $b^4 + 4b^3 + 2b^2 - 4b + 1$

8. $\dfrac{2(2c - 3)}{3(2c + 3)}$ **9.** $\dfrac{p(p - 3)}{(p + 4)(p + 3)}$ **10.** $4y^3 - 2y^2 + y - 3$

Section 9-5. Supplementary Exercises

1. -54 **2.** -30 **3.** 4   **4.** 4   **5.** -21 **6.** -45 **7.** -5 **8.** -22 **9.** $\dfrac{11i\sqrt{3}}{15}$ **10.** $\dfrac{\sqrt{10}}{5}$ **11.** $-i$

12. 1   **13.** i **14.** -1 **15.** $-i$ **16.** 1   **17.** i **18.** -1 **19.** i **20.** $-i$
21–30.   **31.** $-7 + 4i$ **32.** $3 - 2i$ **33.** $5 + 8i$ **34.** $-4 - 6i$ **35.** -5

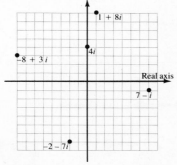

36. $8 + 3i$ **37.** $6 - 7i$ **38.** $-7 - 4i$ **39.** $6i$ **40.** $1 - 5i$ **41.** $\dfrac{-3}{2}$ and 5   **42.** $\dfrac{2}{3}$ and -7 **43.** $\dfrac{9 \pm 7\sqrt{5}}{2}$

44. $-2 \pm 2\sqrt{3}$ **45.** $-6 \pm 5i$ **46.** $-10 \pm 3i$ **47.** $\dfrac{-1 \pm i\sqrt{5}}{2}$ **48.** $\dfrac{-5 \pm i\sqrt{3}}{2}$ **49.** $\dfrac{2 \pm 3\sqrt{2}}{3}$ **50.** $\dfrac{1 \pm 5\sqrt{3}}{2}$

Section 9-6. Practice Exercises

1. -3 and 4   **2.** -4 and 2   **3.** 1 and 3   **4.** 1 and 5   **5.** $\dfrac{-1}{2}$ and 5   **6.** $\dfrac{1}{2}$ and 2

7. $\dfrac{-7}{4}$ and -5 **8.** $\dfrac{-7}{3}$ and -4 **9.** $-4 < u < 2$

10. $-1 < u < 5$ **11.** $v < -5$ or $v > 2$

12. $v < -8$ or $v > 5$ **13.** $1 < t < 5$

14. $2 < t < 8$ **15.** $w < -9$ or $w > \dfrac{-2}{3}$

16. $w < -5$ or $w > \dfrac{-1}{2}$ **17.** $-10 \le a \le 2$

18. $-9 \le a \le 5$ **19.** $b \le -7$ or $b \ge 9$

20. $b \le -8$ or $b \ge 3$ **21.** $-5 \le w \le \dfrac{-1}{2}$

22. $-7 \le w \le \dfrac{-1}{3}$ **23.** $n \le \dfrac{8}{5}$ or $n \ge 2$

24. $n \le \dfrac{5}{6}$ or $n \ge 3$ **25.** $10 < w < 20$ in   **26.** $6 < w < 23$ cm   **27.** $10 < s \le 12$ in

28. $8 < s \le 15$ in   **29.** $10 < b \le 19$ cm   **30.** $15 < b \le 21$ in

Section 9-6. Ten Review Exercises

1. $k - 5$ **2.** $3k + 8$ **3.** $(3k + 8)(k - 5)$ **4.** 8   **5.** 8   **6.** 0   **7.** 0   **8.** $\dfrac{-8}{3}$ and 5

9. $\dfrac{-8}{3} < k < 5$ **10.** $k \le \dfrac{-8}{3}$ or $k \ge 5$

Section 9-6. Supplementary Exercises

1. -6 and 0   **2.** -9 and 0   **3.** 0 and 4   **4.** 0 and 4   **5.** $\dfrac{-5}{2}$ and $\dfrac{4}{3}$ **6.** $\dfrac{-1}{4}$ and $\dfrac{2}{5}$

7. $\dfrac{-7}{4}$ and $\dfrac{7}{4}$ **8.** $\dfrac{-10}{3}$ and $\dfrac{10}{3}$ **9.** $0 \le b \le 5$

10. $b \le \dfrac{-10}{3}$ or $b \ge 0$ **11.** $\dfrac{-1}{4} < k < 0$

12. $\dfrac{-2}{3} < k < \dfrac{1}{5}$ **13.** $t \le \dfrac{5}{2}$ or $t \ge \dfrac{9}{2}$

14. $t \le \dfrac{-10}{3}$ or $t \ge \dfrac{-2}{3}$ **15.** $w \le -6$ or $w \ge 3$

16. $w \leq -2$ or $w \geq 11$ 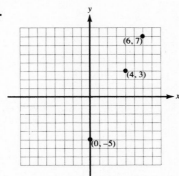 **17.** $-5 < u < -3$

18. $-7 < u < -4$

19 and 20. No real number squared is less than 0 **21 and 22.** All real numbers squared are greater than or equal to 0

23 and 24. Only true for $z = 7$ and $z = \dfrac{-1}{2}$, respectively **25 and 26.** True for all real numbers except $t = -8$ and $t = \dfrac{1}{6}$, respectively

Chapter 9 Review Exercises

1. a. $2x^2 - 3x + 4 = 0$ **b.** $2, -3,$ and 4 **2. a.** $6y^2 - 2y + 8 = 0$ **b.** $6, -2,$ and 8 **3.** $\dfrac{7}{2}$ **4.** -8 and 11

5. 0 and $\dfrac{1}{6}$ **6.** -3 and 5 **7.** -11 and 11 **8.** -1 and -9 **9.** $\dfrac{-5 \pm 2\sqrt{15}}{3}$ **10.** $\dfrac{2 \pm \sqrt{10}}{5}$

11. 13 and -1 **12.** $6 \pm \sqrt{23}$ **13.** $-10 \pm 4\sqrt{5}$ **14.** $3 \pm \sqrt{19}$ **15.** $\dfrac{5 \pm \sqrt{29}}{2}$ **16.** $\dfrac{3}{2}$ and $\dfrac{-1}{2}$

17. $10 \pm \sqrt{58}$ **18.** $\dfrac{\pm\sqrt{21}}{3}$ **19.** $\dfrac{-1 \pm \sqrt{5}}{2}$ **20.** $\dfrac{-3}{2}$ and $\dfrac{5}{4}$ **21.** -1 and 6 **22.** 0 and -3 **23.** $\pm 2\sqrt{6}$

24. $\dfrac{1}{3}$ and -1 **25.** 20, 21, and 29 cm **26.** 6 hr for Fred; 9 hr for Chris

27. The numbers are $\dfrac{5 + 5\sqrt{5}}{2}$ and $\dfrac{-5 + 5\sqrt{5}}{2}$, or $\dfrac{5 - 5\sqrt{5}}{2}$ and $\dfrac{-5 - 5\sqrt{5}}{2}$ **28.** 6 and 8 in **29.** $\pm 5i$ **30.** $\pm 4i$

31. $\dfrac{-2 \pm 4i}{3}$ **32.** $\dfrac{5 + i\sqrt{10}}{2}$ **33.** $\dfrac{-1 \pm i\sqrt{3}}{2}$ **34.** $-1 \pm i\sqrt{2}$ **35.** $\dfrac{-5 \pm i\sqrt{47}}{4}$ **36.** $\dfrac{-5 \pm i\sqrt{3}}{2}$

37. $-5 < t < \dfrac{3}{2}$ **38.** $\dfrac{-1}{3} < t < 4$

39. $x \leq -6$ or $x \geq -1$ **40.** $x \leq -4$ or $x \geq 3$

Chapter 10 Relations and Functions

Section 10-1. Practice Exercises

1. a. $(0, -5), (4, 3), (6, 7)$ **b.** **2. a.** $(-1, 8), (2, -1), (4, -7)$

b. 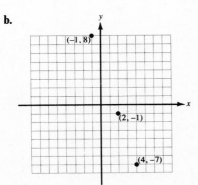 **3. a.** $(-4, -6), (0, -3), (8, 3)$ **b.**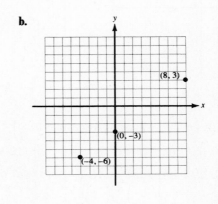

4. a. $(-4, -6), (0, 4), (-2, -1)$ **b.**

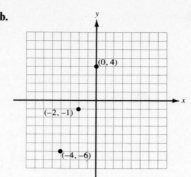

5. a. $(-3, 5), (0, 5), (6, 5)$

b.

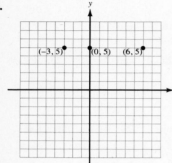

6. a. $(-7, -3), (0, -3), (7, -3)$ **b.**

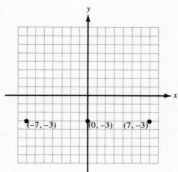

7. a. $\left(\frac{5}{2}, -5\right), \left(\frac{5}{2}, 0\right), \left(\frac{5}{2}, 5\right)$ **b.**

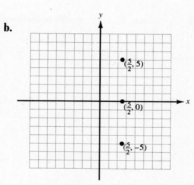

8. a. $(-2, -3), (-2, 0), (-2, 3)$

b.

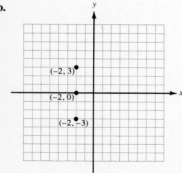

9. $(-2, -2), (0, 0), \left(6, \frac{2}{3}\right)$ **10.** $(-1, 6), (5, 0), \left(10, \frac{1}{2}\right)$ **11.** $(-3, 5), (0, -4), (3, 5), (5, 21)$

12. $(-2, 0), (0, 8), (2, 0), (4, -24)$ **13.** $(3, 1), (7, 3), (15, 5), (27, 7)$ **14.** $(-1, 0), (0, 1), (3, 2), (35, 6)$

15. $(-2, -6), (0, 2), (2, 10), (3, 29)$ **16.** $(-2, -10), (0, 0), (2, 10), (4, 116)$ **17.** $\left(-3, \frac{4}{5}\right), (-1, 0), (0, -1), \left(2, \frac{3}{5}\right)$

18. $(-2, 0), (0, 1), (2, 0), \left(4, \dfrac{-3}{5}\right)$ **19.** $(-2, 8), (0, 0), (2, 8), (4, 63)$ **20.** $(-1, -7), (0, 0), (1, 1), (2, -16)$

21. a. $-2, -1, 0, 1,$ and 2 **b.** $3, 4, 5, 6,$ and 7 **22. a.** $-2, -1, 0, 1,$ and 2 **b.** $-7, -5, -3, -1,$ and 1

23. a. $-4, -2, 0, 2,$ and 4 **b.** $-5, -1, \dfrac{-1}{5}, \dfrac{1}{7},$ and 1 **24. a.** $-4, -1, 0, 1,$ and 4 **b.** $\dfrac{-8}{3}, \dfrac{-2}{3}, 0, \dfrac{2}{3},$ and $\dfrac{8}{3}$

25. a. $5, 6, 8, 9,$ and 12 **b.** $0, 1, \sqrt{3}, 2,$ and $\sqrt{7}$ **26. a.** $0, 1, 4, 5,$ and 12 **b.** $1, \sqrt{3}, 3, \sqrt{11},$ and 5

27. All x, all y **28.** All x, 2 **29.** All x; all y, $y \geq -4$ **30.** All x, $x \leq 6$; all y **31.** All x, all y **32.** All x, all y

33. All x, $x \geq -4$; all y, $y \geq -2$ **34.** All x, $x \geq -4$; all y, $y \geq -3$ **35.** All x; all y, $y \geq -3$ **36.** All x, $x \geq -6$; all y

37. Function **38.** Function **39.** Not a function **40.** Not a function **41.** Function **42.** Not a function

43. Function **44.** Not a function **45.** Not a function **46.** Not a function **47.** Function **48.** Function

49. a. -17 **b.** -5 **c.** 3 **50. a.** 13 **b.** 4 **c.** 8 **51. a.** -22 **b.** 2 **c.** 8 **52. a.** 0 **b.** -1 **c.** -5

53. a. 0 **b.** 9 **c.** 5 **54. a.** 1 **b.** 2 **c.** $\sqrt{6}$ **55. a.** 3 **b.** $\sqrt{6}$ **c.** 2 **56. a.** 11 **b.** 10 **c.** $\dfrac{28}{3}$

Section 10-1. Ten Review Exercises

1. $\dfrac{3}{2}$ and 3 **2.** 4 **3.** No solution **4.** 16 **5.** $(3, -2)$ **6.** $(x - 8)(x + 5)$ **7.** $3t(t - 4)$ **8.** $(5u + 1)^2$

9. $(2v + 5)(4v^2 - 10v + 25)$ **10.** $(2m - 1)(3n + 2)$

Section 10-1. Supplementary Exercises

1. a. (Jim, Jason), (Jim, Jacob), (Judy, Jason), and (Judy, Jacob) **b.** Yes **c.** No

2. a. (Sam, Carla), (Sam, Woody), (Rebecca, Carla), and (Rebecca, Woody) **b.** Yes **c.** No

3. a. (Glenn, Patricia), (Jerry, Patricia) **b.** Yes **c.** Yes **4. a.** (Larry, Eric), (Samantha, Eric) **b.** Yes **c.** Yes

5. a. (Fred, Ed), (Fred, Mark), (Fred, Tammie) **b.** Yes **c.** No **6. a.** (Bunty, Martha), (Bunty, Clyde) **b.** Yes

c. No **7. a.** (Ted's, roast beef), (Ted's, hamburgers), (Ted's, meat loaf), (Melba's, chicken), (Melba's, ribs), and (Melba's, hamburgers) **b.** Ted's and Melba's **c.** roast beef, hamburgers, meat loaf, chicken, and ribs **d.** No

8. a. (Jill and Jack's, salami), (Jill and Jack's, pastrami), (Jill and Jack's, ham), (Clair's, tuna), (Clair's, ham), and (Clair's, bologna) **b.** Jill and Jack's, and Clair's **c.** salami, pastrami, ham, tuna, and bologna **d.** No

9. a. (Smokie's, cola), (Smokie's, orange soda), (Tom's, coffee), and (Tom's, tea) **b.** Smokie's and Tom's **c.** cola, orange soda, coffee, and tea **d.** No **10. a.** (Red Top Eatery, cherry pie), (Red Top Eatery, apple turnovers), (Fast Stop Beanery, chocolate cake), and (Fast Stop Beanery, donuts) **b.** Red Top Eatery and Fast Stop Beanery **c.** Cherry pie, apple turnovers, chocolate cake, and donuts **d.** No **11.** $(0, 2), (4, 12),$ and $(6, 5)$

12. $(-1, 12), (2, 5),$ and $(4, 3)$ **13.** $(-4, 2), (-4, 5), (0, 7), (0, -1), (8, 0)$ and $(8, 1)$ **14.** $(5, 2), (5, 10), (7, 10), (9, 2),$ and $(9, 10)$

15. a. All x, $x \neq 0$ **b.** -1 and 1 **c.** Yes **16. a.** All x, $x \neq -2$ and 2 **b.** $-4, 0,$ and 4 **c.** Yes **17. a.** All x

b. $y < -2$ or $y \geq 2$ **c.** Yes **18. a.** All x **b.** $y \leq -4$ or $y > 0$ **c.** Yes **19. a.** All x **b.** $y \geq -6$

c. No **20. a.** All x **b.** $y \leq 4$ **c.** No **21. a.** $(2, 0)$ and $(2, 4)$ **b.** $(4, -2)$ and $(4, 6)$ **c.** $(1, 1)$

22. a. $(2, -1)$ and $(2, -5)$ **b.** $(4, 1)$ and $(4, -7)$ **c.** $(4, 1)$ **23. a.** $(-3, -1)$ and $(-3, 1)$

b. $(0, -2)$ and $(0, 2)$ **c.** $(4, -3)$ **24. a.** $(-3, -2)$ and $(-3, 0)$ **b.** $(0, -3)$ and $(0, 1)$ **c.** $(-4, -1)$

25. a. $(-2, 3)$ **b.** $(0, 2)$ **c.** $(-6, 5)$ **26. a.** $(-6, -4)$ **b.** $(0, -2)$ **c.** $(3, -1)$

Section 10-2. Practice Exercises

1. $f(x) = \dfrac{2}{3}x - \dfrac{16}{3}$ **2.** $f(x) = x + \dfrac{2}{5}$ **3.** $f(x) = \dfrac{-7}{4}x + 3$ **4.** $f(x) = \dfrac{-1}{3}x - 5$ **5.** $f(x) = -x - \dfrac{9}{4}$

6. $f(x) = 2x + 9$ **7.** $f(x) = 6$ **8.** $f(x) = -12$

9.

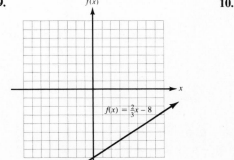

10.

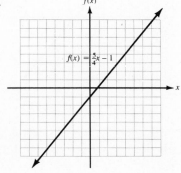

11.

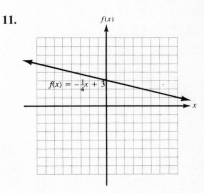

12.
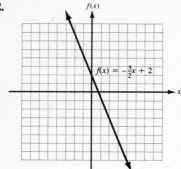
$f(x) = -\frac{5}{2}x + 2$

13.
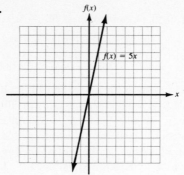
$f(x) = 5x$

14.
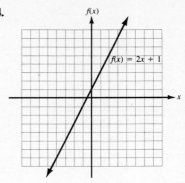
$f(x) = 2x + 1$

15.
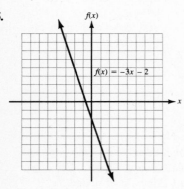
$f(x) = -3x - 2$

16.
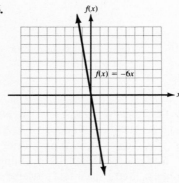
$f(x) = -6x$

17.
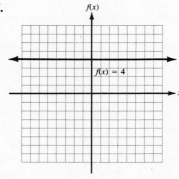
$f(x) = 4$

18.
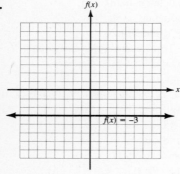
$f(x) = -3$

19.
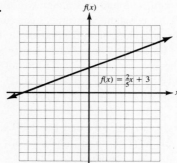
$f(x) = \frac{2}{5}x + 3$

20.
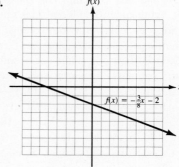
$f(x) = -\frac{3}{8}x - 2$

21.
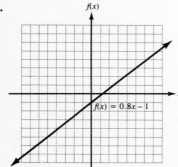
$f(x) = 0.8x - 1$

22.
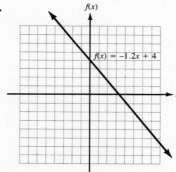
$f(x) = -1.2x + 4$

23. a. -1 **b.** -2 **c.** $x = -1$ **24. a.** 2 **b.** 12 **c.** $x = 2$ **25. a.** 2 **b.** 0 **c.** $x = 2$
26. a. -3 **b.** 0 **c.** $x = -3$ **27. a.** 0 **b.** 4 **c.** $x = 0$ **28. a.** 0 **b.** -6 **c.** $x = 0$ **29. a.** 2 **b.** 7
c. $x = 2$ **30. a.** -4 **b.** -22 **c.** $x = -4$

31.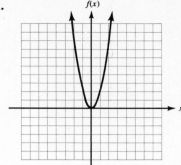

32.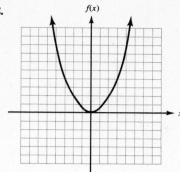

33.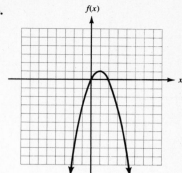

34.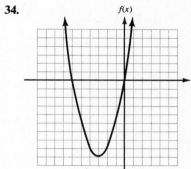

35.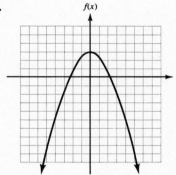

36.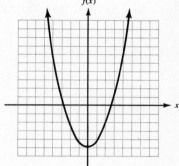

37.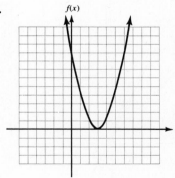

38.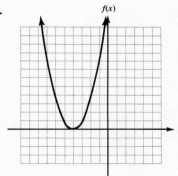

39.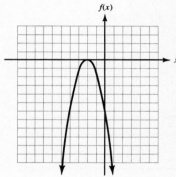

40.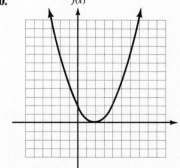

41.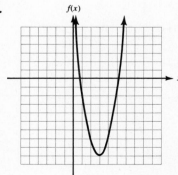

42.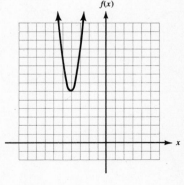

43.

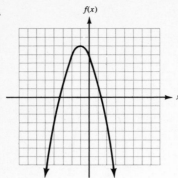

44.

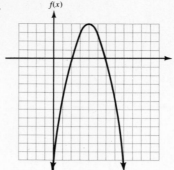

Section 10-2. Ten Review Exercises

1. $\dfrac{-3t^2}{u}$ **2.** $-3a^4b^3$ **3.** $-8x^3y^6z^9$ **4.** $\dfrac{-2k^2}{3}$ **5.** $\dfrac{-m^6}{2}$ **6.** $2\sqrt{30t}$ **7.** $\dfrac{\sqrt{2u}}{6}$ **8.** $3\sqrt{2}-3$ **9.** $\dfrac{-5+\sqrt{10}}{3}$

10. $4\sqrt{6x}$

Section 10-2. Supplementary Exercises

1.

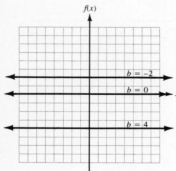

2.

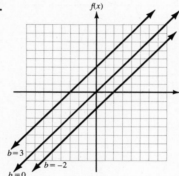

3.

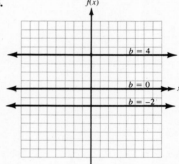

4.

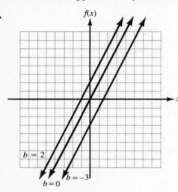

5.

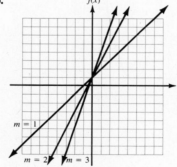

6.

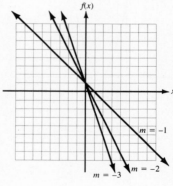

7.

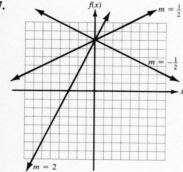

8.

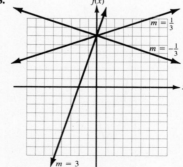

9. a. $V(4, -6)$ **b.** $x = 4$ **c.** Positive   **10. a.** $V(-8, -10)$ **b.** $x = -8$ **c.** Positive   **11. a.** $V(-4, 6)$

b. $x = -4$ **c.** Negative   **12. a.** $V(7, -1)$ **b.** $x = 7$ **c.** Negative   **13. a.**

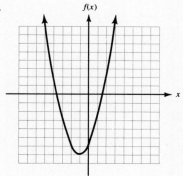

b. 2 and 3   **c.** -3 and -2 **d.** $\pm\sqrt{6}$ **e.** 2.5   **f.** -2.5 **14. a.**   **b.** 3 and 4

c. -4 and -3 **d.** $\pm 2\sqrt{3}$ **e.** 3.5   **f.** -3.5 **15. a.**   **b.** 1 and 2

c. -4 and -3 **d.** $-1 \pm \sqrt{7}$ **e.** 1.6   **f.** -3.6 **16. a.**   **b.** 5 and 6

c. -2 and -1 **d.** $2 \pm \sqrt{10}$ **e.** 5.2   **f.** -1.2

17. a.

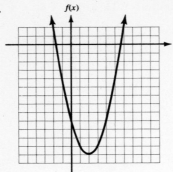

b. 5 and 6   **c.** −2 and −1   **d.** $2 \pm \sqrt{13}$ **e.** 5.6   **f.** −1.6

18. a.

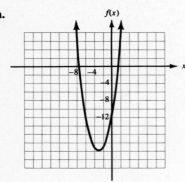

b. 1 and 2   **c.** −8 and −7   **d.** $-3 \pm 2\sqrt{5}$ **e.** 1.5   **f.** −7.5

Section 10-3. Practice Exercises

1.

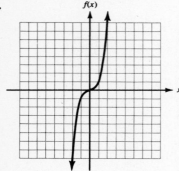

2.

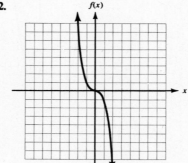

3.

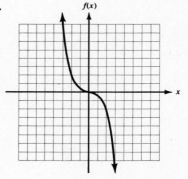

4.

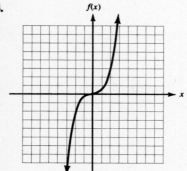

5.

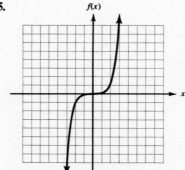

6.

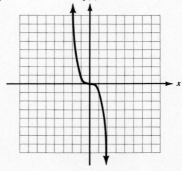

7.
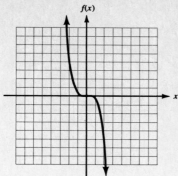

8.
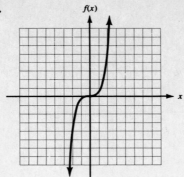

9.
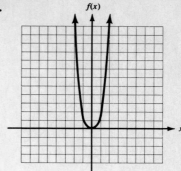

10.
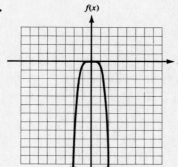

11.
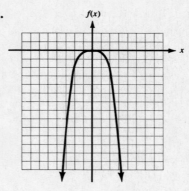

12.
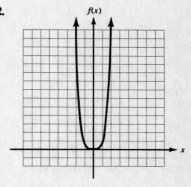

13.
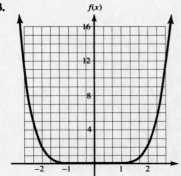

14.
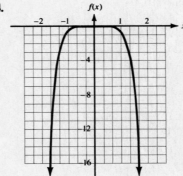

15.
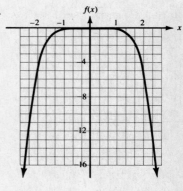

16.
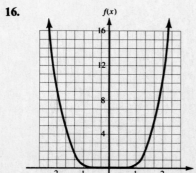

17.
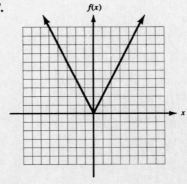

18.
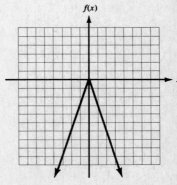

19.

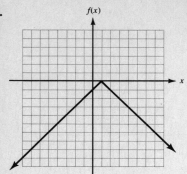

20.

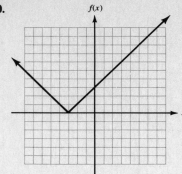

21.

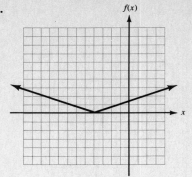

22.

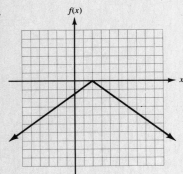

23.

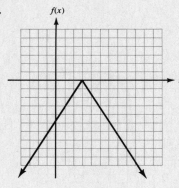

24.

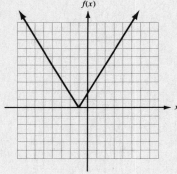

Section 10-3. Ten Review Exercises

1. $2t - 8$ **2.** $16u + 32$ **3.** $4x^2 - 41xy - 143y^2$ **4.** $4p^2 - 12p + 2pq - 6q + q^2 + 9$ **5.** $3m^4 + 8m^3 + m^2 + 4m + 4$

6. $2n^3 + n^2 - 3n + 7$ **7.** $\dfrac{(t-3)(t+3)}{(t-5)(t-1)}$ **8.** $\dfrac{4u-1}{4u+3}$ **9.** $\dfrac{3a-2b}{(3a-b)(2a-b)}$ **10.** $\dfrac{x-1}{x+1}$

Section 10-3. Supplementary Exercises

1.

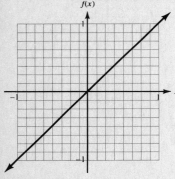

2.

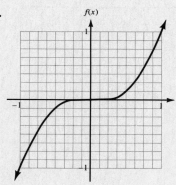

3.

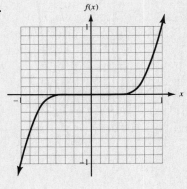

4. The "bend" gets sharper.

5.

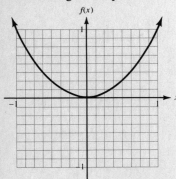

6.

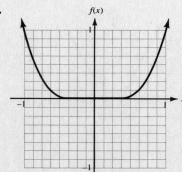

7.

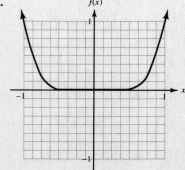

8. The "cup" gets flatter.

9.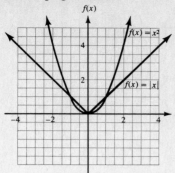

10.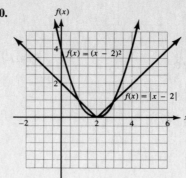

11.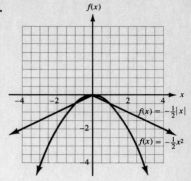

12.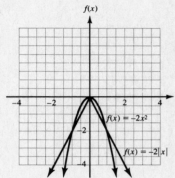

13.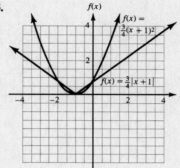

14.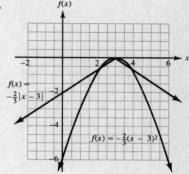

Chapter 10 Review Exercises

1. a. $(-5, 20)$ **b.** $\left(\frac{1}{2}, 9\right)$ **c.** $(6, -2)$ **2. a.** $(1, -3)$ **b.** $(-4, -3)$ **c.** $(2, 0)$ **3. a.** $(-3, 0)$

b. $(4, -11)$ and $(4, 13)$ **c.** $(8, 13)$ and $(4, 13)$ **4. a.** $(-4, 1)$ **b.** $(4, 3)$ **c.** $(-1, 2)$ **5. a.** 6, 4, 3, and -2

b. $-1, 0,$ and 5   **6. a.** $-6 \le x \le 6$ **b.** $-5 \le y \le 5$ **7. a.** Function   **b.** Function   **c.** Not a function

8. a. Function   **b.** Not a function   **9. a.** 0   **b.** -91 **10. a.** 3   **b.** 0   **11.** $f(x) = \frac{2}{3}x + \frac{22}{3}$ **12.** $f(x) = \frac{3}{2}x - 4$

13.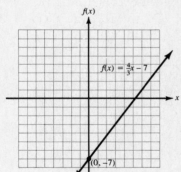

14.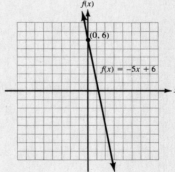

15. $f(x) = 2x^2 + 7x - 33$ **16.** $f(x) = 2x^2 + 12x + 19$ **17. a.** 0   **b.** -8 **c.** $x = 0$ **d.**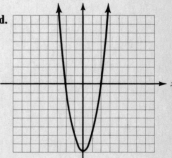

18. a. $\dfrac{3}{2}$ **b.** 1   **c.** $x = \dfrac{3}{2}$ **d.**

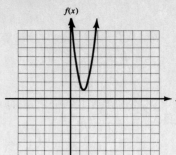

19. a. 3   **b.** 0   **c.** $x = 3$

d.

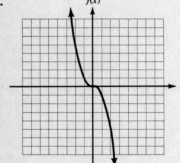

20. a. 1   **b.** 4   **c.** $x = 1$ **d.**

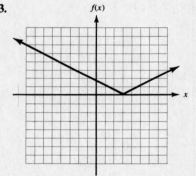

21.

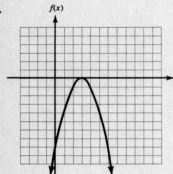

22.

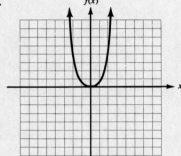

23.

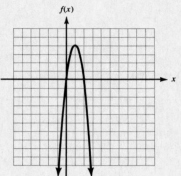

24.

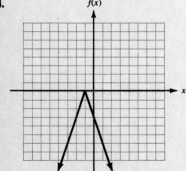

Solutions and Answers
to Summary Exercises

Section 1-1. Summary Exercises

1. $1 \cdot 7 = 7; 2 \cdot 7 = 14; 3 \cdot 7 = 21; 4 \cdot 7 = 28; 5 \cdot 7 = 35$
  7, 14, 21, 28, 35

2. $1 \cdot 36 = 36$; thus 1 and 36
  $2 \cdot 18 = 36$; thus 2 and 18
  $3 \cdot 12 = 36$; thus 3 and 12
  $4 \cdot 9 = 36$; thus 4 and 9
  $6 \cdot 6 = 36$; thus 6
  1, 2, 3, 4, 6, 9, 12, 18, 36

3. $7.3 = \dfrac{73}{10}$
  $\dfrac{73}{10}$

4. a. Any negative is less than any positive.
  $<$
  b. -32 is less negative.
  $>$
  c. $\dfrac{1}{2}$ is less than $\dfrac{4}{5}$, and whole number parts are equal.
  $<$

5. $= 22 - 12 + 17$
  $= 10 + 17$
  $= 27$
  27

6. $= \dfrac{3 \cdot 8 \cdot 14}{4 \cdot 7 \cdot 15}$

  $= \dfrac{1 \cdot 2 \cdot 2}{1 \cdot 1 \cdot 5} = \dfrac{4}{5}$

  $\dfrac{4}{5}$

7. $= 72 - 36 \div 9 - 8$
  $= 72 - 4 - 8$
  $= 68 - 8$
  $= 60$
  60

8. $= 1 + 128 + 5 - 9$
  $= 129 + 5 - 9$
  $= 134 - 9$
  $= 125$
  125

Section 1-2. Summary Exercises

1. $= 4 + (-15)$
  $= -11$
  -11

2. $= 27 - 11$
  $= 16$
  16

3. $= 3 + (-13)$
  $= -10$
  -10

4. $= 43 - 22$
  $= 21$
  21

5. $= -32 + 73 - (92 - 49)$
  $= -32 + 73 - 43$
  $= 41 - 43$
  $= -2$
  -2

6. $= -15 + (39 - 39)$
  $= -15 + 0$
  $= -15$
  -15

7. $= 34 + (17 + (35 - (-6))) - 30$
  $= 34 + (17 + 41) - 30$
  $= 34 + 58 - 30$
  $= 62$
  62

8. $163 - 4 = 159$
  $159 - 5 = 154$
  $154 + 2 = 156$
  $156 - 4 = 152$
  $152 + 1 = 153$
  153 pounds

Section 1-3. Summary Exercises

1. $= 882$
  882

2. $= -7$
  -7

3. $= 60 + (-56)$
  $= 4$
  4

4. $= -17 + 45 - (-4)$
  $= 28 - (-4)$
  $= 32$
  32

5. $= \dfrac{-41 + (-90) + 14}{-5 - (-44)}$

  $= \dfrac{-117}{39}$

  -3

6. $= 14 - 34 + 8 - (-27)$
  $= -20 + 8 - (-27)$
  $= -12 - (-27)$
  $= 15$
  15

7. $= -375 \div 75$
  $= -5$
  -5

8. $\dfrac{-7}{22} \cdot \dfrac{8}{8} = \dfrac{-56}{176}$

  $\dfrac{-56}{176}$

Section 1-4. Summary Exercises

1. $= 4 + 6 \cdot 2 - 27$
$= 4 + 12 - 27$
$= 16 - 27$
$= -11$
-11

2. $= 2(7) - 4(-1)$
$= 14 - (-4)$
$= 18$
18

3. $= 12 \div 3 \cdot 5 - 16$
$= 4 \cdot 5 - 16$
$= 20 - 16$
$= 4$
4

4. $= \dfrac{2 \cdot 64 + 4 \cdot 8}{16}$
$= \dfrac{128 + 32}{16} = \dfrac{160}{16} = 10$
10

5. $= \dfrac{9 \cdot 3 - 16 + 2}{9 + 4}$
$= \dfrac{27 - 16 + 2}{13} = \dfrac{13}{13} = 1$
1

6. $= \{12^2 \div (4)(-2)\} + (7)(-3)$
$= \{144 \div (4)(-2)\} + (-21)$
$= \{36(-2)\} + (-21)$
$= -72 + (-21) = -93$
-93

7. $= [64][100] + (0)$
$= 6400$
6400

8. $= -3[2(7) - 12]$
$= -3[14 - 12]$
$= -6$
-6

Section 1-5. Summary Exercises

1. (Interchange terms.)
$10.7 + (-8.5)$

2. (Interchange factors.)
$\dfrac{-8}{3} \cdot \dfrac{7}{10}$

3. (Regroup the middle factor.)
$-10 \cdot \left(\dfrac{4}{9} \cdot 27\right)$

4. (Regroup the middle term.)
$(-26 + 26) + 89$

5. a. 7.4
  b. Write 4.7 as $\dfrac{47}{10}$.
  $\dfrac{10}{47}$

6. $= 45 \cdot \dfrac{-7}{9} + 45 \cdot \dfrac{2}{5}$
$= -35 + 18 = -17$
  a. $45 \cdot \dfrac{-7}{9} + 45 \cdot \dfrac{2}{5}$
  b. -17

7. $= 0.49(62 + 38)$
$= 0.49(100)$
$= 49$
  a. $0.49(62 + 38)$
  b. 49

8. (0 is added to $4 + 9$.)
Identity element for addition

9. (-15 is multiplied by 1.)
Identity element for multiplication

10. (Middle term has been regrouped.)
Associative property of addition

11. (The 2.8 has been distributed.)
Distributive property of multiplication over addition

12. (Middle factor has been regrouped.)
Associative property of multiplication

13. (The sum is 0.)
Additive inverse

14. (The product is 1.)
Multiplicative inverse

15. (The common factor has been removed.)
Distributive property of multiplication over addition

Section 1-6. Summary Exercises

1. a. -6
  $x^2 y$
  b. $\left(\text{Write as } \dfrac{3}{5}a^4.\right)$
    (i) $\dfrac{3}{5}$
    (ii) a^4
  c. (Write as $1 \cdot pq$.)
    (i) 1
    (ii) pq

2. $= (5 - 4 + 7)x$
$= 8x$
$8x$

3. $= (-6m + 12m) + (4 - 11)$
$= 6m - 7$
$6m - 7$

4. $= (b^2 - 12b^2) + (-3b - b) + (7 + 15)$
$= -11b^2 + (-4b) + 22$
$-11b^2 - 4b + 22$

5. $= (6 + 1)x^2 y + (-9 + 11)xy^2$
$= 7x^2 y + 2xy^2$
$7x^2 y + 2xy^2$

6. $= -30x + 20 + 7x - 49$
$= -23x - 29$
$-23x - 29$

7. $= 8x^2 + 56xy + 8y^2 - 6x^2 + 2xy - 12y^2$
$= 2x^2 + 58xy - 4y^2$
$2x^2 + 58xy - 4y^2$

8. $= -6c^2 d + 12cd^2 - 9cd - 25c^2 d + 35cd^2 - 5cd$
$= -31c^2 d + 47cd^2 - 14cd$
$-31c^2 d + 47cd^2 - 14cd$

Section 1-7. Summary Exercises

1. $5(10)^2 - 15$
$= 500 - 15$
$= 485$
485

2. $(-3)^2 - 3(-3) + 6$
$= 9 - (-9) + 6$
$= 24$
24

3. $-9(6) + 3(-5)$
$= -54 + (-15)$
$= -69$
-69

4. $(4^2 - 9)^2 + (4^2 - 9) + 4$
$= 7^2 + 7 + 4$
$= 49 + 7 + 4 = 60$
60

5. $11[2(-6) + 3(5 - 2(-6))]$
$= 11[-12 + 3(17)]$
$= 11[39] = 429$
429

6. $\dfrac{(-12)^2 - 5(-12)(3)}{3(-12) - 15(3)} = \dfrac{144 - (-180)}{-36 - 45} = \dfrac{324}{-81} = -4$
-4

7. a.

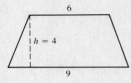

b. $A = \dfrac{1}{2}(4)(6 + 9)$
$= 2(15)$
$= 30$
30 in^2

8. $A = \$500(1 + 0.02)^8$
$= \$500(1.17165937)$
$\$585.83$

Section 1-8. Summary Exercises

1. "Sum of x and 3" is $(x + 3)$.
"Sum of y and 5" is $(y + 5)$.
". . . is decreased by . . ." is $(x + 3) - (y + 5)$.
$(x + 3) - (y + 5)$

2. "Product of two and x" is $2x$.
". . . is increased by 11" is $2x + 11$.
$2x + 11$

3. "3 more than x" is $(x + 3)$.
"6 less than y" is $(y - 6)$.
". . . is multiplied by" is $(x + 3)(y - 6)$.
$(x + 3)(y - 6)$

4. "k reduced by 1" is $k - 1$.
". . . is divided by the square of x" is $\dfrac{k - 1}{x^2}$.
$\dfrac{k - 1}{x^2}$

5. "y minus 6" is $(y - 6)$.
". . . is multiplied by one-half of x" is $\dfrac{1}{2}x(y - 6)$.
$\dfrac{1}{2}x(y - 6)$

6. "The cube of the sum of a and b" is $(a + b)^3$.
". . . is divided by the square of c" is $\dfrac{(a + b)^3}{c^2}$.
$\dfrac{(a + b)^3}{c^2}$

7. "The sum of k-squared and 3" is $(k^2 + 3)$.
". . . is divided by 2" is $\dfrac{k^2 + 3}{2}$.
$\dfrac{k^2 + 3}{2}$

8. "The quotient of 3 and x" is $\dfrac{3}{x}$.
". . . is subtracted from . . ." is $3x - \dfrac{3}{x}$.
$3x - \dfrac{3}{x}$

Section 2-1. Summary Exercises

1. $5(4) - 3(-2) = (-4)(-1) + (-3)(10)$
$20 - (-6) = 4 + (-30)$
$26 = -26$, false
Not true

2. $6(-3) - 15 = -23 + 2(-3)$
$-18 - 15 = -23 + (-6)$
$-33 = -29$, false
Not a root

3. $4(3) - 5 - 3 = 13 - 2(3) - 3$
$12 - 5 - 3 = 13 - 6 - 3$
$4 = 4$, true
Yes, a root

4. $2\left(\dfrac{25}{4}\right) - 20 = -2\left(\dfrac{25}{4}\right) + 5$

$\dfrac{25}{2} - \dfrac{40}{2} = \dfrac{-25}{2} + \dfrac{10}{2}$

$\dfrac{-15}{2} = \dfrac{-15}{2}$, true

Yes, a root

5. $-168 + 20 = -24 - 76$, false
$-112 + 20 = -16 - 76$, true
$-56 + 20 = -8 - 76$, false
$0 + 20 = 0 - 76$, false
-4

6. $9 + 3(-3) = -3 + 3$, true
$1 + 3(-1) = -1 + 3$, false
$1 + 3(1) = 1 + 3$, true
$9 + 3(3) = 3 + 3$, false
-3 and 1

7. In A: $9(3) - 36 = 3(3) - 18$
$27 - 36 = 9 - 18$, true
Yes, equivalent

8. a. $= 5x + 5 - 4x + x - 5$
$= 2x$
$2x$

b. In A: $5(-2 + 1) - 4(-2) = -2 - 5$
$5(-1) - (-8) = -7$
$3 = -7$, false
Not equivalent

Section 2-2. Summary Exercises

1. $7w = 63$
$w = 9$
The root is 9.

2. $t + 13 = -5$
$t = -18$
The root is -18.

3. $8a \cdot \dfrac{1}{8} = -56 \cdot \dfrac{1}{8}$
$a = -7$
The root is -7.

4. $\dfrac{-2}{3}w \cdot \dfrac{-3}{2} = 36 \cdot \dfrac{-3}{2}$
$w = -54$
The root is -54.

5. $30 = 5x$
$6 = x$
The root is 6.

6. $-17 = 4z + 19$
$-36 = 4z$
$-9 = z$
The root is -9.

7. $11 + 12a = -10$
$12a = -21$
$a = \dfrac{-21}{12} = \dfrac{-7}{4}$
The root is $\dfrac{-7}{4}$.

8. $\dfrac{4x}{3} = 8$
$x = 8 \cdot \dfrac{3}{4}$
$x = 6$
The root is 6.

Section 2-3. Summary Exercises

1. Let n be the number.
$4n + 30 = 3n - (-25)$
$n + 30 = 25$
$n = -5$
The number is -5.

2. $350x + 225(x + 3) = 3,550$
$350x + 225x + 675 = 3,550$
$575x = 2,875$
$x = 5$
$x + 3 = 8$
5 large mugs, 8 small mugs

3. Let w be the width
$3w - 1$ is then the length
$2w + 2(3w - 1) = 158$
$8w - 2 = 158$
$8w = 160$
$w = 20$
$3w - 1 = 3(20) - 1$
$= 59$
width is 20 cm; length is 59

4.

Car	Truck
Rate = 60 mph	Rate = x mph
Time = 4 hours	Time = 4 hours
Distance = (60)(4) miles	Distance = $4x$ miles

$(60)(4) = 4x + 12$
$228 = 4x$
$57 = x$
Rate of truck is 57 mph.

5. Let t be the age of the white bike.
$t + 6$ is then the age of the blue bike.
Since the blue bike is four times as old as the white bike;
$t + 6 = 4t$
$6 = 3t$
$2 = t$
$8 = t + 6$.
White bike is 2 years old; blue bike is 8 years old.

Section 2-4. Summary Exercises

1. $-5 = 4y + 7$
$-12 = 4y$
$-3 = y$
The root is -3.

2. $6x - 12 + 21 = 5x - 55$
$6x + 9 = 5x - 55$
$x + 9 = -55$
$x = -64$
The root is -64.

3. $-4w + 6 = 6 - 4w$
$6 = 6$, true
All real numbers are roots.
R

4. $17y - 7y - 14 = 8y - 6 - 7$
$10y - 14 = 8y - 13$
$2y - 14 = -13$
$2y = 1$
$y = \dfrac{1}{2}$
The root is $\dfrac{1}{2}$.

5. $5p + 4 = 7p + 7 - 2p$
     $4 = 7$, false
   No roots.

6. $6y - 12 = 0$
     $6y = 12$
     $y = 2$
   The root is 2.

7. Let n, $(n + 2)$, and $(n + 4)$ be the integers.
   $3(n + 4) - 2n = -3$
   $3n + 12 - 2n = -3$
   $n = -15$; $n + 2 = -13$; $n + 4 = -11$
   The integers are -15, -13, -11.

8. Let x be the value of A.
   $2x - 10°$ is the value of B.
   $2(x + 5°)$ is the value of C.
   $x + (2x - 10°) + 2(x + 5°) = 180°$
   $5x = 180°$
   $x = 36°$, $2x - 10° = 62°$, $2(x + 5°) = 82°$
   A is $36°$, B is $62°$, C is $82°$

Section 2-5. Summary Exercises

1. a. $(42 \text{ mph}) \cdot t = 126 \text{ miles}$
     $t = \dfrac{126}{42} = 3 \text{ hours}$
     3 hours
   b. $11:00 \text{ A.M.} - 3 \text{ hours} = 8:00 \text{ A.M.}$
     $8:00 \text{ A.M.}$

2. $106 = 2(31) + 2w$
     $44 = 2w$
     $22 = w$
     22 yards wide

3. $57,600 = 300w$
     $192 = w$
     192 feet wide

4. $\dfrac{PV}{nT} = \dfrac{nRT}{nT}$
     $R = \dfrac{PV}{nT}$

5. $2FR = w(R - r)$
     $w = \dfrac{2FR}{R - r}$
     $w = \dfrac{2FR}{R - r}$

6. a. $5z = 8 - 3x + 2y$
     $z = \dfrac{8 - 3x + 2y}{5}$
     $z = \dfrac{8 - 3x + 2y}{5}$
   b. $z = \dfrac{8 - 3(-5) + 2(6)}{5}$
     $z = \dfrac{35}{5} = 7$
     7

7. a. $4w = P - 4l - 4h$
     $w = \dfrac{P - 4l - 4h}{4}$
     $w = \dfrac{P - 4l - 4h}{4}$
   b. $w = \dfrac{60 - 4(7) - 4(3)}{4}$
     $w = \dfrac{20}{4} = 5$
     5

8. $3y + 6 = ay - 10$
     $16 = ay - 3y$
     $16 = y(a - 3)$
     $y = \dfrac{16}{a - 3}$
     $y = \dfrac{16}{a - 3}$

Section 2-6. Summary Exercises

1. $3t \le 45$
     $t \le 15$
     $t \le 15$

2. $-5x > 15$
     $x < -3$
     $x < -3$

3. $3x + 21 \ge 4x - 3$
     $-x \ge -24$
     $x \le 24$
     $x \le 24$

4. $\dfrac{-3}{2}w < 9$
     $w > -6$
     $w > -6$

5. $-17y < 34$
     $y > -2$
     $y > -2$

6. $6x + 2 - 5x - 4 > 2$
     $x - 2 > 2$
     $x > 4$
     $x > 4$

7. $3x - 1 \le x - 3x + 14$
     $3x - 1 \le -2x + 14$
     $5x \le 15$
     $x \le 3$
     $x \le 3$

8. Let n be the numbers
     $5n \le n + 8$
     $4n \le 8$
     $n \le 2$
     All numbers less than or equal to 2

Section 3-1. Summary Exercises

1. a. Binomial

   b. Polynomial of five terms

2. Degree of $4x^2$ is 2.

Degree of $2xy$ is $1 + 1 = 2$.

Degree of y^2 is 2.

Degree of 9 is 0.

a. 2, 2, 2, 0

b. Highest degree is 2.

3. Degree of $-9r^2s^3$ is $2 + 3 = 5$.

Degree of $3rs^2$ is $1 + 2 = 3$.

Degree of $5s$ is 1.

Degree of 1 is 0.

a. 5, 3, 1, 0

b. Highest degree is 5.

4. Write $3b^2 - 18b^3 - 27b$

as $3b^2 + (-18b^3) + (-27b)$.

$-18b^3 + 3b^2 - 27b$

5. $7x^5 - 3x^4 + 6x^3 - x^2 + 8$

6. a. $6a^3 + 60a^2b + 150ab^2 - 25$

   b. $150ab^2 + 60a^2b + 6a^3 - 25$

7. $3 + k = 7$

$k = 4$

4

8. a. x^2y^2 and $6xy^3$ have degree 4.

Therefore, $4x^3y^k$ must be the term with degree 5.

Yes

   b. $3 + k = 5$

$k = 2$

2

Section 3-2. Summary Exercises

1. $= (6x^2 - 12x^2) + (3x + 7x) + (-10 + 14)$

$= -6x^2 + 10x + 4$

$-6x^2 + 10x + 4$

2. $= (-7x^2y + 10x^2y) + (3xy + 9xy) + (-4xy^2 - xy^2)$

$= 3x^2y + 12xy - 5xy^2$

$3x^2y + 12xy - 5xy^2$

3. $= (7z^3 - 7z^3) + 3z^2 + 4z + (5 - 5)$

$= 3z^2 + 4z$

$3z^2 + 4z$

4. $= 4p^2 + 8p - 7 - p^2 + 2p - 6$

$= (4p^2 - p^2) + (8p + 2p) + (-7 - 6)$

$= 3p^2 + 10p - 13$

$3p^2 + 10p - 13$

5. $= a^3 - b^3 + a^2 + ab + b^2 - a^3 - ab$

$= (a^3 - a^3) + a^2 + (ab - ab) + b^2 - b^3$

$= a^2 + b^2 - b^3$

$a^2 + b^2 - b^3$

6. $= m^2n^2 + 4m^2n - 3mn^2 - 6 - 5m^2n^2 - m^2n + 3mn^2 - 10$

$= (m^2n^2 - 5m^2n^2) + (4m^2n - m^2n) + (-3mn^2 + 3mn^2) + (-6 - 10)$

$= -4m^2n^2 + 3m^2n - 16$

$-4m^2n^2 + 3m^2n - 16$

7. $= y^2 + 6y + 9 + 2y^2 - 11y + 5 - 3y^2 + 5y - 12$

$= (y^2 + 2y^2 - 3y^2) + (6y - 11y + 5y) + (9 + 5 - 12)$

$= 2$

2

8. $= -3t^2 + 12t - 24 + 10t^2 + 2t - 6$

$= 7t^2 + 14t - 30$

$7t^2 + 14t - 30$

Section 3-3. Summary Exercises

1. $= w^{4+1+5}$

$= w^{10}$

w^{10}

2. $= (-6)(-12)x^{2+3} \cdot y^{1+2}$

$= 72x^5y^3$

$72x^5y^3$

3. $= p^{15}q^6$

$p^{15}q^6$

4. $= (-5)^3(a^4)^3(b)^3$

$= -125a^{12}b^3$

$-125a^{12}b^3$

5. $= (-5a^2b)(a^{12}b^3)(9a^2b^4)$

$= (-5 \cdot 9)(a^2 \cdot a^{12} \cdot a^2)(b \cdot b^3 \cdot b^4)$

$= -45a^{16}b^8$

$-45a^{16}b^8$

6. $= \dfrac{(2m^4n)^4}{(p^3)^4}$

$= \dfrac{16m^{16}n^4}{p^{12}}$

$\dfrac{16m^{16}n^4}{p^{12}}$

7. $= (z^2)(3z) - (11z)(3z) + (10)(3z)$

$= 3z^3 - 33z^2 + 30z$

$3z^3 - 33z^2 + 30z$

8. $= \dfrac{-cd^2}{3} \cdot 33c^2 + \dfrac{-cd^2}{3} \cdot 24cd - \dfrac{-cd^2}{3} \cdot 12d^2$

$= -11c^3d^2 - 8c^2d^3 + 4cd^4$

$-11c^3d^2 - 8c^2d^3 + 4cd^4$

Section 3-4. Summary Exercises

1. $= (x)(3x) + (x)(5) + (7)(3x) + (7)(5)$

$= 3x^2 + 5x + 21x + 35$

$= 3x^2 + 26x + 35$

$3x^2 + 26x + 35$

2. $= (4x)(2x) + (4x)(4) - 9(2x) - 9(4)$

$= 8x^2 + 16x - 18x - 36$

$= 8x^2 - 2x - 36$

$8x^2 - 2x - 36$

3. $= (5a^2)(3a) + (5a^2)(-1) + 2(3a) + 2(-1)$

$= 15a^3 - 5a^2 + 6a - 2$

$15a^3 - 5a^2 + 6a - 2$

4. $x^2 - 2x + 5$
$3x + 2$
$\overline{3x^3 - 6x^2 + 15x}$
$2x^2 - 4x + 10$
$\overline{3x^3 - 4x^2 + 11x + 10}$
$3x^3 - 4x^2 + 11x + 10$

5. $z^2 - 4z + 5$
$z - 3$
$\overline{z^3 - 4z^2 + 5z}$
$- 3z^2 + 12z - 15$
$\overline{z^3 - 7z^2 + 17z - 15}$
$z^3 - 7z^2 + 17z - 15$

6. $x^2 + 3x - 1$
$x^2 - 3x + 1$
$\overline{x^4 + 3x^3 - x^2}$
$- 3x^3 - 9x^2 + 3x$
$x^2 + 3x - 1$
$\overline{x^4 - 9x^2 + 6x - 1}$
$x^4 - 9x^2 + 6x - 1$

7. $7w^3 + 3w + 2$
$2w - 5$
$\overline{14w^4 + 6w^2 + 4w}$
$- 35w^3 - 15w - 10$
$\overline{14w^4 - 35w^3 + 6w^2 - 11w - 10}$
$14w^4 - 35w^3 + 6w^2 - 11w - 10$

8. a. $x(x + 13)$ or $x^2 + 13x$
b. $11(11 + 13) = 264$ cm^2
c. $2x + 2(x + 13)$ or $4x + 26$
d. $4(11) + 26 = 70$ cm

Section 3-5. Summary Exercises

1. $= x^2 + 8x + 10x + 80$
$= x^2 + 18x + 80$
$x^2 + 18x + 80$

2. $= 6x^2 - 15x + 16x - 40$
$= 6x^2 + x - 40$
$6x^2 + x - 40$

3. $= p^3 + p^2q + 2pq + 2q^2$
$p^3 + p^2q + 2pq + 2q^2$

4. $= 15a - 6a^2 + 35 - 14a$
$= -6a^2 + a + 35$
$-6a^2 + a + 35$

5. $= (3y^2)^2 - (4)^2$
$= 9y^4 - 16$
$9y^4 - 16$

6. $= (5x)^2 - (2)^2$
$= 25x^2 - 4$
$25x^2 - 4$

7. $= (7x)^2 - 2(7x)(3) + (3)^2$
$= 49x^2 - 42x + 9$
$49x^2 - 42x + 9$

8. $= 3t(28t^2 + 3t - 135)$
$= 84t^3 + 9t^2 - 405t$
$84t^3 + 9t^2 - 405t$

Section 3-6. Summary Exercises

1. $= 6^{4-3} \cdot a^{3-1} \cdot b^{5-2}$
$= 6a^2b^3$
$6a^2b^3$

2. $= \dfrac{x^6y^9 \cdot xy^2}{x^3y^3}$
$= \dfrac{x^7y^{11}}{x^3y^3}$
$= x^4y^8$
x^4y^8

3. $= m^{8-5} \cdot n^{4-4} \cdot p^{10-1}$
$= m^3p^9$
m^3p^9

4. $= \dfrac{16}{5} \cdot s^{9-3} \cdot t^{3-2} \cdot u^5$
$= \dfrac{16}{5}s^6tu^5$
$\dfrac{16}{5}s^6tu^5$

5. $= \dfrac{35a^6}{7a^2} - \dfrac{21a^3}{7a^2} + \dfrac{49a^2}{7a^2}$
$= 5a^4 - 3a + 7$
$5a^4 - 3a + 7$

6.
$$\begin{array}{r}x - 10 \\ x - 3 \overline{)\, x^2 - 13x + 30}\\ (-)\ \underline{x^2 - 3x}\\ - 10x + 30\\ (-)\ \underline{- 10x + 30}\end{array}$$
$x - 10$

7.
$$\begin{array}{r}4y^3 - y \\ y + 5 \overline{)\, 4y^4 + 20y^3 - y^2 - 5y + 3}\\ (-)\ \underline{4y^4 + 20y^3}\\ - y^2 - 5y + 3\\ (-)\ \underline{- y^2 - 5y}\\ 3\end{array}$$
$4y^3 - y + \dfrac{3}{y + 5}$

8.
$$\begin{array}{r}x^2 + 4x + 16 \\ x - 4 \overline{)\, x^3 + 0x^2 + 0x - 64}\\ (-)\ \underline{x^3 - 4x^2}\\ 4x^2 + 0x - 64\\ (-)\ \underline{4x^2 - 16x}\\ 16x - 64\\ (-)\ \underline{16x - 64}\end{array}$$

$x^2 + 4x + 16$

Section 4-1. Summary Exercises

1. $= 6(2x^2) + 6(x) - 6(3)$
$= 6(2x^2 + x - 3)$
$6(2x^2 + x - 3)$

2. $= z^2(5z^2) - z^2(3z) + z^2(4)$
$= z^2(5z^2 - 3z + 4)$
$z^2(5z^2 - 3z + 4)$

3. $= a^2b^2(8a) - a^2b^2(3) + a^2b^2(5b)$
$= a^2b^2(8a - 3 + 5b)$
$a^2b^2(8a + 5b - 3)$

4. $= 5x(3x^2) - 5x(4x) + 5x(2)$
$= 5x(3x^2 - 4x + 2)$
$5x(3x^2 - 4x + 2)$

5. $= 6a(5a) - 6a(16) + 6a(a^2) + 6a(3a^3)$
$= 6a(5a - 16 + a^2 + 3a^3)$
$6a(3a^3 + a^2 + 5a - 16)$

6. $= 5x^2y(-1) + 5x^2y(9y) - 5x^2y(5y^2)$
$= 5x^2y(-1 + 9y - 5y^2)$
$5x^2y(-5y^2 + 9y - 1),$
or $-5x^2y(5y^2 - 9y + 1)$

7. $= 4j^2k(5j^2k^2) - 4j^2k(6jk) + 4j^2k(9)$
   $= 4j^2k(5j^2k^2 - 6jk + 9)$
   $4j^2k(5j^2k^2 - 6jk + 9)$

8. $= 6s(-2s^2) - 6s(8s^3) + 6s(3s^4) + 6s(1) - 6s(4s)$
   $= 6s(-2s^2 - 8s^3 + 3s^4 + 1 - 4s)$
   $6s(3s^4 - 8s^3 - 2s^2 - 4s + 1)$

Section 4-2. Summary Exercises

1. $= (x - 5)(x + 8)$ ⟨ $-5 \cdot 8 = -40$
   $-5 + 8 = 3$
   $(x - 5)(x + 8)$

2. $= (z + 2)(z - 25)$ ⟨ $2(-25) = -50$
   $2 + (-25) = -23$
   $(z + 2)(z - 25)$

3. $= (y + 1)(y + 16)$ ⟨ $1 \cdot 16 = 16$
   $1 + 16 = 17$
   $(y + 1)(y + 16)$

4. There do not exist two integers m and n such that
   $m \cdot n = 7$ and $m + n = -14$
   Cannot be factored.

5. $= (m + 3n)(m + 9n)$ ⟨ $3 \cdot 9 = 27$
   $3 + 9 = 12$
   $(m + 3n)(m + 9n)$

6. $= y(y^2 - 17y + 42)$
   $= y(y - 3)(y - 14)$ ⟨ $3 \cdot 14 = 42$
   $3 + 14 = 17$
   $y(y - 3)(y - 14)$

7. $= -ab(a^2 - 15a + 26)$
   $= -ab(a - 2)(a - 13)$ ⟨ $2 \cdot 13 = 26$
   $2 + 13 = 15$
   $-ab(a - 2)(a - 13)$

8. $= 2x(x^2 + 20x + 64)$
   $= 2x(x + 4)(x + 16)$ ⟨ $4 \cdot 16 = 64$
   $4 + 16 = 20$
   $2x(x + 4)(x + 16)$

Section 4-3. Summary Exercises

1. $= 2s^2(8s + 7) + 3(8s + 7)$
   $= (8s + 7)(2s^2 + 3)$
   $(8s + 7)(2s^2 + 3)$

2. $= 2x(7y + 2) + 7(7y + 2)$
   $= (7y + 2)(2x + 7)$
   $(7y + 2)(2x + 7)$

3. $= 36x^2 + 24x + 15xy + 10y$
   $= 12x(3x + 2) + 5y(3x + 2)$
   $= (3x + 2)(12x + 5y)$
   $(3x + 2)(12x + 5y)$

4. $= 18x^2 + 3x + 42xy + 7y$
   $= 3x(6x + 1) + 7y(6x + 1)$
   $= (6x + 1)(3x + 7y)$
   $(6x + 1)(3x + 7y)$

5. $\quad 2x(4y + 3) - 7(4y + 3)$
   $= (4y + 3)(2x - 7)$
   $(4y + 3)(2x - 7)$

6. $= 2a(b - 2c) - 7(b - 2c)$
   $= (b - 2c)(2a - 7)$
   $(b - 2c)(2a - 7)$

7. $= 10x^2(6x + 1) - 3y(6x + 1)$
   $= (6x + 1)(10x^2 - 3y)$
   $(6x + 1)(10x^2 - 3y)$

8. $= 2pr(20pr + 15p + 24qr + 18q)$
   $= 2pr[5p(4r + 3) + 6q(4r + 3)]$
   $2pr(4r + 3)(5p + 6q)$

Section 4-4. Summary Exercises

1. $= 16w^2 - 36w + 12w - 27$
   $= 4w(4w - 9) + 3(4w - 9)$
   $= (4w - 9)(4w + 3)$
   $(4w - 9)(4w + 3)$

2. $= 15x^2 - 3x + 35x - 7$
   $= 3x(5x - 1) + 7(5x - 1)$
   $= (5x - 1)(3x + 7)$
   $(5x - 1)(3x + 7)$

3. $= 2a^2 - 2a - 7a + 7$
   $= 2a(a - 1) - 7(a - 1)$
   $= (a - 1)(2a - 7)$
   $(a - 1)(2a - 7)$

4. $= 15y^2 - 10y - 18y + 12$
   $= 5y(3y - 2) - 6(3y - 2)$
   $= (3y - 2)(5y - 6)$
   $(3y - 2)(5y - 6)$

5. $= 2t^2 - 9t + 6t - 27$
   $= t(2t - 9) + 3(2t - 9)$
   $= (2t - 9)(t + 3)$
   $(2t - 9)(t + 3)$

6. $= 8u^2 + 14uv - 15v^2$
   $= 8u^2 + 20uv - 6uv - 15v^2$
   $= 4u(2u + 5v) - 3v(2u + 5v)$
   $= (2u + 5v)(4u - 3v)$
   $(2u + 5v)(4u - 3v)$

7. $= 2a(5a^2 + 13a - 6)$
   $= 2a(5a^2 + 15a - 2a - 6)$
   $= 2a[5a(a + 3) - 2(a + 3)]$
   $= 2a(a + 3)(5a - 2)$
   $2a(a + 3)(5a - 2)$

8. $= 2m^2(12m^2 - 19m - 10)$
   $= 2m^2(12m^2 - 24m + 5m - 10)$
   $= 2m^2[12m(m - 2) + 5(m - 2)]$
   $= 2m^2(m - 2)(12m + 5)$
   $2m^2(m - 2)(12m + 5)$

Section 4-5. Summary Exercises

1. $= (2x)^2 - (11)^2$
$= (2x + 11)(2x - 11)$
   $(2x + 11)(2x - 11)$

2. $= (8t)^2 - (9)^2$
$= (8t + 9)(8t - 9)$
   $(8t + 9)(8t - 9)$

3. $= (a)^2 - 2(a)(15) + (15)^2$
$= (a - 15)^2$
   $(a - 15)^2$

4. $= (3x)^2 + 2(3x)(14) + (14)^2$
$= (3x + 14)^2$
   $(3x + 14)^2$

5. $= 3(4a^2 - 4ab + b^2)$
$= 3[(2a)^2 - 2(2a)(b) + (b)^2]$
$= 3(2a - b)^2$
   $3(2a - b)^2$

6. $= 2(16a^2 + 24ab + 9b^2)$
$= 2[(4a)^2 + 2(4a)(3b) + (3b)^2]$
$= 2(4a + 3b)^2$
   $2(4a + 3b)^2$

7. $= (2x)^3 + (3)^3$
$= (2x + 3)((2x)^2 - (2x)(3x) + (3)^2)$
$= (2x + 3)(4x^2 - 6x + 9)$
   $(2x + 3)(4x^2 - 6x + 9)$

8. $= 5b(125a^3 - b^3)$
$= 5b((5a)^3 - (b)^3)$
$= 5b(5a - b)((5a)^2 + (5a)(b) + (b)^2)$
$= 5b(5a - b)(25a^2 + 5ab + b^2)$
   $5b(5a - b)(25a^2 + 5ab + b^2)$

Section 4-6. Summary Exercises

1. $x - 8 = 0$ or $3x + 7 = 0$
     $x = 8$ $3x = -7$
                      $x = \dfrac{-7}{3}$

$8, \dfrac{-7}{3}$

2. $4x = 0$ or $x + 20 = 0$
   $x = 0$ $x = -20$
$0, -20$

3. $(a - 10)(a + 3) = 0$
$a - 10 = 0$ or $a + 3 = 0$
   $a = 10$ $a = -3$
$-3, 10$

4. $y^2 - 5y = 0$
$y(y - 5) = 0$
$y = 0$ or $y - 5 = 0$
                   $y = 5$
$0, 5$

5. $6b^2 + 5b + 1 = 0$
$(3b + 1)(2b + 1) = 0$
$3b + 1 = 0$ or $2b + 1 = 0$
   $3b = -1$ $2b = -1$
     $b = \dfrac{-1}{3}$ $b = \dfrac{-1}{2}$

$\dfrac{-1}{3}, \dfrac{-1}{2}$

6. $8k^2 - 6k + 1 = 0$
$(4k - 1)(2k - 1) = 0$
$4k - 1 = 0$ or $2k - 1 = 0$
   $4k = 1$ $2k = 1$
     $k = \dfrac{1}{4}$ $k = \dfrac{1}{2}$

$\dfrac{1}{4}, \dfrac{1}{2}$

7. $(3z + 4)(3z - 4) = 0$
$3z + 4 = 0$ or $3z - 4 = 0$
   $3z = -4$ $3z = 4$
    $z = \dfrac{-4}{3}$ $z = \dfrac{4}{3}$

$\dfrac{-4}{3}, \dfrac{4}{3}$

8. $4w^2 - 36w + 81 = 0$
$(2w - 9)(2w - 9) = 0$
$2w - 9 = 0$
   $2w = 9$
    $w = \dfrac{9}{2}$

$\dfrac{9}{2}$

Section 4-7. Summary Exercises

1. Let n be one integer.
$19 - n$ is then the other integer.
$n(19 - n) = 84$
$19n - n^2 = 84$
      $0 = n^2 - 19n + 84$
      $0 = (n - 7)(n - 12)$
$n - 7 = 0$ or $n - 12 = 0$
   $n = 7$ $n = 12$
$19 - n = 12$ $19 - n = 7$
The integers are 7 and 12.

2. Let x be the length of the shorter leg.
$x + 6$ is the length of the longer leg.
$x^2 + (x + 6)^2 = 30^2$
$x^2 + x^2 + 12x + 36 = 900$
   $2x^2 + 12x - 864 = 0$
   $2(x^2 + 6x - 432) = 0$
   $2(x - 18)(x + 24) = 0$

$2 = 0$, false   or $x - 18 = 0$ or $x + 24 = 0$
                        $x = 18$ $x = -24$, reject
                  $x + 6 = 24$

a. 18 centimeters
b. 24 centimeters

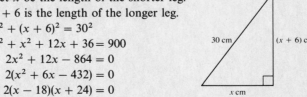

3. Let w be the width of the rectangle.
$w + 12$ is then the length.
$$w(w + 12) = 160$$
$$w^2 + 12w - 160 = 0$$
$$(w - 8)(w + 20) = 0$$
$$w - 8 = 0 \quad \text{or} \quad w + 20 = 0$$
$$w = 8 \qquad\qquad w = -20, \text{ reject}$$
$$w + 12 = 20$$
a. 8 inches
b. 20 inches

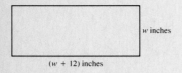

w inches

$(w + 12)$ inches

4. Let b be the length of the base.
$\frac{1}{2}b$ is then the height.
$$\frac{1}{2} \cdot b \cdot \frac{1}{2}b = 25$$
$$\frac{1}{4}b^2 = 25$$
$$b^2 = 100$$
$$b^2 - 100 = 0$$
$$(b + 10)(b - 10) = 0$$
$$b + 10 = 0 \quad \text{or} \quad b - 10 = 0$$
$$b = -10, \text{ reject} \qquad b = 10$$
$$\frac{1}{2}b = 5$$
a. 10 meters
b. 5 meters

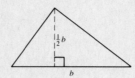

Section 5-1. Summary Exercises

1. $\dfrac{(-3)^2 + 9}{(-3)^2 + (-3) - 3} = \dfrac{18}{3} = 6$

6

2. $\dfrac{2(3) - 2}{2(3)^2 - 3(3)(2) + 2^2} = \dfrac{4}{4} = 1$

1

3. $\dfrac{(-2)^2 - 6^2 + 5^2}{(-2)^3 + 5^3} = \dfrac{-7}{117}$

$\dfrac{-7}{117}$

4. $p^2 - 5p = 0$
$p(p - 5) = 0$
$p = 0 \quad \text{or} \quad p - 5 = 0$
$p = 5$

0 and 5

5. $a^2 - b^2 = 0$
$(a + b)(a - b) = 0$
$a + b = 0 \quad \text{or} \quad a - b = 0$
$a = -b \qquad\qquad a = b$
$a \neq -b$ and $a \neq b$

6. $\dfrac{2c(3c + 8)}{(3c - 8)(3c + 8)}$
$= \dfrac{2c}{3c - 8}$

$\dfrac{2c}{3c - 8}$

7. $\dfrac{2(m - 2)(m - 2)}{6(m + 2)(m - 2)}$
$= \dfrac{m - 2}{3(m + 2)}$

$\dfrac{m - 2}{3(m + 2)}$

8. $\dfrac{(3u - 5)(9u^2 + 15u + 25)}{(3u - 5)(2u + 3v)}$
$= \dfrac{9u^2 + 15u + 25}{2u + 3v}$

$\dfrac{9u^2 + 15u + 25}{2u + 3v}$

Section 5-2. Summary Exercises

1. $\dfrac{1}{4^3} \cdot \dfrac{1}{x^2} \cdot y^5 = \dfrac{y^5}{64x^2}$

$\dfrac{y^5}{64x^2}$

2. $\dfrac{10}{\frac{1}{p^3}} = 10p^3$

$10p^3$

3. $\left(\dfrac{3}{s}\right) = \dfrac{9}{s^2}$

$\dfrac{9}{s^2}$

4. w^{-11+7}
$= w^{-4}$
$= \dfrac{1}{w^4}$

$\dfrac{1}{w^4}$

5. $\dfrac{1}{t^{-6-(-10)}} = \dfrac{1}{t^4}$

$\dfrac{1}{t^4}$

6. $9^{-2} \cdot m^{(-3)(-2)}$
$= \dfrac{m^6}{81}$

$\dfrac{m^6}{81}$

7. $\dfrac{(b^{-5})^{-2}}{5^{-2} \cdot (a^{-1})^{-2}} = \dfrac{25b^{10}}{a^2}$

$\dfrac{25b^{10}}{a^2}$

8. a. 2.75×10^7
b. 0.00063

Section 5-3. Summary Exercises

1. $\dfrac{9 \cdot 4}{10 \cdot 3} \cdot \dfrac{x}{x^2} \cdot \dfrac{y^2}{y} = \dfrac{6y}{5x}$

$\dfrac{6y}{5x}$

2. $\dfrac{30}{x^2 y^2} \cdot \dfrac{xy}{18}$

$= \dfrac{30}{18} \cdot \dfrac{x}{x^2} \cdot \dfrac{y}{y^2} = \dfrac{5}{3} \cdot \dfrac{1}{x} \cdot \dfrac{1}{y} = \dfrac{5}{3xy}$

$\dfrac{5}{3xy}$

3. $\dfrac{(2x + 1)(x - 3) \cdot 2x^4}{(x - 3)(x - 3) \cdot x(2x + 1)} = \dfrac{2x^3}{x - 3}$

$\dfrac{2x^3}{x - 3}$

4. $\dfrac{(a + b)(a + b)(2a + b)}{2ab(2a + b)(a + b)}$

$= \dfrac{a + b}{2ab}$

$\dfrac{a + b}{2ab}$

5. $\dfrac{(2a + b)(a + b)(a^2 + ab + b^2)}{(a - b)(a^2 + ab + b^2)(2a + b)(a - b)}$

$= \dfrac{a + b}{(a - b)(a - b)}$

$\dfrac{a + b}{(a - b)(a - b)}$

6. $\dfrac{y(x - y^2)(x^2 + y)(2x + 3y)}{2x(x^2 + y)(x - y^2)(2x + 3y)}$

$= \dfrac{y}{2x}$

$\dfrac{y}{2x}$

7. $\dfrac{(8x^2 + 18x - 5)(10x^2 + 11x - 6)(4x + 5)}{(10x^2 - 9x + 2)(8x^2 + 22x + 15)(4x - 1)}$

$= \dfrac{(4x - 1)(2x + 5)(5x - 2)(2x + 3)(4x + 5)}{(5x - 2)(2x - 1)(4x + 5)(2x + 3)(4x - 1)} = \dfrac{2x + 5}{2x - 1}$

$\dfrac{2x + 5}{2x - 1}$

8. $\dfrac{3n^2(m^2 - 4n^2)(2m)(10m^2 n - 5mn^2)}{(6m^2 - 3mn)(m^2 n + 2mn^2)(45m - 15n)(mn)}$

$= \dfrac{3n^2(m + 2n)(m - 2n)(2m)(5mn)(2m - n)}{3m(2m - n)(mn)(m + 2n)(15)(3m - n)(mn)} = \dfrac{2n(m - 2n)}{3m(3m - n)}$

$\dfrac{2n(m - 2n)}{3m(3m - n)}$

Section 5-4. Summary Exercises

1. $\dfrac{3a^3 + 9a^2}{4(a + 3)} = \dfrac{3a^2(a + 3)}{4(a + 3)} = \dfrac{3a^2}{4}$

$\dfrac{3a^2}{4}$

2. $\dfrac{5y + 2 - y + 4}{2y + 3} = \dfrac{4y + 6}{2y + 3} = \dfrac{2(2y + 3)}{(2y + 3)} = 2$

2

3. $\dfrac{a + 4 - a}{a + 2} = \dfrac{4}{a + 2}$

$\dfrac{4}{a + 2}$

4. $\dfrac{x + y + 3x - y}{x^2 + y^2} = \dfrac{4x}{x^2 + y^2}$

$\dfrac{4x}{x^2 + y^2}$

5. $\dfrac{x^2 + 1 + x - 3}{x^2 - 3x + 2}$

$= \dfrac{x^2 + x - 2}{x^2 - 3x + 2} = \dfrac{(x - 1)(x + 2)}{(x - 1)(x - 2)} = \dfrac{x + 2}{x - 2}$

$\dfrac{x + 2}{x - 2}$

6. $\dfrac{5m^2 + 2mn - m^2 - 2mn - n^2}{2m^2 + 3mn - 2n^2}$

$= \dfrac{4m^2 - n^2}{2m^2 + 3mn - 2n^2} = \dfrac{(2m + n)(2m - n)}{(2m - n)(m + 2n)} = \dfrac{2m + n}{m + 2n}$

$\dfrac{2m + n}{m + 2n}$

7. $\dfrac{y^2 - 4y + 3(y - 2)}{3y^2 + 5y - 2}$

$= \dfrac{y^2 - y - 6}{3y^2 + 5y - 2} = \dfrac{(y - 3)(y + 2)}{(3y - 1)(y + 2)} = \dfrac{y - 3}{3y - 1}$

$\dfrac{y - 3}{3y - 1}$

8. $\dfrac{3a^2}{a^2 + 5a + 6} + \dfrac{2(a + 3)(a)(a - 2)}{(a + 2)(a - 2)(a + 3)(a + 3)} + \dfrac{2 + 5a}{a^2 + 5a + 6}$

$= \dfrac{3a^2 + 2a + 2 + 5a}{(a + 2)(a + 3)} = \dfrac{(3a + 1)(a + 2)}{(a + 2)(a + 3)} = \dfrac{3a + 1}{a + 3}$

$\dfrac{3a + 1}{a + 3}$

Section 5-5. Summary Exercises

1. The LCM of 6 and 15 is 30.
The LCM of m^2 and m is m^2.
The LCM of n and n is n.
$30m^2n$

2. $x^2 - 7x = x(x - 7)$
$2x^2 - 98 = 2(x + 7)(x - 7)$
$5x + 35 = 5(x + 7)$
$10x(x + 7)(x - 7)$

3. The LCM is $24a^2b$.
$$\frac{5}{12a^2b} \cdot \frac{2}{2} - \frac{1}{8ab} \cdot \frac{3a}{3a}$$
$$= \frac{10 - 3a}{24a^2b}$$
$$\frac{10 - 3a}{24a^2b}$$

4. The LCM is $6(y - 3)$.
$6y - 18 = 6(y - 3)$
$6 - 2y = -2(y - 3)$
$$\frac{5y}{6(y - 3)} + \frac{-3y}{2(y - 3)} \cdot \frac{3}{3}$$
$$= \frac{5y - 9y}{6(y - 3)} = \frac{-4y}{6(y - 3)}$$
$$\frac{-2y}{3(y - 3)} \text{ or } \frac{2y}{3(3 - y)}$$

5. The LCM is $(b - 3)(b + 3)^2$.
$$\frac{2}{(b + 3)(b + 3)} + \frac{2}{(b + 3)(b - 3)}$$
$$= \frac{2(b - 3) + 2(b + 3)}{(b - 3)(b + 3)^2} = \frac{4b}{(b - 3)(b + 3)^2}$$
$$\frac{4b}{(b - 3)(b + 3)^2}$$

6. The LCM is $3(p + 2)(p - 1)$.
$$\frac{2p + 5}{(p + 2)(p - 1)} + \frac{4p + 9}{3(p + 2)}$$
$$= \frac{3(2p + 5) + (4p + 9)(p - 1)}{3(p + 2)(p - 1)} = \frac{6p + 15 + 4p^2 + 5p - 9}{3(p + 2)(p - 1)} = \frac{(4p + 3)(p + 2)}{3(p + 2)(p - 1)}$$
$$\frac{4p + 3}{3(p - 1)}$$

7.
$$\frac{z}{z + 7} + \frac{z}{z - 9} - \frac{z^2 + 4z + 27}{(z + 7)(z - 9)}$$
$$= \frac{z(z - 9) + z(z + 7) - (z^2 + 4z + 27)}{(z + 7)(z - 9)} = \frac{z^2 - 6z - 27}{(z + 7)(z - 9)} = \frac{(z + 3)(z - 9)}{(z + 7)(z - 9)}$$
$$\frac{z + 3}{z + 7}$$

8.
$$\frac{2}{t} - \frac{1}{t + 5} + \frac{(t - 4)(t + 2)(t - 7)(t + 6)}{(t + 5)(t + 2)(t - 7)(t - 4)}$$
$$= \frac{2}{t} \cdot \frac{t + 5}{t + 5} - \frac{1}{t + 5} \cdot \frac{t}{t} + \frac{t + 6}{t + 5} \cdot \frac{t}{t}$$
$$= \frac{2t + 10 - t + t^2 + 6t}{t(t + 5)}$$
$$= \frac{(t + 2)(t + 5)}{t(t + 5)}$$
$$\frac{t + 2}{t}$$

Section 5-6. Summary Exercises

1.
$$\frac{6x^2 + 7x - 3}{x^2} \div \frac{9x^2 + 3x - 2}{x^2}$$
$$= \frac{(3x - 1)(2x + 3) \cdot x^2}{x^2 \cdot (3x + 2)(3x - 1)} = \frac{2x + 3}{3x + 2}$$
$$\frac{2x + 3}{3x + 2}$$

2.
$$\frac{3y - 1}{y} \div \frac{9y^2 - 1}{y^2}$$
$$= \frac{(3y - 1) \cdot y^2}{y \cdot (3y + 1)(3y - 1)} = \frac{y}{3y + 1}$$
$$\frac{y}{3y + 1}$$

3. $\dfrac{a(a + b) - b(a - b)}{(a - b)(a + b)} \div \dfrac{a(a - b) + b(a + b)}{(a + b)(a - b)} = \dfrac{(a^2 + b^2)(a + b)(a - b)}{(a - b)(a + b)(a^2 + b^2)}$
$$1$$

4. $\left(\dfrac{t - \frac{9}{t}}{t - 7 + \frac{12}{t}}\right)\dfrac{t}{t} = \dfrac{t^2 - 9}{t^2 - 7t + 12}$
$$= \frac{(t + 3)(t - 3)}{(t - 4)(t - 3)}$$
$$\frac{t + 3}{t - 4}$$

5. $\left(\dfrac{\dfrac{1}{p} - \dfrac{2}{p^2} - \dfrac{3}{p^3}}{1 - \dfrac{9}{p^2}}\right) \cdot \dfrac{p^3}{p^3} = \dfrac{p^2 - 2p - 3}{p^3 - 9p}$

$= \dfrac{(p-3)(p+1)}{p(p-3)(p+3)}$

$\dfrac{p+1}{p(p+3)}$

6. $\left(\dfrac{\dfrac{m}{n} + \dfrac{n}{5}}{\dfrac{mn}{10}}\right) \cdot \dfrac{10n}{10n} = \dfrac{10m + 2n^2}{mn^2}$

$\dfrac{10m + 2n^2}{mn^2}$

7. $\left(\dfrac{1 + \dfrac{1}{x+1}}{\dfrac{1}{2} + \dfrac{1}{x}}\right) \cdot \dfrac{2x(x+1)}{2x(x+1)}$

$= \dfrac{2x(x+1) + 2x}{x(x+1) + 2(x+1)}$

$= \dfrac{2x^2 + 4x}{x^2 + 3x + 2} = \dfrac{2x(x+2)}{(x+1)(x+2)}$

$\dfrac{2x}{x+1}$

8. $= \dfrac{a + \dfrac{3}{a}}{a^2 - \dfrac{9}{a^2}} = \dfrac{a^2 + 3}{a} \div \dfrac{a^4 - 9}{a^2}$

$= \dfrac{(a^2 + 3) \cdot a^2}{a \cdot (a^2 + 3)(a^2 - 3)}$

$\dfrac{a}{a^2 - 3}$

Section 5-7. Summary Exercises

1. The LCM is 6.

$6\left(\dfrac{2x+1}{3} - 1\right) = \dfrac{x}{2} \cdot 6$

$2(2x+1) - 6 = 3x$

$x = 4$

4

2. The LCM is $12y$.

$12y\left(\dfrac{1}{3} + \dfrac{y+1}{4y}\right) = \left(\dfrac{1}{3y} - \dfrac{y+5}{6y}\right)12y$

$4y + 3y + 3 = 4 - 2y - 10$

$9y = -9$

$y = -1$

-1

3. The LCM is $(2a+1)(a+4)(a-2)$

$\text{LCM}\left(\dfrac{4}{(2a+1)(a+4)} = \dfrac{1}{(2a+1)(a-2)}\right)$

$4(a-2) = a + 4$

$3a = 12$

$a = 4$

4

4. The LCM is $b(b+1)(b-1)$.

$\text{LCM}\left(\dfrac{4b}{b(b+1)} + \dfrac{3}{b(b-1)} = \dfrac{4}{b}\right)$

$4b(b-1) + 3(b+1) = 4(b+1)(b-1)$

$4b^2 - 4b + 3b + 3 = 4b^2 - 4$

$-b = -7$

$b = 7$

7

5. The LCM is $(u-3)(u-2)$.

$\text{LCM}\left(\dfrac{u-1}{u-3} = \dfrac{2u}{u-2} + \dfrac{4}{(u-3)(u-2)}\right)$

$(u-1)(u-2) = 2u(u-3) + 4$

$0 = u^2 - 3u + 2$

$0 = (u-2)(u-1)$

$u = 2, \text{reject}$ or $u = 1$

1

6. The LCM is $(v+2)(v-3)$.

$\text{LCM}\left(\dfrac{v+1}{v+2} + \dfrac{-1(v+1)}{v-3} = \dfrac{5}{(v+2)(v-3)}\right)$

$(v+1)(v-3) - (v+1)(v+2) = 5$

$-5v = 10$

$v = -2, \text{reject}$

No solutions

7. The LCM is $(5-a)$.

$\text{LCM}\left(\dfrac{5}{5-a} - \dfrac{a^2}{5-a} = -2\right)$

$5 - a^2 = -2(5-a)$

$0 = a^2 + 2a - 15$

$0 = (a+5)(a-3)$

$a = -5$ or $a = 3$

-5 and 3

8. The LCM is $(x-2)$.

$\text{LCM}\left(\dfrac{x}{x-2} = 2 - \dfrac{-1}{x-2}\right)$

$x = 2(x-2) + 1$

$x = 2x - 4 + 1$

$3 = x$

3

Section 5-8. Summary Exercises

1. $12t = 924$

$t = 77$

77

2. $4(u+3) = 36$

$4u + 12 = 36$

$4u = 24$

$u = 6$

6

3. $3x^2 - 3x = 2x^2 + 3x$

$x^2 - 6x = 0$

$x(x-6) = 0$

$x = 0$ or $x = 6$

Reject 0.

6

4. $5y^2 + 2y = 30y - 15$

$5y^2 - 28y + 15 = 0$

$(5y-3)(y-5) = 0$

$5y - 3 = 0$ or $y - 5 = 0$

$y = \dfrac{3}{5}$ $y = 5$

$\dfrac{3}{5}$ and 5

5. $\dfrac{3 \text{ ounces}}{260 \text{ calories}} = \dfrac{12 \text{ ounces}}{x}$

$3x = 3,120$

$x = 1,040$

1,040 calories

6. $\dfrac{5 \text{ avocados}}{\$1.98} = \dfrac{45 \text{ avocados}}{x}$

$5x = 89.10$

$x = 17.82$

$17.82

7. $\dfrac{2 \text{ voted}}{5 \text{ registered}} = \dfrac{246 \text{ voted}}{x}$

$2x = 1,230$

$x = 615$

615 registered voters

8. $\dfrac{a}{x} = \dfrac{b}{y} = \dfrac{c}{z}$

$\dfrac{4}{x} = \dfrac{6}{10}$

$6x = 40$

$x = \dfrac{40}{6} = \dfrac{20}{3}$

$x = \dfrac{20}{3}$

$\dfrac{6}{10} = \dfrac{c}{18}$

$10c = 108$

$c = \dfrac{108}{10} = \dfrac{54}{5}$

$c = \dfrac{54}{5}$

Section 5-9. Summary Exercises

1. Let n be the number.

$\dfrac{9 + n}{13 + (n - 1)} = \dfrac{4}{5}$

$5(9 + n) = 4(12 + n)$

$45 + 5n = 48 + 4n$

$n = 3$

The number is 3.

2. Let n be the number.

$n - \dfrac{3}{n} = \dfrac{23}{6}$

$6n\left(n - \dfrac{3}{n}\right) = \dfrac{23}{6} \cdot 6n$

$6n^2 - 18 = 23n$

$6n^2 - 23n - 18 = 0$

$(3n + 2)(2n - 9) = 0$

$3n + 2 = 0 \quad$ or $\quad 2n - 9 = 0$

$n = \dfrac{-2}{3} \qquad\qquad n = \dfrac{9}{2}$

The number is $\dfrac{-2}{3}$ or $\dfrac{9}{2}$.

3.

Machine	Rate	Time	Tub
Faucets	$\dfrac{1}{8}$	t	$t \cdot \dfrac{1}{8}$
Drain	$\dfrac{-1}{12}$	t	$t \cdot \dfrac{-1}{12}$

Let t be the time.

$24\left(\dfrac{t}{8} - \dfrac{t}{12}\right) = 1 \cdot 24$

$3t - 2t = 24$

$t = 24$

24 minutes

4.

	Distance	Rate	Time
With current	$\dfrac{1}{3}$	$r + \dfrac{1}{2}$	t
Against current	$\dfrac{1}{5}$	$r - \dfrac{1}{2}$	t

Let r be the rate with no current.

$\dfrac{\frac{1}{3}}{r + \frac{1}{2}} = \dfrac{\frac{1}{5}}{r - \frac{1}{2}}$

The LCM is 30.

$30\left(\dfrac{1}{3}r - \dfrac{1}{6}\right) = \left(\dfrac{1}{5}r + \dfrac{1}{10}\right)30$

$10r - 5 = 6r + 3$

$4r = 8$

$r = 2$

2 mph

Section 6-1. Summary Exercises

1. $3(5) - 4(2) = 9$
$15 - 8 = 9$
$7 = 9$, false
No

2. $4(-4) + 3(9) = 11$
$-16 + 27 = 11$
$11 = 11$, true
Yes

3. $0 - 7(-3) = 21$
$0 - (-21) = 21$
$21 = 21$, true
Yes

4. $2(6) - 3\left(\dfrac{-1}{3}\right) = 13$
$12 - (-1) = 13$
$13 = 13$, true
Yes

5. $-2 + v = 9$
$v = 11$
$(-2, 11)$

6. $b = \dfrac{1}{2} \cdot 4 - 3$
$b = 2 - 3$
$b = -1$
$(4, -1)$

7. $\dfrac{3}{2} = \dfrac{5}{2}m - 1$
$3 = 5m - 2$
$5 = 5m$
$1 = m$
$\left(1, \dfrac{3}{2}\right)$

8. $(-5)^2 + 2s = 33$
$25 + 2s = 33$
$2s = 8$
$s = 4$
$(-5, 4)$

Section 6-2. Summary Exercises

1. $A(-4, 2)$
$B(7, -3)$
$C(3, 6)$
$D(0, -5)$

2.

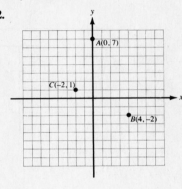

3. a. $x < 0$ and $y < 0$
III
b. $x > 0$ and $y > 0$
I
c. $x < 0$ and $y > 0$
II
d. $x > 0$ and $y < 0$
IV

4. $3y = 2x + 15$
$-2x + 3y = 15$
$2x - 3y = -15$
$2x - 3y = -15$

5. $4y = -x$
$y = \dfrac{-1}{4}x$

x	y
0	0
4	-1
-4	1

6. $y = -2x + 3$

x	y
0	3
1	1
-1	5

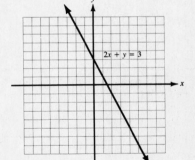

7. $-4y = -5x + 20$
$y = \dfrac{5}{4}x - 5$

x	y
0	-5
4	0
8	5

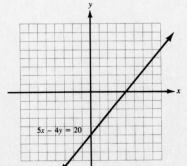

8.

x	y
-5	-3
0	-3
5	-3

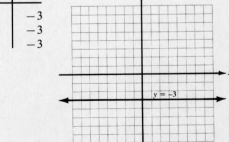

Section 6-3. Summary Exercises

1. $m = \dfrac{7-1}{3-(-2)} = \dfrac{6}{5}$
$$\dfrac{6}{5}$$

2. $m = \dfrac{2-(-2)}{-6-6} = \dfrac{4}{-12} = \dfrac{-1}{3}$
$$\dfrac{-1}{3}$$

3. $m = \dfrac{0-5}{0-(-4)} = \dfrac{-5}{4}$
$$\dfrac{-5}{4}$$

4. $m = \dfrac{6-6}{-4-(-7)} = \dfrac{0}{3} = 0$
$$0$$

5. a. $\dfrac{7-7}{0-(-3)} = \dfrac{0}{3}$ Horizontal line.
Horizontal line

b. $\dfrac{-2-10}{4-4} = \dfrac{-12}{0}$ Vertical line.
Vertical line

6. a. Let $y = 0$: $3x = 18$
$$x = 6$$
$$6$$

b. Let $x = 0$: $-6y = 18$
$$y = -3$$
$$-3$$

7. If $x = 0$,
$$-2y = 6$$
$$y = -3$$
$P(0, -3)$ is on the line.

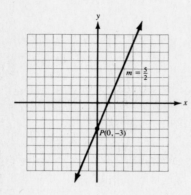

8. If $x = 1$,
$$2 - 3y = -1$$
$$-3y = -3$$
$$y = 1$$
$P(1, 1)$ is on the line.

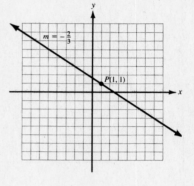

Section 6-4. Summary Exercises

1. $-5y = -x - 15$
$$y = \dfrac{1}{5}x + 3$$
a. $y = \dfrac{1}{5}x + 3$
b. $\dfrac{1}{5}$
c. 3

2. $5y = -6x + 35$
$$y = \dfrac{-6}{5}x + 7$$
a. $y = \dfrac{-6}{5}x + 7$
b. $\dfrac{-6}{5}$
c. 7

3. $y = \dfrac{-3}{5}x;\ m_1 = \dfrac{-3}{5}$
$$3x + 5y = 15$$
$$5y = -3x + 15$$
$$y = \dfrac{-3}{5}x + 3;\ m_2 = \dfrac{-3}{5}$$
Parallel

4. $-4y = -7x + 28$
$$y = \dfrac{7}{4}x - 7;\ m_1 = \dfrac{7}{4}$$
$$7y = -4x - 14 \qquad \dfrac{7}{4} \cdot \dfrac{-4}{7} = -1$$
$$y = \dfrac{-4}{7}x - 2;\ m_2 = \dfrac{-4}{7}$$
Perpendicular

5. Use $(0, -3)$ and $m = \frac{2}{3}$.

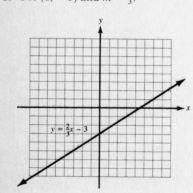

6. $3y = -4x - 12$
$$y = \dfrac{-4}{3}x - 4$$
Use $(0, -4)$ and $m = \frac{-4}{3}$.

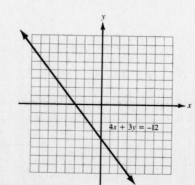

7. $6x + 4 = 4y - 8$
$$4y = 6x + 12$$
$$y = \dfrac{3}{2}x + 3$$
Use $(0, 3)$ and $m = \frac{3}{2}$.

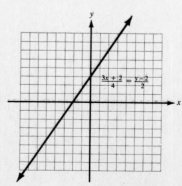

8. Vertical line through $(-6, 0)$

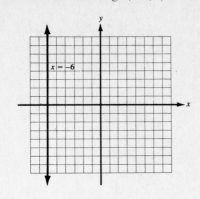

Section 6-5. Summary Exercises

1. a. $y - 4 = \dfrac{-6}{7}(x - (-3))$

$y - 4 = \dfrac{-6}{7}(x + 3)$

b. $7y - 28 = -6x - 18$
$6x + 7y = 10$
$6x + 7y = 10$

2. a. $y - 0 = \dfrac{1}{2}(x - 0)$

$y - 0 = \dfrac{1}{2}(x - 0)$

b. $2y = x$
$x - 2y = 0$
$x - 2y = 0$

3. $m = \dfrac{7 - (-1)}{-3 - 2} = \dfrac{-8}{5}$

a. $y - 7 = \dfrac{-8}{5}(x - (-3))$

$y - 7 = \dfrac{-8}{5}(x + 3)$

b. $5y - 35 = -8x - 24$
$8x + 5y = 11$
$8x + 5y = 11$

4. $m = \dfrac{-7 - 0}{-7 - (-3)} = \dfrac{-7}{-4} = \dfrac{7}{4}$

a. $y - (-7) = \dfrac{7}{4}(x - (-7))$

$y + 7 = \dfrac{7}{4}(x + 7)$

b. $4y + 28 = 7x + 49$
$7x - 4y = -21$
$7x - 4y = -21$

5. $3y = -x + 6$

$y = \dfrac{-1}{3}x + 2; m = \dfrac{-1}{3}$

a. $y - (-2) = \dfrac{-1}{3}(x - (-4))$

$y + 2 = \dfrac{-1}{3}(x + 4)$

b. $3y + 6 = -x - 4$
$x + 3y = -10$
$x + 3y = -10$

6. $-5y = -6x$

$y = \dfrac{6}{5}x; m = \dfrac{6}{5}$

a. $y - (-4) = \dfrac{-5}{6}(x - 9)$

$y + 4 = \dfrac{-5}{6}(x - 9)$

b. $6y + 24 = -5x + 45$
$5x + 6y = 21$
$5x + 6y = 21$

7. For a horizontal line, $m = 0$.

a. $y - (-5) = 0(x - 11)$
$y + 5 = 0$

b. $y = -5$
$y = -5$

8. For vertical lines, the slope is undefined.
With $x_1 = -13, x = -13$.
a. Not possible
b. $x = -13$

Section 6-6. Summary Exercises

1. $1 > \dfrac{-1}{2}(-6) - 1$

$1 > 3 - 1$, false
No

2. $2(3) + (-5) \leq 3$
$6 + (-5) \leq 3$, true
Yes

3. $-6 \geq -4$, false
No

4. $-y = -4x + 4$
$y = 4x - 4$
$m = 4$ and
$b = -4$

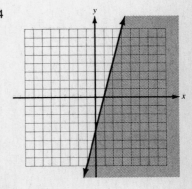

5. $2y = -6x + 6$
$y = -3x + 3$
$m = -3$ and $b = 3$

6. $-y = -x + 2$
$y = x - 2$
$m = 1$ and $b = -2$

7. $-5y = -2x - 1$
$y = \dfrac{2}{5}x + \dfrac{1}{5}$

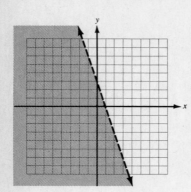
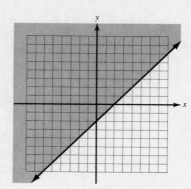
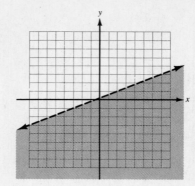

8. $P_1(0, 3)$ and $P_2(4, 6)$ are on the boundary

$m = \dfrac{6-3}{4-0} = \dfrac{3}{4}$ Pick $(0, 0)$ as a test point.

$3(0) - 4(0) \boxed{} - 12$

$y - 3 = \dfrac{3}{4}(x - 0)$ $0 \boxed{>} - 12$

An inequality is $3x - 4y > -12$.

$4y - 12 = 3x$
$3x - 4y = -12$

Section 7-1. Summary Exercises

1. "Four times a number x is three more than $\frac{1}{2}$ a number y."

$4x = 3 + \dfrac{1}{2}y$

"The sum of x and y is 12."

(1) $4x = 3 + \dfrac{1}{2}y$

(2) $x + y = 12$

2. (1) $3(2) + 3(1) = 9$
$6 + 3 = 9$, true
Thus $(2, 1)$ is a solution.
Yes

(2) $3(2) - 6(1) = 0$
$6 - 6 = 0$, true

3. (1) $3(4) - 2(9) = -6$
$12 - 18 = -6$, true
Thus $(4, 9)$ is a solution.
Yes

(2) $9(4) + 20(9) = 216$
$36 + 180 = 216$, true

4. (1) $10 \cdot \dfrac{1}{2} + 9 \cdot \dfrac{1}{3} = 8$
$5 + 3 = 8$, true
Thus $\left(\dfrac{1}{2}, \dfrac{1}{3}\right)$ is a solution.
Yes

(2) $12 \cdot \dfrac{1}{2} - 18 \cdot \dfrac{1}{3} = 0$
$6 - 6 = 0$, true

5. (1) $3(7) - 5y = 6$
$-5y = -15$
$y = 3$
$(7, 3)$

(2) $7 - 2(3) = 1$
$7 - 6 = 1$, true

6. (1) $24a - 15\left(\dfrac{-7}{3}\right) = 35$
$24a - (-35) = 35$
$24a = 0$
$a = 0$
$\left(0, \dfrac{-7}{3}\right)$

(2) $5 \cdot 0 - 3 \cdot \dfrac{-7}{3} = 7$
$0 - (-7) = 7$, true

7. (1) $2(5) + 10q = 5$
$10q = -5$
$q = \dfrac{-1}{2}$
$\left(5, \dfrac{-1}{2}\right)$

(2) $5 - 4\left(\dfrac{-1}{2}\right) = 7$
$5 - (-2) = 7$
$7 = 7$, true

8. (1) $2(7) - 2(5) + 4 = 8$
$14 - 10 + 4 = 8$, true
(2) $7 + 3(5) - 3(4) = 10$
$7 + 15 - 12 = 10$, true
(3) $3(7) - 5(5) + 4(4) = 12$
$21 - 25 + 16 = 12$, true
Yes

Section 7-2. Summary Exercises

1. (1)

x	y
0	−5
3	−2
5	0

(2)

x	y
0	3
3	2
6	1

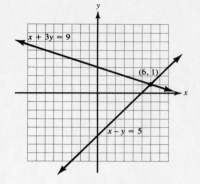

2. (1)

x	y
−2	4
0	0
2	−4

(2)

x	y
−1	−6
0	−4
1	−2

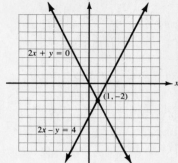

3. (1)

x	y
−3	1
0	2
3	3

(2)

x	y
−3	−1
0	0
3	1

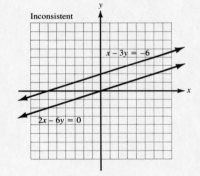

4. (1)

x	y
−3	−5
0	−1
3	3

(2)

x	y
−3	1
0	2
3	3

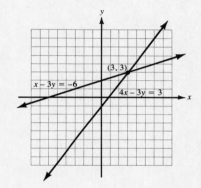

5. (1)

x	y
−5	6
0	2
5	−2

(2)

x	y
−5	6
0	2
5	−2

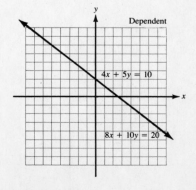

6. (1)

x	y
−5	0
−3	2
0	5

(2)

x	y
−4	6
0	5
4	4

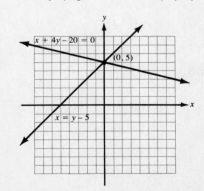

7. (1)

x	y
−5	6
0	1
5	−4

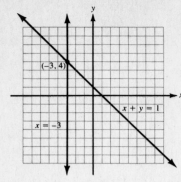

(2)

x	y
−3	5
−3	0
−3	−5

8. (1) $-5y = -2x + 70$

$y = \dfrac{2}{5}x - 14$

(2) $10y = 4x + 20$

$y = \dfrac{2}{5}x + 2$

$m_1 = m_2$ and $b_1 \neq b_2$

Thus the system is inconsistent.

Section 7-3. Summary Exercises

1. (1) $7x - 9(-4) = 64$
$7x - (-36) = 64$
$7x = 28$
$x = 4$
$(4, -4)$

2. (2) $5(2s) - 3s = 7$
$10s - 3s = 7$
$10s - 3s = 7$
$7s = 7$
$s = 1$
(1) $\quad r = 2(1) = 2$
$(2, 1)$

3. (2) $4x + 3(x - 1) = 18$
$4x + 3x - 3 = 18$
$7x = 21$
$x = 3$
(1) $\quad y = 3 - 1 = 2$
$(3, 2)$

4. (1) $n = -2m - 8$
(2) $3m - 5(-2m - 8) = 14$
$3m + 10m + 40 = 14$
$13m = -26$
$m = -2$
(1) $n = -2(-2) - 8$
$= 4 - 8$
$= -4$
$(-2, -4)$

5. (1) $x = 5 - y$
(2) $4(5 - y) + 5y = 28$
$20 - 4y + 5y = 28$
$y = 8$
(1) $\quad x = 5 - 8 = -3$
$(-3, 8)$

6. (1) $4a - 4a = 3$
$0 = 3$, false
Inconsistent

7. (2) $s = 3t + 1$
(1) $3(3t + 1) - 3 = 9t$
$9t + 3 - 3 = 9t$
$3 - 3 = 0$, true
Dependent

8. (1) $c = 4d + 10$
(2) $5(4d + 10) + 8d = 1$
$20d + 50 + 8d = 1$
$28d = -49$
$d = \dfrac{-49}{28} = \dfrac{-7}{4}$

(1) $c = 4 \cdot \dfrac{-7}{4} + 10$

$= 3$

$\left(3, \dfrac{-7}{4}\right)$

Section 7-4. Summary Exercises

1. (1) $4m - 2n = -11$
(2) $\quad\quad\quad 2n = \quad 19\,(+)$
$\overline{4m \quad\quad = \quad 8}$
$m = 2$
(2) $2n = 19$
$n = \dfrac{19}{2}$
$\left(2, \dfrac{19}{2}\right)$

2. (1) $x + y = \quad 2$
(2) $\quad x - y = -6\,(+)$
$\overline{2x \quad\quad = -4}$
$x = -2$
(1) $-2 + y = 2$
$y = 4$
$(-2, 4)$

3. Multiply (2) by (-1):
(1) $\quad s - 2t = \quad 11$
(2) $\underline{-s + 3t = -18\,(+)}$
$t = \quad -7$
(1) $s - 2(-7) = 11$
$s - (-14) = 11$
$s = -3$
$(-3, -7)$

4. Multiply (1) by 2:
(1) $\quad 8a - 2b = \quad 6$
(2) $\underline{5a + 2b = 20\,(+)}$
$13a \quad\quad = 26$
$a = 2$
(1) $4(2) - b = 3$
$-b = -5$
$b = 5$
$(2, 5)$

5. Multiply (1) by 2 and (2) by (-3):

(1) $\quad 4p - 6q = 36$

(2) $\underline{-21p + 6q = 15\,(+)}$

$\qquad -17p \quad\;\; = 51$

$\qquad\qquad\quad p = -3$

(1) $2(-3) - 3q = 18$

$\qquad\quad -3q = 24$

$\qquad\qquad q = -8$

$(-3, -8)$

6. Multiply (1) by (-5):

(1) $-5y + 10x = -30$

(2) $\underline{\;\;5y - 10x = \quad 1\,(+)}$

$\qquad\qquad\quad 0 = -29,\text{ false}$

Inconsistent

7. Multiply (2) by (-2):

(1) $\quad 8a + 20b = \quad 6$

(2) $\underline{-8a - 20b = -6\,(+)}$

$\qquad\qquad\quad 0 = \quad 0,\text{ true}$

Dependent

8. Multiply (1) by 2 and (2) by (-3):

(1) $\quad 12m - 24n = 14$

(2) $\underline{-12m - 27n = \quad 3\,(+)}$

$\qquad\quad -51n = 17$

$\qquad\qquad\qquad n = \dfrac{-17}{51} = \dfrac{-1}{3}$

(2) $4m + 9 \cdot \dfrac{-1}{3} = -1$

$\qquad\qquad 4m = 2$

$\qquad\qquad\; m = \dfrac{1}{2}$

$\left(\dfrac{1}{2}, \dfrac{-1}{3}\right)$

Section 7-5. Summary Exercises

1. Let x be the smaller number.

Let y be the larger number.

(1) $x + y = 9$

(2) $2y - 3x = 93$

Multiply (1) by 3 and add to (2):

(1) $3y + 3x = \quad 27$

(2) $\underline{2y - 3x = \quad 93\,(+)}$

$\quad 5y \qquad\;\; = 120$

$\qquad\qquad y = 24,\text{ the larger number}$

$\qquad\qquad x = 9 - 24 = -15,\text{ the smaller number}$

The numbers are -15 and 24.

2. Let w be the width.

Let l be the length.

(1) $2w + 2l = 24$

(2) $\quad l - w = 1$

Solve (2) for l and substitute in (1):

(2) $l = w + 1$

(1) $2w + 2(w + 1) = 24$

$\qquad 2w + 2w + 2 = 24$

$\qquad\qquad\qquad 4w = 22$

$\qquad\qquad\qquad\;\; w = 5.5\text{ ft}$

$\qquad\qquad\qquad\;\;\; l = 5.5 + 1 = 6.5\text{ ft}$

The width is 5.5 ft and the length is 6.5 ft.

3.

Items	Quantity	Value
Dimes	x	$0.10x$
Quarters	y	$0.25y$

(1) $x + y = 118$

(2) $0.10x + 0.25y = 18.55$

(1) $x = 118 - y$

(2) $0.10(118 - y) + 0.25y = 18.55$

$\qquad 11.8 - 0.10y + 0.25y = 18.55$

$\qquad\qquad\qquad\quad 0.15y = 6.75$

$\qquad\qquad\qquad\qquad\;\; y = 45\text{ quarters}$

$\qquad\qquad\qquad\qquad\;\; x = 73\text{ dimes}$

73 dimes and 45 quarters

4.

	Rate	Time	Distance
Train	45 mph	x hr	$45x$ mi
Bus	60 mph	y hr	$60y$ mi

(1) $x - y = 0.5$

(2) $\quad 45x = 60y$

(1) $x = y + 0.5$

(2) $45(y + 0.5) = 60y$

$\qquad 45y + 22.5 = 60y$

$\qquad\qquad\quad 1.5 = y$

If $y = 1.5$ hr, then distance $= 1.5(60)$.

$\qquad\qquad\qquad\qquad\qquad = 90$ mi

90 mi from Sacramento

5. Let x be Judy's age now.

Let y be Jason's age now.

(1) $x = y + 9$

(2) $x + 10 = 2(y + 10) - 9$

Substitute $y + 9$ for x in (2):

(2) $y + 9 + 10 = 2y + 20 - 9$

$\qquad y + 19 = 2y + 11$

$\qquad\qquad\; 8 = y$

$\qquad\qquad\; x = 8 + 9 = 17$

Judy is 17 and Jason is 8.

Section 8-1. Summary Exercises

1. a. $(-17)^2 = 289$, yes   **2.** $\sqrt{81} = 9$, since $9^2 = 81$ **3.** $-\sqrt{\dfrac{25}{16}} = \dfrac{-5}{4}$, since $\left(\dfrac{-5}{4}\right)^2 = \dfrac{25}{16}$
   b. $(-21)^2 = 441$, no   9
   c. $17^2 = 289$, yes $\dfrac{-5}{4}$
   d. $21^2 = 441$, no
   a and c

4. $\sqrt{0.49} = 0.7$, since $0.7^2 = 0.49$ **5.** $\sqrt{\dfrac{256}{101}}$ is irrational, because 101 is not a perfect square.   **6.** $\sqrt{89} \approx 9.434$
        0.7   Irrational   9.434

7. $\sqrt{45.6} = 6.75277\ldots \approx 6.753$ **8.** $\dfrac{\sqrt{8}+\sqrt{12}}{\sqrt{20}} = \dfrac{2.8284\ldots + 3.4641\ldots}{4.4721\ldots} = \dfrac{6.2925\ldots}{4.4721\ldots} = 1.4070\ldots \approx 1.407$
        6.753   1.407

Section 8-2. Summary Exercises

1. $\sqrt{45} \cdot \sqrt{5} = \sqrt{45 \cdot 5} = \sqrt{225} = 15$ **2.** $\sqrt{7} \cdot \sqrt{14} = \sqrt{7 \cdot 14} = \sqrt{98 \cdot 2} = 7\sqrt{2}$ **3.** $\sqrt{529} = 23$, since $23^2 = 529$
        15 $7\sqrt{2}$ 23

4. $\sqrt{363} = \sqrt{121 \cdot 3} = 11\sqrt{3}$ **5.** $\dfrac{-\sqrt{162}}{\sqrt{2}} = -\sqrt{\dfrac{162}{2}} = -\sqrt{81} = -9$ **6.** $\dfrac{\sqrt{350}}{\sqrt{14}} = \sqrt{\dfrac{350}{14}} = \sqrt{25} = 5$
        $11\sqrt{3}$ -9 5

7. $\sqrt{\dfrac{5}{36}} = \dfrac{\sqrt{5}}{\sqrt{36}} = \dfrac{\sqrt{5}}{6}$ **8.** $\sqrt{\dfrac{17}{3}} = \sqrt{\dfrac{17}{3} \cdot \dfrac{3}{3}} = \dfrac{\sqrt{51}}{\sqrt{9}} = \dfrac{\sqrt{51}}{3}$
        $\dfrac{\sqrt{5}}{6}$ $\dfrac{\sqrt{51}}{3}$

Section 8-3. Summary Exercises

1. $\sqrt{11} - 3\sqrt{7} + 6\sqrt{11} + 14\sqrt{7} = (1+6)\sqrt{11} + (14-3)\sqrt{7} = 7\sqrt{11} + 11\sqrt{7}$
                                      $7\sqrt{11} + 11\sqrt{7}$

2. $2\sqrt{108} + \sqrt{27} = 2(6\sqrt{3}) + 3\sqrt{3} = (12+3)\sqrt{3} = 15\sqrt{3}$
                    $15\sqrt{3}$

3. $\sqrt{243t} - \sqrt{300t} + 3\sqrt{3t} = 9\sqrt{3t} - 10\sqrt{3t} + 3\sqrt{3t} = (9+3-10)\sqrt{3t} = 2\sqrt{3t}$
                    $2\sqrt{3t}$

4. $10\sqrt{10x} - 12\sqrt{3y} - \sqrt{75y} + \sqrt{160x} = 10\sqrt{10x} - 12\sqrt{3y} - 5\sqrt{3y} + 4\sqrt{10x} = 14\sqrt{10x} - 17\sqrt{3y}$
                         $14\sqrt{10x} - 17\sqrt{3y}$

5. $\dfrac{9+\sqrt{45}}{3} = \dfrac{9+3\sqrt{5}}{3} = \dfrac{3(3+\sqrt{5})}{3} = 3+\sqrt{5}$ **6.** $\sqrt{14}(3-\sqrt{7}) = 3\sqrt{14} - \sqrt{98} = 3\sqrt{14} - 7\sqrt{2}$
                $3+\sqrt{5}$ $3\sqrt{14} - 7\sqrt{2}$

7. $(8+5\sqrt{2})(3-2\sqrt{2}) = 24 - 16\sqrt{2} + 15\sqrt{2} - 10(2) = 4 - \sqrt{2}$ **8.** $(9-3\sqrt{5})(9+3\sqrt{5}) = 9^2 - (3\sqrt{5})^2 = 81 - 45 = 36$
                $4-\sqrt{2}$ 36

Section 8-4. Summary Exercises

1. $-9\sqrt{90} = -9\sqrt{9 \cdot 10} = -9(3\sqrt{10}) = -27\sqrt{10}$ **2.** $\dfrac{2\sqrt{147}}{21} = \dfrac{2\sqrt{49 \cdot 3}}{21} = \dfrac{14\sqrt{3}}{21} = \dfrac{2\sqrt{3}}{3}$ **3.** $\sqrt{\dfrac{8}{7}} = \dfrac{\sqrt{8}}{\sqrt{7}} \cdot \dfrac{\sqrt{7}}{\sqrt{7}} = \dfrac{\sqrt{56}}{7} = \dfrac{2\sqrt{14}}{7}$
        $-27\sqrt{10}$ $\dfrac{2\sqrt{3}}{3}$ $\dfrac{2\sqrt{14}}{7}$

4. $\dfrac{\sqrt{45b}}{9} + \sqrt{20b} - \sqrt{\dfrac{b}{5}} = \dfrac{3\sqrt{5b}}{9} + 2\sqrt{5b} - \dfrac{\sqrt{5b}}{5}$

$\qquad\qquad = \dfrac{5\sqrt{5b} + 30\sqrt{5b} - 3\sqrt{5b}}{15}$ $\quad$ (LCM of denominators is 15.)

$\qquad\qquad = \dfrac{32\sqrt{5b}}{15}$

$\qquad\qquad \dfrac{32\sqrt{5b}}{15}$

5. $\dfrac{7}{\sqrt{6}} + \sqrt{\dfrac{2}{3}} - \sqrt{\dfrac{3}{2}} = \dfrac{7\sqrt{6}}{6} + \dfrac{\sqrt{6}}{3} - \dfrac{\sqrt{6}}{2}$

$\qquad\qquad = \dfrac{7\sqrt{6} + 2\sqrt{6} - 3\sqrt{6}}{6}$ $\quad$ (LCM of denominators is 6.)

$\qquad\qquad = \dfrac{6\sqrt{6}}{6} = \sqrt{6}$

$\qquad\qquad \sqrt{6}$

6. $\dfrac{-6}{\sqrt{60}} + \dfrac{3\sqrt{15}}{5} = \dfrac{-6}{2\sqrt{15}} \cdot \dfrac{\sqrt{15}}{\sqrt{15}} + \dfrac{3\sqrt{15}}{5} = \dfrac{-3\sqrt{15}}{15} + \dfrac{9\sqrt{15}}{15} = \dfrac{6\sqrt{15}}{15} = \dfrac{2\sqrt{15}}{5}$

$\qquad\qquad \dfrac{2\sqrt{15}}{5}$

7. $\dfrac{4}{2+\sqrt{5}} - \dfrac{1+\sqrt{5}}{2-\sqrt{5}} = \dfrac{4}{2+\sqrt{5}} \cdot \dfrac{2-\sqrt{5}}{2-\sqrt{5}} - \dfrac{1+\sqrt{5}}{2-\sqrt{5}} \cdot \dfrac{2+\sqrt{5}}{2+\sqrt{5}}$

$\qquad\qquad = \dfrac{8 - 4\sqrt{5}}{4-5} - \dfrac{2 + \sqrt{5} + 2\sqrt{5} + 5}{4-5} = \dfrac{8 - 4\sqrt{5} - 7 - 3\sqrt{5}}{-1} = \dfrac{1 - 7\sqrt{5}}{-1}$

$\qquad\qquad 7\sqrt{5} - 1$

8. $\dfrac{-2}{1-3\sqrt{5}} \cdot \dfrac{1+3\sqrt{5}}{1+3\sqrt{5}} = \dfrac{-2(1+3\sqrt{5})}{1-45} = \dfrac{-2(1+3\sqrt{5})}{-44} = \dfrac{1+3\sqrt{5}}{22}$

$\qquad\qquad \dfrac{1+3\sqrt{5}}{22}$

Section 8-5. Summary Exercises

1. $(\sqrt{5a})^2 = 10^2$
$\qquad 5a = 100$
$\qquad\quad a = 20$
Check: $\sqrt{100} = 10$, true
$\quad$ 20

2. $(\sqrt{3u+1})^2 = 5^2$
$\qquad 3u + 1 = 25$
$\qquad\quad 3u = 24$
$\qquad\quad\ u = 8$
Check: $\sqrt{24+1} = 5$, true
$\qquad\quad$ 8

3. $(3\sqrt{z})^2 = (\sqrt{7z+2})^2$
$\qquad 9z = 7z + 2$
$\qquad 2z = 2$
$\qquad\ z = 1$
Check: $3\sqrt{1} = \sqrt{7+2}$, true
$\qquad$ 1

4. $x^2 = (\sqrt{10x - 24})^2$
$\quad x^2 = 10x - 24$
$\quad x^2 - 10x + 24 = 0$
$\quad (x-4)(x-6) = 0$
$\quad x = 4 \qquad$ or $\qquad x = 6$
Check: $4 = \sqrt{40 - 24}$, true
$\qquad\quad 6 = \sqrt{60 - 24}$, true
$\qquad$ 4 and 6

5. $(\sqrt{2n-1})^2 = (n-2)^2$
$\qquad 2n - 1 = n^2 - 4n + 4$
$\qquad\quad 0 = n^2 - 6n + 5$
$\qquad\quad 0 = (n-1)(n-5)$
$\qquad\quad n = 1 \quad$ or $\quad n = 5$
Check: $\sqrt{2-1} + 2 = 1$, false
$\qquad \sqrt{10-1} + 2 = 5$, true
$\quad$ 5

6. $\qquad (p+5)^2 = (\sqrt{2p+13})^2$
$\qquad p^2 + 10p + 25 = 2p + 13$
$\qquad p^2 + 8p + 12 = 0$
$\qquad (p+2)(p+6) = 0$
$\qquad p = -2 \quad$ or $\quad p = -6$
Check: $-2 + 5 = \sqrt{-4+13}$, true
$\qquad\ -6 + 5 = \sqrt{-12+13}$, false
$\quad$ -2

7. $(\sqrt{3y+9})^2 = (2\sqrt{y})^2$
$\qquad 3y + 9 = 4y$
$\qquad\quad 9 = y$
Check: $\sqrt{27+9} = 2\sqrt{9}$
$\qquad\ \sqrt{36} = 2(3)$, true
$\qquad$ 9

8. $25 - 10\sqrt{u} + u = u + 5$ Check: $5 - \sqrt{4} = \sqrt{4 + 5}$

$\qquad\qquad 20 = 10\sqrt{u}$ $3 = \sqrt{9}$, true

$\qquad\qquad 2^2 = (\sqrt{u})^2 \qquad 4$

$\qquad\qquad\quad 4 = u$

Section 9-1. Summary Exercises

1. $6 - 3x - 6 = x^2 - 5x$

$\qquad\quad 0 = x^2 - 2x$

$\qquad\quad 0 = x^2 - 2x$

2. $5x - 8 = 0$ or $9x + 1 = 0$

$\qquad 5x = 8$ or $9x = -1$

$\qquad\quad x = \dfrac{8}{5}$. or $x = \dfrac{-1}{9}$

$\qquad\qquad \dfrac{8}{5}$ and $\dfrac{-1}{9}$

3. $b^2 - 26b + 169 = 0$

$\quad (b - 13)(b - 13) = 0$

$\quad b - 13 = 0$ or $b - 13 = 0$

$\qquad b = 13$ or $b = 13$

$\qquad\qquad 13$

4. $\qquad x^2 + 4x = 21$

$\quad x^2 + 4x - 21 = 0$

$\quad (x + 7)(x - 3) = 0$

$\quad x + 7 = 0$ or $x - 3 = 0$

$\qquad x = -7$ or $x = 3$

$\quad -7$ and 3

5. $x^2 + 2x + 1 = -2x + 46$

$\quad x^2 + 4x - 45 = 0$

$\quad (x + 9)(x - 5) = 0$

$\quad x + 9 = 0$ or $x - 5 = 0$

$\qquad x = -9$ or $x = 5$

$\quad -9$ and 5

6. $\qquad x^2 + 4x = 16 - 2x$

$\quad x^2 + 6x - 16 = 0$

$\quad (x + 8)(x - 2) = 0$

$\quad x + 8 = 0$ or $x - 2 = 0$

$\qquad x = -8$ or $x = 2$

$\quad -8$ and 2

7. Let x be the length of the longer leg.

$x + 2$ is the length of the hypotenuse.

$x - 7$ is the length of the shorter leg.

$\quad (x - 7)^2 + x^2 = (x + 2)^2$ $x = 3$ $x = 15$

$\quad x^2 - 18x + 45 = 0$ $x + 2 = 5$ $x + 2 = 17$

$\quad (x - 3)(x - 15) = 0$ $x - 7 = -4$, reject. $x - 7 = 8$

8, 15, and 17 yd

8. Let t be the time for Juan to paint it alone.

Let $t + 9$ be the time for Scott to paint it alone.

$\dfrac{20}{t} + \dfrac{20}{t + 9} = 1$ $t - 36 = 0$ or $t + 5 = 0$

$20t + 180 + 20t = t^2 + 9t$ $t = 36$ $t = -5$, reject

$\qquad 0 = t^2 - 31t - 180$ $t + 9 = 45$

$\qquad 0 = (t - 36)(t + 5)$ 36 hours for Juan; 45 hours for Scott

Section 9-2. Summary Exercises

1. $s = \sqrt{175}$ or $s = -\sqrt{175}$

$\quad s = 5\sqrt{7}$ or $s = -5\sqrt{7}$

$\quad \pm 5\sqrt{7}$

2. $y^2 = \dfrac{45}{16}$

$\quad y = \pm\sqrt{\dfrac{45}{16}}$

$\quad y = \pm\dfrac{3\sqrt{5}}{4}$

$\quad \pm\dfrac{3\sqrt{5}}{4}$

3. $5x = \pm 10$

$\quad x = \pm\dfrac{10}{5}$

$\quad x = \pm 2$

$\quad \pm 2$

4. $3b = \pm 2$

$\quad b = \pm\dfrac{2}{3}$

$\quad \pm\dfrac{2}{3}$

5. $p - 3 = \pm\sqrt{220}$

$\quad p - 3 = \pm 2\sqrt{55}$

$\quad p = 3 \pm 2\sqrt{55}$

$\quad 3 \pm 2\sqrt{55}$

6. $3t + 7 = \pm\sqrt{8}$

$\quad 3t + 7 = \pm 2\sqrt{2}$

$\quad 3t = -7 \pm 2\sqrt{2}$

$\quad t = \dfrac{-7 \pm 2\sqrt{2}}{3}$

$\quad \dfrac{-7 \pm 2\sqrt{2}}{3}$

7. $(3b - 1)^2 = 289$

$\quad 3b - 1 = \pm 17$

$\quad 3b = 1 \pm 17$

$\quad b = \dfrac{1 \pm 17}{3}$

$\qquad\qquad b = \dfrac{1 + 17}{3} = 6$

$\qquad\qquad b = \dfrac{1 - 17}{3} = \dfrac{-16}{3}$

$\quad \dfrac{-16}{3}$ and 6

8. $(6x + 5)^2 = 50$

$\quad 6x + 5 = \pm\sqrt{50}$

$\quad 6x = -5 \pm 5\sqrt{2}$

$\quad x = \dfrac{-5 \pm 5\sqrt{2}}{6}$

$\quad \dfrac{-5 \pm 5\sqrt{2}}{6}$

Section 9-3. Summary Exercises

1. $x^2 - 8x + 16 = (x-4)^2$

$\dfrac{1}{2} \cdot 8 = 4$

$4^2 = 16$

  a. $x^2 - 8x + 16$

  b. $(x-4)^2$

2. $y^2 + 2y + 1 = 8 + 1$

$(y+1)^2 = 9$

$y + 1 = \pm 3$

$y = -1 \pm 3 \begin{cases} y = -1 + 3 = 2 \\ y = -1 - 3 = -4 \end{cases}$

-4 and 2

3. $z^2 - 6z + 9 = -2 + 9$

$(z-3)^2 = 7$

$z - 3 = \pm\sqrt{7}$

$z = 3 \pm \sqrt{7}$

$3 \pm \sqrt{7}$

4. $p^2 - 24p + 144 = -44 + 144$

$(p-12)^2 = 100$

$p - 12 = \pm 10$

$p = 12 \pm 10 \begin{cases} p = 12 + 10 = 22 \\ p = 12 - 10 = 2 \end{cases}$

2 and 22

5. $s^2 - 20s + 100 = -50 + 100$

$(s-10)^2 = 50$

$s - 10 = \pm 5\sqrt{2}$

$s = 10 \pm 5\sqrt{2}$

$10 \pm 5\sqrt{2}$

6. $\quad b^2 + 12b = -24$

$b^2 + 12b + 36 = -24 + 36$

$(b+6)^2 = 12$

$b + 6 = \pm 2\sqrt{3}$

$b = -6 \pm 2\sqrt{3}$

$-6 \pm 2\sqrt{3}$

7. $x^2 + 4x + \dfrac{1}{4} = 0$

$x^2 + 4x + 4 = \dfrac{-1}{4} + 4$

$(x+2)^2 = \dfrac{15}{4}$

$x + 2 = \pm\sqrt{\dfrac{15}{4}}$

$x = -2 \pm \dfrac{\sqrt{15}}{2} = \dfrac{-4 \pm \sqrt{15}}{2}$

$\dfrac{-4 \pm \sqrt{15}}{2}$

8. $\quad y^2 - 8y + \dfrac{1}{2} = 0$

$y^2 - 8y + 16 = \dfrac{-1}{2} + 16$

$(y-4)^2 = \dfrac{31}{2}$

$y - 4 = \pm\sqrt{\dfrac{31}{2}}$

$y = 4 \pm \dfrac{\sqrt{62}}{2} = \dfrac{8 \pm \sqrt{62}}{2}$

$\dfrac{8 \pm \sqrt{62}}{2}$

Section 9-4. Summary Exercises

1. $6x^2 - 3x - 4 = 0$

a is 6

b is -3

c is -4

2. $x^2 + 10x + 25 = 6 + 10x$

$x^2 + 19 = 0$

a is 1

b is 0

c is 19

3. $y = \dfrac{-4 \pm \sqrt{16 - 4(1)(-9)}}{2(1)}$

$= \dfrac{-4 \pm \sqrt{52}}{2}$

$= \dfrac{-4 \pm 2\sqrt{13}}{2} = \dfrac{2(-2 \pm \sqrt{13})}{2} = -2 \pm \sqrt{13}$

$-2 \pm \sqrt{13}$

4. $z^2 - 3z - 11 = 0$

$z = \dfrac{-(-3) \pm \sqrt{9 - 4(1)(-11)}}{2(1)}$

$= \dfrac{3 \pm \sqrt{53}}{2}$

$\dfrac{3 \pm \sqrt{53}}{2}$

5. $\quad 6p^2 + p - 1 = -5p$

$6p^2 + 6p - 1 = 0$

$p = \dfrac{-6 \pm \sqrt{36 - 4(6)(-1)}}{2(6)}$

$= \dfrac{-6 \pm \sqrt{60}}{12}$

$= \dfrac{-6 \pm 2\sqrt{15}}{12} = \dfrac{2(-3 \pm \sqrt{15})}{12} = \dfrac{-3 \pm \sqrt{15}}{6}$

$\dfrac{-3 \pm \sqrt{15}}{6}$

6. $4x^2 - 12x + 9 = 10 - 8x$

$4x^2 - 4x - 1 = 0$

$$x = \frac{-(-4) \pm \sqrt{16 - 4(4)(-1)}}{2(4)}$$

$$= \frac{4 \pm \sqrt{32}}{8} = \frac{4 \pm 4\sqrt{2}}{8} = \frac{4(1 \pm \sqrt{2})}{8} = \frac{1 \pm \sqrt{2}}{2}$$

$$\frac{1 \pm \sqrt{2}}{2}$$

7. Let x be the smaller.

$x + 10$ is the larger.

$$x(x + 10) = 20$$

$$x^2 + 10x - 20 = 0$$

$$x = \frac{-10 \pm \sqrt{100 + 80}}{2}$$

$$= \frac{-10 \pm \sqrt{180}}{2}$$

$$= \frac{-10 \pm 6\sqrt{5}}{2} = -5 \pm 3\sqrt{5}$$

$$x + 10 = 5 \pm 3\sqrt{5}$$

The larger number is $5 \pm 3\sqrt{5}$

The smaller number is $-5 \pm 3\sqrt{5}$

8. Let h be the height.

$h - 1$ is the base.

$$8 = \frac{1}{2}h(h - 1)$$

$$0 = h^2 - h - 16$$

$$h = \frac{-(-1) \pm \sqrt{1 + 64}}{2}$$

$$= \frac{1 \pm \sqrt{65}}{2}$$

$$h = \frac{1 + \sqrt{65}}{2}$$

$$b = \frac{1 + \sqrt{65}}{2} - 1 = \frac{1 + \sqrt{65} - 2}{2} = \frac{-1 + \sqrt{65}}{2}$$

The height is $\dfrac{1 + \sqrt{65}}{2}$ m

The base is $\dfrac{-1 + \sqrt{65}}{2}$ m

Section 9-5. Summary Exercises

1. $\sqrt{-72} = i\sqrt{72} = 6i\sqrt{2}$

$6i\sqrt{2}$

2. $-\sqrt{\dfrac{-3}{25}} = -i\sqrt{\dfrac{3}{25}} = \dfrac{-i\sqrt{3}}{5}$

$\dfrac{-i\sqrt{3}}{5}$

3. $-13 + \sqrt{-144} = -13 + i\sqrt{144}$

$= -13 + 12i$

$-13 + 12i$

4. $\sqrt{20} - \sqrt{-125} = \sqrt{20} - i\sqrt{125}$

$= 2\sqrt{5} - 5i\sqrt{5}$

$2\sqrt{5} - 5i\sqrt{5}$

5. $9t^2 = -100$

$$t = \frac{-100}{9}$$

$$t = \pm\sqrt{\frac{-100}{9}}$$

$$= \pm\frac{10}{3}i$$

$$\pm\frac{10}{3}i$$

6. $k^2 - 6k + 58 = 0$

$$k = \frac{-(-6) \pm \sqrt{36 - 4(1)(58)}}{2(1)}$$

$$= \frac{6 \pm \sqrt{-196}}{2}$$

$$= \frac{6 \pm 14i}{2}$$

$$= \frac{2(3 \pm 7i)}{2} = 3 \pm 7i$$

$$3 \pm 7i$$

7. $0 = a^2 + 2a + 19$

$a = \dfrac{-2 \pm \sqrt{4 - 4(1)(19)}}{2(1)}$

$= \dfrac{-2 \pm \sqrt{-72}}{2}$

$= \dfrac{-2 \pm 6i\sqrt{2}}{2}$

$= \dfrac{\cancel{2}(-1 \pm 3i\sqrt{2})}{\cancel{2}} = -1 \pm 3i\sqrt{2}$

$\qquad -1 \pm 3i\sqrt{2}$

8. $b = \dfrac{-(-12) \pm \sqrt{144 - 4(9)(11)}}{2(9)}$

$= \dfrac{12 \pm \sqrt{-252}}{18}$

$= \dfrac{12 \pm 6i\sqrt{7}}{18}$

$= \dfrac{6(2 \pm i\sqrt{7})}{\cancel{18}} \atop 3$

$\dfrac{2 \pm i\sqrt{7}}{3}$

Section 9-6. Summary Exercises

1. Region I, use -5: $(-5 - 3)(-5 + 4) < 0$, false
  Region II, use 0: $(0 - 3)(0 + 4) < 0$, true
  Region III, use 4: $(4 - 3)(4 + 4) < 0$, false

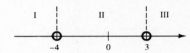

$-4 < z < 3$

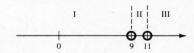

2. Region I, use -3: $(-6 + 5)(-3 - 7) \le 0$, false
  Region II, use 0: $(0 + 5)(0 - 7) \le 0$, true
  Region III, use 8: $(16 + 5)(8 - 7) \le 0$, false

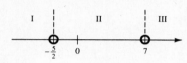

$\dfrac{-5}{2} \le x \le 7$

$(a - 10)(a - 2) > 0$

3. Region I, use -9: $-27(-9 + 8) \le 0$, false
  Region II, use -1: $-3(-1 + 8) \le 0$, true
  Region III, use 1: $3(1 + 8) \le 0$, false

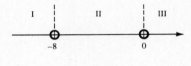

$-8 \le x \le 0$

4. Region I, use 0: $(0 - 10)(0 - 2) > 0$, true
  Region II, use 3: $(3 - 10)(3 - 2) > 0$, false
  Region III, use 11: $(11 - 10)(11 - 2) > 0$, true

$a < 2$ or $a > 10$

5. $b^2 - 20b + 99 < 0$
  $(b - 9)(b - 11) < 0$
   Region I, use 0: $0 + 99 < 0$, false
   Region II, use 10: $100 + 99 < 200$, true
   Region III, use 12: $144 + 99 < 240$, false

$9 < b < 11$

6. $3t^2 + 26t - 9 \le 0$
  $(3t - 1)(t + 9) \le 0$
   Region I, use -10: $-260 \le 9 - 300$, false
   Region II, use 0: $0 \le 9 - 0$, true
   Region III, use 1: $26 \le 9 - 3$, false

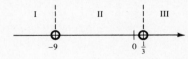

$-9 \le t \le \dfrac{1}{3}$

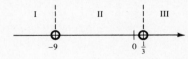

7. $0 > 5y^2 - 18y - 8$
$0 > (5y + 2)(y - 4)$
  Region I, use -1: $-18 + 8 > 5$, false
  Region II, use 0: $0 + 8 > 0$, true
  Region III, use 5: $90 + 8 > 125$, false

8. $25m^2 - 144 \geq 0$
$(5m + 12)(5m - 12) \geq 0$
  Region, I, use -3: $225 \geq 144$, true
  Region II, use 0: $0 \geq 144$, false
  Region III, use 3: $225 \geq 144$, true

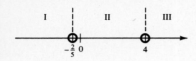

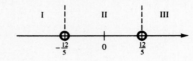

$\dfrac{-2}{5} < y < 4$

$m \leq \dfrac{-12}{5}$ or $m \geq \dfrac{12}{5}$

Section 10-1. Summary Exercises

1. a. For $x = -1$: $y = -2(-1) + 5 = 7$
  For $x = 0$: $y = -2(0) + 5 = 5$
  For $x = 6$: $y = -2(6) + 5 = -7$
  $(-1, 7)$
  $(0, 5)$
  $(6, -7)$

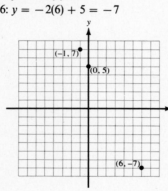

2. If $x = -2$, $y = \dfrac{(-2)^2}{(-2)^2 + 4} = \dfrac{4}{8} = \dfrac{1}{2}$

  If $x = 0$, $y = \dfrac{0}{0 + 4} = 0$

  If $x = 4$, $y = \dfrac{16}{16 + 4} = \dfrac{16}{20} = \dfrac{4}{5}$

  $\dfrac{1}{2}$, 0, and $\dfrac{4}{5}$

3. $-6, 6, -12, 15$, and 0
  24, 19, 0, and 8

4. a. Domain: $-6 \leq x \leq 6$
  b. Range: $-3 \leq y \leq 3$

5. Yes

6. Yes

7. a. $f(2) = 6(4) + \dfrac{1}{2} = 24 + \dfrac{1}{2}$

    $\dfrac{49}{2}$

  b. $f(-1) = 6(1) + \dfrac{1}{-1} = 5$

    5

  c. $f\left(\dfrac{1}{2}\right) = 6\left(\dfrac{1}{4}\right) + \dfrac{1}{\dfrac{1}{2}} = \dfrac{3}{2} + 2 = \dfrac{7}{2}$

    $\dfrac{7}{2}$

8. a. $f(4) = \dfrac{\sqrt{4 - 4}}{16 - 1} = 0$

    0

  b. $f(5) = \dfrac{\sqrt{5 - 4}}{25 - 1} = \dfrac{1}{24}$

    $\dfrac{1}{24}$

  c. $f(8) = \dfrac{\sqrt{8 - 4}}{64 - 1} = \dfrac{2}{63}$

    $\dfrac{2}{63}$

Section 10-2. Summary Exercises

1. $5x + 5y = 3x + 3$
    $5y = -2x + 3$
    $y = \dfrac{-2}{5}x + \dfrac{3}{5}$
    $f(x) = \dfrac{-2}{5}x + \dfrac{3}{5}$

2. $12x + 48 = 15 - 5y$
    $5y = -12x - 33$
    $y = \dfrac{-12}{5}x - \dfrac{33}{5}$
    $f(x) = \dfrac{-12}{5}x - \dfrac{33}{5}$

3.

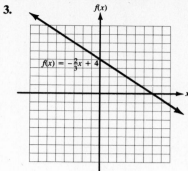

$f(x) = -\frac{2}{3}x + 4$

4.

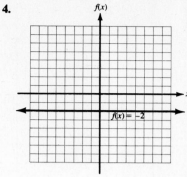

$f(x) = -2$

5. $f(x) = -1 \cdot x^2 + 10x - 15$

a. $\dfrac{-b}{2a}$ becomes $\dfrac{-10}{-2} = 5.$

5

b. $f(5) = -25 + 50 - 15 = 10$

10

c. $x = 5$

6. a. $\dfrac{-b}{2a}$ becomes $\dfrac{12}{4} = 3.$

3

b. $f(3) = 2(9) - 36 + 9$

-9

c. $x = 3$

7. $V(1, -1)$ is the vertex.

$x = 1$ is the axis of symmetry.

Cup opens upward.

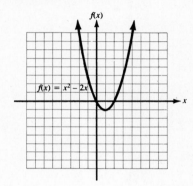

$f(x) = x^2 - 2x$

8. $V(1, -1)$ is the vertex.

$x = 1$ is the axis of symmetry.

Cup opens upward.

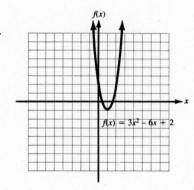

$f(x) = 3x^2 - 6x + 2$

Section 10-3. Summary Exercises

1.

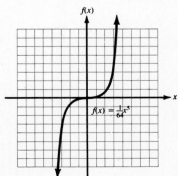

$f(x) = \frac{1}{64}x^5$

2.

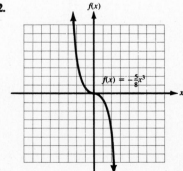

$f(x) = -\frac{5}{8}x^3$

3.

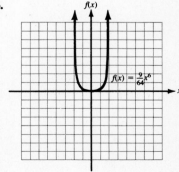

$f(x) = \frac{9}{64}x^6$

4.

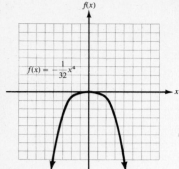

$f(x) = -\dfrac{1}{32}x^4$

5.

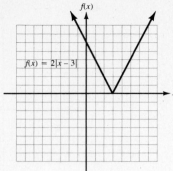

$f(x) = 2|x - 3|$

6.

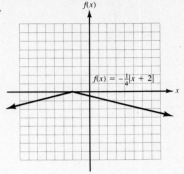

$f(x) = -\dfrac{1}{4}|x + 2|$

Index

A

Absolute value
  function 623
  of a number 6
Addition
  associative property 39
  commutative property 38
  exponents 168
  integers 13, 14
  polynomials 162
  radical expressions 512
  rational expressions 299, 306
  rational numbers. *See* Fraction,
    numerical
Addition method (solving systems) 465
Addition property
  equations 96
  inequalities 131
  square roots 513
Additive inverse 40
Age problems 105, 477
Algebraic expressions 45, 88
Algebraic fraction 273
Angles of a triangle 116
Apparent roots 325
Applied problems
  age 105, 477
  angles of a triangle 116
  areas
    rectangles 178, 252, 568
    trapezoid 54
    triangles 122, 253
  compound interest 55
  consecutive integer 115, 256
  distance-rate-time 56, 104, 476
  elevation 17
  falling object 56
  geometry 103, 473
  guess and test 106
  interest, compound 55
  miles per gallon 333
  mixture 103, 474
  motion 104, 343, 476
  number 102, 134, 250, 471, 568
  paint an area 334
  procedure for solving 102
  Pythagorean theorem 251, 550
  quadratic inequality 583
  rectangle area 178, 252, 568
  right-triangle 251, 550
  similar triangles 334
  statistics 324
  total cost 334

  trapezoid area 54
  trigonometry 323
  volume 179
  wildlife populations 336
  work 341, 550
Approximating
  roots of quadratic equations 568, 569
  square roots 498, 503, 511, 519, 521
Areas
  rectangles 178, 252, 568
  surface 155
  trapezoid 54
  triangles 122, 253
Ascending order 157
Associative property 39
Axes of a graph 378
Axis of symmetry 616

B

Bar 33
Base 30, 283
Binary operation 39
Binomial 154
  difference of
    cubes 239
    squares 185, 237
  FOIL 183
  length-width-area 182
  multiplying 175
    sum and a difference 184
  squaring 185
Boundary 416, 580
Braces 33
Brackets 33
Building up fractions 308

C

Calculator
  approximate radicals 519
  clear/clear all 18
  evaluate polynomials 157
  fix the number of decimals 503
  memory 157
  order of operations 34
  parentheses 278
  $+/-$ 18
  raising to a power 35
  square root 499
Checking solutions
  radical equations 528
  rational equations 325

Coefficient 47
Combining like terms 162
Common
  denominator 299, 306, A-7
  factor 213
  multiples 13
Commutative property 38
Comparing numbers 5
Complement of a set A-3
Completing the square 562
Complex
  fraction 316
  fraction, negative exponents 320
  number 574
  plane 578
Compound
  inequalities in two variables 417
  inequality in one variable 134, 137
  interest 55
Conditional equation 88
Conjugate 523
Consecutive integers 115, 256
Consistent system 450
Coordinates 378
Cube root 526

D

Degree
  no degree 156
  polynomial 156
  term 156
  zero 156
Denominator
  least common (LCM) 306
  rationalizing 522
Dependent
  system 452, 459, 467
  variables 373, 605
Descending order 157
Difference of the squares 237
Direct variation 387
Discriminant 572
Distance-rate-time 56, 104, 476
Distributive property 41
Divine proportion 338
Division
  exponents 190
  fractions A-7
  integers 22
  one 24
  polynomials 189

Division (continued)
  radical expressions 508, 522
  rational expressions 294
  scientific notation 292
  zero 26
Domain 603
Double root 247

E

Element A-1
Elevation problems 17
Elimination method. See Addition method
Empty set A-2
Equal (=) 5
Equality properties
  addition 96
  multiplication 97
Equations. See also Formula
  with all reals as roots 114
  conditional 88
  definition 87
  direct variation 387
  equivalent 89
  versus expressions 89
  finding 106
  fractional 324
  graphing
    absolute value function 623
    linear 381
    linear function 614
    linear system 449
    parabola 615
    power function 621
    quadratic function 617
  with grouping symbols 113
  for horizontal lines 410
  linear 95, 111, 378
  lines
    $ax + by = c$ form 380
    point-slope 407
    slope-intercept 399
  with no roots 113
  radical 527
  rational 324
  solving linear 95, 111
  systems 440
  in two variables 371
  for vertical lines 410
  $x^2 = k$ form 556
Equivalent
  equations 89
  expressions 308
  fractions 25, 309
Exponents
  base 30
  laws 168
  negative 284
  positive integer 30

power of a product 171
power of a quotient 171
power, same base 170
power term 30
product, same base 168
quotient, same base 190
rational 502
simplifying expressions with 32, 53
zero 191, 283
Expression
  algebraic 45, 88
  versus equations 89
  evaluating 53
  mixed 316
  numerical 45
  radical 505
  rational 272
  simplifying. See Simplifying
  variable 52
  terms of 49, 51
Extraneous solutions 530
Extremes 331

F

Factor 47
Factoring
  as an area 235
  common factor 214
  definition 214
  difference of
    cubes 239
    squares 237
  general trinomial
    grouping 229
    trial-and-error 232
  greatest common factor 214
  grouping 224, 229
  negative factor 225
  perfect square trinomial 238
  simple trinomial 218
    cannot be factored 221
    with a common factor 220
  solving equations by 245, 548
  special forms 236
  sum of cubes 239
Falling object 56
FOIL 183
Formula
  definition 120
  evaluating 121
  quadratic 565
  solving for a variable 122
Fraction, algebraic. See Rational
    expression
Fraction, numerical
  adding A-8
  definition A-5
  dividing A-7

equal A-5
least common denominator A-9
multiplying A-6
reciprocal A-7
reduced form A-6
subtracting A-8
Fractional equation 324
Function 605
$f(x)$ notation 606

G

General form of a linear equation 380
General trinomial 224, 229
Geometry problems 103, 473
Golden section 338
Graph
  absolute value function 623
  inequality one variable 131, 580
  linear in two variables
    equation 381
    function 614
    inequality 416
    intercept method 397
    slope and a point 392
    slope and y-intercept 399, 404
    system 449
  number 5
  ordered pair 378
  parabola 615
  power function 621
  quadratic
    function 617
    inequality 579
Graphical method of solving a system 449
Greater than (>) 5
Greater than or equal (≥) 134
Greatest common factor 214
Grouping
  symbols 33
  symbols in equations 113
Guess and test 106

H

Half-plane 415
Horizontal line
  equation 409
  graph 382
  slope 392
Hypotenuse 251, 550

I

i-form 504
Identity element 40
Imaginary unit i 574
Inconsistent 451, 459, 467
Independent system 450

Independent variables 373, 603
Inequality
   one variable 129, 580
   properties
      addition 131
      multiplication 133
   two variables 416
Integers
   adding 13
   consecutive 115, 256
   definition 3
   dividing 22
   multiplying 22
   subtraction 15
Intercepts
   definition 393
   graph using 397
Interest, compound 55
Intersection of sets A-3
Inverse, additive and multiplicative 40
Irrational number 4

L
Leading coefficient 218
Least common multiple (LCM) 307
Less than (<) 5
Less than or equal (≤) 134
Like
   radicals 513
   terms 162
Line of symmetry. *See* Axes of symmetry
Linear
   equation in one variable 95
      solving 111
   equation in two variables 378
      general form 380
      preferred form 380
   function 614
   inequality in one variable 130
   inequality in two variables 416
   system of equations 441
Lines
   horizontal 382
   number 5
   parallel 451
   perpendicular 401
   point-slope 407
   slope 390
   slope-intercept 399
   vertical 382
Literal equation 123

M
Magnitude of a number 6
Mean proportional 338
Means 331
Means-extremes product property 332
Member of a set. *See* Element

Miles-per-gallon problems 333
Mixed expression 316
Mixture problems 103, 474
Monomial 154
Motion problems 104, 343, 476
Multiples 3
Multiplication
   associative property 39
   commutative property 38
   distributive property 41
   exponents 168
   fractions A-6
   integers 22
   one 24
   polynomials 172, 175
      horizontal format 176
      vertical format 177
   property for
      equations 97
      inequalities 133
   radical expressions 505
   rational expressions 292
   scientific notation 291
   zero 26
Multiplicative
   identity 40
   inverse 40

N
Natural numbers 2
Negative
   integer 3
   integer exponent 284
   radicand 500
   square root 497
Null set A-2
Number
   integer 3
   irrational 4
   natural 2
   rational 4
   real 4
   whole 3
Number line
   coordinate of the point 5
   graph of a number 5
   order on 5
Number problems 102, 134, 250, 471, 568
Numerical expressions 45

O
Opposite 16, 25, 40
Order of operations 9, 32, 33
Ordered pair
   graph 378
   solution 371
Ordered triple 377, 443, 463

Ordered four-tuple 463
Origin 378

P
Paint an area 334
Parabola
   axes of symmetry 616
   definition 615
   graph 617
   symmetric points 617
   vertex 616
Parallel lines 451
Parentheses
   definition 33
   equations 113
   expressions 33
Perfect
   cube factors 527
   square 498
   square trinomial 186, 238
Perimeter, rectangle 104, 106
Plotting points
   one dimension 5
   two dimensions 371
Points
   coordinates 5, 378
   plotting 5, 379
Point-slope equation 407
Polynomial
   adding 162
   ascending order 157
   binomials 154
   coefficient 154
   combining like terms 162
   definition 154
   degree 156
   descending order 157
   dividing 189
   evaluating 157
   factoring. *See* Factoring
   monomials 154
   multiplying 175
   subtracting 163
   terms 55
   trinomials 154
Power term 30
Preferred form
   linear equation 380
   radicals 520
Principal square root 497
Problems. *See* Applied problems
Product property of square roots 506
Products. *See* Multiplication
Property
   addition of equality 96
   addition of inequality 131
   associative 39
   commutative 38

Property *(continued)*
   distributive 41
   equality 528
   exponents 168, 190, 286
   identity 40
   inverse 40
   means-extremes 332
   multiplication
      of equality 97
      of inequality 133
      of one 24
      of zero 23
   special of zero 26
   zero product 245, 548
Proportions 331
Pythagorean theorem 251, 550

Q

Quadrant 378
Quadratic
   approximating roots 568, 569
   completing the square 562
   definition 547
   discriminant 572
   equation 245, 547
      complex solutions 575
      solve by factoring 246, 548
   formula 565
   function 615, 617
   graph 617
   inequality 579
   inequality applied problems 583
Quotient property of square roots 508

R

Radical expressions
   addition 512
   approximate 498, 519
   conjugate 523
   definition 505
   division 508, 522
   equations 528
   like 513
   multiplication 505
   not real 500
   preferred form 520
   radicand 497
   rationalizing the denominator 522
   subtraction 512
Radical sign 497
Radicand 497
Range 603
Rational
   equations 324
   exponent 502
   numbers 4
Rational expression
   complex 316

definition 274
dividing 294
multiplying 292
reciprocal 294
restricted values 276
sum with different denominator 309
sum with same denominator 299
Rationalize the denominator 522
Real numbers 4
Reciprocal 40, 97, 284, 294
Rectangle area problems 568
Reduced form 277
Relation 602
Repeating decimal 4
Replacement set 53
Restricted values 276
Right-triangle problems 251, 550
Root
   cube 526
   equations 88, 245, 371
   square 496
Rounding A-13

S

Scientific notation
   converting 288
   definition 288
   dividing 292
   multiplying 291
Set A-1
Shaded half-plane 416
Similar triangles 334
Simple trinomial 218
Simpler form 506
Simplified form
   complex fraction 316
   radical 520
Simplifying
   complex fractions 316
   expressions with grouping symbols 33
   radical expressions 520
   rational expressions 277
Slope
   definition 390
   horizontal line 382
   parallel lines 400
   perpendicular lines 401
   rise/run 392
   slope-intercept form 399
   vertical line 382
Slope-intercept equation 399
Solution
   equation
      linear in one variable 88
      quadratic 245
      radical 528
      rational 325
      system 371

inequalities
   one variable 131
   two variables 417
linear system 442
system of three variables 443
Solution set 371
Solving
   applied problems 102
   equations
      fractional 325
      linear in one variable 94–116
      proportion 332
      quadratic 548, 555, 562, 565
      radical 528
      rational 325
      systems 448, 457, 465
   inequalities
      in one variable 131
      in two variables 417
Special factoring forms 236
Speed (rate) 56, 104, 476
Square
   binomial 185
   difference 238
   sum 238
   trinomial, perfect 238
Square root
   addition property 513
   approximating 498
   calculator 499
   definition 496
   negative
      numbers 500
      roots 497
   principal 497
   product property 505
   quotient property 508
   subtraction property 513
   symbol 496
   theorem 555
Squares, difference 237
Statistics 324
Substitution method (solving systems) 457
Subtraction
   exponents 190
   integers 16
   polynomials 163
   radical expressions 512
   rational expressions 299, 306
   rational numbers. *See* Fraction,
      numerical
Subtraction property of square roots 513
Surface area
   rectangular solid 155
   sphere 155
Symbols
   equal (=) 5
   greater than (>) 5
   greater than or equal (≥) 134
   less than (<) 5

less than or equal ($\leq$) 134
radical 497
Symmetric points 617
System of equations
   consistent 450
   definition 441
   dependent 452
   graph 449
   inconsistent 451
   independent 450
   solution 442
   solving
      addition 465
      graphing 449
      substitution 456

T
Term
   conventional form 48
   definition 46
   degree 156
   like 48
Terminating decimal 4
Test point 417
Translating
   algebraic expressions 60
   equations 102–7
Trapezoid area 54
Trial-and-error factoring 232
Triangle
   angles 116

area 122, 253
perimeter 104, 106
Pythagorean theorem 251, 550
right 251, 550
similar 334
Trigonometry 323
Trinomial
   definition 154
   factoring
      general 229, 232
      simple 218

U
Union of sets A-3
Universal set A-2

V
Variable expression 52
Variables 45
Venn diagrams 637
Vertex 616
Vertical line
   equation 409
   graph 382
   slope 391
Vertical line test 605
Volume
   rectangular solid 155, 179
   sphere 155

W
Whole numbers 3
Word phrases
   addition 60
   division 63
   multiplication 62
   raising to powers 62
   subtraction 61
Word problems. *See* Applied problems
Work problems 341, 550

X
x-intercept 393
xy-coordinate system 378

Y
y-intercept 393

Z
Zero
   add 26
   additive identity 40
   divide 26
   exponent 191, 283
   multiply 26
   special properties 26
Zero-product property 245, 548

A 0
B 1
C 2
D 3
E 4
F 5
G 6
H 7
I 8
J 9